AF664050

Agricultural Salinity Assessment and Management

Kenneth K. Tanji, Editor

Prepared by the
Water Quality Technical Committee
of the Irrigation and Drainage Division
of the American Society of Civil Engineers.

1996 Corrected Edition

United Book Prints

Reprinted in India by:

UNITED BOOK PRINTS
5-B, Bhagat ki Kothhi, Near Police Chowki
New Pali Road, Jodhpur 342 001 (INDIA)
E-mail: info@scientificpub.com

Print : 2020

ISBN: 978-81-72331-26-9

Corrected Edition reprinted in 1996

Printed in India

MANUALS AND REPORTS ON ENGINEERING PRACTICE

(As developed by the ASCE Technical Procedures Committee, July 1930, and revised March 1935, February 1962, and April 1982.)

A manual or report in this series consists of an orderly presentation of facts on a particular subject, supplemented by an analysis of limitations and applications of these facts. It contains information useful to the average engineer in his everyday work, rather than the findings that may be useful only occasionally or rarely. It is not in any sense a "standard," however; nor is it so elementary or so conclusive as to provide a "rule of thumb" for nonengineers.

Furthermore, material in this series, in distinction from a paper (which expresses only one person's observations or opinions), is the work of a committee or group selected to assemble and express information on a specific topic. As often as practicable the committee is under the general direction of one or more of the Technical Divisions and Councils, and the product evolved has been subjected to review by the Executive Committee of that Division or Council. As a step in the process of this review, proposed manuscripts are often brought before the members of the Technical Divisions and Councils for comment, which may serve as the basis for improvement. When published, each work shows the names of the committee by which it was compiled and indicates clearly the several processes through which it has passed in review, in order that its merit may be definitely understood.

In February 1962 (and revised in April 1982) the Board of Direction voted to establish:

> A series entitled 'Manuals and Reports on Engineering Practice,' to include the Manuals published and authorized to date, future Manuals of Professional Practice, and Reports on Engineering Practice. All such Manual or Report material of the Society would have been refereed in a manner approved by the Board Committee on Publications and would be bound, with applicable discussion, in books similar to past Manuals. Numbering would be consecutive and would be a continuation of present Manual numbers. In some cases of reports of joint committees, bypassing of Journal publication may be authorized.

AVAILABLE* MANUALS AND REPORTS OF ENGINEERING PRACTICE

Number

10 Technical Procedure for City Surveys
13 Filtering Materials for Sewage Treatment Plants
14 Accommodation of Utility Plant Within the Rights-of-Way of Urban Streets and Highways
31 Design of Cylindrical Concrete Shell Roofs
33 Cost Control and Accounting for Civil Engineers
34 Definitions of Surveying and Associated Terms
35 A List of Translations of Foreign Literature on Hydraulics
36 Wastewater Treatment Plant Design
37 Design and Construction of Sanitary and Storm Sewers
40 Ground Water Management
41 Plastic Design in Steel—A Guide and Commentary
42 Design of Structures to Resist Nuclear Weapons Effects
44 Report on Highway and Bridge Surveys
45 Consulting Engineering—A Guide for the Engagement of Engineering Services
46 Report on Pipeline Location
47 Selected Abstracts on Structural Applications of Plastics
49 Urban Planning Guide
50 Report on Small Craft Harbors
51 Survey of Current Structural Research
52 Guide for the Design of Steel Transmission Towers
53 Criteria for Maintenance of Multilane Highways
54 Sedimentation Engineering
55 Guide to Employment Conditions for Civil Engineers
56 Subsurface Investigation for Design and Construction of Foundations of Buildings
57 Operation and Maintenance of Irrigation and Drainage Systems
58 Structural Analysis and Design of Nuclear Plant Facilities
59 Computer Pricing Practices
60 Gravity Sanitary Sewer Design and Construction
61 Introductory Manual on Computer Services
62 Existing Sewer Evaluation and Rehabilitation
63 Structural Plastics Design Manual
64 Manual on Engineering Surveying
65 Construction Cost Control
66 Structural Plastics Selection Manual
67 Wind Tunnel Model Studies of Buildings and Structures
68 Aeration—A Wastewater Treatment Process
69 Sulfide in Wastewater Collection and Treatment Systems
70 Evapotranspiration and Irrigation Water Requirements
71 Agricultural Salinity Assessment and Management
72 Design of Steel Transmission Structures
73 Quality in the Constructed Project—a Guide for Owners, Designers, and Constructors

*Numbers 1, 2, 3, 4, 5, 6, 7, 8, 9, 11, 12, 15, 16, 17, 18, 19, 20, 21, 22, 23, 24, 25, 26, 27, 28, 29, 30, 32, 38, 39, 43, and 48 are out of print.

Preface

In July 1984, Fred Hotes and Marvin E. Jensen suggested that ASCE produce a manual on agricultural salinity management. The Water Quality Technical Committee of the Irrigation and Drainage Division of ASCE accepted this challenge and appointed Kenneth K. Tanji of the University of California, Davis, as Project Coordinator to explore the development of such a manual.

After extensive consultations with colleagues in salinity research groups, cooperative extensions, universities, and regulatory and management agencies, Tanji submitted, in July 1986, a prospectus for a salinity manual. He outlined the needs for a manual that integrates contemporary concepts and management practices for agricultural water and salinity problems. He proposed that the manual be produced by an interdisciplinary team of authors and editors and that it contain engineering and scientific aspects as well as environmental, economic, and legal considerations. The principal objective was to prepare a manual that would be a valuable reference to a wide spectrum of users.

In September 1986, the Water Quality Technical Committee submitted a proposal to form a task committee for overseeing the development of the salinity manual. ASCE approved this proposal in February 1987. The task committee appointed an editorial committee in May 1987 to outline chapters, select contributing authors, and review outlines and manuscripts of chapters. In April 1988, as the draft of the manual was being developed, the executive committee of the Irrigation and Drainage Division appointed a peer review committee to independently review the manual. A complete draft of the manual was submitted to the Irrigation and Drainage Division in September 1989 for their approval and to the ASCE Board on Publications in October 1989.

Acknowledgments

Division of Agriculture and Natural Resources, University of California for providing editorial and secretarial assistance and travel expenses for editorial committee meetings.

U.S. Department of Agriculture, Soil Conservation Service for re-drafting all figures and graphs to maintain uniformity.

TABLE OF CONTENTS

Foreword

I. Introduction 1

Chapter 1. The Nature and Extent of Agricultural Salinity Problems 1
Chapter 2. Overview: Diagnosis of Salinity Problems and Selection of Control Practices 18

II. Effects of Salts on Soils 42

Chapter 3. Chemistry of Salt-Affected Soils and Waters 42
Chapter 4. Chemistry of Trace Elements in Soils and Ground Water 64
Chapter 5. Soil Responses to Saline and Sodic Conditions 91

III. Effects of Salts on Plants 113

Chapter 6. Plant Responses to Saline and Sodic Conditions 113
Chapter 7. Deficiencies and Toxicities of Trace Elements 138
Chapter 8. Genetic Approaches for Developing Economic Salt-Tolerant Crops 161

IV. Sampling, Monitoring, and Measurement 186

Chapter 9. Field Sampling and Monitoring of Soil, Water, and Plants 186
Chapter 10. Field and Laboratory Measurements 201

V. Diagnosis of Salt Problems 220

Chapter 11. Irrigation Water Quality Assessments 220
Chapter 12. Leaching Fraction and Control of Root Zone Salinity 237
Chapter 13. Crop Salt Tolerance 262
Chapter 14. Crop Production Functions 305
Chapter 15. Spatial and Temporal Variabilities in Salinity 327

VI. Salinity Management Options 349

Chapter 16. On-Farm Irrigation and Drainage Practices 349
Chapter 17. Management of Dryland Saline Seeps 372
Chapter 18. Project-Level Salinity Management Options 393

VII. Land Reclamation, Treatment, and Disposal Of Drainage Waters 410

Chapter 19. Reclamation of Saline-, Sodic-, and Boron-Affected Soils 410
Chapter 20. Use of Saline Drainage Water for Irrigation 432
Chapter 21. Drainage Water Treatment and Disposal Options 450

VIII. The Use of Models in Salinity Assessment 469

Chapter 22. Steady-State Root Zone Salt Balance 469
Chapter 23. Concepts of Modeling Transient Root Zone Salinity 482
Chapter 24. Conceptual Irrigation Project Hydrosalinity Model 504
Chapter 25. Dynamic Optimization Models for Salinity and Drainage Management 530

IX. Environmental, Legal, and Economic Issues 549

Chapter 26. Institutional and Legal Constraints 549
Chapter 27. Economic Incentives and Environmental Quality 561

X. The Future of Irrigated Agriculture 584

Chapter 28. Irrigated Agriculture: Is it Sustainable? 584

Appendices

A.1 Examples of Preferred (P) and Acceptable (A) Units for General Use 595
A.2 Conversion Factors for SI and non-SI Units 599
A.3 Conversion Factors for Chemical Concentrations 605
A.4 Other Useful Conversion Factors and Formulas 607

Index 608

Foreword

Salinity has plagued irrigated agriculture throughout history. Considerable progress has been made in managing and controlling salinity in irrigated lands on the basis that saline drainage waters would be disposed of somewhere. However, coping with salinity and drainage problems has become much more complex because of increasing environmental constraints on the discharge of drainage waters that contain salts and, in some instances, fertilizer and pesticide residues and naturally-occurring toxic trace elements.

The goal of this ASCE Salinity Manual is to bring together contemporary concepts and management practices addressing agricultural water and soil salinity problems that will be useful to engineers, scientists, practitioners and educators.

The manual consists of 28 chapters, written by 49 contributing authors. The manual is organized into ten sections: Introduction; Effects of Salts on Soils; Effects of Salts on Plants; Sampling, Monitoring, and Measurement; Diagnosis of Salt Problems; Salinity Management Options; Land Reclamation, Treatment and Disposal Of Drainage Waters; The Use of Models in Salinity Assessment; Environmental, Legal and Economic Issues; and The Future of Irrigated Agriculture.

This manual should be a valuable addition to earlier references regarding mitigation of salinity problems in irrigated agriculture, such as USDA Handbook No. 60, "Diagnosis and Improvement of Saline and Alkali Soils," published in 1954, and FAO Irrigation and Drainage Paper 29, "Water Quality for Agriculture," by R. Ayers and D. Westcot (1976, revised in 1985).

PART ONE—INTRODUCTION

CHAPTER 1

NATURE AND EXTENT OF AGRICULTURAL SALINITY[a]

INTRODUCTION

Irrigated agriculture has faced the challenge of sustaining its productivity for centuries. Because of natural hydrological and geochemical factors, as well as irrigation-induced activities, soil and water salinity and associated drainage problems continue to plague agriculture.

The problems have extended far beyond the farmlands, where saline soils and waters impair crop production. Practices based on the presumption that saline drainage waters will somehow, somewhere, be discharged are now being challenged. New and extended regulations on the discharge of nonpoint source pollutants in agricultural drainage waters are expected in the United States.

This chapter presents an overview of the nature of salinity in soils and waters, its extent from global to regional scales, the reactivity of salts and salt flows, and the concerns of agriculture and other sectors of society.

Nature of Agricultural Salt Problem

Salinity Constituents. Salinity is the concentration of dissolved mineral salts present in waters and soils on a unit volume or weight basis. The major solutes comprising dissolved mineral salts are the cations Na, Ca, Mg, and K and the anions Cl, SO_4, HCO_3, CO_3, and NO_3. Other constituents contributing toward salinity in hypersaline waters include B, Sr, Li, SiO_2, Rb, F, Mo, Mn, Ba, and Al.

Salinity Parameters. Salinity is expressed in a number of ways, depending on the method and purpose of the measurements. The salinity constituents listed above are frequently reported in terms of $mol_{(c)}/l$ (equivalents per liter) or mg/l (ppm) for major solutes and μg/l (ppb) for trace elements.

Salinity is often expressed as a lumped parameter, e.g., electrical conductivity (EC), an intensive parameter, total dissolved solids (TDS), an extensive gravimetric measure in mg/l, or total concentration of soluble cations (TSC) and anions (TSA) in $mol_{(c)}/l$. The EC of saline soils and waters are reported in decisiemens per meter (dS/m, which is equiv-

---

[a]Prepared by: Kenneth K. Tanji, Dept. of Land, Air, and Water Resour., Veihmeyer Hall, Univ. of Calif., Davis, CA 95616.

alent to millimhos per cm) and the EC of lesser saline soils and waters in dS/m $\times$ 10^{-3} (or microsiemens/cm, which is equivalent to micromhos per cm).

No exact relationship exists between these measures of lumped salinity parameters, but TDS may be approximated by multiplying EC(dS/m) by a factor of 640 for lesser saline samples and a factor of 800 for hypersaline samples. To obtain TSC or TSA, multiply EC(dS/m) by a factor of 0.1 for $mol_{(c)}/l$ and a factor of 10.0 for $mmol_{(c)}/l$.

Salinity Measurements. The measurement of salinity in waters for EC, TDS, TSC, and TSA is straightforward. In contrast, soil-water contents significantly affect the measurement of salinity as a lumped parameter or dissolved mineral contents in soils. Soil salinity is typically measured (1) In a saturation soil extract; (2) in soil solutions extracted by vacuum in the field; or (3) by electroconductimetric methods. The concentration of salts in the soil solution does not typically change in proportion to change in soil-water contents because the major solute species participate in sink/source mechanisms, such as mineral precipitation and dissolution, cation exchange, and ion association (Tanji et al., 1967). Moreover, salinity is a dynamic property in soil-water systems, as it is highly mobile and related to water content.

Salinity Criterion. The criterion used for salinity hazard depends on the use of the water or soil. For instance, an irrigation water of EC < 0.7 dS/m poses little or no threat to most crops, while EC > 3.0 dS/m may restrict the growth of most crops (Ayers and Westcot 1985). A soil with an EC > 4 dS/m in the soil saturation extract is traditionally characterized as a saline soil (Richards 1954). Soil and water salinity decreases the availability of soil water and leads to reduced germination, growth, and yields.

Furthermore, certain constituents in waters or soils, such as B, may be toxic to plant growth. Others, such as Na, detrimentally affect the soil's physical properties, such as its infiltration rate. Therefore, characterizing a water or soil as saline is relative and may vary widely, since responses by plants and soils to salinity are highly variable.

Sources of Salts. The problem of salinity manifests itself in the environment in a number of ways: saline irrigation and drainage waters, saline and sodic soils, saline ground waters, seawater intrusion, brines from natural salt deposits or geologic formations, and brines from oil and gas fields and mining.

The primary source of salts in waters and soils is the chemical weathering of earth materials, i.e., minerals that are constituents of rocks and soils. Evaporative salinization, e.g., surface evaporation of water and transpiration by plants, and dilution, e.g., rainfall, snowmelt waters, and irrigation waters, affect the level of concentration of dissolved mineral salts. Mineral solubility principally regulates the extent to which salts accumulate or dissolve.

Natural secondary sources of salts include atmospheric deposits of oceanic salts along coastal areas, seawater intrusion into estuaries due to tidal events, seawater intrusion into ground-water basins in coastal

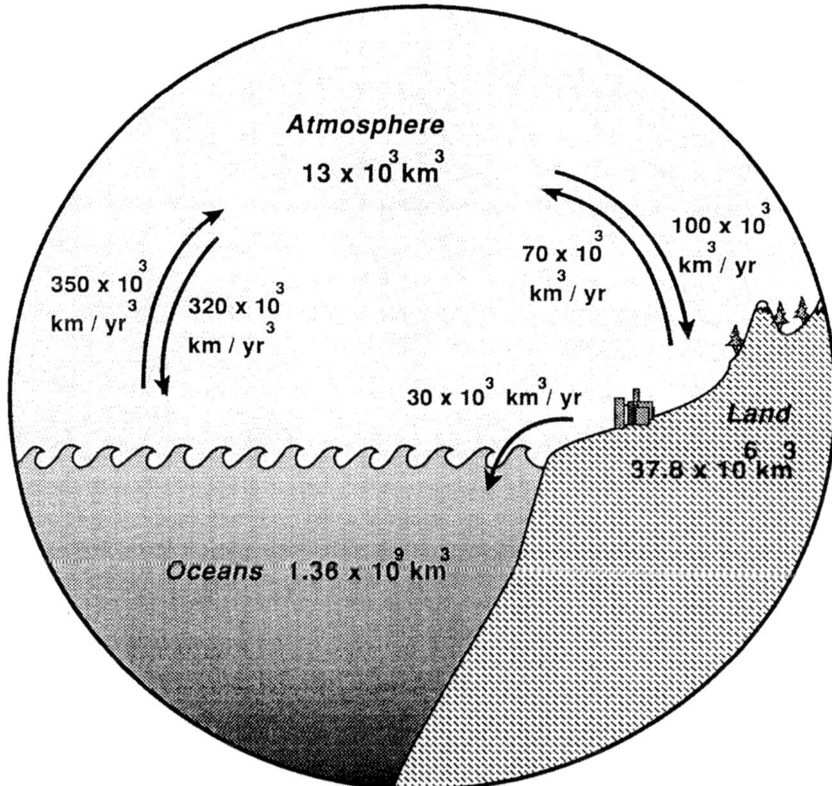

Fig. 1.1 Global Water Balance and Fluxes (Todd 1970)

areas due to overdraft, saline water from rising ground waters, inland saline lakes and playas, and leaching of saline land forms.

Anthropogenic sources of salts include irrigation and drainage waters, soil and water amendments, animals manures and wastes, chemical fertilizers, sewage sludges and effluents, and oil and gas field brines.

Extent of Agricultural Salt Problem

Global Scale. Most of the water in the hydrosphere is salty, and much of the fresh water is frozen. Fig. 1.1 shows that the oceans contain about 97.3% of the water, continents about 2.8%, and the atmosphere about 0.001% (Todd 1970). About 77.2% of the water associated with land is found in ice caps and glaciers and about 22% in ground waters, much of which are economically irretrievable. This leaves only a small percentage of readily manageable fresh water as a resource for water supply.

The world's land surface occupies about 13.2 $\times$ 10^9 ha, no more than 7 $\times$ 10^9 ha of which are arable and only 1.5 $\times$ 10^9 ha of which are cultivated (Massoud 1981). Of the cultivated lands, about 0.34 $\times$ 10^9 ha

Fig. 1.2 Global Distribution of Salt-Affected Soils (Szabolcs 1989)

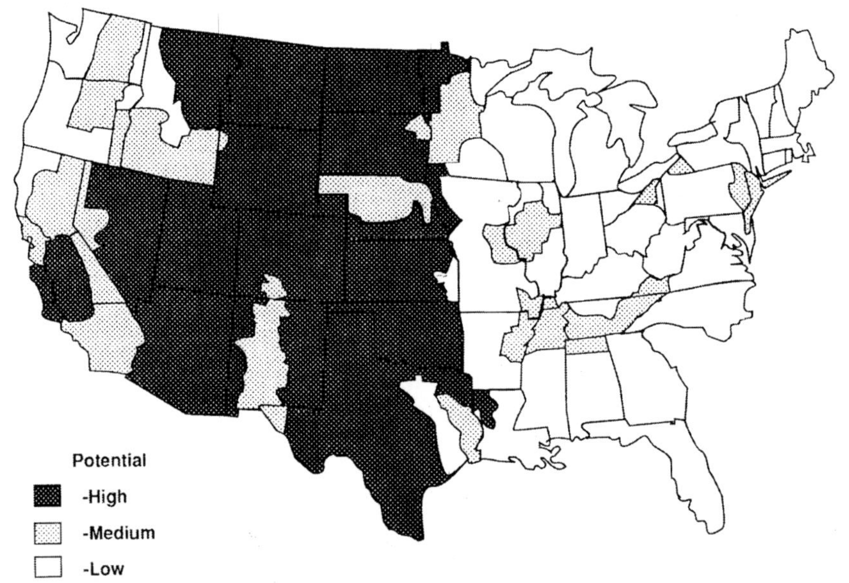

Fig. 1.3 Potential Water and Soil Salinity Problems in U.S. (USDA 1988)

(23%) are saline and another 0.56 × 10^9 ha (37%) are sodic. Fig. 1.2 (Szabolcs 1989) shows that saline and sodic soils cover about 10% of total arable lands and exist in over 100 countries. Salt-affected soils are not limited to semiarid to arid regions. In several other regions, the climate and mobility of salts produce saline waters and soil seasonally.

United States. Fig. 1.3 (U. S. Department of Agriculture 1988) shows areas with the potential for problems with water and soil salinity in the conterminous United States. The map is based on spatial scales of river

TABLE 1.1 Extent of Salinity and Associated Problems by Land Use in California, in Millions of Hectares (Backlund and Hoppes, 1984)

Primary land use (1)	Nonfederal land area (2)	Salinity/ sodicity soils (3)	High water table (4)	Water quality (5)
Irrigated cropland	4.1	1.2	1.1	1.4
Dry cropland	0.7	0.00	0.04	0.04
Grazed land	7.9	0.32	0.16	0.16
Timberland	3.6	0.00	0.00	0.04
Wildlife land	0.5	0.08	0.04	0.08
Urban	2.0	0.04	0.04	0.20
Other	3.6	0.08	0.04	0.12
Total	22.4	1.72	1.42	2.04

basins or watersheds and salt indicators, which include TDS and Cl in waters and salts in surface soils and geologic formations. About 30% of the land in the conterminous United States has a moderate to severe potential for such problems. The western United States, where geologic formations and soils derived from them are salt-affected, has a high potential for problems with salinity.

California. Table 1.1 (Backlund and Hoppes 1984) shows the extent of salinity and associated problems by land use in California. About 1,720,000 ha (29%) of nonfederal land are saline (EC > 4 dS/m) or sodic (exchangeable sodium percentage > 15%), 1,100,000 ha (27%) have a water table at a depth of 1.5 m or less, and 1.4 million ha (34%) have a problem with water quality. Table 1.2 indicates that much of the area affected by salinity, sodicity, a high water table, and problems with water quality is in the San Joaquin Valley.

Reactivity of Salts and Salt Flows

World Water Chemistry. Fig. 1.4 (Gibbs 1970) plots TDS of world waters against the cation ratio (Na to Na + Ca) in mg/l. When the scatter of data points are enclosed, a boomerang-shaped figure emerges. In its center are waters with about 50 to 500 mg/l TDS and cation ratios from near zero to about 0.6. Gibbs indicates that rock dominance or the watershed's geochemical nature primarily influence these waters. As TDS increases, the cation ratio increases, forming the upper leg of the boomerang with the major oceans at its apex. Gibbs points out that waters become saline as rates of evaporative salinization exceed precipitation (rainfall and snowfall) and the cation ratio increases from the selective precipitation of Ca over Na. As TDS decreases, the cation ratio also increases to form the lower leg. Gibbs attributes this trend to the dilution of the waters by the dominance of rainfall or high runoff over evapoconcentration and the presence of sea salts, mainly NaCl, in high-rainfall coastal areas, which result in high cation ratios. A similar diagram emerges when TDS is plotted against the anion ratio (Cl to Cl + HCO_3).

TABLE 1.2 Salinity and Drainage Problems by Major Irrigated Areas in California, in Millions of Hectares (Backlund and Hoppes, 1984)

Location (1)	Irrigated area (2)	Salinity/ sodicity (3)	High water table (4)	Water quality (5)
San Joaquin Valley	2.3	0.89	0.61	0.93
Sacramento Valley	0.85	0.08	0.16	0.12
Imperial Valley	0.20	0.08	0.20	0.20
Other areas	0.77	0.12	0.12	0.12
Total	4.12	1.17	1.09	1.37

It illustrates the three major mechanisms that regulate the chemistry of the world's water: (1) Evapoconcentration; (2) selective mineral precipitation; and (3) rainfall of variable composition.

Evaporative Salinization. Fig. 1.5 (Drever 1982) presents a simplified interpretation of the Hardie-Eugster model for evaporative salinization of waters. The chemical divides lead to several types of hypersaline water. With a water containing the most dominant solute species and subject to evapoconcentration, the first mineral to precipitate in large quantities is calcite ($CaCO_3$). What happens with further evapoconcentration depends on whether the molarity, *m*, of Ca is greater than or less than carbonate alkalinity (Alk). If $2m_{Ca} >$ Alk, the water would be depleted of HCO_3. Further evapoconcentration results in the precipitation of gypsum ($CaSO_4 \cdot 2H_2O$). If $m_{Ca} > m_{SO_4}$, the result would be a Cl type of water (e.g., brines in Death Valley, Calif., or the Carson Sink, Nev.). If $m_{Ca} < m_{SO_4}$, the brine will also contain appreciable amounts of SO_4, e.g., world oceans, the Dead Sea, the Salton Sea in California, and the Great Salt Lake in Utah.

In contrast, if $2m_{Ca} <$ Alk, the water would be depleted of Ca and the next mineral to precipitate would be sepiolite ($MgSi_3O_6(OH)_2$). If the water contains $2m_{Mg} >$ Alk, further evapoconcentration results in a brine similar to world oceans. If $2m_{Mg} <$ Alk, the brine would not accumulate Mg, e.g., Owens Lake in California and Pyramid Lake in Nevada. There is some question of invoking sepiolite formation as the primary control for Mg. Other possible Mg-controlling precipitates include Mg-rich smectite, Mg-rich calcite, or dolomite ($CaMg(CO_3)_2$).

Cyclic Salinization and Dilution. Natural waters are also subjected to dilution by inputs of fresh water. Fig. 1.6 (Tanji 1972) shows classes of precipitates, with evaporites as the most soluble types of minerals and resistates and hydrolyzates as the least soluble. It gives examples of solid phases for each class and the solute species that form or dissolve from these minerals. The precipitation of minerals and the dissolution of minerals are essentially engaged in a tug-of-war that depends on the relative rates of evapoconcentration versus dilution. Hypersaline waters typically would have solute abundances in the order of Na $\geq$ Mg $>$ Ca

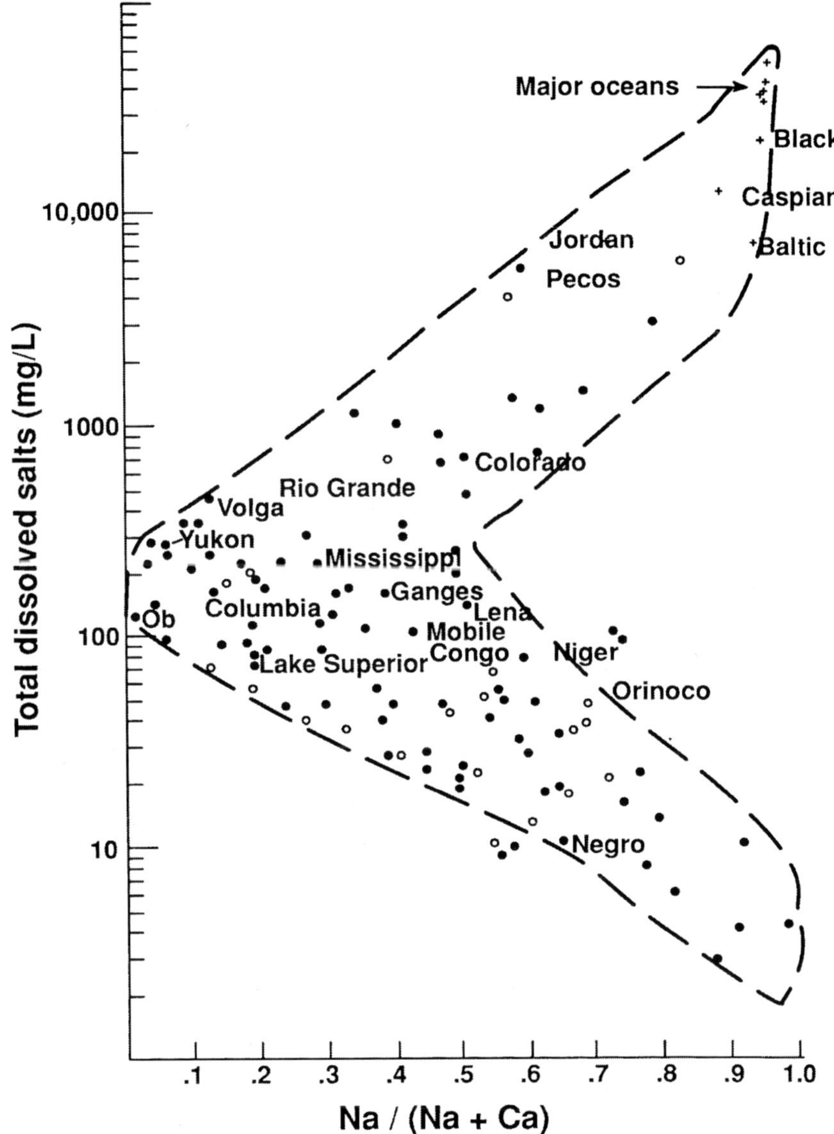

Fig. 1.4 Chemical Characteristics of World Waters (Gibbs 1970)

and $Cl \geq SO_4 > HCO_3$. Fresh waters would typically have nearly the reverse order of $Ca \geq Mg > Na$ and $HCO_3 > Cl \geq SO_4$.

Soil-Water System. Figs. 1.4, 1.5, and 1.6 indicate that mineral solubility principally controls water chemistry and salinity. In a soil-water system, where the ratio of water to soil is considerably smaller than the ratio of water to sediments in streams and lakes, other mechanisms may

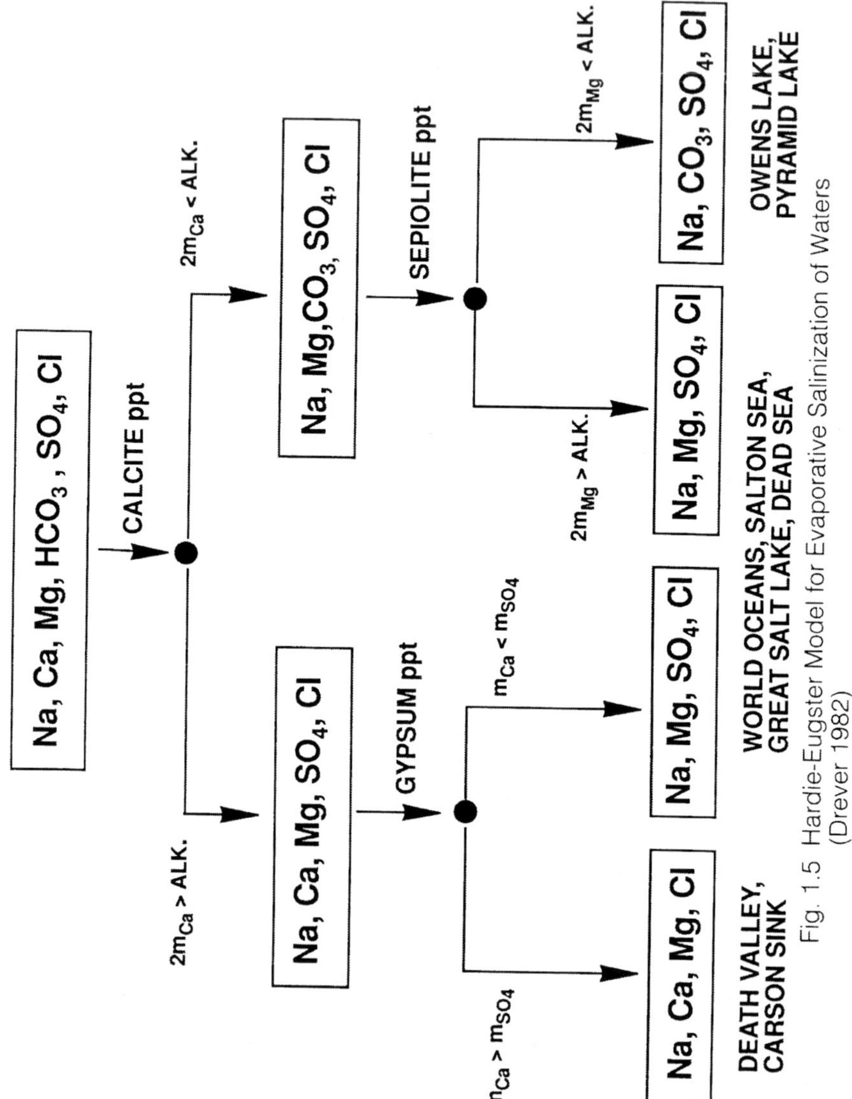

Fig. 1.5 Hardie-Eugster Model for Evaporative Salinization of Waters (Drever 1982)

deposition

First to precipitate ⟷ Last to Precipitate

Last to Dissolve ⟷ First to Dissolve

pickup

	Resistates	Hydrolyzates	Oxidates	Carbonates	Evaporltes
	(silica)	(aluminosilicates)	(oxides, sulfides)	(carbonates)	(chorides, sulfates)
SOLID	quartz, SiO_2	gibbsite, Al_2O_3 kaolinite, $Al_2Si_2O_5(OH)_4$ feldspar, $NaAlSi_3O_5$	hematite, Fe_2O_3 pyrite, FeS_2 manganite, Mn_2O_3	calcite,$CaCO_3$ dolomite, $CaMg(CO_3)_2$	halite, NaCl mirabilite, $Na_2SO_4 \cdot 10H_2O$ gypsum, $CaSO_4 \cdot 2H_2O$ bloedite, $Na_2Mg(SO_4)_2 \cdot 4H_2O$
SOLUTES	SiO_2 $Si(OH)_4$ $SiO(OH)_3^-$	Al^{3+} $Al(OH)^{2+}$ $Al(OH)_4^-$	Fe^{2+}, Mn^{2+} $Fe(OH)^{2+}$ HSO_4^- PO_4^{3-}, HPO_4^{2-}	Ca^{2+}, Mg^{2+} $CaHCO_3^+$ HCO_3^- $CaCO_3^0$	Na^+, Cl^-, Mg^{2+} Ca^{2+}, SO_4^{2-} $CaSO_4^0$, $MgSO_4^0$

Fig. 1.6 Mineral Dissolution-Mineral Precipitation in Waters (Tanji 1972)

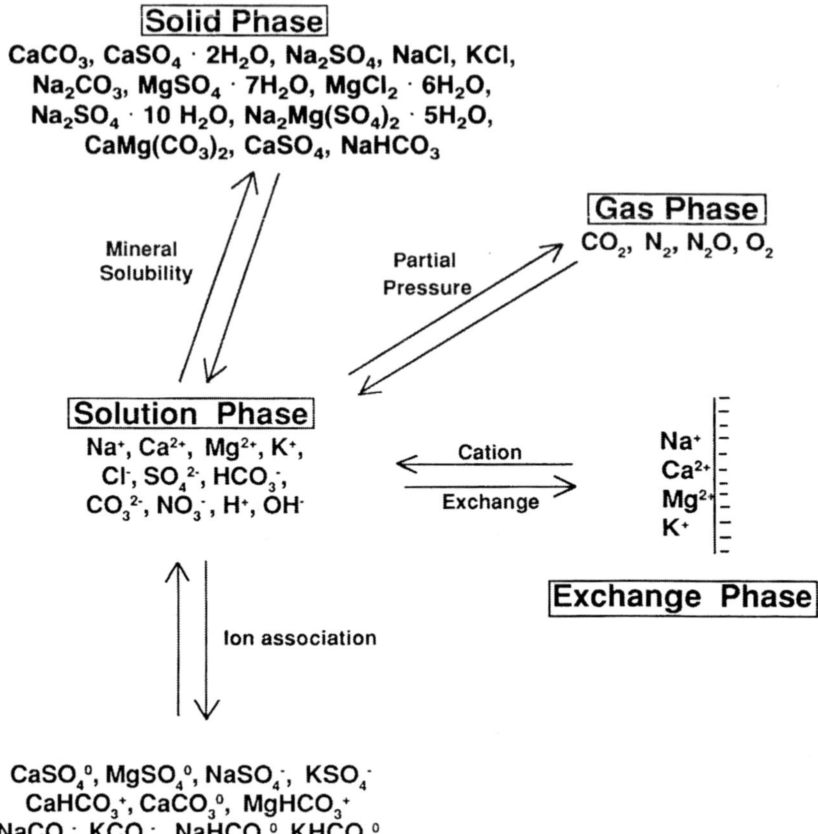

Fig. 1.7 Interactive Chemical Reactions in Soil-Water Systems (Tanji 1972)

also be significant. Fig. 1.7 (Tanji 1972) illustrates the complex chemical interactions that take place between the solution, solid, exchanger, and gas phases. A change in soil-water content by irrigation and rainfall or evapotranspiration by plants will cause the equilibrium to shift due to mineral precipitation or dissolution, association or dissociation of ion pairs, adsorption or desorption of cations, and emission or absorption of gases (Paul et al., 1966). Moreover, the free ions and ion pairs are subject to transport by diffusion and convection, e.g., water movement.

Irrigation Project. Fig. 1.8 (Hornsby 1973) schematically represents water and salt flows in an irrigation project with surface-water supplies. The top left-hand side depicts irrigation water diverted from a river to cropland, the center left depicts irrigated cropland, and the top and bottom left respectively depict the surface and subsurface irrigation return flows. The right-hand side contains water and salt transfers between the interface of irrigation projects and its environs. One way to interpret, synthesize, and simulate water and salt flows in irrigation

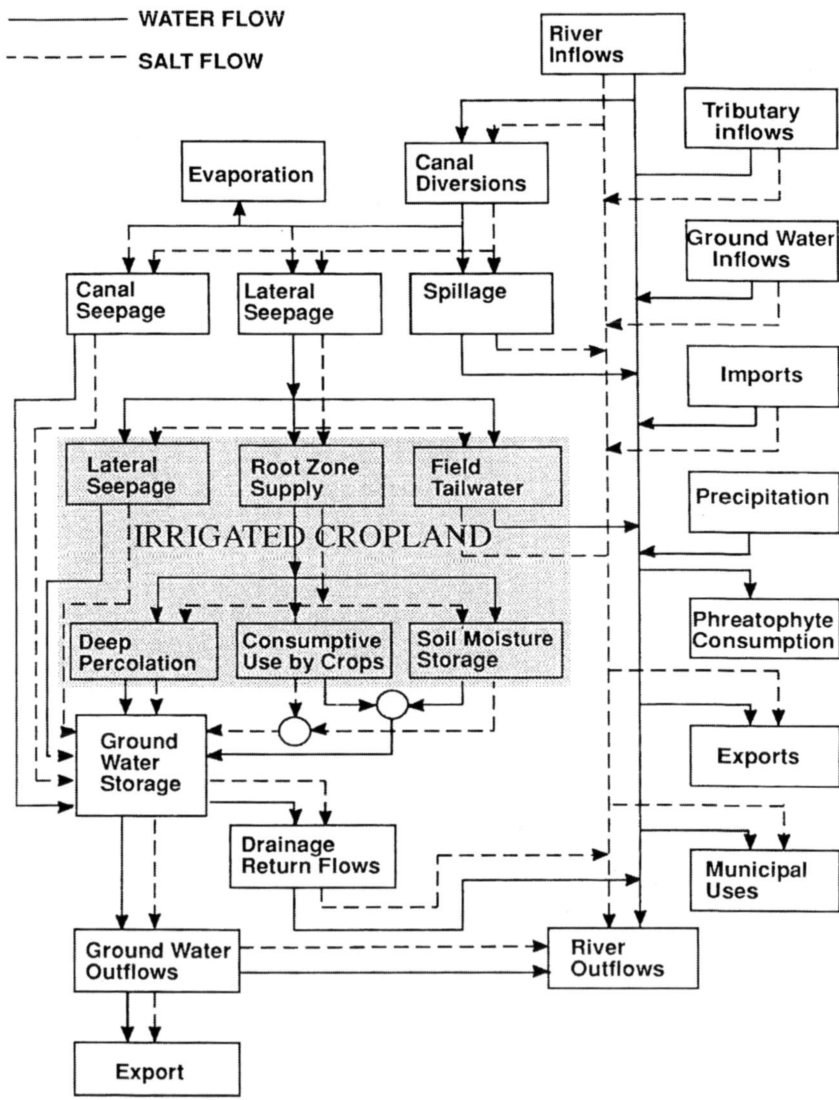

Fig. 1.8 Schematic Representation of Water and Salt Flows in an Irrigation Project with Surface Water Supply (Hornsby 1973)

projects and river basins is hydrosalinity modeling (Tanji 1981). In summary, the behavior of dissolved mineral salts in waters and soils is dynamic and subject to physical, chemical, and biological mechanisms on micro- and macro-levels.

Concerns Over Agricultural Salinity Problems

Historical Perspective. The accumulation of salt in soils and the frequently accompanying problem of drainage has plagued irrigated agri-

TABLE 1.3 Potential Management Options for Agricultural Drainage Problems in the San Joaquin Valley, California (San Joaquin Valley Drainage Program, 1988)

I.	Source Control for Subsurface Drainage, Salinity and Selenium	
	A.	Water Management
	B.	Drainage Management
	C.	Crop Management
	D.	Alternate Land Use
II.	Drainage Water Reuse	
	A.	Agriculture
	B.	Cooling Water for Power Plants
	C.	Solar Ponds for Production of Electricity
	D.	Salt Harvest from Evaporation Ponds
	E.	Aquaculture
	F.	Agroforestry
III.	Drainage Water Treatment for Removal of Selenium	
	A.	Biological Processes
	B.	Physical and Chemical Processes
IV.	Drainage Water Disposal	
	A.	River Discharge With and Without Dilution
	B.	Evaporate Drainage Waters in Ponds
	C.	Deep Well Injection
V.	Fish and Wildlife Measures	
	A.	Protection, Restoration and Enhancement
	B.	Freshwater Substitute for Drainage Water Supplies
VI.	Institutional Changes	
	A.	Water Pricing and Marketing
	B.	Modify or Eliminate Irrigation Subsidies
	C.	Modify Tax Structures and Rates
	D.	Impose Drainage Effluent Fees
	E.	Reallocate Water from Agriculture to Other Uses

culture for centuries. Such accumulation results when plants transpire waters, but leave most of the salts in the soil solution. Over time, salts may concentrate to such an extent that they hinder germination, seedling, and vegetative growth, and the yield and quality of crops.

Historical records for the past 6,000 years reveal that numerous societies based on irrigated agriculture have failed. One of the most highly publicized is ancient Mesopotamia, now Iraq. This once productive land appears to have suffered salt damage from about 2400 B.C. to 1700 B.C. This problem (Jacobsen and Adams 1958) stemmed from a dispute between two Sumerian cities, Umma and Girsu, over land and water rights (a problem still common today). Umma, located upstream from Girsu on the Euphrates River, blocked branch canals that supplied water to Girsu's agriculture. The latter responded by building other canals off the Euphrates to irrigate a large basin. Flooding, seepage, over-irrigation, and siltation resulted in a rising water table, which led to excessive soil salinity (Gelburd 1985). The crop records show that production of wheat was phased out over time and replaced by the more salt-tolerant barley, but the yields of barley gradually declined to 10 bushels per acre. After 1,000 to 1,500 years of successful irrigated agriculture, the Sumerian civilization declined. Numerous references to canal building in Sumeria from the third millennium B.C. are available, but no record of drainage canals being built to sustain agriculture (a common present-day occurrence) exists. Similar problems have developed in more recent times in the Indus Plain region, which includes parts of modern-day India and Pakistan, where irrigation began about 2,000 years ago by the Harappa civilization. Serious salinity and drainage problems did not develop until recently, within the last 150 years or so (Taylor 1965).

Less well known are the historical salt and drainage problems in North and South America. The inhabitants of Viru Valley on the coast of Peru developed an irrigation system between 400 B.C. and the time of Christ (Willey 1953). By 800 A.D., the population of the Viru Valley reached its peak and from 1200 A.D. dramatically decreased. Evidence shows that people relocated from the previously densely-settled valley bottoms to the upper narrows of the valley. Historians attribute this relocation to increasing soil salinity and rising water tables from lack of drainage (Armillas 1961). To say that the sole cause of relocation of the people of Viru Valley in the 13th century was due to soil salinity may not be entirely accurate, but to ignore this fact is also unacceptable to historians.

A second documented case is the Hohokam Indians, who lived in the Salt River region of what is now Arizona. They practiced a form of flood irrigation, similar to that practiced by the farmers in the Viru Valley, beginning about 300 B.C. This civilization flourished through 900 A.D. Historical records are sketchy until about 1275 A.D., when this region, along with much of the southwestern United States, suffered a drought (Willey 1953). After 1450 A.D., no evidence of the Hohokam civilization exists. Records indicate that waterlogging and salt accumulation in the valley floor caused crop failures. Historians surmise that these problems led the Hohokams to either relocate or starve.

Current Perspective. The recently discovered presence of selenium and other toxic elements in subsurface drainage waters from the west side of California's San Joaquin Valley, coupled with the age-old threat of salinity and high water tables, has heightened our awareness of the difficulties in sustaining irrigated agriculture (Tanji et al., 1986). Before the 1950s, only salinity and boron were of major concern in agricultural drainage. During the 1960s, more and more attention was paid to nitrates. In the 1970s and the early part of the 1980s, pesticides became the major problem. Now, selenium and other toxic elements are of major concern.

Eugene Hilgard first identified the growing problem of drainage and alkaline soils in the San Joaquin Valley around 1877. He concluded in 1889 that subsurface drainage was needed for waterlogged, saline, and sodic soils (Kelley 1951). W. W. Mackie in 1905 first investigated land reclamation with an experimental tile-drainage system at Kearney Park, near Fresno, in the San Joaquin Valley (Kelley 1951). Tile-drainage systems were installed beginning in the early 1950s in the west side of the San Joaquin Valley.

Since irrigation began about 1850, surface and subsurface drainage waters have been discharged into the San Joaquin River system. Federal and state agencies envisioned a 450-km master drain to serve the west side of the valley. However, by 1975, the U.S. Bureau of Reclamation had built only the upper 137-km reach of the San Luis Drain, terminating at Kesterson Reservoir, a flow-regulating reservoir in the master drain plan. Restrictions in the federal budget due to the Vietnam War delayed completion of the lower reaches of the drain. In 1979, the San Joaquin Valley Drainage Program recommended that the master drain be completed; however, in 1981, uncertainties about the potential environmental and health effects of the discharge of San Luis Drain into the Sacramento-San Joaquin Delta and the San Francisco Bay halted further construction.

Meanwhile, the San Luis Drain conveyed saline subsurface drainage waters from a 17,000-ha area affected by drainage to Kesterson Reservoir, which served as a terminal evaporation pond. From about 1981 to 1986, the average annual flow of tile effluents into the San Luis Drain was about 85,200 ha-cm, collected from about 2,000 ha of tile-drained fields and 1,200 ha drained by the open-joint collector system.

In 1982, selenium toxicity of fish and, in 1983, deformed and dead waterfowl were found in the Kesterson Reservoir. The selenium, averaging 300 μg/l (ppb) in the drain water, came from the Moreno shale, a geologic formation in the Coast Range Mountains that contributed the parent material for some of the soils formed in the west side of the San Joaquin Valley. The discovery of selenium toxicosis of waterfowl at Kesterson Reservoir and the presence of selenium in irrigation return flows in the western United States emphasized the need for agriculture to be concerned about edge-of-field effects on the environment. Discharge of the effluents into the San Luis Drain was halted in June 1986.

Massive state and federal investigations have been underway since 1985 on how to contain and dissipate selenium in the Kesterson Reservoir and develop management options to mitigate drainage problems in the San Joaquin Valley (Letey et al., 1987). An irrigation-drainage study in seven Western states where selenium toxicity is suspected has been underway since 1986 (Sylvester et al., 1988). The San Joaquin Valley Drainage Program drafted planning alternatives to solve agricultural drainage and drainage-related problems in the San Joaquin Valley. This program's goals are to minimize potential health risks, protect reasonable and beneficial uses of waters, sustain productivity of farmlands, and protect and restore fish and wildlife.

Table 1.3 outlines alternative management options under investigation (San Joaquin Valley Drainage Program 1988). A consensus exists among all interested parties that source control for subsurface drainage, salinity, and selenium is the most important option. Some potential exists for the reuse of drainage water. Various drainage water treatment processes to remove selenium are available, but they are too expensive for agriculture to bear the entire cost. Limited potential exists for drainage-water disposal in the San Joaquin River, but not all drainage districts have a drainage outlet to the river. Evaporation ponds are currently being used, but some appear to be producing symptoms of toxicity in waterfowl. Nonagricultural sectors of society strongly support the use of fresh water for fish and wildlife. Some of the proposed institutional changes and other technical options will be needed to solve the drainage problem. Drainage-water disposal through a master drain into the San Francisco Bay Delta system or into the Pacific Ocean is not being considered by the San Joaquin Valley Drainage Program because it is perceived as politically infeasible.

Contemporary salt problems exist elsewhere in the United States, e.g., the upper Colorado River Basin (see Chapter 18) and the northern Great Plains (see Chapter 17).

Dilemma over Agricultural Salinity and Drainage Problems

As mentioned earlier, the history of agriculture has shown that irrigated agriculture cannot survive in perpetuity without adequate salt balance and drainage. The length of time that irrigated agriculture can survive without adequate drainage depends on hydrogeology and water management. The accumulation of salt in soils depends on the salinity of the applied waters, the salinity of the native soil, and the rate at which salts are leached out of the root zone. Waterlogging in the root zone depends on the presence of restricting soil and substrata layers and the vadose region's capacity for deep percolation. If restricting layers exist close to the surface, waterlogging occurs and the accumulation of salt associated with waterlogged soils develops within a comparatively short time, such as decades. If no restricting layers exist and the vadose region has a large capacity, irrigation may be practiced for a long time, such as centuries, before problems with surface drainage arise.

The technology to sustain irrigated agriculture exists, but its use is limited by a number of factors, such as lack of economic incentives for irrigators; lack of education on best management practices; the high cost of improving structures, pressurized irrigation systems, and drainage systems; institutional constraints, such as water rights and water transfers; and the effects of irrigation return flows on the environment.

Since irrigated agriculture operates in an open system and even the best source control measures still result in drainage water and salt, fertilizer, and pesticide residues, ultimate salt "sinks," such as the oceans or inland closed basins, are needed.

In semiarid to arid climates, irrigated agriculture requires leaching and drainage. In the past, it was assumed that residuals from agriculture, especially saline drainage waters, would be disposed of somewhere, somehow. Now other water users have demanded that agriculture minimize the degradation of water quality. Federal, state, and local regulations on nonpoint sources of pollutants are being promulgated.

Irrigated agriculture needs to strive for a balance between maintaining agricultural productivity and protecting our natural resources. To this end, this manual on "Agricultural Salinity Assessment and Management" serves as a reference to help sustain irrigated agriculture.

REFERENCES

Armillas, P. 1961. "Land use in pre-Columbian America," pp. 255–276, *A History of Land Use in Arid Regions,* L. D. Stamp (ed.) UNESCO Arid Zone Research, 17, Paris, France.

Ayers, R.S., and Westcot, D.W. 1985. "Water Quality for Agriculture." FAO Irrigation and Drainage Paper 29 (Rev. 1), Rome, Italy, 174 p.

Backlund, V.L., and Hoppes, R.R. 1984. "Status of soil salinity in California." *Calif. Agric.* 38(10):8–9.

Drever, J.I. 1982. *The Geochemistry of Natural Waters,* 2nd ed., Prentice Hall, New Jersey, 437 p.

Gelburd, D.E. 1985. "Managing salinity lessons from the past." *Journal of Soil and Water Conservation* 40(4):329–331.

Gibbs, R. J. 1970. "Mechanisms controlling world water chemistry." *Science* 170:1088–1090.

Hornsby, A.G. 1973. "Prediction modeling for salinity control in irrigation return flows," EPA-R2-73-168, U.S. Envir. Protection Agency, 55 p.

Jacobsen, T., and Adams, R.A. 1958. "Salt and silt in ancient Mesopotamian agriculture." *Science* 128:1251–1258.

Kelley, W.P. 1951. *Alkali Soils: Their Formation, Properties and Reclamation.* Reinhold Pub. Corp., N.Y., 176 p.

Letey, J., Roberts, C., Penberth, M., and Vasek, C. 1987. An agricultural dilemma: Drainage water and toxic disposal in the San Joaquin Valley. Special Publication 3319, Div. of Agric. and Natural Resour., Univ. of Calif., 56 p.

Massoud, F.I. 1981. "Salt affected soils at a global scale and concepts for control." FAO Land and Water Development Div., Technical Paper, Rome, Italy, 21 p.

Paul, J.L., Tanji, K.K., and Anderson, W.D. 1966. "Estimating soil and saturation extract composition by a computer method." *Soil Sci. Society of America Proc.* 30:15–17.

Richards, L.A., ed. 1954. *Diagnosis and Improvement of Saline and Alkali Soils.* USDA Handbook No. 60, Wash. D.C., 160 p.

San Joaquin Valley Drainage Program. 1988. "Preliminary planning alternatives for solving agricultural drainage and drainage related problems in the San Joaquin Valley," Draft Report, Sacramento, California, pp. 1–1 to 5–16.

Sylvester, M.A., Deason, J.P., Feltz, H.R., and Engberg, R.A. 1988. "Preliminary results of the Department of Interior's Irrigation Drainage Studies," pp. 665–677, *Planning Now for Irrigation and Drainage in the 21st Century,* Proc. of Conference, D. L. Hay, ed., ASCE, Irrig. and Drainage Div.

Szabolcs, I. 1989. *Salt-Affected Soils.* CRC Press, Inc. Boca Raton, Fla., 274 p.

Tanji, K.K., Dutt, G.R., Paul, J.L., and Doneen, L.D. 1967. "Quality of percolating waters. II. A Computer method for predicting salt concentrations at variable moisture contents. *Hilgardia* 38(9):307–318.

Tanji, K.K. 1972. Lecture notes from the course, *Chemistry of the Hydrosphere,* University of Calif., Davis, Calif.

Tanji, K.K. 1981. "River basin hydrosalinity modeling." *Agricultural Water Management* 4:207–225.

Tanji, K., Läuchli, A. and Meyer, J. 1986. "Selenium in the San Joaquin Valley." *Environment* 28(6):6–11, 34–39.

Taylor, A.C. 1965. "Water, history and the Indus Plain." *Natural History* 24(5):40–49.

Todd, D.E. 1970. *The Water Encyclopedia.* Water Information Center, Port Washington, N.Y. 559 p.

USDA. 1988. "Water quality eduction and technical assistance plan." USDA-Soil Conservation Service and USDA-Extension Service report, 20 p + app.

Willey, G.R. 1953. "Prehistoric settlement patterns in the Viru Valley, Peru." Smithsonian Institute, Bureau of American Ethnology, Bulletin 155, Wash. D.C., 453 p.

CHAPTER 2

OVERVIEW: DIAGNOSIS OF SALINITY PROBLEMS AND SELECTION OF CONTROL PRACTICES[a]

INTRODUCTION

This chapter summarizes the principal adverse effects of salts on soils and plants and describes a methodology for diagnosing the nature and cause of salt problems, assessing reclaimability, and selecting appropriate reclamation and management practices.

Other chapters in this manual discuss salt effects, salinity-measurement techniques, diagnostic criteria, and salinity-control practices in more detail. They are referenced as appropriate.

POTENTIAL ADVERSE EFFECTS OF SALTS ON SOILS AND PLANTS—BRIEF SUMMARY

The development of effective salinity control practices requires an understanding of cause and effect. Toward that end, this section describes briefly the principal effects of salts on soils and plants.

Effects of Salts on Soils

The suitability of a soil for cropping depends greatly on the degree to which it conducts water and air (permeability) and on physical properties that control the friability of the seedbed (tilth). Poor permeability and tilth often are major problems in irrigated lands. Saline soils generally have normal physical properties. However, in sodic soils, physicochemical reactions cause the slaking of aggregates and the swelling and dispersion of clay minerals, leading to reduced permeability and poor tilth.

To understand how the poor physical properties of sodic soils are developed, the interaction between the negatively-charged colloidal clay particles in the soil and the exchangeable cation envelopes that are associated with those solid particles must be understood. The interactions between these exchangeable cations and their counterparts in the external bulk solution are also important. The envelope of exchangeable cations are subject to two opposing processes: 1) They are attracted to the negatively-charged clay surface in proportion to their charge; and 2) they tend to diffuse from the surface of the clay, where their concentration is higher, and into the bulk of the solution, where their concentra-

---

[a]Prepared by: J.D. Rhoades, U.S. Salinity Lab., 4500 Glenwood Dr., Riverside, CA 92501

tion is lower. The two opposing processes result in an approximately exponential decrease in cation concentration with distance from the clay surface out to the bulk solution. Divalent cations, like calcium and magnesium, are attracted to the surface with a force twice as great as sodium and other monovalent cations. In soils containing high proportions of Ca and Mg relative to Na or high concentrations of total dissolved electrolytes (i.e., saline soils), the exchangeable cation envelope, therefore, is compressed toward the clay surface. Consequently, the repulsion forces between the like-charged envelopes decrease, and the particles can approach closely enough to permit their cohesion into aggregates. The packing of aggregates yields a matrix of greater pore size compared to that of individual particles. Hence, the permeability and tilth of aggregated soils are better than those of non-aggregated soils.

The repulsion between clay particles also allows more solution to be imbibed between them. This phenomenon is referred to as swelling. Because of the plate-like shape and parallel orientation of clays, such swelling reduces the size of inter-aggregate pore spaces in the soil, reducing permeability. Swelling is particularly important in soils that contain expandable phyllosilicate minerals, e.g., smectites, and have ESP values of greater than about 15. Dispersion, the release of individual clay platelets from aggregates, and slaking, the breakdown of aggregates into subaggregate assemblages, can occur at ESP values of lower than 15 if the electrolyte concentration is low. Dispersed platelets or slaked subaggregate assemblages can lodge in pore interstices, reducing permeability. Slaking also decreases porosity, which leads to crusting, reduced permeability, and poor tilth.

Water entering the soil must pass through the soil surface, so the combined properties of water and soil affect the water-entry rate. Thus, soil permeability and tilth problems must be evaluated in terms of both the salinity of the infiltrating water and the exchangeable sodium percentage (or its equivalent Sodium Adsorption Ratio (SAR) value) of the topsoil. Representative threshold values of SAR (~ESP) and the electrical conductivity of infiltrating water for maintenance of soil permeability and tilth are given in Chapter 11.

The addition of gypsum to the soil surface or to irrigation water can help to avoid or possibly even alleviate problems with reduced infiltration rate and seedling emergence. Gypsum increases the electrolyte concentration of the infiltrating water and supplies divalent calcium to replace sodium.

In summary, soil solutions with high salt concentrations and low SAR values are conducive to good soil physical properties. Conversely, low electrolyte concentrations and relatively high proportions of sodium salts, i.e., high SAR values, adversely affect permeability and tilth.

For more specific information on the effects of exchangeable sodium, electrolyte concentration, pH, and exchangeable Mg and K on the permeability, infiltration, and crusting of soils, see Chapters 3, 4, 5, and 11.

Effects of Salts on Plants

Excess salinity within the root zone reduces plant growth rate. The hypothesis that seems to fit observations best asserts that excess salt reduces plant growth, primarily because it increases the energy that the plant must expend to acquire water from the soil and make the biochemical adjustments necessary to survive. This energy is diverted from the processes that lead to growth and yield, including cell enlargement and the synthesis of metabolites and structural compounds.

Typically, growth is suppressed when a threshold value of salinity is exceeded. This threshold value depends on the crop, external environmental factors such as temperature, relative humidity, or wind speed, and the water-supplying potential of the root zone. Suppression of growth increases as salinity increases until the plant dies. The salt tolerance of many crops can be expressed as follows (Maas and Hoffman, 1977):

$$Y_r = 100 - b\,(\mathrm{EC}_e - a) \qquad (2.1)$$

where Y_r = the percentage of the yield of the crop grown under saline conditions relative to that obtained under nonsaline, but otherwise comparable conditions; a = the threshold level of soil salinity at which yield decreases begin; and b = the percentage yield loss per increase of salinity in excess of a, and the unit of soil salinity ($\mathrm{EC_e}$) is dS/m.

Eq. 2.1 assumes that crops respond primarily to the osmotic potential of the soil solution. Effects of specific ions or elements must be considered separately. They are generally secondary in importance.

Complete compilations of data on crop tolerances to salinity and specific ions and elements are given in Chapter 13. Such salt-tolerance data cannot provide accurate, quantitative values of crop yield losses from salinity for every situation. Actual response to salinity varies with many factors, including climate, soil conditions, agronomic practices, irrigation management, crop variety, stage of growth, and salt composition.

Most plants are relatively salt-tolerant during germination and more sensitive during seedling emergence and early stages of seedling growth. Hence, it is imperative to keep salinity in the seedbed low after germination.

Differences in salt tolerance often occur between different varieties of a given species. Rootstocks affect the salt tolerances of tree and vine crops because they regulate the uptake and translocation of potentially toxic salts, such as sodium and chloride, to the shoots.

Salt tolerance depends on the method of irrigation and its frequency. As water becomes limiting, plants experience stresses from low matric potential, as well as low osmotic potential. The available salt tolerance data apply most directly to crops irrigated by furrow and flood irrigation with conventional irrigation management. Salt concentrations within irrigated soils profiles change constantly. They may differ several fold over the depth of the profile. The plant is most responsive to salinity in that part of the root zone where maximum water uptake occurs.

Sprinkler-irrigated crops are potentially subject to additional damage by foliar salt uptake and burn from spray contact of the foliage. Susceptibility to foliar salt injury depends on leaf characteristics and rate of absorption. It does not correlate with general salt tolerance. The degree of injury caused by spray effects depends on the weather and water stress. Relatively little information is available for predicting yield losses caused by foliar salt uptake and sprinkler spray effects (see Chapter 13).

Climate is a major factor affecting salt tolerance. Most crops can tolerate greater salt stress if the weather is cool and humid than if it is hot and dry. Yield is reduced more by salinity when humidity is low. Ozone decreases the yield of crops more under nonsaline than saline conditions. Thus, the effects of ozone increase the apparent salt tolerance of oxidant-sensitive crops.

While the primary effect of soil salinity on herbaceous crops is to retard growth, as discussed above, certain constituents often present in saline waters are toxic to particular crops. Boron, when present in the soil solution at concentrations of only a few parts per million, is highly toxic to many crops. Boron toxicities can be described in terms of a threshold value and yield-decrement slope parameters, as for salinity. Sodium and chloride may accumulate in the tissues of woody crops over time to toxic levels that produce foliar burn. Tolerances of a wide variety of plants to toxic elements are given in Chapter 13. In sodic soils, calcium deficiencies may occur, as well as deficiencies of other nutrients, such as P, Zn, and Fe. These conditions can be moderated though the use of certain amendments, such as gypsum and sulfuric acid. For more information on the amelioration of such soils, see Chapter 15.

Crops grown on infertile soil may seem more salt-tolerant than those grown with adequate fertility, because fertility is the primary factor limiting growth. The addition of extra fertilizer will not alleviate growth inhibition by salinity.

For more information on how salinity affects the physiology and biochemistry of plants, see Chapters 6, 7, 8, and 13.

DIAGNOSING SALT PROBLEMS

This section provides a methodology for diagnosing salt-related problems. Some guidelines are given for diagnosing problems caused by salinity, permeability and crusting, toxicity, nutritional imbalances, and drainage. Because salinity is usually the dominant problem in field soils, it should be considered first. If it is found to be dominant, then there is no need to proceed with diagnoses of permeability/crusting or toxicity/nutrition problems, especially since the latter two conditions may be altered by the leaching or drainage undertaken to alleviate the excess salinity. If salinity is not a problem, then permeability/crusting should be considered next, followed by toxicity/nutrition. Following leaching, toxicity or permeability problems that may have been overridden in the presence of salinity can appear. Subsequent evaluations of these problems may be needed.

Visual observations of soils, crops, topography, elevation of water in surface drainages, and plants are rarely sufficient to diagnose a salinity problem conclusively. For example, salinity may reduce the yields of crops by as much as 25% without visible symptoms. Moreover, visual observations may lead to a false diagnosis. For example, the minerals calcite and gypsum are essentially harmless to plants and soil properties, but they can form a white crust on the surface of soils that may be confused with potentially harmful soluble salts. Still, visual information can be useful. It provides clues to the problem and suggests what measurements should be made and where to make them. *In situ* measurements of soil electrical conductivity and on-the-spot measurements of the salinities of irrigation water and drainage water will help to narrow the possibilities and pinpoint the areas to sample.

Final diagnosis must rely on appropriate chemical analyses of plant, soil, and water samples. Before actively correcting problems caused by salt, the cause of a salt problem should be determined and steps taken to eliminate it.

Soil Salinity Problems

Visual Indications. Saline fields often can be identified by the presence of spotty white patches of precipitated salt. Such precipitates usually occur on slightly elevated positions or in unvegetated areas, where water evaporates and leaves the salts behind. Visible salt crusts obviously indicate surface salt accumulation, but they do not provide reliable evidence of high salinity in the root zone.

Recent excavations and open ditch banks more reliably indicate highly saline soil in the root zone. If salinity is high, white salt deposits will appear on the newly-exposed surfaces of excavations within a few hours, as the water evaporates. During germination and establishment of seedlings, only the salinity of the topsoil affects the plant. Bare spots in a cropped field may indicate only that excessive salt occurs in the shallow topsoil or seedbed rather than in the deeper root zone. The vigor of the plants adjacent to these bare spots is usually a good indicator of the depth of excessive soil salinity. Crops growing on saline soils typically show uneven growth and may exhibit other symptoms induced by salinity, such as stunted growth or unusually small leaves of a deep blue-green color. When different crops are grown in a single salt-affected area, a knowledge of which crops show symptoms induced by salinity and which do not can be combined with the crop tolerance values (see Chapter 13) to estimate the magnitude of soil salinity within the root zone of the area. Some caution should be exercised in using crop distribution and growth to diagnose salinity, because water deficiencies, disease, soil infertility, or misapplication of herbicides all may produce similar visual symptoms.

Measurements. Soil salinity can be determined in the field by a number of techniques. Soil samples can be extracted and the EC of their paste extracts, EC_e, measured in the field using field-kit techniques. More conveniently, soil salinity may be determined from measurements of the electrical conductivity of the soil paste itself, EC_p, or measurements of the electrical conductivity of the bulk field soil, EC_a. The latter measurement is particularly suited for reconnaissance because soil samples do not have to be collected and information about soil salinity can be quickly derived on the spot.

To establish the horizontal and vertical distribution of salinity over a field, EC_a data can be collected from a spatially-distributed set of field sampling sites. This information is used to select the areas to be separately evaluated and to determine the number of soil samples necessary to determine the chemical composition of each area. The resulting EC_e data from each area of concern are used with appropriate salt-tolerance criteria to determine whether soil salinity is a problem.

The principles and practices of measuring EC_p and EC_a, and determining EC_e and chemical composition are described in Chapter 10. Sampling guidelines and statistical procedures are described in Chapter 9.

Diagnostic Criteria. To determine whether salinity is excessive for establishment of seedlings, the EC_e of the topsoil is compared with the tolerance of the crop to be grown (see Chapter 13). To determine whether salinity is excessive for the vegetative and reproductive growth of conventionally irrigated, established crops, the mean salinity of the root zone is compared with the corresponding crop salt tolerances. When evaluating salinity of fields that use high-frequency irrigation or drip irrigation, the water-uptake-weighted salinity, i.e., the salinity within each soil depth weighted in proportion to the estimated fraction of root-zone water extracted from it, is used instead of the depth-averaged mean.

The variation of salinity over the area of concern should be determined and included in the diagnosis. To do this, a salt index is computed and compared with the salt tolerance of the crop to be grown. If a salt problem in any part of the cropped area is unacceptable, then the salt index for the entire area is equivalent to the maximum salinity observed in the area. The standard deviation may be computed if at least 10 data are available. If salt problems are allowable in about 15% of the area, the sum of mean and standard deviation may be used as the salt index. If the standard deviation of salinity values in small, e.g., less than 15% in coefficient of variability, then the sample mean should be used as the salt index.

Salinity is diagnosed as a growth-limiting factor, i.e., as the first-order salt-related problem, if the tolerances are exceeded for seedling establishment or vegetative and reproductive growth. The source of this salinity should then be ascertained. If salinity is not limiting, then the possibility that either permeability and crusting or toxicity and nutrition problems exist should be evaluated next.

Potential Causes of Soil Salinity Problems. The potential causes or sources of excess salts include saline irrigation water, inadequate leaching, inadequate drainage, indigenous soil salts, and inundation by and evaporation of salt-laden waters.

If none of the above factors is the cause of excessive salinity, other possibilities should be considered, such as excessively applied fertilizer or manure, or wind-blown saline water spray. Specific methods are not given here for diagnosing such situations, since they are unusual and case-specific.

To evaluate the possibility that saline irrigation water is the cause of soil salinity problems, compare the upper root-zone (0–30-cm depth) salinity and average root-zone salinity are compared with the salinity levels that would likely result from the use of the irrigation water. The latter values are predicted by multiplying the electrical conductivity of the irrigation water, EC_i, by appropriate concentration-factors (F_c) for these depths (see Chapter 11). If rainfall is significant, a rainfall-weighted averaged value of EC_i is used.

The level of salinity (EC_e basis) that should result in the topsoil from irrigation with a water of EC_i is estimated as 0.6 EC_i. The water is too saline and judged to be the cause of seedling emergence problems if the resulting value of 0.6 EC_i (and the observed topsoil EC_e) exceeds the value that is tolerable for germination and seedling establishment. A spotty stand is typically observed under such circumstances.

The level of average root zone salinity that should result from irrigating with a water of EC_i, when leaching is at its upper practical limit, is estimated as 0.8 EC_i. The water is judged to be too saline if 0.8 EC_i and average observed root zone EC_e exceed the average salinity level tolerable by the crop being grown and the salinity distribution indicates a net downward water flux. In this case, either a crop of greater tolerance must be substituted, or a water of lower salinity should be used for irrigation.

This evaluation scheme applies to situations where the soil has been under irrigation for at least three to five years and salts have not accumulated in the root zone due to upward flow of ground water or inundation by salt water. Upward flow of ground water from the water table can be ascertained from the distribution of salinity within the root zone.

To evaluate the possibility that inadequate leaching is the cause of soil salinity problems, the average root zone salinity or the concentration of a specific solute, such as chloride, and its distribution throughout the root zone should be determined. These may be excessive for crop growth, even when low-salinity irrigation water is used. The salinity distribution should be used to determine the direction of net flow of water. That, in turn, can be used to diagnose the adequacy of leaching or the presence of shallow saline water tables.

Leaching is judged insufficient if the EC_e values increase with depth and the ratio of average root zone EC_e to EC_i exceeds (EC_e'/EC_i), where EC_e' is the maximum average root zone salinity without yield reduction for the crop in question (see Chapter 13). The adequacy of leaching can also be evaluated by comparing an estimate of the leaching fraction, L, or the theoretically achievable leaching fraction under present management, LA, with the leaching requirement, L_r. The L_r for a given crop can be obtained from Chapter 12. LA is estimated from the following equation:

$$LA = 1 - \frac{ET_c}{I \cdot T_i} \qquad (2.2)$$

where ET_c = the estimated crop evapotranspiration requirement; and $I \cdot T_i$ = the amount of water infiltrated over the irrigation period. If LA (or L) is less than L_r, leaching is judged inadequate. This problem can be alleviated by increasing L, either by increasing the infiltration rate (I) or the irrigation set period (T_i), or by changing to a crop of lower ET_c. If excessive surface ponding of water is observed following irrigation, the cause of inadequate leaching is an insufficient infiltration rate.

Inadequate drainage due to a high water table can cause an upward net annual flux of water and salt, which may cause excessive salt accumulation in the soil. To determine whether poor drainage is the cause of a soil salinity problem, the field site is visually inspected for abnormally wet soil or weedy areas dominated by water-loving plants. Plants growing in excessively wet areas often have pale, yellowish leaves, which indicate nitrogen or iron deficiencies. In exposed soil profiles, manganese concretions and red or yellow mottled spots associated with ion compounds also indicate poor soil aeration. Also, water that fills an excavated hole without appreciably draining for several days indicates a high water table.

The adequacy of drainage should be diagnosed from measurements of the magnitude and distribution of EC_e above the water table during the cropping season and observations of the water table depth over the same period of time. If these measurements can only be made at one time during the year, then they should be made at the end of or late in the irrigation season, when the net effect of irrigation practices and the water table on the root zone salinity is generally most evident. Problems caused by inadequate drainage often develop progressively through the season, so the late-season condition is reasonable for diagnosis.

Drainage is judged inadequate if: (1) The EC_e pattern with depth indicates that the net annual flux of water is upward into the root zone and the magnitude of the average root-zone EC_e value exceeds that tolerable by the crop in question; or (2) the root zone is inundated for more than a few days or the soil is waterlogged within a distance of approximately one root-zone depth from the bottom of the crop's roots.

An alternative method for evaluating the adequacy of drainage in the presence of a shallow water table is as follows: the leaching fraction being attained is estimated by using Equation 1, the corresponding value of F_c is obtained from the appropriate table in Chapter 11, and the maximum expected average root zone salinity, $EC_i \cdot F_c$ is computed. It may be assumed that drainage is inadequate if the measured value of average root zone EC_e exceeds the value of the quantity $EC_i \cdot F_c$ for the crop being grown. For more discussion of drainage, see Chapter 16.

Indigenous soil salts are more common in dry-land soils than in irrigated soils. The management and reclamation of soils containing indigenous salts differ from the management and reclamation of soils in which salts are derived from external sources.

To discern whether indigenous soil salts are the cause of soil salinity problems, the methods related in Chapter 10 can be used to determine the gypsum content, the EC_e, and the composition of the soluble salts in a soil. Indigenous salts are diagnosed as the likely source of excessive salinity in an irrigated soil when the determined levels of EC_e through the root zone exceed those predicted from the quantity $EC_i \cdot F_c$ and there is no evidence of a shallow water table. Highly soluble, indigenous salts, e.g., sodium and magnesium sulfates, require different abatement strategies from the relatively insoluble types, e.g., gypsum.

If a soil contains enough gypsum to increase the average root zone salinity to a level that is greater than the level that could result only from irrigating with a water in the absence of gypsum, i.e., $EC_i \cdot F_c$, and if this level exceeds the crops' tolerance, than a crop of higher salt tolerance should be selected, since it is generally impractical to leach out the gypsum. However, if the salts present in an excessively saline soil are of the highly soluble type, they should be removed by leaching. The total volume of water required to dissolve these soluble salts (the reclamation requirement) can be determined as described in Chapter 19.

Salinity problems in soils can result from the periodic inundation and subsequent evaporation of salt-laden waters. To determine whether this could be the cause of a soil salinity problem, the local topography and topsoil properties of the site is examined. Evidence of a playa-type topographic feature indicates that this process of salinization has occurred. To the extent that this process is still operative, it should be countered by an appropriate diversion measure (see Chapter 16).

Soil Permeability and Crusting Problems

Sodicity enhances the degradation of the tilth of a surface soil. This is usually manifested in decreased permeability and increased crusting. The loss of permeability may so restrict water infiltration into the root zone that plants become stressed from lack of water. Crusting may so impede or prevent seedling emergence as to reduce crop stand.

Visual Indications. The structural condition and color of the surface soil may be used to judge whether permeability or crusting problems caused by sodicity are likely to occur. Sodic soils are poorly aggregated, crusted, and hard when dry, and sticky and plastic when wet. They often are dark brown or black due to surface coatings of dispersed organic matter. Following rainfall or irrigation of sodic soils, surface water persists even though the underlying soil may be dry. Phenolphthalein indicator solution dropped into the ponded water typically turns pink because of the high alkalinity. The pH of such soils and water is usually greater than 8.5.

Soils subjected to sodicity conditions for very long periods of time and to successive periods of wetting and drying tend to develop distinctive morphological properties. Their clay content is relatively high in the illuvial horizons, and the structure of the upper B horizon is either columnar or prismatic.

The appearance of excessive ponding or runoff during irrigations or periods of rainfall are good visual indicators of permeability problems. The slaked and compacted (parallel orientation) appearance of the topsoil indicates crusting problems.

Measurements. Measurements of soil intake rate and soil strength are sometimes made to diagnose a permeability problem. The ultimate criteria for judging permeability are: 1) Whether or not the soil water reservoir is being adequately replenished by irrigation; and 2) whether excess salts are being leached from the root zone. Thus, to diagnose a permeability problem, the depth of wetting and degree of salt buildup in the root zone resulting from irrigation should be measured. Fortunately, these parameters are much easier to measure than are the intake rate and soil strength.

To determine the extent of soil water replenishment (or depth of wetting), changes in soil water content are measured by a suitable means. The depth of water penetration, a useful indirect measure of soil water replenishment, can be measured by inserting a pointed rod into the soil before and after irrigation. Other methods are described in Chapters 9 and 10.

To determine the degree of leaching and the direction of net water flux, the magnitude and distribution of EC_e within the root zone is measured. Portable insertion probes are especially useful for this. The principles and methods of such techniques are described in Chapter 10.

To diagnose a crusting problem, measurements of the strength of the surface soil could be made in the field, using some appropriate method. However, because the ultimate criterion for judging crust strength is whether or not seedlings emerge adequately, simple observations of emergence success provide a more appropriate method.

Diagnostic Criteria. Permeability is diagnosed as a problem if soil water cannot be adequately replenished by irrigation without excessive ponding or runoff. It is also judged to be a problem if leaching is inadequate to maintain a net downward flux of water through the root zone (as deduced from the distribution of salinity with soil depth). Additional evaluations may be necessary to determine whether the apparent permeability problems are not simply the result of poor irrigation water management.

Crusting is diagnosed as a problem if the strength of the surface soil is determined from visually observed to significantly hinder seedling emergence.

Potential Causes of Permeability Problems and Crusting. Inadequate water infiltration rates usually are due to low soil permeability or topsoil crusting. These, in turn, are caused by excessive exchangeable sodium, excessively low electrolyte concentrations in the infiltrating water, poor intrinsic soil properties, e.g., massive structure, high clay content, high bulk density, lack of sufficient macro-porosity, low organic matter; or compaction of soil resulting from excessive water application rates or improper tillage and traffic operations. The cause of a permeability problem is important to know because the cause must be abated to achieve a permanent solution.

The various causes of inadequate permeability include under-irrigation, excessive runoff, excessive sodicity, and inherently poor soil physical properties. Under-irrigation is diagnosed as the cause of the permeability and crusting problem if the amount of water applied is less than the crop consumptive use. Excessive runoff is diagnosed as the problem if the difference between the amount of water applied and the amount of runoff from the field during an irrigation is less than the crop water requirement. It may be caused by improper field grade or crusted soils. Measurements of the soil grade and observations of aggregation and crust conditions should be made to distinguish whether the excessive runoff is due to improper field grade, poor soil physical conditions, or improper irrigation system design and management. Such requirements are described in Chapters 10 and 16.

To evaluate whether excessive sodicity is the cause of inadequate permeability, surface soil SAR at 0-cm to 15-cm depth is measured, or the adjusted SAR from the EC, SAR, Ca, and HCO_3 concentrations in the irrigation water is calculated. If the SAR of the surface (0-cm to 15-cm depth) soil, or the adjusted SAR of the irrigation water exceeds 10, then the poor permeability probably is caused by excessive exchangeable sodium, especially if the area receives appreciable rainfall or if the electrolyte concentration of the irrigation water is $\sim<3$ meq/liter. Other combinations of soil SAR and irrigation water EC may also be harmful (Chapter 11).

When the EC_i is $\sim<3$ $mmol_c/l$, its harmful effect may be reduced by adding electrolyte to the irrigation water. If the adjusted SAR_i is high (>10), its suitability may be improved by adding a calcium amendment, such as gypsum, to the irrigation water. Such measures should be undertaken only if the addition of ~ 3 $mmol_c/l$ or less of calcium sufficiently improves the SAR-EC relation. If the bicarbonate concentration of the irrigation water, in meq/liter, appreciably exceeds the calcium concentration, the suitability of the water for irrigation may be improved by amending it with enough sulfuric acid to neutralize up to 90% (equivalent basis) of the bicarbonate.

If the EC, SAR, Ca, and HCO_3 of the irrigation water do not indicate high sodium, then the excessive exchangeable sodium might be caused by upward flow of water from a shallow water table. If a shallow ground water with relatively low salinity and sodic-chemistry is present, and if the SAR distribution in the soil decreases with depth, upward flow of water is indicated. If the excessive exchangeable sodium derives from a water table, drainage must be improved before soil reclamation is attempted.

If neither the irrigation water nor the ground water is determined to be the source of excessive exchangeable sodium, then the sodium is indigenous to the soil. Excessive exchangeable sodium must be replaced by reclamation. Methods for reclamation are described in Chapter 19.

If none of the above factors are determined to be the cause of low permeability, then excessive compaction or inherently poor soil physical properties are likely. Some soils may be so high in clay content and so poorly structured that their infiltration rate is too low for normal irrigation practices and cropping needs. Such soils can only be irrigated "down-the-cracks," and irrigations and cultivation must be timed at the proper soil water content. These problems are distinguished from sodic problems (though they may occur simultaneously) by the lack of excessive sodium and the presence of high clay content and massive soil structure. For further discussion of soil structure and other physical properties, see Chapter 5.

Plant Toxicity Problems and Nutritional Imbalance Problems

Visual Indications. Certain crops are more severely affected by sodium, chloride, and boron, or the lack of calcium, than are others (see Chapter 13). When one of the susceptible crops shows stunted growth or other salt-related effects and other kinds of crops nearby do not, a toxicity problem or nutritional imbalance problem, rather than a salinity problem, may exist. Excessive foliar uptake of chloride and sodium may occur with sprinkler irrigation. Therefore, if sprinkled plants show problems and nearby non-sprinkled plants do not, the toxicity is not likely soil-derived.

Plant appearance should be compared with typical toxicity and deficiency symptoms to help diagnose toxicity and calcium deficiency problems. These symptoms are described below and in Chapter 10.

Calcium deficiency generally impairs root growth more than top growth. Affected roots tend to be white and stunted in appearance. The young leaves of new plants are affected before the older leaves, and are often distorted and small with irregular margins and spotted or necrotic areas. There may be dieback at the terminal buds. Alfalfa is a good indicator crop because it is particularly sensitive to calcium deficiency.

Symptoms of excess chloride include burning and firing of leaf tips or margins, bronzing, premature yellowing, abscission of leaves and, less frequently, chlorosis. Smaller leaves and slower growth also are typical. Symptoms of excess sodium include necrotic areas on the tips, margins, or interveinal areas. Incipient injury is indicated by a mottled or chlorotic condition.

Boron toxicity occurs in citrus, avocado, persimmon, and many other species. These plants may develop tip burn or marginal burn of mature leaves accompanied by chlorosis of interveinal tissue. Boron injury to walnut leaves is characterized by marginal burn and brown necrotic areas between the veins. Stone-fruits, apples, and pears are sensitive to boron, but do not accumulate high boron concentrations in their leaves nor do they develop leaf symptoms. Cotton, grapes, potatoes, beans, peas, and several other plants show marginal burning and a cupping of the leaf that results from a restriction of the growth of the margin.

Measurements. Appropriate plant and soil analysis methods are described in Chapter 10. The likelihood of a calcium deficiency, Ca/Mg imbalance, or sodium-, chloride-, or boron-toxicity problem should be evaluated before chemical analyses are undertaken to diagnose other toxicity- or nutrition-related problems. The tolerance tables given in Chapter 13 should be used to evaluate crop susceptibility. Laboratory analyses of plant tissue samples or soil samples for the suspected toxicant should be made to confirm the field diagnosis. If the foliage is wetted by sprinkler irrigation and either sodium or chloride toxicity is suspected, then only plant-tissue analyses are necessary.

Diagnostic Criteria. Specific solute toxicity is diagnosed as a problem if an excessive level of boron, chloride, or sodium is found in the plant or soil samples.

Chloride and sodium are particularly toxic to fruit crops and woody ornamentals. When the leaves of these plants accumulate more than about 0.5% to 1.0% chloride or 0.25% to 0.5% sodium on a dry-weight basis, they develop characteristic leaf-injury symptoms.

Boron is toxic to many plants when it occurs in susceptible tissue at concentrations exceeding approximately 100 mg/kg to 300 mg/kg. Tolerances vary among varieties and rootstocks and are influenced by various growth factors. Specific concentrations in plant tissue associated with toxicity vary with sampling method, crop type, and leaf position. More specific diagnostic criteria for assessing boron, sodium, and chloride toxicities are given in more detail in Chapter 7 and in Reisenauer (1983). Levels of boron and chloride in soil saturation paste extracts associated with toxicity problems for various crops are given in Chapter 13.

Calcium deficiency is an important nutritional imbalance problem associated with some types of salt-affected soils. This problem occurs as a result of either excessive soil sodicity, high pH, or high soluble Mg/Ca ratio in the soil water. Calcium deficiency is diagnosed as a problem if the concentration ratio $Ca^{2+}/Mg^{2+} < 0.5$, or if the absolute concentration of Ca is less than 1 $mmol_c/l$ to 2 $mmol_c/l$ in the soil solution. Calcium deficiency can also be determined from plant tissue tests. Guidelines for various crops are described in Chapter 11.

Potential Causes of Plant Toxicity or Plant Nutritient Imbalances. Plant toxicities are caused by the same factors as salinity problems. Thus, the procedures for their diagnosis are the same as those for diagnosing salinity problems.

When Na or Cl plant toxicity occurs and concentrations of sodium, chloride, or both, in the irrigation water exceed about 5 $mmol_c/l$, foliar uptake is diagnosed as the likely cause.

Calcium deficiency can be due to an inadequate concentration of Ca or excessively low Ca/Mg ratio in the soil solution. If the absolute concentration is too low ($<\sim 2$ $mmol_c/l$), the cause is either that the soil is too alkaline (pH >8.5) or the irrigation water is too sodic (adjusted Ca <2 $mmol_c/l$). Inadequate Ca concentration is diagnosed as a problem if the adjusted Ca value is <2 $mmol_c/l$. Ca/Mg imbalance is diagnosed as a problem if the ratio $[(\text{adjusted Ca})/(Mg_i) \cdot (F_c)]$ is <0.5. Methods for calculating adjusted Ca and F_c are given in Chapters 3 and 11, respectively, and their use in water quality assessment is discussed in Chapter 11.

Deficiencies of other nutrients, such as P, Fe, and Zn, may be evaluated by analyzing soil samples and plant tissues. Procedures for doing so are described in Chapters 7 and 10. Under sodic soil conditions, permeability and crusting conditions are generally more limiting to plant growth than are nutrient deficiencies, and such deficiencies will usually be eliminated by removing the excess sodicity.

ASSESSING RECLAIMABILITY

Once the cause of a salinity problem has been determined, steps should be taken to abate the cause and eliminate the excess salts, toxicants, or sodicity by reclamation. Good drainage is the first requirement for the reclamation of any salt-affected soil. After drainage is provided, excessive salinity and toxic ions can be reduced to acceptable levels by leaching. Some soils, however, have such poor physical properties that leaching is impractical. In sodic soils, for example, amendments may be applied to increase soil permeability and reduce the exchangeable sodium content.

Before any reclamation is undertaken, it should be determined whether leaching, amendments, or special tillage operations are necessary.

Determine Leachability

The feasibility of leaching without amendments may be estimated in two ways. The first method involves plotting the sodium-adsorption-ratio of the saturation extract or the irrigation water against the EC_i, on a figure similar to Fig. 11.3 in Chapter 11. If the plotted point falls into the area of the figure that indicates permeability hazards are unlikely, then leaching without amendments is feasible.

The second method involves leaching packed soil columns with samples of the irrigation water and with 0.05 N $CaCl_2$ solution. Reclamation without amendments is unlikely to be successful if the permeability of the soil leached by irrigation water is very low relative to the case where the same type of soil is leached by $CaCl_2$. This evaluation may also be made in the field, using infiltrometers.

These two methods provide estimates of the reclaimability of a soil without amendments, but should not be relied upon for a final decision. The final evaluation of leachability should be based on the results of field trials.

To set up a field trial, salinity-sensing devices and tensiometers are installed at various depths in the field plot. The plot is then leached, using the same water and method of application as will be used for full-scale irrigation. The changes in salinity and hydraulic gradient are measured as a function of time and volume of water applied (or infiltration). Alternatively, measurements of the rate of water infiltration can be used to predict the soil's reclaimability, as described in Chapter 19. Reclamation by leaching without amendments or special tillage operations is likely to be successful if the rate of leaching or the permeability measured in the field trials is relatively high and sustainable.

If soil permeability is found to be inadequate for reclamation by simple leaching, then the viability of increasing it sufficiently with the use of amendments should be ascertained. Low soil permeability may result from the low EC of the leaching water, excessive ESP, or the inherently poor physical properties of the soil. To determine whether the electrolyte content of the leaching water can be increased by the addition of amendments enough to make leaching practical, the values of EC_i and SAR for the 0-cm to 15-cm soil depth are plotted on Fig. 11.3 of Chapter 11. If adding electrolyte to the leaching water or using another water of higher salinity, such as drainage water, moves the point from the hazard area into the non-hazard area of the figure, reclamation with such amendments or waters may be feasible.

The potential benefit of an amendment or alternative water supply can also be evaluated by leaching a packed soil column with the proposed leachant and determining whether the permeability increases appreciably. This test can also be made in the field, using an infiltrometer. A solution of 0.05 N $CaCl_2$ is recommended for this purpose.

Various soil amendments, e.g., gypsum, $CaCl_2$, H_2SO_4, and high-electrolyte water, should be evaluated to determine the most suitable one for increasing permeability. Laboratory column studies, field infiltrometer studies, and the Watsuit program (Chapter 22) may be used for this evaluation.

The water intake rate of sodic soils is often too low to reclaim the soil at an acceptable rate, especially when gypsum is used as an amendment. Using $CaCl_2$ or H_2SO_4 instead of gypsum usually eliminates this problem.

Another factor affecting the suitability of amendments is their acidity or acid-forming properties. If the soil contains at least 0.5% $CaCO_3$ equivalent, an acid-type amendment can be used. Under some circumstances, the dissolution of $CaCO_3$ can be enhanced by the presence of growing plants (which evolve CO_2), supplying the needed calcium and electrolyte concentration to accomplish reclamation. The final decision on amendment requirement and rates should be based on the results of field trials. For more discussion of amendments and reclamation procedures, see Chapter 19.

Determine Adequacy of Soil Structure and Need for Tillage

If the soil permeability is inadequate when tested using a 0.05 N $CaCl_2$ solution or an acidifying soil amendment, then poor physical properties of the soil are the likely cause of the low permeability. The soil profile may often be sufficiently altered by deep plowing, deep chiseling, shattering, or trenching operations, to achieve an adequate leaching rate. To help to determine whether these methods will succeed for a given soil profile, note the presence or absence of a topsoil crust or permeable materials deeper in the profile that could be brought up by tilling and mixed with permeability-limiting strata. Unfortunately, deep tillage can also bring up saline material into the root zone. This potential hazard must be carefully evaluated by measuring the salinity of the deeper material. If mixing seems like a possible solution, additional chemical and physical analyses of the deeper soils should be made. The final decision on the reclaimability of the soil by tillage and amendments should be based on the results of field trials.

Estimate Amount of Leaching and Time Required for Reclamation

Once it is determined that the soil can be reclaimed by leaching with or without amendments and tillage, the amount of leaching required to reclaim the soil and the amount of time needed for reclamation must be estimated. One method involves installing a succession by depth of salinity-sensing devices in the field and then measuring the change in EC_e by soil depth, time, and depth of infiltrated water. This directly, simply, and accurately determines the reclamation requirements. Such sensors are described in Chapter 10. Other methods for estimating the amount of leaching and time necessary for saline or sodic soil reclamation are given in Chapter 19.

Other Considerations

If the concentrations of chloride, boron, or magnesium in the soil are excessive, then the excessive amounts must be removed by leaching.

Assessing the reclaimability of such soils is similar to the assessment of reclaimability of soils containing excessive salinity.

If nutrient deficiency poses a problem, the methods for assessing reclaimability are similar to those described for assessing the reclaimability of sodic soils, with particular emphasis on pH reduction.

SELECTING SALINITY CONTROL AND MANAGEMENT MEASURES

Soil and water salinization inevitably occur with irrigation. Salinity control measures, therefore, must be implemented if irrigated agriculture is to be sustained over the long-term. Ideally, such measures should be compatible with processes in the natural geohydrologic system.

Management practices offer different levels of control. Some are aimed at controlling salinity within the root zone in a particular field; others control salinity over a much larger scale, such as an irrigation project. The practices should protect offsite environments, such as surface water and ground water. On-farm salinity-management practices consist of agronomic and engineering techniques applied on a field-by-field basis. Management practices applied on a district-wide scale generally consist of structures for water control (both delivery and discharge) and systems for the treatment and disposal of drainage waters.

Usually, no single method suffices to control salinity of an irrigated land. Instead, many practices are combined. The appropriate combination of salinity control methods depends on economic, climatic, social, edaphic, and hydrogeologic factors. Thus, diagnostic procedures cannot be given here for selecting a generic set of control practices. Rather, some important goals, principles, and practices that should be considered in selecting salinity-management practices are reviewed.

Controlling Plant Toxicity and Root Zone Salinity and Sodicity

Management practices for the control of salinity, sodicity, and toxicity in the root zone include:

1. Selecting crops or varieties (rootstocks) that produce satisfactory yields under the existing conditions of salinity, sodicity, and toxicity.
2. Using planting procedures that minimize salt accumulation around the seed.
3. Using land preparation methods that facilitate uniform infiltration of irrigation water.
4. Adopting irrigation procedures that maintain sufficient available soil moisture and cause periodic leaching of the soil profile.
5. Installing and maintaining drainage systems that facilitate leaching, rooting, and trafficability.
6. Using treatments, e.g., chemical amendments and organic matter, to maintain soil permeability and tilth.

Soil properties, water table depth, and the quality of irrigation water also must be considered when selecting a particular management practice.

Growing Suitably Salt-Tolerant Crops. A crop that produces satisfactory yields under present or anticipated conditions of salinity, sodicity, or specific solute concentrations should be selected using the tolerances given in Chapter 13. Barley, sugar beet, cotton, Bermuda grass, Rhodes grass, western wheat grass, bird's-foot trefoil, table beets, kale, asparagus, spinach, and tomato are all highly salt-tolerant. Radish, celery, beans, clovers, and nearly all fruit trees are low-tolerance crops. The maximum expected levels of salinity, sodicity, and specific solute concentration resulting from long-term irrigation can be predicted by using the models described in Chapters 22, 23, and 24.

Minimizing Salt Accumulation in the Seed Bed. Many crops are more sensitive to salinity during early growth. Such crops are particularly susceptible to losses in stand due to high levels of salinity in the seedbed (see Chapter 13). Failure to obtain a satisfactory stand frequently limits the successful cropping of moderately saline soils. Establishment of seedlings in furrow-irrigated row crops is particularly difficult because soluble salts tend to accumulate in raised beds that are wetted by subbing water from the furrow (see Chapter 16).

Practices that facilitate establishment of seedlings in saline soils involve modification of seedbed shape, seed number, and placement and irrigation techniques. To speed germination, the seeds may be presoaked before planting. This is effective, yet seldom used.

As discussed in Chapter 16, salts tend to accumulate near the seed during furrow irrigation. This tendency is greatest in single-row, round-topped beds. Such beds should be avoided under saline conditions. With double-row beds, most of the salt is carried into the center of the bed, leaving the shoulders relatively free of salt for seedling establishment. Sloping beds are best for saline soils, because seedlings can be established on the slope below the zone of salt accumulation. The salt is moved away from the seedlings instead of toward them. Planting in furrows or basins is satisfactory from the standpoint of salinity control, but often unfavorable for the emergence of many row crops because less-favorable crusting and aeration conditions exist in furrows and basins.

Pre-emergence irrigation by sprinklers or drip lines placed close to the seed line may help to keep the soluble salt concentration low in the seedbed. Also, temporary small furrows may be used instead of drip lines. After the seedlings are established, these furrows may be abandoned and new furrows made between the rows. The bed shape may also be altered and sprinkling replaced by furrow irrigation.

The seeding rate should be increased to help to counter the loss of stand associated with germination and emergence problems. Alternatively, seedling transplants can be used to establish stand in saline soils.

Applying Irrigation Water Uniformly and Efficiently. Irrigation management is of prime importance in the control of salinity (see Chapters 12 and 16).

Precise land leveling is necessary to facilitate the uniform application and infiltration of irrigation water and, hence, to control salinity. Barren or poor areas in otherwise productive fields often occur on high spots that do not infiltrate enough water for crop growth or leaching. Usually, lands that have been irrigated for one or two years after initial leveling should be replaned to remove unevenness caused by the settling of fill material. Annual crops should be grown following the first leveling, so that re-planing can be performed before perennial crops are planted.

Flood irrigation is suitable for salinity control if the land is sufficiently level, though soil aeration and crusting problems may occur. Furrow irrigation is well adapted to row crops and land that is too steep for flooding. Aeration and crusting problems are minimal with furrow irrigation, but salts tend to accumulate in the beds, as discussed above. If excess salt accumulates, a periodic change of irrigation method to flooding or sprinkling, along with a switch to grain crops or other more salt-tolerant crops, is advisable. Alternatively, after the seedlings are established, furrow depths can be made shallower or irrigation depths increased, so that the beds will be inundated and leached.

Irrigation by sprinkling allows superior control of the amount and distribution of water. It, therefore, is often used on steep land. There is a tendency to apply too little water for leaching requirements with this method, and leaching of salts beyond the root zone often requires special effort. As explained earlier, crusting is more likely to become a problem with sprinkler irrigation. Another potential hazard of sprinkler-irrigation is foliar uptake of salt and leaf burn due to contact with water. Sprinkler irrigation should be avoided if the water contains excessive levels of sodium and chloride, although sprinkling at night can help in such cases.

Sub-irrigation, in which the water table is maintained close to the soil surface, is not advised when salinity is a problem, unless the water table can be lowered periodically and the accumulated salts leached out by rainfall or by periodic surface applications of water.

Drip irrigation, if properly designed, minimizes salinity and matric stresses because the soil water content is maintained at a high level and the salts are leached to the perimeter of the wetted volume, where rooting activity is minimal. Drip irrigation is usually the method of choice when the water is high in salts, though the high build-up of salts in the fringe of the wetted area may eventually become a problem.

Solid-set, linear-move, and center-pivot sprinkler systems give good control and distribution of water if managed properly. Gravity systems, if designed and operated properly, can also achieve good control. For tree crops, a low-head bubbler system provides excellent control while minimizing pressure requirements.

Laser-controlled precision land leveling allows better areal water distribution and smaller applications. Combined with automation, it has led to high irrigation efficiencies for dead-level, flooded systems. Such systems improve irrigation efficiency and salinity control.

Excessive loss of irrigation water from unlined distribution canals is a major cause of high water tables and saline soils in many irrigation projects (see Chapter 17). Such seepage losses should be reduced by lining the canals with impermeable materials or by compacting the canal wall to achieve low permeability. Closed conduits, rather than open waterways, should be used for laterals wherever possible, because of their more effective off-on control. In addition, they can provide gravitational energy for pressurizing delivery systems or controls. In furrow-irrigated areas, intake uniformity can sometimes be increased and tail water runoff reduced by shortening furrow lengths and using multi-set systems. For some soils, surge irrigation can also be used to improve irrigation uniformity in graded furrows.

It is important to recognize that inefficient irrigation is the major cause of salinity and shallow water tables in most irrigation projects of the world and that the need for artificial drainage can be substantially reduced through improvements in irrigation management. Many delivery systems encourage over-irrigation by supplying water according to fixed schedules, or in fixed amounts, regardless of seasonal variations in crop needs. Ways to improve irrigation efficiency should be sought before drainage capacity is increased.

Irrigating Frequently to Maximize Soil Water Potential. The optimum irrigation scheme provides water continuously at a rate sufficient to keep the soil water content in the root zone within narrow limits, although carefully programmed periods of stress may be desirable to obtain maximum yield with some crops. Cultural practices also may demand periods of dry soil. Thus, the timing and amount of water applications should be carefully controlled to achieve high crop yield and water-use efficiency, especially in saline soils. This calls for water delivery to the field on demand, which, in turn, requires the farmer and the organization that distributes the water to coordinate their efforts closely.

The level of salinity that can be tolerated in the soil depends on the distribution of salinity in the soil profile, the frequency and extent to which the soil water is depleted between irrigations, and the water content of the soil. Plants usually tolerate higher levels of salinity under conditions of low matric stress, such as is achieved with high-frequency irrigation, e.g., drip irrigation. High soil-water salinities occurring in deeper regions of the root zone should be offset by sufficient low-salinity water added to the upper profile. Irrigations should be implemented so as to maximize soil water potential, thereby minimizing the effects of salinity on soil water availability, and to minimize deep percolation, thereby minimizing drainage and pollution problems.

Frequently, the efficiency of on-farm water use is limited by lack of knowledge about when to irrigate. Ideal irrigation management would supply enough water to hold the soil water content near field capacity at planting time and reduce water content at least 50% at harvest and would maintain water within the root zone during the major period of vegetative growth as high as possible without incurring aeration or leaching problems. Under saline conditions, this requires that water be applied in excess of consumptive use.

Irrigation scheduling requires some method of assessing the water availability to the crop with sufficient lead time to provide for a water application before significant plant stress occurs. In addition, the quantities of water needed for replenishing depleted soil moisture and leaching must be estimated.

Both direct and indirect measurements are commonly used to predict the onset of plant water stress and the soil water deficit. Chapters 14 and 16 describe some of these methods in detail.

Providing Leaching and Drainage to Avoid Root Zone Salt Accumulation. Additional water above the amount required by plant transpiration and evaporation must be applied, at least occasionally, to leach out the salt that has accumulated during previous irrigations. This leaching requirement depends on rainfall, the salt content of the irrigation water, and the maximum salt concentration permissible in the soil solution. The leaching requirement should be estimated by using the empirical relations given in Chapter 12, the crop production relations given in Chapter 14, or the Watsuit model described in Chapter 22.

Often, much of the necessary leaching is achieved during preirrigations and early-season irrigations, when soil permeability is at its seasonal maximum. If rainfall is significant, especially between cropping periods, then the necessity of irrigative leaching will be reduced or eliminated.

The control of salinity by leaching is accomplished most easily in permeable, coarse-textured soils. Medium-textured soils, with their greater water-holding capacity, ordinarily present no major problem from the standpoint of salinity control if they have good structure and overlie subsurface materials that facilitate the removal of drainage water. Preventing salt accumulation is most difficult in fine-textured, slowly permeable soil, especially if drainage is poor. For more discussions on leaching requirement, see Chapter 12.

If natural drainage is insufficient, it should be supplemented by drainage installations. For more discussion of drainage and irrigation management for salinity control, see Chapter 16. For a discussion of irrigation requirements, see Chapter 14. For more discussion on project-level management, see Chapter 18.

Controlling Salinity of Water Resources

Irrigated agriculture cannot be sustained without adequate leaching and drainage to prevent excessive salinization of the soil. Yet these processes are the very ones that contribute to the salt loading of our rivers and ground waters. Several approaches are available for minimizing salinity pollution of water resources that result from irrigation activities.

First, irrigation can be eliminated. This should be undertaken in instances where the detrimental effects of irrigating the land outweigh the benefits.

Second, the amount of water lost in seepage and deep percolation can be reduced, lessening the amount of saline water that passes through the soil and substrata.

Third, point sources of drainage-return flow into streams or rivers can be intercepted and diverted to other outlets and uses. For example, saline drainage water can be desalted and reused, disposed of by evaporation in ponds or by injection into some isolated deep aquifer, or used as a water supply for applications in which brackish waters are acceptable.

The second and third approaches are discussed in greater detail below.

Minimize Leaching and Deep Percolation. With irrigation water containing between 0.05 tons and 3.5 tons of salt per 1,000 m^3 and crops requiring annual applications of 6,200 m^3 to 9,300 m^3 water per hectare to meet evapotranspiration, an average of 0.3 tons to 32 tons of salt/ha are added annually to irrigated soils. Clearly, if the volume of water applied for irrigation can be minimized, then the amount of salt added to the soil and the amount that must be removed by leaching will be minimized.

The leaching fraction, L, should also be minimized, to maximize the precipitation of applied Ca, HCO_3, and SO_4 salts as carbonate and gypsum minerals in the soil and to minimize the amount of weathered and dissolved salts picked up from the soil and substrata. By reducing L from 0.3 to 0.1, the amount of salt discharged from the root zone can be reduced by as much as 2 tons/ha to 12 tons/ha annually.

The extent to which leaching can be minimized is limited by the tolerances of plants to increased salinity in the lower part of the root zone. In most irrigation projects, current leaching fractions could be reduced appreciably without harming crops or soils, especially with improved irrigation management (see Chapter 12).

Minimizing leaching may or may not reduce the degradation of the receiving water by salinity. For example, rivers essentially saturated with $CaCO_3$ and gypsum will not benefit from reduced leaching unless salts other than those derived from the diverted water or from soil mineral weathering and dissolution in the root zone are encountered in the drainage flow path. Each situation must be evaluated according to its specific hydrogeologic conditions.

Like surface waters, ground waters receiving irrigation drainage water may or may not benefit from reduced leaching. If no sources of recharge other than irrigation drainage exist, the chemical composition of the ground water will eventually approach that of the drainage water.

Intercept, Isolate, and Reuse Drainage. The ultimate goal for efficient irrigation is to prevent the mixing of highly saline waters and freshwater irrigation supplies and to maximize the utilization of an irrigation water supply, with minimum drainage. If drainage water from one irrigation is acceptable for use by a crop of higher salt tolerance, it should be used again, ideally in the same project. This can be achieved by intercepting the water draining from the root zone, isolating it from higher-quality water supplies and using it to irrigate crops of appropriate salt tolerance. Such reuse of drainage water reduces the salinity of the associated river system or ground water reservoir because the salt load of the drainage return would not be added to it. A saline drainage water may be envisioned as consisting of an usable component, which varies with the plants' salt-tolerance, and an unusable component.

When the growth-limiting factor is salinity, the ultimate fraction of water in a supply that can be used for crop growth is:

$$1 - \frac{EC_{iw}}{EC_m} \tag{2.3}$$

where EC_{iw} = the electrical conductivity (alternatively, salt concentration can be used) of the water supply and EC_m = the maximum electrical conductivity (or salt concentration) that the plant can tolerate in the root zone (in terms of the EC of the soil water). Values of EC_m vary among the crop species. Approximate typical values are: 45 dS/m for highly salt-tolerant crops, such as cotton, sugarbeets, and barley; 30 dS/m for crops of intermediate salt tolerance, such as tomatoes, wheat, and alfalfa; and 15 dS/m for sensitive crops, such as beans, clovers, and onions (Bernstein et al., 1975).

If the usable component of drainage water is allowed to mix with fresh water and the blend is used to irrigate a crop of the same salt tolerance, then the plant can, at best, only consume through transpiration the fresh-water fraction of the mix, and the same "unusable" fraction will pass through the profile again without contributing to transpiration and growth. In the second drainage, more salt may be displaced, or "picked up." Greater flexibility and opportunity for crop production results if the initial drainage water can be intercepted, isolated and prevented from returning to the general water supply. The waters can then be blended or left separate for other uses, such as the irrigation of appropriately salt-tolerant plants. Once the waters are mixed, such alternatives are lost.

Thus, the best strategy for controlling the salinity of water supplies is to intercept drainage waters before they can flow into low-salinity water supplies and to substitute the drainage water for conventional irrigation water after seedling establishment of certain crops in the rotation. When the drainage water is too salty for reuse, it should be discharged to evaporation ponds or other appropriate outlets. This strategy will conserve water, sustain crop production, and minimize the salt loading of fresh-water supplies that occurs by irrigation return flow. It will also reduce the amount of fresh water diverted for irrigation.

The feasibility of this strategy is supported by the following: (1) The maximum possible soil salinity resulting from continuous use of drainage water does not occur during the time such water is only used for a fraction of the time; (2) substantial alleviation of salt build-up resulting from irrigation of salt-tolerant crops with drainage water occurs when salt-sensitive crops are irrigated with fresh water; (3) proper preplant irrigation and careful irrigation management during germination and seedling establishment leaches salts out of the seed area and from shallow soil depths; and (4) data from field experiments to date support the credibility of this cyclic reuse strategy. For more on the reuse of drainage waters for irrigation, see Chapter 20.

Dilution of return flows is often advocated for controlling water salinity and increasing the water supply for irrigation. But, as previously discussed, this concept has serious limitations when its effect on the true volume of usable water for crop transpiration is considered. It is not advocated as a general method of salinity control or water supply enhancement. For more discussion of river basin water and salt flows, salinity control measures, and treatment and disposal options, see Chapters 17, 18, and 21.

Summary. The natural and artificially-induced hydrogeological processes in a given system must be carefully analyzed before management practices for controlling salinity pollution are selected. The relative contributions of root-zone drainage and seepage should be assessed, so that priorities for pollution control can be established. For further discussion of this topic, see Chapters 17 and 18.

IDENTIFYING POTENTIAL SALINITY PROBLEMS

Modeling Methods

Analytical quantitative techniques and numerical computer models are available for predicting soil solution composition, crop response to water applications, and the quantity and quality of subsurface return flows under a variety of water management alternatives.

Deterministic, or process-oriented, models can simulate the physical and chemical processes that occur as water moves through the soil profile. The processes include salt precipitation, mineral weathering, and cation exchange reactions. Stochastic models and optimization models may be used to predict problems that are likely to result from the development of new irrigation projects. Management plans can then be developed to prevent or solve these problems in a more effective and timely manner. Economic models can be used to estimate the damages that may occur as a result of salinization of soils and water. Such economic models can also be used to predict the local, state, regional, and national benefits that result from implementation of a salinity control program.

Area-wide modeling studies can define the need for and potential benefit of salinity control measures. These studies pinpoint the sources and causes of salinity and provide information for selecting the most appropriate control measures. Once the sources of salts are defined, more detailed, local-scale models may help to specify how those sources can be controlled.

For more information on the use of models for salinity evaluation and control, see Chapters 14, 18, 22, 23, 24, and 25.

Monitoring Methods

A viable, permanent irrigated agriculture requires periodic information on soil salinity. Only with this information can the salt balance and water-use efficiency of irrigation projects be assessed. The collection and monitoring of soil salinity is complicated by the spatial variability of salinity data. Typically, numerous samples are needed to characterize an area. Monitoring is also complicated by changes in salinity that occur over time due to changing weather patterns, changing management practices, and movement of the water table.

Representative areas for collecting samples can be established within irrigation projects to minimize the number of samples required. Also, geophysical instruments often provide efficient, fast measurements of parameters, such as soil electrical conductivity.

A network of representative soil salinity monitoring stations should be established in irrigation projects, especially those projects undergoing changes in operation. When EC_e measurements are made at a number of locations, maps of the distribution of EC_e by area and depth can be prepared. Iso-lines of EC_e on such maps delineate the areas of salinity-affected soil. Shallow water tables often are responsible for such salt accumulations and can be associated with them if contour map overlays of the surface elevations and the ground water are prepared. Maps of the net vertical soil-water fluxes can help locate areas of inadequate drainage or salt loading. Techniques and equipment for measuring EC_a and estimating EC_e are described in Chapters 9 and 10.

REFERENCES

Bernstein L., Francois L.E., and Clark, R.A. (1975). "Minimal leaching with varying root depths of alfalfa. *Soil Sci. Soc. Am. Proc.* 39:112–115.

Reisenauer, H.M. (1983). *Soil and plant tissue testing in California*. Bulletin 1879. Div. of Agric. Sci., Univ. of Calif., Davis, Calif.

NOTATION

a = threshold level of soil salinity at which yield decreases begin; b = percentage yield loss per increase of salinity in excess of *a*; E_m = maximum electrical conductivity that plant can tolerate in root zone; EC_a = electrical conductivity of the bulk field soil; EC_e = electrical conductivity of paste extracts of soil samples; EC_i = electrical conductivity of irrigation water; EC_{iw} = electrical conductivity of water supply; EC_p = electrical conductivity of soil paste; ET_c = estimated crop evapotranspiration requirement; F_c = concentration factor; I = infiltration rate; $(I \cdot T_1)$ = amount of water infiltrated over irrigation period; L = leaching fraction; LA = theoretically achievable leaching fraction; L_r = leaching requirement; T_i = irrigation set period; and Y_r = percentage of yield of crop grown under saline conditions relative to that obtained under nonsaline, but otherwise comparable, conditions.

PART TWO—EFFECTS OF SALTS ON SOILS

CHAPTER 3

THE CHEMISTRY OF SALT-AFFECTED SOILS AND WATERS[a]

ORIGIN OF SALT IN SOIL AND WATER

The primary source of salts in soil and waters is the ever-continuing geochemical weathering of rocks that form the upper strata of the earth's continental crust. It represents one step in the geochemical exchange of matter between the land and oceans that has been in existence throughout geologic time.

Weathering

Because most rocks have formed under high temperature and pressure, the constituent minerals usually are thermodynamically unstable when exposed to atmospheric conditions. Weathering is a spontaneous process that transforms primary minerals to other minerals that are more stable at the earth's surface. The reagents involved in geochemical weathering include atmospheric water, oxygen, and carbon dioxide. The biosphere enhances weathering because of its increased levels of CO_2 and organic matter. Organic matter can serve as a reducing agent and a source of organic acids, which promote weathering and cation migration through complexation.

Three types of reactions describe the chemical weathering of rock-forming minerals: congruent dissolution, incongruent dissolution, and reduction-oxidation (redox) reactions.

Congruent Dissolution. In this reaction, the solution products exist in the same proportion as they occur in the mineral, i.e.,

Mineral → soluble species

Examples of congruent weathering processes are:

$$CaCO_3 \text{ (calcite)} + H_2O + CO_2 \rightleftarrows Ca^{2+} + 2HCO_3^-$$
$$SiO_2 \text{ (quartz)} + 2H_2O \rightleftarrows H_4SiO_4 \text{ (silicic acid)}$$

Incongruent Dissolution. In this reaction, part of the mineral dissolves and leaves behind a secondary solid phase (secondary alumino-

---

[a]Prepared by: J. J. Jurinak, Dept. of Soil Sci. and Biometerology, Utah State Univ., Logan, UT 84322; and D. L. Suarez, U.S. Salinity Lab., Riverside, CA 92501.

silicate clay minerals) that differs in composition from the original mineral, i.e.,

$$\underset{\text{(albite)}}{2NaAlSi_3O_8} + 3H_2O \rightleftharpoons \underset{\text{(kaolinite)}}{Al_2Si_2O_5(OH)_4} + \underset{\text{(quartz)}}{4SiO_2} + 2Na^+ + OH^-$$

$$\underset{\text{(orthoclase)}}{3KAlSi_3O_8} + CO_2 + 14H_2O \rightleftharpoons 2K^+ + 2HCO_3^- + 6H_4SiO_4 + \underset{\text{(mica)}}{KAlSi_3O_{10}(OH)_2}$$

Redox Reactions. Changes in the oxidation states of minerals modifies weathering. Redox reactions between ions dissolved in solution and minerals in contact with that solution often influence the pH of the solution, for example:

$$FeS_2 + 15/4\ O_2 + 7/2\ H_2O \rightleftharpoons Fe(OH)_3(s) + 4H^+ + 2SO_4^{2-}$$

In this reaction, the protons produced have a strong local influence on subsequent weathering.

Climate and Landscape Effects

Although weathering is continuous and occurs universally, the intensity and extent of the weathering reactions strongly reflect the influence of climate. The presence of water is most important to the weathering process. It serves as a reactant in mineral transformation and is the medium that transports dissolved and suspended matter from the system. The transport of the weathering products depends on sufficient rainfall to move the soluble salts through the surface soil into the ground water, eventually into rivers, and, ultimately, the oceans. The oceans' chemical composition reflect the constant inflow of salt from the land masses as modified by chemical interaction and evapoconcentration and augmented by volcanic activity.

Salt-affected soils are natural components of arid landscapes. Their presence directly correlates with limited rainfall, i.e., the evapotranspiration greatly exceeds precipitation throughout most of the year. Lack of moisture limits the intensity of the chemical weathering of minerals. It also limits the measurement of the product of weathering (salts), and the secondary minerals formed often are constrained to a localized area. In sub-humid areas, the properties of the parent rock largely dictate the properties of the soils formed. Most arid-zone soils are classified under the order of entisols and aridisols.

Because water serves as the principal vehicle for salts, salinity is closely linked to lowlands or depressions where water drains and accumulates. Salinization is enhanced when restricted soil drainage promotes a high water table and the balance of mineralized ground water is regulated by the evaporation of water, or transpiration, or both, rather than by surface runoff and drainage. Areas of impeded drainage vary in size from a fraction of a hectare to thousands of square kilometers, e.g., Utah's Great Salt Lake Basin and Pakistan's Indus Plain.

Fossil or Secondary Deposits

Throughout geologic time, saline seas have inundated large areas of continents. These submerged areas have subsequently been uplifted. The resulting geologic formations provide only parent material for soils and outcrops and underlying saline strata, both of which are important zones of contact for salt loading of surface and ground water. The secondary deposits (sedimentary rocks) from the weathering of continental rock during inundation are substantial sources of salinity and sodicity. The term "fossil salt" has been used to describe the salinity of these deposits.

A good example of a secondary saline deposit that markedly affects the salinity of a region is the Mancos shale formation in the upper Colorado River Basin. This formation, deposited in the late Cretaceous epoch, reaches a maximum thickness of about 2,000 m (Williams 1975). Mancos shale forms an extensive belt of erodible outcrops and underlying strata in southern Wyoming, eastern Utah, and western Colorado.

Atmospheric Deposition

The atmospheric deposition of salt can be of localized importance. Dry and wet aerosol fallout contributes up to 100 kg/y-ha to 200 kg/y-ha along sea coasts and from about 10 kg/y-ha to 20 kg/y-ha in the interior. The composition of atmospheric salt deposition varies with distance from the source. The salt is predominately a NaCl-type at the coast. It becomes dominated by Ca and SO_4 ions as the air mass moves inland. Atmospheric contribution to the salt load of arid lands is from 10% to 25% of the total yearly contribution of weathering (Bresler et al., 1982). In the overall picture of salinization of soils, the contribution of atmospheric salt is often overlooked, but it is a factor that must be considered, particularly in highly weathered landscapes, which have poor drainage. A case in point is large areas of saline soils in western Australia, whose origin is attributed to long-term salt inputs from rain in combination with limited drainage.

Anthropogenic Activities

Soils made saline by mankind are of major historical and economic importance. Industrialization has increased the atmospheric loading of gaseous nitrogen and sulfur components, both of which increase the salt concentration in the atmosphere and result in acidic fallout, which intensifies the soil mineral weathering rate. Energy-related mining activities have brought to the surface saline and sodic materials that, left in the ground, would have had little effect on the environment.

Irrigation has a dramatic effect on the introduction of salts to the terrestrial system. All irrigation water contains salt in varying amounts and differing types. During evapotranspiration, essentially pure water is used by the plant and the salt that is left behind is added to the salt already present in the rhizosphere. For example, for every 100 mg/l salt in the water, one megagram of salt is added per ha-m of applied water.

If one ha-m of water with a salt concentration of 850 mg/l is applied to a crop during the growing season, then 8.5 megagrams of salts are added. Without salinity management, salts will eventually accumulate in the rhizosphere. Application of irrigation water also greatly increases the rate of mineral weathering. The control of salinity or sodicity in the rhizophere and in surface or ground waters is, therefore, closely associated with soil and water management practices.

MEASUREMENT OF SALINITY

The chemical and physical properties of a salt-affected soil reflect the amount and type of salt present. Although the use of salinity sensors in the field is increasingly common (see Chapter 10), laboratory analysis of aqueous extracts of soil is still the most common technique for assessing salinity and other potential hazards.

One of the earliest methods for determining the amount of dissolved salts is based on evaporating a given volume of water or soil extract and measuring the weight of the residue. The result is called the total dissolved solids (TDS) and the dimensions are mg/l. This method, although still used, has been largely replaced by a more convenient measurement of the electrical resistance of the solution, or its reciprocal, the electrical conductance.

The current-carrying capacity of a solution is proportional to the concentration of ions in the solution. The electrical conductance is measured in reciprocal ohms, i.e., $ohms^{-1}$, also known as mhos. In the Systeme International d'Unites (SI units), mhos are designated as siemens (S).

Measurements of conductivity are made in a cell containing two electrodes of defined geometry. An electric potential is imposed across the electrodes and the resistance of the solution between the electrodes is measured. The results are multiplied by a "cell constant" to correct for variations in cell geometry, allowing all data to be reported as specific conductance (EC), i.e., the conductivity per unit volume (1 cm^3) of solution in siemens (S) per cm.

The siemen is too large for measuring electrical conductance in most natural systems. Hence, the working unit is the millisiemen ($S \times 10^{-3}$) per cm, i.e., mS/cm. Because the basic length in the SI convention is the meter, the preferred dimension for EC is decisiemens per meter (dS/m),

$$\text{dS/m} = \text{mS/cm} = \text{mmhos/cm}$$

where mmhos/cm (millimhos per cm) are the traditional and now-obsolete dimensions for EC.

Saturation Extract

A variety of soil/water ratios can be used to obtain an aqueous extract from a soil sample. Therefore, a standard extraction method must be used if saline-soil chemical data are to be compared. Because the amount of water that a soil holds at saturation (saturation percentage, SP) is related to a number of soil parameters, such as texture, surface area, clay content, and cation exchange capacity (Merrill et al., 1987), one widely-used technique is to obtain an extract by vacuum filtration of a saturated soil paste made with distilled water (U.S. Salinity Laboratory

1954). This extract is then analyzed for electrical conductance (EC_e) and any soluble constituents of interest. Although lower soil-water ratios (1:1, 1:2, etc.) make obtaining an extract easier, they are less related to the field moisture condition than the saturation paste. However, when only relative salinity changes are of interest, the lower soil-water ratios can be used to advantage.

The value of EC is sensitive to the temperature of the solution. EC increases about 2% for each degree increase in solution temperature in the range of 15°C to 35°C. All EC data are normalized to a temperature of 25°C for a valid comparison. A temperature correction to 25°C (EC_{25}) can be approximated by

$$EC_{25} = EC_t - 0.02\,(t-25)\,EC_t \qquad (3.1)$$

where EC_t = the value at temperature t.

Predicting EC From Ionic Composition

Theoretical and empirical approaches have been used to predict the EC of a solution from its composition. An example of the theoretical approach is a model based on Kohlrauch's Law of independent migration of ions (Harned and Owen 1953), in which each ion contributes to the current-carrying ability of an electrolyte solution. The equation used is:

$$EC = \sum_i EC_i = \sum_i \frac{c_i\,(\lambda_o^i - \beta\sqrt{c_i})}{1{,}000} \qquad (3.2)$$

where EC = the specific conductance (dS/m); EC_i = the ionic specific conductance (dS/m); summed over all ith species in solution; c_i = the concentration of the ith ion ($mmol_c$/l); λ_i^o = the ionic equivalent conductance (cm^2 S/mol_c) at infinite dilution; and β is an empirical interactive parameter obtained by plotting the ionic equivalent conductance of the ith ion (λ_i in cm^2 S mol_c^{-1}) versus $(c_i)^{1/2}$. For relatively dilute mixtures, β ranges in value from 2 to 9, with an average of 5.5 (Tanji and Biggar 1972). Values of λ_i^o can be obtained from standard textbooks on electrochemistry and physical chemistry.

A more accurate, but mathematically more complex, model based on the modified Onsanger-Fuoss equation and corrected for ion pair formation (which will be discussed later) gives reasonable agreement (± 8%) between calculated and measured EC up to 15 dS/m (Tanji 1969; Marion and Babcock 1976). Marion and Babcock also developed an empirical approach relating measured specific conductance (EC in dS/m) to total soluble salt concentration (TSS in $mmol_c$/l) and ionic concentration (C in $mmol_c$/l), where C is corrected for ion pairs. In the absence of ion complexation, TSS = C. The derived relationship was determined on a composite data base obtained from soil extracts, river waters, and pure salt solutions. The equations, suitable for most purposes, are:

$$\log C = 0.955 + 1.039 \log EC \qquad r^2 = 0.997 \qquad (3.3)$$

$$\log TSS = 0.990 + 1.055 \log EC \qquad r^2 = 0.993 \qquad (3.4)$$

Equation 3.4 supercedes the former relationship (U.S. Salinity Laboratory 1954):

$$\text{TSS (mmol}_c\text{/l)} = 10 \quad \text{EC(dS/m)}. \tag{3.5}$$

CHEMISTRY OF SALT-AFFECTED SOIL SOLUTION

Major cations in salt-affected soils are Na^+, Ca^{2+}, Mg^{2+}, and, to a lesser extent, K^+. The major anions are Cl^-, SO_4^{2-}, HCO_3^-, NO_3^-, and, at high pH, CO_3^{2-}. The ions HCO_3^- and CO_3^{2-} are reported together as carbonate alkalinity. Other ions that are sometimes present under anaerobic conditions, but neglected from a salinity viewpoint, include NH_4^+, NO_2^-, and organic anions.

Dissolution and precipitation of minerals often determine the composition of the soil solution. The degree of saturation of the soil solution with respect to a particular solid phase can be estimated from the ion activities. The activity concept accounts for the non-ideal behavior of ions in solution. Activities can be calculated from solution concentrations and ionic strength as follows:

$$a_i = \gamma_i m_i \tag{3.6}$$

where γ_i is the activity coefficient and m_i is the molal (moles/kg solvent) concentration of the ith ion. Except for very saline solutions, the assumption that $m_i = M_i$ (moles/liter of solution) is reasonable. The activity coefficients are, in turn, related to the ionic strength, I, defined by:

$$I = 0.5 \sum_i m_i z_i^2 \tag{3.7}$$

where z = the valence of the ith ion. The summation in Equation 3.7 is for all charged species in solution.

An empirical relationship between EC and ionic strength is given by

$$I = 0.0127\ \text{EC} \tag{3.8}$$

where EC is in dS/m at 25°C (Griffin and Jurinak 1973). The activity coefficients for the ith ion is calculated by using the Davies equation (Stumm and Morgan 1981):

$$\log \gamma_i = -0.509\, z_i^2 \left[\frac{\sqrt{I}}{1 + \sqrt{I}} - 0.3I\right] \tag{3.9}$$

This is valid when $I < 0.5$ mol/l and water at 25°C is the solvent. An estimate of the activity coefficients, based on Equation 3.9, can be obtained from the nomograph in Figure 3.1. Individual ion activities can then be estimated from concentration data using Equation 3.6.

Salt tolerance data for crops usually are expressed as a function of osmotic potential, $-\tau_o$, where τ_o = the soil solution's osmotic pressure. The osmotic pressure, in kPa, can be calculated from:

$$\tau_o = 2480 \sum_i m_i \upsilon_i \phi_i \tag{3.10}$$

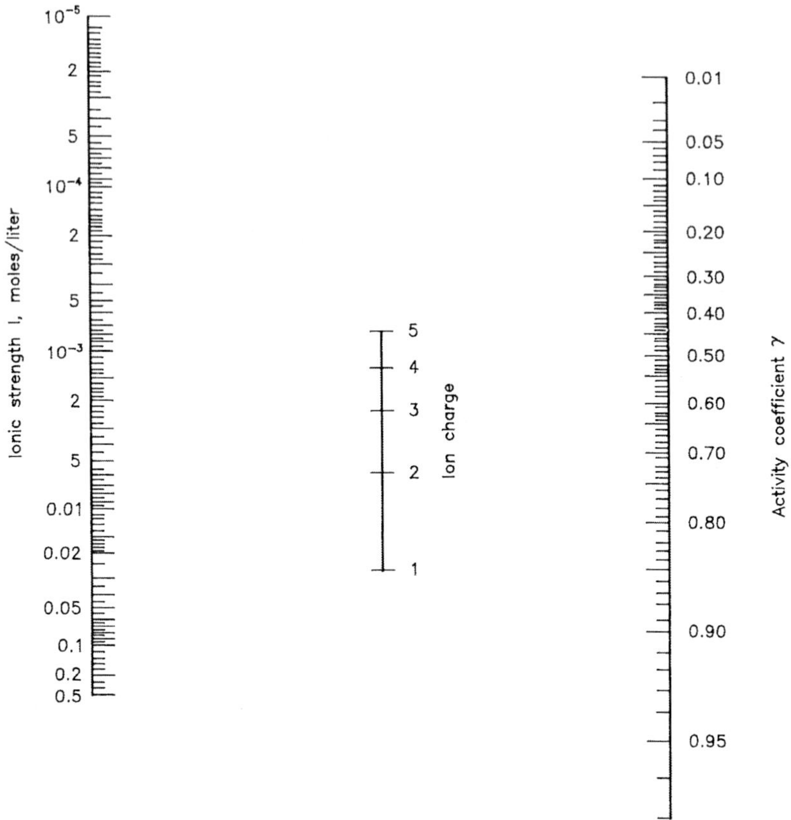

Fig. 3.1 Nomograph for Estimating Ionic Activity Coefficients from the Davies Equation (Butler 1964)

where ϕ = the osmotic coefficient of the ith salt and ν = the stochiometric number of ions yielded by the salt. The ϕ value for each salt can be obtained by using the salt's total concentration (Fig. 3.2).

The approximate relationship between τ_o and EC at 25°C, is:

$$\tau_o \text{ (kPa)} = 40 \text{ EC} \tag{3.11}$$

where EC is in dS/m. The commonly used proportionality constant of 36 is for τ_o data obtained at 0°C (U.S. Salinity Laboratory 1954).

Salt and pH Effect on Chemical Mass Action

The general mass action equation for any reaction is

$$aA + bB = cC + dD \tag{3.12}$$

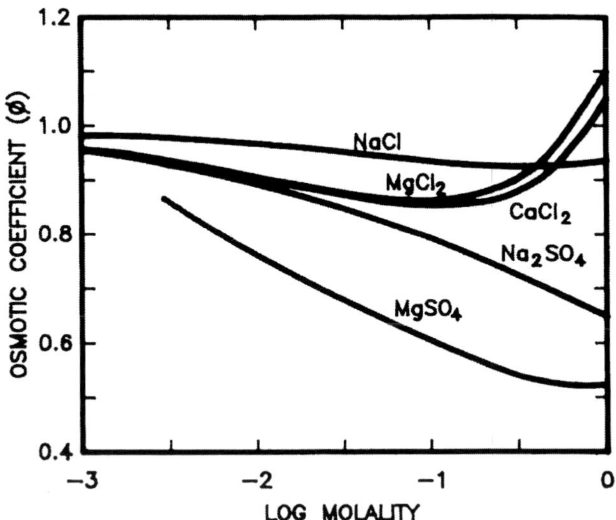

Fig. 3.2 Osmotic Coefficients of Electrolytes at 25°C (Robinson and Stokes 1959)

where the lowercase letters are the stochiometric coefficients and uppercase letters represent chemical symbols.

In Eq. 3.12, the thermodynamic equilibrium constant for the general reaction is

$$K_{eq} = \frac{(c)^c\ (D)^d}{(A)^a\ (B)^b} \qquad (3.13)$$

where ( ) represent activities as defined in Equation 3.6.

The dissolution of carbon dioxide in water is represented by the reactions

$$CO_2 + H_2O \rightleftarrows H_2CO_3 \rightleftarrows H^+ + HCO_3^- \qquad (3.14)$$

Biological production and water content-dependent diffusion usually control the CO_2 concentration in soil air. The CO_2 concentration of the soil solution is, thus, assumed to be independent of water quality or precipitation-dissolution reactions. The above equation shows that an increase in CO_2 results in a production of H^+ and, thus, a reduction in pH.

Calcite ($CaCO_3$), or lime, is commonly found in most arid soils. This mineral is important because much of the soil solution's chemistry can be defined in the context of the CO_2—$CaCO_3$—H_2O system. The dissolution of calcite is represented by

$$CaCO_3 + CO_2 + H_2O \rightleftarrows Ca^{2+} + 2HCO_3^- \qquad (3.15)$$

An increase in aqueous CO_2 concentration shifts the reaction to the right. This increases the Ca^{2+} concentration and alkalinity of the solution. The presence of ions other than Ca^{2+} or HCO_3^- increases the ionic strength, which, in turn, decreases the ionic activity coefficients of Ca^{2+}, HCO_3^-, and CO_3^{2-}. This, known as the ionic strength effect, increases the solubility of $CaCO_3$.

Gypsum ($CaSO_4 \cdot 2H_2O$) is another mineral often found naturally in arid soils. It is frequently added as a soil amendment to reclaim sodic soils and maintain soil permeability. It is moderately soluble and readily precipitated when its solubility is exceeded, such as when irrigating with a high-sulfate water. The dissolution reaction for gypsum is

$$CaSO_4\ 2H_2O \rightleftharpoons Ca^{2+} + SO_4^{2-} + 2H_2O \qquad (3.16)$$

Adding salts that contain ions such as Na^+, Mg^{2+}, Cl^-, and NO_3^- enhances the solubility of gypsum by the ionic strength effect. At constant temperature, the solubility product, K_{sp}, of gypsum is invariant, i.e., assuming the activities of H_2O and solid phase gypsum are unity, the product of the activities $(Ca^{2+}) \cdot (SO_4^{2-})$ is constant, even though the concentration of its constituent ions in solution increase.

Gypsum's solubility is also increased by ion association, i.e., the formation of soluble ion pairs. The concentration of ion pairs can constitute an important percentage of the total ions in a saline solution. For example, in a system of pure gypsum dissolved in water, the concentration of free Ca^{2+} ion is 10 mmol/l, but the total Ca concentration in solution is 15 mmol/l. The difference between the two concentrations is the presence of 5 mmol/l of the $CaSO_4^o$ ion pair in the solution phase. Because the formation of $CaSO_4^o$ requires both Ca^{2+} and SO_4^{2-} ions, gypsum solubility increases as ion association occurs. The formation of ion pairs is prevalent between multiple valence ions. It is also important for monovalent ions at high salinities. Ion association always enhances the solubility of minerals, so much that speciation corrections are needed to accurately estimate the soil-solution composition of salt-affected systems.

In a solution, the presence of excess ions that are common to a mineral phase tend to suppress its solubility (common ion effect). For example, gypsum solubility is greatly reduced in the presence of water containing large amounts of Na_2SO_4 or $MgSO_4$ salt (Tanji 1969b). The addition of gypsum to a solution saturated with respect to $CaCO_3$ increases the Ca ion concentration, causing additional $CaCO_3$ to precipitate and reducing the bulk solution alkalinity.

The solubility of gypsum is not pH-dependent. Adding sulfuric acid to a soil suppresses its solubility due to the common ion effect, rather than the effect of the acidity. Equation 3.15 shows that $CaCO_3$ solubility, on the other hand, is pH-dependent. Addition of acid reduces the alkalinity of the solution and allows $CaCO_3$ to dissolve until equilibrium is re-established.

Chemical Composition of Surface and Ground Waters

The composition of surface and ground water varies greatly in arid regions. Table 3.1 shows data selected from saturation extracts of salt-

TABLE 3.1 Chemical Characteristics of Saturation Extracts of Salt-Affected Soils, Well Waters, and River Waters Analyzed at U.S. Salinity Laboratory

Number of samples (1)			Mean (2)	Median (3)	10% less than (4)	90% less than (5)
(a) mol_c/m^3						
Ca	sat.ext.	139	27.8	10.6	1.4	71.8
	well	115	5.9	3.1	0.8	14.5
	river	68	4.7	3.4	1.0	10.9
Mg	sat. ext.	139	22.2	8.0	1.1	79.1
	well	115	4.3	1.5	0.5	15.3
	river	61	3.6	2.2	0.6	9.2
Na	sat. ext.	139	93.2	53.5	1.8	219
	well	115	15.8	6.6	0.8	44.8
	river	58	7.5	3.7	0.7	18.9
K	sat. ext.	128	1.6	0.5	0.1	2.2
	well	101	0.6	0.1	0.01	0.9
	river	30	0.3	0.2	0.06	0.5
SO_4	sat. ext.	134	400	29.4	3.7	94.1
	well	23	6.7	3.6	0.4	15.4
	river	58	6.7	4.1	0.3	19.0
Cl	sat. ext.	139	95.5	34.8	1.4	281
	well	115	15.0	2.5	0.2	54.2
	river	58	5.9	1.5	0.2	20.7
HCO_3	sat. ext.	139	8.3	3.0	1.2	10.3
	well	115	4.9	4.1	1.9	8.2
	river	58	3.3	3.0	1.7	5.7
(b) (dS/m)						
EC	sat. ext.	134	12.9	8.8	1.1	33.7
	well	115	2.4	1.2	0.3	7.3
	river	58	1.4	0.92	0.3	3.2
(c) $((mol/m^3)^{0.5})$						
SAR	sat. ext.	139	38.5	15.1	0.9	56.1
	well	115	6.6	4.7	0.6	16.5
	river	58	3.3	2.4	0.7	7.4

affected soils, well waters, and surface waters analyzed at the U.S. Salinity Laboratory. The data represent samples obtained throughout the world.

The soil extracts were more saline and more sodic than the well waters and surface waters. This reflects the effects of evapotranspiration

TABLE 3.2 Properties of Common Soil Clays

Clay (1)	Formula (general) (2)	Surface Area ($10^4 m^2/kg$) (3)	CEC ($cmol_c/kg$) (4)
Smectite (Montmorillonite)	$Si_8(Al,Mg)_4O_{20}(OH)_4$	60–80	80–120
Mica (Illite)	$K(Al,Si)_8(Al,Fe,Mg)_4O_{20}(OH)_4$	10–20	20–40
Kaolin (Kaolinite)	$Si_4Al_4O_{10}(OH)_8$	1–2	3–6

and mineral weathering in the soil. With increasing salinity, Na predominates over Ca because Na salts are more soluble than Ca salts. The importance of Na is also reflected in the increasing SAR.

These data also show that Cl is the dominant anion in saline waters, whereas SO_4 is prevalent in dilute solutions. This reflects the influence of gypsum precipitation.

The increase in the Mg/Ca ratio reflects the fact that Mg salts are more soluble than Ca salts. In general, the ratio Ca/HCO_3 is greater than one. As the calcium is removed by precipitation of $CaCO_3$, the Mg/Ca ratio will then increase.

Surface Chemistry

Clay Minerals. Incongruent weathering processes in soils result in the production of soluble salts and, more importantly, the formation of secondary alumino-silicate clay minerals. These secondary clays are what give soils their ion-exchange and adsorptive properties. Because weathering processes differ in response to changing environmental conditions, and because weathering products from one reaction may simultaneously participate as reactants in other, different reactions, a given soil often contains a wide range of clay minerals. For the same reasons, a given clay type typically exhibits a wide range of chemical composition and physical properties.

With their plate-like shape and finely divided state, clay minerals exhibit large specific surface areas. This, coupled with their permanent charge (the result of isomorphous substitution), and their pH-dependent charge, causes the colloidal clay fraction to be the center of chemical activity in the soil. In addition, the colloidal hydrous oxides of iron and aluminum, the stable end-products of weathering, enhance the adsorptive and exchange capacity of soils.

Table 3.2 gives the qualitative unit cell formulae, the range of the cation exchange capacities (CEC), and specific surface areas for three clay minerals commonly found in arid soils. Textbooks on clay mineralogy can be consulted for more information on crystal structure and properties (Grim 1968).

Although inorganic aspects of the soil matrix are emphasized here, the colloidal organic matter (humus) fraction of a soil cannot be overlooked. In some soils, the effect of organic matter is more important than the inorganic-fraction processes.

The Diffuse Double Layer. The diffuse double layer (DDL) model developed from basic electrostatic theory describes electrochemical phenomena at the charged solid-liquid interface. The charge associated with a given surface is viewed as a layer, with an adjacent diffuse layer of oppositely charged ions, known as counter ions, in solution. The layer of counter ions maintains a charge in solution that is equal and opposite to the net charge of the mineral-surface layer. Various forces influence the diffuse nature of the counter ions: electrostatic attraction, which draws them toward the charged surface, repulsion among the ions and thermal energy, which tends to equalize the concentration of ions in the system and draw the counter ions back into the bulk solution.

Applying the DDL model to the clay-mineral surface allows one to predict the distribution of cations (counter ions) and anions (co-ions) at the charged mineral interface. It is an alternate approach to the mass action model (discussed later) for describing the exchange phenomena in soils.

Many physical properties of soils can be modeled as an interaction between DDLs of soil clay particles. The degree and nature of interaction is determined by the effective thickness of the DDL, which can be estimated with the K parameter in units of cm^{-1}.

$$K = \sqrt{\frac{8\pi e^2 z^2 n^o}{D \cdot k \cdot T}} \quad (3.17)$$

where e = the electron charge (coulomb/ion); z = the valence of the counter ion, n^o is the electrolyte concentration in the bulk solution (ion/cm^3); D = the dielectric constant (coulomb/V/cm); k = the Boltzman's constant (V coulomb/K/ion); and T = absolute temperature (°K). The effective thickness of the DDL is 1/K, which has units of cm.

Compression of the DDL (the desired condition) is promoted by; (1) Increasing the valence of the counter ion; (2) increasing the concentration of the bulk solution; or (3) reducing the dielectric constant of the medium. Several excellent references discuss the DDL theory at length (Babcock 1963; Bolt 1979; Bresler et al., 1982).

Anion Exclusion. The DDL model has successfully predicted anion exclusion, the negative adsorption of ions from the clay-water interface. Anions excluded from the double layer must be associated with cations to maintain solution electroneutrality. This increases the apparent concentration of soluble cations in solution.

Fig. 3.3 shows how the concentration of counter ions (n_+) and anions (n_-) vary with distance in the DDL of a negatively-charged particle. Assuming a 1:1 symmetrical salt, e.g., NaCl, in the bulk solution (infinite distance from the surface), the concentration of the electrolyte n^o equals n_+ or n_-. This is represented by the dashed line in the figure. The excess of cations (α_+), which neutralizes both the surface charge and the anions in the DDL, is given by area ABCDEF. The amount of

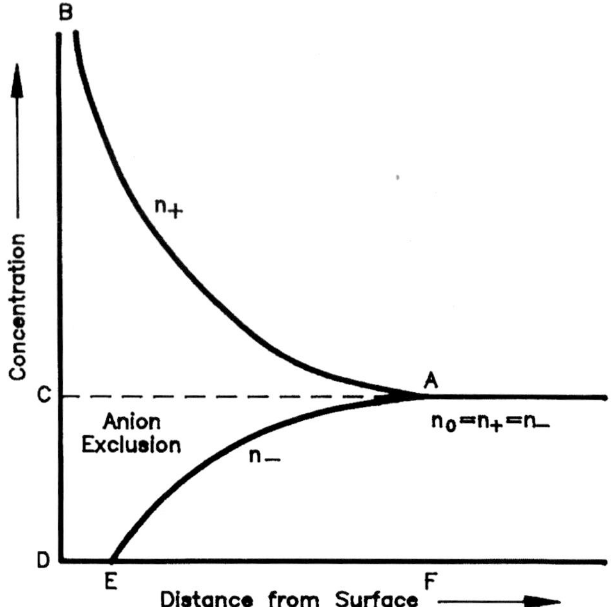

Fig. 3.3 Charge Distribution in Diffuse Double Layer of a Negatively Charged Particle

anions (α_-) in the DDL is given by area *AEF*. Total surface charge (α_{tot}), equivalent to CEC, is given by:

$$\begin{aligned} \text{CEC} &= \alpha_{tot} = \alpha_+ - \alpha_- \\ &= \textit{area ABCDEF} - \textit{area AEF} = \textit{area ABCDE} \end{aligned} \tag{3.18}$$

$$\text{CEC} = \text{surface cation excess} + \text{anion exclusion} \tag{3.19}$$

where anion exclusion is measured by the increase of anion concentration in the contacting solution. The CEC is equal to the surface cation excess (total exchangeable cations) only when anion exclusion is negligible. This occurs at low electrolyte concentrations. Increasing electrolyte concentration compresses the DDL and increases anion exclusion. The correction for anion exclusion becomes significant above approximately 10 dS/m at 25°C.

Bower and Hatcher (1962) reported the effect of anion exclusion correction on exchangeable sodium percentage, ESP. (ESP = NaX · 100/*CEC*, where the units of NaX and CEC are $cmol_c$/kg.) Table 3.3 shows selected data from their study. To correct for anion exclusion, they determined the Cl concentration in the saturation extract. The Cl concentration in the whole soil was determined by extensive leaching. The soluble cation content of the saturation extract was then multiplied by the ratio, $Cl_{leaching}/Cl_{extract}$.

Table 3.3 indicates a marked increase in Na ion preference by the exchanger, ESP, after correction. This underscores the importance of

TABLE 3.3 Anion Exclusion Correction on ESP (Bower and Hatcher, 1962)

			Chloride[a]		Soluble Sodium[a]		Exchangeable Sodium[a]		ESP (percentage)	
Soil (1)	SP (2)	Ec_e (dS/m) (3)	Sat.ext. (4)	Cor. (5)	Uncor. (6)	Cor. (7)	Uncor. (8)	Cor. (9)	Uncor. (10)	Cor. (11)
1	101	9.0	0.078	0.068	0.071	0.062	0.076	0.085	21	23
2	60	67.4	0.406	0.368	0.464	0.420	0.184	0.228	60	75
3	57	32.3	0.074	0.061	0.165	0.135	0.027	0.057	13	27
4	59	17.7	0.033	0.026	0.076	0.059	0.023	0.040	12	21

[a]Chloride, soluble sodium and exchangeable sodium are given in mol_c/kg.

considering anion exclusion in high-salinity environments. Anion exclusion fostered the concept known as the apparent exclusion volume, the hypothetical volume of water near the surface that must be insoluble to anions to account for the bulk solution's anion concentration. When calculating anion exclusion, the mathematically simpler concept of apparent exclusion volume is often used.

Ion Exchange

The ability to predict the distribution of ions between the soil solution and the exchanger phase is vital in the management of salt-affected soils, particularly when dealing with a potential sodic hazard. The complexity of the clay mineral fraction and the multi-ion nature of the soil solution make a rigorous definition of ion distribution in the soil difficult. A common approach to ion exchange is to apply the mass action principle.

Mass Action. The general mass action equation for describing ion exchange in a binary cation system is

$$bAX_a + aB^{b+} = aBX_b + bA^{a+} \tag{3.20}$$

where X = one mole of negative charge on the exchanger; a = the valence of cation A; and b = the valence of cation B. Ion concentrations are in mol/l. When a = b, the exchange is homovalent. When a ≠ b, the exchange is heterovalent. Applying the general exchange equation to Ca/Na exchange, where a = 2 and b = 1, gives

$$CaX_2 + 2Na^+ = 2NaX + Ca^{2+} \tag{3.21}$$

The thermodynamic equilibrium constant is

$$K_{eq} = \frac{(NaX)^2(Ca^{2+})}{(CaX_2)\ (Na^+)^2} \tag{3.22}$$

where ( ) signifies activities in both solution and exchanger phase. The method for calculating activities of ions in solution is well established (Stumm and Morgan 1981), but determining the activities of ions in the exchanger phase is difficult and subject to question. A simple mass action approach, which gives the selectivity coefficient, can be used to estimate the ion distribution between the solution and adsorbed phase.

For Ca/Na exchange (equation 3.21), the selectivity coefficient k_s is:

$$k_s = \frac{[NaX]^2[Ca^{2+}]}{[CaX]\,[Na^+]^2} \qquad (3.23)$$

where [ ] signifies concentrations units in both the solution and exchanger phase. The ion concentration in the exchanger phase is expressed in units of equivalent fraction, using the Gapon convention, or, less frequently, the mole fraction, using the Vanselow convention. These terms are related, but they produce different values for the selectivity coefficient (Sposito 1977).

Although not a true thermodynamic constant, the variability of k_s is small. Using the equivalent fraction, E_i, for the adsorbed ion concentration and mol/l for the solution concentration, Eq. 3.23 becomes

$$k_s = \frac{[E_{Na}]^2[Ca^{2+}]}{[E_{Ca}]\,[Na^+]^2} \qquad (3.24)$$

where E_i = iX/CEC and both iX and CEC have the units of $cmol_c$/kg ($cmol_c$/kg is equivalent to the formerly popular units of meq/100g).

Levy and Hillel (1968) found the k_s value for this exchange reaction to be constant over a wide range of E_{Na} values, i.e., $E_{Na} \approx 0.1$ to 0.7, for typical Israeli soils. The magnitude of the k_s, however, varied between soils.

The Gapon Constant. The Gapon exchange reaction has been widely used in salinity and sodicity studies. The original expression for Ca/Na exchange is

$$Ca_{1/2}X + Na^+ \rightleftarrows NaX + 1/2\ Ca^{2+} \qquad (3.25)$$

where solution concentration is in mol/l and exchange ion concentration is in $cmol_c$/kg. At equilibrium, we can write

$$k_g = \frac{[NaX]\,[Ca^{2+}]^{1/2}}{[Ca_{1/2}X]\,[Na^+]} \qquad (3.26)$$

where k_g = the Gapon selectivity coefficient, often assumed to be constant. The ratio of adsorbed ions is

$$\frac{[NaX]}{[Ca_{1/2}X]} = k_g \frac{[Na^+]}{[Ca^{2+}]^{1/2}} \qquad (3.27)$$

TABLE 3.4 X-Values for Various HCO_3/Ca Mole Ratios and Ionic Strengths (Suarez 1981)

	Ionic Strength (I)[a]														
HCO_3/Ca (1)	0.001 (2)	0.005 (3)	0.01 (4)	0.02 (5)	0.03 (6)	0.04 (7)	0.05 (8)	0.07 (9)	0.1 (10)	0.15 (11)	0.2 (12)	0.25 (13)	0.3 (14)	0.4 (15)	0.5 (16)
0.1	73.4	79.6	84.1	90.0	94.4	97.9	101.0	106.0	112.0	120.0	125.0	130.0	133.0	139.0	144.0
0.2	46.2	50.1	53.0	56.7	59.5	61.7	63.6	66.8	70.5	75.3	78.8	81.7	84.0	87.9	90.4
0.3	35.3	38.2	40.4	43.3	45.4	47.1	48.5	51.0	53.8	57.5	60.1	62.3	64.1	67.0	69.0
0.4	29.1	31.6	33.4	35.7	37.5	38.9	40.1	42.1	44.4	47.4	49.6	51.5	52.9	55.3	57.0
0.5	25.1	27.2	28.8	30.8	32.3	33.5	34.5	36.3	38.3	40.9	42.8	44.3	45.6	47.6	49.1
0.6	22.2	24.1	25.5	27.3	28.6	29.7	30.6	32.1	33.9	36.2	37.9	39.3	40.4	42.2	43.5
0.7	20.1	21.7	23.0	24.6	25.8	26.8	27.6	29.0	30.6	32.7	34.2	35.4	36.4	38.1	39.2
0.8	18.3	19.9	21.0	22.5	23.6	24.5	25.2	26.5	28.0	29.9	31.3	32.4	33.3	34.8	35.9
0.9	17.0	18.4	19.4	20.8	21.8	22.6	23.3	24.5	25.9	27.6	28.9	30.0	30.8	32.2	33.2
1.0	15.8	17.1	18.1	19.4	20.3	21.1	21.8	22.8	24.1	25.8	26.9	27.9	28.9	30.0	30.9
1.5	12.1	13.1	13.8	14.8	15.5	16.1	16.6	17.4	18.4	19.7	20.6	21.3	21.9	22.9	23.6
2.0	9.96	10.8	11.4	12.2	12.8	13.3	13.7	14.4	15.2	16.2	17.0	17.6	18.1	18.9	19.5
2.5	8.58	9.31	9.84	10.5	11.0	11.5	11.8	12.4	13.1	14.0	14.6	15.2	15.6	16.3	16.8
3.0	7.60	8.24	8.71	9.33	9.78	10.1	10.5	11.0	11.6	12.4	13.0	13.4	13.8	14.4	14.9
3.5	6.86	7.44	7.86	8.42	8.82	9.15	9.44	9.91	10.5	11.2	11.7	21.1	12.5	13.0	13.4
4.0	6.27	6.80	7.19	7.70	8.07	8.37	8.63	9.06	9.57	10.2	10.7	11.1	11.4	11.9	12.3
4.5	5.80	6.29	6.65	7.12	7.46	7.74	7.98	8.38	8.85	9.45	9.88	10.2	10.2	11.0	11.3
5.0	5.41	5.86	6.20	6.63	6.96	7.22	7.44	7.81	8.25	8.81	9.21	9.55	9.83	10.3	10.6
6.0	4.79	5.19	5.49	5.87	6.16	6.39	6.59	6.92	7.30	7.80	8.16	8.46	8.70	9.09	9.37
7.0	4.31	4.68	4.95	5.30	5.56	5.77	5.95	6.24	6.59	7.04	7.36	7.63	7.85	8.20	8.45
8.0	3.95	4.29	4.53	4.85	5.09	5.28	5.44	5.71	6.03	6.44	6.74	6.98	7.18	7.50	7.73
9.0	3.65	3.96	4.19	4.48	4.70	4.88	5.03	5.28	5.57	5.95	6.23	6.46	6.64	6.94	7.15
10.0	3.41	3.69	3.90	4.18	4.38	4.55	4.69	4.92	5.20	5.55	5.80	6.02	6.19	6.46	6.66

[a] I = 0.0127 EC (mol/l)

where the units of k_g are $(mol/l)^{-1/2}$. The U.S. Salinity Laboratory (1954) assumed that Mg behaves similarly to Ca in the adsorbed phase and modified the Gapon equation to the following

$$\frac{[NaX]}{[CaX + MgX]} = k_g' \frac{[Na]}{[Ca + Mg]^{1/2}} = k_g' SAR \tag{3.28}$$

where k_g' = the modified Gapon selectivity coefficient $(mmol/l)^{-1/2}$; and SAR = the sodium adsorption ratio, defined as

$$SAR = \frac{[Na]}{[Ca + Mg]^{1/2}} \tag{3.29}$$

where total analytical concentrations are used (mmol/l) with no account of ion association. Because Ca, Mg, and Na are the most common exchangeable cations in arid soils, Eq. 3.29 may be simplified to

$$\frac{[Na]}{CEC - [NaX]} = k_g' \, SAR = ESR \tag{3.30}$$

where ESR is called the exchangeable sodium ratio. In terms of the ESP, Eq. 3.30 becomes

$$\frac{ESP}{100 - ESP} = k_g' \, SAR = ESR \tag{3.31}$$

where ESP = NaX 100/CEC. Because obtaining reliable exchangeable ion data is difficult, the SAR of the soil solution or extract has become the principle parameter for diagnosing sodic hazards in soils.

The value of k_g' is determined by the regression (slope) of the ESR-SAR relationship, which is linear. A value of $k_g' = 0.015$ $(mmol/l)^{-1/2}$ is widely used. It is useful in field studies when the ESR $\leq$ 30. Correspondingly, as a rough approximation in this sodicity range, the value of SAR and ESP are assumed equal (U.S. Salinity Laboratory 1954). Recent studies, however, indicate that numerous factors influence the ESR-SAR couple and the value of k_g' ranges from 0.016 $(mmol/l)^{-1/2}$ to 0.008 $(mmol/l)^{-1/2}$. This suggests that, to be accurate, k_g' should be estimated using data specific to the site under study (Doering and Willis 1980; Jurinak et al., 1984).

Effect of Salinity on Ion Exchange. If the soil solution is concentrated by a factor ΔC, maintaining the same ionic ratios, the initial SAR increased by the factor $(\Delta C)^{1/2}$, i.e.:

$$SAR_{final} = (\Delta C)^{1/2} \cdot SAR_{initial} \tag{3.32}$$

Thus, if the concentration is increased by a factor of three and no chemical precipitation occurs, $SAR_{final} = (3)^{1/2} \cdot SAR_{initial}$. Increasing soil salinity increases the SAR, and if k_g' remains constant, then ESR and ESP must increase.

As an example, assume a soil is irrigated with water of EC = 1.5 dS/m and SAR = 8 $(mmol/l)^{1/2}$. Assuming an evapotranspiration concentration factor of 3 in the root zone, and using Eq. 3.32, the SAR of the soil solution will then be 13.9 $(mmol/l)^{1/2}$. Applying the Gapon convention and assuming that k_g' is constant and equal to 0.015 $(mmol/l)^{-1/2}$, the steady-state ESR of the soil is 0.21 (ESP = 17.4). If evapotranspiration increases the concentration factor to 6, the SAR will then equal 19.6 $(mmol/l)^{1/2}$ and the ESR increases to 0.29 (ESP = 22.5). Thus, if the selectivity coefficient remains the same, an increase in the concentration of the equilibrium solution increases the affinity of the clay fraction for monovalent cations, and dilution increases the affinity for divalent cations.

Ion Demixing. Early theories of cation exchange assumed exchangeable ions were randomly distributed at the clay surface. Recent evidence suggests, however, that random distribution is rare for expanding lattice clays. Instead, divalent ions are preferentially adsorbed in the interlayer spaces and monovalent ions are adsorbed on the edges and planar surfaces. This phenomenon, referred to as ion demixing, affects a soil's response to relatively low amounts of exchangeable sodium (see Chapter 5).

Rhizosphere Chemistry in Salt-Affected Soils

The composition of soil water depends on the composition of irrigation and rain water; chemical reactions, such as dissolution, precipitation, adsorption, and exchange; and the extent to which evapotranspiration concentrates the water. Predicting the chemistry of a soil solution is difficult due to the dynamic nature of the system. For a more convenient prediction assume, though it may be far from true, that steady-state conditions exist.

When assuming steady-state conditions, ion exchange is viewed as a static condition affected only by evapotranspiration and precipitation-dissolution reactions. It is also assumed that Na, Mg, K, and Cl do not undergo chemical reactions. Concentrations of these ions can be calculated by multiplying their initial concentrations in the applied water, C_{iw}, by the ratio of the water consumed to water applied.

Magnesium Carbonates. In the rhizosphere, many soil waters are supersaturated with respect to dolomite, $CaMg(CO_3)_2$, and possibly magnesite, $MgCO_3$. Dolomite is not considered a sink for Mg because of the kinetic constraints on precipitation existing in rhizosphere environments. Although dolomite has approximately the same molar solubility as calcite, the rate at which it dissolves is about 70 times slower. The result is that dolomite is slow to reach equilibrium with a soil solution. Because dolomite does not precipitate readily from the soil solution, whereas calcite does, the Mg/Ca ratio increases as the leaching fraction is reduced.

Magnesite precipitation in soils has not been documented. The solubility of magnesite is in dispute, but pure solutions may attain ion activity products (values of $(Mg)\cdot(CO_3)$) of about 3×10^3 times greater than the solubility product of calcite before hydromagnesite

($MgCO_3 \cdot H_2O$) or nesqehonite ($MgCO_3 \cdot 3H_2O$) precipitate. Before either dolomite or Mg-carbonate minerals form, Mg will likely form either an Mg-silicate mineral, such as sepiolite, $MgSi_3O_6(OH)$, or co-precipitate with Ca and form Mg-calcite.

The Adjusted SAR. The dissolution and precipitation of $CaCO_3$ influences rhizosphere chemistry. Indeed, calcite reaction is generally assumed to dominate the chemistry of the soil solution in arid-zone soils. Studies show that calcite usually supersaturates waters below the root zone and the value of its ion activity product (IAP) equals 10^{-8} (Suarez 1977), i.e.,

$$K' = (Ca^{2+})(CO_3^{2-}) = 10^{-8.0} \tag{3.33}$$

where K' = the apparent solubility product for calcite in the rhizosphere.

However, because HCO_3^- is the dominant carbonate constituent in the soil solution, solving Eq. 3.33 directly is difficult. The equation describing calcite solubility, written in terms of HCO_3^-, is

$$K'' = \frac{[Ca^{2+}]\,[HCO_3^-]^2}{P_{CO_2}} \tag{3.34}$$

When P_{CO_2} is expressed in kilopascals (1 atm = 100 kPa) and a calcite thermodynamic K_{sp} of $10^{-8.47}$ is used, the theoretical value of $K'' = 10^{-8.0}$. If the apparent solubility product of calcite (Eq. 3.33) is used in Eq. 3.34 for predictive purposes, the value of K″ increases to $10^{-7.5}$. In that situation, Eq. 3.33 loses its thermodynamic significance. If computer programs, such as the one by Oster and Rhoades (1975) and others, described at the end of this chapter, are unavailable for calculating equilibrium Ca^{2+} and HCO_3^- activities, a graphical method can be used (Suarez 1982), which involves the following equation

$$[Ca^{2+} - x]\,[HCO_3^- - 2x]^2 = \frac{10^{-7.5} P_{CO_2}}{\gamma_{Ca^{2+}} \gamma^2_{HCO_3^-}} \tag{3.35}$$

where x = the amount of Ca that is precipitated or dissolved.

Another way to solve this equation is to successively approximate x by using input values of $\gamma_{Ca^{2+}}$ and $\gamma_{HCO_3^-}$ from Fig. 3.1. The values of Ca and HCO_3 are those of the irrigation water multiplied by the concentration factor at the depth of interest. For surface environments, P_{CO_2} = 0.1 kPa (0.001 atm) is a reasonable estimate. In the absence of specific data for the rhizosphere, use P_{CO_2} = 1 kPa (0.01 atm) for sandy soils and 5 kPa (0.05 atm) for clay soils. No correction for ion association is made, so the results are only approximate.

The SAR relationship is useful for estimating the ESP. The SAR of soil water, assuming steady-state conditions, is determined from the composition of the irrigation water, after correcting for evapotranspiration and $CaCO_3$ precipitation or dissolution.

Various methods are used to adjust the SAR of waters. The preferred method is to calculate the equilibrium Ca concentration in the rhizosphere (Ca_{eq} in mmol/l) and use that in the following

$$SAR_{adj} = \frac{Na_{iw}F_c}{\sqrt{(Mg_{iw}F_c + Ca_{eq})}} \tag{3.36}$$

where Na_{iw} and Mg_{iw} are concentrations of Na and Mg (mmol/l), respectively, in the applied water. The concentration factor F_c equals $1/L$, in which L = the leaching fraction. Its value depends on the soil depth. For calculating the SAR_{adj} of both the surface soil and irrigation water, $F_c = 1.0$.

The value of Ca_{eq} can be obtained by multiplying Ca and HCO_3 by the appropriate F_c and solving Eq. 3.30. An alternate equation for calculating SAR_{adj} in drainage or soil water is

$$SAR_{adj} = \frac{Na_{iw}F_c}{\sqrt{[Mg_{iw}F_c + 0.215X(P_{CO_2})^{1/3}]}} \tag{3.37}$$

where P_{CO_2} is in kPa, and X values are found in Table 3.4 (Suarez 1981) using the molar HCO_3/Ca ratio and ionic strength values for the irrigation water.

CONCLUSIONS

The composition of the soil solution results from numerous, interdependent, multi-phase chemical interactions, including dissolution and precipitation of minerals, formation of inorganic and organic coordination compounds in solution, adsorption of ions and ligands onto the surfaces of minerals and organic compounds, exchange of ions between clay mineral surfaces and the solution phase, transport of compounds between the gas and liquid phases, and reduction and oxidation of ions and minerals. This list suggests not only the complexity of the system, but the need for conceptual geochemical computer models to predict the composition of the soil solution as affected by man's activities. Geochemical models can be linked with soil-water flow models, plant growth models, and other models to predict solute migration in soils.

Numerous geochemical models are available. An abbreviated list includes MINEQL (Westfall et al., 1976), EQ3/EQ6 (Wolery 1979 and 1983) and MINTEQ (Felmy et al., 1984). They represent the current state-of-the-art in computer simulation of soil-solution chemistry.

REFERENCES

Babcock, K.L. 1963. "Theory of the chemical properties of soil colloidal systems at" equilibrium. *Hilgardia* 34(11). University of Calif., Berkeley, Calif.

Bolt, G.H. 1979. *Soil Chemistry, B. Physico-Chemical Models*. Elsevier Scientific Publishing Co., New York.

Bower, C.A., and Hatcher, J. T. 1962. "Characterization of Salt-Affected Soils with Respect to Sodium." *Soil Science*. 93:275–280.

Bresler, E., McNeal, B.L., and Carter, D.L. 1982. *Saline and Sodic Soils*. Springer-Verlag, New York.

Butler, J.N. 1964. *Ionic Equilibrium, A Mathematical Approach*. Addison-Wesley, Boston.

Doering, E.J., and Willis, W.O. 1980. "Effect of Soil Solution Concentration on Cation Exchange Relationships." *Proc. Inter. Symp. on Salt-Affected Soils. Karnal, India.*

Felmy, A.R., Girvin, D.C., and Jenne, E.A. 1984. MINTEQ: "A computer program for calculating aqueous geochemical equilibria." (NTIS PB84-157148) EPA-600/3-84-032. Nat. Tech. Information Service, Springfield, Virginia.

Griffin, R.A., and Jurinak, J.J. 1973. "Estimation of activity coefficients from the electrical conductivity of natural aquatic systems and soil extracts." *Soil Science*. 116:26–30.

Grim, R.E. 1968. *Clay Mineralogy*. 2nd ed. McGraw-Hill, New York.

Harned, H.S., and Owen, B.B. 1958. *The Physical Chemistry of Electrolyte Solutions*. Reinhold Publ. Corp., New York.

Jurinak, J.J., Amrhein, C., and Wagenet, R.J. 1984. "Sodic Hazard: The Effect of SAR and Salinity in Soils and Overburden Materials." *Soil Science* 137:152–158.

Levy, R., and Hillel, D. 1968. "Thermodynamic equilibrium constants of sodium-calcium exchange in some Israel soils." *Soil Science*. 106:393–398.

Marion, G.M., and Babcock, K.L. 1976. "Predicting Specific Conductance and Salt Concentration of Dilute Aqueous Solution." *Soil Science* 122:181–187.

Merrill, S.D., Deutsch, J.R., and Pole, M.W. 1987. "Saturation percentage." *Reclaiming Mine Soils and Overburden in the Western United States*, Williams, D. R., and Schuman, G. E., eds., Soil Conservation Society America, Ankeny, Iowa.

Oster, J.D., and Rhoades, J.D. 1975. "Calculated drainage water compositions and salt burdens resulting from irrigation with river waters in the Western United States." *J. Environ. Qual.* 4:73–79.

Robinson, R.A., and Stokes, J.J. 1959. *Electrolyte Solutions*. 2nd ed. Butterworth & Co., London.

Sposito, G. 1977. "The Gapon and Vanselow selectivity coefficients." *Soil Sci. Soc. Am. J.* 41:1205–1206.

Stumm, W., and Morgan, J.J. 1981. *Aquatic Chemistry*. 2nd ed. John Wiley and Sons, New York.

Suarez, D.L. 1977. "Ion activity products of calcium carbonate in waters below the root zone." *Soil Sci. Soc. Am. J.* 41:310–315.

Suarez, D.L. 1981. "Relation between pHc and sodium adsorption ratio (SAR) and an alternative method of estimating SAR of soil or drainage waters." *Soil Sci. Soc. Am. J.* 45:469–475.

Suarez, D.L. 1982. "Graphical calculation of ion concentrations in $CaCO_3$ and/or gypsum soil solutions." *J. Environ. Qual.* 11:302–308.

Tanji, K.K. 1969a. "Predicting specific conductance from electrolytic properties and ion association in some aqueous solutions." *Soil Sci. Soc. Am. Proc.* 33:887–889.

Tanji, K.K. 1969b. "Solubility of gypsum in aqueous systems as affected by ion association and ionic strength up to 0.15M and at 25°C. Environ. Sci. & Tech. 3:656–661.

Tanji, K.K., and Biggar, J.W. 1972. "Specific conductance models for natural waters and soil solutions of limited salinity levels. *Water Resource Res.* 8:145–153.

U.S. Salinity Laboratory Staff. 1954. "Diagnosis and Improvement of Saline and Alkali Soils." *U.S. Department of Agriculture Handbook #60.* U.S. Printing Office, Washington, D.C.

Westall, J.C., Zachary, J.L., and Morel, F.M. 1976. "MINEQL: A computer program for calculation of chemical equilibrium composition of aqueous systems." *Tech Note 18.* Dept. Civil Engineering, Mass. Inst. of Tech., Cambridge, Mass.

Williams, J.S. 1975. "The natural salinity of the Colorado River." *Occasional Paper 7,* Utah Water Res. Lab., Logan, Utah.

Wolery, T.J. 1979. "Calculation of chemical equilibrium between aqueous solution and minerals: The EQ3/6 software package." *UCRL-52658.* Lawrence Livermore Lab., University of Calif., Livermore, Calif.

Wolery, T.J. 1983. *EQ3NR, A computer program for geochemical aqueous speciation-solubility calculations.* Lawrence Livermore Lab., Livermore, CA.

NOTATION

a = valence of cation A; b = valence of cation B; a,b,c = stoichiometric coefficients; A,B,C = chemical symbols; c_i = concentration of the *i*th ion; D = dielectric constant; e = electron charge; EC = specific conductance; EC_i = ionic specific conductance; EC_t = value at temperature t; ESR = exchangeable sodium ratio; F_c = concentration factor; k = Boltzmann's constant; k_g = Gapon selectivity coefficient; K' = solubility product for calcite; K_g' = modified Gapon selectivity coefficient; L = leaching fraction; m_i = molal concentration of the *i*th ion; Mg_{iw} = concentration of Mg in applied water; n^0 = electrolyte concentration in the bulk solution; Na_{iw} = concentration of Na in applied water; SAR = sodium adsorption ratio; T = absolute temperature; TSS = total soluble salt concentration; X = one mole of negative charge on the exchanger; x = amount of Ca precipitated or dissolved; z = valence of the counter ion; β = empirical interactive parameter; γ_i = activity coefficient; λ_i^0 = ionic equivalent conductance at infinite dilution; ϕ = osmotic coefficient of the *i*th salt; ( ) = activities; [ ] = concentrations.

CHAPTER 4

CHEMISTRY OF TRACE ELEMENTS IN SOILS AND GROUND WATER[a]

INTRODUCTION

High concentrations of inorganic trace elements in irrigated soils and shallow ground water pose a threat to agricultural production and the health of humans and animals. They do so in three ways: 1) Trace elements can accumulate in plants to levels that cause phytotoxicity; 2) trace elements in plants can adversely affect humans and animals that consume those plants; and 3) trace elements can migrate with seepage through the root zone and into ground water, possibly re-emerging with subsurface drainage in surface waters, thereby affecting wildlife, or with ground water pumped for domestic use, thereby threatening the health of humans.

The objective of this chapter is to provide a framework for evaluating trace-element concentrations in soils and shallow ground water in irrigated agricultural areas. High concentrations of trace elements in soils and ground water in irrigated areas can occur concomitant with soil and ground water salinity and can be affected by the same processes that affect soil and ground water salinity. High trace element concentrations may also occur independent of salinity.

We will describe processes that affect the mobility of trace elements. We have focused on elements designated by the U.S. Environmental Protection Agency (EPA) as priority pollutants (EPA 1986) and on elements that have been documented as pollutants associated with irrigated agriculture. We have divided the trace elements into four categories: (1) The alkali and alkaline-earth metals [barium (Ba) and lithium (Li)]; (2) the transition metals [chromium (Cr), molybdenum (Mo), and vanadium (V)]; (3) the nonmetals [arsenic (As), boron (B), and selenium (Se)]; and (4) the heavy metals [cadmium (Cd), copper (Cu), lead (Pb), mercury (Hg), nickel (Ni), and zinc (Zn)].

PROCESSES AFFECTING TRACE-ELEMENT CONCENTRATIONS

Precipitation and Dissolution

Precipitation and dissolution reactions may govern the activities of trace elements in soil solutions. These reactions generally are described

---

[a]Prepared by: Steven J. Deverel and Roger Fujii, U.S. Geological Survey, Sacramento, Calif.

by a solubility product relation, in which the solid dissolves to form soluble constituents. For example, the solubility of solid barite, $BaSO_4(s)$, may control the activity of Ba^{2+} in the soil solution of arid-zone soils and ground water (Hem 1985). At constant temperature and pressure, a solution in equilibrium with $BaSO_4(s)$ is described by the following equation:

$$BaSO_4\ (s) \rightleftarrows Ba^{2+} + SO_4^{2-} \qquad (4.1)$$

The reduced thermodynamic equilibrium constant expression for this reaction is

$$K_{so} = (Ba^{2+}) \cdot (SO_4^{2-}) \qquad (4.2)$$

where ( ) denote activities, the solubility product constant K_{so} has a value of about 10^{-10} at 25°C and 1 atmosphere pressure (Hem 1985), and H_2O and $BaSO_4(s)$ are assumed to be in their standard states with activities equal to unity.

Mineral phases affecting the activities of some trace elements in soil solutions and ground water can be assessed by calculating mineral saturation indices. The mineral saturation index (SI) is equal to the logarithm value for the quotient of the ion-activity product and mineral solubility product constant, or SI = log (IAP/K_{so}). Negative saturation-index values indicate mineral undersaturation, while positive values indicate supersaturation. Values approaching zero indicate possible thermodynamic mineral equilibrium.

Supersaturation often indicates that kinetic constraints are preventing precipitation. The result of this evaluation alone, however, does not constitute proof that a particular mineral phase is present and affecting the trace-element concentrations, nor that it is absent and not a factor in a water's trace-element chemistry. Mineralogic data can be collected to identify the presence of suspected minerals, thereby providing evidence to support the results of the saturation-index calculations.

Evaluation of saturation indices is limited by the availability and accuracy of mineral thermodynamic data that affect trace-element concentrations. Several computer models that calculate saturation indices based on currently available thermodynamic data exist. Trace element data generally are lacking in these codes. Nordstrom et al. (1979b) and Nordstrom and Ball (1984) review a number of these codes. They all assume that pure solid phases at their standard states either are present or will precipitate. Solid phases with unit activities probably are rare in natural systems (Corey 1981).

The evaluation of mineral saturation indices assumes equilibrium. This may be invalid in some natural systems. In general, the use of equilibrium calculations in mineral aqueous systems is valid only when the residence time of the water greatly exceeds the reaction half-life (Langmuir and Mahoney 1984).

For the case of mineral dissolution (mineral-water equilibria) and crystallization, equilibrium conditions may not be present in many natural systems, depending on the specific reactions and residence times. In contrast, the use of equilibrium calculations for solute-solute or solute-water interactions, such as complexation or acid-base reactions, generally will be valid because of the relatively rapid reaction rates, as long as the solution-theory model is adequate to describe the solution considered and the thermodynamic parameters are known with reasonable accuracy.

Solution-Phase Speciation of Trace Elements

The total (analytical) concentration of any given trace element in the aqueous phase depends on the ionic strength of the solution and the concentrations of other ions with which the trace element forms complexes. For example, the total aqueous concentration of Cd consists of the sum of the concentrations of uncomplexed Cd^{2+} and all the possible complexes formed in solution. This may be expressed mathematically by the following mass-balance equation:

$$[\mathrm{Cd}]_{\mathrm{total}} = [\mathrm{Cd}^{2+}] + [\mathrm{CdSO}_4^0] + [\mathrm{CdNO}_3^+] + [\mathrm{CdCl}^+] + [\mathrm{CdCl}_2^0] + [\mathrm{Cd\text{-}Org}] + \ldots \quad (4.3)$$

where [ ] represents molar or molal concentration and Cd–Org represents Cd complexed by dissolved organic matter.

At constant temperature and pressure, the formation of a soluble complex, for example, $CdCl^+$, can be described by the following mass action equation:

$$\mathrm{Cd}^{2+} + \mathrm{Cl}^- \rightleftarrows \mathrm{CdCl}^+ \quad (4.4)$$

The reduced thermodynamic stability constant expression for this reaction is

$$K_s = \frac{(\mathrm{CdCl}^+)}{(\mathrm{Cd}^{2+})\,(\mathrm{Cl}^-)} \quad (4.5)$$

where ( ) denote activities and H_2O is assumed to be at unit activity. Values of stability constants for inorganic complexes involving trace elements or free energy values from which the stability constants can be calculated are published extensively in the literature (Nordstrom et al. 1979b).

The activity of a single ion can be expressed as $a_i = m_i \gamma_i$, where i = the subscript; a_i = the activity of ion i; m_i = the molal (or, in dilute solutions, molar) concentration of species i; and γ_i is the activity coefficient of species i. For extremely dilute solutions (i.e., approaching infinite dilution), activities are approximated by molar concentrations for all species and γ_i equals unity. As the concentrations of charged dissolved constituents increase, γ_i decreases. This decrease represents the degree of departure from ideality of the ionic properties of ion i.

Individual ion-activity coefficients can be calculated in different ways, depending on the ionic strength of the solution (Table 4.1). At ionic strengths below 0.1 molal, the extended Debye-Hückel limiting

TABLE 4.1 Formulas for Calculating Ion-Activity Coefficients

Extended Debye-Huckel limiting law:

$$\log \gamma_i = -A \cdot z_i^2 \, [\, I^{0.5} / (1 + Ba_o \cdot I^{0.5}) \,]$$

Davies approximation:

$$\log \gamma_i = -A \cdot z_i^2 \, \{ [\, I^{0.5} / (1 + I^{0.5})] - 0.3\, I \}$$

Approximation used by Truesdell and Jones (1974):

$$\log \gamma_i = -A \cdot z_i^2 \, (\{ I^{0.5} / [1 + Ba_o \, (I)^{0.5}] \} + b \cdot I)$$

Ionic strength:

$$I = \tfrac{1}{2} \Sigma \, (m_i \cdot z_i^2)$$

The constants *A* and *B* depend on temperature, and a_o is related to the size of the hydrated radius of the particular ion. Drever (1988) lists the numerical values of these constants.

law is applicable (Stumm and Morgan 1981). The more general Davies equation is valid up to 0.5 molal (or molar). For solutions with ionic strengths approaching one molal, the equation presented by Truesdell and Jones (1974) (see Table 4.1) may be applied.

For solutions of extremely high ionic strengths (greater than 1 molal), the above equations are invalid. Pitzer's equations have been applied successfully to the prediction of solubilities of some evaporite minerals in solutions as concentrated as 20 molal (Pitzer 1979).

Adsorption

Activities of trace elements in soil solutions and natural waters frequently are too low to be controlled by precipitation and dissolution of pure solid phases. Jenne (1976) showed that adsorption reactions occurring at the solid-liquid interface are a significant mechanism controlling trace-element activities in many natural waters.

Adsorption reactions occur mainly in the clay-mineral size fraction of the soil and aquifer material at the surfaces of layered alumino-silicate minerals. The reactive surfaces include amorphous oxides and hydroxides of iron, manganese, and aluminum; organic matter containing weak-acid functional groups; and metal carbonates, such as calcite, $CaCO_3(s)$, and dolomite, $CaMg(CO_3)_2(s)$ (Jenne 1976). These soil components have the highest surface area and, therefore, the greatest contact with soil water. Even though their total concentrations may be relatively low, amorphous oxides, hydroxides, and organic matter occur as coatings on clay-mineral surfaces, thereby exerting considerable control on activities of trace elements in natural waters.

In contrast, colloidal calcite is often found in the clay mineral fraction of arid and semiarid soils. Calcite is an important adsorptive surface for heavy metals. It buffers the soil pH in the alkaline range and decreases heavy metal activities. Jenne (1976) presents an excellent review of these topics.

Adsorption of trace elements can be nonspecific, involving simple electrostatic attractions, or specific, involving coordinate covalent bonding. Trace-element adsorption occurs at surfaces where the nature of the surface charge affects the extent of the reaction. The permanent negative surface charge of 2:1 layered silicates arises from isomorphous substitution of Al^{3+} for silicon (Si^{4+}), substitution of a divalent cation for a trivalent cation (e.g., Mg^{2+} for Al^{3+}), or substitution of a monovalent cation for a divalent cation (e.g., Li^{+} for Mg^{2+}) (Bohn et al. 1979).

The permanent negative charge of layered silicate minerals is balanced by a diffuse cloud of cations. The cations, or counter ions, are electrostatically attracted to the negative surface, thereby increasing their concentration near the surface, which creates a concentration gradient favoring diffusion away from the surface. The opposite is true for anions. The balancing of these two opposing processes determines the distribution of cations and anions near the permanent charge surface, often referred to as the diffuse or electric double layer (Van Olphen 1977). The simplest interpretation of adsorption selectivity of cations is that constant-charge surfaces favor higher valence cations (stronger coulombic attraction) and smaller hydrated radii (closer approach to the surface).

In contrast to permanent-charge sites, charge sites that are pH-dependent exist on broken edges of layered silicate minerals, hydroxylated surfaces of metal oxides and hydroxides, and weak-acid functional groups of organic material. Adsorption reactions at pH-dependent charge sites require the formation of surface complexes and some degree of coordinate covalent bonding, which results in the specificity of the adsorption reactions. Fig. 4.1 depicts reactions that may take place at the hydroxylated surfaces of metal oxides, as suggested by Schindler (1981).

Fig. 4.1a shows the weak-acid nature of the surface hydroxyls. At low pH, the surface hydroxyl will be protonated and positively charged, giving rise to some anion exchange capacity. At higher pH, the surface is negatively charged and can contribute to the soil's cation exchange capacity.

Fig. 4.1b shows reactions for a dissolved trace-metal cation (e.g., Cd^{2+} or Zn^{2+}) forming a surface complex by covalent bonding to the surface, displacing protons. Similarly, ligands can exchange with surface hydroxyls (Fig. 4.1c). The adsorption of a dissolved metal ion also can coordinate a dissolved ligand (Fig. 4.1d), and adsorption of a dissolved ligand can coordinate a dissolved metal ion (Fig. 4.1e). Similar reactions occur at edges of layered silicate minerals (e.g., kaolinite edges) and at weak acid functional groups attached to organic material (e.g., carboxylic acid functional groups).

Hendrickson and Corey (1981) demonstrated the dependence of Cd selectivity on adsorption site coverage for a variety of solids. A continuum of different types of sites exists in natural systems, ranging from extremely specific sites at low trace-element concentrations to non-specific, cation-exchange-type sites at high concentrations. In natural systems where a small proportion of total site coverage by trace metals or trace-element oxyanions is expected, specific adsorption is usually the dominant process controlling the concentrations of many trace metals and trace-element oxyanions.

Trace metals compete with each other for adsorption on minerals

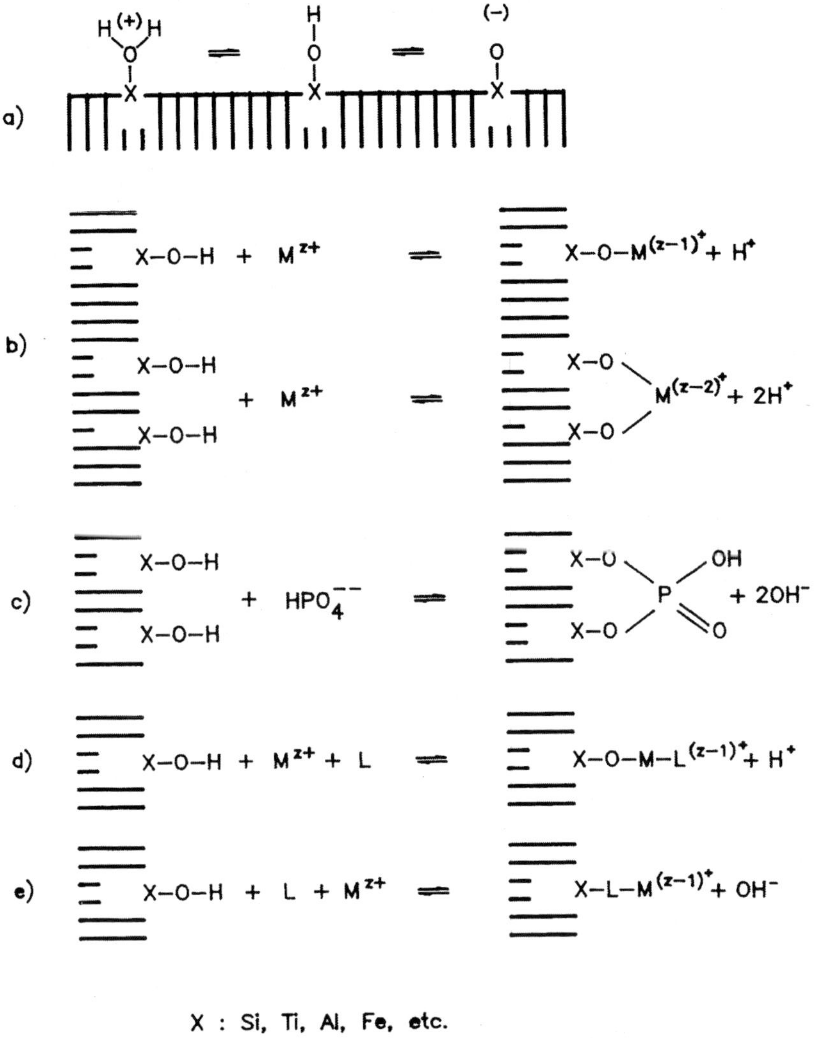

a) $X-OH_2^{(+)} \rightleftharpoons X-O-H \rightleftharpoons X-O^{(-)}$

b) $X-O-H + M^{z+} \rightleftharpoons X-O-M^{(z-1)+} + H^+$

$(X-O-H)_2 + M^{z+} \rightleftharpoons (X-O)_2 M^{(z-2)+} + 2H^+$

c) $(X-O-H)_2 + HPO_4^{--} \rightleftharpoons (X-O)_2 P(OH)=O + 2OH^-$

d) $X-O-H + M^{z+} + L \rightleftharpoons X-O-M-L^{(z-1)+} + H^+$

e) $X-O-H + L + M^{z+} \rightleftharpoons X-L-M^{(z-1)+} + OH^-$

X : Si, Ti, Al, Fe, etc.

Fig. 4.1 Coordination Phenomena at Oxide-Water Interfaces: (a) Acid-Base Reactions at Surface Hydroxyl Groups; (b) Deprotonated Surface Hydroxyls Coordinate with Dissolved Metal Ions; (c) Surface Hydroxyls Are Replaced by Dissolved Ligands; (d) Dissolved Metal Ion Coordinates with Both Deprotonated Surface Hydroxyls and Dissolved Ligands; (e) Dissolved Multidentate Ligand Coordinates with Both *X* and the Dissolved Metal Ion *M* (Schindler 1981)

and soils. Kinniburgh et al. (1976) showed that the selectivity sequence for iron-gel is $Pb^{2+} > Cu^{2+} > Zn^{2+} > Ni^{2+} > Cd^{2+} > Co^{2+} > Sr^{2+} > Mg^{2+}$. For aluminum-gel, the sequence is $Cu^{2+} > Pb^{2+} > Zn^{2+} > Ni^{2+} > Co^{2+} > Cd^{2+} > Mg^{2+} > Sr^{2+}$. Forbes et al. (1976) found trace-metal affinities for goethite to be $Cu^{2+} > Pb^{2+} > Zn^{2+} > Co^{2+} > Cd^{2+}$. Kuo and Mikkelsen (1979) compared Zn^{2+} competition with other trace metals for adsorption on two alkaline soils. They found the adsorption selectivities relative to Zn^{2+} to be: $Hg^{2+} > Cu^{2+} > Fe^{2+} > Mn^{2+}$.

Ligand exchange of anions for surface hydroxyls is the predominant process affecting the specific adsorption of oxyanions on goethite (e.g., Hingston 1981; Goldberg 1985) and probably on soils (Neal et al. 1987a and b; Fujii et al. 1988). Hingston et al. (1968) showed that affinity for anion adsorption on goethite followed the order: phosphate > silicate > selenite > fluoride.

Describing Adsorption Data

Adsorption isotherms describe the relation between the mass of a substance adsorbed (adsorbate) by the solid (adsorbent) and the equilibrium solution concentration (or, more appropriately, activity) supported by the adsorbed phase. These data usually are fit to a model from which adsorption mechanisms may be inferred. Adsorption equations have been incorporated into solute-transport models to estimate the retention and release of an adsorbing species during transport (e.g., Lewis et al. 1986).

One common model used to describe trace-element adsorption data is the Freundlich equation

$$\frac{x}{m} = KC^{1/n} \tag{4.6}$$

where x = the mass of trace element adsorbed; m = the mass of solid phase adsorbent; C = the equilibrium concentration of adsorbate in solution; and K and n are empirically-determined constants.

Many researchers have applied the Freundlich adsorption equation to describe adsorption of trace-metal adsorption onto soils and soil minerals (e.g., Garcia-Miragaya and Page 1976, Bowman et al. 1981).

Another model commonly used to describe trace-element adsorption is the Langmuir equation, which Langmuir (1918) originally developed for adsorption of gases onto solids. When applied to minerals, soils, and sediments, the equation can take the form

$$\frac{x}{m} = \frac{kbC}{(1 + kC)} \tag{4.7}$$

where k = a parameter related to the affinity of the adsorbent for the adsorbate; and b = the adsorption maximum. A linearized form of the equation typically is used to describe sorption data as follows:

$$\frac{C}{x/m} = \left(\frac{1}{kb}\right) + \left(\frac{C}{b}\right) \tag{4.8}$$

TABLE 4.2 Critical Values of pE for Changes in Redox Speciation of Selected Trace Elements at 25°C

Oxidation states (1)	Reduction half-reaction (2)	pE^c_{crit} pH 5 (3)	pH 8 (4)
Mn(III)/Mn(II)[a]	$MnOOH(s) + 3H^+ + e^- \rightleftarrows Mn^{2+} + 2H_2O$	15.9	6.9
Mn(IV)/Mn(II)[a]	$1/2\ MnO_2(s) + 2H^+ + e^- \rightleftarrows 1/2\ Mn^{2+} + H_2O$	14.7	8.7
Cr(VI)/Cr(III)[a]	$1/6\ Cr_2O_7^{2-} + 4/3\ H^+ + e^- \rightleftarrows 1/3\ Cr(OH)_3\ (s) + 1/6\ H_2O(l)$	13.0	8.0
Hg(II)/Hg(I)[a]	$HgO(s) + 2H^+ + e^- \rightleftarrows 1/2\ Hg_2^{2+} + H_2O$	11.8	5.8
Fe(III)/Fe(II)[a]	$Fe(OH)_3(s) + 3H^+ + e^- \rightleftarrows Fe^{2+} + 3H_2O$	8.9	−0.1
Cu(II)/Cu(I)[a]	$CuO(s) + 2H^+ + e^- \rightleftarrows Cu^+ + H_2O$	8.3	2.3
Ag(I)/Ag(0)[a]	$Ag^+ + e^- \rightleftarrows Ag(s)$	5.5	5.5
Mo(VI)/Mo(IV)[a]	$1/18\ MoO_4^{2-} + 1/9\ SO_4^{2-} + 4/3\ H^+ + e^- \rightleftarrows 1/18\ MoS_2(s) + 2/3\ H_2O$	0.5[d]	−3.9[d]
Se(VI)/Se(IV)[b]	$SeO_4^{2-} + 2e^- + 2H^+ \rightleftarrows SeO_3^{2-} + H_2O$	9.5	6.5
Se(VI)/Se(s)[b]	$SeO_4^{2-} + 6e^- + 8H^+ \rightleftarrows Se_{(s)}$ (hexagonal, black) $+ 4H_2O$	6.8	2.9

[a]Sposito (1983)
[b]Adapted from Elrashadi et al. (1987)
[c]Critical activities of trace elements assumed equal to 10^{-8} (Sposito 1983)
[d]$(SO_4^{2-}) = 10^{-3}$ assumed in calculating pE_{crit}

From a plot of *C*/*x*/*m* versus *C*, the *b* and *k* terms can be estimated from the slope and intercept, respectively.

The original derivation of the Langmuir model assumes that only one type of adsorption site exists and that it has constant binding energy (homogeneous surface). However, minerals, soils, and sediments contain surface sites of varying specificity or binding energies. Syers et al. (1973) addressed this problem by using a two-site Langmuir model (high-energy and low-energy sites) to describe the sorption of phosphate by soils. Harter and Baker (1977) and Sposito (1984) addressed other problems associated with using the Langmuir model, such as precipitation of the adsorbate and competition from other adsorbates.

The constant capacitance model (Stumm et al. 1980) assumes and exchange of anionic trace elements for surface hydroxyls on oxide minerals, forming tightly bound inner-sphere complexes. The model can be applied to adsorption of phosphate, silicate, boron, arsenate, and selenite (Goldberg and Sposito 1984a; Goldberg 1985; Goldberg and Glaubig 1985; Goldberg 1986). The model can also describe phosphate adsorption by noncalcareous soils (Goldberg and Sposito 1984b). Anderson and

Rubin (1981), Davis and Hayes (1986), and Sposito (1983, 1984) have reviewed other models based on surface complexation and ligand exchange.

Oxidation-Reduction Processes

Many trace elements exist in more than one oxidation state (e.g., V, Cr, Se, As), and are therefore affected by electron transfer or oxidation-reduction (redox) reactions. Comprehensive reviews of redox reactions in natural systems are presented by Stumm and Morgan (1981), Sposito (1981), and Ponnamperuma (1972).

Most unsaturated soils contain enough oxygen (O_2) to maintain oxidizing conditions. In poorly aerated soils where the supply of O_2 is limited by the rate of diffusion, anoxic micro-sites can develop. Under water-saturated conditions, the O_2 supply rate may be slower than the O_2 demand, and anaerobic or reducing conditions may develop (Reddy and Patrick 1983).

The potential for a substance to accept or donate electrons under equilibrium conditions is represented by standard electrode potentials of reduction half-reactions (*Eh*) relative to the half-reaction for the hydrogen electrode. These reactions involve the transfer of electrons and depend on the activity of electrons in solution, represented by *pE*, the negative logarithm of the aqueous electron activity.

A major electron-donating process in soils is the microbial oxidation of reduced organic carbon. Oxygen is the primary electron acceptor for this reaction. In the absence of O_2, other soil constituents act as the electron acceptor and are reduced. Bohn et al. (1979) lists the principal electron acceptors in soils in the order of their tendency to be reduced, as indicated by the equilibrium potentials of the half-reactions at pH 7: $O_2 > NO_3^- > MnO_2 > FeOOH > SO_4^{2-} > H^+ > (CH_2O)_n$.

Table 4.2 reports critical values of the redox potential for changes in speciation for selected trace elements. Within the range of redox potentials reported for soils (Baas-Becking et al. 1960), changes in oxidation states for these trace elements are expected.

The reduction reactions in Table 4.2 can increase or decrease the concentration of trace elements. For example, reduction of Mn(III) or Mn(IV) to Mn(II) increases the concentration of Mn because Mn^{2+} is more soluble. In contrast, reduction of SeO_4^{2-} to the much-less-soluble elemental selenium decreases Se concentration. Selenate reduction may be the major mechanism immobilizing Se in sediments and deterring its transport to the ground water at the Kesterson National Wildlife Refuge on the west side of the San Joaquin Valley, Calif. The refuge was formerly an impoundment for agricultural drain water with high concentrations of selenium.

Minerals such as MnO_2 and $Fe(OH)_3$ act as sorbents for trace elements; hence, dissolution of these solid phases also can release adsorbed trace elements. During an investigation of As mobility in ground water, Gulens et al. (1979) found decreased adsorption of As by hydrated ferric oxide under reducing conditions. He attributed this to reduction of Fe (III) to the more soluble Fe(II) and subsequent release of As. Conversely, Reddy and Patrick (1977) reported that water-soluble Pb concentrations decreased (from 45 to 22 μg/l at pH 5) with increases in redox potential due to the probable formation of Fe and Mn oxyhydroxides under oxidized conditions, with subsequent adsorption of Pb.

Under sufficiently reducing conditions, sulfate is reduced to sulfide and trace metal sulfides precipitate (Lindsay 1979). Bingham et al. (1976), and Reddy and Patrick (1977) reported a decrease in Cd mobility under reducing conditions due to precipitation of CdS.

Considerable effort has been expended to represent stability or predominance fields for minerals and dissolved constituents in natural systems, as a function of *Eh* (or *pE*) and *pH* (e.g., Garrels and Christ 1965; Lindsay 1979; Drever 1988; Hem 1985). These diagrams represent equilibrium systems and provide critical information on possible equilibrium assemblages of minerals and coexisting water compositions. They also help identify disequilibrium assemblages of minerals and aqueous redox couples (Thorstensen 1984).

Redox reactions in natural systems generally are considered to be at only partial equilibrium because the kinetics usually are slow; therefore, equilibrium is not achieved, owing to lack of effective coupling of redox reactions (Sposito 1983; Thorstensen 1984). The platinum (Pt) electrode has been used extensively for the measurement of redox potential in field situations. However, Pt-electrode measurements do not necessarily represent the redox potential of the system; they may be the result of a single electroactive redox couple or may represent mixed potentials at the electrode surface. Thus, in natural systems, Pt-electrode measurements should be interpreted with great caution (Lindberg and Runnells 1984; Thorstensen 1984).

In natural systems, kinetically limited redox reactions are commonly mediated by microbial organisms. Therefore, the appropriate microbial population must be present to effect a particular reaction. Organisms affect only the kinetics of the reaction and do not alter the thermodynamic constraints (Sposito 1983).

In spite of the problems discussed above, environments exist where dominant redox couples are effectively catalyzed and thermodynamic interpretations may be applied. In flooded soils, for example, reduction of Fe and Mn minerals produces high concentrations of Fe^{2+} and Mn^{2+}, which can dominate the redox potentials of the system and, thus, allow a quantitative thermodynamic description of redox-active ion activities (Bohn et al. 1979; Ponnamperuma 1972; Sposito 1983). Other investigators have reported Nernstian behavior between measured *Eh* and various redox-sensitive elements (e.g., Nordstrom et al. 1979a).

GEOCHEMICAL BEHAVIOR OF TRACE ELEMENTS

The relative importance of the geochemical processes discussed in the preceding sections is determined by an individual element's mode of occurrence. An element's mode of occurence is, in turn, determined primarily by the pH and redox status of soil solutions and ground water, and by the solid phase controlling solubility and distribution. A review of the distribution, sources, and behavior of individual trace elements in

TABLE 4.3 Average Composition in mg/kg of Some Types of Sedimentary Rocks (Horn and Adams 1966)

Element (1)	Resistates (2)	Hydrolyzates (3)	Precipitates (4)
Arsenic	1.0	9.0	1.8
Barium	193	250	30
Boron	90	194	16
Cadmium	0.020	0.22	0.19
Copper	15	45	4.4
Chromium	120	423	7.1
Lead	14	80	16
Lithium	15	46	5.2
Mercury	0.057	0.18	0.068
Molybdenum	0.50	4.2	0.11
Nickel	2.6	29	13
Selenium	0.52	0.60	0.32
Zinc	16	130	16

the near-earth-surface environment is presented here to provide a framework for evaluating potential trace-element pollution problems associated with irrigated agriculture.

Sedimentary rocks and soils are the solid-earth materials that most affect and determine the trace-element chemistry of ground water (Hem 1985). Sedimentary rocks can be divided into four groups, based on their resistance to weathering: resistates, hydrolyzotes, precipitates, and evaporites. Resistates are rocks composed primarily of residual aluminosilicate minerals that are not easily altered chemically by the weathering of parent rock. Hydrolyzates are composed primarily of insoluble metal oxide and aluminosilicate minerals derived from the weathering of parent rock. Precipitates are rocks that are a result of direct precipitation of mineral matter from aqueous solution. Evaporites are rocks composed of minerals deposited during evaporation.

Table 4.3 presents the concentrations of trace elements in sedimentary rocks. Evaporites also affect the composition of some soil solutions and ground water associated with irrigated agriculture. The table excludes evaporites, however, because little is known about their trace-element concentrations (Hem 1985). Trace-element concentrations generally are higher in hydrolyzates than in resistates and precipitates (Table 4.3). This is especially true for those elements that adsorb strongly to metal oxide minerals, e.g., Zn, Ni, Hg, Pb, Cu, Cr, Ba, and As.

Alkali and Alkaline Earth Metals

Barium. Mineralogic sources of Ba include aluminosilicate mineral structures, such as feldspars and micas, where Ba substitutes for K and Ca. Barium, which is present in the +2 valence state, also substitutes for Ca in secondary minerals such as calcite and apatite. In weathering environments, barium solubility seems to be primarily controlled by reactions with mineral surfaces and by the precipitation and dissolution of barite ($BaSO_4$) and witherite ($BaCO_3$) (Rai and Zachara 1984).

The interaction of Ba with mineral surfaces is important in controlling aqueous concentrations. Nonspecific adsorption on constant-charge surfaces exerts substantial control on aqueous Ba concentrations (e.g., Elprince et al. 1980). Barium is also specifically adsorbed onto silicate and aluminum-, manganese-, titanium-, and iron-oxide minerals (Kinniburgh et al., 1976; Posselt et al. 1978). Organic complexing of Ba also affects aqueous concentrations.

Lithium. Although typically present as a monovalent cation, lithium's small ionic and atomic size causes it to behave similarly to the divalent alkaline earth metals, especially Mg^{2+}. Mineralogically, Li occurs in Li-halogen and Li-oxygen bonds, primarily in silicate, phosphate, and halide minerals.

Released from these primary minerals during weathering, Li^+ commonly is removed from solution by incorporation into clays. Among monovalent cations, Li is the most weakly bonded of all the alkali metals. In soils and ground water in which Na and K are present at high concentrations, Li may remain in solution and behave conservatively. Deverel and Millard (1988) found that Li concentrations greater than 100 μg/l were associated with ground-water salinity in $NaSO_{4-}$ dominated shallow ground water beneath the San Joaquin Valley, California.

The environmental toxicity of Li is not well documented; however, it is apparently toxic to plants at concentrations of less than 100 μg/l in irrigation water. Bradford (1963) reported that citrus trees may be damaged by irrigation water containing 60 to 100 μg/l of *Li*.

Transition Metals

Chromium. Chromium exists primarily in the trivalent (Cr^{3+}) and hexavalent (CrO_4^{2-}) forms. The hexavalent form is substantially more toxic than the trivalent form (EPA 1986). Turner and Rust (1971) and Ross et al. (1981) reported reduced plant growth and soil microbial activity due to high concentrations of Cr in soil.

Anthropogenic inputs in the form of sludges, wastewater, and mining activities are the primary sources of high Cr concentrations in soils and ground water. Robertson (1975) has reported naturally occurring CrO_4^{2-} concentrations as high as 200 μg/l in Arizona, which apparently resulted from the dissolution and oxidation of Cr^{3+}-bearing minerals by oxygenated ground water.

The solubility of Cr^{3+} in soil solution and ground water is affected primarily by its interaction with mineral surfaces and co-precipitation with iron-oxide type minerals. Chromite ($FeCr_2O_4$) is stable at low temperatures for most of the water stability region and could be formed as an admixture or co-precipitate with ferrous hydroxide (Hem 1977). Rai and Zachara (1986) identified the solid solution $Cr_xFe_{1-x}(OH)_3$ and $Cr(OH)_3$ (chromium hydroxide) as the solid phases exercising the predominant solubility controls on Cr^{3+} in geologic systems. The amount of Cr^{3+} adsorbed by a Typic Fragiorthod soil increases with pH increasing to 4 (Bartlett and Kimble 1976). Trivalent Cr is also complexed by organic compounds in soil.

Trivalent Cr oxidizes readily to the hexavalent form under conditions prevalent in field soils (Bartlett and James 1979). Organic material and manganese oxides (Bartlett and James 1979; Rai and Zachara 1986) are important electron acceptors for this reaction.

The mobility of CrO_4^{2-} in soil solutions and ground water is controlled primarily by adsorption/desorption reactions at mineral surfaces. Amorphous iron oxide has a high capacity for adsorption of CrO_4^{2-} (Leckie et al. 1980). Davis and Leckie (1980) developed a mathematical model which closely simulated the pH and ionic-strength dependent adsorption of CrO_4^{2-} on aluminum oxide. Their approach and the use of the constant capacitance model (Stumm et al. 1980; Rai and Zachara 1986) underscores the need to consider the supporting electrolyte and competing ions. James and Bartlett (1983) and Stollenwerk and Grove (1985a) used Langmuir isotherms to describe adsorption of CrO_4^{2-} onto soils and aquifer materials. The hexavalent form can be reduced to Cr^{3+} in soils (James and Bartlett 1983) and in water samples acidified for preservation (Stollenwerk and Grove 1985b).

Molybdenum. Molybdenum occurs in oxidation states from $^{+}3$ to $^{+}6$. Under acidic reducing conditions, the Mo(IV) oxidation state dominates and usually precipitates as a sulfide. Under oxidized conditions, the Mo(VI) oxidation state dominates, primarily as MoO_4^{2-}. Because of its oxyanion form, it commonly behaves conservatively in alkaline, oxidized environments. Molybdenum can substitute for aluminium, iron, and titanium in aluminosilicate or metal oxides, and is often associated with organic materials in sedimentary rocks and soils.

Mineral surface reactions and interactions with organic matter are the predominant processes controlling the mobility of Mo in soils and ground water (Jarrel et al. 1980). Adsorption and incorporation into manganese, iron, and aluminum oxides are predominant mechanisms controlling Mo solubility in oxidized environments (Hem 1977; Reyes and Jurinak 1967). Co-precipitation and adsorption of Mo with and onto manganese and iron oxides decreases with increasing pH greater than 5 (Chan and Riley 1966; Reyes and Jurinak 1967).

The EPA (1985) has proposed a drinking-water criterion for molybdenum of 0.1 mg/l. Little evidence exists of human Mo toxicity from drinking water, and few data are available on chronic Mo toxicity. Concentrations of Mo in soils can cause problems if vegetation high in Mo is consumed by livestock (Jarrel et al. 1980).

Vanadium. Vanadium (V) exists in four different geochemical environments: sulfide deposits, oxidized sulfide ores, aluminosilicates, and iron-oxide deposits (Wedepohl 1972). Vanadium can exist in three oxidation states: +3, +4, and +5 (Hem 1977). The trivalent form of vanadium is incorporated into sulfide minerals, and it often replaces Al in aluminosilicate structures and Fe in iron-oxide minerals.

In oxidized deposits, V is present in the tri-, tetra- and pentavalent forms. Evans and Garrels (1958) identified a V mineral weathering sequence spanning the valence states from +3 to +5 and associated with oxidized uranium deposits in the Colorado Plateau. Hem (1977) presents evidence that ferric and ferrous vanadates may control the solubility of V in natural waters. Interactions with mineral surfaces substantially influence aqueous V activities in acidic environments. McBride (1979) and Taylor and Giles (1970) described the adsorption of vanadium on montmorillonite and Mn- and Fe^{3+}-oxides.

Because it tends to exist in oxyanion forms, V commonly is mobile in alkaline, oxidized aqueous environments. Deverel and Millard (1988) reported V concentrations to be significantly correlated with specific conductance data in shallow ground water in the San Joaquin Valley, California. Anthropogenic sources of V in soils and ground water include industrial wastes, sewage sludge, fossil-fuel byproducts, and mining spoils (National Academy of Sciences 1974; Mattigod and Page 1983). Because of its mobility in oxidized environments, V from these sources may represent a potential pollution hazard. Although tolerance levels for aquatic life have not been established, incidences of industrial V toxicosis, primarily from airborne V, have been documented (National Academy of Sciences 1974).

Nonmetals

Arsenic. Because of its carcinogenic effects on human health, detectable concentrations (ng/l levels) or arsenic (As) in water consumed by humans are undesirable (EPA 1986). The predominant factors governing As levels in natural waters are the redox state of the water, aqueous-mineral interactions, and biochemical transformations.

Arsenic exists in several oxidation states. In the tri- and pentavalent forms, it can be incorporated into aluminosilicates and titanium- and iron-oxide minerals. In the more reduced −3 valence state, As forms sulfides such as arsenopyrite (FeAsS), realgar (AsS), and orpiment (As_2S_3).

Adsorption and coprecipitation of arsenic by clay minerals and metal oxides are the predominant solubility controls in oxidized environments (Hem 1977; Pierce and Moore 1980; Leckie et al. 1980). Leckie et al. (1980) demonstrated that arsenate adsorption on amorphous iron oxyhydroxide ($Fe(OH)_3$) decreased at pH values greater than 8.

Welch et al. (1988) described geochemical environments in the western United States where arsenic may be present in the ground water. High arsenic concentrations are found in ground water in alluvial basins derived from neutral to acidic extrusive and volcanic rocks. Ground water in gold mining areas and geothermal areas where arsenosulfide minerals are present generally has high arsenic concentrations. The application of arsenical pesticides is another source of arsenic pollution in ground water and soils (Hem 1985).

Biogeochemical transformations of arsenic include biomethylation to produce soluble methylated arsenic compounds and mono- or dimethyl arsine gases (Welch et al. 1988; Lemmo et al. 1983; Faust et al. 1983).

Boron. Boron generally behaves conservatively in natural waters. It occurs primarily as the undissociated $B(OH)_3$ at $pH < 9$ and as the $B(OH)_4^-$

at pH > 9. Primary and secondary aluminosilicate minerals often contain boron, as it substitutes for Si and Al. In semiarid and arid environments, boron may substantially threaten irrigated agriculture due to its phytotoxic effects and conservative behavior. Evidence exists that bioaccumulation to levels toxic to waterfowl can occur (U.S. Fish and Wildlife Service 1986). Deverel and Millard (1988) reported boron concentrations that were highly correlated with ground water salinity in the San Joaquin Valley, Calif.

Boron concentrations can be affected substantially by interaction with mineral surfaces. Biggar and Fireman (1960) attributed the removal of boron from solution as precipitation of solid phases. Most researchers have described the B-soil association as one of adsorption or desorption. Goldberg and Glaubig (1986b) and Rai and Zachara (1984) have summarized research in this area.

Elrashidi and O'Connor (1982) and Biggar and Fireman (1960) used the Langmuir and Freulich equations to describe boron adsorption. The change in adsorptive behavior with changing pH cannot be accurately described using this approach. Boron adsorption onto various minerals is strongly pH-dependent, exhibiting maximum adsorption from pH 7 to 10.

Keren et al. (1981) successfully modeled the pH dependence of boron adsorption on soils and montmorillonite. Goldberg and Glaubig (1985, 1986a and 1986b) applied the constant capacitance model to boron adsorption on soils and various minerals from pH 3 to 12. Boron is mainly adsorbed on metal-hydroxy clusters and coatings in soils.

The soluble boron in irrigated soils may regenerate or increase in concentration after reaching a constant low concentration. Peryea et al. (1985) concluded that slow release of boron from adsorption sites and diffusion of boron from small pores to large pores are the mechanisms responsible for the regeneration of boron in the soil solution. See Chapter 19 for a discussion on boron reclamation in soils.

Selenium. Selenium generally is similar to sulfur in its chemical behavior (Lakin 1973). Substantial evidence exists that suggests Se is associated primarily with sulfide minerals present in sedimentary environments (Sindeeva 1964). The mobility of Se in the aqueous phase depends on the aqueous complex formation and redox speciation and reactions at mineral surfaces. High concentrations of Se in ground water and agricultural drain water in several western states primarily result from the high mobility of the SeO_4^{2-} (selenate) ion, the predominant species in alkaline, oxidized environments (Deverel and Millard 1988; Crist 1974; Sylvester, et al. 1988).

Selenate (SeO_4^{2-}) generally is nonadsorptive, especially in the presence of high concentrations of sulfate. Selenite (SeO_3^{2-}), in contrast, generally adsorbs strongly onto mineral surfaces (Balistrieri and Chao 1987; Ryden et al. 1987).

Selenium toxicity in waterfowl at Kesterson Reservoir, the San Joaquin Valley, Calif., resulted from high concentrations of SeO_4^{2-} in shallow ground water (e.g., Presser and Ohlendorf 1987). This selenium contamination of the ground water results from irrigation of soils derived from Cretaceous and Tertiary geologic materials in the California Coast

Range. The semiarid climate resulted in the formation of naturally saline soils in many areas. Irrigation of these soils resulted in the movement of soluble salts, including Se, into the ground water. Partial evaporation of a shallow water table in low-lying areas resulted in even higher salinity and Se concentrations (Deverel and Fujii 1988). Farm drainage systems remove this high-Se ground water in some areas. Drain water in some areas is mixed with nonsaline water and reused for irrigation. In other areas, drain water is mixed with surface runoff and irrigation water and discharged to adjacent wetlands and the San Joaquin River.

Heavy Metals

The geochemical behavior of the heavy metals cadmium, copper, lead, nickel, and zinc generally is similar; therefore, they will be discussed as a group. Mercury will be discussed separately.

Cadmium, Copper, Lead, Nickel, and Zinc. In oxidized sedimentary and secondary mineral environments, the solubility, mobility, and distribution of the heavy metals is affected by adsorption onto and coprecipitation with iron and manganese oxides and oxyhydroxides (e.g., Gibbs 1976; Kinniburgh et al. 1976; Jenne 1968; Quirk and Posner 1976). Organic complexing needs to be considered when evaluating the mobility and solubility of these elements in soils and ground water. Numerous authors have reported that variable proportions of heavy metals are present in organically bound forms in soils and sludge-amended soils. Hickey and Kittrick (1984) used selective extraction techniques to determine that, with the exception of Cu, the organically complexed trace metals represented small (less than 10%), but significant, proportions of the total metal present in the soil. They observed that 28% of the Cu present in the soils studied seemed to be organically complexed. Other studies with sludge-amended soils also demonstrate that Cu is organically complexed to a greater degree than the other metals (e.g., Emmerich et al. 1982a, 1982b).

Modeling and quantitative estimation of organically complexed metals in natural waters commonly is limited by lack of information on the chemical composition of the dissolved organic matter and the appropriate thermodynamic constants. Lerman and Childs (1973), Stumm and Morgan (1981), Davis (1984), Davis and Leckie (1978), and others have attempted to quantify the effect of organic ligands on the aquatic solubility and mobility of various metals. The general effect of organic complexing is to maintain higher concentrations of heavy metals in solution.

Sources of heavy metals in soils and ground water include primarily anthropogenically derived point and nonpoint sources. Point sources include mining residues and irrigation water. Nonpoint sources of metals include sewage sludge, fertilizers, amendments, and pesticides. There is little documentation of high concentrations of heavy metals in agricultural soils and ground water underlying agricultural areas from natural geologic sources. Mattigod and Page (1983) have extensively reviewed the sources and mobility of heavy metals in soils.

The toxicity of heavy metals to aquatic life varies, depending on the hardness of the water. Higher hardness generally increases the total concentration of heavy metals that can be tolerated by aquatic life. Most aquatic life forms have high tolerance levels for Cu, Ni, and Zn and low tolerance levels for Cd and Pb (EPA 1986).

Mercury. Mercury exists in the mercurous (+1), mercuric (+2) and metallic (0) forms in soils and ground water (Hem 1970). Mercury readily forms aqueous organic complexes and commonly is present in methylated forms derived from biological metabolism (Jensen and Jernolov 1969). Chronic toxicity of Hg is at present a dominant environmental issue (EPA 1986).

Sources of Hg in soils and ground water include sewage sludges, mining wastes, atmospheric pollutants, soil parent materials, pesticides, fungicides, and fertilizers. Phosphorous fertilizers in particular are reported to have high Hg concentrations—4 μg/g to 100 μg/g (Anderson 1979). Organic-rich shales, sedimentary clays, and sulfides are the major geologic sources of Hg (Fleischer 1970).

The fluxes and forms of mercury in soils and ground water depend on several factors, including temperature, moisture content, and oxidation and reduction state (e.g., Hem 1985; Benes and Havlik 1979). Soil atmospheric flux can be considerable, with volatilization of inorganic forms and release of methylated forms predominant. Although mercury sometimes can move through the soil profile to shallow ground water and drain water, it usually is substantially removed from solution by adsorption or incorporation into organic matter (Anderson 1979).

In oxidized solutions, the Hg^{2+} ion predominates. The divalent form complexes with Cl^- ion, so aqueous $HgCl_2^o$ will be the predominant dissolved inorganic form in most oxidized waters (Benes and Havlik 1979). Under reducing conditions, Hg exists in the relatively insoluble elemental or sulfide forms, $Hg(SH)_2$ and HgS_2^{2-}. Dismutation of the mercurous form, Hg_2^{2+}, to $Hg0 + Hg^{2+}$ also increases volatility and solubility (Hem 1970).

Organic Hg complexes, including methylated and phenylated forms (Benes and Havlik 1979), comprise a significant part of soluble Hg in soil solutions and ground water. These complexes generally are unstable and easily decomposed or volatilized. Adsorption of Hg^{2+} removes Hg from solution. Adsorption of Hg^{2+} onto Mn oxides (Jenne 1970), Fe oxides (Anderson 1979), and soils and sediments (Benes and Havlik 1979) controls the fate of Hg in aqueous environments.

Caution must be exercised in the collection and preservation of samples for Hg analysis. Adsorption to container walls can be prevented by acidification. Reduction to elemental Hg, which is easily volatilized, can be prevented by the addition of an oxidizing agent such as potassium dichromate or potassium manganate.

EVALUATION OF POLLUTION POTENTIAL IN SOILS

The evaluation of inorganic trace-element concentrations in soils involves consideration of: (1) Loading from various sources; (2) potential uptake by plants; (3) potential leaching to ground water or drain water; and (4) potential movement of trace elements with eroded soils or in surface runoff. Evaluation of uptake by plants requires comparison of trace-element concentrations in the soil with levels known to be associated with harmful levels of plant uptake. Evaluation of all considerations requires that the mode of occurrence of the element or elements be determined and the effects of the various chemical and physical processes be identified.

Soil chemical extractants have been used successfully to predict plant uptake, especially under controlled conditions. However, the use of chemical extractants may be ineffective for predicting plant uptake in uncultivated soils and among native species (Gough et al. 1979). Soil-chemical extraction procedures theoretically dissolve specific phases and minerals in a soil sample. Chao (1984) has written a comprehensive review of selective dissolution techniques. The partitioning of elements such as Se and As, which typically are present as anions and undergo oxidation and reduction reactions in soils, may not be characterized by these procedures (Greubel et al. 1988). Less destructive techniques, such as scanning electron microscope, electron-microprobe energy-dispersive x-ray analysis, and x-ray photoelectron spectroscopy, can provide information about the mode of occurrence of trace elements in soils, but are limited to high detection limits.

Evaluating the spatial distribution of the total concentrations of trace elements in soils in relation to morphological features and hydrologic processes can provide insight about sources, transport processes, and mobility. Tidball et al. (1986a), for example, plotted a spatial interpolation of the results of the analysis of more than 700 samples collected in the western San Joaquin Valley, Calif. These samples were analyzed for more than 30 trace elements. Mobile elements, such as Se, were concentrated in areas where there is less leaching of alluvial material or evapoconcentration from a shallow water table. They also found that less mobile elements, such as Hg and As, tended to be concentrated near the source of the alluvial deposits.

Factor analysis provides a way of examining the results from chemical analyses of a large number of samples containing trace elements (Joreskog et al. 1976) by simplifying large quantities of data into associations with master variables. Controversy arises because of the subjectivity involved in identifying master variables and assigning physical and chemical meanings to them.

Tidball et al. (1986b) used factor analysis to evaluate the results of the sampling described above by plotting the spatially interpolated factor scores. McNeal et al. (1985) determined the mode of occurrence of various trace metals in uncultivated soils of the northern Great Plains by combining partial dissolution techniques (Chao 1984) and factor analysis.

Another method of understanding the areal distribution and spatial variability of soil trace-element concentrations is the low-density, hierarchial, analysis-of-variance approach extensively used by the U.S. Geological Survey Branch of Regional Geochemistry (Tourtelot and Miesch 1975). Geochemical differences among regional subgroups are identified by determining the proportion of the total variability among all samples represented by the subgroups. This division also allows for evaluation of the proportion of the variance attributable to sampling and analytical error.

The comparison of soil trace element concentrations with back-

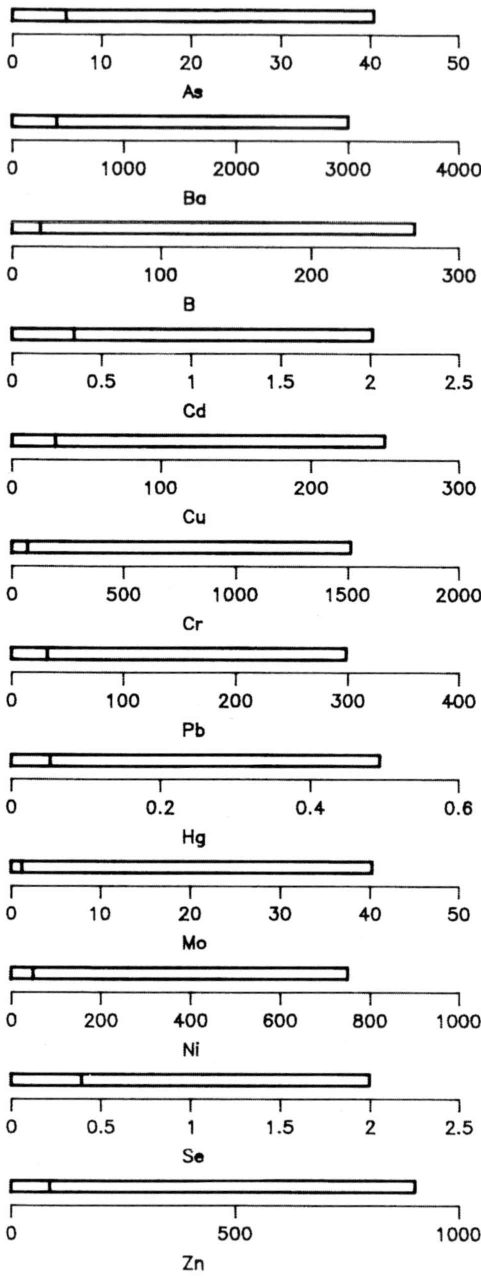

Fig. 4.2 Ranges of Concentrations of Trace Elements in Soils, in mg/kg (Bowen 1979). The Rectangle Represents the Middle 50% of the Data and Vertical Lines Represent the Median.

ground levels needs to be assessed in conjunction with an element's mode of occurrence and mobility. Bowen (1979) has compiled the concentration ranges and median values of trace elements shown in Figure 4.2. Values of total concentrations higher than those listed may indicate that a problem exists. However, the mode of occurrence and mobility of an element in the soil needs to be evaluated before drawing conclusions about potential harmful effects to plants or animals or movement into the ground water.

REFERENCES

Anderson, A. 1979. "Mercury in soils." *The Biogeochemistry of Mercury in the Environment.* J.O. Nriagu, ed. Elsevier, Amsterdam, pp. 79–110.

Anderson, M.A., and Rubin, A.J. 1981. *Adsorption of Inorganics at Solid-Liquid Interfaces.* Ann Arbor Science, Ann Arbor, Michigan.

Balistrieri, L.S., and Chao, T.T. 1987. "Selenium adsorption by goethite." *Soil Sci. Soc. Amer. J.* 51(5):1145–1151.

Bartlett, R.J., and James, B. 1979. "Behavior of chromium in soils: III. Oxidation." *J. Environ. Qual.* 8:31–35.

Bartlett, R.J., and Kimble, J.M. 1976. "Behavior of chromium in soils: 1. trivalent forms." *J. Environ. Qual.* 5:379–383.

Baas-Becking, L.G.M., Kaplan, I.R., and Moore, D. 1960. "Limits of Natural Environment in Terms of pH and Oxidation-Reduction Potentials. *J. Geology* 68:243–284.

Benes and Havlik 1979. "Speciation of mercury in natural waters." *Biogeochemistry of Mercury in the Environment.* J.O. Nriagu, ed. Elsevier, Amsterdam, pp. 175–202.

Biggar, J.W., and Fireman, M. 1960. "Boron adsorption and release by soils." *Soil Sci. Soc. Am. Proc.* 24:115–119.

Bingham, F.T., Page, A.L., Mahler, R.J., and Ganje, T.J. 1976. "Cadmium availability to rice in sludge-amended soil under 'flood' and 'nonflood' culture." *Soil Sci. Soc. Am. J.* 40:715–719.

Bohn, H.L., McNeal, B.L., and O'Connor, G.A. 1979. *Soil Chemistry.* John Wiley and Sons, New York.

Bowen, H.J.M. 1979. *Environmental Chemistry of the Elements.* Academic Press, pp. 32–46.

Bowman, R.S., Essington, M.E., and O'Connor, G.A. 1981. "Soil sorption of nickel: influence of solution composition." *Soil Sci. Soc. Am. Journal* 45:860–865.

Bradford, G.R. 1963. "Lithium survey of California's water resources." *Soil Sci.* 96:77–81.

Chan, K.M., and Riley, J.P. 1966. "The determination of molybdenum in natural waters, silicates and biological materials." *Anal. Chim. Acta* 36:220.

Chao, T.T. 1984. "Use of partial dissolution techniques in geochemical exploration." *J. Geochem. Explor.* 20:101–135.

Crist, M.A. 1974. "Selenium in waters in and adjacent to the Kendrick project," Natrona County, Wyo. U.S. Geol. Survey Water-Supply Paper 2023.

Corey, R.B. 1981. "Adsorption vs. precipitation." *Adsorption of Inorganics at Solid-Liquid Interfaces*. M.A. Anderson, and A.J. Rubin, eds. Ann Arbor Science Publishers, pp. 161–182.

Davis, J.A. 1984. Complexation of trace metals by adsorbed natural organic matter. *Geochimica et Cosmochimica Acta* 48:679–691.

Davis, J.A., and Leckie, J.O. 1978. Effect of adsorbed complexing ligands on trace metal uptake by hydrous oxides." *Env. Sci. Tech.* 12:1309–1315.

_____. 1980. "Surface ionization and complexation at the oxide/water interface. 3. Adsorption of anions." *J. Colloid and Interface Science* 74: 32–43.

Davis, J.A., and Hayes, K.F. 1986. *Geochemical Processes at Mineral Surfaces*. Amer. Chem. Soc., Washington, D.C., p. 2–18.

Deverel, S.J., and Fujii, R. 1988. "Processes affecting the distribution of selenium in shallow ground water of agricultural areas, western San Joaquin Valley, Calif." *Water Resources Research* 24(4):516–524.

Deverel, S.J., and Millard, S.P. 1988. "Distribution and mobility of selenium and other trace elements in shallow ground water of the western San Joaquin Valley, Calif." *Environ. Sci. and Tech.* 22:697–702.

Drever, J.I. 1988. *The Geochemistry of Natural Waters*. 2nd ed. Prentice Hall, New Jersey.

Elprince, A.M., Vanselow, A.D., and Sposito, G. 1980. "Heterovalent ternary cation exchange equilibria: $NH^{4+} - Ba^{+2} - La^{+3}$ exchange on montmorillonite." *Soil Sci. Soc. Am. J.* 44:964–969.

Elrashidi, M.A., Adriano, D.C., Workman, S.M., and Lindsay, W.L. 1987. "Chemical equilibria of selenium in soils: A theoretical development." *Soil Sci.* 144(2):141–152.

Elrashidi, M.A., and O'Connor, G.A. 1982. "Boron sorption and desorption in soils." *Soil Sci. Soc. Am. J.* 46:27–31.

Emmerich, W.E., Lund, L.J., Page, A.L., and Chang, A.C. 1982a. "Solid phase forms of heavy metals on sewage sludge-treated soils." *J. Environ. Qual.* 11:178–181.

_____. 1982b. "Predicted solution phase forms of heavy metals in sewage sludge-treated soils." *J. Environ. Qual.* 11:182–186.

Evans, H.T. and Garrels, R.M. 1958. "Thermodynamic equilibria of vanadium in aqueous systems as applied to the interpretation of the Colorado Plateau ore deposits." *Geochimica et Cosmochimica Acta* 15: 131–149.

Faust, S.D., Winka, A., Belton, T., and Tucker, R. 1983. "Assessment of the chemical and biological significance of arsenical compounds in a heavily contaminated watershed, part II. Analysis and distribution of several arsenical species." *J. Environ. Sci. Health* 18:389–411.

Fleischer, M. 1970. "Summary of the literature on the inorganic geochemistry of mercury," *Mercury in the Environment*. U.S. Geological Survey Professional Paper 713.

Forbes, E.A., Posner, A.M., and Quirk, J.P. 1976. "The specific adsorption of divalent Cd, Co, Cu, Pb, and Zn on goethite." *J. Soil Sci.* 27:154–166.

Fujii, R., Deverel, S.J., and Hatfield, D.B. 1988. "Distribution of selenium in soils of agricultural fields, western San Joaquin Valley, Calif." *Soil Sci. Soc. Am. J.* 52:1274–1283.

Garcia-Miragaya, J., and Page, A.L. 1976. "Influence of ionic strength and inorganic complex formation of the sorption of trace amounts of Cd by montmorillonite." *Soil Sci. Soc. Am. J.* 40:658–663.

Garrels, R.M., and Christ, C.L. 1965. *Solutions, Minerals and Equilibria.* Harper and Row, New York. 450 p.

Gibbs, R.J. 1976. "Transport phases of transition metals in the Amazon and Yukon Rivers." *Geological Society of America Bulletin* 88:829–843.

Goldberg, S. 1985. "Chemical modeling of anion competition on goethite using the constant capacitance model." *Soil Sci. Soc. Amer. J.* 49:851–856.

_____. 1986. "Chemical modeling of arsenate adsorption on aluminum and iron oxide minerals." *Soil Sci. Soc. Amer. J.* 50:1154–1157.

Goldberg, S., and Glaubig, R.A. 1985. "Boron adsorption on aluminum and iron oxide minerals." *Soil Sci. Soc. Amer. J.* 49:1374–1379.

_____. 1986a. "Boron adsorption on California soils." *Soil Sci. Soc. Am. J.* 50:1173–1176.

_____. 1986b. "Boron adsorption and silicon release by the clay minerals kaolinite, montmorillonite, and illite." *Soil Sci. Soc. Am. J.* 50:1442–1448.

Goldberg, S., and Sposito, G. 1984a. "A chemical model of phosphate adsorption by soils: I. Reference oxide minerals." *Soil Sci. Soc. Amer. J.* 48:772–778.

_____. 1984b. "A chemical model of phosphate adsorption by soils: II. Noncalcareous soils." *Soil Sci. Soc. Amer. J.* 48:779–783.

Gough, L.P., Severson, R.C., and McNeal, J.M. 1979. "Extractable and total-soil element concentrations favorable for native plant growth in the northern Great Plains." *Ecology and Coal Resource Development* Vol 2. M.K. Wali, ed. Pergamon Press, New York, pp. 859–869.

Greubel, K.A., Davis, J.A., and Leckie, J.O. 1988. The feasibility of using sequential extraction techniques for As and Se in soils and sediments. *Soil Sci. Soc. Am. J.* 52:390–397.

Gulens, J., Champ, D.R., and Jackson, R.E. 1979. Influence of redox environments on mobility of arsenic in ground water. In: *Chemical Modeling in Aqueous Systems.* E.A. Jenne, ed. Amer. Chem. Soc., Washington, D.C., pp. 81–95.

Harter, R.D., and Baker, D.E. 1977. "Application and misapplication of the Langmuir equation to soil adsorption phenomena." *Soil Sci. Soc. Am. J.* 41:1077–1080.

Hem, J.D. 1970. "Chemical behavior of mercury in aqueous media." *Mercury in the Environment.* U.S. Geological Survey Professional Paper 713.

_____. 1977. "Reactions of metal ions at surfaces of hydrous iron oxide." *Geochimica et Cosmochimica Acta* 41:527–538.

_____. 1985. "Study and interpretation of the chemical characteristics of natural water." U.S. Geological Survey Water Supply Paper 2254.

Hendrickson, L.L., and Corey, R.B. 1981. "Effect of equilibrium metal concentrations on apparent selectivity coefficients of soil complexes." *Soil Sci.* 131:163–171.

Hickey, M.G., and Kittrick, J.A. 1984. "Chemical partitioning of cadmium, copper, nickel, and zinc on soils and sediments containing high levels of heavy metals." *J. Environ. Qual.* 13:372–376.

Hingston, F.J. 1981. "A review of anion adsorption." *Adsorption of Inor-*

ganics at Solid-Liquid Interfaces. M.A. Anderson, and A.J. Rubin, eds. Ann Arbor Science, Ann Arbor, Mich., pp. 51–90.

Hingston, F.J., Posner, A.M., and Quirk, J.P. 1968. "Adsorption of selenite by geothite." *Adsorption of Aqueous Solution.* R.F. Gould, ed. Advances in Chemistry Series, No. 79, Washington, D.C., pp. 82–90.

Horn, M.K., and Adams, J.A.S. 1966. "Computer-derived geochemical balances and element abundances." *Geochimica et Cosmochimica Acta* 30:279–297.

James, B.R., and Bartlett, R.J. 1983. "Behavior of chromium in soils, VI. Interactions between oxidation-reduction and organic complexation." *J. Environ Qual.* 12:173–176.

Jarrel, W.M., Page, A.L., and Elseewi, A.A. 1980. "Molybdenum in the environment." *Residue Reviews* 74:1–42.

Jenne, E.A. 1968. "Controls on Mn, Fe, Co, Ni, Cu, and Zn concentrations in soils and water: the significant role of hydrous Mn and Fe oxides." *Adv. Chem.* 73:337–387.

_____. 1970. Atmospheric and fluvial transport of mercury. In Mercury in the Environment. U.S. Geological Survey Professional Paper 713.

_____. 1976. "Trace element sorption by sediments and soils—sites and processes." *Symposium on Molybdenum in the Environment.* W. Chappell, and K. Peterson, eds., Vol. 2. Marcel-Dekker, Inc., New York, pp. 425–553.

Jensen, S., and Jernolov, A. 1969. "Biological methylation of mercury in aquatic organisms." *Nature* 223:753–754.

Joreskog, K.G., Klovan, J.E., and Reyment, R.A. 1976. *Geological Factor Analysis.* Elsevier Scientific Publishing, New York.

Keren, R., Gast, R.G., and Bar-yosef, B. 1981. "pH-dependent boron adsorption by Na-montmorillonite." *Soil Sci. Soc. Am. J.* 45:45–48.

Kinniburgh, D.G., Jackson, M.L., and Syers, J.K. 1976. "Adsorption of alkaline earth, transition, and heavy metal cations by hydrous oxide gels of iron and aluminum." *Soil Sci. Soc. Am. J.* 40:796–799.

Kuo, S., and Mikkelsen, D.S., 1979. "Zinc adsorption by two alkaline soils." *Soil Sci.* 128:274–279.

Lakin, H.W. 1973. "Selenium in our environment." *Trace Elements in the Environment.* E.L. Kothy, ed. American Chemical Society, Advances in Chemistry Series 123, Washington, D.C., pp. 96–111.

Langmuir, I. 1918. "The adsorption of gases on plane surfaces of glass, mica, and platinum." *J. Am. Chem. Soc.* 40:1361–1403.

Langmuir, D., and Mahoney, J. 1984. "Chemical equilibrium and kinetics of geochemical processes in ground water studies." First Canadian/American conference on hydrogeology. B. Hitchon, and E.I. Wallick, eds. National Water Well Association, Worthington, Ohio, pp. 69–95.

Leckie, J.O., Benjamin, M.M., Hayes, K., Kaufman, G., and Altmann, S. 1980. "Adsorption/coprecipitation of trace elements from water with iron oxy-hydroxide." Electric Power Research Institute Report CS-1513, Stanford Univ., Stanford, Calif.

Lemmo, N.V., Faust, S.D., Belton, T., and Tucker, R. 1983. "Assessment of the chemical and biological significance of arsenical compounds in a heavily contaminated watershed. Part I. The fate and speciation of arsenical compounds in aquatic environments—a literature review. *J. Environ. Sci. Health* 18:335–387.

Lerman, A., and Childs, C.W. 1973. "Metal-organic complexes on natural waters." *Trace Metals and Metal-organic Interactions in Natural Waters*. P.C. Singer, ed. Ann Arbor Science, Ann Arbor, Mich., pp. 201–235.

Lewis, F.M., Voss, C.I., and Rubin, J. 1986. "Numerical simulation of advective-dispersive multisolute transport with sorption, ion-exchange, and equilibrium chemistry. U.S. Geological Survey, Water-Resources Investigations Report 86-4022.

Lindberg, R.D., and Runnells, D.D. 1984. "Ground water redox reactions: An analysis of equilibrium state applied to Eh measurements and geochemical modeling." *Science* 225:925–927.

Lindsay, W.L. 1979. *Chemical Equilibria in Soils*. Wiley-Interscience, New York.

Mattigod S.V., and Page, A.L. 1983. "Assessment of metal pollution in soils." Thorton, I. ed., *Applied Environmental Geochemistry*. Academic Press, London, pp. 355–391.

McBride, M.B. 1979. "Mobility and reactions of VO^{2+} on hydrated smectite surfaces." *Clays and Clay Minerals* 27:91–96.

McNeal, J.M., Severson, R.C., and Gough, L.P. 1985. "The occurrence of extractable elements in soils from the Northern Great Plains." *Soil Sci. Soc. Am. J.* 49:873–881.

National Academy of Sciences. 1974. *Vanadium*. National Research Council, Washington, D.C.

Neal, R.H., Sposito, G., Holtzclaw, K.M. and Traina, S.J. 1987a. "Selenite adsorption on alluvial soils: I. Soil composition and pH effects." *Soil Sci. Soc. Am. J.* 51(5):1161–1165.

_____. 1987b. "Selenite adsorption on alluvial soils: II. Solution composition effects." *Soil Sci. Soc. Am. J.* 51(5):1165–1169.

Nordstrom, D.K., Jenne, E.A., and Ball, J.W. 1979a. "Redox equilibria of iron in acid mine waters." *Chemical Modeling in Aqueous Systems*. E.A. Jenne, ed. Amer. Chem. Soc., Washington, D.C., pp. 51–79.

Nordstrom, D.K., Plummer, L.N., Wigley, T.M.L., Wolery, T.J., Ball, J.W., Jenne, E.A., Bassett, R.L., Crerar, D.A., Florence, T.M., Fritz, B., Hoffman, M., Holdren, Jr, G.R., Lafon, G.M., Mattigod, S.V., McDuff, R.E., Morel, F., Reddy, M.M., Sposito, G., and Thraikill, J. 1979b. "A comparison of computerized chemical models for equilibrium calculations in aqueous systems." *Chemical Modeling in Aqueous Systems*. E.A. Jenne, ed. Amer. Chem. Soc., Washington, D.C., pp. 857–892.

Nordstrom, D.K., and Ball, J.W. 1984. "Chemical models, computer programs and metal complexation in natural waters." *Complexation of Trace Metals in Natural Waters*, C.J.M. Kramer, and J.C. Duinker, eds. Matinus and Nijoff/Dr. W. Jurk, The Hague.

Peryea, F.J., Bingham, F.T., and Rhoades, J.D. 1985. "Mechanisms of boron regeneration." *Soil Sci. Soc. Am. J.* 49:840–843.

Pierce, M.L., and Moore, C.B. 1980. "Adsorption of arsenite on amorphous iron hydroxide from dilute aqueous solution." *Env. Sci. Tech.* 14:214–216.

Pitzer, K.S. 1979. "Theory: ion interaction approach." R.M. Pytkowicz, ed., *Activity Coefficients in Electrolyte Solutions, 1.* CRC Press.

Ponnamperuma, F.N. 1972. "The chemistry of submerged soils." *Advan. Agron.* 24:29–96.

Posselt, H.S., Anderson, F.J., and Weber, W.J. 1978. "Cation sorption on colloidal hydrous manganese dioxide." *Environmental Sci. Tech.* 2: 1087–1093.

Presser, T.S., and Ohlendorf, H.M. 1987. "Biogeochemical cycling of selenium in the San Joaquin Valley, California." *Environmental Management* 11:805–821.

Quirk, J.P., and Posner, A.M. 1976. "Trace element adsorption by soil minerals." *Trace Elements in Soil-Plant-Animal Systems.* D.J.D. Nicholas, and A.R. Egan, eds. Academic Press, New York.

Rai, D., and Zachara, J.M. 1984. "Chemical attentuation rates, coefficients, and constants in leachate migration, a critical review." Electric Power Research Institute Report, EA-3356. Palo Alto, California, Volume 1.

——. 1986. "Geochemical behavior of chromium species." Electric Power Research Institute, Report EA-4544, Palo Alto, Calif.

Reddy, C.N., and Patrick, Jr., W.H. 1977. "Effect of redox potential on the stability of zinc and copper chelates in flooded soils." *Soil Sci. Soc. Am. J.* 41, 729–732.

Reddy, K.R., and Patrick, W.H., Jr. 1983. "Effect of aeration on reactivity and mobility of soil constituents." *Chemical Mobility and Reactivity in Soil Systems.* D.W. Nelson et al. (eds.). Soil Science Society America, Madison, Wisconson, p. 11–33.

Reyes, E.D., and Jurinak, J.J. 1967. "A mechanism of molybdate adsorption in Fe_2O_3. *Soil Sci. Soc. Am. Proc.* 31:637–641.

Robertson, F.N. 1975. "Hexavalent chromium in the ground water in Paradise Valley, Maricopa County, Arizona." *Groundwater* 13:516–527.

Ross, D.S., Sjogren, R.E., and Bartlett, R.J. 1981. "Behavior of chromium in soils: IV. Toxicity to microorganisms." *J. Environ. Qual.* 10: 145–148.

Ryden, J.C., Syers, J.K., and Tillman, R.W. 1987. "Inorganic anion sorption and interaction with phosphate sorption by hydrous ferric oxide gel." *J. Soil Sci.* 38:211–217.

Schindler, P.W. 1981. "Surface complexes at oxide-water interfaces." *Adsorption of Inorganics at Solid-Liquid Interfaces.* M.A. Anderson, and A.J. Rubin, eds. Ann Arbor Science, Ann Arbor, Mich., pp. 1–50.

Sindeeva, N.D. 1964. *Mineralogy and Types of Deposits of Selenium and Tellurium.* Wiley-Interscience, New York, 363 p.

Sposito, G. 1981. *The Thermodynamics of Soil Solutions.* Oxford University Press, New York.

——. 1983. "The chemical forms of trace metals in soils." *Applied Environmental Geochemistry.* I. Thornton, ed. Academic Press, London, pp. 123–170.

——. 1984. *The Surface Chemistry of Soils.* Oxford University Press, New York.

Stollenwerk, K.G., and Grove, D.B. 1985a. "Adsorption and desorption of hexavalent chromium in an alluvial aquifer near Telluride, Colorado." *J. Environ. Qual.* 14:150–155.

_____. 1985b. "Reduction of hexavalent chromium in water samples acidified for preservation." *Journal of Environ. Quality* 14:396–399.

Stumm, W., Kummert, R., and Sigg, L. 1980. "A ligand exchange model for the adsorption of inorganic and organic ligands at hydrous oxide interfaces." *Croatica Chem. Acta* 53:291–312.

Stumm, W., and Morgan, J.J. 1981. *Aquatic Chemistry.* 2nd ed., John Wiley, New York.

Syers, J.K., Browman, M.G., Smillie, G.W., and Corey, R.B. 1973. "Phosphate sorption by soils evaluated by the Langmuir adsorption equation." *Soil Sci. Soc. Amer. Proc.* 37:358–363.

Sylvester, M.A., Deason, J.P., Feltz, H.R. and Engber, R.A. 1988. "Preliminary results of the Department of the Interior Irrigation Drainage Studies." *Proc. on Planning Now for Irrigation and Drainage,* Irrigation and Drainage Div., ASCE, Lincoln, Nebraska, Aug. 1988.

Taylor, R.M., and Giles, J.B. 1970. "The association of vanadium and molybdenum with iron oxides in soils." *J. Soil Sci.* 21:203–215.

Thorstensen, D.C. 1984. "The concept of electron activity and its relation to redox potentials in aqueous geochemical systems." U.S. Geological Survey Open-File Report 84-072.

Tidball, R.R., Grundy, W.D., and Swatzky, D.L. 1986a. "Kriging techniques applied to element distribution in soils of the San Joaquin Valley, California." Proc. HAZTECH International Conference, Denver, Colorado, August 1986.

Tidball, R.R., Severson, R.C., Gent, C.A. and Riddle, G.O. 1986b. "Element associations in soils of the San Joaquin Valley, California." U.S. Geological Survey Open-File Report, 86-583.

Tourtelot, H.A., and Miesch, A.T. 1975. "Sampling designs in environmental geochemistry," Geol. Soc. Amer. Spec. Paper, 155, pp. 107–118.

Truesdell, A.H., and Jones, B.F. 1974. "WATEQ, a computer program for calculating chemical equilibria of natural waters." *Journ. Research U.S. Geol. Survey* 2:233–274.

Turner, M.A., and Rust, R.H. 1971. "Effect of chromium on growth and mineral nutrition of soybeans." *Soil Sci. Soc. Am. Proc.* 35:755–758.

U.S. Environmental Protection Agency. 1985. "National Primary Drinking Water Regulations." *Federal Register* 50,219,46,936–47,022.

_____. 1986. "Quality Criteria for Water," Washington, D.C., 440/5-86-001, 453.

U.S. Fish and Wildlife Service. 1986. "Effects of irrigation drain-water contaminants on wildlife. Annual Report of the Patuxent Wildlife Research Center, Laurel, Maryland.

Van Olphen, H. 1977. *An Introduction to Clay Colloid Chemistry.* John Wiley and Sons, New York.

Wedepohl, K.H. 1972. "Vanadium." *Handbook of Geochemistry,* K.H. Wedepohl, ed. Springer-Verlag, Berlin, p. 23-A-1 to 23-O-1.

Welch, A.H., Lico, M.S., and Hughes, J.L. 1988. "Arsenic in ground water of the western United States." *Groundwater* 26:333–347.

NOTATION

a_i = activity of ion i; b = absorption maximum; C = equilibrium concentration of adsorbate; IAP = ion activity product; k = parameter that reflects affinity of adsorbent for adsorbate; K, n = constants (empirically evaluated); K_s = stability constant; K_{so} = solubility product constant; m = mass of solid phase adsorbent; m_i = molal concentration of species i; SI = saturation index; x = mass of trace element adsorbed; and γ_i = activity coefficient of species i.

CHAPTER 5

SOIL RESPONSE TO SALINE AND SODIC CONDITIONS[a]

INTRODUCTION

Salinity and sodicity affect soil structure, which must be stable for adequate permeability and water infiltration. High sodium levels combined with low soil-water electrical conductivity can lower a soil's permeability and decrease its infiltration capacity through the swelling and dispersion of clays and the slaking of aggregates.

Swelling of clays narrows the conducting pores in soils (McNeal and Coleman 1966; Quirk and Scholfield 1955). Dispersion and deposition of clays in the narrow necks of conducting pores also can reduce soil permeability (Felhendler et al. 1974; Frenkel et al. 1978; Shainberg et al. 1981a,b). Slaking can reduce both the amount of macropores in a soil and the soil's hydraulic conductivity (Abu-Sharar et al. 1987).

This chapter reviews the importance of swelling, dispersion, and slaking, and examines the effect of sodicity and electrolyte concentration on swelling, dispersion, and slaking and the consequent changes in soil hydraulic properties.

A soil in which the primary sand, silt, and clay particles are not bound together into larger units is rare. These structural units are known as aggregates or peds and they are described according to their size, shape, and distinction in the soil profile. They are formed when clay particles interact with each other and with silt- and sand-size particles. Harris et al. (1966) categorize the many possible interactions into four types: (1) Clay-clay interactions, in which a cation forms the linkage between negatively-charged clay faces; (2) clay-edge to clay-face interactions; (3) clay-edge or clay-face to clay-edge or clay-face interactions, in which a polymer forms the linkage; and (4) interactions among uncharged particles linked by silt and organic or inorganic colloids. This basic model continues to be studied, and an extensive literature on soil structure now exists (Emerson 1984; Emerson et al. 1978).

Aggregate stability is the measure of the resistance of aggregates to breakdown. Kemper and Koch (1966) showed this stability to be a function of organic matter, clay, and oxide contents. In subsequent papers, Kemper and Rosenau (1984) and Kemper et al. (1987) showed that cohesion recovery is primarily a physicochemical process. Of particular importance to this review is the fact that aggregate stability and soil permeability to water depend both on soil exchangeable sodium per-

---

[a]Prepared by: I. Shainberg, Inst. of Soil and Water, P.O. Box 6, Bet Dagan, Israel 03-9683-302; and M.J. Singer, Dept. of Land, Air, and Water Resour., Univ. of Calif., Davis, CA. 95616.

centage (ESP) and on the salt concentration of the percolating solution. Emerson (1984) has reviewed the effect of salinity and sodicity on soil structure.

EFFECT OF SOLUTION COMPOSITION ON CLAY SWELLING AND DISPERSION

Colloidal clay is the soil fraction that most decisively determines soil physical behavior. It possesses the greatest specific surface area and is, therefore, most active in physicochemical processes such as swelling and dispersion. Because of its extremely high specific surface area (800 m^2/g), smectite, the dominant clay in semiarid and arid regions, is the most active clay in clay-solution interactions.

This section will describe some theories on the properties of colloidal-clays when in aqueous solution. This information is essential to understand the effects of exchangeable sodium percentage (ESP) and salt concentrations on soil properties.

Sodium and Calcium Distribution in Smectites

When two clay platelets approach each other, their diffuse atmospheres overlap. Work must be done to overcome the electrical repulsion forces between the two positively-charged ionic atmospheres (Bresler et al. 1982; Van Olphen 1977). The greater the compression of the ionic atmosphere toward the clay surface, the smaller the overlap of the atmospheres for a given distance between particles. With an increase in salt concentration and in the valences of the adsorbed ions, the repulsion forces between particles consequently decrease. Adsorbed sodium ions form a diffuse layer and high swelling pressures develop between sodium-montmorillonite platelets. Also, single platelets tend to form and persist in dilute solution (Shainberg and Otoh 1968; Van Olphen 1977).

The low swelling pressure between calcium-clay platelets and the electrical attraction forces between exchangeable calcium and the negative clay surface (forces not considered in the diffuse double-layer theory) prevent the indefinite swelling of calcium-montmorillonite, even in distilled water. Hence, the calcium-platelets aggregate into tactoids or quasi-crystals (Aylmore and Quirk 1959). Each tactoid consists of four to nine clay platelets, which are parallel and 0.9 nm apart. The exchangeable cations form a diffuse double-layer only on the outside of these tactoids, and the clay behaves like a system with a much smaller surface area.

Bresler (1970) pictured a mixture of monovalent and divalent ions randomly mixed throughout the exchange complex, which means that, for any ratio of sodium to calcium, as many sodium ions exist inside the tactoids as outside. Mering and Glaeser (1954) advanced another hypothesis. They suggested that cation de-mixing occurs, so that a few interlayer spaces contain mainly sodium and other mainly calcium ions. Support for de-mixing came from Bar-On et al. (1970), who found that adding a small amount of exchangeable sodium to calcium-saturated

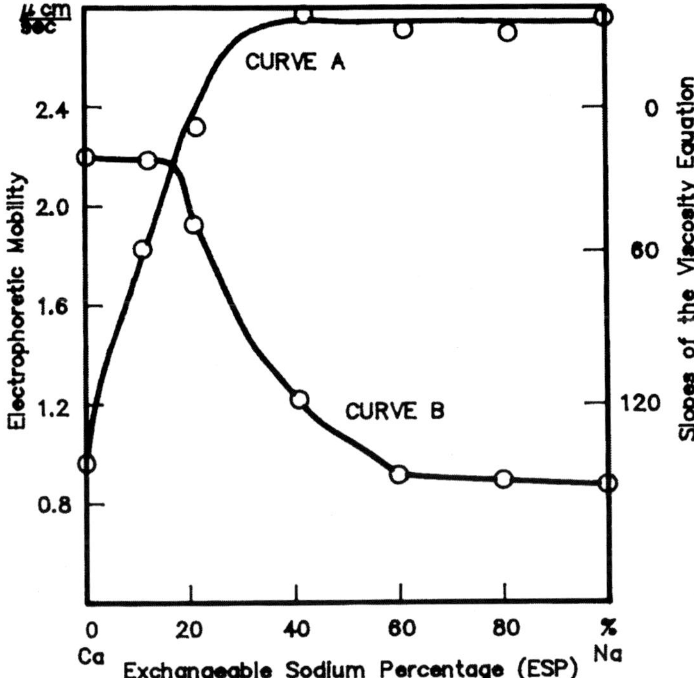

FIG. 5.1 Dependence of Electrophoretic Mobility (Curve A) and Relative Size (Curve B) of Montmorillonite Particles on the Exchangeable Sodium Percentage (ESP) [Relative size is expressed in terms of the slope of Einstein's equation for the viscosity of the suspension (Bar-On et al. 1970)]

clay increases the clay's electrophoretic mobility, but does not increase the size of the calcium-tactoids (Fig. 5.1). At low ESP, the adsorbed sodium concentrates on the external surfaces of the tactoids, increasing their electrophoretic mobility. As ESP increases, sodium then penetrates into and disintegrates the tactoids. Shainberg and Otoh (1968) found that the introduction of greater than 20% sodium into the exchange complex caused tactoid breakdown (Fig. 5.1). The platelets completely separated when sodium replaced 50% to 60% of the adsorbed calcium.

Swelling and Dispersion of Sodium/Calcium Clays

As ESP increases for various consolidation pressures (Fig. 5.2), the volume of water retained by montmorillonite increases. Increases in ESP up to the threshold of 15 only slightly affects clay swelling (Shainberg et al. 1971). These results compare favorably with the work of McNeal and Coleman (1966). They found that an ESP increase above 15 increased swelling very sharply for two montmorillonitic soils.

In a stable clay suspension, dispersed particles collide frequently because of their Brownian motion, but diffuse double-layer repulsion forces separate them again. Subsequently adding salt to a clay suspen-

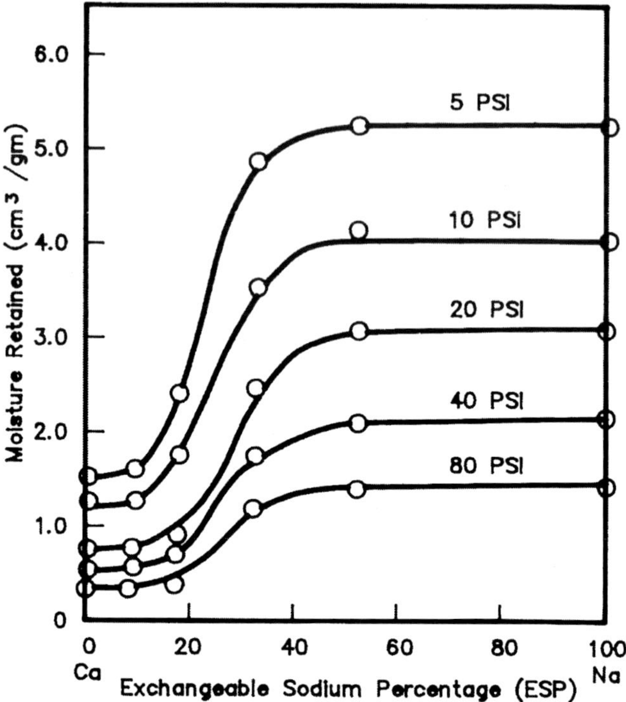

FIG. 5.2 Water Retention as a Function of ESP and Pressure Applied on Montmorillonite (Shainberg et al. 1971)

sion increases the electrolyte concentration and causes the particles to stick together upon collision, forming flocs that settle. The minimum electrolyte concentration that causes flocculation is referred to as the flocculation value (FV). The FV depends on counter-ion valency and clay type. The respective FVs of sodium-montmorillonite and calcium-montmorillonite are 12 mol_c/m^3 NaCl and 0.25 mol_c/m^3 $CaCl_2$ (Van Olphen 1977). Corresponding respective values for sodium- and calcium-illites are 40 mol_c/m^3 to 50 mol_c/m^3 NaCl and 0.25 mol_c/m^3 $CaCl_2$, (Arora and Coleman 1979). Oster et al. (1980) found that a small increase in the ESP of montmorillonite had a considerable effect on FV, similar to that of electrophoretic mobility (Fig. 5.1).

The effect of ESP on FV of calcium-illite suspensions follows a simple linear relationship more closely (Oster et al. 1980). This suggests that de-mixing of the exchangeable ions in sodium/calcium illite is not as pronounced as in sodium/calcium montmorillonite. However, illite of low ESP actually dispersed more easily than montmorillonite of the same ESP. For ESP 5, 10, and 20, the respective FVs for sodium/calcium montmorillonite were 3.0, 4.0, and 7.0 mol_c/m^3; for illite, they were 6, 10, and 18 mol_c/m^3 (Oster et al. 1980). The FV of illite was higher than that of montmorillonite, probably because of the smaller edge-to-face attractive force in sodium-illite (Shainberg and Letey 1984).

Dispersion of Kaolinite

Schofield and Samson (1954) found that a pure sodium-kaolinite flocculated at pH < 7 under conditions that dispersed illite and montmorillonite. They regarded attraction of positively-charged edges to negatively-charged planar surfaces as the cause of flocculation in the absence of salt. This was supported by their observation that the salt-free suspension de-flocculated upon addition of NaOH. At pH values greater than eight, the clay crystals released chloride particles. This indicated that the edges were no longer positively charged. Eliminating the positive edge charge also dispersed the clay.

Schofield and Samson (1954) found that adding small amounts of sodium oxalate, sodium pyrophosphate, sodium polymetaphosphate, sodium alginate, and sodium bentonite also de-flocculated kaolinite suspensions. The negatively-charged polyanions are apparently adsorbed onto the positively charged edges, preventing edge-to-face flocculation. Arora and Coleman (1979) and Frenkel et al. (1978) found kaolinite with montmorillonite impurities to be more dispersive than pure kaolinite. Frenkel et al. (1978) found that adding as little as 2% montmorillonite to kaolinitic soil greatly increased its susceptibility to increased ESP.

Differences of opinion exist in the literature regarding the effect of kaolinite clay on the hydraulic conductivity of soils leached with sodic water. McNeal and Coleman (1966) concluded that the most labile soils were those high in 2:1 layer silicates, especially montmorillonite, and the least labile were those high in kaolinite and sesquioxides. Soil kaolinites containing small amounts of montmorillonite and mica were found to be dispersive (Arora and Coleman 1979; Frenkel et al. 1978; Schofield and Samson 1954). El-Swaify and Swindale (1969) studied the hydraulic conductivities of tropical soils with clay fractions dominated by kaolins, iron oxides, amorphous silicates, and gibbsite, and found that exchangeable sodium had a negligible effect, even in the absence of salinity. McNeal et al. (1968) found that partially removing free iron-oxides greatly reduced the stability of the hydraulic conductivity of tropical Hawaiian soils under high-sodium, low-salt conditions. They concluded that the cementing action of iron oxides prevented dispersion. On the other hand, Elgabaly and Elghamry (1970) found that the hydraulic conductivity of ground and sieved kaolinitic systems decreased rapidly when leached with de-ionized water at ESP greater than or equal to 10. Frenkel et al. (1978) found that, although kaolinitic soil was less sensitive than montmorillonitic soil at low electrolyte concentrations, its hydraulic conductivity was reduced markedly, even at ESP values of 10, when leached with de-ionized water. They concluded that soil behavior depended on the type of minerals present in combination with kaolinite.

EFFECTS OF ELECTROLYTES ON HYDRAULIC CONDUCTIVITY OF SODIC SOIL

Hydraulic conductivity and permeability tend to decrease as ESP increases and salt concentration decreases (Quirk and Schofield 1955; McNeal and Coleman 1966; McNeal et al. 1968). Quirk and Schofield

(1955) postulated that this is caused by clay swelling, clay dispersion and the subsequent plugging of conducting pores by the dispersed clay, and failure of soil aggregates. Permeability can be maintained even at high ESPs if the electrical conductivity (EC) of the percolating water remains above a critical (threshold) level (Quirk and Schofield 1955). The threshold is the salt concentration in the percolating solution that causes a 10% to 15% decrease in soil permeability at a given ESP. When low-salt water is applied, even an ESP value of 5 was sufficient to decrease the hydraulic conductivity by at least two orders of magnitude for 71 Australian soils (McIntyre 1979).

The effect of electrolytes on soil hydraulic conductivity and infiltration rate and the different responses among soils to sodic conditions depend in part on the water-application method. Surfaces disturbed by the beating of raindrops or the stirring action of flowing water are more susceptible to an ESP value of less than 5 because mechanically-dispersed soil is more susceptible to sodicity than soil left to disperse spontaneously (Emerson 1967; Rengasamy et al. 1984).

Most studies on the effect of solution concentration and soil ESP on soil physical and hydraulic properties have used a solution concentration in which the EC exceeds 0.5 dS/m. Consequently, they have focused on soils that have an ESP of greater than 10 (U.S. Salinity Laboratory 1954; McNeal and Coleman 1966; McNeal et al. 1968). Bresler et al. (1982) reviewed these studies. Shainberg and Letey (1984) and Shainberg (1984) reviewed the effect of good-quality water, such as rainwater, on the hydraulic conductivity and hydraulic conductivity of various soils.

A rainy season is typical for many of the arid and semiarid regions where irrigation is practiced. The electrolyte concentration in most irrigation water exceeds 5 mol_c/m^3, (EC = 0.5 dS/m), while the electrolytic concentration in rainwater is less than 1 mol_c/m^3 (EC $<$ 0.1 dS/m) (Shalhevet and Kamburov 1976). Thus, the deleterious effect of exchangeable sodium is more evident during the rainy season than during the irrigation season. Moreover, some deterioration of soil physical properties that occurs because of low ESP during the rainy season tends to be irreversible.

Some data on salt concentration and ESP effects on soil permeability have been obtained from laboratory studies on clay-sand mixtures. These studies involve imperfect models of soils, but they nevertheless provide useful information on the swelling and dispersion of soil clays under various salinity and sodicity conditions.

Salt Effects on Hydraulic Conductivity of Clay-Sand Mixtures

Alperovitch et al. (1985) found that the relative hydraulic conductivity of Wyoming montmorillonite-sand mixtures equilibrated with 10 mol_c/m^3 Na-Ca solutions decreases as ESP increases (see the *y*-intercept of Fig. 5.3). Clay swelling, which reduced pore diameters, caused the hydraulic conductivity to decrease relatively little (0.30) at an SAR of 10 in 10 mol_c/m^3 Cl solution. Smectite dispersion was not expected at this electrolyte concentration, which is well above the clay FV for such ESP values (Oster et al. 1980). The additional large relative hydraulic con-

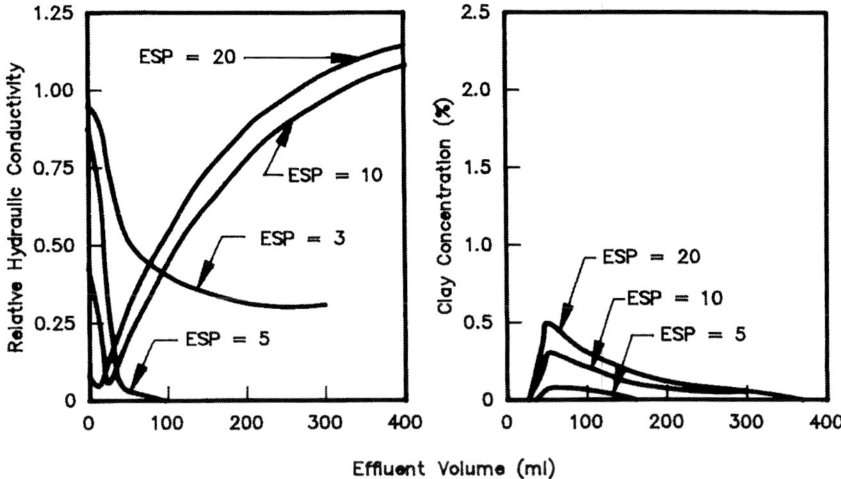

FIG 5.3 Effect of Equilibrium Exchangeable Sodium Percentage on Relative Hydraulic Conductivity and Clay Content in Effluent from a 5% Montmorillonite-95% Sand (0.1 mm to 0.6 mm diameter) Mixture. Y-axis intercept reflects hydraulic conductivity when effluent is changed from 0.01 N Cl^- to distilled water.

ductivity decrease between ESP values of 10 and 20 was mainly due to much-increased swelling at ESP 20, since the electrolyte concentration still exceeded the clay FV.

When de-ionized water replaced the 10 mol_c/m^3 Na-Ca solutions having SAR values of 10 and 20, the relative hydraulic conductivity of the mixture dropped sharply to less than 0.05, then sharply increased. The clay concentration in the effluent peaked at a value equal to about one pore volume before it gradually decreased (Fig. 5.3).

Frenkel et al. (1978) and Pupisky and Shainberg (1979) found similar hydraulic conductivity and clay concentration curves for sandy soils. The large percentage of clay in the effluent suggests that the clay particles had dispersed and moved a long distance. Clay deposition in pore spaces could have led to the initial sharp drop in relative hydraulic conductivity. The subsequent increase in relative hydraulic conductivity occurred for two reasons. First, the clay dispersed in the de-ionized water. Second, the flow regime changed from flow of solution through a sandy clay matrix, in which the sand particles are covered with a skin of clay, to flow of clay suspension through a porous sand matrix. Removing the clay skin from the sand particles increased both the pore radii and the relative hydraulic conductivity. With continued leaching, the increase in pore radii resulted in a relative hydraulic conductivity value of greater than 1.0. At SAR values of 5, the decrease in relative hydraulic conductivity was not followed by an increase, as the clay dispersion was insufficient to change the flow regime (Fig. 5.3).

Clay swelling is a continuous process that proceeds slowly as the solution concentration decreases (Keren and Singer 1988). Dispersion is possible only at solution concentrations below the clay FV. By leaching the mixture with solutions that have electrolyte concentrations below 10 mol_c/m^3, but above the FV (solutions of 5 to 3 mol_c/m^3), slow swelling and pore clogging occurs. Clay doesn't appear in the effluent upon subsequent leaching with de-ionized water because the pores are too narrow for clay movement. However, when de-ionized water replaces the 10 mol_c/m^3 solution immediately, incomplete swelling and dispersion take place. Because the swelling is incomplete, the pores are open and clay movement occurs (Fig. 5.3).

Clay Mineralogy and Exchangeable Potassium Effects on Hydraulic Conductivity

Alperovitch et al. (1985) found that smectites with low charge density, such as Wyoming montmorillonite, swelled and dispersed more than did smectites with high charge density, such as Polkville and Otay smectites. Potassium effects on the hydraulic conductivity of the smectites measured by exchangeable potassium percentage (EPP) also depended on the charge density of the clays (Shainberg et al. 1987). The effect of EPP values of less than 20 were negligible on the hydraulic conductivity of high charge-density smectite. The effect on hydraulic conductivity of low charge-density smectites from Wyoming and Belle Fourche, South Dakota, was similar to the dispersive effect of low ESP values. Potassium ion's low hydration energy, coupled with the strong electrostatic attraction forces between smectite platelets with high charge density, account for potassium's inefficiency as a dispersing agent.

Effect of Electrolyte Solutions on Hydraulic Conductivity of Soil

McNeal and Coleman (1966) and others extended the threshold concentration approach of Quirk and Schofield (1955) to a large number of soils. McNeal and Coleman maintained the electrolyte concentrations of the percolating solutions above 3 mol_c/m^3. Each soil responded differently to the same combination of salt concentration and ESP; thus, each has a unique threshold concentration (Fig. 5.4). For solutions of ECs less than 0.3 dS/m, the soil properties that appeared to most affect soil permeability at differing sodic and electrolyte concentrations were: clay content, clay mineralogy, iron oxide or aluminum oxide content, organic matter content, and bulk density. Shainberg and Letey (1984) reviewed the effects of these factors in detail. When the salinity of the soil solution exceeds 3 mol_c/m^3, soil susceptibility to sodic conditions increased as: (1) Clay content increased (McNeal et al. 1968; Frenkel et al. 1978); (2) the amount of swelling clay, especially montmorillonite, increased (McNeal and Coleman 1966); and (3) the soil bulk density increased (Frenkel et al. 1978). Soils rich in sesquioxides appear to be less susceptible to sodic conditions (McNeal et al. 1968).

Hydraulic conductivities, even at low exchangeable sodium percentages, may significantly decrease when leached with de-ionized water to simulate rainwater (Felhendler et al. 1974; Frenkel et al. 1978; McIntyre 1979; and others). This sensitivity of sodic soils to solutions of low electrolyte concentration led Shainberg et al. (1981a, 1981b) to hypothesize that the rate of salt release in sodic soils by mineral dissolution

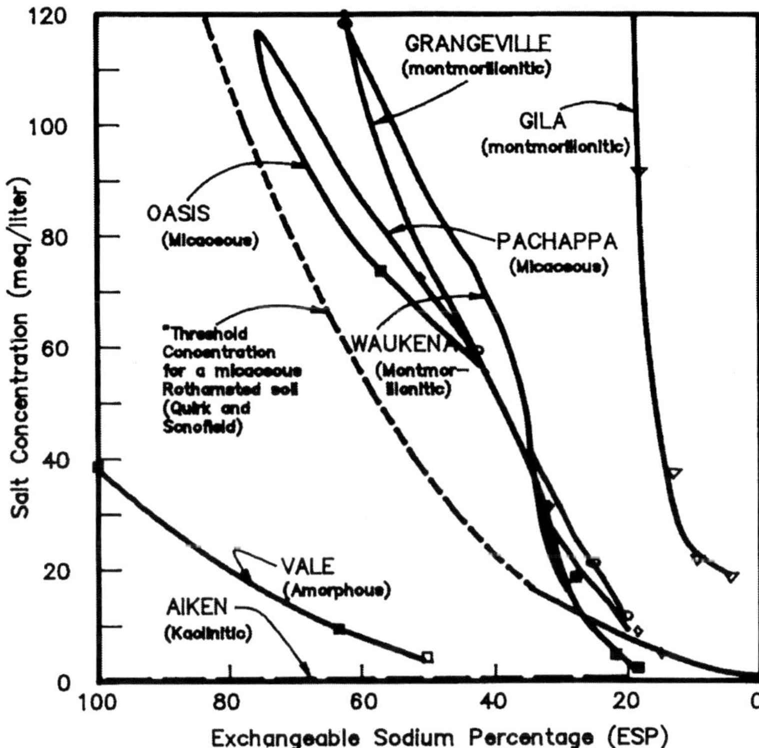

FIG. 5.4 Combinations of Salt Concentration and ESP Required to Produce a 25% Reduction in Hydraulic Conductivity for Selected Soils (McNeal and Coleman 1966)

is a major factor in their susceptibility to decreases in hydraulic conductivity when leached with low-electrolyte water. Relatively unweathered soils may release 3 to 5 mol_c/m^3 of calcium and magnesium into the solution by dissolution of plagioclase, K-feldspars, hornblende, and other minerals (Rhoades et al. 1968). The solution composition of a calcareous soil at a given ESP, when placed in contact with de-ionized water, ranges from 2 to 3 mol_c/m^3 (Shainberg et al. 1981b).

Shainberg et al. (1981a) postulated that sodic soils containing minerals that readily release soluble electrolytes will not easily disperse when leached with de-ionized water at moderate value of ESP. This is because sodic soils maintain a sufficiently high salt concentration in the soil solution (3 mol_c/m^3) to prevent clay dispersion. Conversely, salt concentrations in the soil solutions of soils that do not contain readily weatherable minerals will be below the FV, and the soils will be more susceptible to clay dispersion. This mechanism explains the response to sodic conditions of the two soils in Fig. 5.5. The more chemically stable Netanya sandy loam was also more susceptible to sodic conditions when leached with de-ionized water. This hypothesis was further supported by measurements of clay dispersivity, hydraulic conductivity, and min-

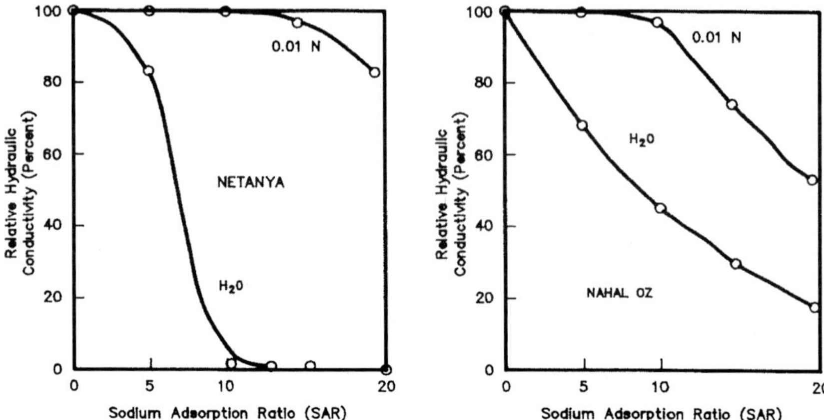

FIG. 5.5 Hydraulic Conductivity of a Sandy Loam (Netanya) and a Silt Loam (Nahal Oz) Soil as a Function of the SAR and Salt Concentration of the Leaching Solutions (Felhendler et al. 1974)

eral-weathering properties of three California soils (Shainberg et al. 1981b).

Effect of pH and Exchangeable Magnesium on Hydraulic Conductivity of Soils

Peele (1936) showed that permeability and aggregate size of a Cecil soil decreased as lime was added. Arora and Coleman (1979) found that increasing the pH of a Georgia kaolinite from 7 to 8.3 made the soil more dispersive than any of the other samples, including smectites, illites, and vermiculites. Suarez et al. (1984) found lower hydraulic conductivity of montmorillonitic and kaolinitic soils at pH 9 than at pH 6. They concluded that differences in hydraulic conductivity resulted primarily from changes in dispersive forces, because changes in hydraulic conductivity were generally consistent with measurements of clay dispersion.

The sensitivity of soil hydraulic conductivity to pH depends on the quantity of variable-charge minerals and organic matter present in the soil (Suarez et al. 1984). Soils with large amounts of variable charge are the most susceptible to pH effects. Changes in pH affect the edge-charge on clays and the surface charge of variable-charge minerals, such as iron and aluminum oxides. At low pH, edge-to-face bonding may occur, as well as bonding of positive iron and aluminum oxides to negative clay surfaces (Van Olphen 1977). This type of bonding hinders dispersion and, thus, results in optimum hydraulic conductivity. With increasing pH, both edge-to-face clay bonding and Fe and Al bonding to clays decrease.

Exchangeable magnesium's effect on the hydraulic conductivity of sodic soil is unclear. The U.S. Salinity Laboratory staff (1954) grouped calcium and magnesium together as similar ions that are beneficial for developing and maintaining soil structure. However, McNeal et al.

(1968) showed that mixed sodium/magnesium soils developed lower hydraulic conductivities than did sodium/calcium soils under similar conditions. Both McNeal et al. (1968) and Chi et al. (1977) made a distinction between the direct (specific) effect of exchangeable magnesium in causing decreases in hydraulic conductivity and the lessened ability of magnesium in irrigation waters (relative to calcium) to prevent the accumulation of exchangeable sodium in the soil. They showed that magnesium affects soils dominated by illite, but not montmorillonitic soils.

More recently, Alperovitch et al. (1981) demonstrated that exchangeable magnesium does not affect the hydraulic conductivity and clay dispersion of calcareous soils. In well-weathered soils that do not contain $CaCO_3$, exchangeable magnesium decreased both hydraulic conductivity and clay dispersion when the sodium/magnesium soils were leached with de-ionized water. Magnesium-saturated soil clays are chemically more stable and do not release electrolytes into the solution (Alperovitch et al. 1981); they, thus, disperse more easily when leached with de-ionized water. Exchangeable magnesium's presence enhances the dissolution of $CaCO_3$ in calcareous soils, and the electrolytes prevent the dispersion of clay and losses of hydraulic conductivity in sodium/magnesium calcareous soils. Alperovitch et al. (1986) reached similar conclusions when they studied clay dispersivity and hydraulic conductivity properties of three California soils as a function of ESP and electrolyte concentration for sodium/calcium and sodium/magnesium systems. They also concluded that soils that were relatively chemically stable were the most sensitive to low ESP and to the effects of exchangeable magnesium when leached with de-ionized water. Such results agreed with the hypothesis that the susceptibility of soils to decreases in hydraulic conductivity produced by exchangeable sodium and exchangeable magnesium depends on their rates of mineral dissolution. Exchangeable magnesium reduces the dissolution rates of noncalcareous soils and increases soil susceptibility to ESP under conditions of low electrolyte concentration.

EFFECT OF ESP AND ELECTROLYTE CONCENTRATION ON INFILTRATION RATE

Infiltration rate is the volume flux of water flowing into the profile per unit of soil surface area. In general, initial soil infiltration capacity is high, particularly when the soil is initially dry, but tends to decrease monotonically until it asymptotically approaches a constant rate—the steady state or final infiltration rate. If water delivery rate to the surface is less than the soil infiltration capacity, water infiltrates as fast as it is delivered and the supply rate determines the infiltration rate. When the delivery rate exceeds the soil infiltration capacity, the latter determines the actual infiltration rate and the process becomes surface-controlled.

In soils with stable surface structures, decreases in infiltration capacity result from the decrease in matric-suction gradient that inevitably occurs as infiltration proceeds (Baver et al. 1972). In the initial wetting, the surface saturates, but the soil below remains dry and the matric suction gradient is steep. As the wetted zone deepens, this gradient is gradually reduced.

Decreases in the infiltration capacity of a soil from an initially high rate can also result when the soil's structure deteriorates and a surface seal forms. When a seal with very low hydraulic conductivity forms at the soil surface, the seal determines the infiltration rate of the soil (Baver et al. 1972).

Crust formation in soils exposed to the beating action of falling water droplets is due to two mechanisms (Agassi et al. 1981; Kazman et al. 1983): (1) Breakdown of the soil aggregates on the soil surface and compaction of the uppermost soil layer; and (2) physicochemical dispersion of soil clays. Breakdown of the soil aggregates reduces the average pore-size of the surface layer. These factors produce a thin ($\approx$0.1 mm) skin seal at the soil surface (McIntyre 1958). Dispersed soil clays can then migrate into the soil with infiltrating water and clog the pores immediately beneath the surface (McIntyre 1958; Agassi et al. 1981; Kazman et al. 1983).

Effect of ESP and Electrolyte Concentration

Kazman et al. (1983) studied the effect of ESP on infiltration rate and crust formation for four soils of varying texture, clay mineralogy, and $CaCO_3$ content, using de-ionized water in a rainfall simulator. In each soil, infiltration rate was highly sensitive to low ESP. For instance, an ESP of 2.2 was enough to decrease final infiltration rate of the non-calcareous sandy loam by 70% (Fig. 5.6). Also, as soil ESP increased, the depth of rain needed to reach the final infiltration rate decreased (Fig. 5.6).

The raindrop-impact energy is the same for all of the experiments described in Fig. 5.6. Thus, differences in infiltration rate curves for the various soil samples are due to the chemical dispersion caused by sodicity. It is evident that the infiltration rate is much more sensitive to soil ESP than is the hydraulic conductivity. The higher susceptibility of the soil surface to low levels of ESP is explained by three factors (Oster and Schroer 1979; Kazman et al. 1983): (1) The mechanical impact of the raindrops, which enhances dispersion; (2) the free movement of clay at the air-soil interface; and (3) the low electrolyte concentration in the applied de-ionized water.

Agassi et al. (1981) found that the infiltration rate of a soil increased markedly as the electrical conductivity of water applied with a rainfall simulator increased in the range between the value for de-ionized water and 5.6 dS/m (Fig. 5.7). The effect of salt concentration on both the final infiltration rate and the infiltration-rate curves is similar to the effect of soil sodicity on the infiltration-rate curves (Figs. 5.6 and 5.7). When solutions of high concentration are applied, the drop impact energy is the main force causing aggregate breakdown and creating a low-permeability, compacted layer at the soil surface. This compacted layer is formed at a relatively slow rate. Final infiltration rate values maintained at 8 mm/h to 15 mm/h. When low-salinity water is applied to the soil, chemical dispersion of the soil clays can occur, even at low ESP. The dispersed clay particles can be washed into the soil with the infiltrating

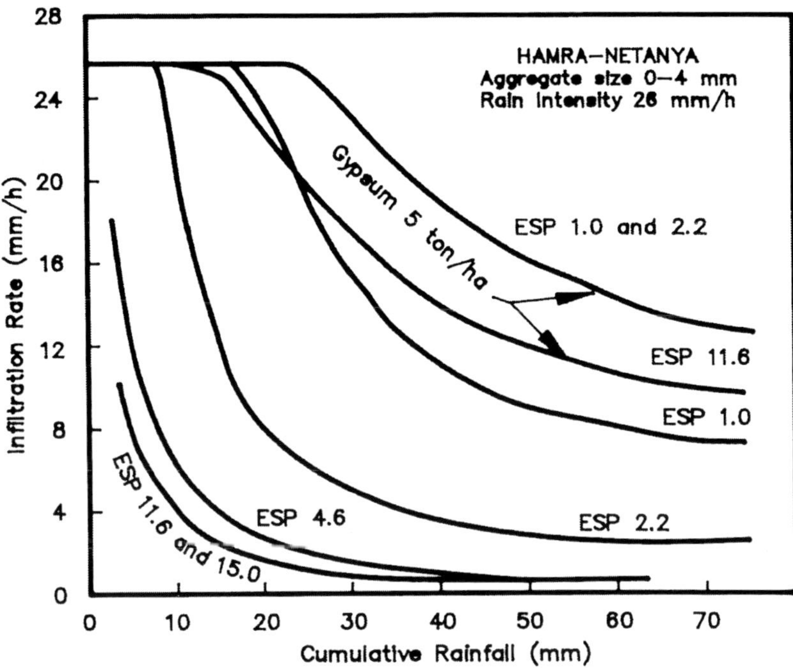

FIG. 5.6 Effect of Soil ESP and Gypsum on the Infiltration Rate of a Sandy Loam (Netanya) soil (Kazman et al. 1983)

water, and the pores immediately beneath the surface can very rapidly become clogged.

Effect of Low-Energy Rainfall

Agassi et al. (1985) simulated a fog-type rain with drops of less than 0.1 mm in size and a maximum velocity of 0.1 m/sec. The kinetic energy of the fog was less than 0.01 joules/mm m^2, simulating field conditions when the soil is protected from rain impact. In high-energy rain, the infiltration rate of the Haploxeralf (silt loam) soil was susceptible to low ESP and the electrolyte concentration of the applied water (Fig. 5.8). In low-energy rain, however, only when de-ionized water (EC = 0.01 dS/m) was applied to soil with an ESP of 6.5 did the infiltration rate gradually decrease (Fig. 5.9). In silt loam with an ESP of 21, only the saline water maintained the initial infiltration rate. When dilute solutions were applied, the infiltration rate dropped gradually. The infiltration rate in the fog experiments changed very gradually, and the final infiltration rate in the high-energy rain was much lower than the final infiltration rate in the low-energy rain. Soils exposed to high-energy rain are much more sensitive to low ESP and electrolyte concentration of the applied water than are soils exposed to low-energy rain.

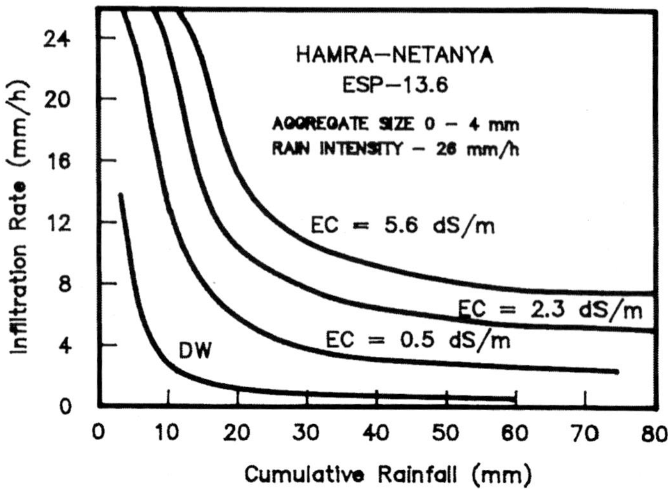

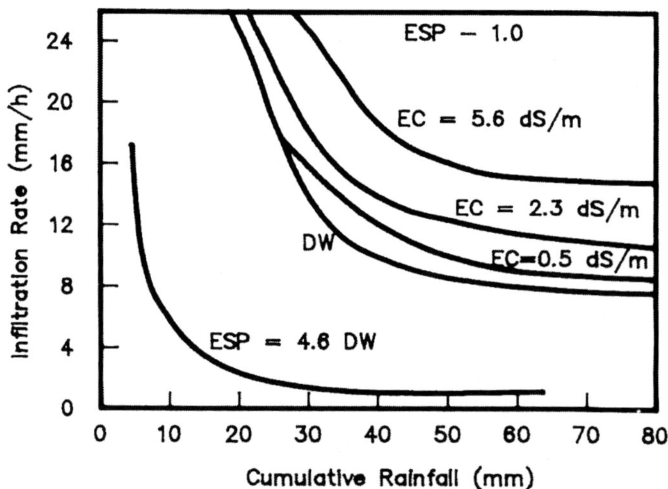

FIG. 5.7 Effect of Electrolyte Concentration in Rain Simulation Experiments on Infiltration Rate of a Sandy Loam Soil (Agassi et al. 1981)

EFFECT OF ELECTROLYTE CONCENTRATION ON DEPOSITIONAL CRUST

Slow water penetration is a major problem in crop production on many surface-irrigated soils. In some areas under surface irrigation, water penetration at the start of the irrigation season is high. The rate of water penetration decreases as the number of irrigations increases be cause depositional crusts form at the soil surface (Oster et al. 1984;

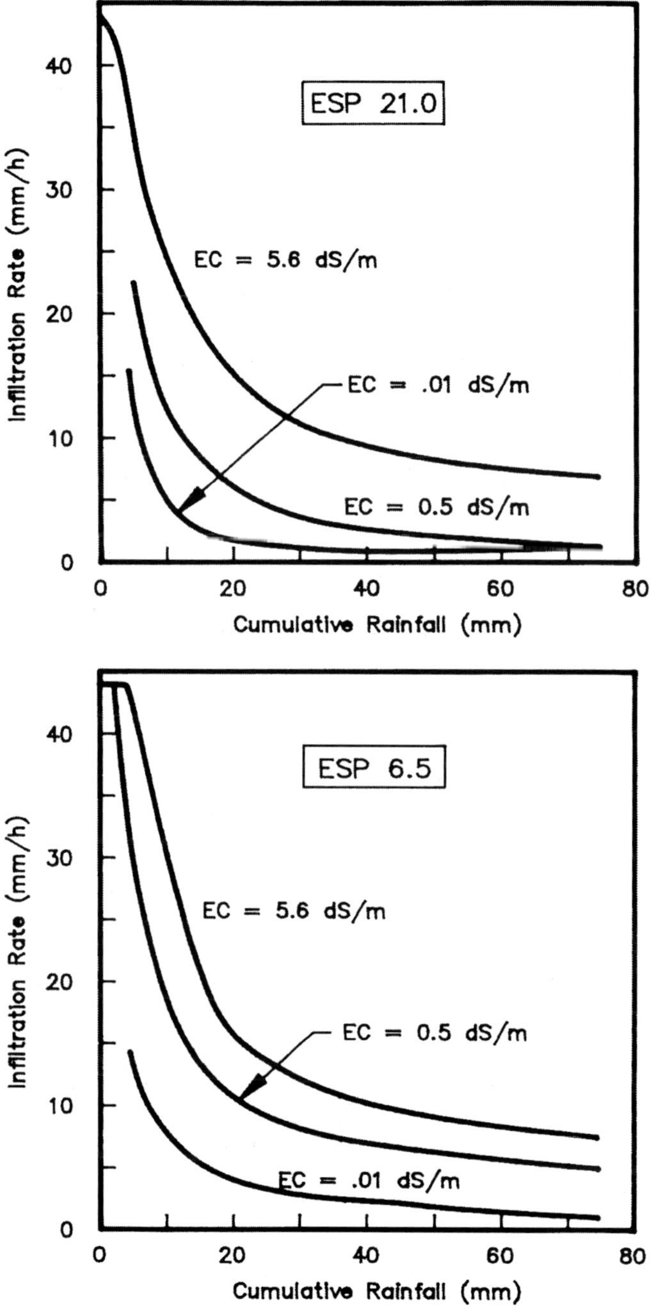

FIG. 5.8 Effect of Electrolyte Concentration on Infiltration Rate of a Haploxeralf (Loess) at Two ESP Levels: High-Energy Rain (Agassi et al. 1985)

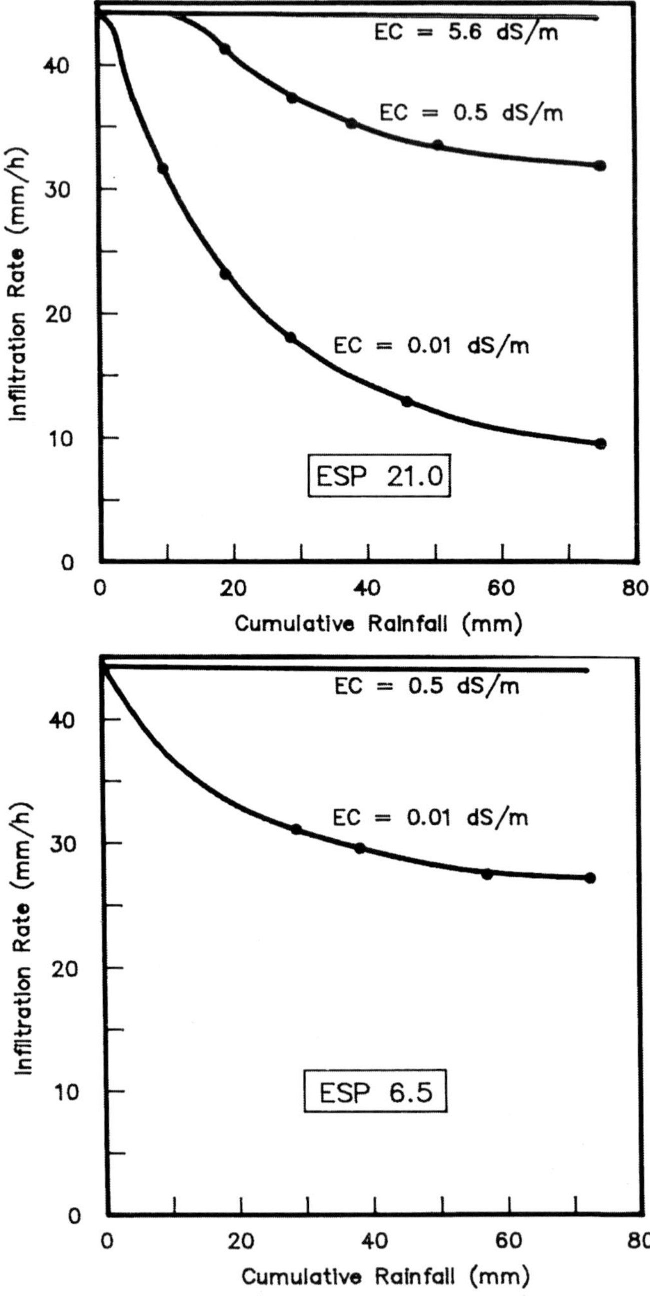

FIG. 5.9 Effect of Electrolyte Concentration on Infiltration Rate of a Haploxeralf (Loess) at Two ESP Levels: Low-Energy Rain (Agassi et al. 1985)

Shainberg and Singer 1985). These crusts form when soil sediments in water remain at the soil surface as the water infiltrates. Sediments in running water originate from the erosion of soil materials, for example, when water flowing in furrows erodes the sides of the furrow during irrigation.

Shainberg and Singer (1985) produced depositional crusts in the laboratory and measured the effects of electrolyte concentration on infiltration rate. They found that the depositional crust's hydraulic conductivity was two to three orders of magnitude lower than that of the underlying bulk soil when the EC of the applied suspension was less than 0.3 dS/m. When the EC of the water exceeded 0.5 dS/m, the crust hydraulic conductivity was only one order of magnitude lower than that of the bulk soil. The high susceptibility of the depositional crust to the electrolytic concentration in the irrigation water was similar to that of structural crusts (Agassi et al. 1981). Whereas the soil's hydraulic conductivity was not affected by the electrolyte concentration of the applied suspension, the hydraulic conductivity of the depositional crust was very sensitive to electrolytic concentration. Shainberg and Singer (1985) and Southard et al. (1988) assumed that the dispersion-flocculation status of the sediments in the suspension determined the hydraulic properties of the depositional crust formed. Depositional crusts formed from flocculated particles had an open and permeable structure. Dispersed particles formed a dense depositional crust in which the particles were oriented parallel to the soil surface. Oster et al. (1980) found that the FV of smectites and illites with ESP of 5.0 was approximately 3 mol_c/m^3. Thus, Shainberg and Singer (1985) suggested that the hydraulic properties of depositional crusts in equilibrium with more dilute solutions is low, while the permeability of the crusts in equilibrium with more concentrated solutions is high.

PREDICTING PROBLEMS

Because the effects of salinity and sodicity on soils are interrelated, no single value for either suffices to predict the degree to which a soil will structurally degrade. As has been shown, the type of clay, the presence of such binding materials as organic matter and oxides, and the amount of clay also influence the response of soils to saline and sodic conditions. Several measures of aggregate stability exist and the methods for measuring salinity and sodicity are well-known. Yet few reliable guidelines have been developed for predicting the development or deterioration of a structural or depositional crust.

Based on this review of the literature, one guideline is that soils with between 15% and 30% clay and less than 1.5% organic carbon have a high potential for soil sealing and crusting. This potential increases if the electrical conductivity of irrigation water is low (e.g., less than 0.2 dS/m). This electrical conductivity is below the flocculation value of the common soil clays at even low soil sodium contents. At such low EC (high-quality water), the potential for clay dispersion and aggregate breakdown increases. As the soil ESP increases, the electrical conductivity of the soil solution must also increase for the clays to remain flocculated.

In soils with an ESP greater than 5, the potential for clay dispersion, structural breakdown, and seal or crust formation increases. Under such conditions, amendments to reduce sodicity and increase soil-solution electrical conductivity are needed. In either high or low sodicity conditions, a relative measure of aggregate stability is often helpful. Several methods of assessing stability are available.

A simple dispersion test developed by Emerson (1967) and modified by Loveday and Pyle (1973) is widely used to assess soil stability. To conduct the test, aggregates are placed in a large volume of distilled water and the clay concentration measured in suspension far removed from the disintegrating aggregates. A distinction is usually made between spontaneous dispersion, where the aggregates in the suspension remain unagitated (Yousaf et al. 1987; Ali et al. 1987), and mechanical dispersion, whereby the aggregates are shaken in water (Rengasamy et al. 1984).

Emerson (1967) and Rengasamy et al. (1984) clearly demonstrated the importance of mechanical stress in influencing dispersion. Rengasamy et al. analyzed 138 soil samples and showed that nearly 80% of the samples dispersed after mechanical shaking. Spontaneous dispersion occurred in only 20% of the samples. These were found to have equilibrium SAR values greater than 3. Under mechanical stress, even samples with SAR values less than 3 dispersed. The electrolyte concentration controlled both mechanical and spontaneous dispersion.

The extent to which soil particles disperse after gentle shaking qualitatively indicates how a soil is likely to behave in the field when it is subjected to raindrop impact without the protection of plant canopy (Rengasamy et al. 1984; Kemper and Rosenau 1986). The remolding test used by Emerson (1967) and others may simulate the effect of the mechanical shear stress that occurs during cultivation when wet.

Knowledge of aggregate stability does not help to distinguish between slow water penetration due to compaction, the presence of natural low-permeability layers in the soil, such as clay pans and hardpans, and surface crusting. Examining the surface to determine visually the presence or absence of a crust is the simplest way to determine whether crusting is the major problem. Using penetrometers or measuring soil bulk density at various depths can be useful in determining if pre-existing density is the major problem. Perhaps the surest method for determining if crusting is a problem and whether low EC water is contributing to that problem is to use a double-ring infiltrometer to measure hydraulic conductivity before and after disturbing the soil surface. One must be cautious in interpreting these data, however, as high soil variability often is reflected in double-ring infiltrometer results.

One can also add gypsum or some other soluble salt to the soil surface before measuring hydraulic conductivity. If the disturbed surface hydraulic conductivity or hydraulic conductivity with gypsum is significantly higher than hydraulic conductivity without disturbance or additions of salt, then crusting is likely to be the major problem.

REFERENCES

Abu-Sharar, T.M., Bingham, F.T., and Rhoades, J.D. (1987). "Stability of soil aggregates as affected by electrolyte concentration and composition." *Soil Sci. Soc. Am. J.:* 51, 309–314.

Agassi, M., Morin, J., and Shainberg, I. (1985). "The effect of water drop impact energy and water salinity on the infiltration rate of sodic soils." *Soil Sci. Soc. Am. J.,* 49:186–190.

Agassi, M., Shainberg, I., and Morin, J. (1981). "Effect of electrolyte concentration and soil sodicity on infiltration rate and crust formation." *Soil Sci. Soc. Am. J.,* 45:848–851.

Ali, O.M., Yousaf, M., and Rhoades, J.D. (1987). "Effect of exchangeable cation and electrolyte concentration on mineralogy of clay dispersed from aggregates." *Soil Sci. Soc. Am. J.,* 51:896–900.

Alperovitch, N., Shainberg, I., and Keren, R. (1981). "Specific effect of magnesium on the hydraulic conductivity of sodic soils." *J. Soil Sci.,* 32:543–554.

Alperovitch, N., Shainberg, I., Keren, R., and Singer, M.J. (1985). "Effect of clay mineralogy, Al and Fe oxides on hydraulic conductivity of clay-sand mixtures." *Clays and Clay Minerals,* 33:443–450.

Alperovitch, N., Shainberg, I., and Rhoades, J.D. 1986. "Effect of mineral weathering on the response of sodic soils to exchangeable magnesium." *Soil Sci. Soc. Am. J.,* 50:901–904.

Arora, H.S., and Coleman, N.T. (1979). "The influence of electrolyte concentrations on flocculation of clay suspensions." *Soil Sci.,* 127:134–139.

Aylmore, L.A.G., and Quirk, J.P. (1959). "Swelling of clay-water systems." *Nature,* 183:1752–1753.

Bar-On, P., Shainberg, I., and Michaeli, I. (1970). "The electrophoretic mobility of Na/Ca montmorillonite particles." *J. Colloid and Interface Sci.,* 33:471–472.

Baver, L.D., Gardner, W.H., and Gardner, W.R. (1972). *Soil Physics.* 4th ed. John Wiley & Sons, New York, N.Y.

Bresler, E. (1970). "Numerical solution of the equation for interacting diffuse layer in mixed ionic system with non-symmetrical electrolytes." *J. Colloid Interface Sci.,* 3, 278–283.

Bresler, E., McNeal, B.L., and Carter, D.L. (1982). *Saline and Sodic Soils.* Springer-Verlag, New York, N.Y.

Chi, C.L., Emerson, W.W., and Lewis, D.G. (1977). "Exchangeable Ca, Mg, and Na and the dispersion of illite in water. I. Characterization of illites and exchange reactions." *Aust. J. of Soil Res.,* 15:243–253.

Elgabaly, M.M., and Elghamry, W.M. (1970). "Water permeability and stability of kaolinite systems as influenced by adsorbed cation ratio." *Soil Sci.,* 110:107–110.

El-Swaify, S.A., and Swindale, L.D. (1969). "Hydraulic conductivity of some tropical soils as a guide to irrigation water quality." *Trans. of the 9th Intern. Cong. Soil Sci. (Adelaide)* 1:381–389.

Emerson, W.W. (1967). "A classification of soil aggregates based on their coherence in water." *Aust. J. Soil Res.,* 5:47–57.

Emerson, W.W. (1984). "Soil structure in saline and sodic soils." Shainberg, I., and Shalhevet, J., eds. *Soil Salinity Under Irrigation*. Springer Verlag, Berlin. pp. 65–76.

Emerson, W.W., Bond, R.D., and Dexter, A.D., eds. (1978). *Modification of soil structure*. John Wiley and Sons. New York.

Felhendler, R., Shainberg, I., and Frenkel, H. (1974). "Dispersion and hydraulic conductivity of soils in mixed solution." *Trans. of the 10th Intern. Cong. of Soil Sci. (Moscow)* 1:103–12. Moscow: Nauka Publ. House.

Frenkel, H., Goertzen, J.O., and Rhoades, J.D. (1978). "Effects of clay type and content, exchangeable sodium percentage, and electrolyte concentration on clay dispersion and soil hydraulic conductivity." *Soil Sci. Soc. Am. J.*, 42:32–39.

Harris, R.F., Chesters, G., and Allen, O.N. (1986). "Dynamics of soil aggregation." A.G. Norman, Ed. *Adv. Agron.*, 18:107–169.

Kazman, Z., Shainberg, I., and Gal, M. (1983). "Effect of low levels of exchangeable Na and applied phosphogypsum on the infiltration rate of various soils." *Soil Sci.*, 135:184–192.

Kemper, W., and Koch, E.J. (1966). "Aggregate stability of soils from western U.S. and Canada." *USDA Tech. Bull.* 1355. US Govt. Printing Office, Washington, DC.

Kemper, W., and Rosenau, R.C. (1984). "Soil cohesion as affected by time and water content." *Soil Sci. Soc. Am. J.*, 48:1001–1006.

Kemper, W., and Rosenau, R.C. (1986). "Aggregate stability and size distribution." A. Klute, ed. *Methods of soil analysis*, Part I. 2nd ed. Agronomy, 9, 435–442.

Kemper, W.D., Rosenau, R.C., and Dexter, A.R. (1987). "Cohesion development in disrupted soils as affected by clay and organic matter content and temperature." *Soil Sci. Soc. Am. J.*, 51:860–867.

Keren, R., and Singer, M.J. (1988). "Effect of low electrolyte concentration on hydraulic conductivity of Na/Ca-montmorillonite-sand systems." *Soil Sci. Soc. Am. J.*, 52:368–373.

Loveday, J., and Pyle, J.C. (1973). "The Emerson dispersion test and its relationship to hydraulic conductivity." Commonwealth Scientific and Industrial Research Organization Div. Soil Tech. Paper 15.

McIntyre, D.S. (1958). "Permeability measurements of soil crusts formed by raindrop impact." *Soil Sci.*, 58,185–189.

McIntyre, D.S. (1979). "Exchangeable sodium, subplasticity and hydraulic conductivity of some Australian soils." *Aust. J. Soil Res.*, 17, 115–120.

McNeal, B.L., and Coleman, N.T. (1966). "Effect of solution composition on soil hydraulic conductivity." *Soil Sci. Soc. Am. Proc.*, 30, 308–312.

McNeal, B.L., Layfield, D.A., Norvell, W.A., and Rhoades, J.D. (1968). "Factors influencing hydraulic conductivity of soils in the presence of mixed-salt solutions." *Soil Sci. Soc. Am. Proc.*, 32, 187–190.

Mering, J., and Glaeser, R. (1954). "On the role of the valency of exchangeable cations in montmorillonite." *Bull. Soc. Fr. Miner. Cristallogr.*, 77, 519–530.

Oster, J.D., and Schroer, F.W. (1979). "Infiltration as influenced by irrigation water quality." *Soil Sci. Soc. Am. J.*, 43, 444–447.

Oster, J.D., Shainberg, I., and Wood, J.D. (1980). "Flocculation value and gel structure of Na/Ca montmorillonite and illite suspensions." *Soil Sci. Soc. Am. J.*, 44, 955–959.

Oster, J.D. and Singer, M.J., et al. (1984). "Water penetration problems in California soils." LAWR Publication 10011, University of California, Davis, CA. 94 p.

Peele, T.C. (1936). "The effect of calcium on the erodibility of soils." *Soil Sci. Soc. Am. Proc.*, 1,47–58.

Pupisky, H., and Shainberg, I. (1979). "Salt effect on the hydraulic conductivity of a sandy soil." *Soil Sci. Soc. Amer. J.*, 43,429–433.

Quirk, J.P., and Schofield, R.K. (1955). "The effect of electrolyte concentration on soil permeability." *J. Soil Sci.*, 6, 163–178.

Rengasamy, P., Greene, R.S.B., Ford, G.W., and Mehanni, A.H. (1984). "Identification of dispersive behavior and the management of red brown earths." *Aust. J. Soil Res.*, 22:413–432.

Rhoades, J.D., Krueger, D.B., and Reed, M.J. (1968). "The effect of soil-mineral weathering on the sodium hazard of irrigation waters." *Soil Sci. Soc. Am. Proc.* 32:643–647.

Schofield, R.K., and Samson, H.R. (1954). "Flocculation of kaolinite due to the attraction of oppositely charged crystal faces." *Faraday Soc. Disc.*, 18:135–145.

Shainberg, I. (1984). "The effect of electrolyte concentration on the hydraulic properties of sodic soils." In: Shainberg, I. and Shalhevet, J., ed. *Soil Salinity under Irrigation*. Springer-Verlag, Berlin, pp. 49–64.

Shainberg, I., Bresler, E., and Klausner, Y. 1971. "Studies on Na/Ca montmorillonite systems, I: The swelling pressure." *Soil Science*, 111: 214–219.

Shainberg, I., Keren, R., Alperovitch, N., and Goldstein, D. (1987). "Effect of exchangeable K on the hydraulic conductivity of smectite-sand mixtures." *Clays and Clay Minerals*, 35:305–310.

Shainberg, I., and Letey, J. (1984). "Response of soils to sodic and saline conditions." *Hilgardia*, 52:1–57.

Shainberg, I., and Otoh H. (1968). "Size and shape of montmorillonite particles saturated with Na/Ca ions." *Isr. J. Chem.*, 6:251–259.

Shainberg, I., Rhoades, J.D., and Prather, R.J. (1981a). "Effect of low electrolyte concentration on clay dispersion and hydraulic conductivity of a sodic soil." *Soil Sci. Soc. Am. J.*, 45:273–277.

Shainberg, I., Rhoades, J.D., Suarez, D.L., and Prather, R.J. (1981b). "Effect of mineral weathering on clay dispersion and hydraulic conductivity of sodic soils." *Soil Sci. Soc. Am. J.*, 45:287–291.

Shainberg, I., and Singer, M.J. (1985). "Effect of electrolyte concentration on the hydraulic properties of depositional crusts." *Soil Sci. Soc. Am. J.*, 49:1260–1263.

Shalhevet, J., and Kamburov, J. (1976). "Irrigation and salinity worldwide survey." *Int. Comm. Irrig. Drainage.*, New Delhi, India.

Southard, R.J., Shainberg, I., and Singer, M.J. (1988). "Influence of electrolyte concentration on the micromorphology of artificial depositional crusts." *Soil Sci.*, 145:278–288.

Suarez, D.L., Rhoades, J.D., Lavado, R., and Grieve, C.M. (1984). "Effect of pH on saturated hydraulic conductivity and soil dispersion." *Soil Sci. Soc. Am. J.*, 48:50–55.

U.S. Salinity Laboratory Staff. (1954). *Diagnosis and Improvement of Saline and Alkali Soils.* USDA Agric. Handbook No. 60.

Van Olphen, H. (1977). *An Introduction to clay colloid chemistry.* 2nd ed. John Wiley & Sons. New York.

Yousaf, M., Ali, O.M., and Rhoades, J.D. (1987). "Clay dispersion and hydraulic conductivity of some salt-affected arid land soils." *Soil Sci. Soc. Am. J.*, 51:905–07.

NOTATION

Term	Abbreviation	SI unit	Other unit
cation exchange capacity	CEC	$cmol_c/kg$	me/100g
electrical conductivity	EC	dS /m	mmho/cm
electrolyte concentration		mol_c/m^3	me/l
exchangeable potassium percentage	EPP	%	%
exchangeable sodium percentage	ESP	%	%
flocculation value	FV	mol_c/m^3	me/l
infiltration rate		mm/h	in/h
interlayer space		nm	Angstroms
kinetic energy		Joules mm m^2	
sodium absorption ratio	SAR		
specific surface		m^2/g	m^2/g
velocity		m/sec	ft/sec

PART THREE—EFFECTS OF SALTS ON PLANTS

CHAPTER 6

PLANT RESPONSES TO SALINE AND SODIC CONDITIONS[a]

INTRODUCTION

Many plants, the glycophytes or non-halophytes, are ill-equipped to cope with the stresses of saline and sodic conditions, while other plants, the halophytes, thrive under them. Stress here refers to a condition that diverges from the normal range that a given plant encounters to such an extent as to prevent the plant from expressing fully its genetic potential for growth, development, and reproduction. Because salinity and sodicity stresses are quantitively expressed as concentrations, activities, and similar units, no sharp dividing line exists between saline and sodic stress and lack of stress. There is instead a continuum from the absence of stress to very intense stress. Neither is there an absolute distinction between salt tolerance and salt sensitivity, even among glycophytes. Even genotypes within a species differ markedly in this regard. The tolerance or sensitivity of a given plant is indicated by the point or range in the continuum of stress where the plant shows signs of being adversely affected. This depends not only on the intensity of stress but on the chemical composition of the medium, temperature, light intensity, relative humidity, and many other factors.

Most crop plants, except for the coconut and date palm, are glycophytes or non-halophytes. In this chapter, the emphasis is on crop species, but halophytes will be discussed as appropriate. Mechanisms of salt tolerance in halophytes have been reviewed by Flowers (1985), Flowers et al. (1977), Waisel (1972), and Wyn Jones (1981), among others. For a discussion of genotypic differences in salt tolerance, see Epstein (1985) and Chapter 8.

SALINITY AND SODICITY

In the context of salinity responses of plants, salinity refers to concentrations of soluble salts so high as to affect significantly the colligative properties of the solution to which the roots are exposed, specifically by reducing its osmotic potential. (For tabulations of the colligative properties of solutions, refer to Wyn Jones and Gorham 1983.) In

---

[a]Prepared by: André Läuchli and Emanuel Epstein, Dept. of Land, Air, and Water Resour., Univ. of Calif., Davis, CA 95616.

the agricultural context, a soil is considered saline if the electrical conductivity of the saturation extract exceeds 4 dS/m at 25° C and the percentage of the cation exchange capacity of the soil occupied by sodium is less than 15 (Richards 1954). The value of 4 dS/m corresponds to about 40 meq/l salt. As discussed below, however, no well-defined limit exists between saline and nonsaline soils. As has been and will be discussed, many factors other than the total salinity influence the responses of plants to salinity.

Sodicity refers to Na^+ ions in particular in the absence of the osmotic component characteristic of salinity. Sodicity is estimated by several criteria. The exchangeable sodium percentage (ESP) and the sodium absorption ratio (SAR) are the most common. The former quantity represents the percentage of the cation exchange complex of the soil occupied by sodium ions. If that percentage exceeds 15, the soil is considered sodic (Richards 1954). For estimating the sodium hazard to plants, the sodium adsorption ratio is often preferred:

$$\mathrm{SAR} = \frac{\mathrm{Na^+}}{\sqrt{\mathrm{Ca^{2+}} + \mathrm{Mg^{2+}}}} \tag{6.1}$$

where all concentrations are molarities. Values of the sodium adsorption ratio, SAR, that exceed 15 indicate sodicity. However, no single value adequately indicates a line of demarcation between sodic and non-sodic soils. According to Northcote and Skene (1972), soils with an ESP of 6 or more are sodic and those with an ESP of 15 or greater are strongly sodic.

The distinction between saline and sodic conditions notwithstanding, salinity and saline are frequently used as generic terms that cover both conditions, the emphasis being on salinity. This convention will be followed here.

PRINCIPAL RESPONSES OF PLANTS TO SALINITY

Fig. 6.1 is a simplified scheme of the effects of salinity and sodicity on plants. The effects of salinity on plants are due to one or the other of two properties of saline media, but usually both are implicated. Salinity depresses the external water potential, a physical effect, and specific ions may have chemical effects. The left and right parts of Fig. 6.1 suggest that some effects of salinity on plants are not deleterious. With few exceptions, however, saline conditions harm crops.

Depression of the external osmotic potential by high salt concentrations tends to narrow the gap between the external and internal water potentials. At high salinities, the external osmotic potential may be depressed below that of the cell water potential, resulting in osmotic desiccation. But, even in less extreme situations, the water availability to the plant will at least initially tend to be lessened. The reduction in the osmotic potential of the medium has long been held to be the prime cause, directly or indirectly, of the adverse effects of salinity on plant growth (Maas and Nieman 1978).

Changes in the colligative properties of a solution, such as the depression of the water potential, depend on the number of the solute entities causing them. Specific ions, however, as well as contributing to the depression of the osmotic potential, influence the response of plants to salinity. As suggested in Fig. 6.1, these specific ion effects may be categorized under three headings.

First, high concentrations of a given ion may cause disorders in mineral nutrition. For example, high sodium concentrations may cause deficiencies of other elements, such as potassium or calcium. Second, certain ions, such as chloride, may have toxic effects, which may not always be clearly distinguishable from deficiencies. Third, there may be specific ion effects that promote the growth or qualitative features of the plant.

Sodicity is due to high activities of Na^{+} ions, relative to those of Ca^{2+} and Mg^{2+}. Sodicity may, therefore, have two of the effects discussed in connection with salinity: disturbances of mineral nutrition and toxicity. Sodicity will not elicit strictly osmotic effects. For detailed considerations of salinity and sodicity, refer to Chapter 3.

As already mentioned, salinity may also positively affect plant growth and composition. It may promote the growth of halophytes or enhance the yield or quality of crops (Pasternak 1987). Moderate salinity may increase the yields of cotton (Pasternak et al. 1979), raise the concentration of total dissolved solids in tomatoes, a quality feature (Rush and Epstein 1981), and increase the freezing tolerance of citrus (Syvertsen and Yelenosky 1988).

MECHANISMS OF RESPONSE

Osmotic Effects

If the osmotic potential of the medium became lower than that of the plant's cells, the latter would suffer osmotic desiccation. To survive, the plant must adjust osmotically, i.e., in a medium of a given, high salinity, it must build up even higher internal solute concentrations. This can be effected by absorption of ions from the medium, or synthesis of organic compounds, or both.

The salt-accumulating halophytes are adapted by absorbing salt from the medium and using it as a major internal osmoticum (Flowers et al. 1977). However, salt in plant cells can be dangerous. Much evidence (Greenway and Munns 1980; Munns et al. 1983; Wyn Jones 1981) indicates that high salt concentrations in the cytoplasm damage enzymes and organelles. Salt from the medium apparently serves as an osmoticum in that large fraction of the total cell volume, the vacuole. In the cytoplasm, the function of osmoregulation is served mainly by organic solutes synthesized by the plant (Wyn Jones and Gorham 1983); refer to the left side of Fig. 6.2. Thus, organic osmolytes are used to a large extent in only a small fraction of the total cell volume. The tonoplast must transport salt into the vacuole, build up a high concentration of the salt there, and prevent any substantial leakage of organic osmolytes from the cytoplasm into the vacuole.

The glycophytes or non-halophytes are unable to cause the sharp asymmetrical intracellular compartmentation of inorganic and organic

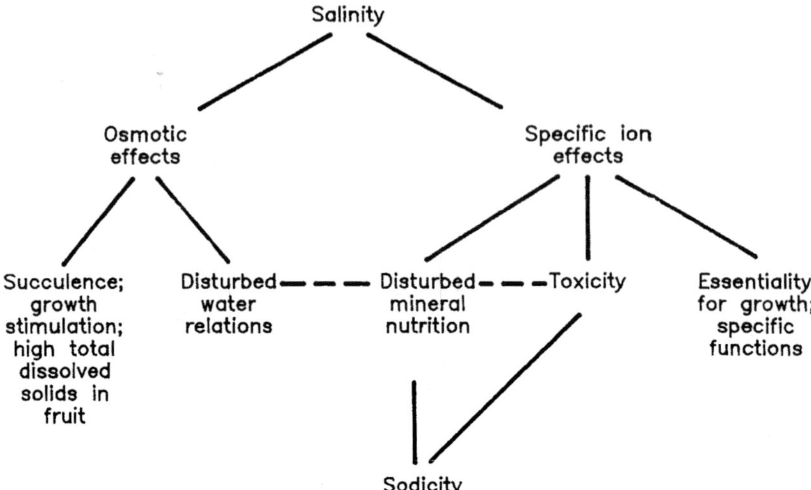

FIG. 6.1. Effects of Salinity and Sodicity on Plants

solutes that is central in the adaptation of salt-absorbing halophytes to their saline habitats. Exposed to moderate salinities, glycophytes tend to exclude salt and to sequester what salt they absorb in the roots and stems, thus minimizing the exposure of the leaf cells and, hence, the photosynthetic apparatus, to salt. This aspect of salt tolerance is discussed more fully below in the "Integration in the Whole Plant" section. Thus, the glycophytes regulate ion fluxes less effectively at the cellular level than do halophytes, but partition ions more effectively at the level of the organ and tissue (Fig. 6.2, right side).

Unable to absorb major quantities of external ions for osmoregulation, glycophytes, to exist in a saline medium, must synthesize organic osmolytes to a larger extent than plants that can use salt itself as a major osmoticum (Fig. 6.2). Many such organic solutes have indeed been identified (Table 6.1). The term "compatible solutes" is often applied to these organic osmolytes because of their presumed compatibility with cytoplasmic entities and processes. There is evidence, however, that not all of these compatible solutes are, in fact, wholly benign (Wyn Jones 1984).

The synthesis and transport of organic osmolytes, the transport of inorganic ions, and the compartmentation of organic and inorganic solutes discussed above all require the expenditure of metabolic energy, i.e., of photosynthate (Epstein 1980). Stavarek and Rains (1985) compared the performance of alfalfa cells selected for salt tolerance in cell culture with that of unselected (non-tolerant) cells. At 170 mM NaCl, the unselected cells hardly grew. The selected (salt-tolerant) cells at the same salinity differed from those in the control treatment (no salt) only in having a lag period of a few days. Thereafter, their growth (dry weight gain) paralleled that of the cells in the control treatment. Measurements of carbohydrate metabolism of the salt-tolerant selected cells showed maintenance costs to be relatively low when the cells grew under saline conditions. The maintenance costs of the unselected cells

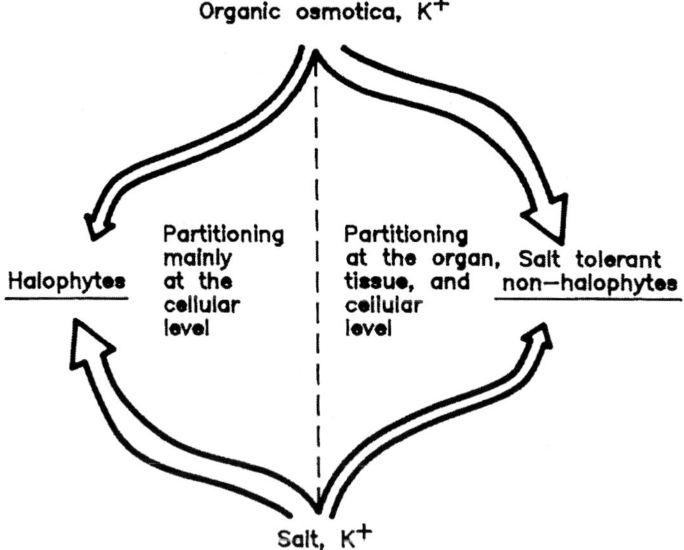

FIG. 6.2 Salt and Organic Solutes as Osmotica in Halophytes and Salt-Tolerant Non-Halophytes (Epstein 1983)

increased with salinity. Even selected cells, however, must divert energy to the processes of ion transport, synthesis of organic osmotica, and compartmentation. These costs result in diminished growth per unit of substrate (sugars) utilized (Stavarek and Rains 1985; Yeo 1983). Plants vary greatly in the adjustment of their energy economy to the presence of salt (Schwarz and Gale 1981). Maintenance respiration rates usually increase at moderate salinities. What is "moderate," however, depends on the salt tolerance of the plant. Fig. 6.3 outlines the main energetics of a salt-stressed cell. Raven (1985) concluded that the use of compatible organic solutes to generate a given internal osmolarity may involve an energy cost one order of magnitude higher than that incurred by the use of NaCl or KCl.

Specific Ion Effects: Nutrition

A universal feature of salinity is the presence of high concentrations or activities of certain ionic species, sodium and chloride in many salt-affected soils (Epstein and Rains 1987; Szabolcs 1989). The ratios of these ions to others may be quite high and may cause deficiencies of nutrient elements present at much lower concentrations. For example, in the saline environments in which sodium predominates over potassium, the plant's paramount nutritional requirement is potassium in adequate amounts (Rains and Epstein 1967). A high degree of selectivity of the potassium transport mechanism was demonstrated in experiments with excised leaf tissue of the mangrove, *avicennia marina*. Even so, this species may show evidence of salt-induced potassium deficiency (Ball et al. 1987).

Saline conditions may inhibit the absorption of nitrate. Thus, in

TABLE 6.1 Organic Osmolytes of Higher Plants and their Taxonomic Distribution[a]

Class of compounds (1)	Major compounds (2)	Taxonomic distribution (3)
Amino acids	Proline	Juncaginaceae Asteraceae Gramineae Aizoaceae Meliaceae Acanthaceae
Methylamines	Glycinebetaine	Chenopodiaceae Amaranthaceae Asteraceae Solanaceae Gramineae Avicenniaceae Sterculiaceae Malvaceae Acanthaceae
	Prolinebetaine	Labiatae Capparidaceae Leguminoseae
	B-alaninebetaine	Plumbaginaceae
Polyhydric alcohols	D-sorbitol	Plantaginaceae Rosaceae
	D-mannitol	Combretaceae Myrsinaceae Rubiaceae
Mono- and disaccharides	Sucrose, glucose, fructose	Malvaceae
Cyclitols	D-pinitol	Leguminoseae Rhizophoraceae Caryophyllaceae
	L-quebrachitol	Euphorbiaceae
	Myo-inositol	Solanaceae (*Lycopersicon esculentum*)
Tertiary sulfonium compounds	b-dimethylsulphonio propionate	Asteraceae Gramineae

[a]Gorham et al. (1985); Sacher and Staples (1985); Wyn Jones (1984); Yancey et al. (1982); and Yopp, J.H. (1989) "Salt and osmolytes" (manuscript).

short-term experiments (up to 12 hours) with barley seedlings, Aslam et al. (1984) found that SO_4^{2-} and, to a greater extent, Cl^- diminish the rate of NO_3^- absorption, the degree of inhibition being 83% at 0.2 M NaCl. The

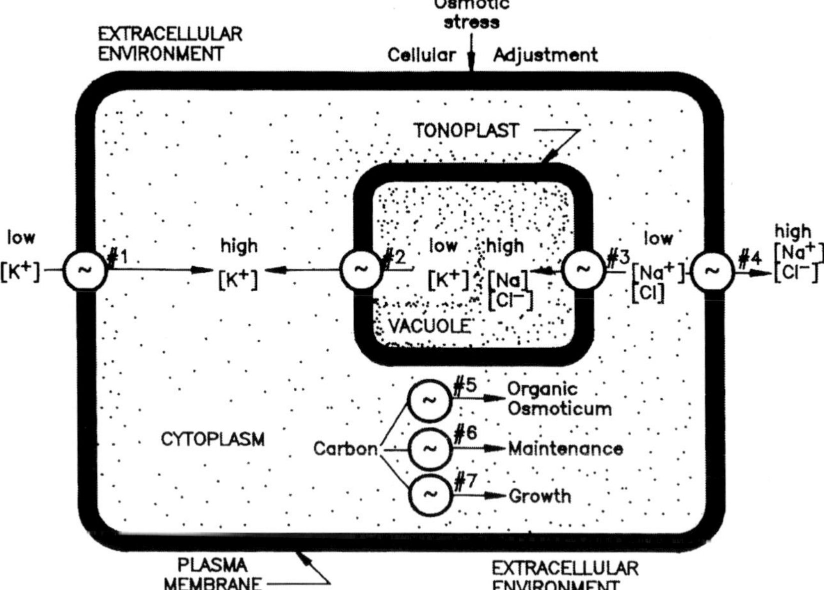

FIG. 6.3. Simplified Model of Vacuolated Plant Cell Exposed to Saline Stress (Rains 1987). [The salinity in the extracellular environment is relatively high in [Na] and [Cl] and low in [K]. The cell has two compartments: the cytoplasm and vacuole. Various processes are depicted which involve expenditure of metabolic energy (~). Processes 1 through 4 involve transport of ions between intracellular compartments and the external environment. Processes 5 through 7 represent metabolism of carbon into various products or processes.]

identity of the cation (N^+ or K^+) had little effect. Salt did not affect subsequent NO_3^- reduction. The inhibition of NO_3^- absorption by salt (0.2 M NaCl) was apparent within one minute of imposing the stress; recovery depended on a number of metabolic processes (Klobus et al. 1988).

Sodium ions have been shown to cause disturbances in calcium nutrition. Nutritional disorders involving other elements may be linked to the effects of salinity on the transport and metabolism of calcium. When external calcium concentrations are high, they may mitigate the effects of salinity. High ratios of sodium to calcium in the medium tend to be deleterious. Inadequate concentrations of Ca^{2+} may adversely affect membrane function and growth within minutes (Epstein 1961; Läuchli and Epstein 1970; Cramer et al. 1988). Different genotypes may have widely different responses. Examples of salinity-calcium interactions follow.

When LaHaye and Epstein (1969) grew highly salt-sensitive bean plants, *phaseolus vulgaris,* in solution cultures at 50 mM NaCl and $CaSO_4$ concentrations of 1 mM or less, NaCl impaired the growth of the plants during the seven days of the experiment. Adding Ca^{2+} at concentra tions of 3 or 10 mM completely protected the plants from the adverse effects of NaCl. Similar findings were obtained with bean plants grown to maturity (LaHaye and Epstein 1971). Effects of Ca^{2+} on the performance of the plants under saline conditions have since been extensively investigated, especially its role in the integrity and unimpaired function of plant cell membranes and its displacement from membranes by Na^+ ions present at high concentrations (Cramer et al. 1985, 1986; Lynch et al. 1987).

Elzam and Epstein (1969) compared two species of wheat grass that were grown in nutrient solutions salinized with NaCl at concentrations ranging up to 500 mM. The salt concentrations that severely affected their growth were 5.0 mM for the salt-sensitive species of wheat grass and 100 mM for the salt-tolerant species. In each species, the calcium concentration in the roots dropped precipitiously at the same concentrations of salt, suggesting a causal relationship between the salt-induced calcium loss and the failure of the plants.

In the experiments just described, the osmolarity of the nutrient solutions increased with increasing salinities, as did the Na^+/Ca^{2+} ratios. Maas and Grieve (1987) compared the effects of exposing corn, *Zea mays*, to isosmotic solutions salinized at various Na^+/Ca^{2+} ratios. At a high ratio (34.6/1 on a molar basis), the plants suffered from calcium deficiency. When that ratio was 5.7/1 or less, no calcium deficiency occurred. In similar experiments with corn, Plaut and Grieve (1988) found that at progressively lower Na^+/Ca^{2+} ratios in the media, the photosynthetic rate (CO_2 fixation) and water-use efficiency declined, the former possibly elicited in part by a calcium-induced magnesium deficiency.

The instances given above of specific ion effects on aspects of mineral nutrition are part of a large body of information obtained over roughly the last 25 years. As a result of these and related researches, the opinion prevailing earlier, that salinity effects were more or less nonspecific and due to the total salinity as often measured by the specific electric conductance of the ambient solution, is being re-evaluated. Now, a much keener appreciation exists of the importance of the ionic composition of the medium. It is no longer possible to refer to the "osmotic school" and the "specific-ion school" (Bernstein 1975). The toxicity effects discussed below reinforce that view.

Specific Ion Effects: Toxicity

In a number of instances, one of two kinds of evidence for specific ion toxicity may be observed. One is that moderate concentrations of sodium, chloride, sulfate or other ions reduce growth or cause specific injury. The other is that isosmotic solutions of different compositions may elicit very different responses. As might be expected, genotypes may differ in these responses even within a species.

Toxicities due to even moderate concentrations of some ions of salt-affected soils are most common in woody plants. Bernstein (1965) shows color photographs of severe leaf injury due to sodium or chloride salts in several fruit crops. These crops have little ability to exclude sodium or chloride from their leaves and are long-lived; hence, they often suffer toxicities at even moderate soil salinities.

Salinity caused leaf injury identified as phosphate toxicity to soybean plants in experiments conducted by Grattan and Maas (1984, 1988). The extent to which salinity caused such leaf injury depended on the concentration of phosphate, the Ca^{2+}/Na^{+} ratio, and the variety. The mechanism by which moderate salinity accompanied by high phosphate concentrations elicits phosphate toxicity is as yet poorly understood (Treeby and van Steveninck 1988).

An example from experiments with wheat, *Triticum aestivum*, shows the usefulness of isosmotic solutions of different ionic composition, as well as the importance of the genotype. Kingsbury and Epstein (1986) found that the growth of a salt-sensitive line of wheat was adversely affected by nutrient solutions containing high concentrations of Na^{+} ($>$ 100 mM), but not by isosmotic solutions without Na^{+} at elevated concentrations. The growth of a line selected for salt tolerance was unimpaired by any of the solutions used. Termaat and Munns (1986) also used isosmotic solutions to investigate osmotic versus specific ion effects.

INTEGRATION IN THE WHOLE PLANT

Salt Responses During Plant Development

Salt sensitivity changes considerably during the development of a plant. Three developmental stages can be distinguished with respect to salt tolerance or salt sensitivity: germination, vegetative growth, and reproductive growth. Considerable evidence indicates that vegetative growth in crop species is particularly salt-sensitive. A more complex response pattern emerges for germination. Many salt-sensitive plant species germinate readily in the presence of high concentrations of salt. This has been documented for corn, *Zea mays,* (Maas et al. 1983); kenaf, *Hibiscus cannabinus,* (Curtis and Läuchli 1985); and tomato, *Lycopersicon esculentum,* (Kurth et al. 1986b). The salt tolerance of triticale lines at germination and emergence varied according to the genotype, and adding calcium significantly improved the performance of one of the lines (Noryln and Epstein 1984). The more salt-tolerant cotton, *Gossypium hirsutum* (Shannon and Francois 1977; Kent and Läuchli 1985), and sugar beet, *Beta vugaris*(Ayers and Hayward 1948), are more salt-sensitive during germination than during later stages of development. Some halophytes, whose vegetative growth is often stimulated by salinity, do not appear to be salt-tolerant during germination (Ungar 1978; Jefferies 1988).

Results from a long-term field study on the reuse of saline drainage water for crop irrigation in the San Joaquin Valley provide a similar pattern of germination response to salinity (Rains et al. 1987). In the fourth year of reuse, irrigation of cotton with water having a salinity of 9,000 mg/l reduced plant density by as much as 70%. However, the more salt-sensitive safflower, grown in the third year of this study, appeared less salt-sensitive during germination.

While vegetative growth in non-halophytes generally is salt-sensitive, some crop plants become dramatically more salt-tolerant during reproductive development. In the aforementioned field study on the reuse of saline drainage water, cotton yield was only slightly reduced at

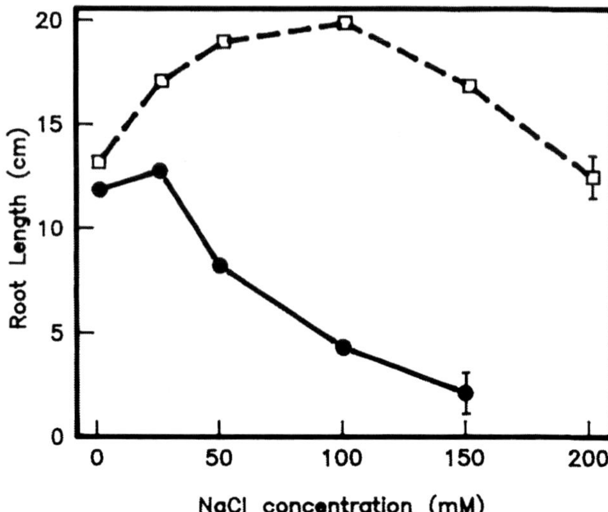

FIG. 6.4. Root Growth of Cotton Seedlings: Effects of NaCl and $CaCl_2$. [Root growth is expressed as tap root length of 6-day-old seedlings. 0.4 mM Ca^{2+} (—); 10 mM Ca^{2+} (– – –). Error bars represent 95% confidence intervals (Cramer et al. 1986)]

the highest salinity level in the irrigation water despite significant reductions in vegetative growth and development (Rains et al. 1987 and Rains' unpublished results). Mechanisms by which the cotton plant becomes more salt-tolerant during its later stages of development are unknown. Developmental shifts in hormonal balance may contribute to the increased salt tolerance during reproductive growth.

Developmental shifts in relative salt tolerance also vary according to the genotype. This has been documented in barley by Lynch et al. (1982). The barley cultivar most sensitive to early salinization, cv. Briggs, proved more resistant at later stages of development, while another cultivar showed the opposite shift in salt tolerance.

Salt Sensitivity During Vegetative Growth

What causes vegetative growth in non-halophytes to be salt-sensitive? First, shoot growth is usually more affected by salinity than is root growth, as demonstrated with barley by Delane et al. (1982) and with sorghum by Weimberg et al. (1984). As has already been discussed, calcium plays a special role in specific ion effects. A wide range of NaCl concentrations does not inhibit the root growth of cotton when supplemental calcium is added (Cramer et al. 1986; Fig. 6.4). Shoot growth, however, remained salt-sensitive (Kent and Läuchli 1985). Adding calcium to the medium apparently favors cell elongation over radial cell expansion and maintains high rates of cell production (Kurth et al. 1986a). At low calcium concentrations, however, root cell elongation was very sensitive to salt; the roots became shorter and thicker, and cell production was inhibited. Adding calcium after adding NaCl only par-

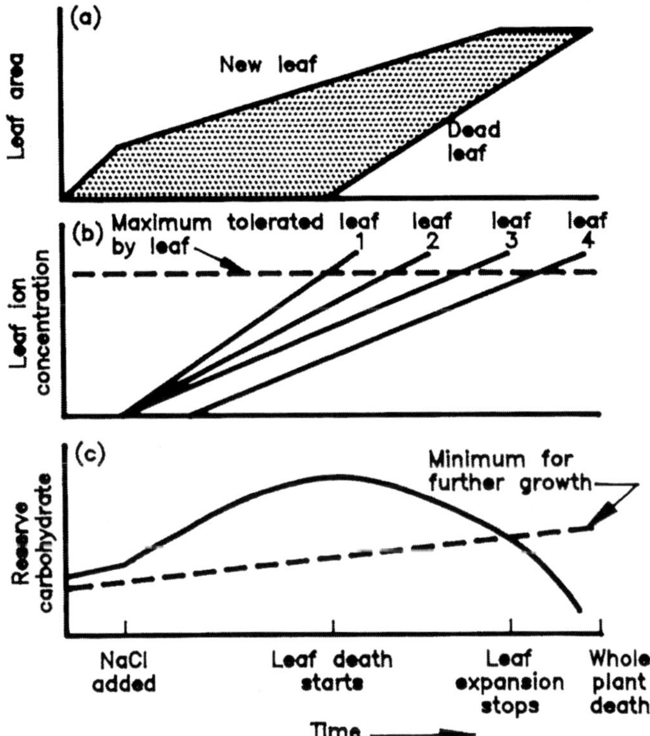

FIG. 6.5. Changes with Time Following Exposure to NaCl: (a) Total Leaf Area; (b) Leaf Mean Ion Concentration; and (c) Total Reserve Carbohydrate Content of a Plant. [The time scale varies with plant species and external NaCl concentration (Munns and Termaat 1986]

tially restored the root growth of corn, but adding calcium before adding NaCl completely protected root growth from the inhibiting effects of salt (Cramer et al. 1988).

Solomon et al. (1986 and unpublished results) found similar effects in pea roots. In a salt-tolerant line of *Citrus sinensis,* salt inhibited cell growth when calcium was removed from the growth medium (Ben-Hayyim and Kochba 1983). Maximal cell production required an increase in calcium for each increase in NaCl (Ben-Hayyim et al. 1987).

In contrast to the results with cotton (Kent and Läuchli 1985) and similar results with rice (Yeo and Flowers 1985), supplemental calcium can partly overcome the effects of NaCl on the shoot growth of beans (LaHaye and Epstein 1969, 1971), barley (Ward et al. 1986), corn (Maas and Grieve 1987), and sorghum (Grieve and Maas 1988). Overall, shoot growth in nonhalophytes is an indicator of salt stress.

What component of shoot growth is inhibited and by what mechanism does this inhibition occur? Reduced shoot growth caused by salinity is normally documented best by a reduced leaf area and stunted shoots. Final leaf size, governed by the number and size of the cells in leaf tissues, depends on cell division and cell enlargement. According to Kriedemann (1986), the early formative stages of shoots and leaves are governed by cell division and are relatively insensitive to drought and salinity but appear to be limited by the supply of assimilates. Leaf expansion is sensitive to drought and salinity. Kriedemann's results agree with those by Papp et al. (1983), who found that the rate of leaf extension in sugar beet fell more or less linearly with increasing salt concentration in the medium, but leaf initiation rate was virtually unaffected. In the salt-sensitive bean plant, salinity suppressed the enlargement and division of cells in the leaves (Nieman 1965). Salt reduced the leaf area of kenaf by reducing the size and number of its epidermal cells (Curtis and Laüchli 1987). This implies that the sensitivity of cell division to salt contributes significantly to the reduction of leaf area after salinization of the latter two plant species.

Munns and Termaat (1986) developed the following hypothesis to explain the relationship between salt sensitivity of leaf growth and the overall sensitivity of a plant to salt (Fig. 6.5). Expansion of leaves initially decreases after exposure to salt and continues at this decreased rate for some time (Fig. 6.5a). Ion concentrations in the leaves build up until toxic concentrations are reached, first in the oldest (leaf 1) and subsequently in the youngest leaves (Fig. 6.5b). Once the rate of leaf death is faster than the rate of leaf expansion, the amount of reserve carbohydrates in the plant will fall in proportion to the reduction in leaf area (Fig. 6.5c). However, the amount of carbohydrate needed for continued growth may increase, particularly because of continuous root growth leading to increased root:shoot ratios. Eventually, the leaves cannot produce enough carbohydrates from photosynthesis to sustain growth of the whole plant (Fig. 6.4c), expansion of leaf area ceases, and, in extreme cases, the plant dies. The time scale of this plant response would vary inversely with the level of salinity in the medium and with the genotype. In the long-term, leaf expansion depends on the avoidance of excessive salt build-up in the leaves and on new leaf production exceeding the rate of death of the old leaves (Munns and Termaat 1986). Regulation of salt balance in the leaves and maintenance of some minimum total leaf area for photosynthesis and production of carbohydrates appear crucial for continued plant growth under salt stress.

Cheeseman (1988) considered the control of sodium acquisition and allocation in plants on the one hand and carbon allocation and use on the other to be critical components of salt tolerance. Photosynthesis per se, however, does not appear to determine plant growth under salt stress for barley (Munns et al. 1982; Rawson et al. 1988), sugar beet (Papp et al. 1983), sunflower (Rawson and Munns 1984), and kenaf (Curtis and Läuchli 1986; Curtis et al. 1988).

Water Relations. Could a water deficit in the leaves limit shoot growth following salinization? Termaat et al. (1985) tested whether leaf turgor limits shoot growth in salt-stressed plants by growing plants in a saline soil within a pressure chamber and applying a pressure to the soil that raised the shoot water potential by the same amount as the pressure applied to the root. Experiments using barley, wheat, Egyptian clover,

TABLE 6.2 Na^+ and Cl^- Concentrations in Leaves of Crop Cultivars Differing in Salt Sensitivity (μmol/g dry wt.)

	NaCl Treatment (mM)	Cultivar				References
		Sensitive		Less Sensitive		
		Na^+	Cl^-	Na^+	Cl^-	
	(1)	(2)	(3)	(4)	(5)	(6)
Avocado	20	220	370	3	300	Downton (1978)
Grapevine[1]	50	—	518	—	30	Bernstein et al. (1969)
Corn (shoot)	50	143	763	363	871	Schubert and Läuchli (1986)
Soybean	100	540	860	170	290	Läuchli and Wieneke (1979)

[1]Thompson Seedless (scion) on root stocks "Cardinal" (sensitive) and "Salt Creek" (less sensitive).

white clover, and white lupin indicated that increased leaf water potential did not improve growth under saline conditions (Munns and Termaat 1986). These results could mean that leaf water deficit does not limit shoot growth in non-halophytes grown at high salinity. Thiel et al. (1988) determined the cellular turgor in leaves of salinized barley seedlings by using the micro-pressure probe method described by Zimmerman et al. (1980) and correlated the turgor values with leaf elongation rates and values of bulk leaf osmotic potential. Adding NaCl to the medium caused the leaf elongation rate to decline immediately, followed by gradual recovery, the extent of which depended on the NaCl concentration. Salt stress reduced cellular turgor in expanding and expanded leaf tissue. The turgor of expanding tissue recovered quickly as a result of osmotic adjustment, but decreased turgor persisted in expanded tissue. These results indicate that salinity reduces leaf elongation through osmotic effects on turgor in the short term, but the sustained component of reduction in leaf elongation is not caused by a leaf water deficit in the long term. The latter part of this conclusion agrees with Munns and Termaat (1986). Further, reduced leaf growth may indirectly increase total solute concentration in the leaves, which contributes to osmotic adjustment unless solutes build up to toxic concentrations in specific cell compartments of the leaves.

Regulation of Salt Balance in Leaves. Regulation of salt balance in the leaves is significant for the salt tolerance of the whole plant. Flowers and Yeo (1986) concluded that for salt-sensitive species, such as rice, sodium supply to the expanding leaves exceeds the demand for osmotic adjustment at high salt stress, leading to excessive sodium concentrations in the leaves. Sodium could accumulate in the cell walls of leaves, as proposed by Oertli (1968) and Munns and Passioura (1984), or reach excessive intracellular concentrations, leading to ion toxicity. Salt accumulation in leaf cell walls could cause loss of turgor and then dehydration of the cells. Such a loss of turgor has been demonstrated in expanded barley leaves (Thiel et al. 1988). It is also well established that salt concentrations are highest in the oldest leaves of non-halophytes (Greenway and Munns 1980). Yeo and Flowers (1982) suggested that lower salt concentrations in younger leaves may partly be due to exclusion of specific ions from the xylem vessels that supply them. In rice, the evidence clearly indicates that, upon salinization, sodium moves first to the older leaves, which in effect protects the younger leaves. This partial exclusion of sodium from the young leaves appears to correlate positively with the salt tolerance of some rice cultivars relative to the germ plasm base (Yeo and Flowers 1986).

The question arises as to how significant salt exclusion from the shoot is as a mechanism of salt tolerance in non-halophytes. First, the capacity of salt exclusion in a plant has its limits. Above a threshold level of salt in the external medium, the capacity becomes saturated and the salt exclusion mechanism breaks down, leading to high rates of transport of Na^+, or Cl^-, or both, to the shoot (Elzam and Epstein 1969; Läuchli 1984). Second, even under conditions where salt exclusion from the shoot does operate, it can be accompanied by the development of a water deficit in the leaves because of inadequate osmotic adjustment (Läuchli and Epstein 1984). Table 6.2 gives examples of four crop species that have cultivars differing in salt tolerance, and the possible correlation with salt exclusion. In avocado, *Persea americana,* a salt-sensitive cultivar does not efficiently exclude Na^+ from the shoot. In soybean, the sensitive cultivar excludes neither Na^+ nor Cl^- efficiently from the shoot. The grapevine scion, Thompson Seedless, excludes Cl^- much more effectively on rootstock "Salt Creek" than on "Cardinal." In corn, however, the more tolerant cultivar of two studied excludes Na^+ poorly (Table 6.2). Läuchli (1984) has cited other examples where higher salt tolerance did not correspond with a higher degree of Na^+ or Cl^- exclusion. Thus, salt exclusion is of limited usefulness as a mechanism of salt tolerance in the whole plant.

In irrigated agriculture under field conditions, surface irrigation, particularly on poorly drained land, can lead to an oxygen deficiency in the root zone of the soil (Meek et al. 1983). Since oxygen deficiency alone adversely affects the growth of most dryland species (Jackson and Drew 1984), the combined effects of salinity and oxygen deficiency would be expected to be particularly damaging. In corn, oxygen deficiency in the root medium under saline conditions led to a breakdown of Na^+ exclusion from the shoots and simultaneously inhibited K^+ transport (Drew and Läuchli 1985). Oxygen deficiency in the root zone interacts with salinity relative to salt exclusion from the shoot in solution culture (Drew et al. 1988) and under field conditions (Barrett-Lennard 1986).

Salt transport to the shoot is clearly regulated by and integrated with growth in a relatively salt-tolerant plant (Cheeseman and Wickens 1986). In the more salt-sensitive crop plants, salt transport to the leaves appears less well regulated, although LaHaye and Epstein (1971) found that in bean plants grown at a Na^+/Ca^{2+} ratio of 5:1 the ratio of total sodium to total plant weight remained constant during the entire six weeks of the experiment. Munns and Termaat (1986) concluded that salt toxicity within the shoot does not limit growth. Munns et al. (1988)

found in dividing and expanding barley leaf tissue that the Na^+ and Cl^- concentrations and the rates of growth and development did not consistently relate. This indicates that local concentrations of Na^+ or Cl^- in the growing tissues do not control shoot growth (Munns et al. 1988). Cheeseman (1988) reviewed considerable evidence that intracellular Na^+ compartmentation in leaf cells may not be as important as previously considered. For example, Seeman and Critchley (1985) found little difference in cytoplasm (plus chloroplast) versus vacuolar Na^+ and Cl^- levels in bean leaves at growth-inhibiting salt concentrations.

Does export of Na^+ and Cl^- from leaves through the phloem effectively help to maintain a salt balance in the leaves of a salinized plant? Läuchli (1984) and Flowers and Yeo (1986) concluded that little evidence exists to support this. Munns et al. (1986) confirmed this conclusion with an experiment that involved salinized barley. They found phloem export of Na^+ and Cl^- from leaves to be only about 10% of the xylem import of those ions.

Chemical Signals in Root-Shoot Interactions. What other components of root-shoot interaction might control shoot growth under saline conditions? Communication between root and shoot is critical for the controlled growth and development of the plant. Such communication might occur in the form of a chemical signal that is transmitted to the shoot at the time of root exposure to salinity. Nutrients and hormones are candidates for such chemical signals. Termaat and Munns (1986) raised the possibility that growth may be NaCl-limited by a reduced rate of transport of an essential nutrient ion to the shoot. Läuchli (1986) briefly discussed the evidence for interaction of salt stress with the acquisition of potassium, calcium, nitrate, and phosphate in plants. Calcium may be the chemical signal from root to shoot that contributes to the control of shoot growth, perhaps in conjunction with a plant hormone. Maas and Grieve (1987) demonstrated that NaCl stress in corn caused calcium deficiency in the leaves; these effects could be partly overcome by decreased Na^+/Ca^{2+} ratios in the medium. Shoot meristems may be deficient of calcium in salinized plants (Grieve and Maas 1988). Lynch and Läuchli (1985) found calcium transport to the shoot of barley to be sensitive even to mild NaCl stress. Increasing NaCl concentrations led to the accumulation of sodium and the depletion of calcium in the region of the third leaf of barley (Lynch et al. 1988). Grieve and Maas (1988) attributed sodium tolerance of the sorghum genotype "Hegari" to the efficient transport of calcium to the developing blades. Salt-induced growth inhibition in barley, corn, and sorghum is associated with inhibited Ca-transport to developing leaves. Evidence on lettuce supports this conclusion (Lazof and Läuchli 1988). Sodium-induced calcium deficiency had been reported for a variety of crop species (Chang and Dregne 1955). Excessive transport of sodium to expanding leaf tissue probably exacerbates the reduced supply of calcium to these regions. Calcium in the cell wall has been implicated in numerous aspects of cell expansion (Hanson 1984). Sodium and calcium interact very strongly in cell walls (Stassart et al. 1981; Zid and Grignon 1985). Excessive accumulation of sodium in leaf cell walls could displace functional calcium from exchange sites of cell walls and the plasma membrane, possibly causing membrane dysfunction and inhibition of cellulose biosynthesis in growing cell walls (Cramer et al. 1985; Zhong and Läuchli 1988). Thus, high Na^+/Ca^{2+} ratios in expanding leaf tissue could interfere with turgor maintenance.

Munns and Termaat (1986) considered the possible involvement of hormones as signals in root-shoot interactions of salt-stressed plants. One possibility is the cytokinins, since they are synthesized in the root and transported to the shoot, and their production appears to be inhibited by salt stress. However, Munns and Termaat (1986) could not find any indication that cytokinins are involved in carrying the message from root to shoot. Earlier, Prisco and O'Leary (1973) had not found evidence that growth inhibition by salinity is the consequence of impaired metabolism or transport of cytokinins in the plant. More likely, salinity-induced growth inhibition may be due to effects on hormonal balance. On the other hand, cytokinin appeared to prevent salt-induced necrosis in tobacco leaves (Benzioni et al. 1974).

Abscisic acid (ABA) is another candidate for a signal to the shoot. ABA increased in concentration in salt-stressed roots, but only after severe salt stress, when growth rates declined before root water potential decreased (Lachno 1983). Results by LaRosa et al. (1985) indicated that ABA is involved in adapting cultured tobacco cells to salts, but not to a water deficit elicited by nonionic osmotic solutes. Other data showed that ABA regulates the synthesis of a new protein in the presence of NaCl (Singh et al. 1987). Evidence shows that Ca^{2+} plays an important part in the regulation of leaf stomata by ABA (DeSilva et al. 1986). This indicates that any disturbance in the supply of Ca^{2+} to the leaves may interfere with the regulatory functions of ABA.

Yet another possibility for involvement of a plant hormone in root-shoot interaction during salt stress is ethylene. The production of ethylene was inhibited by NaCl in mung beans, but stimulated by $CaCl_2$ (Chrominsky et al. 1986). In tomato and pepper roots and shoots, salt treatment increased internal ethylene levels, but the calcium concentration in the medium was not given (El-Beltagy et al. 1979). Bangerth (1979) found that calcium deficiency stimulates ethylene production. Jeschke and Wolf (1988) demonstrated leaf epinasty in castor bean in response to NaCl treatment. Epinastic leaf growth of salt-stressed tomatoes was correlated with increased ethylene production in the shoot (El-Abd et al. 1988). Production of ethylene was a trait in rice cultivars (Khan et al. 1987) and tomato genotypes (Jones and El-Beltagy 1989) and correlated well with salt tolerance at the seedling stage of the two species. Research is necessary to determine the role of the ethylene hormonal system in inhibition of shoot growth by salt. Research is also necessary to determine how the postulated salt-ethylene interaction relates to the hypothesis that long-term, salt-induced inhibition of leaf growth involves a disturbance in the supply of calcium to growing leaves.

SUMMARY

Saline conditions are stressful to many plants, including most crops, for two principal reasons: First, they depress the external water potential, in effect making water less readily available to the plant (an osmotic effect); Second, sodium, chloride, and other ions may disturb mineral nutrition or cause toxicities (specific ion effects). Sodicity, i.e., high ratios of sodium to the sum of calcium and magnesium not accompanied by high concentrations of salts in solution, may have the effects identified as specific ion effects, without the osmotic feature of interference with normal water relations. The osmotic effects of salinity are dealt with by two mechanisms: absorption of salt from the medium or synthesis of organic solutes. Usually both are involved.

Highly salt-tolerant plants, or halophytes, tend to absorb salts from the medium and sequester it in the vacuoles, while organic compatible solutes serve the function of osmotic adjustment in the cytoplasm. Most crop plants are salt-sensitive, i.e., glycophytes. They tend to exclude sodium and choride from the shoots and, especially, the leaves. Hence, they have to rely more heavily than do halophytes on the synthesis of organic osmolytes. Disturbances in various aspects of mineral nutrition caused by the presence in the medium of ions, such as sodium and chloride, and by toxicities are referred to collectively as specific ion effects. The most common specific ion effects are salt-induced deficiencies of nutrient elements. Calcium at elevated concentrations often mitigates the adverse effects of salinity.

The integration of responses in the whole plant is critical for the survival, growth, and development of plants under saline conditions. Salt sensitivity changes during the development of the plant. Root growth is often less affected by salinity than shoot growth, particularly when supplemental calcium is provided. In the shoot, reduction in leaf area is most prominent under salinity. Leaf elongation appears to be reduced in the short term by osmotic effects on turgor. In the long term, regulation of salt balance in the leaves is significant, but shoot growth is probably not controlled by local concentrations of sodium and chloride. Shoot growth may be regulated by a chemical signal or signals transmitted from the root to the shoot upon exposure to salinity. Salt-induced inhibition of leaf growth in the long-term may be related to a disturbance in the supply of calcium and possibly other nutrients to growing leaves. The plant hormones, ethylene and abscisic acid, may be involved in this root-shoot interaction.

REFERENCES

Aslam, M., Huffaker, R.C. and Rains, D.W. (1984) "Early effects of salinity on nitrate assimilation in barley seedlings." *Plant Physiol.* 76: 321–325.

Ayers, D.A., and Hayward, H.E. (1948). "A method for measuring the effects of soil salinity on seed germination with observations on several crop plants." *Soil Sci. Soc. Am. Proc.* 13:224–226.

Ball, M.C., Chow, W.S. and Anderson, J.M. (1987). "Salinity-induced potassium deficiency causes loss of functional photosystem II in leaves of the grey mangrove, *avicennia marina,* through depletion of the atrazine-binding polypeptide." *Aust. J. Plant Physiol.* 14:351–361.

Bangerth, F. (1979). "Calcium-related physiological disorders of plants." *Ann. Rev. Phytopathol.* 17:97–122.

Barrett-Lennard, E.G. (1986). "Effects of waterlogging on the growth and NaCl uptake by vascular plants under saline conditions." *Reclamation and Revegetation Research.* 5:245–261.

Ben-Hayyim, G., Kafkafi, U., and Ganmore-Neumann, R. (1987). "Role of internal potassium in maintaining growth of cultured citrus cells on increasing NaCl and $CaCl_2$ concentrations." *Plant Physiol.* 85: 434–439.

Ben-Hayyim, G., and Kochba, J. (1983). "Aspects of tolerance in a NaCl-selected stable cell line of *citrus sinensis.*" *Plant Physiol.* 72:685–690.

Benzioni, A., Mizrahi, Y., and Richmond, A.E. (1974). "Effect of kinetin on plant response to salinity." *New Phytol.* 73:315–319.

Bernstein, L. (1965). "Salt tolerance of fruit crops." United States Department of Agriculture, Agriculture Information Bulletin No. 292.

Bernstein, L. (1975). "Effects of salinity and sodicity on plant growth." *Ann. Rev. Phytopathol.* 13:295–312.

Bernstein, L., Ehlig, C.F., and Clark, R.A. (1969). "Effect of grape root stocks on chloride accumulation in leaves." *J. Amer. Soc. Hort. Sci.* 94(6):584–590.

Chang, C.W., and Dregne, H.E. (1955). "Effect of exchangeable sodium on soil properties and on growth and cation content of alfalfa and cotton." *Soil Sci. Soc. Am. Proc.* 19:29–55.

Cheeseman, J.M. (1988). "Mechanisms of salinity tolerance in plants." *Plant Physiol.* 87:547–550.

Cheeseman, J.M. and Wickens, L.K. (1986). "Control of Na^+ and K^+ transport in *spergularia marina* iii. Relationship between ion uptake and growth at moderate salinity." *Plant Physiol.* 67:15–22.

Chrominski, A., Khan, M.A., Weber, D.J., and Smith, B.N. (1986). "Ethylene and ethane production in response to salinity stress." *Plant, Cell and Environment.* 9:687–691.

Cramer, G.R., Läuchli, A. and Polito, V.S. (1985). "Displacement of Ca^{2+} from the plasmalemma of root cells. A primary response to salt stress?" *Plant Physiol.* 79:207–211.

Cramer, G.R., Läuchli, A. and Epstein, E. (1986). "Effects of NaCl and $CaCl_2$ on ion activities in complex nutrient solutions and root growth of cotton." *Plant Physiol.* 81:792–797.

Cramer, G.R., Epstein, E., and Läuchli, A. (1988). "Kinetics of root elongation of maize in response to short-term exposure to NaCl and elevated calcium concentration." *J. Expt. Bot.* 39:1513–1522.

Curtis, P.S., and Läuchli, A. (1985). "Responses of kenaf to salt stress: germination and vegetative growth." *Crop Sci.* 25:944–949.

Curtis, P.S., and Läuchli, A. (1986). "The role of leaf-area development and photosynthetic capacity in determining growth of kenaf under moderate salt stress." *Aust. J. Plant Physiol.* 13:553–565.

Curtis, P.S., and Läuchli, A. (1987). "The effect of moderate salt stress on leaf anatomy in hibiscus cannabinus (kenaf) and its relation to leaf area." *Amer. J. Bot.* 74:538–542.

Curtis, P.S., Zhong, H.L., Läuchli, A., and Pearcy, R.W. (1988). "Carbohydrate availability, respiration and the growth of kenaf (hibiscus cannabinus) under moderate salt stress." *Amer. J. Bot.* 75: 1293–1297.

Delane, R., Greenway, H., Munns, R., and Gibbs, J. (1982). "Ion concentration and carbohydrate status of the elongating leaf tissue of hordeum vulgare growing at high external NaCl. I. Relationship between solute concentration and growth." *J. Exp. Bot.* 33:557–573.

DeSilva, D.L.R., Cox, R.C., Hetherington, A.M. and Mansfield, T.A. (1986). "The role of abscisic acid and calcium in determining the behavior of adaxial and abaxial stomata." *New Phytol.* 104:41–51.

Downton, W.J.S. (1978). "Growth and flowering in salt-stressed avocado trees." *Aust. J. Agric. Res.* 29:523–534.

Drew, M.C., and Läuchli, A. (1985). "Oxygen-dependent exclusion of sodium ions from shoots by roots of *Zea mays* (cv Pioneer 3906) in relation to salinity damage." *Plant Physiol.* 79:171–176.

Drew, M.C., Guenther, J., and Läuchli, A. (1988). "The combined effects of salinity and root anoxia on growth and net Na^+ and K^+ accumulation in zea mays grown in solution culture." *Ann. Bot.* 61: 41–53.

El-Abd, S., Jones, R.A. and El-Beltagy, A.S. (1988). "Production and action of ethylene in salinity-stressed tomato seedlings." *Egypt J. Hort. Sci.* 15:171–176.

El-Beltagy, A.S., Khalifa, M.M., and Hall, M.H. (1979). "Salinity in relation to ethylene." *Egypt. J. Hort.* 6:269–271.

Elzam, O.E. and Epstein, E. (1969). "Salt relations of two grass species differing in salt tolerance. I. Growth and salt content at different salt concentrations." *Agrochimica* 13:187–195.

Epstein, E. (1961). "The essential role of calcium in selective cation transport by plant cells." *Plant Physiol.* 36:437–444.

Epstein, E. (1980). "Responses of plants to saline environments." In: *Genetic Engineering of Osmoregulation: Impact on Plant Productivity for Food, Chemicals, and Energy.* D.W. Rains, R.C. Valentine and A. Hollaender, eds. Plenum Press, New York, pp. 7–21.

Epstein, E. (1983). "Crops tolerant of salinity and other mineral stresses." In: *Better Crops for Food.* Ciba Foundation Symposium 97. J. Nugent and M. O'Connor, eds. Pitman, London, pp. 61–82.

Epstein, E. (1985). "Salt-tolerant crops: origins, development, and prospects of the concept." *Plant and Soil* 89:187–198.

Epstein, E. and Rains, D.W. (1987). "Advances in salt tolerance." *Plant and Soil* 99:17–29.

Flowers, T.J. (1985). "Physiology of halophytes." *Plant and Soil* 89:41–56.

Flowers, T.J., Troke, P.F. and Yeo, A.R. (1977). "The mechanism of salt tolerance in halophytes." *Ann. Rev. Plant Physiol.* 28:89–121.

Flowers, T.J., and Yeo, A.R. (1986). "Ion relations of plants under drought and salinity." *Aust. J. Plant Physiol.* 13:75–91.

Gorham, J., Wyn Jones, R.G. and McDonnell, E. (1985). "Some mechanisms of salt tolerance in crop plants." *Plant and Soil* 89:15–40.

Grattan, S.R. and Maas, E.V. (1984). "Interactive effects of salinity and substrate phosphate on soybean." *Agron. J.* 76:668–676.
Grattan, S.R., and Maas, E.V. (1988). "Effect of salinity on phosphate accumulation and injury in soybean. I. Influence of $CaCl_2$/NaCl ratios." *Plant and Soil* 105:25–32.
Greenway, H. and Munns, R. (1980). "Mechanisms of salt tolerance in nonhalophytes." *Ann. Rev. Plant Physiol.* 31:149–190.
Grieve, C.M., and Maas, E.V. (1988). "Differential effects of sodium/calcium ratios on sorghum genotypes." *Crop Sci.* 28:659–665.
Hanson, J.B. (1984). "The functions of calcium in plant nutrition." *Advances in Plant Nutrition,* Vol. 1, Tinker, P.B., and Läuchli, A., eds. Praeger, New York, pp. 149–208.
Jackson, M.B., and Drew, M.C. (1984). "Effects of flooding on growth and metabolism of herbaceous plants." *Flooding and Plant Growth.* T.T. Kozlowski, ed. Academic Press, Orlando, Fla. pp. 47–128.
Jefferies, R.L. (1988). "Halophytes: unity and diversity in response to salinity." *International Symposium on Physiological Ecology of Aquatic Plants,* Aarhus, Denmark.
Jeschke, W.D., and Wolf, O. (1988). "Effect of salinity on growth, development, ion distribution, and ion translocation in castor bean (*Ricinus communis* L.)." *J. Plant Physiol.* 132:45–53.
Jones, R.A., and El-Beltagy, A.S. (1989). "Epinasty promoted by salinity or ethylene is an indicator of salt-sensitivity in tomatoes." *Plant, Cell and Environment.* 12:813–817.
Kent, L.M., and Läuchli, A. (1985). "Germination and seedling growth of cotton: salinity-calcium interactions." *Plant, Cell and Environment.* 8:155–159.
Khan, A.A., Akbar, M., and Seshu, D.V. (1987). "Ethylene as an indicator of salt tolerance in rice." *Crop Sci.* 27:1242–1247.
Kingsbury, R.W. and Epstein, E. (1986). "Salt sensitivity in wheat. A case for specific ion toxicity." *Plant Physiol.* 80:651–654.
Klobus, G., Ward, M.R., and Huffaker, R.C. (1988). "Characteristics of injury and recovery of net No_3^- transport of barley seedlings from treatment of NaCl." *Plant Physiol.* 87:878–882.
Kriedemann, P.E. (1986). "Stomatal and photosynthetic limitations to leaf growth." *Aust. J. Plant Physiol.* 13:15–31.
Kurth, E., Cramer, G.R., Läuchli, A., and Epstein, E. (1986a). "Effects of NaCl and $CaCl_2$ on cell enlargement and cell production in cotton roots." *Plant Physiol.* 82:1102–1106.
Kurth, E., Jensen, A. and Epstein, E. (1986b). "Resistance of fully imbibed tomato seeds to very high salinities." *Plant, Cell and Environment* 9:667–676.
Lachno, D.R. (1983). "Abscisic acid and indole-3-acetic acid in maize roots subject to water, salt and mechanical stress." *Growth Regulators in Root Development.* Jackson, M.B., and Stead, A.D., eds. British Plant Growth Regulator Group Monograph, No. 10, pp. 37–54.
LaHaye, P.A. and Epstein, E. (1969). "Salt toleration by plants: enhancement with calcium." *Science.* 166:395–396.

LaHaye, P.A. and E. Epstein. (1971). "Calcium and salt toleration by bean plants." *Physiol. Plant.* 25:213–218.

LaRosa, P.C., Handa, A.K., Hasegawa, P.M. and Bressan, R.A. (1985). "Abscisic acid accelerates adaptation of cultured tobacco cells to salt." *Plant Physiol.* 79:138–142.

Läuchli, A. (1984). "Salt exclusion: an adaptation of legumes for crops and pastures under saline conditions." *Salinity Tolerance in Plants—Strategies for Crop Improvement*. Staples, R.C. and Toenniessen, G.H., eds. Wiley, New York, pp. 171–187.

Läuchli, A. (1986). "Responses and adaptations of crops to salinity." *Acta Hort.* 190:243–246.

Läuchli, A. and Epstein, E. (1970). "Transport of potassium and rubidium in plant roots. The significance of calcium." *Plant Physiol.* 45:639–641.

Läuchli, A. and Epstein, E. (1984). "Mechanisms of salt tolerance in plants." *Calif. Agric.* 38(10):18–20.

Läuchli, A., and Wieneke, J. (1979). "Studies on growth and distribution of Na^+, K^+ and Cl^- in soybean varieites differing in salt tolerance." *Z. Pflanzenernähr. Bodenkde.* 142:3–13.

Lazof, D., and Läuchli, A. (1988). "Effect of NaCl salinization on nutrient supply to growing lettuce leaves." *Plant Physiol. Suppl.* 86:57.

Lynch, J., Cramer, G.C. and Läuchli, A. (1987). "Salinity reduces membrane-associated calcium in corn root protoplasts." *Plant Physiol.* 83: 390–394.

Lynch, J., E. Epstein, E., and Läuchli, A. (1982). "Na^+-K^+ relationships in salt-stressed barley." *Plant Nutrition 1982, Proc. 9th Int. Plant Nutrition Colloq.*, Warwick, England, Vol. 1, Scaife, A., ed. Commonwealth Agricultural Bureaux, Slough, U.K., pp. 347–352.

Lynch, J., and Läuchli, A. (1985). "Salt stress disturbs the calcium nutrition of barley (Hordeum vulgare L.). *New Phytol.* 99:345–354.

Lynch, J., Thiel, G., and Läuchli, A. (1988). "Effects of salinity on the extensibility and Ca availability in the expanding region of growing barley leaves." *Botanica Acta* 101:355–361.

Maas, E.V. and Grieve, C.M. (1987). "Sodium-induced calcium deficiency in salt-stressed corn." *Plant, Cell and Environment* 10: 559–564.

Maas, E.V., Hoffman, G.J., Chaba, G.D., Poss, J.A. and Shannon, M.C. (1983). "Salt sensitivity of corn at various growth stages." *Irrig. Sci.* 4:45–57.

Maas, E.V. and Nieman, R.H. (1978). "Physiology of plant tolerance to salinity." In: *Crop Tolerance to Suboptimal Land Conditions.* G.A. Jung, ed. American Society of Agronomy, Madison, pp. 277–299.

Meek, B.D., Ehlig, C.F., Stolzy, L.H., Graham, L.E. (1983). "Furrow and trickle irrigation: effects on soil oxygen and ethylene and tomato yield." *Soil Sci. Soc. Am. J.* 47:631–635.

Munns, R. and Termaat, A. (1986). "Whole plant responses to salinity." *Aust. J. Plant Physiol.* 13:143–160.

Munns, R., Greenway, H., Delane, R., and Gibbs, J. (1982). "Ion concentrations and carbohydrate status of the elongating leaf tissue of

Hordeum vulgare growing at high external NaCl. II. Cause of the growth reduction." *J. Exp. Bot.* 33:574–583.

Munns, R., Greenway, H., and Kirst, G.O. (1983). "Halotolerant eukaryotes." In: *Encylopedia of Plant Physiology New Series.* Vol. 12C.Physiological Plant Ecology III. O.L. Lange, P.S. Nobel, C.B. Osmond, and H. Ziegler, eds., Springer-Verlag, Berlin, pp. 59–135.

Munns, R. and Passioura, J.B. (1984). "Effect of prolonged exposure to NaCl on the osmotic pressure of leaf xylem sap from intact, transpiring barley plants." *Aust. J. Plant Physiol.* 11:497–507.

Munns, R., Fisher, D.B., and Tonnet, M.L. (1986). "Na^+ and Cl^- transport in the phloem from leaves of NaCl-treated barley." *Aust. J. Plant Physiol.* 13:757–766.

Munns, R., Gardner, P.A., Tonnet, M.L., and Rawson, H.M. (1988). "Growth and development in NaCl-treated plants. II. Do Na^+ or Cl^- concentrations in dividing or expanding tissues determine growth in barley?". *Aust. J. Plant Physiol.* 11:497–507.

Nieman, R.H. (1965). "Expansion of bean leaves and its suppression by salinity." *Plant Physiol.* 40:156–161.

Norlyn, J., and Epstein, E. (1984). "Variability in salt tolerance of four triticale lines at germination and emergence." *Crop Sci.* 24:1090–1092.

Northcote, K.H., and Skene, J.K.M. (1972). "Australian Soils with Saline and Sodic Properties." *Commonwealth Scientific and Industrial Research Organization,* Soil Publ. No. 27.

Oertli, J.J. (1968). "Extracellular salt accumulation, a possible mechanism of salt injury in plants." *Agrochimica* 12:461–469.

Papp, J.C., Ball, M.C. and Terry, N. (1983). "A comparative study of the effects of NaCl salinity on respiration, photosynthesis and leaf extension growth in *Beta vulgaris* (sugar beet)." *Plant, Cell and Environment* 6:675–677.

Pasternak, D. (1987). "Salt tolerance and crop production—a comprehensive approach." *Ann. Rev. Phytophathol.* 25:271–291.

Pasternak, D., Twersky, M. and De Malach, Y. (1979). "Salt resistance in agricultural crops." *Stress Physiology in Crop Plants.* H. Mussel and R.C. Staples, eds. John Wiley and Sons Inc., New York, pp. 127–142.

Plaut, Z. and Grieve, C.M. (1988). "Photosynthesis of salt-stressed maize as influenced by Ca:Na ratios in the nutrient solution." *Plant and Soil.* 105:283–286.

Prisco, J.T., and O'Leary, J.W. (1973). "The effects of humidity and cytokinin on growth and water relations of salt-stressed bean plants." *Plant and Soil.* 39:263–276.

Rains, D.W. (1987). "Salinity stress-energy cost of plant cells exposed to salinity." *Models in Plant Physiology and Biochemistry.* Vol. III. D.W. Newman and K.G. Wilson, eds. CRC Press, Inc., Boca Raton, pp. 91–93.

Rains, D.W. and Epstein, E. (1967). "Preferential absorption of potassium by leaf tissue of the mangrove, *Avicennia marina:* an aspect of halophytic competence in coping with salt." *Aust. J. Bio. Sci.* 20:847–857.

Rains, D.W., Goyal, S., Weyrauch, R., and Läuchli, A. (1987). "Saline drainage water reuse in a cotton rotation system." *Cal. Agric.* 41(9–10):24–26.

Raven, J.A. (1985). "Regulation of pH and generation of osmolarity in vascular plants: a cost-benefit analysis in relation to efficiency of use of energy, nitrogen and water." *New Phytol.* 101:25–77.

Rawson, H.M., Long, M.J., and Munns, R. (1988). "Growth and development in NaCl-treated plants I. Leaf Na^+ and Cl^- concentrations do not determine gas exchange of leaf blades in barley." *Aust. J. Plant Physiol.* 15:519–527.

Rawson, H.M., and Munns, R. (1984). "Leaf expansion in sunflower as influenced by salinity and short-term changes in carbon fixation." *Plant, Cell and Environment.* 7:207–213.

Richards, L.A., ed. (1954). *Diagnosis and Improvement of Saline and Alkali Soils.* Agriculture Handbook No. 60. United States Department of Agriculture, Washington, D.C.

Rush, D.W. and Epstein, E. (1981). "Breeding and selection for salt tolerance by the incorporation of wild germplasm into a domestic tomato." *J. Amer. Soc. Hort. Sci.* 106:699–704.

Sacher, R.F. and Staples, R.C. (1985). "Inositol and sugars in adaptation of tomato to salt." *Plant Physiol.* 77:206–210.

Schubert, S., and Läuchli, A. (1986). "Na^+ exclusion, H^+ release, and growth of two different maize cultivars under NaCl salinity." *J. Plant Physiol.* 126:145–154.

Schwarz, M. and Gale, J. (1981). "Maintenance respiration and carbon balance of plants at low levels of sodium chloride salinity." *J. Exp. Bot.* 32:933–941.

Seeman, J.R., and Critchley, C. (1985). "Effects of salt stress on the growth, ion content, stomatal behavior and photosynthetic capacity of a salt sensitive species, *Phaseolus vulgaris* L." *Planta.* 164:151–162.

Shannon, M.C., and Francois, L.E. (1977). "Influence of seed pretreatments on salt tolerance of cotton during germination." *Agron. J.* 69: 619–622.

Singh, N.K., LaRosa, P.C., Handa, A.K., Hasegawa, P.M. and Bressan, R.A. (1987). "Hormonal regulation of protein synthesis associated with salt tolerance in plant cells." *Proc. Natl. Acad. Sci. USA* 84:739–743.

Solomon, M., Gedalovich, E., Mayer, A.M., and Poljakoff-Mayber, A. (1986). "Changes induced by salinity to the anatomy and morphology of excised pea roots in culture." *Ann. Bot.* 57:811–818.

Stassart, J.M., Neirinckx, L., and Dejaegere, R. (1981). "The interactions between monovalent cations and calcium during their adsorption on isolated cell walls and absorption by intact barley roots." *Ann. Bot.* 47:647–652.

Stavarek, S.J. and Rains, D.W. (1985). "Effect of salinity on growth and maintenance costs of plant cells." *Cellular and Molecular Biology of Plant Stress.* J.L. Key and T. Kosuge, eds. Alan R. Liss, Inc., New York, pp. 129–143.

Syvertsen, J.P. and Yelenosky, G. (1988). "Salinity can enhance freeze tolerance of citrus rootstock seedlings by modifying growth, water relations, and mineral nutrition." *J. Amer. Soc. Hort. Sci.* 113:889–893.

Szabolcs, I. (1989). *Salt-Affected Soils*. CRC Press, Boca Raton, Fla.

Termaat, A. and Munns, R. (1986). "Use of concentrated macronutrient solutions to separate osmotic from NaCl-specific effects on plant growth." *Aust. J. Plant Physiol.* 13:509–522.

Termaat, A., Passioura, J.B. and Munns, R. (1985). "Shoot turgor does not limit shoot growth of NaCl affected wheat and barley." *Plant Physiol.* 77:869–872.

Thiel, G., Lynch, J., and Läuchli, A. (1988). "Short-term effects of salinity stress on the turgor and elongation of growing barley leaves." *J. Plant Physiol.* 132:38–44.

Treeby, M.T. and van Steveninck, R.F.M. (1988). "Effects of salinity and phosphate on ion distribution in lupin leaflets." *Physiol. Plant.* 73: 317–322.

Ungar, I.A. (1978). "Halophyte seed germination." *Bot. Rev.* 44:233–264.

Waisel, Y. (1972). *Biology of Halophytes*. Academic Press, New York and London.

Ward, M.R., Aslam, M., and Huffaker, R.C. (1986). "Enhancement of nitrate uptake and growth of barley seedlings by calcium under saline conditions." *Plant Physiol.* 80:520–524.

Weimberg, R., Lerner, H.R., and Poljakoff-Mayber, A. (1984). "Changes in growth and water-soluble solute concentrations in *Sorghum bicolor* stressed with sodium and potassium salts." *Physiol. Plant.* 62:472–480.

Wyn Jones, R.G. (1981). "Salt tolerance." *Physiological Processes Limiting Plant Productivity*. C.B. Johnson, ed. Butterworths, London, pp. 271–292.

Wyn Jones, R.G. (1984). "Phytochemical aspects of osmotic adaptation." *Phytochemical Adaptation to Stress*. B.N. Timmermann, C. Steelink, and F.A. Loewus, eds. Plenum Press, New York and London, pp. 55–78.

Wyn Jones, R.G. and Gorham, J. (1983). "Osmoregulation." *Encyclopedia of Plant Physiology New Series*. Vol. 12C. Physiological Plant Ecology III. O.L. Lange, P.S. Nobel, C.B. Osmond, and H. Ziegler, eds., Springer-Verlag, Berlin, pp. 35–58.

Yancey, P.H., Clark, M.E., Hand, S.C., Bowlus, R.D., and Somerio, G.N. (1982). "Living with water stress: evolution of osmolyte systems." *Science* 217:1214–1222.

Yeo, A.R. (1983). "Salinity resistance: physiology and prices." *Physiol. Plant.* 58:214–222.

Yeo, A.R., and Flowers, T.J. (1982). "Accumulation and localization of sodium ions within the shoots of rice (*Oryza sativa*) varieties differing in salinity resistance." *Physiol. Plant.* 56:343–348.

Yeo, A.R., and Flowers, T.J. (1985). "The absence of an effect of the Na/Ca ratio on sodium chloride uptake by rice (*Oryza sativa* L.)." *New Phytol.* 99:81–90.

Yeo, A.R., and Flowers, T.J. (1986). "Salinity resistance in rice (*Oryza sativa* L.) and a pyramiding approach to breeding varieties for saline soils." *Aust. J. Plant Physiol.* 13:161–173.

Zhong, H., and Läuchli, A. (1988). "Incorporation of [14C]-glucose into cell wall polysaccharides of cotton roots: effects of NaCl and $CaCl_2$." *Plant Physiol.* 88:511–514.
Zid, E., and Grignon, C. (1985). "Sodium-calcium interactions in leaves of *Citrus auranitum* grown in the presence of NaCl." *Physiol. Veg.* 23:895–903.
Zimmermann, U., Hüsken, D., and Schulze, E.D. (1980). "Direct turgor measurement in individual leaf cells of *Tradescantia virginiana.*" *Planta* 149:445–453.

CHAPTER 7

DEFICIENCIES AND TOXICITIES OF TRACE ELEMENTS[a]

INTRODUCTION

"Trace element" is used in the geochemical and biochemical literature to refer to a group of otherwise unrelated chemical elements that occur in nature at low concentrations. Their concentrations in different natural environments vary widely. Iron, for example, is the fourth most abundant element in soils, yet its concentration in terrestrial plants rarely exceeds 200 mg/kg.

Many trace elements are biologically beneficial at very low concentrations, but become toxic or otherwise detrimental to the health of organisms and plants at low to moderate concentrations. Long-term exposure at sublethal concentrations to certain trace elements may result in chronic biological effects. Some trace elements can accumulate to potentially harmful levels in soils and sediments through chemical immobilization, in plants through bioconcentration, or via a food chain through biomagnification.

Other terms for chemical elements that affect the chemistry or biology of the natural environment at low concentrations include trace metals, trace inorganics, micro-elements, and heavy metals. Heavy metal should be used only for those elements with densities greater than 5.0 g/cm^3. Trace metal obviously is inappropriate for the metalloids, such as boron and arsenic.

The trace elements that can be translocated from soils through the human and animal food chain are cadmium (Cd), molybdenum (Mo), selenium (Se), and cobalt (Co). Arsenic (As), boron (B), cadmium (Cd), copper (Cu), lithium (Li), nickel (Ni), manganese (Mn), and zinc (Zn) can be toxic to plants. Elements that are common in the natural environment, have no known physiological function and are potentially harmful to plants and animals include mercury (Hg), lead (Pb), chromium (Cr), antimony (Sb), beryllium (Be), silver (Ag), and tellurium (Te). They usually occur in soils at low concentrations and, if introduced into soils, convert to chemical forms not readily absorbed by plants.

Trace elements that are essential for the growth of higher plants are B, Cu, iron (Fe), Mn, Mo, Zn, sodium (Na), and chlorine (Cl). Cobalt, Se, and silicon (Si) are considered essential for some species.

In saline soils, trace elements most likely to be deficient for plant growth are Cu, Fe, Mn, and Zn. Boron deficiencies, common to soils in

---

[a]Prepared by: A.L. Page, A.C. Chang, Dept. of Soil and Envir. Sci., Univ. of Calif., Riverside, CA 92521; and D.C. Adriano, Biogeochemistry Div., Savannah River Ecology Lab., Aiken, SC 29801.

temperate, semi-humid, and humid regions, are rarely a problem for agricultural production in saline soils. Since most trace elements have low solubility in neutral and alkaline soils, trace element toxicities to plants rarely occur. More common is the uptake of certain trace elements by plants in amounts great enough to be toxic to the domestic animals and wildlife that consume the plants. Numerous cases of livestock poisoning caused by elevated levels of Mo and Se in forage have been reported worldwide.

This chapter will review trace elements in soils and soil factors that influence their availability and accumulation by crops. Particular attention will be paid to trace elements that are important in managing saline soils.

FACTORS INFLUENCING TRACE ELEMENT DEFICIENCIES AND TOXICITIES OF PLANTS

Plants subjected to nutritional deficiencies or exposed to phytotoxic substances usually experience reduced growth, or exhibit visual symptoms of injuries, or both. If the adverse conditions worsen, dieback and then death of the plant will occur. As the effects of nutritional deficiencies and phytotoxicities gradually intensify, criteria based on observation are needed to define deficiency and toxicity thresholds of an element in the growing medium. For agronomic plants, the crop yield is the ultimate measurement used most commonly to set such criteria. Also, visual symptoms of plant damage often help to define deficiencies and phytotoxicities to plants, especially perennials.

Agronomists have long studied nutrient deficiencies and phytotoxicity. Fig. 7.1 shows the relationship between the concentration of an essential chemical element in the soil and plant growth as measured by the yield. If the concentration of a micronutrient element in the soil fails to reach a threshold (D), the plant's growth is inhibited. As the elemental content of the soil increases, a critical concentration (T) will be reached beyond which growth is again depressed due to toxicity (Maas and Hoffman 1977; Bingham et al. 1985; Burton et al. 1983; Davis and Carlton-Smith 1984). The D and T values, if known, allow the status of this element in the soil to be assessed. With a slight modification ($D = 0$), the same model can describe the effects of phytotoxic elements on the yield. Small amounts of trace elements satisfy the plant's nutritional requirements, i.e., D is small. Compared with the major elements, the concentration range for optimum growth ($T - D$) is consistenly narrow. Extracting the soil with an appropriate chemical reagent reveals the soil's elemental concentrations, which help to indicate the element's availability to plants (Lindsay and Norvell 1978).

The crop's yield is a nonspecific measurement that can be affected by factors unrelated to nutritional deficiency or phytotoxicity. The leaf tissue concentration of elements is specifically related to the plant's uptake of the elements and relatively independent of other factors (Burton et al. 1983; Davis and Carlton-Smith 1984). The same model (Fig. 7.1) can represent the relationship between leaf concentration of an element and

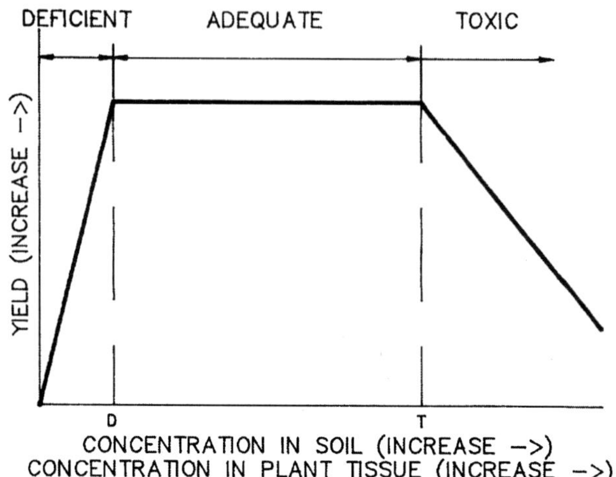

FIG. 7.1 Diagnosis of Trace Element Deficiencies and Toxicities of Crops

the growth of the plant, with the elemental concentration of soil on the x-axis replaced by the elemental concentration of plant tissue. Burton et al. (1983) reported that when the metal concentration is plotted on a log scale, the decrease of crop and yield beyond T and the metal concentrations in plant tissue follow a straight line. Although the threshold concentrations expressed by the elemental content in plant tissue do not vary as much as those expressed in soils, the threshold levels do vary according to the species.

Nutritional deficiencies and elemental toxicities of crops are related to a number of soil and plant factors. The physical, mineralogical, chemical, and biological properties of soils govern the concentration of an element in the soil solution that, along with the rate at which the element can be replenished, governs the element's bioavailability. Species respond differently to the same concentration of an element in the soil solution. Hence, soil and plant factors that are specific to a site must be evaluated to diagnose soils for trace element deficiencies and toxicities.

Soil Factors

As the primary medium in which plants obtain nutrients and extract water, the soil significantly influences plant growth. The soil is a porous medium made up of mixtures of weathered mineral fragments, organic debris, and microorganisms. Air and water fill voids between the granular particles. Most important to plant nutrition and soil fertility is the extent to which soil can store, transmit, and retain water. The rate at which applied water enters the soil (infiltration), the rate at which it passes through the soil profile (leaching), and the soil's water content strongly influence the chemical forms and concentrations of trace elements that are present.

Chemical reactions in the soil can be described as a heterogeneous system with various solid phases existing more or less in equilibrium with chemical species in the soil solution and soil atmosphere. Influx and efflux of water, fluctuation of the soil's water content, the plant's absorption of nutrients and excretion of metabolites, and microbial transformations constantly change this equilibrium. Because soil is a porous medium and biological activity is present, true chemical equilibrium is seldom, if ever, attained. Therefore, the chemical forms and concentrations in soil solutions are not easily, if ever, accurately predicted.

Chemical characteristics of soils, such as cation, anion, and molecular adsorption, oxidation/reduction potential, precipitation, and soil pH and buffering capacity influence trace elements in the soil solution. Adsorption of cations, anions, and molecules by the solids is fundamental to chemical activities in soils. The extent of adsorption is related to the kind and amount of clay, organic matter, and amorphous oxides of Al, Fe, and Mn. The adsorption removes trace elements from the soil solution. The ability of soils to adsorb trace elements that occur in cationic or molecular form generally increases as their contents of clay and organic matter increase. The adsorption of trace elements occurring in anionic form tends to increase as the soil's content of amorphous oxides of Fe, Al, and Mn increases.

Chemical precipitation in soils results in the formation of sparingly soluble compounds and the removal of ions from the soil solution. Precipitation, a nucleation process, differs mechanistically from adsorption, a surface phenomenon. Because both remove ions from the solution, distinguishing the two is difficult (Sposito 1986). As most soil systems are undersaturated with solubilities of known inorganic solid compounds of many trace elements, including Cd, Cu, Ni, and Zn, the solubility of trace elements in soils is thought to be governed by either adsorption or the formation of a mixed solid. In saline soils, because concentrations of inorganic ligands, like those of chloride, sulfate, bicarbonate, and carbonate, are orders of magnitude greater than those of trace elements, ion association that results in ion pairing of trace elements with these ligands is the rule, not the exception (Page et al. 1981). Ion association markedly affects the adsorption, precipitation, and concentration of trace elements in the soil solution.

The moisture content of the soil governs the oxygen level, which governs the reduction/oxidation potential of the soil.. The redox potential of soils ranges from approximately -400 mV for a flooded soil with strongly reduced environment to $+700$ mV for a well-drained soil with oxidized environment. Under reducing conditions, most trace elements, including Cu, Ni, Cd, Pb, Hg, Se, and Zn, have a very low solubility due mainly to the formation of sulfides and selenides. Because crops, with the exception of rice, cannot tolerate reduced soil conditions for an extended period of time, the redox potential's effect on the chemistry of trace elements is transitory. It depends on the rate at which the products of reduction are re-oxidized after the system returns to oxidative conditions. The elements Fe and Mn, however, are exceptions. They are more soluble in their reduced forms than in oxidized forms. The concentration of hydrogen ion (pH) in the soil solution is related to all of the previously discussed soil properties. Each, along with the hydrogen ion concentration, is interdependent.

The surface charge on an inorganic solid in soil comprises a permanent component and a pH-dependent component. The permanent charge is created when an ion in the crystal structure of the soil mineral is substituted by an ion of about equal size and lesser valence, resulting in a charge imbalance. The association of hydrogen ions or dissociation of hydroxyl ions from inorganic and organic surfaces in aqueous solution creates positively charged surface sites capable of adsorbing negatively charged trace elements. Hydrogen and hydroxyl ions *per se* are adsorbed on inorganic and organic surfaces. They are also involved in the dissolution and precipitation of minerals and amorphous solids in soils. Through these types of reactions, the hydrogen ion concentration (pH) of the soil solution-more precisely, the activity of hydrogen ions-affects the concentrations of trace elements. The solubility of the cationic trace elements decreases with increasing pH, while the solubility of the anionic trace elements increases with increasing pH. Therefore, deficiencies of Cu, Fe, Mn, and Zn, rather than toxicities of these elements, commonly exist in saline soil. Phytotoxicities caused by excessive amounts of Al, Cd, Mn, and Ni in acid soils essentially do not exist in alkaline soils. Solubilities of the anionic trace elements increase with increasing pH, but seldom reach levels that are toxic to plants. The increased solubility and availability of the anionic Mo and Se species in alkaline soils may pose a problem, though, because certain plants absorb these elements in amounts that are harmful to animals that eat the plants.

Boron is probably the most troublesome trace element in managing saline and alkaline soils. While low concentrations of boron are essential for the growth of plants, it becomes phytotoxic at concentrations only slightly higher than that required for optimum growth. In soil solutions, boron occurs mainly as undissociated boric acid [$B(OH)_3$] and is much more soluble than the other trace elements of concern, such as As, Cd, Ni, and Cu. Lithium, like boron, is also very soluble. Because lithium rarely occurs in soils and because plants have a relatively high tolerance to it, Li toxicity is observed in only a few crops (Bradford 1966).

Plant Factors

Plants differ in their ability to tolerate trace elements and in their ability to absorb and accumulate trace elements. Table 7.1 summarizes the concentrations of essential and non-essential trace elements in selected crops grown in major producing regions throughout the United States (Wolnick et al. 1983a, 1983b, 1985). This data set is representative of concentrations in crops entering the marketplace, as plant samples for the study were collected from fields of commercial growers throughout the United States. Concentrations of Cu, Fe, Mn, Mo, and Zn in various species varied by factors of about 5 (1.8 mg/kg to 11 mg/kg), 70 (3 mg/kg to 200 mg/kg), 12 (7 mg/kg to 81 mg/kg), 6 (0.098 mg/kg to 6.65 mg/kg), and 3 (15 mg/kg to 46 mg/kg), respectively. Similar patterns of variations among crops are observed for the non-essential trace elements—Cd, Ni, Pb, and Se.

Data presented in Table 7.1 represent crops grown on many different soils. The crops obviously contain different amounts of trace elements and are chemically, physically, and morphologically different.

TABLE 7.1 Median Concentrations of Essential and Non-Essential Trace Elements in Raw Crops Grown in Major Producing Areas of the United States (Wolnick et al. 1983a and b, 1985)

Crop species (1)	Number of observations (2)	Trace Element Concentration (mg/kg, oven-dry weight)								
		Cu (3)	Fe (4)	Mn (5)	Mo (6)	Zn (7)	Cd (8)	Ni (9)	Pb (10)	Se (11)
Lettuce	150	6.1	57	31	0.25	46	0.435	—	0.19	0.039
Spinach	104	8.5	200	81	0.22	43	0.80	1.1	0.53	—
Tomato	231	11	48	15	0.30	22	0.22	0.84	0.027	—
Wheat	288	4.9	36	43	0.43	29	0.036	—	0.02	0.19
Sweet corn	268	1.8	18	7	0.16	25	0.008	0.26	0.009	0.014
Soybean	322	13	71	27	—	45	0.045	4.8	0.036	0.082
Rice	166	2.1	3	11	0.65	15	0.005	0.26	0.005	—
Carrot	207	4.7	27	12	0.098	20	0.16	0.41	0.055	—
Potato	297	4.4	20	7	0.19	15	0.14	—	0.025	0.013
Onion	228	3.6	13	9	0.14	16	0.09	0.32	0.038	—
Peanut	320	8.3	20	18	0.28	31	0.068	1.5	0.040	0.040

Even among crops grown on the same soil, concentrations of trace elements in plant tissue vary. Table 7.2 shows that the Zn and Cd in the leaf tissue of 12 plant species grown in the greenhouse on the same soils vary. Data accumulated by Davis and Carlton-Smith (1980) showed comparable variations in the concentrations of Cd, Cu, Ni, Pb, and Zn in the leaves and edible parts of 39 commercial crops grown on two soils.

For a given species, concentrations of trace elements in various parts of the plants and among different cultivars also vary. Generally, concentrations of Cd, Cu, and Zn in the leaves of a plant exceed concentrations in grains, edible roots, or fruits (Davis and Carlton-Smith 1980) (Table 7.3). Logan and Chaney (1983) compiled the differences in the uptake of Cd and Zn among different cultivars of corn, soybean, potato, lettuce, and wheat.

A single biological mechanism apparently does not cause the differences in nutrient absorption among species and within cultivars. Genetically controlled features of plants, morphological and anatomical differences between plants, and physiology of the ion transport mechanism may all be responsible (Chang et al. 1982). When a plant takes up more cations than anions from the soil solution to satisfy the charge balance, hydrogen ions are effluxed by the roots. The opposite effect is observed when the plant extracts more anions than cations: the liberation of acids or bases by plant roots changes the pH of the soil immediately adjacent to the root, thereby changing the solubility and availability of trace elements (Pierre et al. 1970; Pierre and Banwart 1973; Jarvis and Robson 1983). Plants also exude, through roots, organic ligands that can complex trace elements in soils and alter the availability of those elements to the crop.

TABLE 7.2 Cadmium and Zinc in Leaves of Various Plants Grown in Two Soils (Kim et al. 1988). (Concentration Measurements in mg/kg on Dry Weight)

	Gazos soil		Domino soil	
Plant (1)	**Cd (2)**	**Zn (3)**	**Cd (4)**	**Zn (5)**
Lettuce (leaf)	6.1	52	6.0	76
Lettuce (head)	5.6	54	6.1	84
Swiss chard	2.0	40	5.7	250
Tomato	5.0	22	7.2	68
Pepper	16.9	179	9.0	243
Broccoli	1.2	14	1.8	76
Corn	2.0	63	5.8	130
Wheat	1.2	34	0.48	90
Radish	4.0	56	6.7	168
Potato	3.3	21	2.6	24
Carrot	6.0	51	5.9	91
Beet	3.0	52	7.8	286

TABLE 7.3 Concentrations of Trace Elements in Leaves and Edible Parts of Various Crops (Davis and Carlton-Smith 1980) (Measurements of Trace Elements in mg/kg on Dry Weight)

Crop (1)	**Tissue (2)**	**Cd (3)**	**Cu (4)**	**Zn (5)**
Sweet corn	Leaf	2.0	7.2	109
	Grain	0.8	3.7	37
Wheat	Leaf	1.0	19	66
	Grain	0.9	6.3	50
Radish	Leaf	1.4	11	84
	Root	0.5	4	44
Carrot	Leaf	1.5	8.6	109
	Root	0.6	2.8	53
Tomato	Leaf	8.3	11	22
	Fruit	1.0	6.6	12

METHODS FOR DIAGNOSING TRACE ELEMENT DEFICIENCIES AND TOXICITIES

Plant tissues can be assayed or soil extracts analyzed to diagnose whether a plant is experiencing deficiencies or toxicities of trace elements. These data are compared against criteria that define ranges of concentrations related to deficiencies and toxicities. Through this relationship, the concentrations of a trace element in plant tissue or a soil extract may be related to the soil's ability to supply the plant with adequate amounts for optimum yield and the nutritional needs of consumers, be they human, livestock, or wildlife. A plant's response to trace elements is affected by many factors specific to the species and the soil and sometimes by synergistic or antagonistic chemical and biochemical interactions at the root-soil interface. These factors complicate the diagnosis of trace element deficiencies and toxicities.

Soil Methods

A soil's ability to supply trace elements needed by plants is related not only to the chemical forms and solubility of the element, but also to a series of reactions. The rate at which the plant extracts trace elements and the rate at which other trace-element pools in the soil system replenish the chemical form of the element determine the total plant uptake. Such uptake depends on the root's rate of growth. Though sufficient amounts exist in the soil, the rate of release must be fast enough to supply the trace element to avoid symptoms of deficiencies. Ideally, the soil extractant used to indicate the availability of the trace element would account for contributions of all applicable physical, chemical, and biological processes in the soil related to the uptake of trace elements. Actually, no static extraction process can simulate these dynamic processes. Thus, the use of trace element extractants to evaluate the soil's deficiency and toxicity potential is derived from observations on the amounts extracted and the corresponding plant responses (yield or crop quality). Therefore, levels of trace elements associated with deficiency or toxicity for one crop may not indicate a similar response to another crop. Calibration of the criteria for conditions specific to the site and the crop is advisable.

Extractants used to diagnose trace element deficiencies and toxicities of crops include water, weak acids, strong acids, buffered and neutral salt solutions, and solutions of organic complexing compounds. Table 7.4 lists the common ones. They are more frequently used to diagnose deficiencies than toxicities. They often are specific to a species or group of species, e.g., sensitive species or tolerant species, and a particular type of soil.

In the western United States, two similar reagents, DTPA-TEA and DTPA-NH_4HCO_3, are most often used. In the rest of the nation, HCl and a mixed reagent containing HCl and H_2SO_4 are commonly used. Other methods involve the use of aqueous solutions of other organic complexing compounds (EDTA and EDDHA), weak acids (HOAc and HOOCCOOH), and buffered and unbuffered salts (NH_4OAc, NH_4NO_3, and $MgCl_2$).

DTPA-TEA ($CaCl_2$). Lindsay and Norvell (1969) used DTPA-TEA to extract zinc from soils and distinguish the soils deficient in Zn from those with adequate Zn. This method can also be used to evaluate the plant nutritonal status of Zn in soil cropped with corn (Brown et al. 1971; Lauer 1971; Alley et al. 1972; Lindsay and Norvell 1978). Other investigators have shown that the amount of DTPA-TEA extractable Zn of a soil linearly

TABLE 7.4 Reagents Used to Extract Plant-Available Trace Elements From Soils

Reagent (1)	Trace elements (2)	References (3)
0.005 M DTPA-0.1 M TEA-0.01 M $CaCl_2$	Cd, Cu, Fe, Mn, Zn	Bingham et al. (1975), Lindsay and Norvell (1978)
0.1 M HCl	Cd, Ni, Zn	Nelson et al. (1959), Mishra and Pande (1974), Burau (1983)
0.01 M EDTA-1M $(NH_4)_2CO_3$ adjusted to pH = 8.6	Cu, Mn, Zn	Trierweiler and Lindsay (1969), Dolar et al. (1971)
0.05 M HCl-0.05 M H_2SO_4	As, Cd, Cu, Ni, Zn	Perkins (1970), Korcak and Fanning (1978), Woolson et al. (1971)
1 M NH_4OAc at pH = 7	Cd, Zn	John et al. (1972), Haq et al. (1980)
0.005 M DTPA-1 M NH_4HCO_3 at pH = 7.6	Cd, Fe, Mn, Mo, Ni, Pb, Se, Zn	Soltanpour and Workman (1980, 1981), Soltanpour (1985)
Distilled water/0.01 M $CaCl_2$	B, Se	Bingham (1982), Berger and Truog (1939), Baker and Mortensen (1966), Soltanpour and Workman (1980, 1981)
0.5 M $NaHCO_3$	As	Woolson et al. (1971)
$(COOH)_2$ + $(COONH_4)_2$ at pH = 3.3	Mo	Grigg (1953), Walsh et al. (1951)

correlates with the concentration of Zn in the plant (Walsh et al. 1972; Evans et al. 1974; Korcak and Fanning 1978; Valdares et al. 1983). The same DTPA-TEA soil tests were also used successfully to indicate whether soils cropped with sorghum were deficient in Fe (Lindsay and Norvell 1978; de Boer and Reinsenauer 1973; Soltanpour et al. 1976).

When amounts of the following trace elements adequate for plant growth are present in soils, the concentrations extracted with DTPA-TEA from soils significantly correlate with the concentrations at which they occur in various species: Cu and Mn (Lindsay and Norvell 1978); Cd, Cu, and Ni (Korcak and Fanning 1978); Cd and Pb (Singh and Narwal 1984); Cd and Ni (Valdares et al. 1983); and Cd (Bingham et al. 1975; Latterell et al. 1978; Street et al. 1977). It can be reasonably assumed that this correlation extends to deficient and toxic ranges in the

TABLE 7.5 Levels of Trace Elements Extracted from Soils with NH_4HCO_3-DTPA, Indicative of Possible Deficient, Adequate, or Phytotoxic Levels for Crops (Soltanpour 1985) (Measurements in mg/kg)

	Extractable level indicative of:		
Element (1)	**Deficiency (2)**	**Adequacy (3)**	**Phytotoxicity (4)**
Cu	0–0.2	0.3–0.5	>0.5
Fe	0–3.0	3.1–5.0	>5.0
Mn	0–0.5	0.6–1.0	>1.0
Zn	0–0.2	0.3–0.5	>0.5

soils. Levels of DTPA-TEA extractable Zn, Fe, Mn, and Cu associated with possible deficiencies to crops are 0.8 mg/kg, 4.5 mg/kg, 1.0 mg/kg, and 0.2 mg/kg, respectively (Lindsay and Norvell 1978).

In a greenhouse study, Bingham et al. (1975) correlated concentrations of Zn extracted from soils with DTPA-TEA with 25% reductions in yield of a number of crops, including carrot, radish, turnip, spinach, curly cress, cabbage, zucchini, squash, tomato, wheat, soybean, and field bean. Cadmium concentrations extracted with DTPA-TEA and associated with the reductions in the yield were not the same for the crops tested. They varied from 2.4 mg to 102 mg DTPA-TEA extractable Cd per kilogram of soil for spinach and cabbage, respectively. Logan and Chanel (1983) have pointed out that the levels of extractable metals associated with reductions in the yield grown under greenhouse conditions do not necessarily represent what is observed in field-grown crops.

DTPA-TEA (NH_4HCO_3 versus DTPA-TEA ($CaCl_2$). Soltanpour and Schwab (1977) modified the DTPA-TEA method to include the extraction of anions of N, P, S, Mo, Se, and As. The change involved replacing $CaCl_2$ with NH_4HCO_3 as the background electrolyte. The extracting reagent consisted of 0.005 M DTPA and 1 M NH_4HCO_3 adjusted to pH 7.6. This method is currently used by Colorado State University to evaluate trace element deficiencies and toxicities of soils (Soltanpour and Workman 1981). Soltanpour (1985) has used DTPA-NH_4HCO_3 extraction to detect potential phytotoxicities due to B, Mo, Pb, Cd, Se, and As and to diagnose the plant nutrient status of NO_3-N, P, K, S, Zn, Fe, Mn, and Cu in soils. Table 7.5 shows the concentrations of selected DTPA-NH_4HCO_3-extractable trace elements in soils that indicate a deficient and toxic condition. The threshold levels of Zn, Fe, Mn, and Cu suggested by Soltanpour (1985) correspond closely to those determined by Lindsay and Norvell (1978). The phytotoxicity threshold of DTPA-NH_4HCO_3 extractable B, Mo, Pb, Se and As, based on either limited data or extrapolations, must be considered tentative.

HCl and HCl + H_2SO_4. Before the mid-1960s, methods were not sensitive enough to detect certain trace elements dissolved in water or neutral electrolytes. Researchers experimented with the use of strong acid extractants to help determine whether trace elements in a soil would be deficient or toxic to plants. Wear and Evans (1968) reported a strong linear correlation between the amount of Zn extracted from 12 soils by both 0.1 N HCl and 0.025 N H_2SO_4 + 0.05 N HCl and the amount of Zn in corn and sorghum grown on these soils. However, they observed no yield response to the Zn added to soils. The amounts of Zn extracted by acids failed to illustrate the usefulness of acid extractants in identifying Zn-deficient soils. In an earlier study, Wear and Sommers (1948) reported a good correlation between 0.1 N HCl extractable Zn of soil and Zn- deficiency symptoms in corn. Brown et al. (1971) reported positive responses of corn to applications of Zn in 66% of the 92 California soils that contained less than 5 mg/l of 0.1 M HCl extractable Zn. Singh and Narwal (1984) reported that 0.1 N HCl extractable Zn, Co, and Ni of soils did not adequately predict the soil's Zn, Co, and Ni availability to rape fodder. However, they significantly correlated the Cd and Pb contents of rape fodder with the concentration of the 0.1 N HCl extractable Cd and Pb in the soils. Burau (1983) showed a significant correlation between the concentration of 2 M HCl extractable soil Cd and the concentration of Cd in 19 crops grown on these soils. Lagerwerff (1971) and Jones et al. (1975) showed that Cd extracted from the soil with 1 M HCl correlated with the concentration in the plant.

Haq and Miller (1972) reported that the contents of Zn in corn grown on 80 soils in Canada related with the amount of Zn extracted by 0.05 M HCl + 0.0125 M H_2SO_4. Korcak and Fanning (1978) observed statistically significant correlations between the concentrations of Cd, Cu, Ni (one exception), and Zn in corn tissue and the amounts of these elements extracted from three limed and unlimed soils by 0.05 M HCl + 0.0125 M H_2SO_4. The extent of correlation increased when the data were separated into the limed and the unlimed soil groups.

The mixed acid reagent can also be used to diagnose other trace elements in soils. Woolson et al. (1971) demonstrated that yields of corn grown in acid soils (pH <6.5) were best correlated with the logarithm of the concentration of soil arsenic extracted by the HCl and H_2SO_4 mixture. As with other trace elements, the phytoxicity and plant availability of arsenic in soils depends on the species.

Other Extractants. A number of other reagents have been used to indicate levels of soil metals associated with deficiency or toxicity to crops. In all cases, the reagent was apparently specific to the crop, the soil, or the element. In the United Kingdom, 0.5 M HOAc and 0.05 M ETDA are commonly used to extract Cd, Cu, Pb, Ni, and Zn in soils and set the toxicity thresholds (Purves 1985). Solutions of neutral and buffered salts have also been used.

Cadmium extracted from soils with 1 M NH_4OAc and 1 M NH_4NO_3, respectively, correlated with concentrations of Cd in radish and lettuce (John et al. 1972) and in radish grown in the tested soils (Symeonides and McRae 1977). Molybdenum extracted from soils by Tamm's Reagent (ammonium oxalate-oxalic acid) appeared to correspond well with levels of Mo in the harvested forage crops (Pierzneski and Jacobs 1986a, 1986b).

Hot-water-soluble boron is used most frequently to assess levels of

TABLE 7.6 Minimum Concentrations of Hot-Water Soluble Boron Associated with Optimum Yield for Various Crops (Keren and Bingham 1985)

0.1 mg B/kg (1)	0.1 mg B/kg–0.5 mg B/kg (2)	0.5 mg B/kg–1.0 mg B/kg (3)
Barley	Tobacco	Apple
Oat	Tomato	Alfalfa
Wheat	Lettuce	Clover
Corn	Peach	Red beet
Soybean	Pear	Turnip
Pea	Cherry	Mustard
Bean	Olive	Asparagus
Strawberry	Sweet potato	Celery
Potato	Peanut	
Flax	Carrot	

boron in soils. Berger and Truog (1939) first developed the method, which Baker and Mortensen (1966) modified to use an aqueous solution of 0.1% $CaCl_2 \cdot 2H_2O$. The $CaCl_2$ electrolyte flocculates the soil particles and facilitates filtration. As with other trace elements, crops differ in their responses to boron. Certain crops grow well in soils where the level of boron extracted with hot water is less than 0.1 mg/kg, while others suffer from a deficiency of boron if the level of hot water-soluble-boron falls below 1 mg/kg (Table 7.6). For crops sensitive to boron, the margin of safety between levels of B in soils associated with deficiencies and toxicities to crops usually is narrow.

Crops grown on saline soils more commonly suffer from boron toxicities than from boron deficiencies. Both the soil and irrigation water can contribute to excessive boron in the growing medium. Ayres and Westcot (1976) classified the relative tolerances of a wide variety of crops to concentrations of B in irrigation water. This study was based on work by Eaton (1944) and on visual symptoms of plants grown in sand cultures irrigated with nutrient solutions containing various concentrations of boron. Sensitive crops show reduced yield or symptoms of injury when concentrations of B in the soil solution exceed 0.3 mg/l, while tolerant crops withstand concentrations of B in the soil solution of 4 mg/l without showing any symptoms. Maas (1984) developed a more realistic classification scheme based on the threshold concentration, which is the maximum concentration of B in the soil solution (point T, Fig. 7.1) that does not restrict the yield. Table 7.7 presents boron tolerances of crops determined by Maas (1984) according to the classification systems of Maas (1984). The relative boron tolerance of crops determined by Maas (1984) based on the concentrations of boron in the soil solution corresponded to the relative boron tolerances of crops to B in irrigation water developed by Ayers and Westcot (1976). Sensitive crops in Maas' classification (Table 7.7) correspond to sensitive crops in Ayres and Westcot's classification based on B concentration in irrigation water (0.4 mg/l to 1.0 mg/l). Those crops in the moderately sensitive and moderately tolerant categories

TABLE 7.7 Relative Tolerance of Crops to Boron in Soil Solutions at Saturated Water Content (Maas 1984)

Sensitive		Moderately sensitive	Moderately tolerant	Tolerant	Very tolerant
(0.05 mg/l) (1)	(0.75 mg/l) (2)	(1.0 mg/l) (3)	(2.0 mg/l) (4)	(4 mg/l) (5)	(6 mg/l) (6)
Avocado	Garlic	Broccoli	*Lettuce*	Tomato	Sorghum (6 mg/l)
Grapefruit	Sweet potato	Red pepper	*Cabbage*	*Alfalfa*	
Orange	Wheat	*Pea*	*Celery*	*Purple Vetch*	Cotton (10 mg/l)
Apricot	Sunflower	Carrot	Turnip	*Parsley*	
Peach	*Mung bean*	Radish	*Bluegrass*	Beet, red	*Asparagus*
Cherry	*Sesame*	Potato	Barley	Sugarbeet	
Plum	*Lupine*	Cucumber	Oats		
Persimmon	*Strawberry*		Corn		
Fig (kadota)	*Kidney bean*		*Tobacco*		
Grape	*Lima bean*		*Mustard*		
Walnut	Peanut		*Sweet clover*		
Pecan			Squash		
Cowpea			*Muskmelon*		
Onion			Cauliflower		
(0.75 mg/l)	(1.0 mg/l)	(2.0 mg/l)	(4 mg/l)	(6 mg/l)	(15 mg/l)

Tolerance is based on permissible concentrations in soil water without yield reduction of either the vegetative growth or the harvested crop. Species for which yield reduction is based on vegetative growth are in italics. Within each column, crop tolerance to boron increases downward.

(Table 7.7) based on the concentration of B in the soil solution would correspond to the semi-tolerant class (1 mg/l to 2 mg/l), and those in the tolerant and very tolerant categories (Table 7.7) would correspond to the tolerant class (2 mg/l to 4 mg/l) for irrigation waters. Keren and Bingham (1984) and Gupta et al. (1985) presented additional information in review articles on the relative tolerances of crops to concentrations of B in irrigation waters and factors influencing these concentrations.

Bingham et al. (1985) refined the Maas (1984) approach of defining a threshold boron concentration, below which no reduction in crop yield due to B toxicity occurs. They also defined the yield reductions of crops when soil B exceeded the threshold concentration. The Bingham extension of the Maas approach for which crop data are available can be used to predict the percent reduction in yield associated with a given concentration of B in the soil solution.

The $NaHCO_3$ reagent is commonly used on neutral and alkaline soils to assess the availability of phosphorus to crops (Olson and Sommers 1982). Because the chemistry of phosphorus is somewhat similar to the chemistry of arsenic, soil tests for available P could reasonably be expected to apply to available As. Woolson et al. (1971) demonstrated that yields of corn correlated with the logarithm of the concentration of As extracted from neutral and alkaline soils by 0.5 M $NaHCO_3$.

Plant Methods

The nutritional deficiencies and toxicities of plants can be diagnosed according to visual symptoms and plant tissue analysis. Leaves of plants affected by nutrient deficiency or toxicity are frequently characterized by necrosis (tip burn or marginal scorch), chlorosis (turning yellow in color), and abscission (premature dropping). These symptoms may appear most frequently on young, old, or mature leaves, depending on whether the toxic trace element is readily translocated. The chlorosis can occur uniformly over an entire leaf surface, as often occurs with Fe and N deficiency, or be restricted to the inter-veinal regions of the leaf, such as occurs with Mn and Mg deficiency. Bould et al. (1984), Scaife and Turner (1984), and Winsor and Adams (1987) have described symptoms associated with deficiencies and toxicities of trace elements for a wide variety of crops. Unfortunately, visual symptoms caused by the deficiency or toxicity of one element often are similar to those of another element. Therefore, an elemental analysis of the leaf tissue is needed to confirm the cause.

The analysis of plant tissue has long been used to diagnose deficiencies and toxicities of trace elements in plants. A number of factors determine the threshold concentrations that indicate deficiencies or toxicities. The trace elements in plants vary according to the species and cultivar, the part of the plant sampled, its stage of development, the position of the tissue on fruiting versus non-fruiting branches, and the substrate. Therefore, to use the trace elements in plant tissue to diagnose nutrient deficiencies and phytotoxicities, the condition under which the threshold levels were derived must be adhered to strictly. Chapman (1966) comprehensively reviews the diagnostic criteria for major and trace elements.

The concentration of a trace element over which a deficient or toxic condition is present varies, so ranges are usually reported (Chapman 1966; Walsh and Beaton 1973; Bould et al. 1984; Scaife and Turner 1984; and Winsor and Adams 1987). Data presented in Table 7.8 show levels of trace elements in leaves of plants that indicate deficient or toxic conditions. The concentrations presented in Table 7.8 are generalized from data in various publications. They should be regarded merely as a guide for screening toxic or deficient conditions. Table 7.9 summarizes concentration ranges of trace elements associated with both normal and phytotoxic levels in forage crops. Among the forage species, trace element concentrations vary by a factor of as much as 10 and concentrations associated with phytotoxicity vary by a factor of two or more.

TABLE 7.8 Threshold Concentrations of Trace Elements in Leaves of Crops Indicative of Possible Deficiency or Phytotoxicity (Measurements in mg/kg on Dry Weight)

Element (1)	Deficiency[a] (2)	Phytotoxicity[b] (3)
As	NE[c]	—
B	20	100
Cd	NE	10
Co	0.2	—
Cu	5	25
Fe	30	—
Mn	15	100
Mo	0.1	100
Ni	NE	50
Se	NE	50
Zn	15	200

[a]Deficiency possible if concentration is less than listed.
[b]Phytotoxicity is possible if concentration is greater than listed.
[c]NE = non-essential element.

TABLE 7.9 Concentration of Trace Elements Commonly Observed in Forage Crops (Logan and Chaney 1983) (Measurements in mg/kg on Dry Weight)

Element (1)	Typical (2)	Phytotoxic (3)
As	0.01–1.0	3–10
B	7–75	75
Cd	0.1–1.0	5–700
Co	0.01–0.3	25–100
Cu	3–20	25–40
Mo	0.1–3.0	100
Ni	0.1–5.0	50–100
Se	0.1–2.0	100
Zn	15–150	500–1,000

ACCUMULATION OF TRACE ELEMENTS IN VEGETATION TO LEVELS POTENTIALLY HARMFUL TO ANIMALS

Plants can grow at normal or near-normal rates but accumulate enough of certain trace elements to cause acute toxicity or chronic metabolic imbalances in consumers of the plants.

The trace elements that can be absorbed by plants from soils in amounts high enough to be toxic or cause metabolic imbalance in consumers include Cd, Se, Mo, and possibly Be and Co (Allaway 1968). Arsenic, Cr, Fe, Pb, and Hg are toxic to animals and humans but are not absorbed by terrestrial plants in sufficient concentrations to be harmful to consumers. Chaney (1980, 1983) described the translocation of trace elements from soils to plants in terms of the soil-plant barrier, which refers to the inclusion or exclusion of trace elements in the food chain. The barrier protects the food chain from potentially toxic trace elements when one or more of the following occur: (1) Low solubility in the soil prevents uptake; (2) insolubility of an element in the fibrous roots prevents translocation to the edible plant tissues; or (3) phytotoxicity of the trace element kills the plant before the concentrations of the element in the edible plant tissues reach concentrations injurious to animals. Although the soil-plant barrier effectively protects consumers of the plants against harmful accumulations of trace elements via the root absorption process, other ways of ingesting contaminated soil, dust particles, or vegetation may cause harmful trace-element intake. For example, infants and toddlers in an inner-city area, where soils and dusts contain large amounts of Pb, may inadvertently ingest high amounts of Pb through the hand-to-mouth route (National Academy of Sciences 1980).

Cadmium

Ample evidence demonstrates that crops grown on Cd-contaminated soils contain elevated concentrations of Cd. If a human population regularly consumes Cd-enriched foods over decades, the Cd will accumulate in the people's bodies and adversely affect the health of susceptible individuals (Foulkes 1986; Friberg et al. 1985; Chaney 1984; Tsuchiya 1978). The most widely publicized example of chronic Cd poisoning occurred in Japan, where wastewater from a Zn mining and smelting complex contaminated the rice grown on fields downstream (Kobayashi 1978; Asami 1981; Nogawa and Ishizaki 1979). Most of the cases of Cd poisoning occurred in multiparous farm women of above middle age who also showed symptoms of Ca, Zn, and Fe deficiency. The concentration of Cd in the locally grown rice and the number of years of residence in the area also strongly related to incidence of the disease. No other documented cases exist of chronic Cd poisoning of humans or animals from foods grown on Cd-contaminated soils.

Cadmium contents of agricultural soils normally are less than 1 mg/kg. Some soils in irrigated regions may have naturally high levels of Cd (Lund et al. 1981; Burau 1983), but chronic Cd poisoning is unlikely, unless the food grown on these soils makes up a substantial part of a consumer's regular diet over an extended period of time.

Molybdenum

Molybdenum, essential in the diet of livestock at low concentrations, is toxic at higher concentrations (Allaway 1968). Ruminant animals are apparently more sensitive to such toxicity. The amount of Mo that an animal can tolerate depends on the levels of Cu and SO_4 in the diet. If the level of Cu is low, as low as 5 mg/kg of Mo in the forage may adversely affect the health of ruminant animals. Because plants can tolerate considerably larger concentrations of Mo than 5 mg/kg in their plant tissue, forage crops grown on soils high in Mo may become unfit as livestock feed without showing any symptoms of injury. The Mo toxicity to livestock is primarily a nutritional deficiency of Cu, and the ambient sulfate level markedly affects the interaction between Cd and Mo (Underwood 1977).

In soils, Mo availability to crops increases as the pH of the soil increases. The Cu availability tends to decrease as pH of the soil increases. Consequently, toxic levels of Mo are more common in forages grown on alkaline soils than in those grown on acid soils.

Selenium

Selenium is essential to livestock at low concentrations and toxic at slightly higher concentrations. Soils from many regions in the world produce forage with amounts of Se that are nutritionally inadequate for animals. In a few isolated areas, however, concentrations of Se are high enough to produce forage with amounts of Se that would be toxic to animals (National Academy of Sciences 1983). Selenium toxicity in humans due to environmental exposure has also been reported (Yang et al. 1983).

For livestock, the margin between concentrations of Se in forage that cause dietary deficiency and those that cause toxicity is quite narrow. According to Allaway (1968), concentrations of Se required to prevent Se deficiency range from 0.04 mg/kg to 0.20 mg/kg of Se on a dry weight basis (DW). If concentrations in forage exceed 4 mg/kg to 5 mg/kg (DW) and this forage makes up a substantial part of the diet, Se toxicity is likely to occur. Allaway (1968) suggests that the most desirable level of Se in forage is between 0.1 mg/kg and 1.0 mg/kg.

Recently, in California's Central Valley, high concentrations of Se in drainage from irrigated fields have been linked to deaths and deformities of waterfowl and the elimination of fish from reservoirs receiving the drainage water (Presser and Ohlendorf 1988). The problem is caused by irrigated agriculture of soils derived from selenium-laden sediments of marine origin. As the irrigation water leaches through the soils, selenium is brought into solution and transported through the drainage network to a terminal reservoir, where evaporation further concentrates the Se.

Cobalt

Cattle have been injured when fed diets spiked with levels of soluble salts of Co that exceed 10 mg/kg. Concentrations of Co in forages are normally from 0.01 mg/kg to 0.3 mg/kg. Some crops may grow well at as much as 50 mg/kg to 100 mg/kg (Logan and Chaney 1983). Although Co is apparently an element that can penetrate the soil-plant barrier, no instances of animals being injured from excessive plant-absorbed Co in their diets have been reported (Logan and Chaney 1983).

SUMMARY

Agronomists have long labored to find ways to diagnose the nutrient deficiencies and phytotoxicities of plants. Many have focused on determining the thresholds of an element in the growing medium that result in nutritional deficiencies or phytotoxicities. Two approaches are commonly used: (1) Relate the concentration of the element in soil to the yield—and therefore the degree of health—of a plant; and (2) examine the plant's foliage for symptoms of nutritional deficiency or phytotoxicity.

The amount of a trace element in a soil is measured by extracting the element with an appropriate chemical reagent and determining the concentration of the element in the extract. Since the physical and chemical properties of the soil influence the results, the deficiency and phytotoxicity thresholds of an element may vary widely in different soils.

A plant's performance can also be indicated by assaying the amount of an element that has accumulated in its tissue. The elemental content of the tissue usually depends solely on the availability of the element in the soil. To determine the elemental concentrations in the tissue that result in nutritional deficiency and phytotoxicity, relate them to the yield or other indicators of growth. Criteria based on the elemental concentration in plant tissue suffer from physiological differences of different species and cultivars and uncertainties introduced by the climate and cultivation practices.

The techniques for diagnosing trace element deficiencies and phytotoxicities in plants grown in saline soils are the same as for plants grown in nonsaline soils. However, the high electrolyte concentration in saline soil solutions, the neutral and alkaline soil chemical conditions, the soil's susceptibility to desiccation, and the salinity stress experienced by the plants influence the diagnostic criteria. In our opinion:

1. Boron is the trace element most likely to be phytotoxic in saline soils.
2. Plants grown on saline soils often experience deficiencies of Cu, Fe, Mn, and Zn because Cu, Fe, Mn, Zn, and other micronutrients have little solubility in saline and alkaline soils.
3. Plants can tolerate high levels of Se and Mo in soils, but many forage crops grown in saline soils may absorb enough Se and Mo to be toxic to consumers.
4. Levels of cadmium in soils are generally low and unlikely to cause problems in crop production, but cadmium introduced into the soil by man may accumulate to toxic amounts in food crops.

REFERENCES

Allaway, W.H. 1968. "Agronomic controls over the environmental cycling of trace elements." *Ad. Agron.* 20:235–274.

Alley, M.M., Martens, D.C., Schnoppinger, M.G. Jr., and Hawkings, G.W. 1972. "Field calibration of soil tests for available Zn." *Soil Sci. Soc. Am. Proc.* 36:621–623.

Asami, T. 1981. "Maximum allowable limits of heavy metals in rice and soil." *Heavy Metal Pollution in Soils of Japan*. K. Kitagishi and I. Yamane, eds. Japan Scientific Societies Press, Tokyo. 302 pp.

Ayres, R.S., and Westcot, D.W. 1976. "Water quality for irrigation."

Irrigation and Drainage Paper 29. Food and Agriculture Organization of the United Nations, Rome, Italy.

Baker, A.S., and Mortensen, W.P. 1966. "Residual effect of single borate applications on western Washington soils." *Soil Sci.* 102:173–179.

Berger, K.C., and Truog, E. 1939. "Boron determination in soils and plants." *Ind. Eng. Chem. Anal. Ed.* 11:540–545.

Bingham, F.T. 1982. "Boron." *Methods of Soil Analysis, Part 2. Chemical and Microbiological Properties*. 2nd ed. A.L. Page, ed. American Soc. of Agronomy, Madison, Wisc. pp. 431–447.

Bingham, F.T., Page, A.L., Mahler, R.J., and Ganje, T.J. 1975. "Growth and cadmium accumulation of plants grown on a soil treated with a cadmium-enriched sewage sludge." *J. Environ. Qual.* 4:207–211.

Bingham, F.T., Strong, J.E., Rhoades, J.D., and Keren, R. 1985. "An application of the Maas-Hoffman salinity response model for boron toxicity." *Soil Sci. Soc. Am. J.* 49:672–674.

Bould, C., Hewitt, E.J., and Needham, P. 1984. *Diagnosis of Mineral Disorders in Plants. Volume 1: Principles.* Chemical Publishing, New York. 170 pp.

Bradford, G.R. 1966. "Lithium." *Diagnostic Criteria for Plants and Soils,* H. D. Chapman, ed. University of Calif., Div. of Agricultural Sciences, Berkeley, Calif. pp. 218–224.

Brown, A.L., Quick, J., and Eddings, J.L. 1971. "A comparison of analytical methods for soil Zn." *Soil Sci. Soc. Am. Proc.* 35:105–107.

Burau, R.G. 1983. "National and local dietary impact of Cd in south coastal California soils." *Ecotoxicology and Environmental Safety.* 7:53–57.

Burton, K.W., Morgan, E., and Roig, A. 1983. "The influence of heavy metals upon growth of sitka-spruce in South Wales forests. I. Upper critical and foliar critical concentration." *Plant and Soil.* 73:327–336.

Chaney, R.L. 1980. "Health risks associated with toxic metals in municipal sludge." *Sludge-Health Risks of Land Application.* G. Bitton, B.L. Damron, G.T. Edds, and J.M. Davison, eds. Ann Arbor Science Publishers, Inc., Ann Arbor, Mich. pp. 59–83.

Chaney, R.L. 1983. "Fate of toxic substances in sludge applied to cropland." Proceedings of the International Symposium on Land Application of Sewage Sludge. Oct. 13–15, 1982. Tokyo, Japan. pp. 259–324.

Chaney, R.L. 1984. "Potential effects of sludge-borne heavy metals and toxic organics on soils, plants, and animals, and related regulatory guidelines." *Proc. Pan American Health Organization Workshop on the International, Transportation, Utilization or Disposal of Sewage Sludge,* pp. 1–56.

Chang, A.C., Page, A.L., Foster, K.W., and Jones, T.E. 1982. "A comparison of cadmium and zinc accumulation by four cultivars of barley grown in sludge-amended soils." *J. Environ. Qual.* 11:409–412.

Chapman, H.D., ed. 1966. *Diagnostic Criteria for Plants and Soils.* University of Calif., Div. of Agric. Sci., Berkeley, Calif. 793 pp.

Davis, R.D., and Carlton-Smith, C. 1980. "Crops as indicators of the significance of contamination of soil by heavy metals." *Technical Report TR140,* Water Research Centre, Stevenage, U.K. 43 pp.

Davis, R.D., and Carlton-Smith, C.H. 1984. "An investigation into the

phytotoxicity of zinc, copper and nickel using sewage sludge of controlled metal content." *Environ. Pollut.* (Ser. B) 8:163–185.
de Boer, G.J., and Reisenauer, H.M. 1973. "DTPA as an extractant of available soil iron." *Commun. Soil Sci. Plant Anal.* 4(2):121–128.
Dolar, S.G., Kenney, D.R., and Walsh, L.M. 1971. "Availability of Cu, Zn, and Mn in soils. III. Predictability of plant intake." *J. Sci. Food Agr.* 22:282–286.
Eaton, F.M. 1944. "Deficiency, toxicity and accumulation of boron in plants." *J. Agr. Res.* 69:237–279.
Evans, C.E., Wear, J.I., Hajek, B.F., and Cope, J.T. Jr. 1974. "The relationship of soil Zn removed by three extractants to Zn uptake by corn and sorghum in medium- to fine-textured soils." *Comm. Soil Sci. Plant Anal.* 5:105–113.
Foulkes, E.C. 1986. "Handbook of experimental pharmacology," Vol. 80: Cadmium. Springer-Verlag, New York. 400 pp.
Friberg, Lars, Elinder, Gustaf-Carl,Kjellstrom, T., and Nordberg, Gunnar F. 1985. *Cadmium and health: A toxicological and epidemiological appraisal. Volume 1. Exposure, dose and metabolism.* CRC Press, Inc., Boca Raton, Fla. 209 pp.
Grigg, J.L. 1953. "Determination of available soil molybdenum." *N. Z. Soil News.* Molybdenum Symposium, Soil Bureau, Wellington N 1, New Zealand. pp 37–40.
Gupta, U.C., Jame, Y.W., Campbell, C.A., Leyshon, A.J., and Nicholaichuk, W. 1985. "Boron toxicity and deficiency: A review." *Can. J. Soil Sci.* 65:381–409.
Haw, A.U. and Miller, M.H. 1972. "Prediction of available soil Zn, Cu and Mn using chemical extractants." *Agron. J.* 64:779–782.
Haq, A.W., Bates, T.E., and Soon, Y.K. 1980. "Comparison of extractants for plant-available zinc, cadmium, nickel, and copper in contaminated soils." *Soil Sci. Soc. Am. J.* 44:772–777.
Jarvis, S.C., and Robson, A.D. 1983. "A comparison of the cation-anion balance of 10 cultivars of Trifolium subterraneum L. and their effects on soil acidity." *Plant Aoil* 75:235–244.
John, M.K., VanLaerhoven, C.J., and Chauh, H.H. 1972. "Factors affecting plant uptake and phytotoxicity of cadmium added to soils." *Environ. Sci. Tech.* 6:1005–1009.
Jones, R.L., Hinesly, T.D., Ziegler, E.L., and Tyler, J.J. 1975. "Cadmium and Zn contents of corn leaf and grain produced by sludge-amended soil." *J. Environ. Qual.* 4:509–514.
Keren, R., and Bingham, F.T. 1984. "Boron in water, soils, and plants." *Advances in Soil Science.* Vol. 1. Springer-Verlag, New York. pp. 228–275.
Kim, S.J., Chang, A.C., Page, A.L., and Warneke, J.E. 1988. "Relative concentrations of cadmium and zinc in tissues of selected food plants grown on sludge-treated soil." *J. Environ. Qual.* 17:568–573.
Kobayashi, J. 1978. "Pollution by cadmium and the itai-itai disease in Japan." *Toxicity of Heavy Metals in the Environment.* F.W. Oehme, ed. Marcel Dekker, Inc., New York. pp. 199–260.
Korcak, R.F., and Fanning, D.S. 1978. "Extractability of cadmium, copper, nickel, and zinc by double acid versus DTPA and plant content at excessive soil levels." *J. Environ. Qual.* 7:506–512.

Lagerwerff, J.V. 1971. "Uptake of Cd, Pb, and Zn by radish from soil and air." *Soil Sci.* 111:129–133.

Latterell, J.J., Dowdy, R.H., and Larson, W.E. 1978. "Correlation of extractable metals and metal uptake of snap beans grown on soil amended with sewage sludge." *J. Environ. Qual.* 7:435–440.

Lauer, D.A. 1971. "Evaluation of plant-available Zn by the DTPA soil test, 0.1 N HCl extraction, and labile Zn measurements." Ph.D. thesis, Department of Agronomy, Colorado State University, Fort Collins. Univ. Microfilms, Ann Arbor, Mich. (Diss. Abstr. 31, 6157B).

Lindsay, W.L., and Norvell, W.A. 1969. "Development of a DTPA micronutrient soil test." *Agron. Abstr.* p. 84.

Lindsay, W.L. and Norvell, W.A. 1978. "Development of a DTPA soil test for Zn, Fe, Mn and Cu." *Soil Sci. Soc. Am. J.* 42:421–428.

Logan, T.J., and Chaney, R.L. 1983. "Utilization of municipal wastewater and sludge on land-Metals." *Proc. of the Workshop on Utilization of Municipal Wastewater and Sludge on Land.* A.L. Page et al., eds. University of Calif. Riverside, Calif. pp. 235–326.

Lund, L.J., Betty, E.E., Page, A.L., Elliott, R.A. 1981. "Occurrence of naturally high Cd levels in soil and its accumulation by vegetation." *J. Environ. Qual.* 10:551–556.

Maas, E.V. 1984. "Salt tolerance of plants." *Handbook of Plant Science.* Christie, B.R., ed. CRC Press, Cleveland, Ohio.

Maas, E.V., Hoffman, G.J. 1977. "Crop salt tolerance—Current assessment." *J. Irrig. Drainage Div. ASCE* 103(IR2):115–134.

Mishra, S.G., and Pande, P. 1974. "Evaluation of suitable extractant for available nickel in soils." *Plant and Soil.* 41:697–700.

National Academy of Science. 1983. *Selenium in nutrition.* Revised Edition. National Academy Press, Washington, D.C.

National Academy of Sciences. 1980. *Lead in the human environment.* National Academy of Science, Washington, D.C. 525 pp.

Nelson, J.L., Brown, L.C., and Viets, F.G. Jr. 1959. "A method for assessing zinc status of soils using acid-extractable zinc and 'titratable alkalinity' values." *Soil Sci.* 88:275–283.

Nogawa, K., and Ishizaki, A. 1979. "A comparison between Cd in rice and renal effects among inhibitants of the Jinzu River basin." *Environ. Res.* 18:410–420.

Olson, S., and Sommers, L.E. 1982. "Phosphorus." *Methods of Soil Analysis. Part 2. Chemical and Microbiological Properties.* wnd ed. A.L. Page et al., ed. American Soc. of Agronomy, Madison, Wisc. pp. 403–427.

Page, A.L., Chang, A.C., Sposito, G., and Mattigod, S. 1981. "Trace elements in wastewater: Their effects on plant growth and composition and their behavior in soils." *Modeling Wastewater Renovation: Land treatment.* I.K. Iskandar, ed. John Wiley and Sons, New York. pp. 182–222.

Perkins, H.F. 1970. "A rapid method of evaluating the zinc status of Coastal Plain soils." *Commun. in Soil Sci. Plant Anal.* 5:115–121.

Pierzyneski, G.M., and Jacobs, L.W. 1986a. "Extractability and plant availability of Mo from inorganic and sewage sludge sources." *J. Environ. Qual.* 15:323–326.

Pierzynski, G.M., and Jacobs, L.W. 1986b. "Molybdenum accumulation by corn and soybeans from a Mo-rich sewage sludge." *J. Environ. Qual.* 15:94–98.

Pierre, W.H., and Banwart, W.L. 1973. "Excess-base and excess-base/nitrogen ratio of various crop species and parts of plants." *Agron. J.* 65:91–96.

Pierre, W.H., Meisinger, J., and Birchett, J.R. 1970. "Cation-anion balance in crops as a factor in determining the effect of nitrogen fertilizers on soil acidity." *Agron. J.* 62:106–111.

Presser, T.S., and Ohlendorf, H.M. 1988. "Biogeochemical cycling of selenium in the San Joaquin Valley, California, USA." *Environ. Management* 11:805–821.

Purves, David. 1985. *Trace-Element Contamination of the Environment*. Elsevier, New York. 243 pp.

Scaife, Alan, and Turner, Mary. 1984. *Diagnosis of mineral disorders in plants. Volume 2: Vegetables*. Chemical Publishing Co., New York, 96 pp.

Singh, B.R., and Narwal, R.P. 1984. "Plant availability of heavy metals in a sludge treated soil: II. Metal extractability compared with plant metal uptake." *J. Environ. Qual.* 13:344–349.

Soltanpour, P.N. 1985. "Use of ammonium bicarbonate DTPA soil test to evaluate elemental availability and toxicity." *J. Plant Nutr.* 8:323–338.

Soltanpour, P.N., and Schwab, A.P. 1977. "A new soil test for simultaneous extraction of macro- and micro-nutrients in alkaline soils." *Commun. Soil Sci. Plant Anal.* 8(3):195–207.

Soltanpour, P.N., and Workman, S.M. 1980. "Use of NH_4HCO_3-DTPA soil test to assess availability and toxicity of selenium to alfalfa plants." *Commun. in Soil Sci. Plant Anal.* 11:1147–1156.

Soltanpour, P.N., and Workman, S.M. 1981. "Soil-testing methods used at Colorado State University Soil Testing Lab for evaluation of fertility, salinity, sodicity, and trace element toxicity." Colo. State Univ. Expt. Sta., Fort Collins, Tech. Bull. 142.

Soltanpour, P.N., Khan, A., and Lindsay, W.L. 1976. "Factors affecting DTPA-extractable Zn, Fe, Mn and Cu from soils." *commun. Soil Sci. Plant Anal.* 7:797–821.

Sposito, Garrison. 1986. "Distinguishing adsorption from surface precipitation." *Geochemical Processes at Mineral Surfaces*. James A. Davis and Kum F. Harges, eds. ACS Symposium Series No. 323. Amer. Chem. Soc., Washington, D.C. pp. 217–228.

Street, J.J., Lindsay, W.L., and Sabey, B.R. 1977. "Solubility and plant uptake of Cd in soils amended with Cd and sewage sludge." *J. Environ. Qual.* 6:72–77.

Symeonides, C. and McRae, S.G. 1977. "The assessment of plant-available Cd in soils." *J. Environ. Qual.* 6:120–123.

Trierweiler, J.F. and Lindsay, W.L. 1969. "EDTA-ammonium carbonate soil test for zinc." *Soil Sci. Soc. Amer. Proc.* 33:49–54.

Tsuchiya, K., ed. 1978. *Cadmium Studies in Japan: A Review*. Elsevier/North-Holland Biomedical Press, New York. 376 pp.

Underwood, E.J. 1977. *Trace elements in Human and Animal Nutrition*. 4th ed. Academic Press, San Francisco.

Valdares, J.M.A.S., Gal, M., Mingelgrin, U., and Page, A.L. 1983.

"Some heavy metals in soils treated with sewage sludge, their effects on yield, and their uptake by plants." *J. Environ. Qual.* 12:49–57.

Walsh, Leo M., and Beaton, James D., eds. 1973. Soil testing and plant analysis. Soil Science Society of America, Madison, Wisc. 491 pp.

Walsh, L.M., Steevens, D.R., Seibel, H.D., and Weis, G.G. 1972. "Effects of high rates of Zn on several crops grown on an irrigated Plainfield sand." *Comm. Soil Sci. Plant Anal.* 3:187–195.

Walsh, T., Nunan, M., and O'Moore, L.B. 1951. "The importance of molybdenum in relation to some cropping practices and livestock problems under Irish conditions." *J. Ireland Dep. Agr.* 48:32–43.

Wear, J.I., and Evans, C.E. 1968. "Relationship of Zn uptake by corn and sorghum to soil Zn measured by three extractants." *Soil Sci. Soc. Amer. Proc.* 32:543–546.

Wear, J.I., and Sommers, A.L. 1948. "Acid extractable Zn of soils in relation to the occurrence of Zn deficiency symptoms of corn. A method of analysis." *Soil Sci. Soc. Amer. Proc.* 12:143–144.

Winsor, G., and Adams, P. 1987. *Diagnosis of Mineral Disorders in Plants, Volume 3: Glasshouse Crops.* Chemical Publishing Co., New York. 168 pp.

Wolnick, K.A., Fricke, F.L., Capar, S.G., Braude, G.C., Meyer, M.W., Satzger, R.D., and Bonnin, E. 1983a. "Elements in major raw agricultural crops in the United States. 1. Cadmium and lead in lettuce, peanuts, potatoes, soybeans, sweet corn, and wheat." *J. Agric. Food Chem.* 31:1240–1244.

Wolnick, K.A., Fricke, F.L., Capar, S.G., Braude, G.C., Meyer, M.W., Satzger, R.D., and Bonnin, E. 1983b. "Elements in major raw agricultural crops in the United States. 2. Other elements in lettuce, peanuts, potatoes, soybeans, sweet corn and wheat." *J. Agric. Food Chem.* 31: 1244–1249.

Wolnick, K.A., Fricke, F.L., Capar, S.G., Braude, G.C., Meyer, M.W., Satzger, R.D., and Bonnin, E. 1985. "Elements in major raw agricultural crops in the United States. 3. Cadmium, lead, and eleven other elements in carrots, field corn, onions, rice, spinach, and tomatoes." *J. Agric. Food Chem.* 33:807–811.

Woolson, E.A., Axley, J.H., and Kearney, P.C. 1971. "Correlation between available soil arsenic, estimated by six methods, and response of corn (Zea mays L.)." *Soil Sci. Soc. Am. Proc.* 35:101–105.

Yang, G.Q., Wang, S., Zhou, R., and Sun, R. 1983. "Endemic selenium intoxication of humans in China." *Amer. J. Clin. Nutr.* 37:872–875.

CHAPTER 8

GENETIC APPROACHES FOR DEVELOPING ECONOMIC SALT-TOLERANT CROPS[a]

INTRODUCTION

A wide range of salt tolerance exists in the plant kingdom. Some species are extreme halophytes and ecologically competitive only under saline conditions. Others, the glycophytes, are salt-sensitive plants that grow well under salt-free conditions. This diversity of salt tolerance has evolved over millions of years.

For several thousands of years, domesticated crop species typically have been provided with optimum conditions to obtain high yields. This has selected against those characteristics that help plants adapt to extreme environments (Rosielle and Hamblin 1981). Thus, a narrower range of tolerance to salt and other environmental stresses exists among crops than for wild species. Indeed, many crops are moderately to highly sensitive to salinity. This chapter focuses on the salt tolerances of crops and the improvements that can be made through conventional plant breeding strategies or the application of new approaches in molecular biology.

Historically, salinization of a crop production area has resulted in the replacement of salt-sensitive species with more tolerant ones. Methods that leached salts from soils and management practices that avoided salinity problems had yet to be developed (Marr 1967). A few highly salt-tolerant crops, such as sugar beet, barley, cotton, asparagus, and sugarcane, could substitute for more sensitive species. Such substitution is still practiced in the world's saline growing areas. However, climatic, nutritional, and economic needs often call for a greater diversity of salt-tolerant crop species than is available. Only recently has man become interested in improving salt tolerance in cultivated plants.

Improving the salt tolerance of sensitive species does more than improve the yield. It gives the farmer greater flexibility for crop selection. Additionally, relatively small increases in salt tolerance may greatly reduce the leaching requirement in areas that need to be reclaimed by leaching. This is because the efficiency of leaching as measured by percent decrease in soil salinity per unit water applied decreases (Hoffman 1980). Improved salt tolerance in sensitive species would allow more extensive use of brackish water; this is especially important where water is expensive or scarce. These potential advantages draw attention to the

---

[a]Prepared by: M.C. Shannon, U.S. Salinity Lab., 4500 Glenwood Dr., Riverside, Calif. 92501; and C.L. Noble, Irrig. and Salinity Res. Inst., Ferguson Rd., Private Bag, Tatura, Victoria 3616, Australia.

TABLE 8.1 Proceedings and Volumes Dealing with Development of Environmental Stress Tolerance and Salt Tolerance in Crops (Epstein, 1985)

Title (1)	Reference (2)
Plant Adaptation to Mineral Stress in Problem Soils	Wright (1976)
Crop Tolerance to Suboptimal Land Conditions	Jung (1978)
Proceedings of the Workshop/Seminar on Membrane Biophysics and Salt and Salt Tolerance in Plants	Qureshi et al. (1981)
The Biosaline Concept: an Approach to the Utilization of Underexploited Resources	Hollaender et al. (1979)
Genetic Engineering of Osmoregulation: Impact on Plant Productivity for Food, Chemicals, and Energy	Rains et al. (1980)
A Conference on Biosalinity: the Problem of Salinity in Agriculture	Rains (1981)
Breeding Plants for Less Favorable Environments	Christiansen and Lewis (1982)
Biosaline Research: a Look to the Future	San Pietro (1982)
Salinity Tolerance in Plants: Strategies for Crop Improvement	Staples and Toenniessen (1984)
Applications of Genetic Engineering to Crop Improvement	Collins and Petolino (1984)
Biosalinity in Action: Bioproduction with Saline Water	Pasternak and San Pietro (1985)
Plant Growth, Drought, and Salinity	Turner and Passiouri (1986)
Plant Breeding for Stress Environments	Blum (1988)

fact that too little progress has been made in improving the salt tolerance of crops.

SELECTIVE BREEDING FOR SALT TOLERANCE

Breeding for improved salt tolerance has been reviewed a number of times (Nieman and Shannon 1976; Gupta 1978; Pasternak et al. 1979; Norlyn 1980; Ramage 1980; Duvick et al. 1981; Shannon 1982, 1984, 1985; Noble 1983; Blum 1984, 1988; Downton 1984; Jones and Qualset 1984; Rana 1985; Tal 1985; Epstein and Rains 1987; Jones 1987). Epstein (1985) tabulated many of the reviews, chapters, and volumes on the genetic control of mineral nutrition and ion relations in plants. These are important resources to those interested in the breeding and genetic aspects of salt tolerance. They may also be useful in finding references on drought and other environmental stresses. Table 8.1 lists some of the books and volumes that review breeding for salt tolerance and other mineral stresses.

Despite great interest and effort and the proposal of several screening and selection schemes (Greenway and Andrew 1962; Dewey 1962; Beatty and Ehlig 1973; Epstein et al. 1980; Shannon et al. 1983; Sykes 1985), limited numbers of varieties selected for improved salt tolerance exist. They include:

1. Arsola 1–18, a cold-hardy and highly salt-tolerant variety of avocado developed by crossing a salt-tolerant West Indian variety with a cold-hardy Mexican variety (Cooper 1951).
2. Nebraska 10, a variety of wheatgrass (*Agropyron desertorum*) selected as a salt-tolerant strain in artificially salinized field trials (Dewey 1962).
3. Giza 159, a variety of rice still used widely in Egypt and selected from a cross between the salt-tolerant variety Agami 1 and the locally-adapted Giza 14. Giza 159 was found to be superior in both yield and salt tolerance (Gad El Hak 1966).
4. Giza 160, recently developed from a cross between Giza 159 and the salt-tolerant variety Reiho to further improve yield in salt-affected areas (M. El Azizi, personal communication). It was released in 1984 on a limited basis but found to be sensitive to two new blast races.
5. Edkway, a highly salt-tolerant variety of tomato developed by growers along the northern coast of Egypt (Hassan and Desouki 1982; A. El-Beltagy, personal communication), selected from an unknown source by regional farmers who grow their crops on sand dunes, using 4 dS/m to 6 dS/m drainage water for irrigation. Even though tomato is highly self-pollinating and, therefore, has limited variability within a variety, enough variation was present for the selection of salt tolerance under high selection pressure.
6. AZ-Germ Salt 1, a variety of alfalfa selected for high germination rates from Mesa Sirsa after five cycles of recurrent selection, using selection pressures as high as 99% in laboratory and greenhouse salt tolerance tests (Dobrenz et al. 1983).
7. AZ-Germ Salt 2, a subsequent selection with an even higher salt tolerance at germination, which is due to be released soon (A. Dobrenz, personal communication). The original selection has been used extensively as a salt-tolerant germplasm source in other breeding programs.
8. Saltol, a salt-tolerant red fescue collected as a naturally selected ecotype from a tidal marsh and used to revegetate roadsides affected by de-icing salts (Cordukes 1981).
9. BG 84-3, a more productive and salt-tolerant variety of muskmelon selected from salt-tolerant, locally-adapted varieties of melon in Israel (S. Mendlinger, personal communication).
10. Shad El-Dokki, a variety of muskmelon developed in Egypt that has superior yields under saline conditions compared to other locally adapted varieties (M. El. Doweiny, personal communication).

11. Arizona 8601, a variety of maize germ plasm developed from a natural selection program using saline irrigation water (2000 mg/l to 4000 mg/l total soluble salts) over a period of 20 years. It reportedly has better yields than Pioneer 3183 in the production of silage (+38%) and grain (+26%) under the conditions at which it was selected (Day 1987).

A limited number of released varieties specifically bred for improved salt tolerance exist, but efforts to breed such varieties are rapidly expanding. Many studies that may result in more salt-tolerant crops in the near future are underway. The International Rice Research Institute (IRRI) screened over 10,000 accessions of rice for salt tolerance in greenhouse tests and identified many promising varieties (IRRI 1977). Thousands of accessions of wheat and barley throughout the world have been extensively screened for salt tolerance (Kingsbury and Epstein 1984). Some efforts involve crossing cultivated species with closely related wild species to improve the variability for salt tolerance.

The cultivated tomato, *Lycopersicon esculentum,* has been hybridized with several salt-tolerant wild species to improve its tolerance. A salt-tolerant line was developed by back-crossing *L. cheesmanii* to the cultivated parent (Frederickson and Epstein 1975; Rush and Epstein 1976, 1981a). No commercial release is yet available. Another wild tomato species, *L. pennellii,* has been hybridized and backcrossed to the New Yorker variety for nine generations and selected for cold tolerance and earliness. Many breeding lines from this cross had a higher salt tolerance than that of the cultivated parent (Sacher et al. 1982). In Israel, a variety of processing tomato with improved salt tolerance was selected by back-crossing a cultivated processing variety with *L. pennellii.* Four cycles of recurrent selection produced a line with high salt tolerance and superior quality characteristics (D. Pasternak and Y. Saranga, personal communication).

In other species, wild germplasm and hybridizations between cultivated varieties are used to establish variability from which to select for improved salt tolerance. For instance, crosses between the salt-tolerant Rungpur citrus line and Rubidoux trifoliate orange produce progeny with a range of salt tolerance (Furr et al. 1963, Furr and Ream 1968). Intravarietal selection for salt tolerance within lettuce has been reported in salinized field plots (Shannon 1980). Similar selections have been made within a variety of white clover, *Trifolium repens,* grown under controlled conditions in the greenhouse (Noble and Shannon 1988). Many examples exist in which salt tolerance has been developed inadvertently in varieties selected for high yield in naturally saline environments. Some varieties of wheat, barley, cotton, and rice developed for high yield in saline regions of Pakistan, India, Egypt, and the United States have a better salt tolerance than varieties developed in nonsaline areas (Akbar et al. 1972, Bernal et al. 1974, Kingbury and Epstein 1986).

Considerable research has quantified the salt tolerance of various crop species (Francois and Maas 1978, 1985; Maas and Hoffman 1977; Maas 1985, 1987), but data are usually based on comparisons among only a few varieties for many species. Some studies on a range of varieties revealed that salt tolerance varied widely according to the species. Others showed limited differences. In many cases, only a small portion of the existing germplasm base has been adequately tested. Many wild progenitors of cultivated species have not been examined at all.

Selection Procedures

Salinity is a term that includes soils and waters with high concentrations of dissolved salts of many kinds, any of which may critically limit plant growth. Saline soils may be sodic or acidic and cover a wide range of soil types and moisture conditions. Genotypes that show similar salt tolerance in one environment may differ in response in a different environment. Rana (1985) notes that crops adapted to alkali soils are usually tolerant of non-alkaline saline soils but that the converse is not true. Salt tolerance in saline-drained conditions differs from that in saline-waterlogged conditions. Waterlogging in the root zone combined with salinity can be extremely hazardous. The loss of needed energy from oxidative metabolism disrupts energy-dependent ion uptake and exclusion and extrusion processes in root membranes. Salinity and waterlogging in the root zone greatly increase the uptake of salt (West 1978; West and Taylor 1984). Other important environmental factors that significantly interact with salinity include temperature, wind, humidity, light, and pollution, which Chapters 6 and 13 cover in detail.

Prerequisites and Limitations. One requirement in breeding for salt tolerance is that genetic variations exist for that character in the gene pool. Such variation may be between individuals, varieties, or species, if some degree of sexual compatibility exists so that genes may be transferred from one individual to another. Also, the salt tolerance character must be identifiable in segregating generations, or information about the numbers of genes involved and their heritability must exist. For example, Hunt (1965) found salt tolerance in *Agropyron intermedium* to be a heritable trait that could be selected based on growth in mature plants. Norlyn (1980) found that salt tolerance in barley was heritable, but that genetic control was complex.

The breeding procedure depends on the pattern of inheritance (qualitative or quantitative), the number of genes with major effects, and the nature of gene action. The heritability and inheritance of salt tolerance in a species determines the selection intensity and the number of selection cycles necessary. The lack of genetic variation and a poor understanding of genetic control often present the greatest barriers to improving salt tolerance.

The multigenic, quantitative nature of the salt tolerance character imposes several constraints on breeding. Ramage (1980) emphasized the need for a favorable background genotype. He suggested the use of recurrent selection for cross-pollinated species and self-pollinated species with male sterility systems that can be genetically maniupulated. Using male sterility systems reduces the work involved during the recombination phase of the recurrent selection procedure with self-pollinators. Jones and Qualset (1984) assert that plant growth attributes must be measured throughout the growth period so that particularly salt-sensitive stages of growth can be identified.

Identifying a quantitative character is difficult at best. The interactions between salinity and other environmental stresses complicate assessments that use yield or growth as an index of tolerance. Other indices have been proposed. Survival after massive doses of salinity has been used as a criterion. Other criteria will be discussed below. Although selection programs using some the the criteria have had limited success, no given protocol has been widely adopted or applied.

The salt responses of various crops often depend on the type of soil and other environmental factors (Levitt 1972). Rana (1985) noted that major obstacles to successful breeding for salt tolerance include the complexity of soils and environmental factors, the identification of clear-cut varietal differences, and the lack of efficient criteria for rating the salt tolerance of individuals in large segregating populations. An inadequate understanding of how plants physiologically integrate and respond to salt throughout their life cycles may be viewed as another factor that limits successful breeding (Shannon 1985). For example, the method of irrigation affects crop rooting patterns. This, in turn, can affect how the crop responds to salinity (Bernstein and Francois 1973). Though such bits and pieces of knowledge are known about salinity-environmental interactions, an integrated model is currently unavailable.

Salt Tolerance Measurement. Salt tolerance is measured as a complex function of yield decline across a range of salt concentrations (Maas and Hoffman 1977; van Genuchten and Hoffman 1984). Measuring salt tolerance for purposes of screening has inherent complications. For one, the tolerance of single plants in a segregating population cannot be assessed, because information must be collected on a relative basis. How well a plant grows under saline conditions depends on its salt tolerance and vigor. The performance of genetically segregating material under control and saline conditions can be compared only if progeny lines with some degree of homozygosity are first established. This requires time and labor. Also, high mean yield and phenotypic stability across environments correlate negatively (Finlay and Wilkinson 1963; Frey 1964). Selection for salt tolerance under the wrong conditions or using the wrong genetic material can result in low-yielding selections (Richards 1983).

Screening cannot be conducted efficiently across the whole range of salinities, so the critical level or levels must be chosen. Considering salt response as a function of a simple threshold-and-slope model (Maas and Hoffman 1977; Chapter 13) may simplify the process, but it also involves some difficulties. The threshold (the salt concentration at which yield begins to decline) is sensitive to environmental interaction and depends on accurate salinity measurements and the method by which they are integrated over the area of the plot, depth, and time. Consequently, a degree of error exists in evaluating the slope at salt concentrations near the threshold. At high salt concentrations, the slope tends to lessen, resulting in added uncertainty at this part of the curve. For agronomic purposes, salt tolerance at high salinities has little economic importance. Physiological and genetic factors that contribute to growth at high salt concentrations proportionally relate more to survival than to high yields.

The fundamental selection criteria in plant breeding are the mean yield and the yield stability across environments. When the genetic correlation for yields across environments is highly negative and genetic variance in the stress environment is less than in the non-stress environment, selection for productivity normally increases yields in both environments (Rosielle and Hamblin 1981). If some capacity exists for selection under a particular stress environment, i.e., genetic variance is high compared to that under non-stress, tolerance might be improved without also decreasing yield in the non-stress environment.

Field Screening. Selection for tolerance in the field is difficult because the spatial and temporal variations in the salinity of soil solutions close to the root zone are difficult to assess. The ability of plants to extract water preferentially from the most nonsaline portions of their root zones further complicates this assessment. To compensate, use very large populations, judiciously select a breeding strategy and experimental design, and precisely control irrigation in artifically salinized plots. Nevertheless, a low selection efficiency is to be expected even under ideal conditions, and the efforts will be very expensive. The advantages of field screening are that space poses no limits on replication and population sizes and that environmental conditions more nearly approximate conditions in the field.

Screening under Controlled Conditions. A screening procedure that is more reliable and efficient than field techniques is needed. An initial consideration would be to control the applied salinity and extraneous environmental factors by using greenhouses and environmental chambers. In these circumstances, mean yield becomes an unmanageable selection index due to space limitations. Jones and Qualset (1984) have suggested that field selection under saline conditions be used to screen out unpromising entries. The type of breeding and selection program must be chosen based on field applicability and greenhouse limitations. The following steps must be taken:

1. Collect available information on the salt tolerance of the crop, including relative salt tolerance range, potential variability among cultivated varieties and closely related species, and sensitivities to specific ions and environmental interactions.
2. Identify the stage of growth that limits productivity and determine whether economical management techniques can overcome the limitation. For example, if stand establishment is the limitation, determine whether it could best be overcome by improved bed preparation (to move dissolved salts away from the seed), by timely irrigation with high- or medium-quality water, or by a closer seed or plant spacing.
3. Determine whether other locally adapted varieties are more tolerant at this stage of growth.
4. Develop a screening procedure for the sensitive stage of growth based on average salt concentration and composition of the soil water during this stage and the environmental conditions during the period of salt damage in the field. The criterion must relate to mean yield response in the field.

5. Conduct field experiments at an early stage to prove the relationship of the criteria selected to the desired field characteristics.

6. Evaluate a range of varieties and introductions to determine genetic variance for the desired character. Include controls to separate genetic from environmental effects under saline and nonsaline conditions. Use the collected data to determine whether intravarietal selection will be effective. If genetic variance is low or a greater degree of tolerance is needed, evaluate related, wild species and hybrid lines.

Indices of Salt Tolerance. Criteria that have been used to evaluate salt tolerance in species include germination, emergence, shoot dry weight, root weight, shoot number, leaf damage, flowering, and seed or fruit yield. Often tolerance for one criterion is unrelated to another. Ayers et al. (1952) found that salinity decreased shoot dry weight in barley and wheat more than seed production. Salinity often affects the growth of shoots more than the growth of roots. In muskmelons, salt tolerance decreased in the following order: total vegetative dry weight > total vine yield > fruit yield > marketable yield (Shannon and Francois 1978). Consequently, the degree of salt tolerance between and within species can vary according to the criteria used for evaluation.

Since salt tolerance varies with ontogeny and development, it cannot be assumed that relative salt tolerance rankings developed during one stage of growth will correlate well with salt tolerance at other stages. Intervarietal variation in salt tolerance during germination in a number of cases varies with tolerance at other stages. Rates of germination as measured in saline filter paper or gels cannot be applied in the field, since many seeds that germinate under these conditions would be too weak to break a soil crust and establish themselves as viable plants. The rate of emergence usually is a better screening criterion. However, seedling vigor may be a discriminating factor in saline soils that form hard crusts.

Due to the differences in salt tolerance between stage of growth, Epstein et al. (1980) selected for tolerance by imposing salt stress over the entire growth cycle. However, if a constant salt concentration is used in this strategy, the degree of selection pressure varies with the stage of growth. For some species, independent selection at more than one stage may be appropriate. Lines with optimal tolerance at each stage of growth could be developed. Then, a crossing program could be designed to combine the lines into a single variety.

For plants, such as sugar beets, that are most sensitive to salinity during germination, selection for salt tolerance may be needed at only one stage of growth (Bernstein and Hayward 1958). For many crops, though, little information is available about salt sensitivities during development. Salt tolerance at specific stages of growth may sometimes be used to advantage. For instance, Shannon and Francois (1978), Mizrahi and Pasternak (1985), and Mizrahi et al. (1988) have shown that applying moderate salinity during the development of fruit can change the partitioning of photosynthates and improve soluble solids in melon and tomato. The small decreases in yield that can result from salinity are offset by the higher, more marketable quality of the fruit.

Whatever approach is used in breeding for salt tolerance, field evaluation must play an integral role. Although selection for tolerance can be based simply on a plant's response to external salt, factors such as soil structure can significantly modify that response. Indeed, in some cases, tolerance to poor soil structure may be as important as tolerance to high external salt concentrations (Levitt 1972, Rana 1985).

Some grain crops, such as sorghum, wheat, and barley, are extremely insensitive to relatively high concentrations of saline water applied during or just before anthesis (Maas et al. 1986, Maas and Poss 1988). Such tolerance could be exploited by substituting brackish water for irrigation water during the later stages of growth. This strategy has been used successfully for field and vegetable crops (Rhoades 1986, Grattan et al. 1987). Genetic variation for salt tolerance at specific stages of growth has been inadequately examined.

Screening for Physiological Factors Related to Salt Tolerance

Salt tolerance and salt resistance are often used synonymously. Another frequently used term is salt avoidance. In a given situation, the appropriate term to use depends on the plant's physiological response to salt stress. Avoidance means that growth is temporarily postponed or interrupted (Levitt 1972). Tolerance usually means that the plant can limit the effects of stress and continue to grow, although perhaps at a lower rate. Resistance usually applies to biotic rather than abiotic stresses and refers to active opposition in a biological sense. Since plant responses to salinity most often involve mechanisms for enduring the salt stress, salt tolerance is usually the most apt term.

Our understanding of the salt tolerance of a plant can be increased by understanding how growth and developmental processes are linked and finding the factor most limited by salt stress in this chain of events. The mechanisms for salt tolerance often depend on the anatomical and physiological complexity of the plant. This makes it difficult to find ways to increase salt tolerance greatly.

Several investigators have demonstrated salt tolerance mechanisms based on factors thought to limit growth under saline conditions, including ion accumulation (Rush and Epstein 1976, 1981b; Tal and Shannon 1983), ion exclusion (Abel 1969; Noble et al., 1984), compatible solute production (Grumet and Hanson 1986; Wyn Jones et al. 1977), pyramiding characters (Pasternak 1987; Yeo and Flowers 1983), late maturation (Bernal et al. 1974), and pollen sterility (Akbar and Yabuno 1977; Akbar et al. 1972). Pyramiding characters refers to building salt tolerance in an additive manner based on strengthening tolerance within lines that are already highly salt-tolerant. This technique can be used even when the physiological basis of salt tolerance is unknown (Ramage 1980). This chapter will examine a few physiological responses to salt that have been considered as selection criteria.

Ion Restriction. Flowers et al. (1977) and Harvey (1985) attributed salt sensitivity in some crops to the failure of plants to keep Na^+ and Cl^- out of the transpiration stream and, consequently, the cytoplasm of the aerial parts. Under salt stress, a plant must absorb nutrients and restrict the uptake of toxic ions at lower water potentials than usual. Munns and

Termaat (1986) divided salt stress into short- and long-term effects. Short-term effects take place over a matter of days. They involve decreased shoot growth, possibly as a result of the root response to water deficit. Long-term effects take place over weeks and result in maximum salt loads in fully expanded leaves and a reduction in photosynthetic activity. Flowers and Yeo (1986) noted that salt damage in the leaves of sensitive species can result from excess apoplastic ion concentrations or ion toxicity effects on metabolic processes in the symplast.

Plants that limit uptake of toxic ions and maintain normal ranges of nutrient ions could be more salt-tolerant than those that do not. Selective ion uptake mechanisms that can differentiate between chemically similar ions, such as Na^+ and K^+, could have adaptive value. These mechanisms probably operate in the membranes of tissues and various organelles throughout the plant (Bliss et al., 1984; Kuiper 1968). Breeding for efficient nutrient uptake or low ion accumulation under salt stress may be one of the simplest ways to improve salt tolerance in sensitive varieties of some species. This also may be accomplished by finding tolerance to the toxicity of a specific ion associated with salt stress.

Munns et al. (1988) concluded that high salt concentrations in the phloem of the salt-sensitive *Lupinus albus* do not directly relate to reduced growth or leaf injury but are a symptom of disrupted regulation of ion transport properties in the root. The passive root membrane filtration system used by mangroves undoubtedly is the most efficient system for restricting salt uptake. The grey mangrove, *Avicennia marina*, can exclude 90% of the salt in the medium surrounding its roots (Burchett et al. 1984). It has maximal growth at 25% seawater. Other species of mangrove can survive salt concentrations two to three times that of seawater (Clough 1984). The mangrove's system is unique; it is not found in other plants. Most species limit the uptake of salt into the transpiration stream by membrane-mediated compartmentalization in organelles (vacuoles) or tissues (see Chapters 6 and 13). Some species can rid themselves of ions through ion sequestering organelles (salt glands) or by storing salt in the root, old leaves, petioles, stems, or tracheids (Jacoby 1964). Salt restriction from the cytoplasm is incomplete, however. A plant will eventually succumb if its rate of growth is not high enough to prevent its salt storage sinks from filling up completely, allowing salt to then accumulate in sensitive tissues (Flowers and Yeo 1986). The genetic variations that may occur in each of these systems are numerous. Little research on the extent of that variation has been conducted.

Genetic differences result in differences in selective ion transport between varieties (Vose 1963; Epstein and Jefferies 1964). Gene differences may affect salt tolerance, as well as nutrient supply. Sometimes interactive effects exist. Salt sensitivities in barley (Greenway 1965), soybean (Abel and Mackenzie 1964), citrus (Cooper et al. 1951, Cooper and Gorton 1952) and grape (Bernstein et al. 1969; Downton 1977, 1985) relate to the plant's inability to regulate the transport of Cl^- adequately from the root to the shoot. In some plants, such as soybean, a single gene may be responsible. In others, such as citrus, multiple genes are involved and the response is complex. Distinguishing whether toxic effects are due to Na^+ or Cl^- sometimes is difficult, as concentrations of these ions often increase concomitantly. Woody species seem sensitive to Cl^-. Varieties of beans and corn that better regulate accumulation of Na^+ in the shoot can tolerate low salinity better than those which cannot (Richter and Marschner 1973; Jacoby and Ratner 1974). This, however, is a generalization. For example, trifoliate orange (*Poncirus trifoilata*) restricts the uptake of Na^+ at low salinities (Walker 1986). Each species should be examined individually to determine ion toxicity thresholds, ion restriction ability, and the capacity for ion selectivity.

Hannon and Barber (1972) found that salt-tolerant clones of *Festuca rubra* and *Agrostis stolonifera* restricted the shoot uptake of both Na^+ and Cl^-. Populations of *A. stolonifera* collected from maritime habitats had lower concentrations of Na^+ and higher concentrations of K^+ in the shoots than inland populations grown at the same salinity. Often, specific ecotypes have been found to be tolerant to salinity. Another closely related species, *A. tenuis Sibth.*, was sensitive to high concentrations of Mg^{2+} in seawater. Tolerant accessions of tall wheatgrass (*Elytrigia pontica*) limited the uptake of Na^+ and Cl^- into shoots more effectively than sensitive accessions (Shannon 1978). Hybridization between tolerant lines yielded progeny with improved tolerance. However, improvement in salt tolerance at this level does not correlate with differences in ion uptake or osmotic regulation (Weimberg and Shannon 1988).

Dvorak and his colleagues (1985, 1986) transferred salt tolerance from *Elytrigia pontica* and *E. elongata* to wheat by hybridizing the two grasses to wheat. Some of the derivatives, which varied in chromosome composition, had a better salt tolerance than wheat. Furthermore, amphidiploids showed tolerance to Na^+, K^+, Mg^{2+}, Cl^-, and SO_4^{2-}, as well as to sea water. Gorham et al. (1987) located genes that control K/Na discrimination in wheat on the long arm of chromosome 4D by using conventional genetic manipulation of chromosomes and chromosome fragments. They used disomic D-substitution lines of tetraploid wheat previously developed (Sears 1966) from crosses between tetraploid and hexaploid wheats with unbalanced A, B, and D genome sets.

The distinction of whether a line is salt-sensitive or salt-tolerant depends on its relationship to the breeding population's average response. Certain gene differences increase sensitivity to a specific ion or salt effect. These genotypes should be differentiated from those that confer greater salt tolerance than is found in the mean population. An example of salt sensitivity is the Jackson variety of soybean (Abel and MacKenzie 1964; Abel 1969). Similarly, an introduction of wheat, PI 94341, was found to be more sensitive to Na^+ than PI 178704 (Kingsbury et al. 1984; Kingsbury and Epstein 1986).

Ion Accumulation. Although restriction of ions to roots or shoots is one of the most frequently reported differences between salt-tolerant and sensitive varieties, halophytes take up very high concentrations of ions to cope with saline environments (Flowers et al. 1977). The accumulation of salt is thought to reduce the requirements for increased wall extensibility, leaf thickness, and water permeability that might otherwise be needed to maintain growth and turgor at low soil-water potentials. The wild tomato species, *Lycopersicon cheesmanii*, is thought to be more salt tolerant than the cultivated species as a result of its capacity to accumulate ions (Rush and Epstein 1981b). The aforementioned salt-tolerant Edkawy tomato also accumulates higher concentrations of Na^+ in leaf tissues than do the more salt-sensitive *L. esculentum* cultivars (Hashim et al. 1986).

As with salt restriction, salt accumulation within tissues is well-regulated and sequestered from cytosolic compartments containing the salt-sensitive metabolic machinery of the cell. Some halophytes can rid themselves of toxic ions through specialized organs, such as salt glands and bladders (Levitt 1972; Schirmer and Breckle 1982). In both glycophytes and halophytes, salt may preferentially accumulate in vacuoles, interstitial compartments, stems, and old leaves. The physical and genetic factors that influence ion compartmentation and distribution within plants are mostly unknown. Only a few species, such as sugar beet, are true halophytes. Transferring halophytism into glycophytic species may be impractical. However, developing wild halophytes into new and useful salt-tolerant crops appears promising.

Osmotic Adjustment. Many reviews of the physiological bases for salt effects have attempted to classify the various components of stress into osmotic, nutritional, and ionic effects. Specific tolerance characters may be associated with each of these components, and certain crops may be more sensitive to some of these stress components than others. For example, many temperate fruit crops suffer losses in growth and yield at external salt concentrations too low to exert significant water (osmotic) stress (Downton 1978). The relative sensitivity to these stress components may vary with external salt concentrations.

Osmotic adjustment is a decrease in plant osmotic potential through an increase in solute content (or a decrease in water content), which is in response to a decrease in external water potential. Morgan (1977) has noted that wheat genotypes have substantially different capacities for osmoregulation. However, some controversy exists as to whether any osmoregulation occurs in higher plants (Munns and Termaat 1986).

High humidities improve the tolerance of corn, bean, onion, radish, and barley, but not of cotton, wheat, and red beet (Gale et al. 1967, Hoffman et al. 1971, Hoffman and Rawlins 1971, Hoffman and Jobes 1978, Prisco and O'Leary 1973). This indicates that certain crops may benefit from selection pressures that improve their capacity to adjust osmotically or maintain more favorable water relations under salt stress (Tal and Gardi 1976; Shannon et al. 1987). Generally, poor osmotic adjustment leads to loss of turgor and closure of stomata, followed by reduced gas exchange and photosynthesis. Loss of turgor can also harm cell division and elongation.

The accumulation of organic solutes, such as sugars, proline, and glycinebetaine, improves salt tolerance by contributing to osmotic balance and preserving enzyme activity in the presence of toxic ions (Greenway and Munns 1980; Grumet et al. 1985; Tal et al. 1979). Rathert (1984) has noted that salinity causes greater increases of sucrose in the leaves of salt-sensitive species than in tolerant species. He has suggested that leaf sucrose and starch concentrations could be used as a selective index in screening for improved salt tolerance. High betaine genotypes of barley maintained lower solute potentials than near-isoline, low-betaine genotypes grown at the same salinities (Grumet and Hanson 1986). This also suggests that betaine could be used as a selection index for improved salt tolerance. However, these characters alone may prove to be inadequate criteria if other salt-tolerant characteristics are not maintained.

Other mechanisms that can prevent turgor loss through better water efficiency include increased leaf resistance, i.e., fewer stomates, increased mesophyll resistance, and increased cuticle thickness; a higher ratio of roots to shoots, and osmotic adaptation through ion accumulation and the production of compatible solutes. Each of these strategies can be accomplished in various ways, as is evident by the diversity within the plant kingdom. Likewise, most strategies are associated with some aspect detrimental to the maximum production of crops.

Most measurements of water relations are insufficiently accurate to screen for salt tolerance. Advances in instrumentation and a more thorough understanding of water relations mechanisms may eventually improve the breeder's ability to select genotypes. Characters available within the breeding population, the range of the genetic variability of each, and the energy cost associated with the character all need to be addressed.

Bioenergetics. Saline conditions force the plant to divert some of the energy expended for growth into increased maintenance. Thus, the efficiency with which the plant uses its resources of energy may be an important aspect of salt tolerance. Salinization of the rooting medium initially increases rates of respiration in many species (Nieman 1962, Livne and Levin 1967), but total respiration then decreases along with decreased photosynthesis. The latter directly relates to stomata closure induced by water stress (Schwartz and Gale 1981, 1983, 1984). Energy costs for cell maintenance increases with salinity. Active transport processes, which maintain ion compartmentation and fuel protein synthesis, are two of the most costly energy sinks in a plant (Penning de Vries 1975). Increased maintenance costs during salinity stress are directed toward ion regulation (compartmentation and transport of toxic ions), osmotic adjustment, and maintenance of membrane integrity (Yeo 1983). In certain cases, salt may induce *de novo* synthesis of proteins or lipids that replace or alter transport properties of membranes (Kuiper 1968; Bliss et al. 1984).

The physiological adaptations and adjustments to salinity that are most advantageous to the plant in terms of energy efficiency and the maintenance of high growth potential are unknown. Evidence indicates that previously stressed plants are more adapted to subsequent stresses. The combined stress from salinity and drought may be less detrimental to plant growth and carbon gain than the sum of the separate effects. Although salinity reduced the rates of leaf expansion in well-irrigated leaves of wheat maize and sorghum, it also permitted leaf expansion to continue at lower leaf-water potentials than under the stress of drought alone (Sepaskhah and Boersma 1979; Stark and Jarrell 1980; Richardson and McCree 1985). Evidence exists that stress from drought develops more slowly in salinized plants and that they can adjust osmotically to lower leaf water potentials (Richardson and McCree 1985).

The energy costs and consequences of the various physiological mechanisms of salt tolerance need to be investigated much more to select the most efficient strategies for a particular crop.

New Technologies for the Development of Salt-Tolerant Crops

Tissue Culture. The possibility of increasing salt tolerance by selecting undifferentiated cells in tissue culture has been explored for many species (Dix and Street 1975; Nabors et al. 1975, 1980; Croughan et al. 1978; Ben-Hayyim and Kochba 1983; Ben-Hayyim et al. 1985; Rangan and Vasil 1983; Pandey and Ganapathy 1984). Tissue cultures have great potential because large populations of cells can be maintained in precisely controlled environments. Mutagenesis, selection, haploid production, somatic hybridization, and transformation are relatively straightforward tissue-culture techniques. Major limitations are posed by the inability to regenerate many important species from cultured cells and the lack of understanding of the relationship between cellular and whole-plant response (Petolino and Collins 1984). The possibility of epigenetic change, or physiological adaptation, presents another limitation, as cells that physiologically adapt to gradually increasing concentrations of salt may be mistaken for cells in which genetic changes have occurred. Characteristics that can be selected from cell cultures include the ability to regulate or partition ions more efficiently, the production of organic solutes to improve osmoregulation under salt stress, and improved metabolic efficiency (Hasegawa et al. 1980; Ben-Hayyim and Kochba 1983, 1985; Binzel et al. 1983; Watad et al. 1983; Stavarek and Rains 1984).

Somaclonal variation (Larkin and Scowcroft 1981), a spontaneous variation arising from cultured cells, has both helped and hindered efforts to regenerate salt-tolerant clones. Somaclonal variation can cause loss of cellular tolerance during the regeneration process. Some investigators have advocated its use to induce variability in germplasm with little inherent variation in salt tolerance.

Molecular Biology. Recent advances in molecular biology have broadened the possibilities for manipulating genes at the level of cell cultures and higher units of organization. New or improved technologies for monoclonal antibodies, endonuclease digestion, cDNA libraries, purified protein isolation, and rapid amino-acid sequencing have made gene identification, isolation, and transformation realities. The use of these techniques to develop salt-tolerant crops is hampered by the fact that salt tolerance is controlled by complex multigenic traits and often is a composite response of the integrated biological system. With current technology, traits that involve whole-plant responses cannot be engineered. However, molecular biology can be used to monitor environmental effects at the level of the genome.

Stress-induced changes in gene expression can occur at the level of transcription or post-translation processing and transport. Ramagopal

(1987a) found that salt stress reduced total mRNA in barley roots by 20 to 30%. The stress induced 21 new RNA species and inhibited others, and unique changes occurred in salt-tolerant and salt-sensitive varieties. Proteins translated *in vivo* and characterized by 2-D gel electrophoresis produced 11 new proteins, not all of which corresponded to the newly-induced mRNA (Ramagopal 1987b). Salt-induced protein induction differed in root, shoot, and callus tissues (Ramagopal 1988). These data suggest that some of the expression for salt tolerance may be controlled at the transcription and post-transcription levels.

Salt-induced proteins have been described recently in Nicotiana (Ericson and Alfinito 1984; Singh et al. 1985; King et al. 1986), barley (Hurkman and Tanaka 1987; Ramagopal 1987b), and a hybrid between wheat and *Elytrigia* (Gulick and Dvorak 1987). Prisco (1971) described salt-induced proteins in germinating bean embryos. However, methods for tracing these proteins back to genes were then unavailable. Recently, King et al. (1988) used HPLC to isolate a salt-induced protein from tomato cell cultures. They sequenced the amino-terminal end of the NP 24 protein and synthesized a 29-base oligomer probe. They then probed a cDNA library from the cell culture with the oligomer and identified the DNA sequence specific to the NP 24 protein. Further work led to the development of a polyclonal antibody to NP 24 that was used to immunocytochemically locate the protein as a component of the cytoplasm. An immunologically identical protein, osmotin, is found in dense bodies within vacuoles and as diffused particles throughout the cytoplasm and tonoplast of salt-stressed tobacco cells (Singh et al. 1987).

As new and improved technologies are developed, the molecular approach will increasingly contribute to our knowledge of genetic and physiological events associated with salt stress. Genes for salt tolerance may someday be transferred to salt-sensitive species by direct injection of DNA or through bacterial or viral plasmid vectors. Because the number and location of specific genes affecting salt tolerance is unknown, techniques that transfer substantial portions of genetic material from one species to another may prove to be the most useful, particularly for hybridizing related species. Electrofusion, for example, can overcome natural interspecific barriers to hybridization by fusing cells from species of different families.

Complete transfer of genetic material between species will require a more complete understanding of the way in which genes from different species interact and the ways in which fused cells must be cultured and plants regenerated. Certainly, we are on the threshold of a new and exciting era.

CONCLUSION

Differences between halophytes and glycophytes exemplify the variability in salt tolerance within the higher plants. These differences have evolved over hundreds of thousands of years. It may be unrealistic to expect to develop halophytic characters in some of our most salt-sensitive crops. Nonetheless, if we commit ourselves to developing an understanding of the physiological mechanisms by which plants cope with salt, the energetic efficiencies associated with those mechanisms, and the coordination of salt-stress-related characters, dramatic improvement in the salt tolerance of many crops may be seen within 10 to 15 years.

REFERENCES

Abel, G.H. (1969). "Inheritance of the capacity for chloride inclusion and exclusion in soybeans." *Crop Sci.* 9:697–699.

Abel, G.H., and Mackenzie, A.J. (1964). "Salt tolerance of soybean varieties during germination and later growth." *Crop Sci.* 4:157–160.

Akbar, M., Yabuno, T., and Nakao, S. (1972). "Breeding for salt-resistent varieties of rice, I: Variability for salt tolerance among some rice varieties." *Japan J. Breed.* 22:272–284.

Akbar, M., and Yabuno, T. (1977). "Breeding for salt-resistent varieties of rice, IV: Inheritance of delayed-type panicle sterility induced by salinity." *Japan J. Breed.* 27:237–240.

Ayers, A.D., Brown, J.W., and Wadleigh, L. (1952). "Salt tolerance of barley and wheat in soil plots receiving several salinization regimes." *Argon. J.* 44:307–310.

Beatty, K.D., and Ehlig, C.F. (1973). "A technique for testing and selecting for salt tolerance in sugarbeet." *J. Amer. Soc. Sugar Beet Technol.* 17:295–299.

Ben-Hayyim, G., and Kochba, J. (1982). "Growth characteristics and stability of tolerance of citrus callus cells subjected to NaCl stress." *Plant Sci. Lett.* 27:87–94.

Ben-Hayyim, G., and Kochba, J. (1983). "Aspects of salt tolerance in a NaCl-selected stable cells line of Citrus sinensis." *Plant Physiol.* 72: 685–690.

Ben-Hayyim, G., Spiegel-Roy, P., and Neumann, H. (1985). "Relation between ion accumulation of salt-sensitive and isolated stable salt-tolerant cell lines of Citrus aurantium." *Plant Physiol.* 78:144–148.

Bernal, C.T., Bingham, F.T., and Oertli, J. (1974). "Salt tolerance of Mexican wheat, II: Relation to variable sodium chloride and length of growing season." *Soil Sci. Soc. Am. Proc.* 38:777–780.

Bernstein, L., and Francois, L.E. (1973). "Comparisons of drip, furrow and sprinkler irrigation." *Soil Sci.* 115:73–86.

Bernstein, L., and Hayward, H.E. (1958). "Physiology of salt tolerance." *Annu. Rev. Plant Physiol.* 9:25–46.

Bernstein, L., Ehlig, C.F., and Clark, R.A. (1969). "Effect of grape rootstocks on chloride accumulation in leaves." *J. Amer. Soc. Hortic. Sci.* 94:584–590.

Binzel, M.L., Hasegawa, P.M., Bressan, R.A., Handa, S., and Handa, A.K. (1983). "Adaptive responses of cultured tobacco cells to NaCl." *Plant Physiol.* 72:S136.

Bliss, R.D., Platt-Aloia, K.A., Thompson, W. W. (1984). "Changes in plasmalemma organization in cowpea radicle during imbibition in water and NaCl solutions." *Plant Cell Environ.* 7:606–606.

Blum, A. (1984). "Breeding crop varieties for stress environments." *CRC Critical Rev. Plant Sci.* 2(3):119–237.

Blum, A. (1988). *Plant Breeding for Stress Environments.* CRC Press Inc., Boca Raton, Fla. 223 p.

Burchett, M.D., Field, C.D., and Pulkownik, A. (1984). "Salinity, growth and root respiration in the grey mangrove, Avicennia marina." *Physiol. Plant.* 60:113–118.

Christiansen, M.N., and Lewis, C.F. eds. (1982). *Breeding Plants for Less Favorable Environments.* Wiley-Interscience, New York. 459 p.

Clough, B.F. (1984). "Growth and salt balance of the mangroves Avicennia marina (Forsk.) Vierh. and Rhizophora stylosa Griff. in relation to salinity." *Aust. J. Plant Physiol.* 11:419–430.

Collins, G.B., and Petolino, J.G., eds. (1984). *Applications of Genetic Engineering to Crop Improvement.* Nijhoff/Junk, Dordrecht. 604 p.

Cooper, W.C. (1951). "Salt tolerance of avocados on various rootstocks." Texas Avocado Society Yearbook, 24–28, Weslaco, Tex.

Cooper, W.C., and Gorton B.S. (1952). "Toxicity and accumulation of chloride salts in cirtus on various rootstock." *Amer. Soc. Hortic. Sci. Proc.* 59:143–146.

Cooper, W.C., Gorton, B.S., and Edwards C. (1951). "Salt tolerance of various citrus rootstocks." *Rio Grande Valley Hort. Inst. Proc.* 5:46–52.

Cordukes, W.E. (1981). "Saltol, a salt-tolerant red fescue." *Can J. Plant Sci.* 61:761–764.

Croughan, T.P., Stavarek, S.J., and Rains, D.W. (1978). "Selection of a NaCl tolerant line of cultured alfalfa cells." *Crop Sci.* 18:959–963.

Day, A.D. (1987). "Registration of Arizona 8601 maize germplasm for saline environments." *Crop Sci.* 27:1096.

Dewey, D.R. (1962). "Breeding crested wheatgrass for salt tolerance." *Crop Sci.* 2:403–407.

Dix, P.J., and Street, H.E. (1975). "Sodium chloride-resistant cultured cell lines from Nicotiana sylvestris Spec. & Comes and Capsicum annum L." *Plant Sci. Lett.* 5:231–237.

Dobrenz, A.K., Schonhorst, M.H., Stone, J.E., Thomspon, R.K., Allen, S., and Smith D. (1983). "AZ-Germ Salt-1 nondormant alfalfa germplasm." *Crop Sci.* 23:807.

Downton, W.J.S. (1977). "Photosynthesis in salt stressed grapevines." *Aust. J. Plant Physiol.* 4:183–192.

Downton, W.J.S. (1978). "Growth and flowering in salt stressed avocado trees." *Aust. J. Agric. Res.* 29:523–534.

Downton, W.J.S. (1984). "Salt tolerance of food crops: prospectives for improvements." *CRC Critical Rev. Plant Sci.* 1:183–201.

Downton, W.J.S. (1985). "Growth and mineral composition of the sultana grapevine as influenced by salinity and rootstock." *Aust. J. Agric. Res.* 36:425–434.

Duvick, D.N., Kleese, R.A., and Frey, N.M. (1981). "Breeding for tolerance of nutrient imbalances and constraints to growth in acid, alkaline and saline soils." *J. Plant Nutr.* 4.

Dvorak, J., and Ross, K. (1986). "Expression of tolerance of Na^+, K^+, Mg^{2+}, Cl^-, and SO_4^{2-} ions and sea water in the amphiploid of Triticum aestivum X Elytrigia elongata." *Crop Sci.* 26:658–660.

Dvorak, J., Ross, K., and Medlinger, S. (1985). "Transfer of salt tolerance from elytrigia ponica (Podp.) holub to wheat by the addition of an incomplete elytrigia genome." *Crop Sci.* 25:306–309.

Epstein, E. (1985). "Salt tolerant crops: origins, development, and prospects of the concept." *Plant Soil* 89:187–198.

Epstein, E., and Jefferies, R.L. (1964). "The genetic basis of selective ion transport in plants." *Annu. Rev. Plant Physiol.* 15:169–184.

Epstein, E., Norlyn, J.D., Rush, D.W., Kingsbury, R.W., Kelly, D.B., and Cunningham, G.A. (1980). "Saline culture of crops; a genetic approach." *Science* 210:399–404.

Epstein, E., and Rains, D.W. (1987). *Advances in Salt Tolerance in Plant and Soil* 99:17–29.

Ericson, M.C., and Alfinito, S.H. (1984). "Proteins produced during salt stress in tobacco cell culture." *Plant Physiol.* 74:506–509.

Finlay, K.W., and Wilkinson, G.N. (1963). "The analysis of adaptation in a plant breeding programme." *Aust. J. Agric. Res.* 14:742–754.

Flowers, T.J., Troke, P.F., and Yeo, A.R. (1977). "The mechanism of salt tolerance in halophytes." *Annu. Rev. Plant Physiol.* 28:89–121.

Flowers, T.J., and Yeo, A.R. (1986). "Ion relations of plants under drought and salinity." *Aust. J. Plant Physiol.* 13:75–91.

Francois, L.E., and Maas, E.V. (1978). "Plant responses to salinity: an indexed bibliography." USDA, ARM-W-6, October 1978, 291 p.

Francois, L.E., and Maas, E.V. (1985). "Plant responses to salinity: a supplement to an indexed bibliography." USDA,ARS-24, 174 p.

Fredrickson, D.L., and Epstein, E. (1975). "Genetically controlled differences in salt tolerance of two species of the tomato." *Plant Physiol.* 56 (2 Suppl.), 4.

Frey, K.J. (1964). "Adaptation reaction of oat strains selected under stress and non-stress environmental conditions." *Crop Sci.* 4:55–58.

Furr, J.R., Carpenter, J.B., and Hewitt, A.A. (1963). "Breeding new varieties of citrus fruits and rootstocks for the Southwest." *J. Rio Grande Valley Hort. Soc.* 17:90–107.

Furr, J.R., and Ream, C.L. (1968). "Breeding and testing rootstocks for salt tolerance." *Calif Citrogr.* 54:30–35.

Gad El-Hak, M. (1966). "Breeding a new variety of rice adapted to recently reclaimed saline soils." *Agric. Res. Rev.* (Cairo) 44:1–13.

Gale, J., Kohl, H.C., and Hagan, R.M. (1967). "Changes in the water balance and photosynthesis of onion, bean and cotton plants under saline conditions." *Physiol. Plant.* 20:408–420.

Gorham, J., Hardy, C., Wyn Jones, R.G., Joppa, L.R., and Law, C.N. (1987). "Chromosomal location of the K/Na discrimination character in the D genome of wheat." *Theor. Appl. Genet.* 74:584–588.

Grattan, S.R., Shennan, C., May, D.M., Mitchell, J.P., and Burau, R.G. (1987). "Use of drainage water for irrigation of melons and tomatoes." *Calif. Agric.* 41:24–26.

Greenway, H. (1965). "Plant responses to saline substrates. VII. growth and ion uptake throughout plant development in two varieties of Hordeum vulgare." *Aust. J. Biol. Sci.* 18:763–779.

Greenway, H., and Andrew, W.D. (1962). "A screening technique to predict field behavior of medics in saline soil." *Aust. J. Exp. Agric. Anim. Husb.* 2:234–235.

Greenway, H., and Munns, R. (1980). "Mechanisms of salt tolerance in non-halophytes." *Annu. Rev. Plant Physiol.* 31:149–190.

Grumet, R., and Hanson, A.D. (1986). "Genetic evidence for an osmoregulator function of glycinebetaine accumulation in barley." *Aust. J. Plant Physiol.* 13:353–364.

Grumet, R., Isleib, T.G., and Hanson, A D. (1985). "Genetic control of glycinebetaine level in barley." *Crop Sci.* 25:618–622.

Gulick, P., and Dvorak, J. (1987). "Gene induction and repression by salt treatment in roots of the salinity-sensitive Chinese spring wheat and the salinity-tolerant Chinese spring x elytrigia elongata amphiploid. *Proc. Natl. Acad. Sci. USA* 84:99–103.

Gupta, U.S. (1978). "Genetic and breeding aspects of salt tolerance in plants." *Curr. Agric.* 2(1–2):1–6.

Hannon, N.J., and Barber, H.N. (1972). "The mechanism of salt tolerance in naturally selected populations of grasses." *Search* 3:259–260.

Harvey, D.M.R. (1985). "The effects of salinity on ion concentrations within the root cells of Zea mays L." *Planta* 165:242–248.

Hasegawa, P.M., Bressan, R.A., and Handa, A.K. (1980). "Growth characteristics of NaCl-selected and non-selected cells of Nicotiana tabacum L." *Plant Cell Physiol.* 21:1347–1355.

Hashim, M., Jones, R.A., and A.S. El-Beltagy. (1986). Comparative responses to high salinity between salt-sensitive and salt-tolerant genotypes of the tomato." *Acta Hortic.* 190:533–543.

Hassan, A.A., and Desouki, I. A.M. (1982). "Tomato evaluation and selection for sodium chloride tolerance." *Egypt. J. Hort.* 9:153–162.

Hoffman, G.J. (1980). "Guidelines for reclamation of salt-affected soils." Proc. of Inter-American Salinity and Water Mgmt. Tech. Conf., Juarez, Mexico. Dec. 11–12, pp. 49–64.

Hoffman, G.J., and Jobes, J.A. (1978). "Growth and water relations of cereal crops as influenced by salinity and relative humidity." *Agron. J.* 20:765–769.

Hoffman, G.J., and Rawlins, S.L. (1971). "Growth and water potential of root crops as influenced by salinity and relative humidity." *Agron. J.* 63:877–880.

Hoffman, G.J., Rawlins, S.L., Garber, M.J., and Cullen, E.M. (1971). "Water relations and growth of cotton as influenced by salinity and relative humidity." *Agron. J.* 63:822–826.

Hollaender, A., Aller, J.C., Epstein, E., San Pietro, A., and Zaborsky, O.R., eds. (1979). *The Biosaline Concept: An Approach to the Utilization of Underexploited Resources.* Environ. Sci. Res. Vol. 14, Plenum Press, NY; 391 p.

Hunt, O J. (1965). "Salt tolerance in intermediate wheatgrass." *Crop Sci.* 5:407–409.

Hurkman, W.J., and Tanaka, C.K. (1987). "The effects of salt on the pattern of protein synthesis in barley roots." *Plant Physiol.* 83:517–524.

International Rice Research Institute (1977). "Screening rice for tolerance to mineral stresses." *Int. Rice Res. Inst. Bulletin,* Los Banos, Laguna Philippines, pp. 1–35.

Jacoby, B. (1964). "Function of bean roots and stems in sodium retention." *Plant Physiol.* 39:445–449.

Jacoby, B., and Ratner, A. (1974). "Mechanism of sodium exclusion in bean and corn plants—a reevaluation." *Plant Analysis and Fertilizer Problems* (J. Wehrmann, ed.) pp. 175–184. German Soc. of Plant Nutrition, Hannover.

Jones, R.A. (1987). "Genetic advances in salt tolerance." *Tomato Biotechnology* (D. J. Evans and R. A. Jones, eds.) Alan R. Liss, Inc., New York. pp. 125–137.

Jones, R.A. and Qualset, C.O. (1984). "Breeding crops for environmental stress tolerance." *Applications of Genetic Engineering to Crop Improvement* (G. B. Collins and J. G. Petolino, eds.) M. Nijhoff/W. Junk, Dordrecht. pp. 305–340.

Jung, G.A., ed. (1978). *Crop Tolerance to Suboptimal Land Conditions.* Amer. Soc. Agron., Crop Sci. Soc. Amer., and Soil Sci. Soc. Amer., Madison, Wisc., 343 p.

King, G.J., Hussey, C.E. Jr., and Turner, V.A. (1986). "A protein induced by NaCl in suspension cultures of Nicotiana tabacum accumulates in whole plant roots." *Plant Mol. Biol.* 7:441–449.

King, G.J., Turner, V.A., Hussey, C.E. Jr., Wurtele, E.S., and Lee, M.S. (1988). "Isolation and characterization of a tomato cDNA clone that codes for a salt-induced protein." *Plant Mol. Biol.* 10:401–412.

Kingsbury, R.W. and Epstein, E. (1984). "Selection for salt-resistant spring wheat." *Crop Sci.* 24:310–315.

Kingsbury, R.W., Epstein, E., and Pearcy, R.W. (1984). "Physiological responses to salinity in selected lines of wheat." *Plant Physiol.* 74: 417–423.

Kingsbury, R.W., and Epstein, E. (1986). "Salt sensitivity in wheat. A case for specific ion toxicity." *Plant Physiol.* 80:651–654.

Kuiper, P.J.C. (1968). "Lipids in grape roots in relation to chloride transport." *Plant Physiol.* 43:1367–1371.

Larkin, P.J., and Scowcroft, W.R. (1981). "Somaclonal variation—a novel source of variability from cell cultures for plant improvement." *Theor. Appl. Genet.* 60:197–214.

Lauchli, A. (1976). "Genotypic variation in transport." *Encyclopedia of Plant Physiology. New Series. Vol. 2. Part B: Transport in Plants* (U. Luttge and M. G. Pittman, eds.), Springer-Verlag, Berlin. pp. 372–393.

Levitt, J. (1972). Responses of Plants to Environmental Stresses. Academic Press, New York. 697 p.

Livne, A., and Levin, N. (1967). Tissue respiration and mitochonrial oxidative phosphorylation of NaCl-treated pea seedlings." *Plant Physiol.* 42:407–414.

Maas, E.V. (1985). "Crop tolerance to saline sprinkling waters." *Plant and Soil* 89:273–284.

Maas, E.V. (1987). "Salt tolerance of plants." *Handbook of Plant Science in Agriculture.* B. R. Christie, ed. CRC Press Inc., Vol. 2, pp. 57–75.

Maas, E.V., and Hoffman, G.J. (1977). "Crop salt tolerance—current assessment." *J. Irrig. and Drainage Div., ASCE* 103(IR2), pp. 115–134.

Maas, E.B., Poss, J.A., and Hoffman G.J. (1986). "Salinity sensitivity of sorghum at three growth stages." *Irrig. Sci.* 7:1–11.

Maas, E.V., and Poss, J.A. (1988). "Salt sensitivity of wheat at various growth stages." *Irrig. Sci.* 10:29–40.

Marr, P.D. (1967). "The social context of irrigation." *Irrigation of Agricultural Lands* (R. M. Hagan, H. R. Haise and T. W. Edminster, eds.), Amer. Soc. Agron., Madison, Wisc. pp. 12–23.

Mizrahi, Y., and Pasternak, D. (1985). "Effect of salinity on quality of various agricultural crops." *Biosalinity in Action: Bioproduction with Sa-*

line Water (D. Pasternak and A. San Pietro, eds.), Martinus Nijhoff, Dordrecht. pp. 301–307.

Mizrahi, Y., Taleisnik, E., Kagan-Zur, V., Zohar, Y., Offenbach, R., Matan, E., and Golan, R. (1988). "A saline irrigation regime for improving tomato fruit quality without reducing yield." *J. Amer. Soc. Hort. Sci.* 113(2):202–105.

Morgan, J.M. (1977). "Differences in osmoregulation between wheat genotypes." *Nature* 270:234–235.

Munns, R., and Termaat, A. (1986). "Whole-plant responses to salinity." *Aust. J. Plant Physiol.* 13:143–160.

Munns, R., Tonnet, M.L., Shennan, C., and Gardner, P.A. (1988). "Effect of high external NaCl concentration on ion transport within the shoot of Lupinus albus, II: Ions in phloem sap." *Plant Cell Environ.* 11:283–289.

Nabors, M.W., Daniels, A., Nadolny, L., and Brown, C. (1975). "Sodium chloride-tolerant lines of tobacco cells." *Plant Sci. Lett.* 4: 155–159.

Nabors, M.W., Gibbs, S.E. Bernstein, C.S., and Meis, M.E. (1980). "NaCl-tolerant tobacco plants from cultured cells." *Z. Pflanzenphysiol.* 97:13–17.

Nieman, R.H. (1962). "Some effects of sodium chloride on growth, photosynthesis, and respiration of twelve crop plants." *Bot. Gaz.* 123: 279–285.

Nieman, R.H. and Shannon, M.C. (1976). "Screening plants for salinity tolerance." *Plant Adaptation to Mineral Stress in Problem Soils* (M. J. Wright, ed.) pp. 359–367. A Special Publ. of Cornell Univ. Agric. Exper. Sta., Cornell Univ., Ithaca, NY.

Noble, C.L. (1983). "The potential for breeding salt-tolerant plants." *Proc. R. Soc. Vict.* 95(3):133–138.

Noble, C.L., Holloran, G.M., and West, D.W. (1984). "Identification and selection for salt tolerance in lucerne (Medicago sativa L.)." *Aust. J. Agric. Res.* 35:239–252.

Noble, C.L., and Shannon, M.C. (1988). "Salt tolerance selection of forage legumes using physiological criteria." *Proc. 1st Int. Congress Pl. Physiol.*, Soc. Pl. Physiol. Biochem., Int. Assoc. Plant. Physiol. In press.

Norlyn, J.D. (1980). "Breeding salt tolerant plants." *Genetic Engineering of Osmoregulation* (D.W. Rains, R. C. Valentine and A. Hollander, eds). Basic Life Sci. 14, Plenum Press, New York. pp. 293–309.

Pandey, R., and Ganapathy, P.S. (1984). "Isolation of sodium-tolerant callus line of Cicer arietinum L. cvBG-203" *Plant Cell Reports* 3:45–47.

Pasternak, D. (1987). "Salt tolerance and crop production—a comprehensive approach." *Ann. Rev. Phytopathol.* 25:271–291.

Pasternak, D., and San Pietro, A., eds. (1985). *Biosalinity in Action: Bioproduction with Saline Water.* M. Nijhoff, Dordrecht. 369 p.

Pasternak, D., Twersky, M., and De Malach, Y. (1979). "Salt resistance in agricultural crops." *Stress Physiology in Crop Plants,* (H. W. Mussel, R. C. Staples, eds.) Wiley, New York. pp. 127–142.

Penning De Vries, F.W.T. (1975). "The cost of maintenance processes in plant cells." *Ann. Bot.* 39:77–92.

Petolino, J.F., and Collins, G.B. (1984). "Applications of genetic engineering to crop improvement." *Application of Genetic Engineering to Crop Improvement,* (G.B. Collins, J.H.G. Petolino, eds.) Nijhoff/W. Junk, Dordrecht. pp. 341–390.

Prisco, J.T. (1971). "Polyacrylamide gel electrophoresis of soluble proteins of salt and water stressed embryo-axis of Phaseolus vulgaris L. seeds during germination." *Cienc. Agron.* 1:47–50.

Prisco, J.T., and O'Leary, J.W. (1973). "The effects of humidity and cytokinin on growth and water relations of salt-stressed bean plants." *Plant Soil* 39:263–276.

Qureshi, R.H., Muhammed, S., and Aslam, M., eds. (1981) *Proc. of the Workshop/Seminar on Membrane Biophysics and Development of Salt Tolerant Plants.* Univ. Agric. Press Faisalabad; 252 p.

Rains, D W., ed. (1981). *A Conference on Biosalinity: The Problem of Salinity in Agriculture.* Univ. Calif., Davis, Calif. 216 p.

Rains, D.W., Valentine, R.C., and Hollaender, A., eds. (1980). *Genetic Engineering of Osmoregulation: Impact on Plant Productivity for Food, Chemicals and Energy.* Basic Life Sci., Vol 14. Plenum Press, New York. 381 p.

Ramage, R.T. (1980). "Genetics methods to breed salt tolerance in plants." *Genetic Engineering of Osmoregulation: Impact on Plant Productivity for Food, Chemicals and Energy* (D. W. Rains, R. C. Valentine and A. Hollaender, eds.) pp. 311–318. Basic Life Sci., Vol. 14. Plenum Press, New York.

Ramagopal, S. (1987a). "Differential mRNA transcription during salinity stress in barley." *Proc. Natl. Acad. Sci. USA* 84:94–98.

Ramagopal, S. (1987b). "Salinity stress induced Tissue-specific proteins in barley seedlings." *Plant Physiol.* 84:324–331.

Ramagopal, S. (1988). "Sodium chloride effects on dedifferentiation and protein synthesis in root meristem cultures of two contrasting barley genotypes." *J. Plant Physiol.* 132:245–249.

Rana, R.S. (1985). "Breeding for salt resistance: Concept and strategy." *Internat. J. Trop. Agri.* III(4):236–254.

Rangan, T.S., and Vasil, I.K. (1983). "Sodium chloride tolerant embryogenic cell lines of Pennisetum americanum (L.) K. Schum." *Annu. of Bot.* 52:59–64.

Rathert, G. (1984). "Sucrose and starch content of plant parts as a possible indicator for salt tolerance of crops." *Aust. J. Plant Physiol.* 11: 491–495.

Rhoades, J.D. (1986). "Use of saline water for irrigation." Nat'l Water Res. Inst., Ontario, Canada, special issue bulletin, Water Quality, Burlington, Ont.

Richards, R.A. (1983). "Should selection for yield in saline regions be made on saline or non-saline soils?" *Euphytica* 32:431–438.

Richardson, S.G., and McCree, K.J. (1985). "Carbon balance and water relations of sorghum exposed to salt and water stress." *Plant Physiol.* 79:1015–1020.

Richter, C., and Marschner, N. (1973). "Turnover of potassium in different root zones of corn seedlings." *Z. Pflanzenphysiol.* 70(3):211–221.

Rosielle, A.A., and Hamblin J. (1981). "Theoretical aspects of selection for yield in stress and non-stress environments." *Crop Sci.* 21:943–946.

Rush, D.W., and Epstein, E. (1976). "Genotypic responses to salinity:

Differences between salt-sensitive and salt-tolerant genotypes of the tomato." *Plant Physiol.* 57:162–166.
Rush, D.W., and Epstein, E. (1981a). "Breeding and selection for salt tolerance by the incorporation of wild germplasm into a domesticated tomato." *J. Amer. Soc. Hort. Sci.* 106:669–670.
Rush, D.W., and Epstein, E. (1981b). "Comparative studies on the sodium, potassium and chloride relations of a wild halophytic and a domestic salt-sensitive tomato species." *Plant Physiol.* 68:1308–1313.
Sacher, R.F., Staples, R.C., and Robinson, R.W. (1982). "Saline tolerance in hybrids of Lycopersicon esculentum x Solanum penellii and selected breeding lines." *Biosaline Research: A Look to the Future,* (A. San Pietro, ed.), Plenum Press, New York. pp. 325–336.
San Pietro, A., ed. (1982). *Biosaline Research: A Look to the Future.* Envir. Sci. Res Vol. 23, Plenum Press, New York. 578 p.
Sears, E.R. (1966). "Nullisomic-tetrasomic combinations in hexaploid wheat." *Chromosome Manipulations and Plant Genetics,* (R. Riley, K.R. Lewis, eds.) Oliver and Boyd, Edinburgh. pp. 29–45.
Sepaskhah, A.R., and Boersma, L. (1979). "Elongation of wheat leaves exposed to several levels of matric potential and NaCl induced osmotic potential of soil water." *Agron. J.* 71:848–852.
Schirmer, U., and Breckle, S W. (1982). "The role of bladders for salt removal in some Chenopodiaceae (mainly Atriplex species)." *Contributions to the Ecology of Halophytes,* (D.N. Sen, K.S. Rajpurohit, eds.), Junk, the Hague, the Netherlands. pp. 215–232.
Schwartz, M., and Gale, J. (1981). "Maintenance respiration and carbon balance of plants at low levels of sodium chloride salinity." *J. Exp. Bot.* 32:933–941.
Schwartz, M., and Gale, J. (1983). "The effect of heat and salinity stress on the carbon balance of Xanthium strumarium." Effects of Stress in Photosynthesis, (R. Marcelle et al., eds.), Nijhus and Junk, the Netherlands. pp 325–331.
Schwartz, M., and Gale, J. (1984). "Growth response to salinity at high levels of carbon dioxide." *J. Exp. Bot.* 35:193–196.
Shannon, M.C. (1978). "Testing salt tolerance variability among tall wheatgrass populations." *Agron. J.* 70:719–722.
Shannon, M.C. (1980). "Differences in salt tolerance within Empire lettuce." *J. Amer. Soc. Hort. Sci.* 105(6):944–947.
Shannon, M.C. (1982). "Genetics of salt tolerance: new challenges." *Biosaline Research: A Look Into the Future.* (A. San Pietro, ed.) Plenum Press, New York. pp. 271–282.
Shannon, M.C. (1984). "Salinity Tolerance in Plant Strategies for Crop Improvement." *Breeding Selection and the Genetics of Salt Tolerance.* (N.R. Staples and G.H. Toenniessen, eds.) Wiley-International, N.Y. pp. 231–254.
Shannon, M.C. (1985). "Principles and strategies in breeding for higher salt tolerance." *Plant and Soil* 89:227–241.
Shannon, M.C., and Francois, L.E. (1978). "Salt tolerance of three muskmelon cultivars." *J. Amer. Soc. Hort. Sci.* 103:127–130.
Shannon, M.C., McCreight, C., and Draper, J.H. (1983). "Screening tests for salt tolerance in lettuce." *J. Amer. Soc. Hort. Sci.* 108:225–230.
Shannon, M.C., Gronwald, J.W., and Tal, M. (1987). "Effects of salinity

on growth and accumulation of organic and inorganic ions in cultivated and wild tomato species." *J. Amer. Soc. Hort. Sci.* 112:416–423.

Singh, N.K., Bracker, C.A., Hasegawa, P.M., Handa, A.K. Buckel, S., Hermodson, M.A., Pfankoch, E., Regnier, F.E., and Bressan, R.A. (1987). "Characterization of Osmotin." *Plant Physiol.* 85:529–536.

Singh, N.K., Handa, A.K., Hasegawa, P.M., and Bressan, R.A. (1985). "Proteins associated with adaptation of cultured tobacco cells to NaCl." *Plant Physiol.* 79:126–137.

Staples, R.C., Toenniessen, G.H., eds. (1984). *Salinity Tolerance in Plants: Strategies for Crop Improvement.* Wiley-Interscience, New York. 443 p.

Stark, J.C., and Jarrell, W.M. (1980). "Salinity-induced modifications in the response of maize to water deficits." *Agron. J.* 72:745–748.

Stavarek, S.J., and Rains, D.W. (1984). "The development of tolerance to mineral stress." *Hort.Science* 19:377–382.

Sykes, S.R. (1985). "A glasshouse screening procedure for identifying citrus hybrids which restrict chloride accumulation in shoot tissues." *Aust. J. Agric. Res.* 36:779–789.

Tal, M. (1985). "Genetics of salt tolerance in higher plants: theoretical and practical considerations." *Plant and Soil* 89:199–226.

Tal, M., and Gardi, I. (1976). "Physiology of polyploid plants: water balance in autotetraploid and diploid tomato under low and high salinity." *Physiol. Plant* 38:257–261.

Tal, M., and Shannon, M.C. (1983). "Salt tolerance in the wild relatives of the cultivated tomato: responses of *Lycopersicon esculentum L. cheesmanii, L. peruvianum, solanum pennellii* and F_1 hybrids to high salinity." *Aust. J. Plant Physiol.* 10:109–117.

Tal, M., Katz, A., Heikin, H., and Dehan, K. (1979). "Salt tolerance in the wild relatives of the cultivated tomato: *proline accumulation in lycopersicon esculentum, L. peruvianum* and *solanum pennellii* under NaCl salinity." *New Phytol.* 82:349–355.

Turner, N.C., and Passiouri, J.B., eds. (1986). *Plant Growth, Drought and Salinity.* CSIRO Editorial and Publ., Melbourne. 201 p.

van Genuchten, M.Th., and Hoffman, G.J. (1984). "Analysis of crop salt tolerance data." *Soil Salinity under Irrigation— Process and Management.* (I. Shainberg, J. Shalhevet, eds.), Ecological studies: 51, Springer-Verlag, New York. pp. 258–271.

Vose, P.B. (1963). "Varietal differences in plant nutrition." *Herbage Abst.* 33:1–13.

Walker, R.R. (1986). "Sodium exclusion and potassium-sodium selectivity in salt-treated trifoliate orange (Poncirus trifoliata) and Cleopatra mandarin (Citrus reticulata) plants." *Aust. J. Plant Physiol.* 13:293–303.

Watad, Abd-Elrahem A., Reinhold, L., and Lerner, H.R. (1983). "Comparison between a stable NaCl-selected Nicotiana cell line and the wild type." *Plant Physiol* 73:624–629.

Weimberg, R., and Shannon, M.C. (1988). "Vigor and salt tolerance in three lines of tall wheatgrass." *Physiol. Plant.* 73:232–237.

West, D.W. (1978). "Water use and sodium chloride uptake by apple trees. II. The response to soil oxygen deficiency." *Plant and Soil* 50: 51–65.

West, D.W., and Taylor, J.A. (1984). "Response of six grape cultivars to the combined effects of high salinity and rootzone waterlogging." *J. Amer. Soc. Hort. Sci.* 109:844–851.

Wright, M.J. ed. (1976). *Plant Adaptation to Mineral Stress in Problem Soils.* A Special Publ. of Cornell Univ. Agric. Exper. Sta., Cornell Univ., Ithaca, New York. 420 p.

Wyn Jones, R.G., Storey, R., Leigh, R.A., Ahmad, N., and Pollard A. (1977). "A hypothesis on cytoplasmic osmoregulation." *Regulation of Cell Membrane Activities in Plants* (E. Marre and O. Ciferri, eds.), Elsevier Press, Amsterdam. pp 121–136.

Yeo, A.R. (1983). "Salinity resistance: physiologies and prices." *Physiol. Plant* 58:214–222.

Yeo, A.R., and Flowers, T.J. (1983). "Varietal differences in the toxicity of sodium ions in rice leaves." *Physiol. Plant.* 59:189–195.

PART FOUR—SAMPLING, MONITORING, AND MEASUREMENT

CHAPTER 9

FIELD SAMPLING OF SOIL, WATER, AND PLANTS[a]

INTRODUCTION

Diagnostic information must be obtained to assess, correct, or prevent salinization of agricultural land, a world-wide problem that affects our food and fiber production. This information can be obtained by sampling soil, water, and plants. Objectives of sampling should include:

- Selection of parameters;
- Identification of problem areas;
- Description of the severity of the problem;
- Monitoring of time-dependent changes in salinity; and
- Assessing effects of different management strategies on salinity and crop yield.

To develop a sampling strategy that will achieve these objectives, the following questions must be answered:

- What is the region or time period to be sampled?
- What types of measurements are needed?
- What sampling technique is appropriate for the particular problem?
- How accurate must estimates of the population's parameters be?
- How many observations are needed in the sample to obtain the desired accuracy?
- Should the sample be distributed spatially or temporally?

A sampling strategy involves collecting a set of observations, called a sample, from a population, which is the total of all possible observations. The parameters or characteristics of the population are estimated by statistics—values calculated from the sample. Inferences about characteristics are made from statistics. Examples of statistics are the mean or average value, standard deviation, range, and mode. Two characteristics of an estimator or statistic of a population parameter are its accuracy—a measurement of how close the estimate is to the true value of

---

[a]Prepared by: Blaine R. Hanson and Stephen R. Grattan, Dept. of Land, Air, and Water Resour./Cooperative Extension, Univ. of Calif., Davis, Davis, CA 95616.

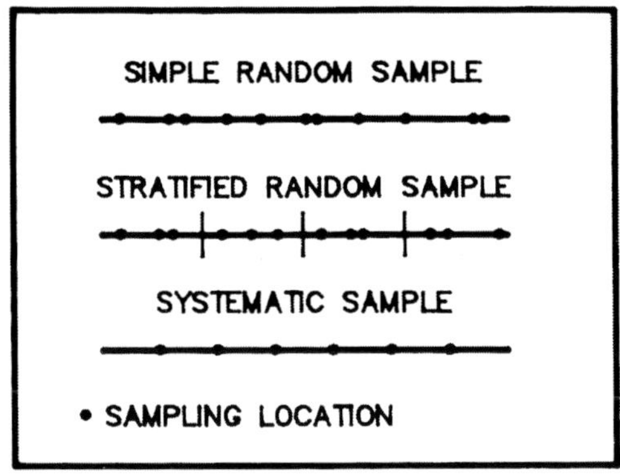

Fig. 9.1 Examples of Different Sampling Techniques

the population parameter—and its precision—the magnitude of the deviations from a mean obtained from repeated sampling.

Due to bias, an estimator can be inaccurate. Such bias can be caused by improperly calibrated instruments, improper analysis procedures, improper sample selection, or improper methods of estimation. Sample selection can cause bias by (Yates 1981) deliberate selection of a sample, selecting samples based on a characteristic correlated with the properties of the sampling area of interest, conscious or unconscious bias in the selection of a sample, substitution of one sampling location with another, and failure to cover all of the sample.

Proper procedures of analysis and estimation and proper methods of sample selection, such as random or systematic methods of selection, must be used to avoid bias.

SAMPLING TECHNIQUES

A number of techniques are used for field sampling (Snedecor and Cochran 1980; Cochran 1977; Yates 1981; Yamane 1967; Peterson and Calvin 1967). Some are described below.

Judgment Sampling

Sampling locations are selected based on the judgement of the sampler (Peterson and Calvin 1967). They may be selected according to such factors as crop growth, visual appearance of soil or plants, and locations of subsurface drains, ditches, and canals. Frequently, samples are taken in a problem area to compare with a few samples from a non-affected area. Samples can be composited to reduce analysis costs.

Simple Random Sampling

Simple random sampling, which is illustrated in Fig. 9.1 and based on the assumption that each sample has an equal chance of being selected, involves randomly selecting observations. A table of random numbers is used to select sampling locations. Peterson and Calvin (1967) suggest establishing a grid or transect throughout a field or plot and then using a table of random numbers to select coordinates for the locations of the observations. Simple random sampling is appropriate for homogeneous populations.

The estimators used for simple random sampling are sample mean:

$$\overline{X} = \frac{\sum X_i}{N} \tag{9.1}$$

standard deviation:

$$s = \left[\frac{\sum (X_i - \overline{X})^2}{N - 1}\right]^{1/2}$$

or

$$s = \left[\frac{N\sum X_i^2 - \left(\sum X_i\right)^2}{N(N - 1)}\right]^{1/2} \tag{9.2}$$

standard error of the mean:

$$s_{\overline{X}} = \frac{s}{\sqrt{N}} \tag{9.3}$$

confidence intervals:

$$\overline{X} \pm (z)\, s_{\overline{X}} \tag{9.4}$$

and coefficient of variation:

$$CV = \frac{100s}{\overline{X}} \tag{9.5}$$

Stratified Random Sampling

This technique, appropriate for areas with heterogeneous populations, consists of subdividing the area of interest into relatively uniform subareas or strata (Fig. 9.1). Each stratum is randomly sampled. Strata need not be of equal size, but they must not overlap. Selection of strata might be based on differences in crop appearance, soil texture, salt accumulation on the soil surface, or management. The number of observations per stratum should be proportional to the percentage of the total area represented by a stratum.

Stratified random sampling can be more accurate than simple random sampling for heterogeneous populations. Since stratified sampling eliminates sampling error between strata, the only sampling error comes from within strata. Miyamoto and Cruz (1986) found that a stratified sample based on soil mapping units in a field greatly reduced the sampling error. Their analysis showed the average coefficient of variation for field-wide transects to be 63%, whereas the average coefficient of variation was 36% for the stratified sample. Soil type accounted for 73% of the variability in the soil salinity. However, for homogeneous populatons, stratified sampling may be no more accurate than simple random sampling.

Estimators used for stratified sampling include

Mean:

$$\overline{X} = \sum W_h \overline{X}_h \tag{9.6}$$

Standard error:

$$s_{\overline{X}} = \left(\frac{\sum W_h^2 s_h^2}{N_h} \right)^{1/2} \tag{9.7}$$

Systematic Sampling

This technique consists of randomly selecting a starting point and then sampling the total area at equal intervals along a grid or a transect (Fig. 9.1). An advantage of this technique is that the area of interest is completely covered, whereas some parts of the field may not be sampled in simple random sampling. However, if a periodicity about equal to the sampling interval exists, a considerable loss of accuracy occurs (Peterson and Calvin 1967).

No reliable way to estimate the standard error of the sample mean exists with systematic sampling, since one observation per interval is normally obtained (Snedecor and Cochran 1980). Peterson and Calvin (1967) and Yates (1981) recommend stratifying the systematic sample with at least two samples per block, assuming that the variation within blocks is the only sampling variation and estimating the sampling error similarly to a stratified random sample. They also suggest obtaining several separate systematic samples from randomly selected starting points, and then calculating the sample mean for each sample. These means are treated as the data from a simple random sample of size equal to the number of separate systematic samples.

CONSIDERATIONS IN DEVELOPING SAMPLING STRATEGIES

Selecting a Sampling Technique

Selecting a sampling technique depends on the objectives and the sources of variance. For salinity problems in relatively small areas, judgment sampling coupled with compositing might be used. Any bias or errors in estimates are likely to be inconsequential. If, however, field-wide salinity needs to be assessed, an unbiased and accurate sampling technique may be required, particularly if monitoring salinity with time is an objective. For this objective, sampling errors from spatial variability of salinity must be minimal to detect differences in salinity.

Common sources of variance are trend, discontinuities, periodicity, and randomness. Trend in soil salinity may result from varying soil texture or varying amounts of infiltrated water in a field. If trend is a major source of variability, stratified random or systematic sampling may be best. Stratified random sampling should be used in areas with discontinuities or relatively abrupt changes in salinity. If random behavior is the major source of variability and the field-wide population appears homogeneous, then simple random sampling would be satisfactory.

Periodicity in soil properties normally is anthropogenic, but it can occur naturally (Webster 1977). If periodicity is substantial, the sampling strategy must account for this source of variance. Sampling in blocks of observations, with the size of a block equal to the period of the frequency, may be needed for relatively high frequencies or small periods. The blocks of observations could be randomly or systematically distributed over the field, depending on the desired sampling technique. Since any periodicity is most likely created, the period or size of blocks might be estimated from information on cultural practices. For example, periodicity in the advance of water across a field could be caused by different rates of infiltration for wheel and nonwheel furrows. Estimates of the mean and standard deviation of advance times might be obtained by sampling in blocks equal to the number of furrows cultivated per pass of the cultivator. The number of wheel and nonwheel furrows sampled in a block should be proportional to the number of wheel and nonwheel furrows in a block.

Spatial Distributions

While population parameters are important, equally important is the spatial distribution of salinity. According to Cameron et al. (1971), use of the mean value may be inappropriate for recommendations, since salinity and toxic elements vary spatially. If distributions are known, adequate treatments can be applied in areas with the more severe problems and the expense of overtreatment avoided in other areas.

Number of Samples

The number of observations needed depends on the variability of the population and the accuracy and confidence level desired in the estimate. A statistical approach for estimating a sample size is (Wadpole and Myers 1978)

$$N = \left[\frac{(z)(s)}{e}\right]^2 = \left[\frac{(z)(CV)}{k}\right]^2 \qquad (9.8)$$

The confidence level is the percent of time that the interval of X ± e contains the true value. For confidence levels of 95%, 90%, and 80%, z = 1.96, 1.64, and 1.28, respectively. Cameron et al. (1971) suggest that e = 0.1 to 0.2 of the sample mean and an 80% confidence level is sufficient for soil recommendations. Thus, the true mean would be expected to be between $\overline{X}\pm0.15\overline{X}$ for 80% of the time for a sample of size N.

Estimating the sample size requires an estimate of the variance or the standard deviation, usually unknown before sampling. However, an analysis of data sets on the variability of soil salinity (Miyamoto and Cruz 1987; Biggar and Nielsen 1985; Hanson and Oster 1986) suggests that a CV = 35 to 45% might first be used in sampling. Subsequent sample sizes could then be adjusted as better estimates of the standard deviation of a specific field are obtained. For example:

> Calculate the number of samples required so that the interval $\overline{X}\pm0.15\overline{X}$ contains the true mean value for 80% of the time with repeated sampling. Assume the coefficient of variation is 40%.

$$N = \left[\frac{(1.28)(0.40)}{0.15}\right]^2 = 11.6, \text{ or collect 12 samples}$$

Once the number of observations has been determined, the sampling interval can be determined by systematically spacing the observations across the region of interest, or by randomly locating sampling locations (simple random and stratified random sampling). A minimum interval must be maintained to satisfy independence of observations.

Independence of Observations

The aforementioned sampling techniques assume independence of samples, i.e., the probability of one value occurring does not depend on the probability of another value. This assumption implies that each observation provides completely new information about the population parameter.

The assumption may be invalid for a given sampling interval due to spatial variability. Observations made relatively close together are more likely to be similar than those made relatively far apart. Thus, over some distance, a soil property may be autocorrelated, and observations taken within that distance are dependent, i.e., each observation does not provide entirely new information about the population parameter.

The distance over which dependence occurs, called the range, is determined with variograms or correlograms (Chapter 15; Shumway 1988; Clark 1979). Dependence exists if observations are autocorrelated. Autocorrelation generally decreases with distance between observations and the variance between observations increases. At distances where no autocorrelation exists, the observations are independent and the variance is maximum.

The sampling interval must satisfy sample independence if classical statistics will be used, e.g., simple random sampling. Sampling at intervals where autocorrelation exists will not improve the estimates of the population parameters but will result in unnecessary sampling.

What is the minimum sampling interval for a given area or field? Most likely, a correlogram or variogram for a particular field will not exist. However, studies of field-wide spatial variability of soil salinity (Miyamoto and Cruz 1987; Gajem et al. 1981; Hajrasuliha et al. 1980; Hanson and Oster 1986) revealed that the range was often less than sampling intervals of 4m to 10m. Thus, the assumption of independence is likely valid for intervals that are practical in terms of cost and time, i.e., intervals of more than 10m.

If a variogram or correlogram is known or can be developed for an area, the standard error can be estimated with fewer observations with a technique called kriging (Clark 1979). McBratney and Webster (1983) reported that estimating the standard error with kriging may require as few as about one-tenth the number of observations as with the classical approach. According to Gutjahr (1984), at least 50 observations are needed for a reliable variogram.

Volume-Variance Relationships

Also to be considered in sampling is the relationship between the volume sampled per observation and the variance or the coefficient of variation. Hawley et al. (1982), Hassan et al. (1983), El-Araby et al. (1987), and Zobeck et al. (1985) have shown that as the volume sampled per observation increases, the variance of the sample decreases. Hanson and Oster (1986) found a coefficient of variation of about 45% when sampling with a four-electrode salinity probe (volume sampled about 90 cm^3) and a CV of about 35% when sampling with an electromagnetic conductivity meter (volume sampled about one cubic meter). These findings imply that the larger the volume sampled per observation (and thus, the smaller the variance), the fewer the number of required observations. For the CV of the four-electrode probe data, 14 observations are necessary. For the electromagnetic conductivity meter, seven are necessary.

Variance-Size of Area Sampled

The number of observations might be expected to increase as the area sampled increases. However, as discussed previously, the number of observations depends on the variance or coefficient of variation. The required number of observations increases with the size of the area sampled only if the CV increases.

Beckett and Webster (1971) concluded that soil variability, as measured by the CV, increases with the size of the area sampled. However, as much as half of the CV present within one ha is already present within a few m^2. Their analysis of data sets showed that relationships between CV and size of area sampled vary greatly. This probably reflects changes in soil properties in the size of the area sampled.

The effect of the area's size on variability depends on the spatial distribution of the soil property. The data in Fig. 9.2a show an increase in the CV with distance. The data in Figs. 9.2c and 9.2d show primarily random behavior, or little change in the CV with distance.

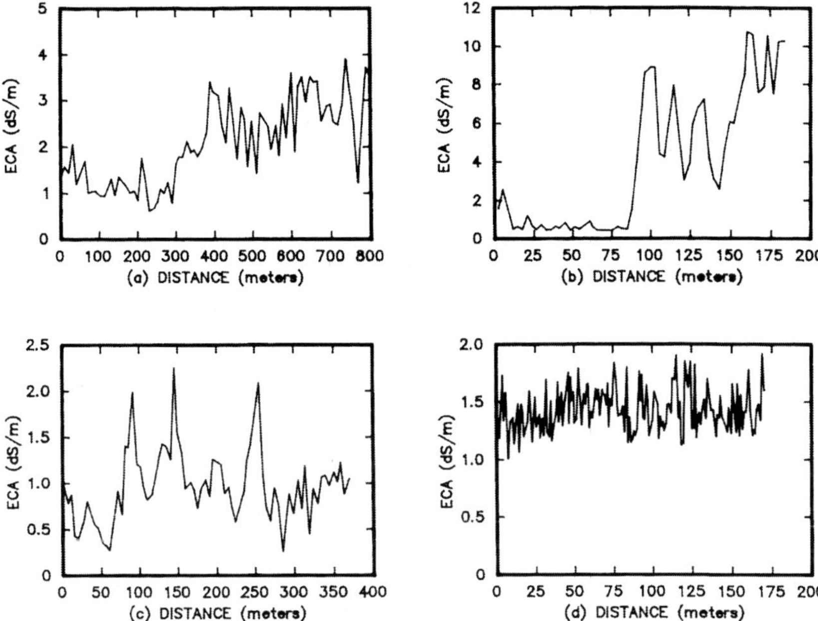

Fig. 9.2 Field-Wide Salinity along Transects for: (a) Site DU; (b) Site FR; (c) Site CR; and (d) Site BO.

Local Variability

Once field-wide locations have been selected, further care is needed in sampling when substantial small-scale variability in soil salinity exists due to patterns of water flow unique to furrow and drip/trickle irrigation systems (Hoffman et al. 1985; Wadleigh and Fireman 1948). Sampling at a particular location must account for this variability. Sampling in the furrow bed would be appropriate for analysis of salinity for achieving germination and stand establishment but inappropriate for analyzing seasonal crop yield-salinity relationships. For drip/trickle systems, sampling midway between emitters would result in different salinity levels than sampling near the emitters. When monitoring time-dependent salinity levels, each observation must be made at the same relative position to the pattern of water flow.

Composite Sampling

Frequently, samples are composited to reduce the cost of sample analysis. Cline (1944) states that composite sampling is valid if:

1. The sampling volume represents a homogeneous population. For heterogeneous populations, knowledge of spatial distributions is important, which a composite sample would fail to provide.
2. Equal amounts of each sampling unit contribute to the composite sample.

TABLE 9.1 Standard Error of the Mean for Several Sampling Strategies as Applied to Example Data for four Transects of Soil Salinity

Site (1)	Simple random (2)	Stratified random (3)	Systematic (4)
DU	0.18	0.11	0.14
BO	0.04	0.03	0.03
CR	0.09	0.10	0.09
FR			
All data	0.94	0.53	0.49
Saline area	0.69	0.44	0.55

3. No interactions occur within the composite sample as a result of the compositing.

4. An unbiased estimate of the mean is the only objective. Composite sampling is unsatisfactory if any other statistic is required.

EVALUATIONS OF SAMPLING STRATEGIES

Which sampling technique is most appropriate where little information is known before sampling? To address this matter, several sampling strategies were evaluated, using data from four transects of soil salinity (Figs. 9.2a–9.2d). Each data set has a different spatial distribution of salinity. Strategies were developed under the assumption that little information exists on field-wide variance and frequency distributions. Sampling techniques evaluated include simple random sampling, stratified random sampling, and systematic sampling (analyzed with stratified random sampling methods). A sample size of 15 observations was determined, using a confidence level of 80%, an error of 15%, and an assumed CV of 45%. This sample size was used for all strategies. Four strata were arbitrarily used for stratified random sampling.

Figs. 9.2a–9.2d show the data sets. The primary source of variability at Site DU (Fig. 9.2a) was a gradual trend in salinity across the field. No surface indications of differences in soil texture or salinity were observed. At Site FR (Fig. 9.2b), a discontinuity in soil salinity appeared between 85m to 90m along the transect. Visible differences in soil texture were observed between areas of low and high salinity. Soil salinity appeared to be somewhat random at Site CR (Fig. 9.2c), but three large spikes occurred at about 100m, 150m, and 250m along the transect. Crop growth was poor at those locations. Fig. 9.2d shows a relatively uniform distribution of salinity at Site BO. Random behavior appears to be the major source of variability.

Results of this analysis (Table 9.1) revealed that the standard error for Sites BO and CR—where random behavior is the major source of variability—was about the same for all sampling techniques. The standard error where other sources of variability contributed substantially was less for the stratified random and systematic sampling techniques.

This suggests that stratified random or systematic sampling should be used when little information is known about the salinity distribution before sampling. Eq. 9.8 and an assumed CV are used to estimate the number of observations. For Sites CR and FR, judgment sampling is used. Areas with relatively high salinity levels experienced poor plant growth. Thus, one strategy would be to sample only those areas with poor growth to determine initial salinity levels and monitor changes in salinity with time. At Site CR, the affected areas were small. Hence, a few composited samples might be sufficient. At Site FR, the affected area was about half the total area sampled. A stratified random or systematic sampling method might be more appropriate (Table 9.1).

SAMPLING PARAMETERS

Soil

Soil chemistry problems are frequently identified by measurements of soil salinity, soluble salts, and toxic materials, such as boron. Other types of measurements that help to develop management strategies for salinity control include water content, water holding capacity of the soil, soil texture, soil infiltration rate, hydraulic conductivity, and nutrients and micronutrients. Chapter 10, Black (1965), Reisenauer (1978), and Chapman and Pratt (1982) cover methods for measuring the various parameters. Some considerations and recent developments in measuring some of these parameters will be discussed here.

Soil Salinity. Several devices that allow rapid field-wide sampling of soil salinity have been developed. They include the four-electrode salinity probe (Rhoades and van Schilfgaarde 1976; Rhoades and Halvorson 1977); the Wenner Array (Rhoades and Ingvalson 1971; Rhoades and Halvorson 1977), and the electromagnetic conductivity meter (Rhoades and Corwin 1981; Corwin and Rhoades 1984). All three instruments measure the bulk electrical conductivity (EC) of the soil.

The four-electrode salinity probe is a probe of small diameter with four electrodes spaced a few centimeters apart along its length of about 150 mm or 6 in. Two of the electrodes create an electric field. The other two measure the electrical resistance of the soil. To use it, the probe is inserted into the soil to the desired depth. The volume sampled is about 90 cm^3.

The Wenner Array consists of placing along the soil surface four equally spaced electrodes, each inserted to a depth of about 40 mm to 50 mm. The depth sampled roughly equals the electrode spacing.

The electromagnetic conductivity meter consists of a transmitter and a receiver coil. The instrument is placed on or held above the surface of the soil. Electrical current introduced into the transmitter coil creates a secondary magnetic field in the soil through magnetic induction. The receiver coil detects the secondary field, the strength of which depends on the electrical conductivity of the soil.

Theoretically, the bulk EC depends on the salinity of the soil solution, soil-water content, porosity, and type and amount of clay in the soil (McNeill 1980). While laboratory studies have shown the bulk EC to depend strongly on water content (Rhoades et al. 1976), field studies have shown little correlation between water content and bulk EC (Hanson and Oster 1986). Calibration curves for a given soil texture are needed to relate the bulk EC to the EC of a saturated extract (Halvorson et al. 1977). Studies on field-wide sampling (Cameron et al. 1981, William and Baker 1982, de Jong et al. 1979, Hanson and Oster 1976) have shown the following results:

1. The four-electrode salinity probe provides good depth resolution, describing the salinity distribution with depth at a particular location. However, a commercially available probe, with a diameter of 28 mm (1.1 in.), was extremely difficult to insert into the soil, even with a pilot hole, and equally difficult to extract. A probe designed by Hanson (1985) with a diameter of 13 mm (0.5 in.) overcame this problem.

2. The Wenner Array provides satisfactory depth resolution over a large volume of soil, using the method of Rhoades and Halvorson (1977). However, the instrument is bulky and requires at least two persons for rapid surveying. Another drawback is that the drying of the soil surface can result in poor electrical continuity between the electrodes and the soil. Due to this problem, Hanson and Oster (1986) found the same coefficient of variation for the Wenner Array and the four-electrode salinity probe, though the array sampled a much larger volume of soil.

3. The electromagnetic conductivity meter (EM-38) provides a way to survey field-wide salinity patterns rapidly. Its effective depth is about 1 m. Deeper sampling can be conducted with other instruments, such as the EM-31. The depth resolution of the instrument is poor, although some information about changes in salinity with depth can be obtained by operating the instrument in both the vertical mode (coils in vertical position) and horizontal mode (coils in horizontal position). (The EM-38 and EM-31 are manufactured by Geonics, Ltd, Mississauga, Ontario, Canada. Mention of these instruments is not an endorsement.)

Gravimetric Sampling. For gravimetric sampling of soil-water content, Hawley et al. (1982) found relatively large coefficients of variation for volumes of soil less than about 50 cm^3 to 60 cm^3 per observation, and little change in the CV for volumes from 50 cm^3 to about 800 cm^3. These results suggest that the volume sampled per observation should be about 50 cm^3 to 100 cm^3 for water content estimates. Larger samples would not improve the estimate of water content. For soil salinity, El-Araby et al. (1987) recommend a sample size of about 500g per observation, based on volume-variance data.

Infiltration Rates. Traditional methods of measuring soil infiltration rates, such as ring infiltrometers, blocked-furrow infiltrometers, and inflow and outflow measurements, usually sample a very small area of a field. They can be highly unreliable due to spatial variability. Elliott and Walker (1980) developed a way to calculate the infiltration rate for furrow or border irrigation systems. Data needed for this method include water advance measured at two locations along the length of the field, border or furrow inflow rate, field slope, run length, surface roughness, and furrow geometry. The advantage of this method is that the entire furrow or border is used as the infiltrometer, thus reducing the variance of infiltration measurements. Hanson et al. (1988), however, found infiltration rates calculated from measurements made in a single furrow to be highly unreliable for cracked soils. Their study showed that measurements must be made in blocks of furrows for these soils.

Water

Measurements made on water samples include salinity, soluble salts, toxic elements, and pH. Other useful measurements for management strategies include flow rates, volumes, and application rates. While strategies for water sampling should be based on the previously discussed techniques, some modifications may be required. Observations could be distributed systematically or randomly. The required number of observations can be calculated from Eq. 9.8, if an estimate of the variance is available. Events responsible for changes in water quality must also be considered in determining sample size and sampling interval. The quality of irrigation water may change slowly during an irrigation season. Thus, infrequent sampling at random or systematic intervals may be sufficient. The quality of subsurface drainage water may vary considerably with irrigation events. Thus, water measurements may be needed just before and just after an irrigation and then at relatively frequent intervals shortly after the irrigation.

Plants

The most important plant parameters to monitor in a saline environment are growth, visual appearance, element concentrations in tissues, and crop quality. All these parameters vary in time and space. Variability can be at the field-wide scale and the local scale, i.e., within a plant.

Elements in plants are usually distributed spatially in predictable patterns in the canopy. For example, chloride and boron are often found in highest concentrations in the margins of older leaves. Consequently, segregation of leaves that vary in physiological age could reduce the variance. Therefore, dividing the plant into strata based on leaf age may be desirable. This suggests a stratified random sampling approach. For practical purposes, judgment sampling can be used to select leaves for the sample within a stratum.

REFERENCES

Beckett, P. H. T., and Webster, R. 1971. "Soil variability: a review." *Soils and Fertilizers* 34:1–15.

Biggar, J. W., and Nielson, D. R. 1985. "Spatial analysis of water table salinity and hydraulic conductivity prior to salinization." *Proceedings of the International Symposium on the Reclamation of Salt-Affected Soils*, Part 1. 13–21 May 1985, Jinan, China.

Black, C. A., ed. (1965). "Methods of Soil Analysis, Part 1." American Society of Agronomy Monograph 9; 770 p.

Cameron, D. R., de Jong, E., Read, D. W. L., and Oosterveld, M. 1981. "Mapping salinity using resistivity and electromagnetic inductive techniques." *Can. J. Soil Sci.* 61:67–78.

Cameron, D. R., Nyborg, M., Toogood, J. A., and Laverty, D. H. 1971. "Accuracy of field sampling for soil tests." *Can. J. Soil. Sci.* 51:165–175.

Chapman, H. D. and Pratt, P. F. 1982. "Methods of analysis for soils, plants, and waters." University of California Publication 4034; 309 p.

Clark I. 1979. *Practical Geostatistics.* Applied Science Publishers Ltd, London. 129 p.

Cline, M. G. 1944. "Principles of soil sampling." *Soil Science* 58:275–288.

Cochran, W. G. 1977. *Sampling Techniques.* John Wiley and Sons. New York. 428 p.

Corwin, D. L., and Rhoades, J. D. 1984. "Measurement of inverted electrical conductivity profiles using electromagnetic induction." *Soil Sci. Soc. of Am. Proc.* 41:966–971.

de Jong, E., Ballantyne, A. K., Cameron, D. R., and Read, D. W. L. 1979. "Measurement of apparent electrical conductivity of soils by an electromagnetic induction probe to aid salinity surveys." *Soil Sci. Soc. of Amer. Proc.* 43:810–812.

El-Araby, A. M., Warrick, A. W., and Tucker, T. C. 1987. "Sample volume effect on determination of soil salinity." *Commun. in Soil Sci. Plant Anal.* 18:593–599.

Elliott, R. L., and Walker, W. R. 1980. "Furrow irrigation infiltration and advance functions." ASAE Paper 80-2075. Presented at the 1980 Amer. Soc. Agr. Engin. Summer Meeting, San Antonio, Tex., June 15–18, 1980.

Gajem, Y. M., Warrick, A. W., and Myers, D. E. 1981. "Spatial dependence of physical properties of a typic torrifluvent soil." *Soil Sci. Soc. of Amer. Proc.* 45:709–715.

Gutjahr, A. 1984. "Spatial variability: geostatistical methods." Soil Spatial Variability, Proceedings of a workshop of the International Soil Science Society and the Soil Science Society of America, Las Vegas, Nev., Nov. 30–Dec 1, 1984.

Hajrasuliha, S., Baniabbassi, N., Metthey, J., and Nielsen, D. R. 1980. "Spatial variability of soil sampling for salinity studies in southwest Iran." *Irrigation Science* 1:197–208.

Halvorson, A. D., Rhoades, J. D., and Reule, C. A. 1977. "Soil salinity—four electrode conductivity relationships for soils of the Northern Great Plains." *Soil Sci. Soc. of Amer. Proc.* 41:966–971.

Hanson, B. R. 1985. Unpublished design.

Hanson, B. R., and Oster, J. D. 1986. "Rapid field-wide assessment of soil salinity." ASAE Paper 86-2063. Presented at the 1986 Amer. Soc. Agr. Engin. Summer Meeting, California Polytechnic Institute, San Luis Obispo, Calif., June 29–July 2, 1986.

Hanson, B. R., Prichard, T. L., Goldhamer, D., and Schulbach, H. 1988. "Evaluating and predicting furrow irrigation system performance." Draft report on Interagency Agreement B55427 between the State Dept. of Water Resour., Office of Water Conservation, and the University of Calif. Cooperative Extension/Dept. of Land, Air, and Water Resources.

Hassan, H. M., Warrick, A. W., Amoozegar-fard, A. 1983. "Sampling volume effects on determining salt in a soil profile." *Soil Sci. Soc. of Amer. Proc.* 47:1265–1267.

Hawley, M. E., McCuen, R. H., and Jackson, T. J. 1982. "Volume-accuracy relationship in soil moisture sampling." *Proc. Am. Soc. of Civil Engr. Irrig. and Drain. Div.* 108(IR1):1–11.

Hoffman, G. J., Shannon, M. C., and Jobes, J. A. 1985. "Influence of rain on soil salinity and lettuce yield." *Drip/Trickle Irrigation in Action,* Proceedings of the Third Drip/Trickle Irrigation Congress, Nov. 18–21, 1985, Fresno, Calif.

McBratney, A. B., and Webster, R. 1983. "How many observations are needed for regional estimation of soil properties?" *Journal of Soil Science* 135:177–183.

McNeill, J. D. 1980. "Electrical conductivity of soils and rocks." Technical Note TN-5. Geonics Limited, Mississauga, Ontario, Canada.

Miyamoto, S., and Cruz, I. 1986. "Spatial variability and soil sampling for salinity and sodicity appraisal in surface-irrigated orchards." *Soil Sci. Soc. of Amer. Proc.* 50:1020–1026.

Miyamoto, S., and Cruz, I. 1987. "Spatial Variability of soil salinity in furrow-irrigated torrifluvents." *Soil Sci. Soc. of Amer. Proc.* 51:1019–1025.

Peterson, R. G., and Calvin, L. D. 1967. "Sampling." *Methods of Soil Analysis, Part 1.* American Society of Agronomy Monograph 9; 770 p.

Reisenauer, H. M. 1978. "Soil and Plant-Tissue Testing in California." University of California Bulletin 1879; 54 p.

Rhoades, J. D., and Halvorsen, A. D. 1977. "Electrical conductivity methods for detecting and delineating saline seeps and measuring salinity in Northern Great Plains soils." ARS W-2. Agricultural Research Service,, USDA; 45 p.

Rhoades, J. D., and Corwin, D. L. 1981. "Determining soil electrical conductivity-depth relations using an inductive electromagnetic soil conductivity meter." *Soil Sci. Soc. of Amer. Proc.* 45:255–260.

Rhoades, J. D., and van Schilfgaarde, J. 1976. "An electrical conductivity probe for determining soil salinity." *Soil Sci. Soc. of Amer. Proc.* 40: 647–651.

Rhoades, J. D., Raats, P. A. C., and Prather, R. J. 1976. "Effects of liquid-phase electrical conductivity, water content, and surface conductivity on bulk soil electrical conductivity." *Soil Sci. Soc. of Amer. Proc.* 40:651–655.

Rhoades, J. D., and Ingvalson, R. D. 1971. "Determining salinity in field soils with soil resistance measurements." *Soil Sci. Soc. of Amer. Proc.* 35:54–60.

Shumway, R. H. 1988. *Applied Statistical Time Series Analysis.* Prentice Hall, New Jersey; 379 p.

Snedecor, G. W., and Cochran, W. G. 1980. *Statistical Methods.* The Iowa State University Press; 507 p.

Wadleigh, C. H., and Fireman, M. 1948. "Salt distribution under furrow and basin irrigation cotton and its effect on water removal." *Soil Sci. Soc. of Amer. Proc.* 13:527–530.

Wadpole, R. E. and Myers, R. H. 1978. *Probability and Statistics for Engineers and Scientists.* MacMillian Publishing Co., New York; 580 p.

Webster, R. 1977. "Spectral analysis of gilgai soil." *Australian Journ. Soil Research* 15:191–204.

William, B. G., and Baker, G. C. 1982. "An electromagnetic induction technique for reconnaissance survey of soil salinity hazards." *Aust. Journ. Soil Research* 20:107–118.

Yamane, T. 1967. *Elementary Sampling Theory.* Prentice-Hall, New Jersey; 405 p.

Yates, F. 1981. *Sampling Methods for Censuses and Surveys.* Charles Griffin and Co. Ltd.; 458 p.

Zobeck, T. M., Fausey, N. R., and Al-Hamdan, N. S. 1985. "Effect of sample cross-sectional area on saturated hydraulic conductivity in two structured clay soils." *Transactions of the ASAE* 28:791–794.

NOTATION

CV = coefficient of variation, expressed as a percent; e = absolute allowable error between the estimate of sample mean and the true value; k = error expressed as a decimal fraction of the mean; N = total number of observations in sample; N_h = number of observations in stratum h; s = sample standard deviation; s^2 = sample variance; $s_{\bar{X}}$ = standard error of the mean; s_h = standard deviation for stratum h; W_h = relative weight of stratum h; $\bar{X}$ = sample mean or average value; X_i = individual observation; $\bar{X}_h$ = mean for stratum h; and z = multiplier, which equals 1.96, 1.64, and 1.28 for confidence levels of 95%, 90%, and 80%, respectively.

CHAPTER 10

FIELD AND LABORATORY MEASUREMENTS[a]

INTRODUCTION

Developing management practices for salt-affected soils and low-quality irrigation waters includes determining the salinity of the soil and the water, the origin of the various salts, and the crops that can be grown. Gathering this information can be very time- and resource-consuming because salt-affected soils tend to be extremely variable with time and space. In this chapter, soil and water sampling methods and the storage of samples will be discussed. References for the more commonly accepted procedures are listed.

REMOTE SENSING

Remote sensing refers to taking observations or measurements with devices that sense and record reflected or emitted energy without touching the subjects of interest. Several such methods for salt-affected soils currently exist.

Aerial Photography. Of the different types of film used for remote sensing, color infrared film has produced the most useful data on how the salinity-induced plant stress varies. With this film, dark-green vegetation produces a bright red image; light-green foliage produces a pink image; barren saline soil produces a white image; non-saline soil a gray, bluish-gray, or greenish-gray image; senescent and salt-stressed crops a reddish-brown to reddish-black image; clear water a very dark-blue to black image, and sediment-laden water a blue image. Thus, plants' responses to salinity, soil, and water can be readily identified.

Modern image analysis systems can digitize infrared photographs, cluster the scene into classes of salinity severity, and produce fractional area estimates that fall into each class. The response of crops to salinity severity usually varies from season to season and from crop to crop and depends mainly on patterns of precipitation and the crop's tolerance to salinity. Infrared photography is also very useful for recording year-to-year and crop-to-crop responses to soil salinity and changes in management practices.

---

[a]Prepared by: C. W. Robbins, USDA Agric. Res. Service, Kimberly, Idaho 83341; and C. L. Wiegand, USDA Agric. Res. Service, Weslaco, TX 78596.

Videography. Video cameras because of their recording ease, immediate playback, real time display, and declining prices are replacing traditional photographic cameras. Manzer and Cooper (1982) first described videography for agricultural applications. Interest has since rapidly increased. Everitt et al. (1987) first used videography to investigate soil salinity. They digitized the video signals and used an image-processing system to distinguish between saline and non-saline areas and to estimate the area of land affected. Considerable progress has been made in applying video sensing to agriculture. The built-in automatic gain controls of most commercial camera outputs make calibrating units of energy or reflectance difficult. Although resolution is lower than with photographic cameras, videography adequately surveys and documents salinity patterns. Hardwired boards are becoming available for rapid video imagery processing on personal computers.

Infrared Thermometry and Imaging. According to Jackson (1982), temperatures of the crop canopy can be measured by infrared thermometers and correlated with the crop water stress index (CWSI) to monitor the water stress caused by soil salinity. Howell et al. (1984) concluded that the CWSI is as sensitive as other methods, provides broader spatial resolution of water stress in plants, and appears to be a practical tool for assessing osmotic stress in the field. Infrared thermometry can also be used to monitor saline seeps, since the temperatures of wet and dry soil differ (Doering et al. 1976).

Infrared thermometers can be hand-held or flown across fields in transects. At midday, dry soil is at least 20°C warmer than the air. Foliage is within about 2°C of the air's temperature. The thermal data provides a vertical view of the bare soil and vegetation, so the infrared thermometer measures the effect of salinity on vegetative cover better than it measures the effect of salinity on canopy temperature (Jackson 1982; Wiegand et al. 1968). Some systems have built-in reference temperature and calibration to provide quantitative temperature data, but most only record relative differences in temperature.

Multispectral Scanners. These systems sense in multiple bands or wavelength intervals (Holter et al. 1970). The visible band (0.4 μm to 0.7 μm) responds to vegetative ground cover as affected by salinity. Reflective infrared (0.75 μm to 1.35 μm) responds to the development of plants expressed as leaf area index. Water-affected bands (1.4 μm to 2.5 μm) contrast living vegetation with dry soil and wet versus dry soil. The thermal bands (5 μm to 20 μm) measure the temperature of the canopy when ground cover is complete and the temperature of plant cover when the growth of plants varies in response to soil salinity. The information from the various bands is complementary and extensively documents the effects of salinity on plant growth.

The costs of operating and processing data are high for multispectral scanners, so they are best suited to gathering data for irrigation districts, river valleys, and other geographic entities of 250 km^2 to 8,000 km^2. They are more cost-effective when the data is used for multiple purposes and several users share the costs.

The thematic mapper of the Landsat satellite series is the best instrument for continental or global surveys. It has the appropriate wavelengths, and makes six observations in each band on each area of the earth's surface every 16 days, cloud cover permitting.

Microwave Sensors. Measurements of microwaves can be related to soil salinity because soil water and solutes affect the natural passive microwave emissions of the soil (Moore et al. 1975). Jackson and O'Neill (1987) report that data on microwaves is currently being applied to salt-affected soils, but results are only qualitative.

MEASUREMENT OF SALINITY

Non-destructive methods for direct measurement of soil salinity include: buried porous electrical conductivity sensors, four-probe electrode systems, electromagnetic induction sensors, and time domain reflectometric electrode systems. These methods measure total salt concentrations in soils but not the concentrations of individual ions. For initially mapping large areas and monitoring changes in the total concentrations of salt with time, they are faster and involve much less labor than traditional methods. They are useful for determining sampling locations and intensity. Other methods must be used for specific ion ratios or concentrations.

Electrical conductivity salinity sensors consist of a pair of small, corrosion-resistant electrodes in a very fine and porous housing of ceramic or glass. The electrodes connect to continual recording monitors or to manually operated, hand-held meters or recorders. The newer sensors contain a thermistor and a portable meter designed to correct the electrical-conductivity readings for variations in temperature. Capillary action draws the soil solution into the porous matrix. The sensors measure changes in electrical conductivity at a given point during a long period of time from a small volume of soil. The calibration can change with time, so periodical recalibration is necessary. The response can take as long as a day or more under dry conditions. This is due to the slow movement of ions into and out of the electrode housing by diffusion (Wood 1978).

A four-electrode system consists of four probes or electrodes placed in a straight line at a predetermined spacing. A constant, alternating electrical current is passed between the outer electrodes. The electrical potential between the inner pair is measured. These systems have been used in various configurations, including four rods pushed into the surface of the soil (Rhoades and Ingvalson 1971), four metal rings incorporated into the surface of a plastic rod to form a probe that can be pushed into an augered hole (Rhoades and van Schilfgaarde 1976), an inexpensive and permanently installed four-electrode unit that monitors the salinity of the soil *in situ* (Rhoades 1979), and round cells for laboratory and lysimeter studies (Gupta and Hanks 1972). To use the round configuration, Gupta and Hanks mounted eight or more electrodes around a cylinder or lysimeter at equal spacings and used four adjacent electrodes simultaneously to make each measurement. Eight separate soil volumes can be measured with each circle of electrodes.

The volume and depth of soil (for which salinity is measured) is determined by the distance between the outer two electrodes. With the probe configuration of Rhoades and van Schilfgaarde (1976), a hole is cored for each sampled site. In dry or stony soils, establishing uniform soil contact with the electrodes may pose a problem. These systems provide quick, easily-obtained measurements of total salinity without the need for taking soil samples to the laboratory.

Salinity measurement by electromagnetic induction consists of inducing current flow in a hand-held primary electromagnet, which, in turn, induces current flow in the primary conductor. The induced current flow in the soil creates a secondary electromagnetic field. The secondary-field strength is proportional to the current flow in the secondary conductor, the soil. Since the current in the soil is a function of the electrical conductivity of the soil, the instrument can be calibrated so that the secondary field strength measured by the instrument provides a measure of the electrical conductivity of the soil. Using this principle, Rhoades and Corwin (1981) developed an inductive electromagnetic soil conductivity meter that could rapidly measure soil salinity. The instrument is placed on the soil surface or at some distance above the surface, making soil sampling and holes for electrodes unnecessary. It can be used to survey salt-affected soils, map saline seeps, and estimate salinity at different depths (Corwin and Rhoades 1984; Rhoades et al. 1989a).

Time domain reflectometry (TDR) is currently being developed for simultaneously measuring volumetric water content and bulk soil electrical conductivity. TDR measures these two parameters *in situ* on a given sample volume for a given point in time. It combines the principle that the relative dielectric constant of a soil is primarily related to its water content and the fact that the propagation velocity of a voltage pulse along the transmission lines is a function of the electrical conductivity of the soil. Soil voids cause reading errors, and the method is inaccurate above a certain electrical conductivity. Dasberg and Dalton (1985) give a detailed explanation and history of the method, and compare it with other methods.

SAMPLING OF SOIL

Once a sampling strategy has been chosen, 0.5 kg to 1.0 kg soil samples are taken from each depth or horizon and chemically analyzed. Larger samples are needed for physical measurements. Care must be taken to avoid contaminating the samples with soil from other depths. The samples should be crushed or broken up, air-dried, passed through a 2.0-mm sieve, thoroughly mixed, and placed in durable, labeled containers. Labels should include the date, the upper and lower depth limits, and the site. The approximate field moisture content, the name of the person who collected the sample, the soil surface conditions, the crop's appearance, history, and yield, the next crop to be grown, and the quality of available irrigation water should also be noted. Samples should be kept dry and free of contamination.

SAMPLING OF WATER

If well water is used for irrigation, samples should be collected after the pumps have run for at least half an hour. This removes standing water from the well casing and the area around the well, ensuring a representative sample. Usually, the well-water quality will not change during the growing season. In a few cases, where the water level is substantially lowered during the pumping season, the quality may change, justifying periodical sampling throughout the irrigation season.

Irrigation water quality remains constant when the water is from large storage reservoirs, but it may vary during the irrigation season when the water comes from small storage systems and stream systems that have annually fluctuating flows or accumulate subsurface irrigation return flow or spring water. Samples should be collected only during the irrigation season and when changes in quality are suspected. Once the annual patterns of quality in a well or stream water are established, samples should be collected only when changes are suspected.

Water samples of 0.2 l to 0.5 l (0.5 to 1.0 pint) are sufficient for analysis. The sample containers should be clean and free of oil, salts, and other chemical contaminants, rinsed with the water to be sampled, and have tight closures. Labels should include the date, time, and location of sampling, the approximate water flow, the crop to be grown, the irrigation method, and the name of the person collecting the sample. The samples should be refrigerated at about 4°C or 40°F and analyzed as soon as possible. They should not be frozen.

WATER TABLE DEPTH AND GROUND WATER QUALITY

To remove excess salts from the root zone, the water table must be deep enough to allow drainage and salt leaching below the root zone. The required depth varies with soil texture, degree of stratification, quality of subsurface and irrigation water, type of crops grown, and application rates and methods of irrigation.

To determine the effects of a water table on soil salinity, it is necessary to determine the depth and salinity of the ground water. When the water table is close enough to the surface to be of concern—usually less than 2.0 m—auger holes, observation wells, or sand points inserted into the saturated zone allow for sampling and monitoring of water depth. With the required equipment, a well can be rapidly installed by jetting water under pressure through the casing of the well and lowering the casing as the cavity is created. To prevent the hole from caving in, slotted pipe or a sand point should be inserted into the hole to below the water table. The wells must not pass through clay or cemented layers below perched water tables. Otherwise, the water level of the well may fall as water drains through the well and into a vadose zone below or rise when artesian water flows into the well from beneath the restrictive layers. Water samples should be collected for chemical analysis. To ensure fresh samples, two or three well volumes should be pumped from each hole before sampling. The same procedures used to store and to analyze well water samples and soil extracts are used to store and analyze these samples.

When a monitoring well is hand-augured or made with a coring machine, soil samples should be kept by depth increments. The samples' texture, structure, and degree of cementation reveal the type and amount of soil stratification above and below the water table. With the methods used for the other soil samples, these samples can be analyzed for various salts and physical properties, such as hydraulic conductivity and texture.

Two alternatives to observation wells are piezometer tubes and tensiometers. A piezometer is an observation well with an open bottom that measures the positive hydraulic head at a given point. A tensiometer is a piezometer closed at the bottom with a porous cup that is used to measure negative hydraulic head at a given point. They are installed similarly to observation wells (Cassell and Klute 1986; Reeve 1986). They can be used to measure the hydraulic head at a given depth and, in the case of clay lens or cemented soil layers, the hydraulic head above, below, and across the layers can be measured. However, they also have some disadvantages. Piezometers are more difficult than observation wells to sample for water. Tensiometers are not used to collect water samples.

The direction of ground-water flow and the hydraulic gradient can be determined if at least four wells or piezometers are inserted in a rectangular pattern over the area of interest to measure the absolute water surface elevation or hydraulic head. These two parameters are useful for determining if an effective drainage system can be installed. How often the water table is measured depends on the aquifer's responsiveness to such events as on-site or up-slope irrigation, precipitation, and the absence or presence of water in irrigation conveyance systems. A log can be maintained during the growing season to determine whether a change in measurement frequency is needed.

CHEMICAL ANALYSES

The chemical species of interest in salt-affected soils include: the soil solution ions, namely, calcium (Ca), magnesium (Mg), sodium (Na), potassium (K), chloride (Cl), sulfate (SO_4), bicarbonate (HCO_3) and carbonate (CO_3); the same ions in the irrigation water; the exchangeable cations, namely, Ca, Mg, Na, and K; and the precipitated salts, namely, calcium carbonate (lime) and calcium sulfate (gypsum). These ions are often referred to as the macroions in salt-affected soils and irrigation waters. Other informative measurements include pH, EC_e, and saturation-paste water content. Other parameters that can be calculated from these data include the sodium adsorption ratio (SAR), the adjusted SAR of the irrigation water (SAR_{adj}), the exchangeable sodium percentage (ESP), and the salt mass per unit volume or area.

Boron (B) concentrations in irrigation waters and soil solutions produce plant injury at much lower concentrations than other ions. Thus, B should be considered more a toxic element than a salt. Because

B toxicity often is associated with salt- and sodium-affected soils, it will be considered here, as well as in Chapters 4 and 7.

The methods of analyzing soil paste, soil water extract, and bulk irrigation and drain water discussed here will be restricted to those appropriate to soils and waters with a pH of 6.5 to 10.0 and concentrations of salt that adversely affect yields. Unless noted otherwise, the same analytical procedure can be used for irrigation waters, drainage waters, and soil extracts. For the reader's convenience, procedures will be referenced and restricted to those that are generally accepted and standardized. Many other research-level procedures exist, but they are inapplicable to data-gathering at the management level.

Saturated Soil Paste Electrical Conductivity. In the past, much of the data on soil salinity was obtained by using a 50 ml cylindrical conductivity cell filled with a saturated soil paste to estimate soluble-salt concentrations. Because of the ease and speed of use, this "Bureau of Soils Cup" was used to map and diagnose salt-affected soils. When Reitemeier and Wilcox (1946) determined that plant responses to soil salinity more nearly correlated with the electrical conductivity values of the saturation paste extract, its use was discontinued. A theoretical relationship between the electrical conductivity of the saturation soil paste and the electrical conductivity of the saturation paste extract has since been developed to overcome the cell's shortcomings. This was done by developing a simple method of determining the volumetric water and volumetric solid contents of the saturation paste, the conductance of the sample surface, and the current pathway of the water in the cell. The cell, an electrical conductivity bridge, a portable balance, and the appropriate graphs can now be used to accurately estimate the electrical conductivity of the saturation paste in the field (Rhoades et al. 1989b).

Saturation Paste Extracts. To apply and interpret data about salinity uniformly from one soil or crop to another, a standardized extraction procedure is needed. Data from soil solution samples with a moisture content that represents field conditions are ideal, but ways to obtain them are presently unavailable. The use of saturation paste extracts is the standard way to measure and interpret soil salinity as a bulk measurement and individual ion concentrations. This is the lowest soil-to-water ratio that can be easily extracted with vacuum, pressure, or centrifugation, while providing a sample of sufficient size to analyze. The water content of a saturation paste is about twice the field capacity for most soils. The concentrations of salt and individual ions have been diluted by about the same ratio between field conditions and the extract for all samples. Exceptions include sandy soils, organic soils, and soils containing gypsum. This standardized extraction procedure enabled the development of crop tolerance tables that can be interpreted across soil-texture classes (U.S. Salinity Laboratory Staff 1954; Rhoades 1982b). Most data on the salt tolerance of a crop are expressed in terms of the electrical conductivity of the saturation paste extract (Bresler et al. 1982; Maas 1986). The relationship between cation solution concentrations and exchangeable cation concentrations in soils, used to predict exchangeable sodium percentage (ESP) from sodium adsorption ratio

(SAR) data, is also based on the dilution ratio of the saturation paste extract (U.S. Salinity Laboratory staff 1954).

Extraction ratios of soil-to-water by weight, e.g., 1:1, 1:2, and 1:5, are often used, especially by commercial soil testing laboratories. These extracts require less skill and time to prepare than saturation paste extracts. Any dilution above field conditions introduces errors in the interpretation of data. The greater the dilution, the greater the deviation between ionic ratios in the sample and the soil solution under field conditions. These errors are associated with mineral dissolution, ion hydrolysis, and changes in exchangeable cation ratios. Soil samples containing gypsum deviate the most because the calcium and sulfate concentration remain near-constant with sample dilution, while the concentrations of other ions decrease with dilution. The standardized relationship between the extract and the conditions of the soil solution in the field for different soils is lost with the use of preset soil-to-water ratios.

Samples to be used for saturation paste extracts should be air-dried, crushed, and passed through a 2-mm sieve. As the name implies, saturation pastes contain enough water to fill voids without leaving excess water on the surface after samples have set. Saturation pastes are prepared by adding distilled water to a soil sample and mixing the two to obtain a glistening surface. The sample should flow together slowly behind a spoon or spatula as it is drawn through the mix. Most saturated pastes will not stick to a clean spatula. When the container is gently tapped on the bench top, the paste's surface should still glisten, but free water should not be standing on the surface. After the container is closed, the sample should be allowed to set for at least an hour and then its consistency rechecked. If water remains on the surface, the sample is too wet and dry soil is added. If the sample has stiffened, more water is added, the sample is allowed to set, and then rechecked. The sample is covered with an airtight lid and allowed to set long enough to establish equilibrium between the soil minerals and the water (at least four hours, but preferably overnight). After equilibrating, the samples are thoroughly remixed and their pH measured with an electrode and pH meter. The pH of the saturation paste extract is generally less meaningful than the pH of the saturation paste for salt-affected soils. The two values are seldom the same for a sample. Existing criteria are based on the saturation paste's pH data.

Take a small (30g to 50g) subsample of the paste at this point. Weigh it, oven-dry it (105°C) overnight, and re-weigh it to determine the saturation moisture content. The moisture content in soils is determined by dividing the water weight by the dry soil weight.

The saturation paste is transferred to a Buchner or similar funnel fitted with a high retention filter paper, and the solution is collected into a bottle or test tube by vacuum extraction. The electrical conductivity (EC_e) is measured immediately and adjusted to 25°C (U.S. Salinity Laboratory Staff 1954). If HCO_3 and CO_3 are of interest, they should be determined at this time. Methods are discussed under procedures for anion analysis. Once the HCO_3 and CO_3 subsamples are removed, lime and gypsum can be prevented from precipitating in the remaining saturation paste extract by diluting the samples in half with distilled water, or adding acid to make the sample 0.1% nitric, sulfuric, or hydrochloric acid, depending on which anions are of interest. The extracts should be stored at 4°C or 40°F (do not allow to freeze) in containers with tight-fitting lids and analyzed for the ions of interest as soon as possible.

Soil Solution Extracts. Suction extraction through ceramic cups and thin-walled hollow tubing can be used to collect samples of the soil solution when the soil water potential is greater than about −30 kPa (−0.3 bar). If the metric potential is lower than this, the water moves toward the cup too slowly to permit the sampling. This method is used immediately after irrigation or from slightly above a water table. If a wastewater disposal field or a high water table is involved, suction extraction through ceramic cups can be used to monitor finite points in plots or fields. If the ceramic cups are not properly cleaned before they are installed, a potential for error exists. Suction-extraction preparation may include forcing 0.1 N HCl through the cups, followed by a liberal volume of distilled water. Timing in relation to irrigation or rainfall needs to remain consistent throughout a study, since the soil volume and solution location (in terms of pore size) changes as soil pores of various sizes drain.

Analysis procedures for saturation paste extracts and samples of irrigation and drain water are the same from here on. Once the liquid samples have been obtained, the solution electrical conductivity (EC) is the simplest analysis to make and the most widely used parameter for evaluating soil and irrigation water salinity (Rhoades 1982). Salt tolerance tables for most irrigated crops and a few ornamental plants have been developed. The electrical conductivity of a saturation paste extract is the only parameter necessary to use these tables to estimate the relative yield decrease caused by the soil solution's soluble salts (Bresler et al. 1982; Maas 1986; Chapter 13).

Saturation paste extracts of sodic soils ($pH > 8.5$ and $EC < 4.0$ dS/m) are usually dark in color due to suspended organic matter, which interferes with most analytical procedures. This material can be removed from the solution by adjusting the pH to 3.0 to 3.5 with a strong acid, then adding a small amount of an aluminum salt, and then centrifuging and filtering it (Robbins 1989).

Cation Analysis. Cation concentrations in irrigation waters, saturation paste extracts, and shallow water table and drainage waters are the second most useful data for diagnosing and managing salt-affected soils. The concentrations and ratios of cations are major factors controlling soil pH and the physical characteristics of salt-affected soils. Flame photometry is the most practical and widely used way to analyze Na and K. Any solid particles should be filtered from the samples and the samples then diluted to the proper concentration ranges. This is usually the only pretreatment needed to determine Na and K. Sodium is usually determined at 589 nm and K at 766.5 nm by flame photometry (Knudsen et al. 1982). Atomic absorption spectrophotometry is the most common way to analyze Ca and Mg in these samples. Elements that form stable oxysalts, such as aluminum, beryllium, phosphorus, silica, titanium, vanadium, and zirconium, reduce their sensitivity in aqueous samples.

These interferences are removed by adding 0.1% to 1.0% lanthanum or strontium chloride to the samples. Calcium is usually measured at 422.7 nm and Mg at 285.2 nm by atomic absorption spectrophotometry (Lanyon and Heald 1982). More recently, procedures involving ion chromatography (IC) have been developed to analyze alkali and alkaline earth metal ions (Nieto and Frankenberger 1985b). Gravimetric and colorimetric methods are available for these cations when instrumentation for the previously mentioned methods are unavailable. These methods are slower, involve more work, and are less sensitive to low concentrations and small differences among samples than the more commonly used methods (Knudsen et al. 1982; Lanyon and Heald 1982).

Anion Analysis. Gravimetric, volumetric, electrometric, and colorimetric methods have been used to determine concentrations of chloride in soil extracts, drainage effluents, and well waters (Adriano and Doner 1982). Ion chromatography has also been used to analyze Cl (Nieto and Frankenberger 1985a). All of these methods, except the gravimetric methods, are easily automated by reliable, available instrumentation. A variety of commercially available automated systems analyze chlorides simply, accurately, and reproducibly. Chloride is usually the predominate anion in salt-affected soils and nonreactive with other components. Due to its chemical inertness, Cl ion is often used as a conservative tracer for calculating leaching fractions and the relative proportions of water from different sources that enter soil water and irrigation systems.

Gravimetric, turbidimetric, titrimetric, and colorimetric methods have been used to determine sulfate concentrations in soil extracts and water samples (Beaton et al. 1968). Problems arise in choosing an extractant and defining what has been extracted and from interferences from suspended organic matter and clay particles in the sample solutions. More recently, IC methods to analyze sulfate have been developed that are free of the usual sulfate interferences, are faster, and will accurately detect sulfate at much lower levels (Adriano and Doner 1982; Nieto and Frankenberger 1985a).

Carbonate and bicarbonate ions are usually determined by titrating aqueous samples to pH 8.4 (carbonate) and then to pH 4.7 (bicarbonate) with a dilute solution of a strong acid, such as HCl or H_2SO_4, to determine carbonate and bicarbonate ions. The end points can be determined colorimetrically, using phenolphthalein for carbonate and methyl orange for bicarbonate, or potentiometrically, with a pH electrode and pH meter. These analyses can be readily performed manually or by automated instrumentation (Rhoades 1982b).

Flow injection analysis procedures and instrumentation have been developed to measure aqueous pH and electrical conductivity and analyze for the cations and anions of interest in saline and sodic soils and waters. This is an automated system for injecting a liquid sample and reagent(s) into a transport fluid. The sample and reagents pass through a reaction chamber and then through a detector. Available detectors include pH, electrical conductivity, calorimetric, turbidimetric, atomic adsorption spectrophotometry, and flame emission. The reagents, reaction chamber, and detectors are interchangeable, and their selection is determined by the parameters to be measured (Ranger 1981).

Exchangeable Cations and Cation Exchange Capacity. To quantitatively determine exchangeable plus soluble cations, the soil is extracted with a salt solution that provides a cation to replace the cations of interest. Neutral 1.0 M ammonium acetate is the generally used solution for sodium and potassium extraction (Knudsen et al. 1982), and 1.0 M sodium acetate adjusted to a pH of 8.2 is generally used for calcium and magnesium extraction. When lime and gypsum are absent, calcium and magnesium can be determined in the ammonium acetate extract. When lime is present, extracting at the higher pH of 8.2 limits lime solubility and overcomes most of the error in extracting calcium (Bower et al. 1952). When gypsum is present, measuring exchangeable calcium is impossible and meaningless. Once the extractable cation values are derived, the individual exchangeable cation values are calculated as the difference between the soluble and the extractable cations (U.S. Salinity Laboratory Staff 1954).

The cation exchange capacity (CEC) can be measured by washing the sodium-saturated samples from the calcium-magnesium analyses with ethanol and then replacing the sodium in the samples with ammonium. The exchangeable sodium concentration in this second washing equals the CEC (U.S. Salinity Laboratory Staff 1954). An independent procedure that uses a sodium acetate-sodium chloride saturating solution and a magnesium nitrate extracting solution can also be used to determine the CEC. This eliminates the ethanol wash cycles. However, the extractable concentrations of Ca and Mg cannot be obtained from this (Rhoades 1982a). The concentrations of calcium, magnesium, sodium, and potassium ion in these extracts are analyzed by the methods used for the water and aqueous extract samples, except the standard solutions are made up in the extracting solutions rather than in water.

Lime and Gypsum. The presence or absence of lime and gypsum in the soil should be considered when making management decisions in salt-affected soils. Soil carbonate content can be estimated from the volume of CO_2 gas evolved from a sample treated with excess HCl, or by the weight loss caused by the CO_2 evolved from samples treated with H_2SO_4 and $FeSO_4$. Gypsum can be semi-quantitatively measured by heating and weighing the water lost from the sample as the gypsum changes to plaster of paris. Gypsum can also be titrated with $BaCl_2$ and the $BaSO_4$ precipitate weighed (Nelson 1982).

Boron. After reviewing several analytical methods for B concentration measurement in water, soil extracts, and plant materials, Bingham (1982) indicated a preference for the azomethine-H method to determine B. He also described a mannitol potentiometric method, a tetrafluoroborate selective electrode method, and a total boron method, as well as hot water, saturation paste, and mannitol-calcium chloride extraction methods.

PLANT SAMPLING AND TISSUE ANALYSIS

The effects of soluble salts and exchangeable sodium on the growth and mineral composition of crops and ornamental plants have been extensively discussed in the literature. Other chapters in this book have also covered some of these effects. Soil salinity or exchangeable sodium ratios do not uniformly affect Ca, Mg, Na, K, N, P, and S uptake ratios, and excesses and deficiencies in different plant species and varieties. For example, fruit and nut trees and woody ornamentals are more sensitive to Cl and Na uptake than field and forage crops (Bernstein and Pearson 1956). Plant maturity and soil nitrate availability also affect salt sensitivity in crops (Bernstein 1974). Pesticide injury and plant pathogens can, under certain conditions, produce changes in mineral balance or physical appearances similar to those caused by excess soluble salts or exchangeable sodium. Plant sampling and tissue analysis may indicate a reduction in yield or quality induced by salinity or exchangeable sodium, but the aforementioned difficulties have made tissue analysis generally ineffective in diagnosing and managing saline or sodium-affected soils. An exception is plant tissue testing for boron (B) toxicity. Boron tolerance and toxicity tables are available for a number of crops and ornamentals (Bresler et al. 1982; U.S. Salinity Laboratory Staff 1954; Wilcox 1960). Gaines and Mitchell (1979) described a tissue analysis procedure used to measure the B status of crops. Plant material was ashed at 600°C for an hour and dissolved in 0.18M H_2SO_4. The B concentration was then measured by the azomethine H procedure.

Plant sampling techniques for salinity considerations include choosing the part of the plant and the stage of the plant's maturity to be sampled. The mineral composition for a particular species is not fixed. The composition of a particular part of a plant also changes with age and the environment. Once the samples are collected, they are prepared for analysis, which usually includes decontaminating, drying, grinding, and ashing or extraction. The procedures vary according to the crop and sampling purpose (Jones and Steyn 1973). After the samples are extracted or ashed and redissolved, the solutions are analyzed by the same procedures that are used for soil extracts and water samples.

PHYSICAL MEASUREMENTS

Saline and sodic soils often exhibit undesirable physical characteristics that also contribute to the poor crop productivity. These may include a shallow water table, low aggregate stability, impervious subsoil layers or lenses, low infiltration rates, low hydraulic conductivities, and variations in horizontal or vertical textures. These physical characteristics also affect plant growth and the effectiveness of reclamation procedures. The following sections briefly review the most common methods to characterize soils physically as they relate to salt-affected soils. The chemical and physical characteristics in salt-affected soils interact strongly. The intensity of these interactions varies with soil texture and clay mineralogy.

Aggregate Stability and Size Distribution. Aggregates are made up of sand, silt, clay and organic particles that adhere to each other and form structural units in the soil matrix that can withstand disruptive forces. The size and number of cleavage planes between aggregates and inter-aggregate air pores greatly influence other physical properties, such as infiltration rates and hydraulic conductivity. Concentrations and kinds of solution ions, soil mineralogy, organic matter, iron oxides, and carbonate coatings largely determine aggregate stability. The disruptive forces of sampling and analysis differ from those encountered in the field, which include cultivation, wheeled traffic, freezing, and thawing. Because of this difference, most investigators use aggregate stability of one aggregate size fraction rather than aggregate size distribution (Kemper and Rosenau 1986). Aggregate stability is the fraction of the aggregates of a selected size that do not disintegrate to smaller particles upon sieving or sedimentation. The electrical conductivity of the water used for wet sieving should be equal to less than 0.01 dS/m, standardized, and specified when aggregate stabilities of sodic soils are reported (Shainberg et al. 1981).

Particle Size Analysis. Spatial variations in soil texture play an important role in the movement and distribution of water and salts in soils. Water will move into and through coarse-textured soils and soil horizons more rapidly than through finer textured clay soils or layers, unless they are highly aggregated or perforated by macropores. Finer textured soils draw water and salts up from a water table by capillary action from a greater depth than do coarser textured soils. The combination of restricted leaching due to low infiltration and percolation rates and greater upward flow of salt in clayey areas perpetuates salinity in areas with the finest soil textures. Clay content and mineralogy also greatly affect specific surface area, cation exchange capacity, and water retention. Consequently, lateral and vertical texture variations and how they relate to patterns of salinity observed in the field tell a lot about processes that cause salt accumulation and aid in reclamation (Wiegand et al. 1966).

Gee and Bauder (1986) summarize ways to pretreat and analyze samples for particle size. Saline and sodic soil procedures are essentially the same as for normal soils, except that saline soils should be leached with distilled water until the leachate EC drops below 1.0 dS/m; chemically pretreated to remove organic matter, iron oxides, and carbonate coatings that inhibit dispersion; and dispersed by turbulent mixing, mechanical shaking, or ultrasound cavitation in an alkaline dilute solution of sodium polyphosphate. Procedures for analyzing particle size in gypsiferous soils have been developed (Hesse 1976), as have procedures for removing gypsum from soil samples (Rivers et al. 1982).

Hydraulic Conductivity. Soil salts are transported as solutes in water, so water must move through the soil. When a shallow water is the source of salinity, the soils must be drained before soil salts can be leached. Hydraulic conductivity of salt-affected soils, which must be measured to determine the rates at which salts can be leached, can be measured in the field with a shallow (saturated) or deep (unsaturated) water table.

Hydraulic conductivity may be limiting at different positions in the soil profile, depending on the circumstances. In sodic soils, low hydraulic conductivity at the surface of the soil may control water movement throughout the profile, or a nearly impermeable layer deeper in the soil can cause a perched water table to develop above which salts accumulate. High clay content can limit hydraulic conductivity throughout the root zone. Amoozegar and Warrick (1986) developed ways to measure the hydraulic conductivity of saturated soils in the field. Green et al. (1986) developed ways to measure the hydraulic conductivity of unsaturated soils in the field. These measurements can also be made in the laboratory on disturbed or undisturbed samples over a range of water contents (Klute and Dirksen 1986).

The most common way to determine the soil's saturated hydraulic conductivity in the presence of a water table is the augered-hole method. A hole is augered into the water table, with care taken to avoid puddling the walls. Water is pumped from the hole several times to ensure that water will flow freely into the cavity. The water level in the hole is allowed to equilibrate, and the water depth in the hole is measured. Water is again removed from the hole and the rate of rise is measured until one half of the original depth is reached. The water-rise rate in the hole, the depth of the hole below the water table, and its radius are the variables used to calculate hydraulic conductivity (Amoozegar and Warrick 1986).

When measuring hydraulic conductivity in the field, predominantly vertical flow during drainage is assumed. Of the four methods described by Green et al. (1986), only the unsteady drainage flux method, often called the instantaneous profile method, is described here. Soil in a 3.6-m square outer plot and a 1.2-m square inner plot, at a site where the water table is deep enough not to inhibit drainage, is thoroughly wetted and covered to prevent evaporation. The hydraulic head profiles are measured with tensiometers installed to different depths, and the water content profiles are measured in close proximity to the tensiometer by neutron scattering. These two measurements are made periodically in close time synchronization. The water fluxes determined from the changes in water content are divided by the hydraulic gradients for the same depth and time intervals to obtain the unsaturated hydraulic conductivity (Green et al. 1986). Watson (1966) summarized data-analysis procedures for this method. The method assumes vertical flow, nonhysteresis of the water content-pressure head relation, and isothermal conditions. It is not suited to soils whose horizons differ greatly in hydraulic conductivity or soils with water tables within about 1.5 m of the deepest depth.

When field measurements are impractical, hydraulic conductivity can be determined in the laboratory. Undisturbed soil cores are preferred to packed columns (McIntyre 1974). Klute and Dirksen (1986) consider cylindrical cores that measure 20 mm to 100 mm in diameter by 50 mm to 250 mm long to be reasonable sizes for laboratory analysis. The number and distribution of the samples should represent the field to be characterized (Chapter 9). The recommended test fluid is degassed 5.0 mM $CaSO_4$ solution, saturated with .2 to .3 g/l thymol to inhibit biological activity. The instantaneous profile method described for the field has also been adapted to the laboratory (Klute and Dirksen 1986).

Infiltration Rates. Getting water to move into and through soils to carry unwanted salts and exchangeable ions from the root zone is usually the primary consideration in reclaiming saline and sodic soils. As with other properties, infiltration rates are seldom uniform in salt-affected fields.

Infiltration-rate measurement procedures have been discussed for ring infiltrometers (Bouwer 1986), sprinkler infiltrometers (Peterson and Bubenzer 1986), and surface irrigation systems (Kincaid 1986). The measurement procedures should correspond to the site's irrigation methods. Infiltration-rate test conditions should duplicate the water quality, soil surface, duration of water application, and other conditions of the system. With all methods, care needs to be taken to avoid puddling the soil surface while applying the water and to avoid clogging the surface with fine particles suspended in water. If a thin surface layer is sodic, affected by water drops, or otherwise suspected of limiting infiltration, then infiltration before and after removal of the thin surface layer is suggested. For nonuniform profiles with layers of different hydraulic conductivity, the final infiltration rate equals the harmonic mean of the hydraulic conductivities of the various layers if the layers become less permeable with depth (Bouwer 1969).

Surface Topography. With natural vegetation and rain-fed (nonirrigated) agriculture, topography usually affects salinity patterns, since the only water normally available for leaching is precipitation. The precipitation in excess of infiltration does not remain on crests and slopes long enough to leach excess salts from the profile (Wiegand et al. 1966). Also, slope changes in irrigated fields can vary the rate of water intake along furrows. In the steeper parts of the field, leaching is insufficient to remove salt. At the same time, leaching may be adequate on the field's more level portion. Grading irrigated fields to uniform slopes and undercutting saline areas on level, non-irrigated cultivated fields has successfully been used to reclaim saline soils (Lyle and Allen 1966). These results demonstrate how important micro-topographical differences are to the salt balance and how valuable measuring the topography of the soil's surface is. In nearly level areas, the topographic effects can be documented visually or photographically by recording standing water or soil-drying patterns after irrigation or rainfall.

Other physical measurements, often made in research studies but seldom used in practical applications, include soil solution and plant sap osmotic potential, matric water potential-water retention curves, field capacity, available water capacity, solute diffusivity, and air permeability. Soil salinity and sodicity affect all of these physical characteristics.

REFERENCES

Adriano, D.C., and Doner, H.E. 1982. "Bromine, chlorine and fluorine." *Methods of Soil Analysis, Part 2: Chemical and Microbiological Properties*. 2nd ed. (A.L. Page, ed.) Am. Soc. Agron., SSSA, Madison, Wisc. pp. 449–483.

Amoozegar, A., and Warrick, A.W. 1986. "Hydraulic conductivity of saturated soils: field methods." *Methods of Soil Analysis: Part 1: Physical and Mineralogical Methods*. 2nd ed. (A. Klute, ed.), Am. Soc. Agron., SSSA, Madison, Wisc., pp. 735–770.

Beaton, J.D., G.R. Burns, and Platou, J. 1968. "Determination of sulphur in soils and plant material." Tech. Bull. 14. The Sulphur Inst., Washington, D.C.

Bernstein, L. 1974. "Crop Growth and Salinity." In: van Schilfgaarde, J., Agronomy 17 *Drainage for Agriculture*. Am. Soc. Agron., Madison, Wisc., pp. 39–89.

Bernstein, L., and Pearson, G.A. 1956. "Influence of exchangeable sodium on the yield and chemical composition of plants: I. green beans, garden beets, clover and alfalfa." *Agron. J.* 82:247–258.

Bingham, F.T. 1982. "Boron." *Methods of Soil Analysis, Part 2: Chemical and Microbiological Properties.* 2nd ed. (A.L. Page, ed.), Am. Soc. Agron., SSSA, Madison, Wisc., pp. 431–447.

Bouwer, Herman. 1969. "Infiltration of water into nonuniform soil." *J. Irrig. Drain. Div., ASCE* 95(IR4):451–462.

Bouwer, Herman. 1986. "Intake rate: Cylinder infiltrometer." *Methods of Soil Analysis, Part 1: Physical and Mineralogical Methods.* 2nd ed. (A. Klute, ed.), Am. Soc. Argron., SSSA, Madison, Wisc., p 825–844.

Bower, C.A., Reitemeier, R.F., and Fireman, M. 1952. "Exchangeable cation analysis of saline and alkali soils." *Soil Sci.* 73:251–261.

Bresler, E., McNeal, B.L., and Carter, D.L. 1982. *Saline and sodic soils.* Springer-Verlag New York, pp. 174–181.

Cassell, D.K., and Klute, A. 1986. "Water potential: Tensiometry." *Methods of Soil Analysis. Part 1: Physical and Mineralogical Methods.* 2nd ed. (A. Klute, ed.), Am. Soc. Argon., SSSA, Madison, Wisc., pp. 563–596.

Corwin, D.L., and Rhoades, J.D. 1984. "Measurement of Inverted Electrical Conductivity Profiles Using Electromagnetic Induction." *Soil Sci. Soc. Am. J.* 48:288–291.

Dasberg, S., and Dalton, F. N. 1985. "Time domain reflectometry field measurements of soil water content and electrical conductivity." *Soil Sci. Soc. Am. J.* 49:293–297.

Doering, E.J., and Sandoval, F.M. 1976. "Hydrology of saline seeps in the northern great plains." *Trans. ASAE.* 19:856–861, 865.

Everitt, J.H., Gerbermann, A.H., Escobar, D.E., Nixon, P.R., and Alaniz, M.A. 1987. "Detecting saline soils with video imagery." Proc. 11th Biennial Workshop on Color Aerial Photography and Videography in the Plant Sciences. Am. Soc. Photog., Falls Church, Va. 318–325.

Gaines, T.P., and Mitchell, G.A. 1979. "Boron determination in plant tissue by the azomethine H method." *Commun. Soil Sci. Plant Anal.* 10:1099–1108.

Gee, G.W., and Bauder, J.W. 1986. "Particle-size analysis." *Methods of Soil Analysis. Part 1: Physical and Mineralogical Methods.* 2nd ed. (A. Klute, ed.), Am. Soc. Agron., SSSA, Madison, Wisc., pp. 383–411.

Green, R.E., Ahuja, L.R., and Chong, S.K. 1986. "Hydraulic conductivity, diffusivity, and sorptivity of unsaturated soils: field methods." *Methods of Soil Analysis. Part 1: Physical and mineralogical methods.* 2nd ed. (A. Klute, ed.), Am. Soc. Agron., SSSA, Madison, Wisc., pp. 771–798.

Gupta, S. C., and Hanks, R. J. 1972. "Influence of water content on electrical conductivity of the soil." *Soil Sci. Soc. Am. Proc.* 36:855–857.

Hesse, P.R. 1976. "Particle size distribution in gypsic soils." *Plant Soil* 44:241–247.

Holter, M.R., Bair, M., Beard, J.L., Limperis, T., and Moore, R.K. 1970. "Imaging with nonphotographic sensors." *Remote Sensing with Special Reference to Agriculture and Foresty.* National Academy of Sciences, Washington, D.C., pp. 73–163.

Howell, T.A., Hatfield, J.L., Rhoades, J.D., and Meron, M. 1984. "Response of cotton water stress indicators to soil salinity." *Irrig. Sci.* 5:25–36.

Jackson, R.D. 1982. "Canopy temperature and crop water stress." *Advances in Irrigation,* (D. Hillel, ed.), Academic Press, New York, pp. 43–85.

Jackson, R.J., and O'Neill, P.E. 1987. "Salinity effects on the microwave emission of soils." *IEEE Trans. Geosci. and Remote Sens.* Vol. GE-25, pp. 214–220.

Jones, J.B. Jr., and Steyn, W.J.A. 1973. "Sampling, handling and analyzing plant tissue samples." In: Walsh, L.M., and Beaton, J.D., eds. *Soil Testing and Plant Analysis.* Soil Sci. Soc. Am. Inc. Madison, Wisc., pp. 249–270.

Kemper, W.D., and Rosenau, R.C. 1986. "Aggregate stability and size distribution." *Methods of Soil Analysis. Part 1: Physical and Mineralogical Methods.* 2nd ed. (A. Klute, ed.), Am. Soc. Agron., SSSA, Madison, Wisc., pp. 425–442.

Kincaid, D.C. 1986. "Intake rate: border and furrow." *Methods of Soil Analysis. Part 1:* SA, Madison, Wisc., pp. 871–887.

Klute, A., and Dirksen, C. 1986. "Hydraulic conductivity and diffusivity: laboratory methods." *Methods of Soil Analysis. Part 1: Physical and Mineralogical Methods.* 2nd ed. (A. Klute, ed.), Am. Soc. Agron., SSSA, Madison, Wisc., pp. 687–734.

Knudsen, D., Peterson, G.A., and Pratt, P.F. 1982. "Lithium, sodium and potassium." *Methods of Soil Analysis: Part 2: Chemical and Microbiological Properties.* 2nd ed. (A.L. Page, ed.) Am. Soc. Agron., SSSA, Madison, Wisc., pp. 225–246.

Lanyon, L.E., and Heald, W.R. 1982. "Magnesium, calcium, strontium, and barium." *Methods of Soil Analysis. Part 2: Chemical and Microbiological Properties.* 2nd ed. (A.L. Page, ed.), Am. Soc. Agron., SSSA, Madison, Wisc., pp. 247–262.

Lyles, L., and Allen, R.R. 1966. "Landforming for leaching of saline soils in a nonirrigated area." *J. Soil and Water Cons.* 21:57–60.

Manzer, F.E., and Cooper, G.R. 1982. "Use of portable videotaping for aerial infrared detection of potato disease." *Plant Disease* 66:665–667.

Maas, E.V. 1986. "Salt Tolerance of Plants." *Appl. Agric. Res.* 1:12–25.

McIntyre, D.S. 1974. "Procuring undisturbed cores for physical measurements." *Methods for Analysis of Irrigated Soils.* (L. Loveday, ed.) Tech. Commun. 54 Commonwealth Bur. Soils, Appendix I, pp. 154–165.

Moore, R.K., Chestnut, L.J., Porcello, L.J., Stevenson, J., and Ulaby, F.T. 1975. "Microwave remote sensors," (Chap. 9). *Manual of Remote Sensing, Vol. I.* (R.G. Reeves, ed.) Am. Soc. Photogram., Falls Church, Va. pp. 399–537.

Nelson, R.E. 1982. "Carbonate and gypsum." *Methods of Soil Analysis.*

Part 2: Chemical and Microbiological Properties. 2nd ed. (A.L. Page, ed.), Am. Soc. Agron., SSSA, Madison, Wisc., pp. 181–197.

Nieto, K.F., and Frankenberger, W.T. Jr. 1985a. "Single column ion chromatography: I. Analysis of inorganic anions in soils." *Soil Sci. Soc. Am. J.* 49:587–592.

Nieto, K.F., and Frankenberger, W.T. Jr. 1985b. "Single column ion chromatography. II. Analysis of ammonium, alkali metals, and alkaline earth cations in soils." *Soil Sci. Soc. Am. J.* 49:592–596.

Peterson, A.E., and Bubenzer, G.D. 1986. "Intake rate: sprinkler infiltrometer." *Methods of Soil Analysis. Part 1: Physical and Mineralogical Methods.* 2nd ed. (A. Klute, ed.), Am. Soc. Agron., SSSA, Madison, Wisc., pp. 845–870.

Ranger, C.B. 1981. "Flow injection analysis: Principles, techniques, applications, design." *Analy. Chem.* 53:20A–29A.

Reeve, R.C. 1986. "Water potential: Piezometry." *Methods of Soil Analysis. Part 1: Physical and Mineralogical Methods.* 2nd ed. (A. Klute, ed.), Am. Soc. Agron., SSSA, Madison, Wisc., pp. 545–561.

Reitemeier, R.F., and Wilcox, L.V. 1946. "A critique of estimating soil solution concentrations from the electrical conductivity of saturated soils." *Soil Sci.* 61:281–293.

Rhoades, J.D. 1979. "Inexpensive four-electrode probe for monitoring soil salinity." *Soil Sci. Soc. Am. J.* 43:817–818.

Rhoades, J.D. 1982a. "Cation exchange capacity." *Methods of Soil Analysis. Part 2: Chemical and Microbiological Properties.* 2nd ed. (A. L. Page, ed.), Am. Soc. Agron., SSSA, Madison, Wisc., pp. 149–157.

Rhoades, J.D. 1982b. "Soluble salts." *Methods of Soil Analysis. Part 2: Chemical and Microbiological Properties.* 2nd ed. (A.L. Page), Am. Soc. Agron., SSSA, Madison, WI, p. 167–179.

Rhoades, J.D., and Corwin, D.L. 1981. "Determining soil electrical conductivity-depth relations using an inductive electromagnetic soil conductivity meter." *Soil Sci. Soc. Am. J.* 45:255–260.

Rhoades, J.D., and Ingvalson, R.D. 1971. "Determining salinity in field soils with soil resistance measurements." *Soil Sci. Soc. Am. Proc.* 35: 54–60.

Rhoades, J.D., Lesch, S.M., Shouse, P.J., and Alves, W.J. 1989a. "New Calibrations for determining soil electrical conductivity-depth relations from electromagnetic measurements." *Soil Sci. Soc. Am. J.* 53: 74–79.

Rhoades, J.D., Manteghi, N.A., Shouse, P.J. and Alves, W.J. 1989b. "Estimating Soil Salinity from Saturated Soil-Paste Electrical Conductivity." *Soil Sci. Soc. Am. J.* (in press).

Rhoades, J.D., and van Schilfgaarde, J. 1976. "An electrical conductivity probe for determining soil salinity." *Soil Sci. Soc. Am. J.* 40:647–651.

Rivers, E.D., Hallmark, C.T., West, L.T., and Drees, L.R. 1982. "A technique for rapid removal of gypsum from soil samples." *Soil Sci. Soc. Am. J.* 46:1338–1340.

Robbins, C.W. 1989. "Sample preparation for determining ions in dark colored sodic soil extracts." *Soil Sci. Soc. Am. J.* 53:721–725.

Shainberg, I., Rhoades, J.D., and Prather, R.J. 1981. "Effect of low electrolyte concentration on clay dispersion and hydraulic conductivity of a sodic soil." *Soil Sci. Soc. Am. J.* 45:273–277.

U.S. Salinity Laboratory Staff. 1954. L.A. Richards, ed. *Diagnosis and Improvement of Saline and Alkali Soils*. U.S. Dept. of Agric. Handbook. No. 60.

Watson, K.K. 1966. "An instantaneous profile method for determining the hydraulic conductivity of unsaturated porous materials." *Water Resour. Res.* 2:709–715.

Wiegand, C.L., Heilman, M.D., and Gerbermann, A.H. 1968. "Detailed plant and soil thermal regime in agronomy." Proc. 5th Int. Symp. Rem. Sens. Environ., pp. 325–342.

Wiegand, C.L., Lyles, L., and Carter, D.L. 1966. "Interspersed salt-affected and unaffected dryland soils of the Lower Rio Grande Valley: II. Occurrence of salinity in relation to infiltration rates and profile characteristics." *Soil Sci. Soc. Am. Proc.* 30:106–110.

Wilcox, L.V. 1960. "Boron injury to plants." U.S. Dept. Agric. Info. Bull. No. 211, 7pp.

Wood, J. D. 1978. "Calibration stability and response time for salinity sensors." *Soil Sci. Soc. Am. J.* 42:248–250.

PART FIVE—DIAGNOSIS OF SALT PROBLEMS

CHAPTER 11

IRRIGATION WATER QUALITY ASSESSMENTS[a]

INTRODUCTION

This chapter discusses the effects of inorganic elements on the quality of irrigation waters. It focuses on salinity, sodicity and the effects of sodicity on soil permeability, major cations and anions, and trace elements.

A meaningful assessment of the quality of water used for irrigation should consider such local factors as the chemical reactivity of constituents dissolved in the water, the soil's chemical and physical properties, climate, and irrigation management practices. It should also consider the effects of irrigation on the quality of agricultural drainage, the effects on humans and animals of chemicals concentrated in harvested plant products, and the economic conditions that determine how much salinity-induced reduction in yield or quality can be tolerated.

To avoid the accumulation of toxic amounts of waterborne substances in the root zone of irrigated lands, the input of those substances to the soil must, over the long term, not exceed the sum of losses from the soil and conversions to unavailable forms. Losses include removal in harvested crops, transport by subsurface drainage, erosion by wind and water, and, for some elements, volatilization of gaseous compounds. Immobile elements, such as As and Cu, often are converted in the root zone to unavailable forms. Conversion processes are less prevalent for readily adsorbed and desorbed elements, such as Na and B, and insignificant for highly mobile elements, such as Cl.

Most soluble constituents, being mobile, can be removed by leaching. Thus, leaching often can be used to adjust the concentrations of soil chemical constituents to accommodate crops.

If the element of interest is immobile under existing soil conditions and if leaching losses are insignificant, then the elemental inputs not removed by plants or converted in the soil to unavailable forms will accumulate as soluble and labile forms. These forms are related as follows:

$$\text{Soluble} \leftrightarrow \text{Labile} \leftrightarrow \text{Residual}$$

The soluble element adsorbs or desorbs into the labile form as the amount in solution increases or decreases. The labile (adsorbed) element

---

[a]Prepared by: Parker F. Pratt and Donald L. Suarez, U.S. Salinity Lab., USDA-ARS, 4500 Glenwood Drive, Riverside, CA 92501.

is transformed to or from the residual (unavailable) form. Only the soluble form is immediately available to the plant. As the soluble element is removed by plant roots, desorption from the labile pool replenishes the soluble pool.

The level of toxicity depends on the amount of the toxic constituent in solution and the capacity of the labile pool. The hazard posed by elements that exist in soluble and labile forms in the soil is that once toxic levels are attained, eliminating or reducing these levels involves removal in harvested crops and conversion to insoluble residual forms, both of which are processes that can take decades, even if inputs of the element cease. These situations can be avoided by ensuring that inputs of potentially toxic elements remain below the levels that are tolerable by the most sensitive crop to be grown and by avoiding crops that will bioaccumulate the elements of concern.

SALINITY

Salinity of an irrigation water is defined as the total sum of dissolved inorganic ions and molecules. The major components of salinity are the cations Ca, Mg, and Na, and the anions Cl, SO_4, and HCO_3. The K and NO_3 ions are usually minor components of the salinity. The effects of these and other minor dissolved constituents, such as boron, are excluded in assessing the salinity of an irrigation water.

Salinity reduces crop growth by reducing the ability of plant roots to absorb water. The soluble ions and molecules reduce the availability of water to a plant, a phenomenon known as the osmotic pressure effect. Water availability in the soil relates to the sum of the matric and osmotic potential. As the water content of the soil decreases, the matric and osmotic potential decrease, i.e., become more negative. Evaporation and transpiration by plants remove almost pure water, leaving behind soluble salts in the soil. Thus, matric and osmotic potential decrease, as the water in the soil decreases.

Salinity is most easily and conveniently measured by determining the electrical conductivity (EC) of the solution. The United States Salinity Laboratory (USSL) showed that the EC of the solution was not very dependent on the type of salt, when the salts were expressed in $mmol_c/l$ (1954). They showed that EC in soil extracts highly correlated with total salts, when the data were expressed in $mmol_c/l$, and that the osmotic potential (OP) is approximately related to EC by the equation $OP = -36 \times EC$, where OP is expressed in kPa and EC in dS/m at 25°C.

The EC is used as an expression of salinity in the irrigation water (EC_{iw}), salinity in the saturation extract (EC_e), and salinity in the soil solution (EC_{ss}). The USSL (1954) developed the saturation extract technique, a way to estimate soil salinity that uses a reference water content. Demineralized water is added to a soil sample until the soil paste glistens and slightly flows when the container is tipped. The soil paste is then filtered under suction. The solution obtained is analyzed for EC_e and soluble constituents. The water content of the saturated paste is roughly twice that of field capacity. Thus, the EC_{ss} at field capacity is about twice that at saturation. The EC_e provides a way to assess the

TABLE 11.1 Relative Solute Concentrations of Soil Water (Field Capacity Basis) Compared to That of Irrigation Water Related to Depth in the Root Zone and Leaching Fraction[a]

Root zone in quarters (1)	V_{cu}[b] (2)	F_c at LF values of: 0.05 (3)	0.10 (4)	0.20 (5)	0.30 (6)	0.4 (7)	0.50 (8)
0	0	1.0	1.0	1.0	1.0	1.0	1.0
1	40	1.61	1.56	1.47	1.39	1.32	1.25
2	70	3.03	2.70	2.27	1.96	1.72	1.54
3	90	7.14	5.26	3.57	2.70	2.17	1.82
4	100	20.00	10.0	5.00	3.33	2.50	2.00
Average F_c[c]		5.58	3.76	2.58	2.06	1.74	1.54

[a]Assuming a water uptake of 0.4, 0.3, 0.2, and 0.1, respectively, from the first through fourth quarters of the root zone.
[b] Cumulative percentage of consumptive use above each indicated depth in the root zone.
[c] The average for each quarter of the root zone divided by four.

salinity of field samples. The relationships among EC_{iw}, EC_e, and EC_{ss} are useful, as a large amount of data on salinity tolerances of crops to EC_e values exists. The saturation extract has the advantage of minimizing salt dissolution, since less water is added than with other dilution-extraction methods, such as the 1-to-5 or the 1-to-10 soil/water ratios.

The salinity of irrigation waters can be assessed by relating EC_{iw}, the leaching fraction (LF), the average EC_{ss} at field capacity (EC_{aw}), EC_e, and the salt tolerance of crops (Rhoades 1984). The EC_{aw} is calculated from EC_{iw} and the concentration factor F_c, which equals 1/LF at the bottom of the root zone, and an assumed distribution of water uptake. It can be assumed that water is removed by evapotranspiration in proportions of 0.40, 0.30, 0.20, and 0.10, from the root zone's first, second, third, and fourth quarters, respectively. Alternatively, an exponential water uptake function can be used. However, the F_c values would not greatly change. Table 11.1 shows that the extent to which soil water is concentrated relative to irrigation water is related to depth in the root zone and LF.

Since the EC_e is about one-half of the EC_{ss}, the F_c values to convert from EC_{iw} to EC_e are 2.79, 1.88, 1.29, 1.03, 0.87, and 0.77 for LF of 0.05, 0.10, 0.20, 0.30, 0.40, and 0.50, respectively. These F_c values can be used to calculate the relationships between EC_e and EC_{iw} for each LF, as presented in Fig. 11.1. The threshold values listed for various crops were derived from the following relationship between yield and EC_e:

$$\text{Yield} = 100 - B\,(EC_e - A)$$

where A = the salinity concentration at which growth depression (threshold) starts, and B = the percent of yield decrease per unit EC_e above the threshold level.

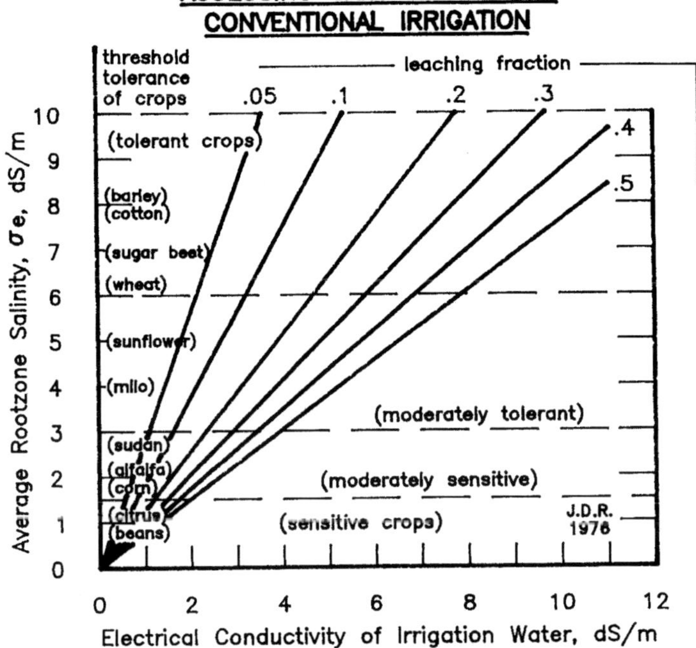

Fig. 11.1 Relationship Between Average Root Zone Salinity (Saturation Extract Basis), Electrical Conductivity of Irrigation Water, and Leaching Fraction to Use for Conditions of Conventional Irrigation Management (Rhoades 1982)

To use Fig. 11.1, determine the EC_{iw} and then estimate the LF that can be obtained for the soil with the available irrigation management system. The point in the figure based on EC_{iw} and LF will indicate crops that can be grown successfully without decreases in yield from salinity. For example, if EC_{iw} is 4.0 and a leaching fraction of 0.20 is expected, only tolerant plants can be grown. If leaching fractions of 0.5 or greater are anticipated, moderately tolerant plants can be grown. If the nature of the soil or water availability is such that only very small leaching fractions are possible, then this water, in which EC_{iw} = 4.0, will reduce yields in even the most tolerant crops. Thus, assessing the effects of salinity as a parameter of water quality depends on the soil, the crops, the water available, the irrigation system, the irrigator's expertise in achieving the needed leaching, and the decrease in yield that can be tolerated.

Fig. 11.1 is based on calculations for conventional irrigation systems, such as furrow or flood irrigation, which allow considerable drying between applications. If high-frequency irrigation systems are used so that development of the matric potential between irrigations is insignificant, the F_c values can be reduced to 1.79, 1.35, 1.03, 0.87, 0.77, and 0.70 for LF of 0.05, 0.10, 0.20, 0.30, 0.40, and 0.50, respectively. Fig. 11.2

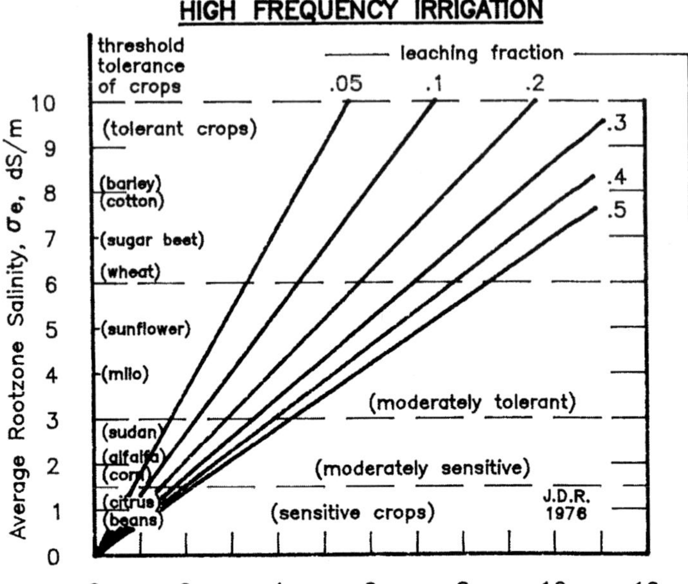

Fig. 11.2 Relationship Between Water-Uptake-Weighted Salinity (Saturation Extract Basis), Electrical Conductivity of Irrigation Water, and Leaching Fraction to Use for Conditions of High-Frequency Irrigation (Rhoades 1981)

shows the relationships between EC_e and EC_{iw} for various LF under high-frequency irrigation. In this case, the utility of the water of any given EC_{iw} increases, as compared to conventional irrigation systems.

The method of assessing the salinity of water described above can be adapted to different sites. The suitability of the water supply can be assessed based on such local conditions as the ease with which the soil can be leached, the salt tolerance of the crops, the irrigation system, the skill of the manager, and the climate. Perhaps the weakest link in this system is the estimation of the LF, which is seldom measured directly, but can be determined by measuring water infiltration and estimating evapotranspiration.

The foliar injury from salts in irrigation waters that are sprinkled on plants depends on the concentrations of the individual ions in the water, the sensitivity of the crop, the frequency of sprinkling, and, to some extent, environmental factors, such as temperature, relative humidity, and water stress of the plants before irrigation. Maas et al. (1982) reported that rates of salt absorption by leaves increased as the frequency of irrigation increased but that a threefold increase in the duration of sprinkling had no measurable effect on salt absorption. Nighttime sprinkling reduces foliar absorption and injury.

Foliar absorption by Na or Cl ions at concentrations of less than 5 mmol/l damages some fruit trees. Other crops can tolerate concentrations of greater than 20 mmol/l. Thus, although an increase in Na or Cl in the water reduces its usefulness for sprinkler systems by reducing the types of crops that can be grown without foliar injury, no concentration limits can be recommended. Also, the degree of injury depends on the irrigation system and how it is operated. Consequently, limits or guidelines for sprinkler irrigation at current levels of knowledge are too arbitrary to be useful.

SODICITY

Sodium hazards of irrigation and soil waters can negatively affect crop production. Unlike the salinity hazard, excessive exchangable Na does not impair the uptake of water by plants but does impair the infiltration of water into the soil. The growth of plants is, thus, affected by an unavailability of soil water or poor aeration.

The reduction in infiltration of water can usually be attributed to surface crusting, the dispersion and migration of clay into the soil pores, and the swelling of expandable clays. All of these phenomena relate to the distance of charge neutralization for soil particles, predominantly clay, but also oxides in more weathered landscapes.

The hydrated exchangeable cations neutralize the net negative charge on clays. The distance of charge neutralization (the double-layer thickness) depends on the cation valence, hydration energy, and ion concentration in solution. Divalent cations, such as Ca and Mg, neutralize the surface charge in relatively short distances, even at low concentrations. Particles are repulsed when the charge is neutralized too far from the surface and the electrostatic repulsion between particles exceeds the attractive (van der Waals) forces. In contrast to Ca and Mg, the exchangeable Na ion has a much larger, double-layer thickness and requires high concentrations in solution before particle aggregation and swelling are reduced. Consider Ca and Mg as stabilizing ions and Na as a destabilizing ion in regard to the soil structure.

The sodicity of a soil is given by the exchangeable Na percentage, ESP, which is the percentage of the exchangeable charge neutralized by Na. The ESP of a soil can be estimated from the SAR of the water, i.e., $ESP = 1.475\ SAR/(1 + 0.0147\ SAR)$, based on a set of data from soils in the western United States (USSL 1954, Chapter 3). The ESP value alone is insufficient for predicting soil stability. Soil structure depends on many other factors, including soil salinity, tillage, mineralogy, organic matter, and pH.

The sodic-hazard potential of a water is often evaluated from the SAR and salinity. At the same SAR, the dispersion potential of a dilute water exceeds that of a more saline water. Various authors have developed stability lines related to concentration and SAR. Fig. 11.3 shows the guidelines of Rhoades (1982) and Quirk and Schofield (1955) represented as solid and dashed lines, respectively. Rhoades based his primarily on experience and data from arid soils in California. Quirk and Schofield (1955) based theirs on a non-calcareous soil in England. In

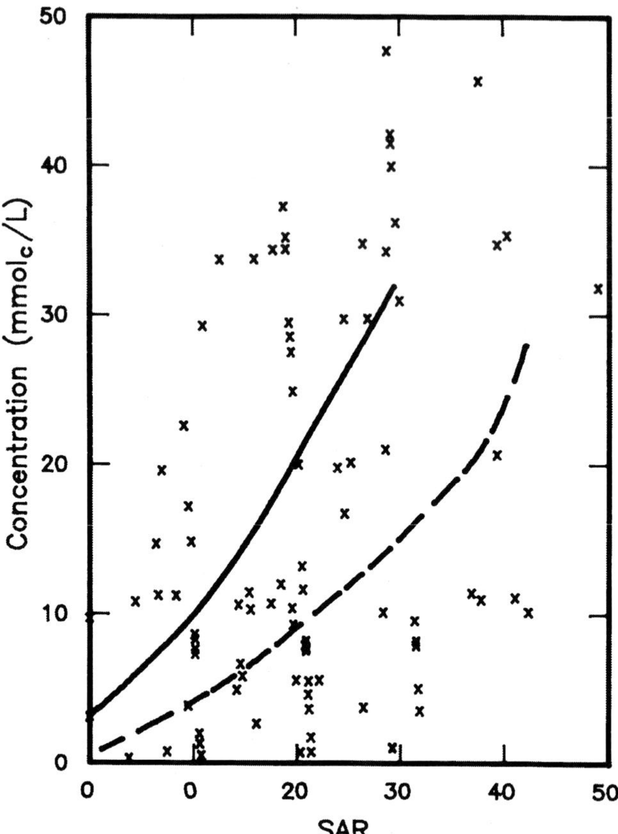

Fig. 11.3 Relationship Between SAR and Solute Concentration (in mmol/l) At Which a 25% Reduction in Soil Hydraulic Conductivity was Observed (The data were obtained from laboratory studies of packed soil columns containing arid-land soils. The dashed and solid lines are guideline values recommended by Quirk and Schofield (1955) and Rhoades (1982), respectively.)

each instance, the region below the line represents unstable permeability, and the region above it represents stable permeability.

Fig. 11.3 also shows the concentration and SAR values at which a 25% reduction in saturated hydraulic conductivity took place in packed laboratory soil columns from all available published data from arid soils. A general relationship cannot be predicted because soils greatly differ, but a good SAR versus concentration relationship for a set of soils from a region or locality is possible. For all arid soils examined, decreasing salinity, or increasing sodicity, or both, decreases soil stability.

Differences among soils shown in Fig. 11.3 are at least partly due to different experimental procedures used by different researchers, such as column packing, flow rates, and saturation methods. However, variations in clay mineralogy, clay content, organic matter, oxide content, and pH likely account for most of the variation.

For specific groups of soils, several researchers have demonstrated that soil stability correlates well with organic matter, or oxide content, or both. Soils with a very high amount of organic matter and oxides, such as some tropical soils, show little or no loss of hydraulic conductivity, even when saturated with Na and equilibrated to very low levels of salinity. Since the effects of variables other than salinity, SAR, and their interactions have not been quantified, this graph offers only an approximate guideline. It is recommended that the effects of a water be tested directly on the soil of interest with column leaching studies, tests of aggregate stability, or tests of flocculation after the soil has been dispersed in a test tube.

The SAR value calculated from the analyses of a surface water usually represents the SAR of the irrigation water on the surface of the soil. This is not the case for ground waters that are equilibrated with a much higher PCO_2 and, thus, are lower in pH. Exposure of the ground water to atmospheric CO_2 conditions, which occurs with sprinkler irrigation or conveyance through open canals, raises the pH and could cause calcite precipitation. The adjusted SAR of the irrigation water is a correction for $CaCO_3$ solubility (see Chapter 3). In such instances, assume a PCO_2 of 0.1 kPa of the soil surface and adjust the SAR as described in Chapter 3. The adjusted SAR can also be used to estimate the SAR in or below the root zone by correcting for precipitation and assuming no ion exchange. The concentration factor (1/ leaching fraction), the PCO_2, and the concentration of irrigation water are needed. If specific PCO_2 data in the root zone are unavailable, the values of 1 kPa and 5 kPa can be used for sandy and clay soils, respectively.

High pH values, i.e., pH > 8.5, indicate waters with an excess of alkalinity over Ca. They usually pose a sodicity hazard. Waters with pH values below 8.5 can also have high alkalinity, depending on PCO_2. For ground waters, samples are aerated or shaken until the water is equilibrated with ambient CO_2 levels, and then the pH re-measured. If pH $>$ 8.5, then $HCO_3 > Ca$ for all waters undersaturated or saturated with calcite. The higher the pH, the greater the imbalance. Imbalances in HCO_3 and Ca can also exist in waters with a pH of less than 8.5 if they are dilute waters that are undersaturated with calcite, such as surface waters from snowmelt. The very low EC of the water (0.1 dS/m) compounds their sodium hazard. High pH (pH > 9.0) directly and adversely affects infiltration (Suarez et al. 1984), as well as limiting Ca concentrations and increasing the SAR.

The sodicity hazard of an irrigation water also depends on the management system used. Dispersing the soil at the surface requires inputs of energy and a weakening of chemical bonding. Irrigation by sprinkler will increase the infiltration problems associated with irrigation waters. Due to the greater likelihood of surface crusting, sprinkler irrigation may be unsuitable for waters that tend to cause dispersion or swelling. Drip or surface systems will produce less physical disruption with such waters. This is a very important consideration when using waters with a potential dispersion or swelling hazard.

High pH values in irrigation waters may cause nutritional and infiltration problems and, thus, need to be amended to reduce the alkalinity. The composition of the divalent ion component only slightly affects the stability of soils at a particular ESP value, with Ca only slightly more stable than Mg (McNeal et al. 1968). The greater selectivity of most soils for Ca as compared to Mg means that the Mg-Na system has a higher ESP than does the Ca-Na system at the same SAR values. This, combined with the high pH that frequently occurs in low Ca systems, may account for why Mg seems deleterious to infiltration as compared to Ca.

Weathering of Ca minerals, especially $CaCO_3$, decreases SAR and increases electrolyte concentration. The response of sodic soil in arid areas to rain or waters of very low salinity seems related to its weathering potential (Shainberg et al. 1981). The more stable soils appear to maintain higher electrolyte levels than unstable soils do. Tropical Hawaiian soils also appear much more stable than arid soils at comparable SAR and salinity levels, possibly due to their high oxide content (McNeal et al. 1968). Moderate amounts of organic matter also increase the stability of a soil (Kemper and Koch 1966; Dong et al. 1983).

The potential hazard of reduced water infiltration is partly related to the intensity and timing of rainfall in a region. Rainfall, generally < 0.06 dS/m, is relatively pure water. When it infiltrates the soil, the salinity of surface soil can decrease rapidly, but the soil may remain at almost the same ESP. As a result, the potential for dispersion by rainfall is especially high if the ESP of the soil is high. Rainfall, as with sprinkler irrigation, contributes dispersive energy.

Surface (flood, furrow, or drip) irrigation also can cause particles to migrate and result in sealing of the soil surface. However, the inputs of energy are less than with sprinkler irrigation, and lower-quality water can be better tolerated. In areas with little rainfall, such as California's Imperial Valley, these phenomena are generally neglected. In areas with appreciable seasonal rainfall (> 200 mm), surface amendments, such as gypsum, can be applied to maintain the electrolyte concentrations above values for dispersion and swelling and to allow the water to infiltrate well. If rainfall occurs throughout the year, tillage or repeated gypsum applications may be needed.

IONIC BALANCES

Ca, Mg, and K are the major cations needed for plant nutrition. Minor requirements include Fe, Mn, Zn, and Cu. Plants generally tolerate widely varying concentrations of the major cations, including Na, which is not required for plant growth. The Ca requirement of a plant is generally low, i.e., 0.7 mmol/l to 1.5 mmol/l. It appears to depend on the presence of other ions. The Ca requirement may be related to ion competition and, thus, is better expressed in terms of ion ratios. High Mg/Ca ratios in solution may result in Ca deficiencies in plants, despite high absolute Ca concentrations. Carter et al. (1979) observed reduced growth in barley, starting at Mg/Ca ratios of 1.0, independent of salinity or absolute Ca concentrations. Calcium requirements are also greater at low pH than at high pH (Marschner 1986). Guidelines for specific cation ratios cannot be established because cultivars respond in widely varied ways to cation composition.

Concentrations of Mg and K generally are high enough in irrigation waters to prevent deficiency symptoms in plants. The micronutrient cations of Fe, Mn, Cu, and Zn are virtually absent from most irrigation waters. The soil generally supplies these nutrients. The irrigation water limits the availability of micronutrients if the water causes the pH to increase. Specific cation toxicity takes place with excess Na, predominantly in citrus and stone fruits.

For anions, specific toxicities occur, rather than ionic imbalances. Although most plants tolerate high Cl concentrations, woody species and some grape rootstocks do not. High levels of NO_3, often associated with saline waters, may narrow the selection of crops suitable for irrigation. Nitrogen is one of the essential elements for the growth of plants. Optimal growth requires 2% to 5% N on a dry weight basis, depending on the species, developmental stage, organ to be optimized, and ultimate use of the plant or parts (Marschner 1986). High levels of NO_3 during early growth enhance shoot elongation which, in cereals, increases susceptibility to lodging. High substrate levels of NO_3 increase total N in ryegrass, but decrease carbohydrates and increase cellulose content. Such high levels of NO_3, i.e., > 1, to 2% by dry weight, can be toxic to grazing animals. High NO_3 concentrations can cause excessive vegetative growth and reduce production of fruits and other harvested products. Excess N reduces the production of fruits in some varieties of tomatoes and reduces the sugar content and increases impurities in sugar beets. Due to such problems, Ayers and Westcott (1985) indicate that the restriction on the use of water increases as the NO_3–N concentration increases from 5 mg/l to 30 mg/l.

BORON

Boron is essential for and potentially toxic to plants. Boron deficiencies take place in the microgram/l concentration range in soil solutions. Toxicities take place at concentrations above a few mg/l for most plants. In the toxicity range, plants respond to B in the soil solution (B_s) rather than to B adsorbed on soil particles. Hence, solution and sand-culture data are used to evaluate the response of plants to B.

Bingham et al. (1985) and Francois (1984) demonstrated that yield decreases to B toxicity can be fitted to a two-parameter model for salt tolerance (Maas and Hoffman 1977). The expression for this model is

$$Y = 100 - m(x-A)$$

where Y = relative yield; m = the decrease in yield per unit increase in B concentration; A = the maximum concentration of B that does not reduce yield (threshold); and x = the B concentration in the nutrient, sand culture, or soil solution.

Early recommendations on B tolerance of plants were largely based

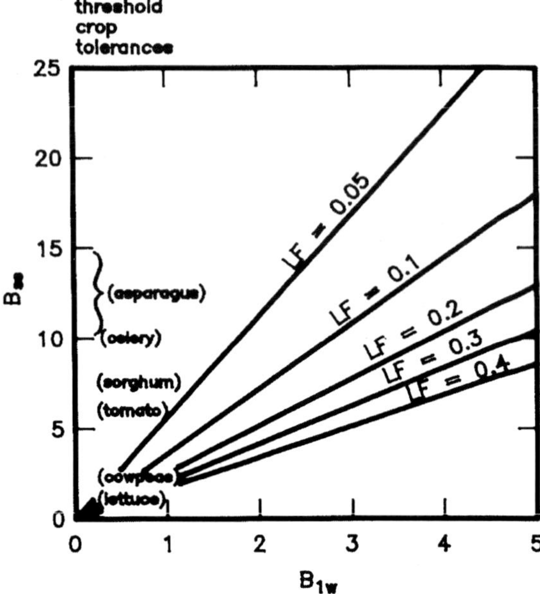

Fig. 11.4 Relationship Between Mean B_s in The Root Zone and B_{iw} for Several LF Values (The threshold for asparagus is given as a range because research to establish a more definite value has yet to be conducted.)

on visual symptoms. Francois (1984) showed that visual symptoms of B toxicity do not correlate with the yield of marketable product.

Decreases in yield from boron toxicity depend on the tolerance of the crop to boron and on the B_s, which depends on the concentration of boron in the irrigation water (B_{iw}), the leaching fraction (LF), and the departure from a steady-state relationship between adsorbed B and B_s. At steady-state input and output of boron from the root zone, the mean B_s is related to B_{iw} and the LF, as shown in Fig. 11.4. In that figure, B_s is plotted against B_{iw} for various values of LF, using mean concentration factors of 5.6, 3.8, 2.6, 2.1, and 1.7 for LF values of 0.05, 0.1, 0.2, 0.3, and 0.4, respectively (Rhoades 1982). These mean concentration factors were calculated based on the assumption that water extraction is 40%, 30%, 20%, and 10%, from the root zone's first, second, third, and fourth fractions, respectively.

Bingham et al. (1988) used the salinity model developed by Rhoades (1982) to predict the B_s to B_{iw} ratio in a three-year lysimeter experiment for which near-steady-state conditions had been achieved. Measured and predicted values (Fig. 11.5) agreed reasonably well, indicating that Rhoades' model (Fig. 11.4) can assess the use of boron-containing waters for irrigation.

Since boron is adsorbed onto and released from the surfaces of soil particles, soil solutions are buffered against rapid changes in B concentration. If the boron in irrigation water is increased, boron is adsorbed,

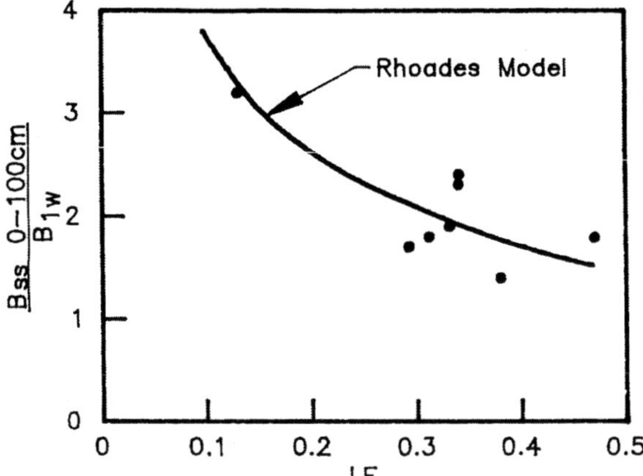

Fig. 11.5 Relationship Between Ratio of B_s in 0 cm to 100 cm Depth of Lysimeters to B_{iw} and LF for an Experiment in Which Boron Input and Output Was Approaching Steady-State (Bingham et al. 1988).

resulting in a smaller increase in the solute boron concentration than the increase to irrigation water. The time required to reach a steady-state concentration of boron depends on the increased B concentration, the amount of water used, the leaching fraction, and the sorption capacity of the soil volume of the root zone. Jame et al. (1981) reported that the time ranged from three to 150 years. Three years was adequate for a sandy soil that can adsorb little B and has been treated with a solution of B at 10 mg/l, and 150 years was required for a clay loam soil that could adsorb a lot of B and was treated with a solution of 0.1 mg/l boron. If boron in the irrigation water is decreased, the soil releases B and time is needed to reduce the B_s. Also, the volume of low-boron water needed to reduce the B_s from toxic to non-toxic levels is two to three times greater than is needed for a comparable reduction in Cl.

The ratio of concentration in the soil solution at field capacity to the concentration in the saturation extract is approximately two for anions not adsorbed or precipitated, such as chloride. However, the ratio is less than two for boron, because the adsorption of boron on the soil surface depends on the concentration. When evapotranspiration decreases the water content and concentrates the soil solution, B is adsorbed. The concentration factor decreases as the adsorption or buffer capacity of the soil increases. Jame et al. (1982) reported that this ratio ranged from 1.0 to 1.8, depending on the concentration of B and the adsorption capacity of the soil. Consequently, the B concentration in the saturation extract does not indicate B toxicity under field conditions well.

TRACE ELEMENTS

Trace elements are those that occur in waters and soil solutions at concentrations of less than a few mg/l, with most concentrations in the

TABLE 11.2 Recommended Maximum Concentrations of 15 Trace Elements in Irrigation Waters for Long-Term Protection of Plants and Animals

Element (1)	Recommended maximum concentration[a] (mg/l) (2)	Comments (3)
Arsenic	0.10	This guideline will protect sensitive crops grown on sandy soils. Higher concentrations can be tolerated by some crops for short periods when grown in fine-textured soils.
Beryllium	0.10	Toxicities to plants have been reported at concentrations as low as 0.5 mg/l in nutrient solutions and at levels in the soil greater than 4% of the cation-exchange capacity.
Cadmium	0.01	Concentrations equal to or less than 0.01 mg/l will require 50 years or more to exceed the recommended maximum Cd loading rate. Removal in crops and by leaching will partially compensate and perhaps allow use of the water indefinitely.
Chromium	0.10	Toxicity in nutrient solutions has been observed at a concentration of 0.50 mg/l and in soil cultures at a rate of 120 kg/ha. Toxicity depends on the form of Cr existing in the water and soil and on soil reactions.
Cobalt	0.05	A concentration of 0.10 mg/l is near the toxic threshold for many plants grown in nutrient solution. Toxicity varies, depending on type of crop and soil chemistry.
Copper	0.20	Concentrations of 0.1 mg/l to 1.0 mg/l in nutrient solutions have been found to be toxic to plants, but soil reactions usually precipitate or adsorb Cu so that soluble Cu does not readily accumulate.
Fluoride	1.0	This concentration is designed to protect crops grown in acid soils. Neutral and alkaline soils usually inactivate F, so higher concentrations can be tolerated.

(continued)

TABLE 11.2 Recommended Maximum Concentrations of 15 Trace Elements in Irrigation Waters for Long-Term Protection of Plants and Animals (Continued)

Element (1)	Recommended maximum concentration[a] (mg/l) (2)	Comments (3)
Lead	5.0	Plants are relatively tolerant to Pb, and soils effectively sorb or precipitate it. Toxicity to animals typically is caused not by Pb absorption from soils, but by aerial deposition of lead on foliage of pasture and forage plants.
Lithium	2.5[b]	Most crops are tolerant to Li up to 5 mg/l in nutrient solutions. Citrus, however, is highly sensitive to Li. Lithium is a highly mobile cation that will leach from soils over an extended period of time.
Manganese	0.20	Some crops show Mn toxicities at a fraction of a mg/l in nutrient solution, but typical soil pH and oxidation-reduction potentials control Mn in the soil solution so that the Mn concentration of irrigation water is relatively unimportant.
Molybdenum	0.01	This concentration is below phytoxic level but is recommended to protect animals from molybdosis because of excess Mo in forages.
Nickel	0.20	Many plants show toxicity at Ni concentrations of 0.5 mg/l to 1.0 mg/l. Toxicity of this element decreases with increase in pH, so acid soils are the most sensitive.
Selenium	0.02	This guideline will protect livestock from selenosis because of Se in forage. Selenium absorption by plants is greatly inhibited by SO_4, so the guideline for this element can be increased for gypsiferous soils and waters.
Vanadium	0.10	Toxicity to some plants has been recorded at V concentrations above 0.5 mg/l.

(continued)

TABLE 11.2 Recommended Maximum Concentrations of 15 Trace Elements in Irrigation Waters for Long-Term Protection of Plants and Animals (Continued)

Element (1)	Recommended maximum concentration[a] (mg/l) (2)	Comments (3)
Zinc	0.50	A number of plants show Zn toxicity at concentration of 1 mg/l in nutrient solution, but soils have a large capacity to precipitate this element. This guideline is designed to provide protection for acid sandy soils. Neutral and alkaline soils can accept much larger concentrations without developing toxicities.

[a] Loading rates in kg/ha-yr can be calculated from the relationship that 1 mg/l in the water gives 10 kg/ha-yr when water is used at a rate of 10,000 m^3/ha-yr.
[b] For citrus, the maximum recommended concentration is 0.075 mg/l.

μg/l range. Some are essential for plants and animals, but all can become toxic to plants and animals at elevated concentrations or doses.

Virtually no experiments have yet been conducted to determine the criteria for quality for trace elements in irrigation water. Hence, guidelines have been based on results from sand, solution, and pot cultures, field trials with applications of chemicals, laboratory studies of chemical reactions, and animal feeding and grazing trials.

Table 11.2 presents the recommended maximum concentrations of 15 trace elements. They are designed to protect the most sensitive crops and animals from toxicities when the most vulnerable soils are irrigated. These concentrations should be considered as guidelines, but not as criteria for water quality. If sufficient knowledge becomes available to show that these concentrations can be exceeded without adversely affecting soils, crops, and animals, then new guidelines can be established. For example, for the irrigated lands of the west side of California's San Joaquin Valley, Pratt et al. (1988) recommended that the guideline for Se in the selenate form be increased to 0.10 mg/l and the guideline for Mo increased to 0.05 mg/l. The conditions included alkaline, fine-textured soils; saline drainage waters, which need high leaching fractions to prevent reduced yields; and drainage waters dominated by SO_4 anions, which inhibit the absorption of Se and Mo by plants.

Other water-quality guidelines list the elements Al, Fe, Sn, Ti, and W (NAS 1973; Ayers and Westcott 1985), but limits for these elements have little meaning. If certain soil conditions develop, e.g., low pH for Al and highly reduced, water-logged conditions for Fe, these elements can become toxic to plants due to the dissolution of Al or Fe from soil solids. Aerated soils with pH values above 5.5 will inactivate the Al and

Fe in irrigation waters. Guidelines for Sn (tin), Ti (titanium), and W (tungsten) cannot be made due to insufficient information.

REFERENCES

Ayers, R. S., and Westcot, D. W. 1985. "Water quality for agriculture." FAO Irrigation and Drainage Paper 29, Rev. 1. FAO, United Nations, Rome. 174 pp.

Bingham, F. T., J. E. Strong, J. D. Rhoades, and R. Keren. 1985. "An application of the Maas-Hoffman salinity response model for boron toxicity."*Soil Sci. Soc. Am. J.* 49: 672–674.

Bingham, F. T., J. D. Rhoades, and J. E. Strong. 1988. "The effect of irrigation water management on the distribution of boron in lysimeters under wheat-sorghum cropping." *J. Environ. Qual.* (Submitted).

Carter, M. R., G. R. Webster, and R. R. Cains. 1979. "Calcium deficiency in some solonetzic soils of Alberta." *J. of Soil Sci.* 30: 161–174.

Dong, A., G. Chesters, and G. V. Simsiman. 1983. "Soil dispersibility." *Soil Sci.* 136: 208–212.

Francois, L. E. 1984. "Effect of excess boron on tomato yield, fruit size, and vegetative growth." *J. Amer. Soc. Hort. Sci.* 109(3): 322–324.

Jame, Y. W., W. Nicholaichuk, A. J. Leyshon, and C. A. Campbell. 1982. "Boron concentration in the soil solution under irrigation: A theoretical analysis." *Can. J. Soil Sci.* 62: 461–471.

Kemper, W. D., and E. J. Koch. 1966. "Aggregate stability of soils from western United States and Canada." U.S. Dept. of Agr. Tech. Bulletin 1335. U.S. Gov. Printing Office, Washington, D.C.

Maas, E. V., and G. J. Hoffman. 1977. "Crop salt tolerance—current assessment." *J. Irrig. Drainage Div., ASCE* 103(IR2): 115–134.

Maas, E. V., R. A. Clark, and L. E. Francois. 1982. "Sprinkler-induced foliar injury to pepper plants: Effects of irrigation frequency duration, and water composition." *Irrig. Sci.* 3: 101–109.

Marschner, Horst. 1986. "Mineral nutrition of higher plants." Academic Press, London. 674 pp.

McNeal, B. L., D. A. Layfield, W. A. Norvell, and J. D. Rhoades. 1968. "Factors influencing hydraulic conductivity of soils in the presence of mixed-salt solutions." *Soil Sci. Soc. Amer. Proc.* 32: 187–190.

National Academy of Sciences. 1973. *Water Quality Criteria, 1972*. Environmental Studies Board Committee on Water Quality Criteria. U.S. Government Printing Office, Washington, D.C. 20402. 594 pp.

Pratt, P. F., Albasel, N., Joseph, H., and Resco, C. 1988. "Trace element guidelines for irrigation waters in the San Joaquin Valley. Report to the Calif. Water Resour. Control Board from the Dept. of Soil and Envir. Sci., Univ. of Calif., Jan. 1988. 120 pp.

Quirk, J. P., and Schofield, R. K. 1955. "The effect of electrolyte concentration on soil permeability." *J. Soil Sci.* 6: 163–178.

Rhoades, J. D. 1982. "Reclamation and management of salt-affected soils after drainage." Proc. of the First Annual Western Provincial Conf. Rationalization of Water and Soil Res. and Management. Lethbridge, Alberta, Can., Nov. 27–Dec. 2; pp. 123–197.

Rhoades, J. D. 1984. "Using saline waters for irrigation." Scientific Re views on Arid Zone Research. Scientific Publishers, Jodhpur India. 2:233–264.

Shainberg, I., Rhoades, J. D., Suarez, D. L., and Prather, R. J. 1981. "Effect of mineral weathering on clay dispersion and hydraulic conductivity of sodic soils." *Soil Sci. Soc. Am. J.* 45: 287–291.

Suarez, D. L., Rhoades, J. D., Lavado, R., and Grieve, C. M. 1984. "Effect of pH on saturated hydraulic conductivity and soil dispersion." *Soil Sci. Soc. Am. J.* 48: 50–55.

U.S. Salinity Laboratory Staff. 1954. *Diagnosis and improvement of saline and alkali soils.* U.S. Dept. Agr. Handbook 60, 160 pp.

NOTATION

A = salinity concentration at which growth depression (threshold) starts, or maximum concentration of boron that does not reduce yield; B = percent of yield decrease per unit EC_e above the threshold level; B_{iw} = concentration of boron in irrigation water; B_s = boron in soil solution; EC = electrical conductivity; EC_{aw} = average EC_{ss} at field capacity; EC_e = electrical conductivity of soil saturation extract; EC_{iw} = electrical conductivity of irrigation water; EC_{ss} = electrical conductivity of soil solution; ESP = exchangeable sodium percentage; F_c = concentration factor; LF = leaching fraction; m = decrease in yield per unit increase in boron concentration; OP = osmotic potential; x = boron concentration in the nutrient, sand culture, or soil solution; and Y = relative yield.

CHAPTER 12

LEACHING FRACTION AND ROOT ZONE SALINITY CONTROL[a]

INTRODUCTION

Successful water management for salinity control depends on adequate leaching, which takes place whenever irrigation and rainfall exceed the soil's capacity to store infiltrated water within the crop's root zone. In humid regions, rainfall results in enough leaching to flush salt from the root zone. In subhumid regions, irrigation water that exceeds the crop's water requirements must be applied to assure adequate leaching. Depending on the salinity control needed, leaching may occur continuously or at intervals of a few weeks to a few years.

The crop's water requirement and salinity control must be prime considerations in places where salinity poses a hazard. Proper irrigation restores the soil's water deficit without a wasteful and potentially harmful excess. Crops need water from irrigation and rainfall to control soil salinity by inducing drainage. Over time, leaching must remove enough salt to prevent it from accumulating in the root zone beyond the crop's tolerance level. Chemical reactions in the soil affect the amount of salt leached, which may be greater than, equal to, or less than the amount of salt added by irrigation water.

WATER AND SALT BALANCE

The amount of irrigation water needed to meet the crop's water requirement can be calculated from a water balance of the crop root zone. The major flows of water into the crop's root zone are irrigation, rainfall, and upward flow from the ground water. The depths of each are expressed in equations as D_i, D_r, and D_g, respectively. Water flows out due to evaporation, transpiration, and drainage. Their equivalent depths are represented in equations as D_e, D_t, and D_d, respectively. The difference between the water flowing in and the water flowing out must equal the change in storage, the depth of which is expressed as D_s in equations. The water balance equation for storage change is as follows:

$$D_s = D_r + D_g + D_i - D_e - D_t - D_d \qquad (12.1)$$

The equation for the change in salt storage, S_s, is the following:

---

[a]Prepared by: Glenn J. Hoffman, Dept. of Agric. Engrg., 223 L. W. Chase Hall, Univ. of Nebraska, Lincoln, NE 68583

$$S_s = D_rC_r + D_gC_g + D_iC_i + S_m + S_f - D_dC_d - S_p - S_c \quad (12.2)$$

where C = salt concentration; and the subscripts r, i, g, and d designate rain, irrigation, upward flow from ground water, and drainage, respectively; S_m = the salt dissolved from minerals in soil; S_f = the salt added to soil as a fertilizer or amendment; S_p = the salt precipitated; and S_c = the salt removed in the harvested crop.

If $D_r + D_g + D_i$ is less than $D_e + D_t$ in Eq. 12.1, the soil-water deficit is met by extraction from soil storage and reduced drainage. With time, the deficit becomes zero. As D_s is depleted, the soil dries, which reduces D_e, and the crop becomes water-stressed, which reduces D_t. Initially, these processes bring water loss from the root zone in balance with the water supply at zero drainage. However, without drainage, salt stored in the root zone concentrates in the remaining stored water. As soil salinity increases, the osmotic stress of the plant increases, which further reduces transpiration. If salts continue to increase in concentration, the plant dies. If the plant is salt-tolerant, transpiration may be reduced to the extent that excess water is again present in the profile and drainage commences. This drainage, in turn, carries salt out of the root zone. The steady leaching fraction that results is the absolute minimum leaching fraction at which the crop can extract water from a saline root zone. This leaching fraction, however, is far less than that needed to prevent a reduced yield.

When a shallow water table exists, deficiencies in $D_i + D_r$ may be offset by D_g. If flow is upward from the ground water, drainage is zero and salt will not be exported from the root zone. This situation cannot continue indefinitely. In the field, upward flow and drainage may take place alternately during the year. Typically, drainage takes place in the winter and early in the irrigation season, when the water requirements of the crop are low and rainfall or irrigation water applications are high. Upward flow takes place late in the irrigation season, when water requirements are high and rainfall and applications of irrigation water are insufficient. If upward flow continues and sufficient leaching does not take place, soil salinity will ultimately reduce the crop's water consumption to the point that the crop dies. Temporary use of water from soil storage beyond that normally removed between irrigations or from shallow ground water is a useful strategy for managing water. But, over the long-term and where salinity is a hazard, a net downward flow of water through the root zone is needed to sustain crop productivity.

Rarely will conditions controlling the water that flows into and out of the root zone last long enough for a true steady state to exist. As a result, the amount of salt in water stored in the root zone fluctuates continually. The goal of water management is to maintain this fluctuation within limits that neither allow excess drainage nor reduce the growth of crops.

Rainfall

The concentration of salt in rainfall (C_r) varies according to distance from the ocean, topography, direction of the wind, intensity of rainfall, and geographical distribution of the storm. The annual deposits of salt from rainfall has been estimated at 100 kg/ha to 200 kg/ha near the sea coast and 10 kg/ha to 20 kg/ha in the continental interiors (Downes 1961; Cope 1958; Yaalon 1963). Although small, these deposits can add up to sizable amounts of salts in areas with low rainfall after several decades. In irrigated areas, the salt applied in the irrigation water normally far exceeds the salt contribution from rainfall. Thus, D_rC_r can normally be assumed to be zero.

Mineral Weathering

Soils in arid and semi-arid regions, except for ancient land masses, such as in parts of Australia, are relatively unweathered. Unweathered minerals provide plant nutrients, but are also a source of soil salinity (S_m). Rhoades et al. (1968) have shown that increases in salt content of 200 mg/l to 300 mg/l are common when arid-land soil solutions remain in contact with relatively unweathered soil minerals for substantial periods of time. The amount of salt dissolved under such conditions depends on the level of carbon dioxide in the soil profile. The partial pressure of carbon dioxide can reach 10% or more when oxygen is consumed and carbon dioxide released during soil respiration (Bohn et al. 1979).

Studies using various simulated irrigation waters from the western United States (Rhoades et al. 1973, 1974) showed that the dissolution of primary minerals is most important when the irrigation water's salt content is low—less than 100 mg/l to 200 mg/l—or when the leaching fraction is at least 0.25. For example, irrigation with water from California's Feather River, which has a salt content of 60 mg/l, results in more salt in the drain water due to weathering than due to the salt content of the irrigation water (Rhoades et al. 1974). The major concern about mineral weathering is the sodicity hazard of a relatively nonsaline irrigation water. For salt-affected soils, mineral weathering is seldom a significant part of salt balance computations.

Salt Precipitation

As indicated in Eq. 12.2, the salt balance is affected by precipitation reactions involving slightly soluble salts, such as gypsum, carbonates, and silicate minerals (S_p). Consequently, the amount of salt leached below the root zone may be less than that applied, as was demonstrated in a three-year lysimeter study (Rhoades et al. 1974). When irrigation waters have a concentration of salt greater than 100 mg/l to 200 mg/l and if leaching fractions are less than 0.25, some salts precipitate in the root zone and become stored in the soil profile. When irrigation waters have a moderate amount of salt, such as the 800 mg/l that occurs in the Colorado River's lower reaches, and leaching fractions are below 0.25, salts precipitated in the soil profile exceed the amount weathered.

Fig. 12.1 shows the relative amounts of salt that may chemically precipitate or become soluble in water due to weathering for various types of irrigation water as a function of leaching fraction. All irrigation waters illustrated have concentrations of salt of above 500 mg/l. Thus, mineral weathering does not exceed chemical precipitation, except for some waters at a leaching fraction of 0.2 and above. At low leaching fractions (L – 0.1), 20% or more of the salt in irrigation water precipi-

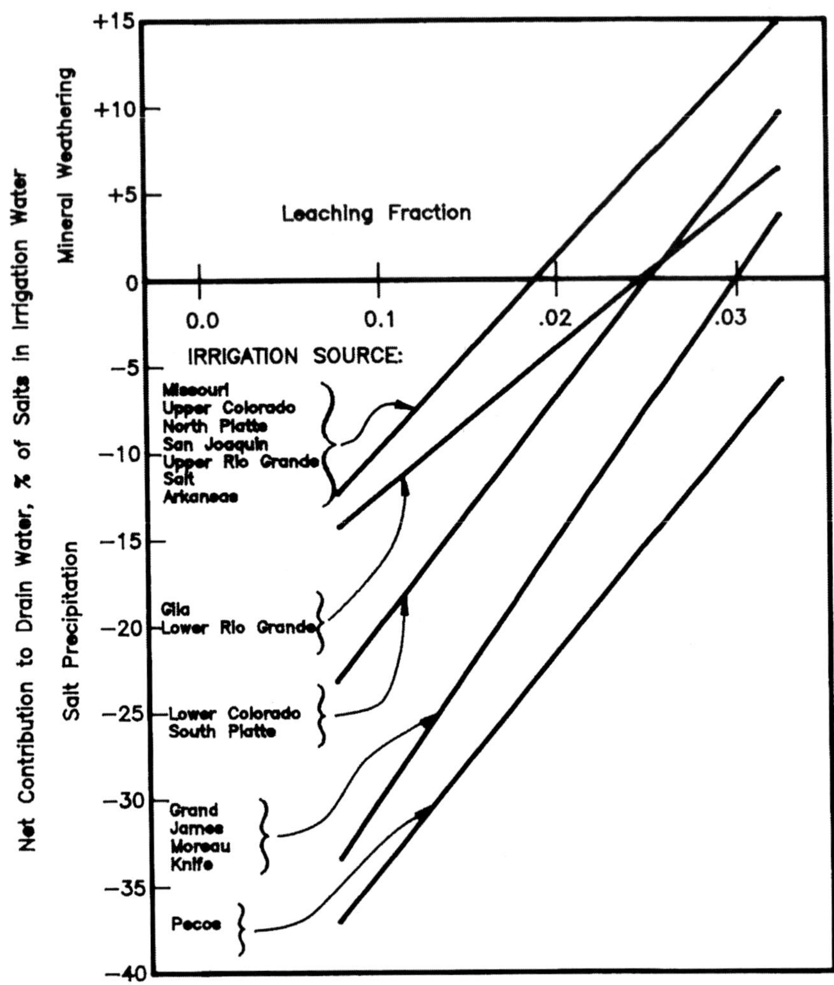

Fig. 12.1 Net Contribution of Mineral Weathering and Salt Precipitation to Salt Content of Drainage Water, Expressed as a Percent of the Salt Applied from Various River Waters (Each line represents an average of the percentages for calcareous and noncalcareous Pachappa sandy loam Soil.) (Rhoades et al. 1974)

tates and is not contained in the drainage water. Consequently, salt precipitation is a significant part of calculating the salt balance when the leaching fraction is low.

Salt Removal by Crops

Salt removed by agronomic crops (S_c) is insufficient to maintain salt balance. The average amount of salt contained in mature crops of alfalfa, barley, corn silage, sudangrass, and sweet clover grown in Texas' Rio

Grande area was 3.6% (Lyerly and Longenecker 1962). At intermediate levels of salinity, Chapman (1966) reported that the salt content of alfalfa, corn, and sorghum was about 3% of the dry tissue weight. In one study, water from the Pecos River with an electrical conductivity, EC_i of 3.3 dS/m was applied to alfalfa grown on sandy loam soil. The amount of salt removed in the harvests was 3% of the dry forage for leaching fractions varying from 0.1 to 0.3 (Rhoades et al. 1974). Francois (1981) reported a salt content of 3% to 4% for alfalfa grown under saline conditions.

Under most agricultural conditions where salinity is a concern, salt removal by crops can be ignored in the salt balance equation. A simple calculation of the degree of salt removal by alfalfa demonstrates this. Plants that are very efficient in removing salt from saline soils, such as sea-blithe (*Suaeda fruticosa*), remove less than 3 Mg/ha with each harvest (Chaudri et al. 1964).

Fertilizer Salts

The upper limit of recommended fertilizer applications for crops, such as corn, is about 250 kg/ha of nitrogen (Kearney et al 1980). If the nitrogen is applied as ammonium sulphate (21% N, 73% SO_4), the amount of fertilizer applied may be as high as 1.2 Mg/ha. When only the sulphate contributes to the salt load, 0.9 Mg/ha is the upper limit. If this amount of fertilizer were added to corn irrigated by 750 mm of water with 800 mg/l of salt, the fertilizer's contribution of salt would be 15% of the amount added by irrigation. The amount of fertilizer in this example is considered excessive for many crops.

SOIL SALINITY WITHOUT LEACHING

High levels of salinity in the lower portion of the root zone has a minor influence on yield if the upper portion is maintained at a relatively low level of salinity (Bernstein and Francois 1973; Bingham and Garber 1970). Invesigators have reported that plants compensate for reduced uptake of water from a zone of highly saline soil by increasing uptake from a zone low in salinity (Shalhevet and Bernstein 1968; Lunin and Gallantin 1965; Eaton 1941; Wadleigh et al. 1947). Although this takes place without significantly reducing yield, there is general concern about how much salt can be stored in the root zone before leaching is needed.

An example of the effect of virtually no drainage on soil salinity is southwest Australia (Peck et al. 1981). Fig. 12.2 illustrates the chloride concentration and the downward velocity of the soil solution for one location in a Mediterranean climate with an annual rainfall of 800 mm on a native Eucalypt forest. Chloride concentration increased most at a soil depth of 7 m, where the downward velocity of the soil solution equaled 0.04% of the annual rainfall, or 0.3 mm/yr. Below 7 m, chloride decreased linearly to less than 2,000 mg/l just above the water table at a depth of 17 m.

The deeper the soil, the greater the capacity to store salt with

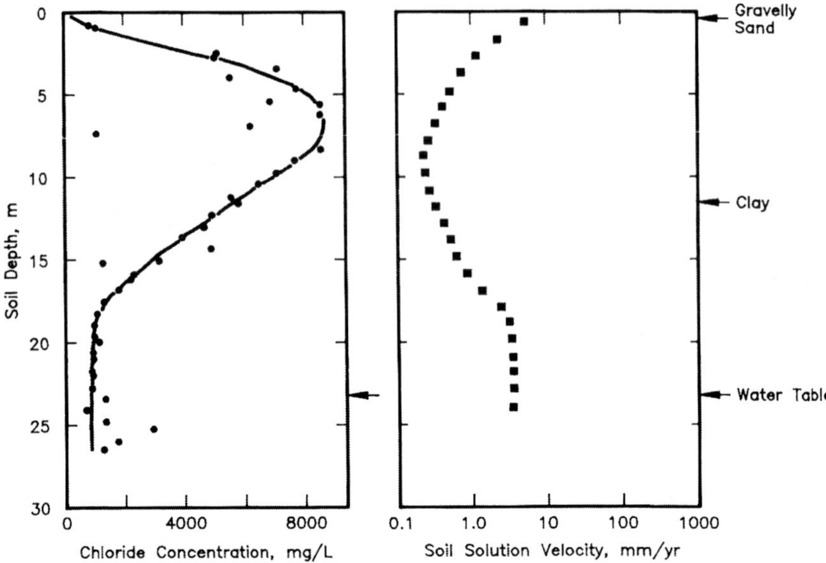

Fig. 12.2 Soil Chloride Concentration and the Downward Rate of Soil-Water Movement as a Function of Soil Depth in Poorly-Drained Soils of Southwest Australia (Peck et al. 1981)

minimal yield reduction. One of the first studies of the effect of no leaching involved alfalfa grown in a greenhouse and irrigated by water with an electrical conductivity (EC_i) of 1 dS/m. The plants were grown without leaching in sandy loam soil with depths of 0.6 m, 1.2 m or 1.8 m for periods of nine, 14, and 20 months, respectively (Francois 1981). Yield was reduced less than 25%, yet 14 mg/ha, 30 mg/ha, and 45 Mg/ha of salt was stored in the lower portions of the three different soil-profile depths. Drastic reductions in yield took place when the salt began to build up in the upper portion of the root zone. This study demonstrated that regardless of soil depth alfalfa can be grown for a considerable period of time without leaching, if the upper part of the root zone is maintained at a low level of salinity.

The Broadview Irrigation District on the west side of the California's San Joaquin Valley is a well-documented example of the effect of accumulating soil salinity on a large scale (Wichelns et al. 1988). The district, made up of 4,000 ha of field crops, has had adequate imported surface water for irrigation since 1957. The water contained approximately 300 mg/l of salt (EC_i of 0.5 dS/m). To facilitate leaching, subsurface drains were installed on more than 80% of the irrigated land. Annual rainfall averages 150 mm. The district had no drainage outlet until 1983, so it had to blend its surface runoff and subsurface drainage with irrigation supply water. The ratio of drain water to fresh water increased from near zero in the early 1960's to about half in the early 1980's, when the mean salt content of the drainage water was about 2,800 mg/l. Although the fields were leached, the salts were reapplied to the fields. Thus, no disposal of salts took place. As illustrated in Table 12.1, crop

TABLE 12.1 Change in the Area and Yield of Cotton and Tomato in Broadview Water District Compared to Fresno County

	1968 to 1972		1978 to 1982		Change (%)	
(1)	Broadview (2)	Fresno County (3)	Broadview (4)	Fresno County (5)	Broadview (6)	Fresno County (7)
Cotton						
Area (ha)	715	76,000	2,130	167,000	+200	+140
Lint Yield (kg/ha)	1,190	1,190	1,305	1,190	+10	0
Tomato						
Area (ha)	605	8,150	170	16,300	−70	+100
Fruit Yield (Mg/ha)	72	50	59	64	−18	+28

selection switched to salt-tolerant crops, such as cotton, to maintain yield. The amount and yield of more salt-sensitive crops, such as tomatoes, dropped drastically as soil salinity increased over time.

LEACHING REQUIREMENT

The amount of leaching needed to maintain a viable irrigated agriculture depends on the salt content of the irrigation water, soil, and ground water; the salt tolerance of the crop; the climate; and soil and water management. The only economical way to control soil salinity is to ensure a net downward flow of water through the root zone to a suitable disposal site. If leaching is inadequate, harmful amounts of salt can accumulate within a few cropping seasons.

In its simplest form, eq. 12.2 reduces to $D_iC_i + D_rC_r - D_dC_d = 0$, with D = depth of irrigation; C_i = the salt content of irrigation water; D_r = the depth of rainfall; C_r = the salt content of rainfall; D_d = the depth of drainage; and C_d = the salt content of drain water. This requires the upward movement of salt to be insignificant and the quantity $(S_m + S_f) - (S_p + S_c)$ to be zero. S_m = the amount of salt dissolved from soil minerals; S_f = the amount of salt added to soil as a fertilizer or an amendment; S_p = the amount of soil precipitated in the soil profile; and S_c = the amount of salt removed by agronomic crops. S_s (salt storage) equals zero at steady-state conditions. In this simple form, Eq. 12.2 can be written as

$$L = \frac{D_d}{D_a} = \frac{C_a}{C_d} = \frac{EC_a}{EC_d} \tag{12.3}$$

in which L = the leaching fraction; D_d = depth of drainage; D_a = depth of applied water (irrigation plus rainfall); C_a = the salt content of applied water; C_d = the salt content of drain water; EC_a = the electrical conductivity of applied water, and EC_d – the electrical conductivity of drainage water. The importance of this concept is greater than its accuracy. Under the assumed conditions, this equation states that the amount of salt added as irrigation water must equal the amount drained to maintain salt balance.

Eq. 12.3, as well as other steady-state equations, applies only to salt constituents that remain dissolved. The electrical conductivity (EC) of a water reliably indicates total salt concentration (U.S. Salinity Laboratory 1954), and EC can replace C in Eq. 12.3 with reasonable accuracy, particularly when weathering and precipitation of salts are insignificant.

The relationship between leaching fraction and root-zone salinity is important in managing soil salinity. The minimum leaching fraction that a crop can endure without yield reduction is the leaching requirement. Its equation is expressed as follows:

$$L_r = \frac{D_d^*}{D_a} = \frac{C_a}{C_d^*} = \frac{EC_a}{EC_d^*} \tag{12.4}$$

in which L_r = the leaching requirement; D_d = the depth of drainage; D_a = the depth of applied water; C_a = the salt content of applied water; C_d = the salt content of drain water; EC_a = the electrical conductivity of applied water; and EC_d the electrical conductivity of drainage waters. The superscript * distinguishes required from actual values. Because electrical conductivity (EC) is easily measured and almost linearly related to salt concentration, the leaching requirement (L_r) is usually expressed as a ratio of electrical conductivity (EC).

Several models have been proposed to relate C_d^* to some readily available value of soil salinity that indicates the crop's leaching requirement. Hoffman and van Genuchten (1983) determined the linearly averaged, mean root zone salinity by solving the continuity equation for one-dimensional vertical flow of water through soil, assuming an exponential soil water uptake function (Raats 1974). Steady-state mass balance was assumed so the Eq. 12.3 could be used to introduce salt concentration into the calculations. The linearly averaged salt concentration of the root zone ($\overline{C}$) as a ratio of the salt concentration of the irrigation water (C_i) is

$$\frac{\overline{C}}{C_i} = \frac{1}{L} + \frac{\delta}{ZL} \ln\left[L + (1-L)e^{-Z/\delta}\right] \tag{12.5}$$

L = the leaching fraction; Z = the depth of the root zone; and δ = an empirical constant set to 0.2Z. Fig. 12.3 illustrates this relationship.

Plants adjust osmotically as soil salinity increases (Maas and Nieman 1978). In salt-tolerance trials, this adjustment prevents yield loss until soil salinity surpasses the salt-tolerance threshold. Since salt-tolerance trials usually are designed to maintain leaching fractions approaching 0.5, the osmotic adjustment consistent with no loss of yield can be estimated by the mean soil salinity at 50% leaching, as given in Fig. 12.3. As a first approximation, the leaching requirement can be expressed as

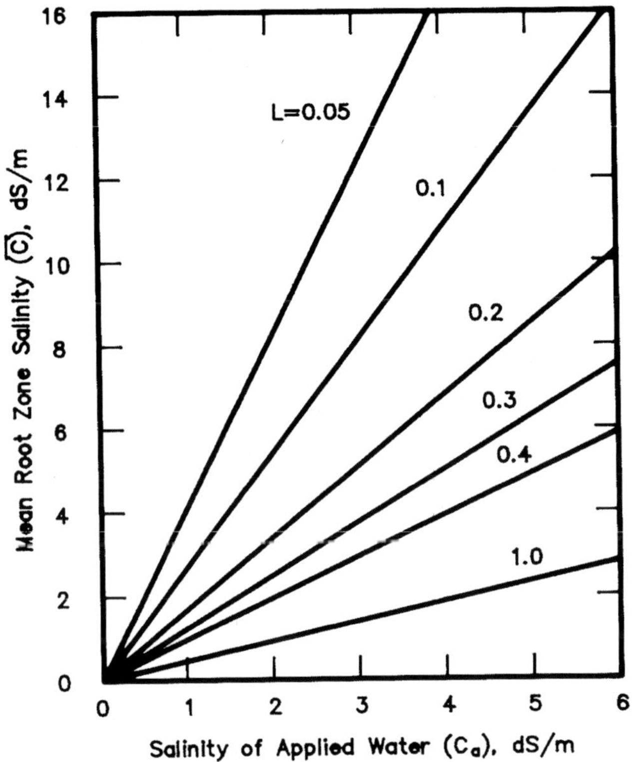

Fig. 12.3 Mean Root Zone Soil Salinity as a Function of Applied Water and Leaching Fraction, *L* (Hoffman and van Genuchten 1983)

a function of the threshold value by reducing $\overline{C}$ at any given L by $\overline{C}$ at 50% leaching. Fig. 12.4 illustrates this relationship among the salinity of the applied water, the salt-tolerance threshold of the crop, and the leaching requirement. Hoffman (1985) compared calculated leaching requirements from this and other experimental models. Of the four models tested, the one presented here agrees well with the measured values throughout the range of L_r of agricultural interest.

The following exemplifies how to estimate the leaching requirement. Assume tomatoes are to be grown in an arid region and irrigated by water with an EC_i of 3 dS/m. Assume rainfall to be insignificant. According to Chapter 13, the salt-tolerance threshold (C_t) of tomatoes is 2.5 dS/m. From Fig. 12.4, using C_a (salt in applied water) of 3 dS/m and C_t of 2.5 dS/m, the leaching requirement (L_r) equals 0.2. To calculate the salinity of the applied water when rainfall is significant, the following equation is used: $C_aD_a = C_rD_r + C_iD_i$. The variable C is salt content, in this case EC, and D is depth. The subscripts a, r, and i indicate applied, rain, and irrigation water, respectively. An iterative process must be used, since D_i is unknown until L_r is known. As an illustration, assume evapotranspiration (D_e) is 750 mm and rainfall (D_r) is 150 mm. Begin

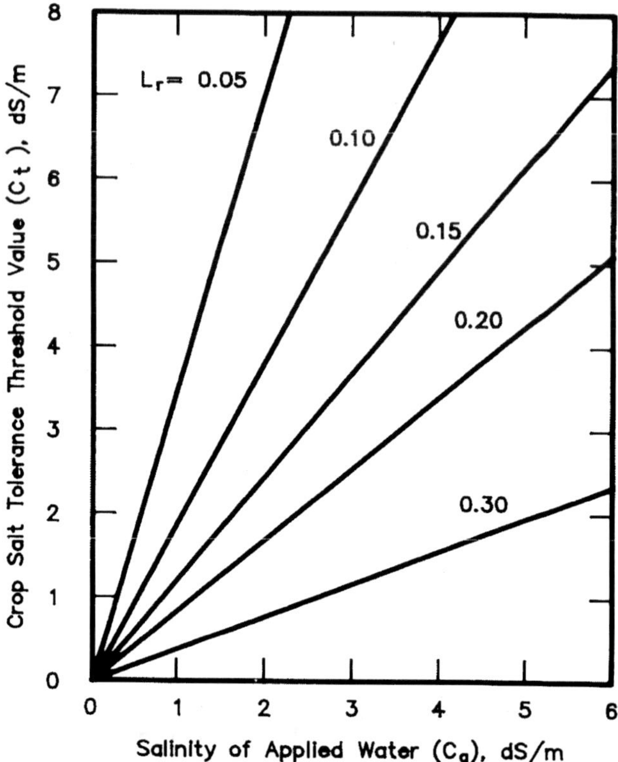

Fig. 12.4 Leaching Requirement (L_r) as a Function of the Salinity of the Applied Water and Salt-Tolerance Threshold Value (Hoffman and van Genuchten 1983)

calculations by assuming that D_i is 900 mm and $C_r = 0$. Because $D_a = D_r + D_i$, then $C_a = C_iD_i/(D_r + D_i) = 3(900)/(150 + 900) = 2.6$. With $C_a = 2.6$, L_r for tomato is 0.18 and $D_i = D_e + D_t + D_d = 750 + 0.18\ D_i$ or $D_i = 915$ mm. This is sufficiently close to the assumed value of 900, making further iterations unnecessary. With this rainfall, L_r for tomatoes is 0.18.

Accounting for nonuniformity of irrigations to estimate L_r has not been addressed to date. If the leaching requirement is not met everywhere in the field, saline soil will develop wherever evapotranspiration plus the leaching requirement is not met. Whether to apply copious amounts of water to assure that the leaching requirement is met throughout the field or to accept some reduction in yield in parts of the field rather than over-irrigate most of the field must be determined.

Extensive advancements in irrigation technology and management are needed to approach the goal of achieving the leaching requirements. Present irrigation practices in many areas are inefficient and inadvertently provide excessive leaching. This is costly. It leads to a loss of water, energy, and nutrients, deteriorates the quality of water, and increases the need for drainage facilities. Consequently, knowing the leaching requirements of crops and striving to attain them is vital.

EFFECT OF SHALLOW GROUND WATER

The upward movement of saline ground water and its subsequent evaporation at the surface of the soil adds to the salinization of soils. To minimize the rate at which salt accumulates and, thus, reduce the salinity hazard, attempts to lower the water table by drainage are usually made. The effects of the depth of the water table and the properties of the soil on the rate of upward movement must be known to determine the depth at which to maintain the water table. This information is also needed to estimate the amount of ground water available to plants from upward movement.

Starting from saturation, the drying rate of the surface of the soils will at first be limited by the atmospheric evaporative conditions. When the surface becomes dry enough, the evaporation rate will be limited by the rate of water movement to the surface in the liquid phase. As the soil dries further, vapor movement is possible, but relatively unimportant, particularly since diurnal fluctuations in temperature may cause the vapor movement to reverse directions.

When a shallow water table exists, upward flow becomes important. Gardner and Fireman (1958) studied how the rate of upward flow relates with the depth to the water table. This study verified the steady-state solutions proposed by Gardner (1957), who based his solutions on the relation between hydraulic conductivity (k) and soil matric potential (suction, S) of the form

$$k = \frac{a}{S^n + b} \tag{12.6}$$

where a, n, and b are constants. For many soils, values of n equal to 2 or 3 fit best with experimental data. For Chino clay, $k = 1100 / (S^2 + 565)$ cm per day, where S is in millibars. For Pachappa sandy loam, $k = 32$ $(S^3 \times 10^{-4} + 2.6)$. Fig. 12.5 gives the theoretical maximum rate of upward flow from the water table for these two soils as a function of the depth to the water table. Two maximum rates of flow must be considered: the potential rate of evaporation from the surface of the soil that atmospheric conditions dictate and the maximum rate at which water can be transmitted upward from the water table. Obviously, the lesser of these limits upward movement. Excluding a shallow water table and humid conditions, the water-transmitting properties of the soil most often limits upward flow.

The depth at which a water table should be maintained to keep upward flow to a minimum can be determined from this type of analysis. Lowering the water table from the surface to a depth of about 1 m would be of little benefit in most soils. Upward flow at these shallow depths could exceed 2.5 mm per day for clay soils (Fig. 12.5) and be even greater for coarser textured soils, depending on the atmospheric evap-

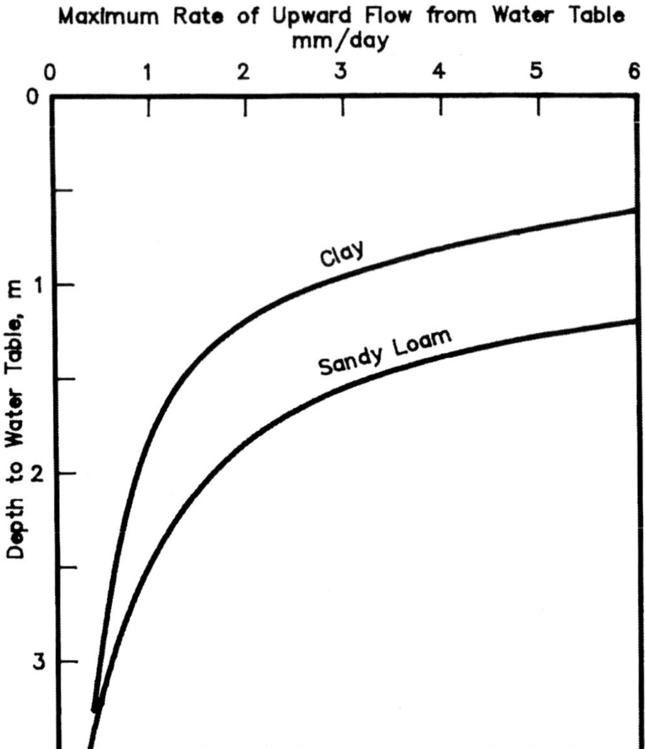

Fig. 12.5 Maximum Theoretical Rate of Upward Water Flow for Chino Clay and Pachappa Sandy Loam as a Function of the Depth of the Water Table (Gardner and Fireman 1958)

orative demand. As the water table is lowered below 1 m, the soil's hydraulic properties and depth limit the rate of upward flow (Fig. 12.5). Lowering the water table from 1.2 m to 3.0 m in Pachappa sandy loam decreases upward flow by a factor of 10. When the water table is at 2.5 m, further lowering reduces upward flow only slightly. Upward movement and evaporation of water from the surface of the soil is possible even with a water table that has a depth of 10 m. Harmful amounts of soluble salts could slowly accumulate if the ground water is sufficiently saline and rainfall and irrigation amounts are sufficiently low. These results, verified by field observations, have lead to the installation of most subsurface drainage systems at depths of 1.5 m to 2.5 m, wherever salinity poses a hazard.

The distribution of salts in the soil profile above the water table depends on the ground water's depth and salt content, the amount of applied water and its salt content and the water uptake pattern of the crop's root system. In a lysimeter study in Texas, researchers studied soil salinity profiles in a Willacy fine sandy loam above a shallow water table (Namken et al. 1969). The study consisted of two moisture treat-

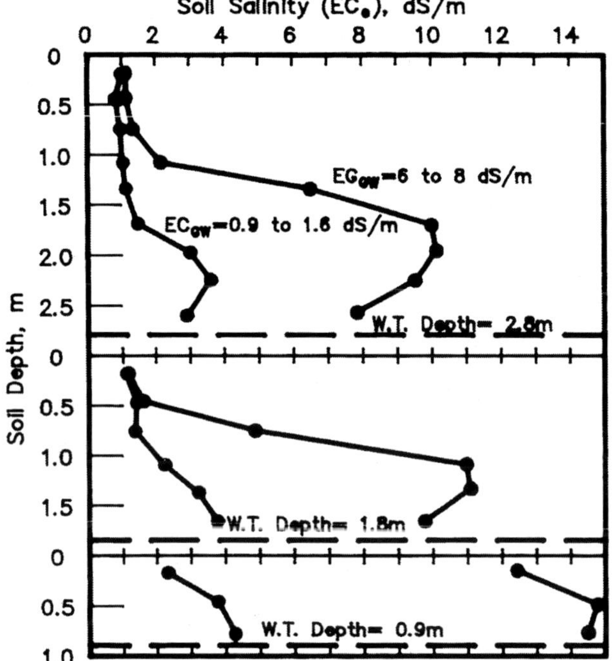

Fig. 12.6 Soil Salinity Distribution for Different Salt Contents and Depths of Ground Water (Namken et al. 1969)

ments and three depths of water tables. Because differences in soil salinity between the moisture treatments were small, Fig. 12.6 illustrates only the influence of the ground water's depth and salt content. During the study's first year, the ground water had a level of salinity (EC_{GW}) of 6 dS/m to 8 dS/m. During its last three years, the EC_{GW} ranged from 0.9 dS/m to 1.6 dS/m. The cotton crop took 57%, 38%, and 28% of the water used when the water table was at depths of 0.9 m, 1.8 m, and 2.8 m, respectively. When the water table was 1.8 m deep or lower, the upper half of the profile remained nonsaline, while the lower half became salinized. When the depth was 0.9 m, the ground water's level of salinity influenced the entire profile.

Water supplied to a crop by capillary rise from shallow ground water can be an important resource. Benefits of using this water include reduced irrigation, lower production costs, the movement of a more moderate amount of ground water to deeper aquifers, and a decrease in the amount of ground water that needs to be disposed through subsurface drainage systems.

Cotton grown on a loam soil in the San Joaquin Valley with a water table located 2.0 m to 2.5 m below the surface received at least 60% of its evapotranspiration from shallow ground water with an EC of 6 dS/m (Wallender et al. 1979). The fewer the irrigations, the more the ground water contributed to evapotranspiration. However, the yields of lint were reduced. Fig. 12.7 illustrates how cotton's use of ground water

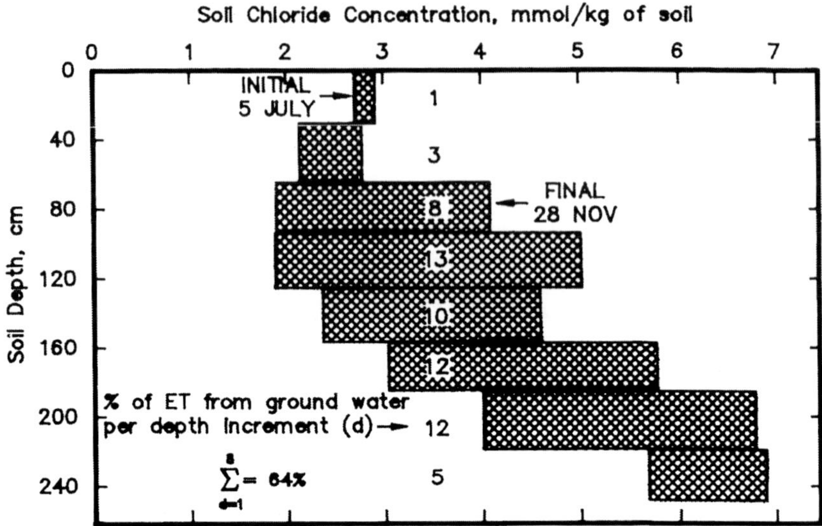

Fig. 12.7 Seasonal Change in Soil Chloride Levels as a Function of Soil Depth and Evapotranspiration (in percent) from Ground Water (Wallender et al 1979)

affected soil salinity. Concentrations of soil chloride from early in the irrigation season (July 5) are compared with concentrations after harvest (November 28). The equivalent depth of water used in evapotranspiration from the ground water equaled 362 mm. This was based on concentrations of soil chloride and the concentration of chloride in the ground water (17.4 mol/m^3) and soil bulk density. The amount agreed with the contribution of ground water based on the soil profile's water budget.

The relationship between the crop's use of water and the ground water's depth and salt content is not well understood. Several experiments have been conducted, but generalizations are difficult to make. Some of the most consistent data have been obtained with cotton (Fig. 12.8). The relationship between cotton's use of water from the ground water and the depth of the water table for soils ranging from clay to clay loam is based on results from field experiments on the west side of California's San Joaquin Valley. The data presented are from three different studies (Grimes et al. 1984; Hanson and Kite 1984; Ayars and Schoneman 1986). The relationship for sandy loam is based on a lysimeter study in Texas (Namken et al. 1969). Results indicate that cotton's uptake of ground water is not reduced measurably until EC_{GW} exceeds 12 dS/m. Use of ground water by alfalfa and corn varies from 15% to 60% of the total seasonal use, but the data are too inconsistent to establish a relationship. Use of ground water by alfalfa from a water table with a depth of 0.6 m in the Grand Valley of Colorado (Kruse et al. 1986) varied from 46% to 94% of the total seasonal use in two different years, when EC_{GW} equaled 0.7 dS/m. It varied from 23% to 91% of the total seasonal use between years, when EC_{GW} equaled 6 dS/m. In the same study,

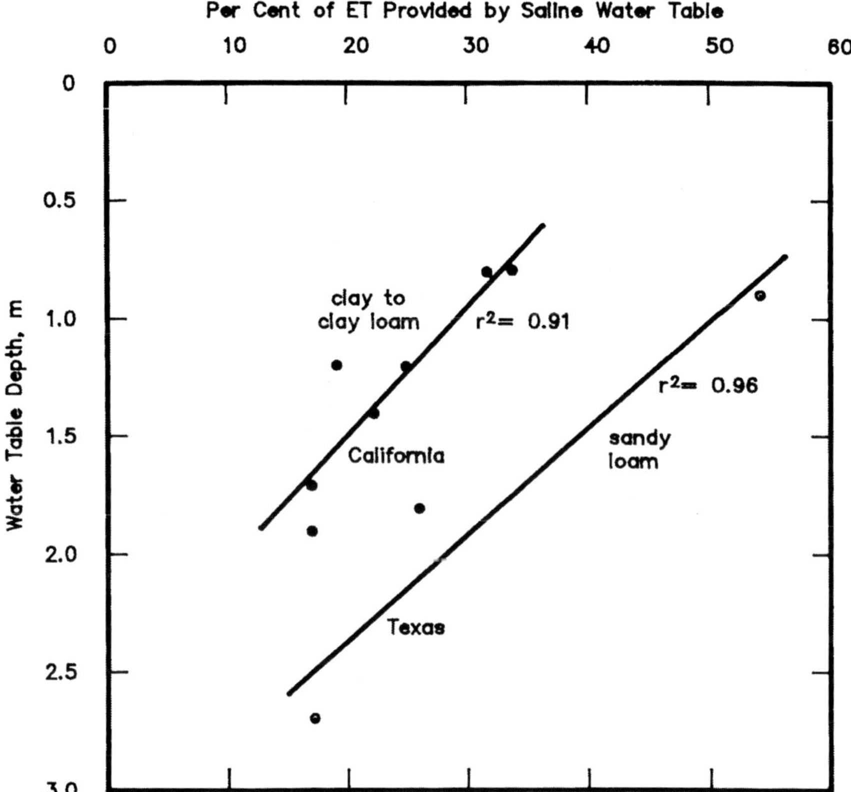

Fig. 12.8 Contribution of Shallow, Saline Ground Water to Evapotranspiration (ET) of Cotton as a Function of Soil Type and Depth to Water Table

Kruse et al. (1985) reported that corn obtained 52% to 68% of its seasonal water requirement when the water table was 0.6 m deep and obtained 25% to 32% of its seasonal water requirement when the water table was 1 m deep. The proportion of use remained unaffected when EC_{GW} varied from 0.7 dS/m to 6 dS/m.

Soils with a shallow water table frequently depress yields due to reduced soil aeration and inhibited root extension. If the shallow ground water is saline, yields may be further reduced. Subsurface drainage benefits crop production in salt-affected soils, but few long-term drainage experiments have been conducted that quantify increased yields and reduced salinity. El-Mowelhi et al. (1988) undertook one such experiment in the Nile Delta of Egypt from 1976 to 1986. Soil salinity to a depth of 1.5 m was reduced from an average of 5.3 dS/m to 2.2 dS/m after one year of drainage (Fig. 12.9) without additional water being applied beyond the normal irrigation amounts and rainfall. For three crop rotations, subsurface drains spaced 20 m apart and placed 1.5 m deep in clay soil increased the yield of cotton and rice by 100% and the yield of wheat and Berseem clover by 50%.

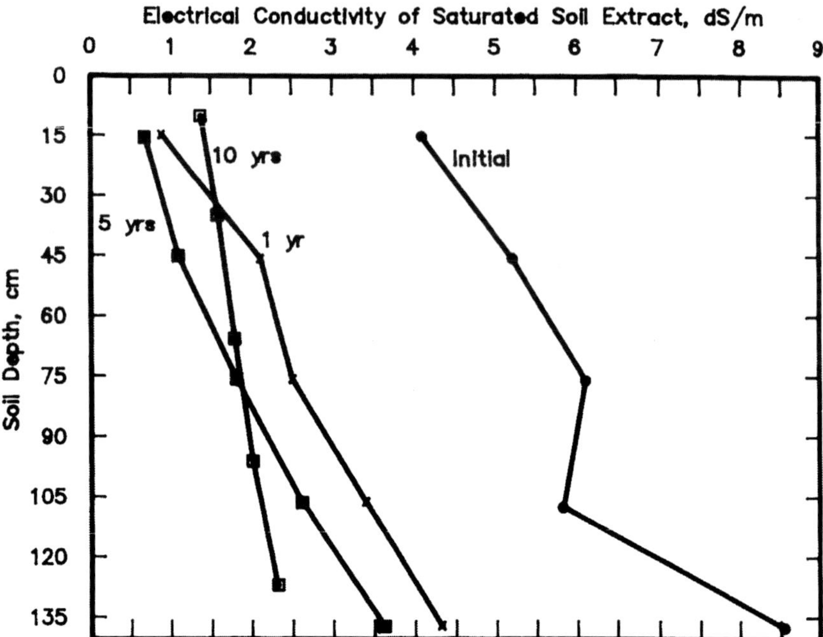

Fig. 12.9 Influence Over Time of Subsurface Drainage on Average Soil Salinity Profile (El-Mowelhi et al. 1988)

INTEGRATION OF SOIL SALINITY BY CROPS

In the field, the distribution of salts is neither uniform nor constant. To apply the results from experiments on the salt tolerance of crops, the response of the plant to salinity, which varies according to time and the soil depth, must be known. The following sections will cover that very topic.

Integration with Soil Depth

Hoffman et al. (1983) conducted a field experiment to establish the salt tolerance of corn in the Sacramento-San Joaquin Delta of California. They tested two irrigation methods. One consisted of mini-sprinklers, each with a wetted diameter of about 4 m, spaced 1.5 m apart along laterals in every other row of corn. Water was applied uniformly to achieve about 50% leaching. Fig. 12.10 illustrates the resulting soil salinity profile for the nonsaline treatment (EC_i equaled 0.2 dS/m) and a saline treatment (EC_i equaled 6 dS/m). The figure gives the values of soil salinity for measurements from soil samples, soil water samples extracted by vacuum through suction cups, and the monitoring of direct-burial, four-electrode salinity probes (Rhoades 1979). Fig. 12.10 also gives the composite values from these three techniques. The linear-averages of the composite values through the root zone is 1.9 dS/m for the nonsaline treatment and 7.3 dS/m for the saline treatment.

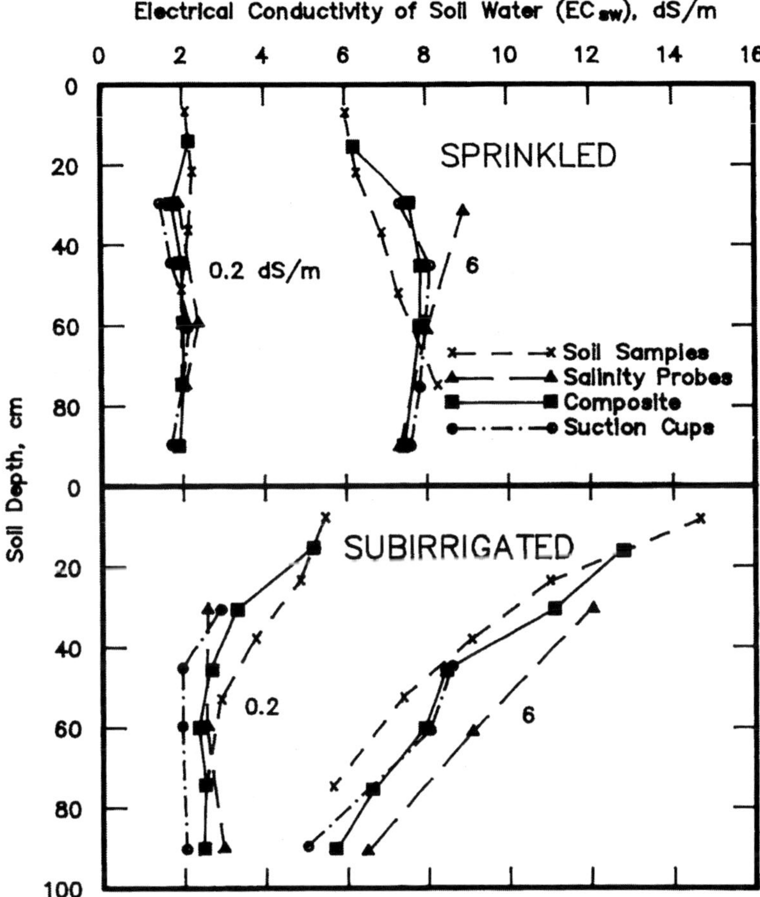

Fig. 12.10 Time-Weighted Averages of Electrical Conductivity of Soil Water from Suction Cups, Salinity Probes, and Soil Samples and Composite Averages for 0.2 dS/m and 6 dS/m Saline Water Applied to Corn, Sacramento-San Joaquin Delta, 1981 (Hoffman et al. 1983)

The second method was subirrigation, which is the one most commonly used method in the Delta. It consisted of ditches, spaced every 16 rows of corn, dug approximately 15 cm wide and 60-cm deep by a trencher each year in mid-June. Irrigation water applied in the ditches moved horizontally and vertically through the soil profile, raising the shallow water table to about 15 cm from the surface of the soil. Fig. 12.10 gives the salinity profiles for the same treatments as for the sprinkled plots. These profiles are representative of those expected in situations with no irrigation; low rainfall; and shallow, saline ground water. The linearly averaged values for the composite salinity profiles were 3.0 dS/m for the 0.2 dS/m treatment and 8.6 dS/m for the 6 dS/m treatment. When the linearly averaged values for these treatments and other levels

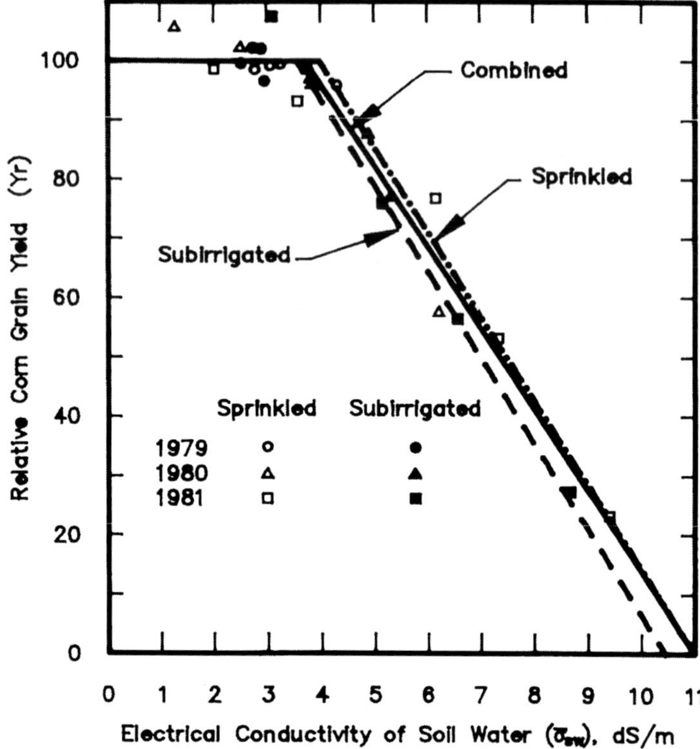

Fig. 12.11 Relative Grain Yield of Corn Grown in the Sacramento-San Joaquin Delta as a Function of Soil Salinity for Sprinkler Irrigation and Subirrigation Water Application Methods (Hoffman et al. 1983)

of salinity tested during the three-year experiment are plotted, the salt tolerance response curves for the sprinkled and subirrigated treatments do not differ statistically (Fig. 12.11). This suggests that plants respond to a linear average of the salinity values through the root zone and that the salt tolerance coefficients apply where the salinity distribution is not uniform with the depth of the soil. These results on organic soil for corn agree with results on field plots in Israel with mineral soil for peanuts (Shalhevet et al. 1969) and tomatoes (Shalhevet and Yaron 1973).

Integration over Time

Soil salinity is typically monitored at the beginning and the end of the crop's growing season, and the mean soil salinity is determined by averaging the values. In experiments, soil salinity is normally monitored more often and by a combination of soil sampling, vacuum extraction of soil water, and various devices that measure salinity. The integration of soil salinity over time is difficult because sensitivity varies from one stage of growth to the next for some crops. Cereal crops seem particularly variable. Results indicate that corn, for example, is most sensitive during the vegetative stage (Maas et al. 1983). Although soil salinity delayed the emergence of seedlings of corn, salinity of up to EC_{SW} of 9 dS/m did not reduce the emergence of seedlings after six days. Increasing the salinity of the irrigation water to 9 dS/m at the tasseling or grain-filling stages did not decrease the yield of corn ear or grain significantly below that achieved where salinity was constant throughout the growing season.

Bernstein and Pearson (1954) compared the influence of a constant level of salinity with cycles of slowly increasing and then abruptly decreasing levels of salinity. They reported that peppers responded to the seasonal mean salinity, while tomatoes were more affected by periods of high salinity. Meiri and Poljakoff-Mayber (1970) noted that the relative leaf area of bean from different salinity experiments was linear with salinity. The response to mean seasonal salinity is probably a reasonable estimate unless salinity during the season ranges both lower and higher than the salt tolerance threshold for the crop or unless salinity occasionally exceeds the range over which linear salt-tolerance response is observed, as probably was the case for tomatoes.

Evaluating the response of perennial crops to salinity over time is more complex than evaluating the response of annuals. This is primarily due to the extended length of time during which the yield of a perennial crop may be affected by soil salinity. With this increased time period come problems of how to compensate for dormant periods, drastic weather changes, such as monsoons and winter rains, and large changes in atmospheric evaporative demand. Deciduous fruit trees exemplify a perennial crop whose response to salinity over time is difficult to assess.

Hoffman et al. (1989) assessed the response of 20-year-old Santa Rosa plum trees in California's San Joaquin Valley to soil salinity. The experiment involved the use of irrigation water with six levels of salinity (EC_i of 0.3 dS/m to 8 dS/m). The water was applied through two mini-sprinklers for each tree to apply published measurements of evapotranspiration (seasonal ET of 1,030 mm) and the desired leaching fraction (0.3). Fig. 12.2 presents soil salinity profiles before the irrigation season (February or March) and during its second half (June to October) for three treatments during the study's first three years. When the experiment began in 1984, all of the treatments had the same low level of salinity before the irrigation season. Winter rainfall that followed the 1984 irrigation season leached soil profiles to below 75 cm before the 1985 irrigation season. The same leaching took place before the 1986 season. The 8 dS/m treatment resulted in such severe salinity damage by the end of 1985 that nonsaline water was applied to that treatment in 1986. This accounts for the low salt content during the second half of the 1986 irrigation season. Soil salinity was relatively uniform with the depth of the soil (Fig. 12.12). Thus, regardless of the integration process used to account for variability with depth, the resulting average soil salinity would be close to a simple average in the increments of depth sampled.

Soil salinity over time, however, changed significantly, as Fig. 12.13 illustrates. The salinity level rises quickly after irrigation begins

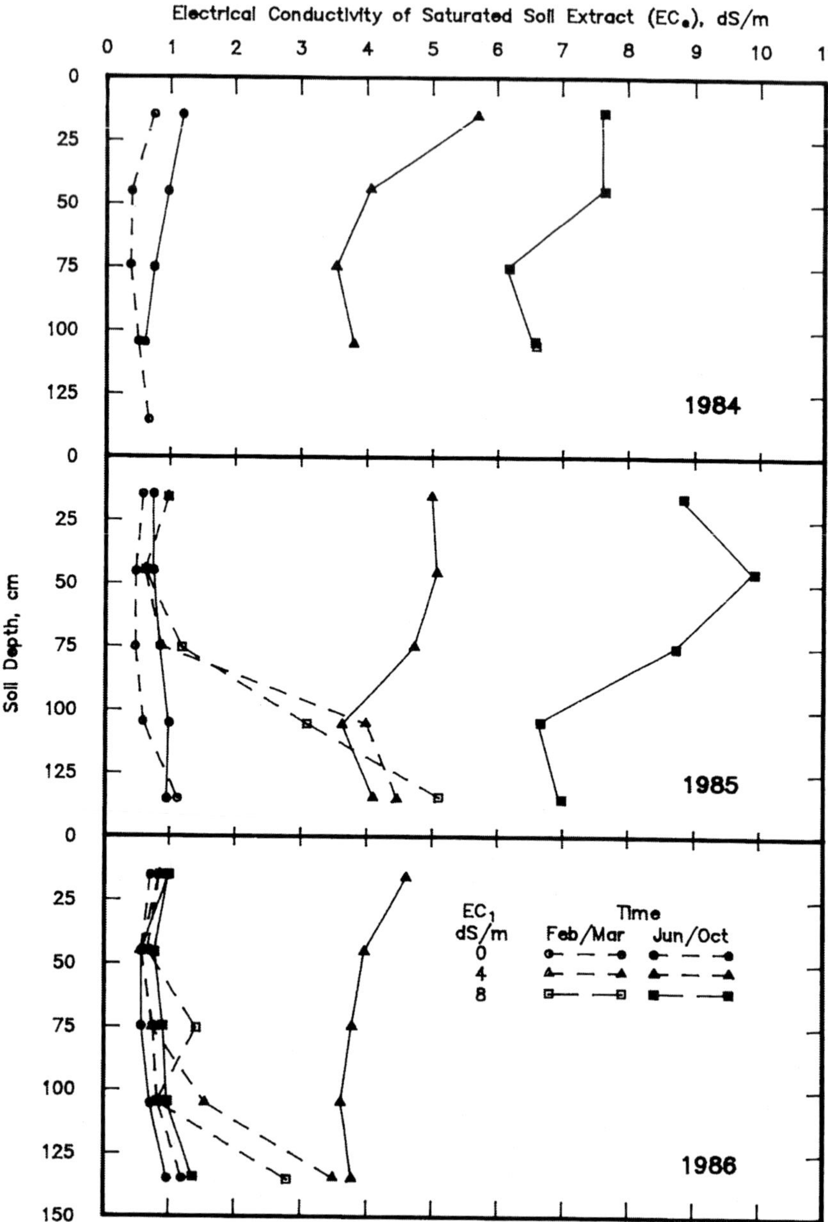

Fig. 12.12 Soil Salinity Profiles During a Salt-Tolerance Experiment on Plum Trees (During 1986, non-saline water (0 dS/m) was applied to the 8 dS/m treatment.) (Hoffman et al. 1989)

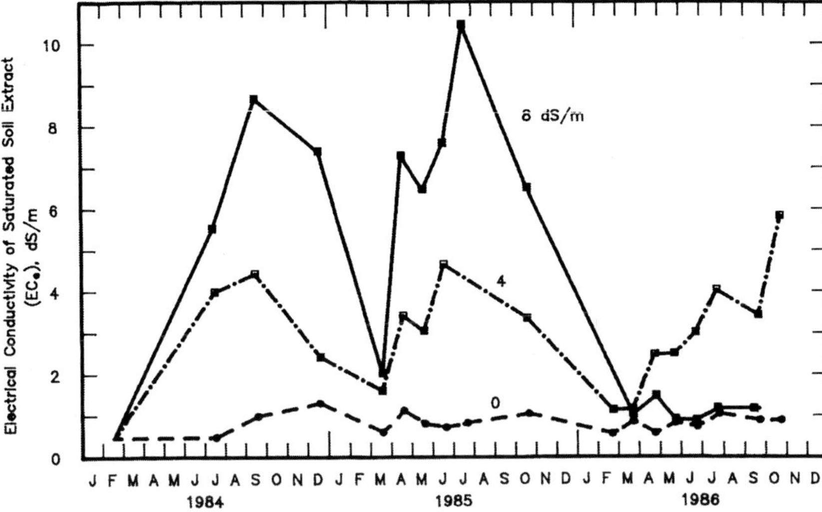

Fig. 12.13 Mean Root Zone Salinity over Time for Plum Trees Irrigated with Water of Electrical Conductivities 0, 4, and 8 dS/m (Hoffman et al. 1989)

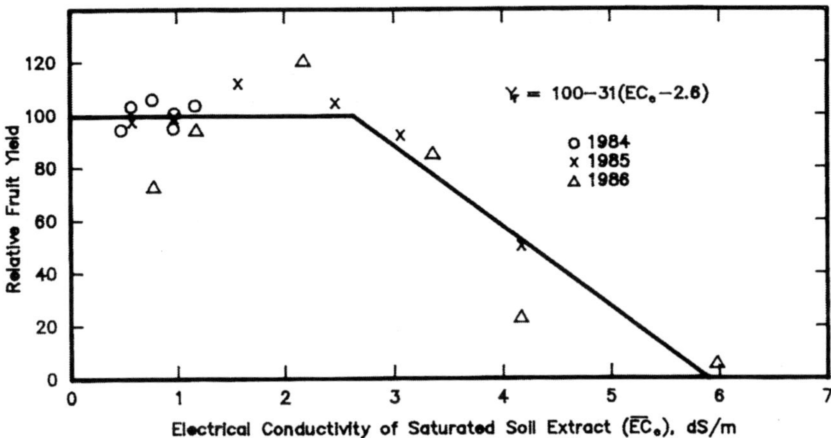

Fig. 12.14 Salt Tolerance of Mature Plum Trees Based on Integrating Soil Salinity over a Two-Year Period (Hoffman et al. 1989)

and drops rapidly due to leaching induced by winter rainfall. Time-integrated values of soil salinity were determined from data similar to that presented in Fig. 12.13 to develop a salt-tolerance curve, as proposed by Maas and Hoffman (1977). To account for salinity's influence on shoot growth, which attributes to bud formation the year before harvest, soil salinity measurements were integrated over the two years before each harvest. Excluded were the months from November to March, when the trees were dormant. Data on flower formation, fruit set, and budwood development can be analyzed to establish a more accurate time frame for integration. The response of mature plum trees to soil salinity is based on the results of the first three years of the field trial (Fig. 12.14). According to these results, soil salinity can apparently be integrated over two years for plum trees. The proper period of time undoubtedly depends on the crop and its environment.

SUMMARY

Salinity always threatens agriculture in arid or coastal environments. However, management strategies for using saline soil and waters to produce crops have been improved immeasurably by knowledge and experience gained over the past century. The basic premise that leaching is essential remains true. The gap between the leaching requirement and the leaching achieved on most irrigated land is being narrowed. As our ability to match crop water requirements with water applications improves throughout each field, our ability to minimize excess drainage will improve proportionately. The ultimate goal is to acquire the skills and knowledge necessary to use as efficiently as possible all available irrigation waters. Achieving this objective will minimize the amount of drain water requiring disposal or treatment.

REFERENCES

Ayars, J. E., and Schoneman, R. A. 1986. "Use of saline water from a shallow water table by cotton." *Transactions of ASAE* 29: 1674–1678.

Bernstein, L., and Francois, L. E. 1973. "Leaching requirement studies: sensitivity of alfalfa to salinity of irrigation and drainage waters." *Soil Sci. Soc. Amer. Proc.* 37: 931–943.

Bernstein, L., and Pearson, G. A. 1954. "Influence of integrated moisture stress achieved by varying the osmotic pressure of culture solutions on growth of tomato and pepper plants." *Soil Science* 77: 355–368.

Bingham, F. T., and Garber, M. J. 1970. "Zonal salinization of the root system with NaCl and boron in relation to growth and water uptake of corn plants." *Soil Sci. Soc. Amer. Proc.* 34: 122–126.

Bohn, H. L., McNeal, B. L., and O'Connor, G. A. 1979. *Soil Chemistry.* Wiley-Interscience, New York. 329 pp.

Chapman, H. D. 1966. "Diagnostic criteria for plants and soils." Div. Agric. Sci., Univ. of Calif., Oakland.

Chaudri, I. I., Shak, B. H., N. Nagro, and Mallick, I. A. 1964. "Investigations on the role of Suaeda Fruiticosa Forsk in the reclamation of saline and alkaline soils in West Pakistan Plains." *Plant and Soil* 21: 1–7.

Cope, F. 1958. "Catchment salting in Victoria." *Soil Conservation Authority Victoria Bull.* 1:1–88.

Downes, R. G. 1961. "Soil salinity in non-irrigated arable and pastoral land as the result of unbalance of the hydrologic cycle." Teheran Symposium "Salinity Problems in the Arid Zones," UNESCO, pp. 105–110.

Eaton, F. M. 1941. Water uptake and root growth as influenced by inequalities in the concentration of the substrate. *Plant Physiol.* 16: 545–564.

El-Mowelhi, N., El-Bershamgy, A., Hoffman, G. J., and Chang, A. C. 1988. "Enhancement of crop yields from subsurface drains with various envelopes." *Agricultural Water Management* 15: 131–140.

Francois, L. E. 1981. "Alfalfa management under saline conditions with zero leaching." *Agronomy J.* 73: 1042–1046.

Gardner, W. R. 1957. "Some steady-state solutions of the unsaturated moisture flow equation with application to evaporation from a water table." *Soil Science* 85(4): 228–232.

Gardner, W. R., and Fireman, M. 1958. "Laboratory studies of evaporation from soil columns in the presence of a water table." *Soil Science* 85(5): 244–249.

Grimes, D. W., Sharma, R. L., and Henderson, D. W. 1984. "Developing the resource potential of a shallow water table." Calif. Water Resour. Center., Univ. of Calif. Davis. Contribution No. 188. 39 p.

Hanson, B. R., and Kite, S. W. 1984. "Irrigation scheduling under saline high water tables." *Transactions of ASAE* 27: 1430–1434.

Hoffman, G. J. 1985. "Drainage required to manage salinity." *J. Irrig. and Drainage Div., ASCE* 111: 199–206.

Hoffman, G. J., Catlin, P. B., Mead, R. M., Johnson, R. S., Francois, L. E., and Goldhamer, D. 1989. "Yield and foliar injury responses of mature plum trees to salinity." *Irrigation Science* 10: 215–229.

Hoffman, G. J., Maas, E. V., Prichard, T. L., and Meyer, J. L. 1983. "Salt tolerance of corn in the Sacramento-San Joaquin Delta of California." *Irrigation Sci.* 4:31–44.

Hoffman, G. J., and van Genuchten, M. Th. 1983. "Water management for salinity control." *Limitations to Efficient Water Use in Crop Production,* Chapter 2C. H. Taylor, W. Jordan, and T. Sinclair, eds. Amer. Soc. Agron. Monograph. pp. 73–85.

Kearney, T. E., Kegel, F. R., Ou, J. P., Smith, J. D., and Ingelbretsen, K. H. 1980. "Field corn production in California." Univ. of Calif. Oakland, Div. Agric. Sci. Leaflet 21163. 9 pp.

Kruse, E. G., Yoder, R. E., Cuevas, D. L., and Champion, D. F. 1986. "Alfalfa water use from high, saline water tables." Amer. Soc. Agric. Engr., Paper No. 86-2597, December, Chicago, Ill., 16 p.

Kruse, E. G., Young, D. A., and Champion, D. F. 1985. "Effects of saline water tables on corn irrigation." Proc., ASCE Specialty Conference, San Antonio, Tex., pp. 444–453.

Lyerly, P. J., and Longenecker, D. E. 1962. "Salinity control in irrigation agriculture." Texas Agr. Exp. Sta. Bulletin 876. 20 p.

Lunin, J., and Gallantin, M. H. 1965. "Zonal salinization of the root system in relation to plant growth." *Soil Sci. Soc. Amer. Proc.* 29: 608–612.

Maas, E. V., and Hoffman, G. J. 1977. "Crop salt tolerance—current assessment." *Journal Irrigation and Drainage Div., ASCE* 103(IR2): 115–134.

Maas, E. V., Hoffman, G. J., Chaba, G. D., Poss, J. A., and Shannon, M. C. 1983. "Salt sensitivity of corn at various growth stages." *Irrigation Science* 4: 45–57.

Maas, E. V., Nieman, R. H. 1978. "Physiology of plant tolerance to salinity." G. A. Jung ed., Crop Tolerance to Suboptimal Land Conditions, Chapter 13, ASA Spec. Publ. 32: 277–299.

Meiri, A., and Poljakoff-Mayber, A. 1970. "Effect of various salinity regimes on growth, leaf expansion, and transpiration rate of bean plants." *Soil Science* 109: 26–34.

Namken, L. N., Wiegand, C. L., and Brown, R. G. 1969. "Water use by cotton from low and moderately saline static water tables." *Agronomy Journal* 61: 305–310.

Peck, A. J., Johnston, C. D., and Williamson, D. R. 1981. "Analyses of solute distributions in deeply weathered soils." *Agric. Water Management* 4: 83–102.

Raats, P. A. C. 1974. "Steady flows of water and salt in uniform soil profiles with plant roots." *Soil Sci. Soc. Amer. Proc.* 38: 717–722.

Rhoades, J. D. 1979. "Inexpensive four-electrode probe for monitoring soil salinity." *Soil Science Soc. of Amer. J.* 43: 817–818.

Rhoades, J. D., Krueger, D. B., and Reed, M. J. 1968. "The effect of soil-mineral weathering on the sodium hazard of irrigation waters." *Soil Sci. Soc. Amer. Proc.* 32: 643–647.

Rhoades, J. D., Ingvalson, R. D., Tucker, J. M., and Clark, M. 1973. "Salts in irrigation drainage waters: I.Effects of irrigation water composition, leaching fraction, and time of year on the salt compositions of irrigation drainage waters." *Soil Sci. Soc. Amer. Proc.* 37: 770–774.

Rhoades, J. D., Oster, J. D., Ingvalson, R. D., Tucker, J. M., and Clark, M. 1974. "Minimizing the salt burdens of irrigation drainage water." *Jour. Environ. Quality* 3: 311–316.

Shalhevet, J., and Bernstein, L. 1968. "Effects of vertically heterogeneous soil salinity on plant growth and water uptake." *Soil Sci.* 106: 85–93.

Shalhevet, J., Reiniger, P., and Shimshi, D. 1969. "Peanut response to uniform and non-uniform soil salinity." *Agronomy Journal* 61: 384–387.

Shalhevet, J., and Yaron, B. 1973. "Effect of soil and water salinity on tomato growth." *Plant and Soil* 39: 285–292.

U.S. Salinity Laboratory Staff. 1954. "Diagnosis and improvement of saline and alkali soils." USDA Agric. Handbook 60, 160 p.

Wadleigh, C. H., Gauch, H. G., and Strong, D. G. 1947. "Root penetration and moisture extraction in saline soil by crop plants." *Soil Sci.* 63: 341–349.

Wallender, W. W., Grimes, D. W., Henderson, D. W., and Stromberg, L. K. 1979. "Estimating the contribution of a perched water table to the seasonal evapotranspiration of cotton." *Agronomy Journal* 71: 1056–1060.

Wichelns, D., Nelson, D., and Weaver, T. March 1988. "Farm-level analyses of irrigated crop production in areas with salinity and drainage problems." San Joaquin Valley Drainage Program. U.S. Bur. of Recl., Sacramento, Calif. 79 pp.

Yaalon, D. H. 1963. "The origin and accumulation of salts in ground-water and in soils of Israel." Bull. Research Council Israel 11G: 105–131.

NOTATION

a, b, n = constants determined by experimental data; C = salt concentration; $\overline{C}$ = linearly-averaged salt concentration of root zone; C_a = salt content of applied water; C_d = salt content of drain water; C_r = salt concentration of the irrigation water; d = drainage; D_a = depth of applied water (irrigation plus rainfall); D_d = depth of flow of water out of the crop's root zone due to drainage; D_e = depth of flow of water out of the crop's root zone due to evaporation; D_g = depth of flow of water from ground water into the crop's root zone; D_i = depth of flow of water from irrigation into the crop's root zone; D_r = depth of flow of water from rainfall into the crop's root zone; D_t = depth of flow of water out of the crop's root zone due to transpiration; EC_a = electrical conductivity of applied water; EC_d = electrical conductivity of drainage water; EC_{GW} = electrical conductivity of ground water; EC_i = electrical conductivity of irrigation water; EC_{SW} = electrical conductivity of soil water; ET = evapotranspiration; g = upward flow from ground water; i = irrigation; k = hydraulic conductivity; L = leaching fraction; L_r = leaching requirement; r = rain; s_c = salt removed in the harvested crop; s_f = salt added to soil as fertilizer or amendment; s_m = salt dissolved from minerals in soil; s_p = salt precipitated; S = soil matric potential; S_s = change in salt storage; Z = depth of root zone; δ = empirical constant set to 0.22; and * = required values.

CHAPTER 13

CROP SALT TOLERANCE[a]

INTRODUCTION

The salt tolerance of a plant can be defined as the plant's capacity to endure the effects of excess salt in the medium of root growth. Implicit in this definition is the idea that a plant can withstand a precise amount of salt without adverse effects. In truth, the salt tolerance of a plant is not an exact value. It depends on many factors, conditions, and limits. First, the salt or salts involved must be specified. Second, the conditions in which the crop is growing, which cause tolerance to vary greatly, must be specified. Third, the age and variety of a plant, which also affects the plant's tolerance, must be specified. To be useful to the farmer, salt tolerance must be defined in terms of the reduction in growth or yield caused by specified concentrations of salt. Of course, the factors mentioned also influence the degree of reduction in growth. Although a plant's capacity to endure salt cannot be stated in absolute terms, relative crop responses to known concentrations of salt under certain conditions can be predicted. This chapter focuses on crop tolerance and the various factors that affect it.

Salinity

When an ion exists in the soil solution at a concentration that exceeds the amount needed for optimum growth, it may become toxic to the plant. Different levels of ions have different toxic levels. Concentrations of chloride of up to 200 mol/m^3 or more may be tolerated by some plants (Maas 1986), while as little as 0.2 mol/m^3 of boron is toxic to some plants (Eaton 1944). Salinity, when used to refer to soil, denotes an excess of salts derived from alkali and alkaline earth metals, primarily Na^+, Ca^{2+}, and Mg^{2+}. The predominant anions are usually Cl^-, SO_4^{2-}, and HCO_3^-. A salt-affected soil is one that contains enough soluble salts to hamper growth of the crop. The proportion of Na^+ to Ca^{2+} and Mg^{2+} that soils contain further differentiate them, i.e., they are either saline, sodic, or saline-sodic. (See Chapter 5.) The relative concentrations of these ions can vary greatly among soils. Their effects on plants, particularly at extreme ratios, can also vary. Generally, however, plants respond similarly to salinity over a fairly wide range of combinations of salt. In this chapter, salt tolerance refers to the capacity of a crop to grow on a saline soil as defined in Chapter 5. Tolerance to specific ions or elements will be considered separately.

---

[a]Prepared by: E. V. Maas, U.S. Salinity Lab., 4500 Glenwood Dr., Riverside, CA 92501.

Criteria for Salt Tolerance

The salt tolerance of a crop is appraised based on one of the following: its ability to survive on saline soils, the reductions in growth or reductions in yield at different levels of salinity, or its growth or yield when grown on saline soil compared to its growth or yield when grown on a nonsaline soil. Plant survival, an important ecological criterion, is of little value for evaluating tolerance at commercial production levels. The capability of plants to survive at extreme salinities often does not correlate with reductions in yield found at more moderate salinities. Actual yield responses to salinity are perhaps the most useful to a farmer, but differences in yield may be caused by different environmental factors, e.g., soil moisture, soil fertility, insects, and plant diseases. Furthermore, it is difficult to compare crops, since yields of different crops are not expressed in comparable units. To overcome this problem, yields can be expressed on a relative basis. Relative yield is the yield of a crop grown under saline conditions expressed as a fraction of that achieved under nonsaline, but otherwise comparable, conditions.

Source of Salt-Tolerance Data

Earlier publications on this subject were updated to compile the data on crop tolerances to salinity and specific ions and elements presented in this chapter (Maas and Hoffman 1977, Maas 1987). Generally, only those papers reporting on both the salinity of the root zone and the yield of the crop were used. In the case of some tree and vine crops, only the responses of vegetative growth were available. Data on ornamentals are based on salt injury and appearance, rather than plant growth. The literature covered in this review can be found in a bibliography compiled by Francois and Maas (1978, 1985).

ENVIRONMENTAL AND EDAPHIC FACTORS INFLUENCING SALT TOLERANCE

A plant's ability to tolerate salinity or specific ions is a function of many other conditions. The reliability of salt-tolerance data depends on whether the interaction between salinity and various conditions of soil, water, and climate influence yield reduction. Other environmental stresses may limit the yield, but they increase, decrease, or leave unaffected the crop's salt tolerance. Therefore, the effects of any interacting factor must be compared based on relative crop yield. Fig. 13.1 shows the types of interactions that illustrate this point. If the response to salinity is proportionately the same when a factor is limiting as when it is adequate, i.e., the absolute yields in both cases are decreased by the same percentage, the relative tolerance would be the same (Type A). If salinity decreases yields by the same absolute amounts for adequate and limiting conditions, a crop may appear relatively less tolerant under the suboptimal condition (Type B). However, if yield is severely limited by some suboptimal condition, e.g., inadequate soil fertility, a crop may appear relatively more tolerant than if it were grown with adequate fertility (Type C) because the effects of salinity on absolute yield cannot be as great as when fertility is adequate. Fig. 13.1 shows idealized yield response relationships to represent only the range over which yield is linearly responsive to salinity. Actual salinity response curves usually follow a sigmoidal relationship.

A relative yield versus salinity response curve usefully expresses salt tolerance if reductions in relative yield are independent of differences in actual yield resulting from differences in the conditions of the soil, properties of the soil, irrigation practices, climate, and other variables. Some of the factors that can affect salt tolerance are discussed below.

Soil Fertility

In irrigated agriculture, fields are usually fertilized to achieve maximum productivity. However, fertilizer applications are sometimes inadequate or even omitted because of cost or unavailability. Crops grown on soils of low fertility may seem more salt-tolerant than those grown with adequate fertility, because fertility and not salinity primarily limits the growth of plants. Proper fertilizer applications would increase yields regardless of the salinity of the soil, but proportionately more if the soil were nonsaline. The effects of salinity and nutritional stresses tend to add up, according to Bernstein et al. (1974). When salinity and soil infertility similarly limit yield, decreasing salinity or increasing fertility gives similar benefits. However, if yields are reduced markedly by one more than the other, alleviating the most severe condition will increase yield more than alleviating the less severe condition would.

Unless salinity causes specific nutritional imbalances (see the "Plant Tolerance to Specific Ions and Elements" section of this chapter), fertilizer applications exceeding that required under nonsaline conditions have rarely helped to alleviate the inhibition of growth by salinity. Most studies indicate that excess applications of N, P, and K do not affect or reduce salt tolerance. However, Ravikovitch and Yoles (1971) found that N, or P, or both, seemed to increase the salt tolerance of millet and clover.

Physical Condition of the Soil

Other suboptimal conditions of the soil can also affect the salt tolerance of crops. For example, soils with a poor structure or impermeable layers could restrict the growth of roots, as well as influence the distribution of water and salt in the soil.

Flooded or poorly drained soils can cause poor aeration of the soil, which may affect the plant's response to salt stress. Studies have shown that low levels of oxygen interact with salinity to affect the shoot growth of tomatoes (Aubertin et al. 1968) and the germination of wheat (Aceves-N et al. 1975). Inadequate drainage that results in a shallow water table could also influence the salt tolerance of crops. Plants could extract water directly from this source and, depending on the quality of water, respond much differently than expected from the level of salinity in the soil.

Salt Distribution in the Profile

Except in experiments, soil salinity is seldom constant with time or uniform in space. Depending on the extent of leaching and drainage, the

A. UNAFFECTED SALT TOLERANCE

Optimal
Suboptimal
Absolute Yield
Optimal & Suboptimal
Relative Yield
Salinity

B. DECREASED SALT TOLERANCE

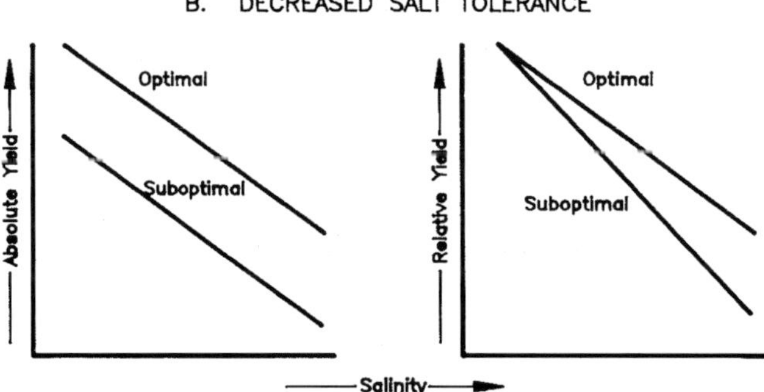

C. INCREASED SALT TOLERANCE

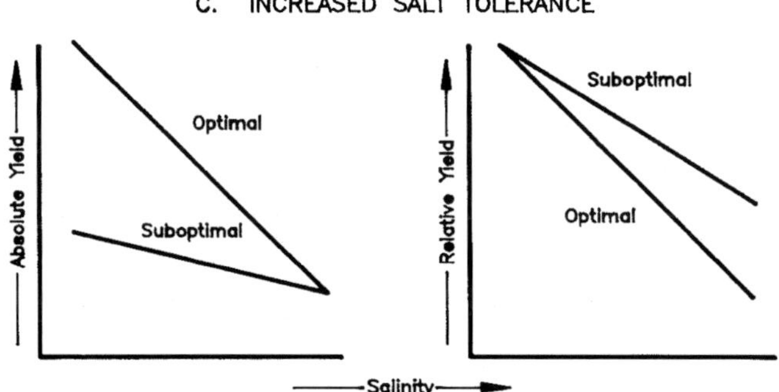

Fig. 13.1 Types of Interactions Between Salinity and Other Variables, Such as Soil Fertility on Absolute and Relative Yields (Relative yields under nonsaline conditions are plotted as 100% for both fertility levels.) (Bernstein et al. 1974); (a) Independent Effects of Salinity and Fertility at Optimal and Suboptimal Levels of Fertility; (b) Decreased Salt Tolerance at Suboptimal Levels of Fertility; and (c) Increased Salt Tolerance at Suboptimal Levels of Fertility

salt distribution may be uniform in the soil and change relatively little with depth, or may be highly irregular and range from concentrations of salt near the surface of the soil that are equal to approximately that of the irrigation water to concentrations at the bottom of the root zone that are many times greater. If a saline water table exists within a meter of the surface, salts may be transported upward by capillary flow. In this case, soil salinity may be inverted, with the highest concentrations at the surface. Salinity also increases between irrigations with the evaporation of soil water. Consequently, the plant must respond to a number of different and continually changing salt concentrations.

Most evidence indicates that plant growth most closely relates to the salinity of the soil water in that part of the root zone where the most of the water uptake takes place. The tolerance of plants to salt, therefore, should be related to salinity integrated over time and measured where roots absorb most of the water.

Irrigation Methods

The salt tolerance of crops also depends on the type and frequency of irrigation. As the soil water content decreases between irrigations, the concentration of salt increases. Consequently, plants are exposed to increasingly saline water with time between irrigations. If water becomes limiting, plants may experience matric stress, as well as osmotic stress. Hoffman et al. (1983) found in a three-year study that salinity caused yields of tall fescue to drop significantly below that expected when the matric potential was decreased.

Plants respond differently to saline waters, depending on the method of irrigation used. Bernstein and Francois (1973) found that irrigating with saline water (2,450 mg/l TDS) reduced the yields of pepper by different amounts, depending on whether a drip, furrow, or sprinkler irrigation system was used. Sprinkler irrigation caused the greatest reduction. Though foliar absorption of salt was discounted as a factor, this was likely an important cause of plant injury. (The "Plant Tolerance to Saline Sprinkling Waters" section of this chapter covers this in more detail.)

Climate

Climate probably influences the response of plants to salinity as much as, if not more than, any other factor. Most crops can tolerate greater salt stress if the weather is cool and humid than if it is hot and dry. Studies on several crops, including alfalfa, bean, beet, carrot, cotton, onion, squash, strawberry clover, salt grass, and tomato, have shown that salinity decreased yields more when these crops were grown at higher temperatures (Ahi and Powers 1938; Magistad et al. 1943; Hoffman and Rawlins 1970). Yields of many crops are also decreased more by salinity as relative humidity decreases. Experiments indicate that barley, bean, corn, cotton, onion, and radish were more sensitive to salt at low humidity than at high humidity. The tolerances of beet and wheat were not greatly affected by humidity (Hoffman and Rawlins 1970, 1971; Hoffman et al. 1971; Nieman and Poulsen 1967).

Many crops are sensitive to air pollutants (Heck et al. 1982).

Ozone, a major pollutant of air, decreases the yield of some crops more under nonsaline than saline conditions (Bytnerowicz and Taylor 1983; Hoffman et al. 1973, 1975; Maas et al. 1973; Ogata and Maas 1973). This strong interaction between the effects of ozone and salinity increases the salt tolerance of oxidant-sensitive crops.

BIOLOGICAL FACTORS INFLUENCING SALT TOLERANCE

Stage of Growth

The sensitivity of crops to soil salinity often changes from one stage of growth to the next. Most plants are tolerant during germination but become sensitive during the stages of emergence and early seedling growth. Plant stand can be seriously affected at this stage. The higher concentration of salts at the soil surface caused by evaporation furthers this effect. The juvenile roots of emerging plants are exposed to greater stress than indicated by the usual sampling methods to measure soil salinity.

Data on the salt tolerance of crops during emergence and seedling growth are extremely limited. Salt stress may delay emergence, but does not affect the final emergence for most crops if concentrations of salt remain at or below the tolerance threshold for mature yields. The tolerance of crop seedlings grown under actual or simulated field conditions has yet to be systematically evaluated. Greenhouse experiments on corn and wheat indicate that dry matter yields of three-week-old plants are reduced by concentrations of salt that are lower than the salinity thresholds found for grain production (Maas et al. 1983; Maas and Poss 1989a).

A sensitive stage in cereal crops is the early reproductive stage. If plants are stressed at the spike or panicle differentiation stage, seed yield is significantly reduced. From the booting stage on, plants appear to become more tolerant, although grain yield may be reduced if plants are stressed during pollination and fertilization (Pearson and Bernstein 1959). However, Kaddah et al. (1973, 1975) did not confirm this. Recent studies indicate that corn (Maas et al. 1983), sorghum (Maas et al. 1986), wheat (Maas and Poss 1989a), and cowpea (Maas and Poss 1989b) are most sensitive during the vegetative and early reproductive stages and least sensitive during the grain-filling stage. Increased tolerance with age has also been observed in asparagus, a perennial that was much more tolerant after the first year's growth (Francois 1987).

Varieties and Rootstocks

Salt tolerance data for most species of crops are based on tests of one or two varieties or cultivars. Tolerance of varieties within species may differ significantly. Perhaps because most commercially grown cultivars are developed under nonsaline conditions and not bred to endure salt stress, their relative tolerances to salinity are often similar and differences are difficult to measure. Nevertheless, significant differences among varieties have been observed for some species, including Bermuda grass, bromegrass, creeping bent grass, bird's-foot trefoil, berseem clover, lettuce, barley, wheat, rice, and soybean. (See Chapter 8.) Whether the differences reflect differences in salt tolerance or differences in adapting to the climate or nutritional conditions is unknown, however.

The ability of different rootstocks to tolerate soil salinity is an important factor in rating fruit tree and vine crops. Their tolerance seems to be related primarily to the capability of different rootstocks to regulate the uptake of Na^+, or Cl^-, or both.

PLANT TOLERANCE TO SOIL SALINITY

Yield Response Functions

The salt tolerance of a crop can best be described by plotting the relative yield as a continuous function of soil salinity. For most crops, this response function follows a sigmoidal relationship. However, some crops may die before salinity reduces the seed or fruit yields to zero, thus eliminating the bottom half of the sigmoidal curve. Maas and Hoffman (1977) proposed that this response curve could be represented by two linear lines, one a tolerance plateau with a slope of zero and the other a concentration-dependent line whose slope indicates the yield reduction per unit increase in salinity. Fig. 13.2 shows the two-piece model fitted to actual data for grain yield from an experiment on corn. The point at which the two lines intersect designates the threshold, i.e., the maximum soil salinity that does not reduce yield below that achieved under nonsaline conditions. This two-piece linear response function provides a reasonably good fit for commercially acceptable yields, when plotted against the electrical conductivity of the saturated-soil extract (EC_e). EC_e is the traditional soil salinity measurement with units of decisiemens per meter (1 dS/m = 1 mmho/cm). For soil salinities exceeding the threshold of any given crop, relative yield (Y_r) can be estimated with the following equation:

$$Y_r = 100 - b\,(EC_e - a) \qquad (13.1)$$

where a = the salinity threshold expressed in dS/m; b = the slope expressed in % per dS/m; and EC_e is the mean electrical conductivity of a saturated-soil extract taken from the root zone.

The two-piece linear response function is also reasonably accurate when salinity is expressed in terms of the osmotic potential of the soil solution at field capacity (OP_{fc}). When the OP_{fc} is known, yield responses can be determined as a function of the osmotic stress experienced by plants. For osmotic potentials exceeding a crop's threshold

$$Y_r = 100 - B\,(OP_{fc} - A) \qquad (13.2)$$

where A = the salinity threshold expressed in bars; B = the slope expressed in % per bar; and OP_{fc} = the osmotic potential of the soil water extracted from the root zone at field capacity. Eq. 13.1, like Eq. 13.2, is linear, though OP_{fc} is not a linear function of EC_e. The deviation

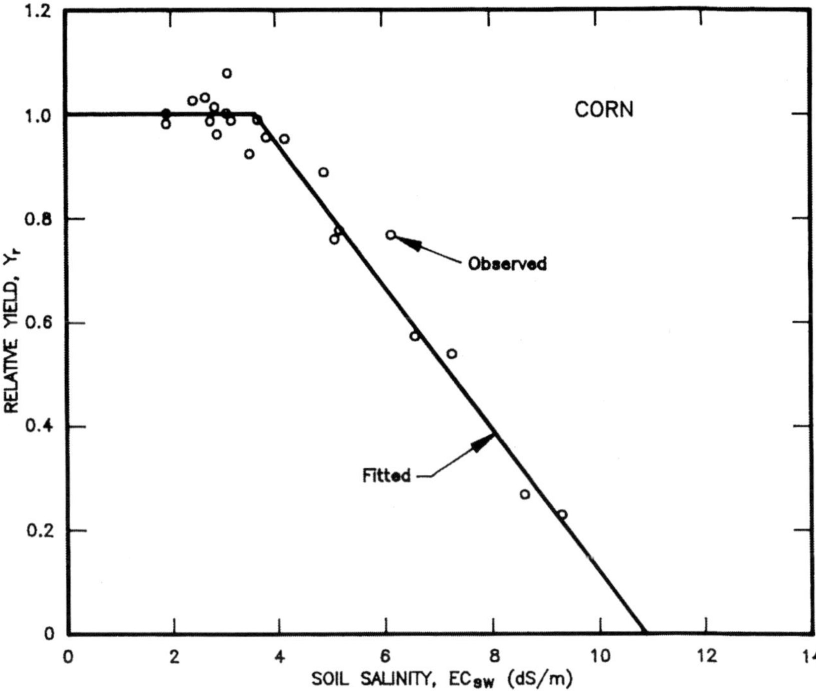

Fig. 13.2 The Piece-Wise Linear Response Function Fitted to Actual Yield Data Obtained from Corn (Hoffman et al. 1983)

from linearity is small, however, and relative yields calculated from Eq. 13.2 are within 1 to 2% of those calculated from Eq. 13.1. The salt-tolerance data in the subsequent sections are given in terms of EC_e. The parameters of the threshold (A) and slope (B) can be determined in terms of OP_{fc} from the EC_e data with the following relationships:

$$A = -0.725\,a^{1.06} \tag{13.3}$$

$$B = \frac{100}{-0.725\left(\dfrac{100 + ab}{b}\right)^{1.06} - A} \tag{13.4}$$

These equations are based on the relationship, $OP_{fc} = -0.725\ EC_e^{1.06}$, from Fig. 6 of the U.S. Department of Agriculture Handbook No. 60 (USSL 1954), after converting osmotic pressure in atmospheres at 0 °C to osmotic potential in bars at 25 °C. The soluble salt concentration in the soil water at field capacity is assumed to be twice that of the saturated-soil extract.

The threshold and slope concept provides general guidelines about salt tolerance for crop management decisions. Farmers need to know the levels of soil salinity that begin to reduce yield and how much yield will be reduced at levels above the threshold. However, more precise plant response functions would be advantageous for crop simulation modeling. Several nonlinear models that more accurately describe the sigmoidal growth response of plants to salinity exist (van Genuchten and Hoffman 1984). The computer programs for these models were developed and documented by van Genuchten (1983). One of these models takes the form:

$$Y_r = \frac{Y_m}{\left[1 + \left(\frac{c}{c_{50}}\right)^p\right]} \qquad (13.5)$$

where Y_m = the yield under nonsaline conditions; c = the average salinity of the root zone; c_{50} = the average salinity of the root zone that reduces yield by 50%; and p = an empirical constant. Values for the root zone's salinity, c and c_{50}, can be expressed in terms of either EC_e or OP_{fc}. Like the two-piece linear model, this sigmoidal model requires two parameters to describe the response curve, the values of c_{50} and p.

Salt Tolerances of Herbaceous Crops

Table 13.1 lists values for the threshold and slope of 69 crops in terms of EC_e. Most of the data were obtained from crops grown under conditions that simulated recommended cultural and management practices for commercial production. They show the relative tolerances of different crops grown under different non-standardized conditions. The data apply only where crops are exposed to fairly uniform salinities from the late seedling stage to maturity. Where crops have particularly sensitive stages, limits of tolerance are given in the footnotes.

The data in Table 13.1 apply to soils where chloride is the predominant anion. Since $CaSO_4$ is dissolved in preparing saturated-soil extracts, the EC_e of gypsiferous soils will range from 1 to 3 dS/m higher than that of non-gypsiferous soils with the same soil water conductivity at field capacity (Bernstein 1962). Therefore, plants grown on gypsiferous soils will tolerate an EC_e of approximately 2 dS/m higher than those listed in Table 13.1. The last column provides a qualitative rating of salt tolerance that can be used to categorize crops. Figure 13.3 illustrates the limits of these categories. Some crops are listed with only a qualitative rating because insufficient data are available to calculate the threshold and the slope.

Salt Tolerance of Woody Crops

Determining the salt tolerance of trees, vines, and other woody crops is complicated, since specific ion toxicities cause additional detrimental effects. The leaves of many perennial woody species are susceptible to injury by the toxic accumulation of Cl^-, or Na^+, or both, in the leaves. Different cultivars and rootstocks absorb Cl^- and Na^+ at different rates, so tolerance can vary considerably within a species. Tolerances to these specific ions will be discussed below. In the absence of specific-ion effects, the tolerance of woody crops can be expressed similarly to the tolerance of herbaceous crops, i.e., as a function of the concentration

TABLE 13.1a Salt Tolerance of Herbaceous Crops.[a]—Fiber, Grain and Special Crops

Crop		Electrical conductivity of saturated-soil extract			
Common name (1)	Botanical name[b] (2)	Threshold[c] dS/m (3)	Slope % per dS/m (4)	Rating[d] (5)	References (6)
Barley[e]	*Hordeum vulgare*	8.0	5.0	T	Maas and Hoffman (1977)
Bean	*Phaseolus vulgaris*	1.0	19.0	S	Maas and Hoffman (1977)
Broad bean	*Vicia faba*	1.6	9.6	MS	Maas and Hoffman (1977)
Corn[f]	*Zea Mays*	1.7	12.0	MS	Maas and Hoffman (1977)
Cotton	*Gossypium hirsutum*	7.7	5.2	T	Maas and Hoffman (1977)
Cowpea	*Vigna unguiculata*	4.9	12.0	MT	West and Francois (1982)
Flax	*Linum usitatissimum*	1.7	12.0	MS	Maas and Hoffman (1977)
Guar	*Cyamopsis tetragonoloba*	8.8	17.0	T	Francois et al. (1990)
Kenaf	*Hibiscus cannabinus*			MT	Francois (1988c)
Millet, foxtail	*Setaria italica*			MS	Maas and Hoffman (1977)
Oats	*Avena sativa*			MT*	
Peanut	*Arachis hypogaea*	3.2	29.0	MS	Maas and Hoffman (1977)
Rice, paddy	*Oryza sativa*	3.0[g]	12.0[g]	S	Maas and Hoffman (1977)
Rye	*Secale cereale*	11.4	10.8	T	Francois et al. (1989a)
Safflower	*Carthamus tinctorius*			MT	Maas and Hoffman (1977)
Sesame[j]	*Sesamum indicum*			S	Yousif et al. (1972)
Sorghum	*Sorghum bicolor*	6.8	16.0	MT	Francois et al. (1984)
Soybean	*Glycine max*	5.0	20.0	MT	Maas and Hoffman (1977)

(continued)

TABLE 13.1a Salt Tolerance of Herbaceous Crops.[a]—Fiber, Grain Special Crops (Continued)

Crop		Electrical conductivity of saturated-soil extract			
Common name (1)	Botanical name[b] (2)	Threshold[c] dS/m (3)	Slope % per dS/m (4)	Rating[d] (5)	References (6)
Sugar beet[h]	*Beta vulgaris*	7.0	5.9	T	Maas and Hoffman (1977)
Sugar-cane	*Saccharum officinarum*	1.7	5.9	MS	Maas and Hoffman (1977)
Sunflower	*Helianthus annuus*			MS*	
Triticale	*X Triti-cosecale*	6.1	2.5	T	Francois et al. (1988)
Wheat	*Triticum aestivum*	6.0	7.1	MT	Maas and Hoffman (1977)
Wheat (semi-dwarf)[i]	*T. aestivum*	8.6	3.0	T	Francois et al. (1986)
Wheat, durum	*T. turgidum*	5.9	3.8	T	Francois et al. (1986)

[a]These data serve only as a guideline to relative tolerances among crops. Absolute tolerances vary, depending on climate, soil conditions, and cultural practices.

[b]Botanical and common names follow the convention of Hortus Third (Liberty Hyde Bailey Hortorium Staff 1976), where possible.

[c]In gypsiferous soils, plants will tolerate EC_es about 2 dS/m higher than indicated.

[d]Ratings are defined by the boundaries in Fig. 13.3. Ratings with an * are estimates. For references, consult the indexed bibliography by Francois and Maas (1978, 1985).

[e]Less tolerant during seedling stage, EC_e at this stage should not exceed 4 dS/m or 5 dS/m.

[f]Grain and forage yields of DeKalb XL-75[e] grown on an organic muck soil decreased about 26% per dS/m above a threshold of 1.9 dS/m (Hoffman et al. 1983).

[g]Because paddy rice is grown under flooded conditions, values refer to the electrical conductivity of the soil water while plants are submerged. Less tolerant during seedling stage.

[h]Sensitive during germination and emergence, EC_e should not exceed 3 dS/m.

[i]Data from one cultivar, "Probred."

[j]Sesame cultivars, Sesaco 7 and 8, may be more tolerant than indicated by the S rating (Francois 1988b).

TABLE 13.1b Salt Tolerance of Herbaceous Crops[a]—Grasses and Forage Crops

Crop		Electrical conductivity of saturated-soil extract			
Common name (1)	Botanical name[b] (2)	Threshold[c] dS/m (3)	slope % per dS/m (4)	Rating[d] (5)	References (6)
Alfalfa	*Medicago sativa*	2.0	7.3	MS	Maas and Hoffman (1977)
Alkali grass, Nuttall	*Puccinellia airoides*			T*	
Alkali sacaton	*Sporobolus airoides*			T*	
Barley (forage)[e]	*Hordeum vulgare*	6.0	7.1	MT	Maas and Hoffman (1977)
Bentgrass	*Agrostis stolonifera palustris*			MS	Maas and Hoffman (1977)
Bermuda grass[h]	*Cynodon Dactylon*	6.9	6.4	T	Maas and Hoffman (1977)
Bluestem, Angleton	*Dichanthium aristatum*			MS*	
Brome, mountain	*Bromus marginatus*			MT*	
Brome, smooth	*B. inermis*			MS	Maas and Hoffman (1977)
Buffelgrass	*Cenchrus ciliaris*			MS*	
Burnet	*Poterium Sanguisorba*			MS*	
Canary grass, reed	*Phalaris arundinacea*			MT	Maas and Hoffman (1977)
Clover, alsike	*Trifolium hybridum*	1.5	12.0	MS	Maas and Hoffman (1977)
Clover, Berseem	*T. alexandrinum*	1.5	5.7	MS	Maas and Hoffman (1977)
Clover, Hubam	*Melilotus alba*			MT*	
Clover, ladino	*Trifolium repens*	1.5	12.0	MS	Maas and Hoffman (1977)
Clover, red	*T. pratense*	1.5	12.0	MS	Maas and Hoffman (1977)
Clover, strawberry	*T. fragiferum*	1.5	12.0	MS	Maas and Hoffman (1977)

(continued)

TABLE 13.1b Salt Tolerance of Herbaceous Crops[a]—Grasses and Forage Crops (Continued)

Crop		Electrical conductivity of saturated-soil extract			
Common name (1)	Botanical name[b] (2)	Threshold[c] dS/m (3)	slope % per dS/m (4)	Rating[d] (5)	References (6)
Clover, sweet	*Melilotus*			MT*	
Clover, white Dutch	*Trifolium repens*			MS*	
Corn (forage)[f]	*Zea Mays*	1.8	7.4	MS	Maas and Hoffman (1977)
Cowpea (forage)	*Vigna unguiculata*	2.5	11.0	MS	West and Francois (1982)
Dallis grass	*Paspalum dilatatum*			MS*	
Fescue, tall	*Festuca elatior*	3.9	5.3	MT	Maas and Hoffman (1977)
Fescue, meadow	*F. pratensis*			MT*	
Foxtail, meadow	*Alopecurus pratensis*	1.5	9.6	MS	Maas and Hoffman (1977)
Grama, blue	*Bouteloua gracilis*			MS*	
Harding grass	*Phalaris tuberosa*	4.6	7.6	MT	Maas and Hoffman (1977)
Kallar grass	*Diplachne fusca*			T*	
Love grass[i]	*Eragrostis* sp.	2.0	8.4	MS	Maas and Hoffman (1977)
Milkvetch, Cicer	*Astragalus cicer*			MS*	
Oat grass, tall	*Arrhenatherum, Danthonia*			MS*	
Oats (forage)	*Avena sativa*			MS*	
Orchard grass	*Dactylis glomerata*	1.5	6.2	MS	Maas and Hoffman (1977)
Panic grass, blue	*Panicum antidotale*			MT*	
Rape	*Brassica napus*			MT*	
Rescue grass	*Bromus unioloides*			MT*	

(continued)

TABLE 13.1b Salt Tolerance of Herbaceous Crops[a]—Grasses and Forage Crops (Continued)

Crop		Electrical conductivity of saturated-soil extract			
Common name (1)	Botanical name[b] (2)	Threshold[c] dS/m (3)	slope % per dS/m (4)	Rating[d] (5)	References (6)
Rhodes grass	*Chloris Gayana*			MT	Maas and Hoffman (1977)
Rye (forage)	*Secale cereale*			MS*	
Ryegrass, Italian	*Lolium italicum multiflorum*			MT*	
Ryegrass, perennial	*L. perenne*	5.6	7.6	MT	Maas and Hoffman (1977)
Salt grass, desert	*Distichlis stricta*			T*	
Sesbania	*Sesbania exaltata*	2.3	7.0	MS	Maas and Hoffman (1977)
Sirato	*Macroptilium atropurpureum*			MS	Maas and Fisher (1986)
Sphaerophysa	*Sphaerophysa salsula*	2.2	7.0	MS	Francois and Bernstein (1964)
Sundan grass	*Sorghum sudanense*	2.8	4.3	MT	Maas and Hoffman (1977)
Timothy	*Phleum pratense*			MS*	
Trefoil, big	*Lotus uliginosus*	2.3	19.0	MS	Maas and Hoffman (1977)
Trefoil, narrowleaf bird's foot	*L. corniculatus tenuifolium*	5.0	10.0	MT	Maas and Hoffman (1977)
Trefoil, broadleaf bird's foot[l]	*L. corniculatus arvenis*			MT	Maas and Hoffman (1977)
Vetch, common	*Vicia angustifolia*	3.0	11.0	MS	Maas and Hoffman (1977)
Wheat (forage)[g]	*Triticum aestivum*	4.5	2.6	MT	Francois et al. (1986)
Wheat, Durum (forage)	*T. turgidum*	2.1	2.5	MT	Francois et al. (1986)

(continued)

TABLE 13.1b Salt Tolerance of Herbaceous Crops[a]—Grasses and Forage Crops (Continued)

Crop		Electrical conductivity of saturated-soil extract			
Common name (1)	Botanical name[b] (2)	Threshold[c] dS/m (3)	slope % per dS/m (4)	Rating[d] (5)	References (6)
Wheat grass, standard crested	*Agropyron sibiricum*	3.5	4.0	MT	Maas and Hoffman (1977)
Wheat grass, fairway crested	*A. cristatum*	7.5	6.9	T	Maas and Hoffman (1977)
Wheat grass, intermediate	*A. intermedium*			MT*	
Wheat grass, slender	*A. trachycaulum*			MT	Maas and Hoffman (1977)
Wheat grass, tall	*A. elongatum*	7.5	4.2	T	Maas and Hoffman (1977)
Wheat grass, western	*A. Smithii*			MT*	
Wild rye, Altai	*Elymus angustus*			T	Maas and Hoffman (1977)
Wild rye, beardless	*E. triticoides*	2.7	6.0	MT	Maas and Hoffman (1977)
Wild rye, Canadian	*E. canadensis*			MT*	
Wild rye, Russian	*E. Junceus*			T	Maas and Hoffman (1977)

[a]These data serve only as a guideline to relative tolerances among crops. Absolute tolerances vary, depending on climate, soil conditions, and cultural practices.
[b]Botanical and common names follow the convention of Hortus Third (Liberty Hyde Bailey Hortorium Staff 1976) where possible.
[c]In gypsiferous soils, plants will tolerate EC_e about 2 dS/m higher than indicated.
[d]Ratings are defined by the boundaries in Fig. 13.3. Ratings with an * are estimates. For references, consult the indexed bibliography by Francois and Maas (1978, 1985).

TABLE 13.1b (Continued)

[e]Less tolerant during seedling stage, EC_e at this stage should not exceed 4 dS/m or 5 dS/m.
[f]Grain and forage yields of DeKalb XL-75[e] grown on an organic muck soil decreased about 26% per dS/m above a threshold of 1.9 dS/m (Hoffman et al. 1983).
[g]Data from one cultivar, "Probred."
[h]Average of several varieties. Suwannee and Coastal are about 20% more tolerant, and common and Greenfield are about 20% less tolerant than the average.
[i]Average for Boer, Wilman, Sand, and Weeping cultivars. Lehmann seems about 50% more tolerant.

TABLE 13.1c Salt Tolerance of Herbaceous Crops[a]—Vegetables and Fruit Crops

Crop		Electrical conductivity of saturated-soil extract			
Common name (1)	Botanical name[b] (2)	Threshold[c] dS/m (3)	slope % per dS/m (4)	Rating[d] (5)	References (6)
Artichoke	*Helianthus tuberosus*			MT*	
Asparagus	*Asparagus officinalis*	4.1	2.0	T	Francois (1987)
Bean	*Phaseolus vulgaris*	1.0	19.0	S	Maas and Hoffman (1977)
Beet, red[e]	*Beta vulgaris*	4.0	9.0	MT	Maas and Hoffman (1977)
Broccoli	*Brassica oleracea botrytis*	2.8	9.2	MS	Maas and Hoffman (1977)
Brussel sprouts	*B. oleracea gemmifera*			MS*	
Cabbage	*B. oleracea capitata*	1.8	9.7	MS	Maas and Hoffman (1977)
Carrot	*Daucus carota*	1.0	14.0	S	Maas and Hoffman (1977)
Cauliflower	*Brassica oleracea botrytis*			MS*	
Celery	*Apium graveolens*	1.8	6.2	MS	Francois and West (1982)

(continued)

TABLE 13.1c Salt Tolerance of Herbaceous Crops[a]—Vegetables and Fruit Crops (Continued)

Crop		Electrical conductivity of saturated-soil extract			
Common name (1)	Botanical name[b] (2)	Threshold[c] dS/m (3)	slope % per dS/m (4)	Rating[d] (5)	References (6)
Corn, sweet	*Zea Mays*	1.7	12.0	MS	Maas and Hoffman (1977)
Cucumber	*Cucumis sativus*	2.5	13.0	MS	Maas and Hoffman (1977)
Eggplant	*Solanum Melongena esculentum*	1.1	6.9	MS	Heuer et al. (1986)
Kale	*Brassica oleracea acephala*			MS*	
Kohlrabi	*B. oleracea gongylode*			MS*	
Lettuce	*Lactuca sativa*	1.3	13.0	MS	Maas and Hoffman (1977)
Muskmelon	*Cucumis Melo*			MS	Shannon and Francois (1978)
Okra	*Abelmoschus esculentus*			S	Maas and Hoffman (1977)
Onion	*Allium Cepa*	1.2	16.0	S	Maas and Hoffman (1977)
Parsnip	*Pastinaca sativa*			S*	
Pea	*Pisum sativum*			S*	
Pepper	*Capsicum annum*	1.5	14.0	MS	Maas and Hoffman (1977)
Potato	*Solanum tuberosum*	1.7	12.0	MS	Maas and Hoffman (1977)
Pumpkin	*Cucurbita Pepo Pepo*			MS*	
Radish	*Raphanus sativus*	1.2	13.0	MS	Maas and Hoffman (1977)
Spinach	*Spinacia oleracea*	2.0	7.6	MS	Maas and Hoffman (1977)
Squash, scallop	*Cucurbita Pepo Melopepo*	3.2	16.0	MS	Francois (1985)

(continued)

TABLE 13.1c Salt Tolerance of Herbaceous Crops[a]—Vegetables and Fruit Crops (Continued)

Crop		Electrical conductivity of saturated-soil extract			
Common name (1)	Botanical name[b] (2)	Threshold[c] dS/m (3)	slope % per dS/m (4)	Rating[d] (5)	References (6)
Squash, zucchini	*C. Pepo Melopepo*	4.7	9.4	MT	Francois (1985)
Strawberry	*Fragaria* sp.	1	33	S	Maas and Hoffman (1977)
Sweet potato	*Ipomoea Batatas*	1.5	11	MS	Maas and Hoffman (1977)
Tomato	*Lycopersicon Lycopersicum*	2.5	9.9	MS	Maas and Hoffman (1977)
Turnip	*Brassica Rapa*	0.9	9	MS	Francois (1984a)
Watermelon	*Citrullus lanatus*			MS*	

[a]These data serve only as a guideline to relative tolerances among crops. Absolute tolerances vary, depending on climate, soil conditions, and cultural practices.

[b]Botanical and common names follow the convention of Hortus Third (Liberty Hyde Bailey Hortorium Staff 1976) where possible.

[c]In gypsiferous soils, plants will tolerate EC_e about 2 dS/m higher than indicated.

[d]Ratings are defined by the boundaries in Fig. 13.3. Ratings with an * are estimates. For references, consult the indexed bibliography by Francois and Maas (1978, 1985).

[e]Sensitive during germination and emergence, EC_e should not exceed 3 dS/m.

of total soluble salts or the osmotic potential of the soil solution. This might also be the case for cultivars and rootstocks that restrict the uptake of Cl^- and Na^+. The data in Table 13.2 for the salt tolerance of woody crops are reasonably accurate in the absence of specific-ion toxicities. Because of the cost and time involved in obtaining fruit yields, tolerances of several crops are based on vegetative growth (see tabular footnotes, Table 13.2). In contrast to other kinds of crops, most woody fruit and nut crops tend to be salt-sensitive, even in the absence of specific-ion effects. Only the date palm is relatively salt-tolerant, and the olive and a few others are believed to be moderately tolerant.

TABLE 13.2 Salt Tolerance of Woody Crops[a]

Crop		Electrical conductivity of saturated-soil extract			
Common name (1)	Botanical name[b] (2)	Threshold[c] dS/m (3)	Slope % per dS/m (4)	Rating[d] (5)	References (6)
Almond[e]	*Prunus duclis*	1.5	19.0	S	Maas and Hoffman (1977)
Apple	*Malus sylvestris*			S	Maas and Hoffman (1977)
Apricot[e]	*Prunus armeniaca*	1.6	24.0	S	Maas and Hoffman (1977)
					(continued)
Avocado[e]	Persea americana			S	Maas and Hoffman (1977)
Blackberry	*Rubus* sp.	1.5	22.0	S	Maas and Hoffman (1977)
Boysenberry	*Rubus ursinus*	1.5	22.0	S	Maas and Hoffman (1977)
Castorbean	*Ricinus communis*			MS*	
Cherimoya	*Annona Cherimola*			S*	
Cherry, sweet	*Prunus avium*			S*	
Cherry, sand	*P. Besseyi*			S*	
Currant	*Ribes* sp.			S*	
Date palm	*Phoenix dactylifera*	4.0	3.6	T	Maas and Hoffman (1977)
Fig	*Ficus carica*			MT*	
Gooseberry	*Ribes* sp.			S*	
Grape[e]	*Vitis* sp.	1.5	9.6	MS	Maas and Hoffman (1977)
Grapefruit[e]	*Citrus paradisi*	1.8	16.0	S	Maas and Hoffman (1977)
Guayule	*Parthenium argentatum*	8.7	11.6	T	Maas et al. (1988)
Jojoba[e]	*Simmondsia chinensis*			T	Yermanos et al. (1967)
Jujube	*Ziziphus jujuba*			MT*	
Lemon[e]	*Citrus Limon*			S	Maas and Hoffman (1977)

(continued)

TABLE 13.2 Salt Tolerance of Woody Crops[a] (Continued)

Crop		Electrical conductivity of saturated-soil extract			
Common name (1)	Botanical name[b] (2)	Threshold[c] dS/m (3)	Slope % per dS/m (4)	Rating[d] (5)	References (6)
Lime	*C. aurantifolia*			S*	
Loquat	*Eriobotrya japonica*			S*	
Mango	*Mangifera indica*			S*	
Olive	*Olea europaea*			MT	Maas and Hoffman (1977)
Orange	*Citrus sinensis*	1.7	16.0	S	Maas and Hoffman (1977)
Papaya[e]	*Carica papaya*			MT	Siegel (1982)
Passion fruit	*Passiflora edulis*			S*	
Peach	*Prunus Persica*	1.7	21.0	S	Maas and Hoffman (1977)
Pear	*Pyrus communis*			S*	
Persimmon	*Diospyros virginiana*			S*	
Pineapple	*Ananus comosus*			MT*	
Plum; Prune[e]	*Prunus domestica*	1.5	18.0	S	Hoffman et al. (1989)
Pomegranate	*Punica granatum*			MT*	
Pummelo	*Citrus maxima*			S*	
Raspberry	*Rubus idaeus*			S	Maas and Hoffman (1977)
Rose apple	*Syzygium jambos*			S*	
Sapote, white	*Casimiroa edulis*			S*	
Tangerine	*Citrus reticulata*			S*	

[a]These data are applicable when rootstocks are used that do not accumulate Na^+ or Cl^- rapidly, or when these ions do not predominate in the soil.

TABLE 13.2 (Continued)

[b]Botanical and common names follow the convention of Hortus Third (Libery Hyde Bailey Hortorium Staff 1976) where possible.
[c]In gypsiferous soils, plants will tolerate EC_e about 2 dS/m higher than indicated.
[d]Ratings are defined by the boundaries in Fig. 13.3. Ratings with an * are estimates. For references, consult the indexed bibliography by Francois and Maas (1978, 1985).
[e]Tolerance is based on growth rather than yield.

Salt Tolerance of Ornamentals

To establish permissible levels of salinity for ornamentals, the aesthetic characteristics of the plant are more important than its growth. The loss or injury of leaves due to salt stress is unacceptable for ornamentals, even if their growth remains unaffected. A 50% reduction in growth might be acceptable, perhaps even desirable, if the plants appeared healthy and attractive. Table 13.3 gives limits of salt tolerance for 49 species of ornamental shrubs, trees, and ground cover grown in Riverside, California. The data indicate the maximum permissible EC_e for an acceptable appearance. Some woody ornamentals, such as rose, Burford holly, and cotoneaster, are very salt-sensitive and exhibit the characteristic symptoms of leaf burn associated with an accumulation of excess Cl^- or Na^+.

PLANT TOLERANCE TO SALINE SPRINKLING WATERS

Sprinkler-irrigated crops can suffer additional salt damage when their foliage is wetted by saline water because the leaves accumulate the salts directly from the impinging solution. Some species become severely injured and lose their leaves. The degree of injury relates to the concentration of salt in the leaves, but the weather and water stress can influence the onset of injury. For instance, leaves may contain toxic levels of Na^+ or Cl^- for several weeks without showing any injury, but the first occurence of hot, dry weather will cause leaf necrosis. Consequently, no guidelines exist for correlating the injury of leaves with concentrations of salt in the leaves.

Foliar Salt Absorption and Injury

The leaves of many plants readily absorb Na^+, Ca^{2+}, and Cl^- when sprinkled with irrigation waters that contain these elements (Maas 1985). The susceptibility to foliar injury varies considerably among species. It depends more on leaf characteristics and the leaves' rate of absorption than on tolerance to soil salinity. Deciduous fruit trees (almond, apricot, and plum) and citrus experience considerable injury to leaves and defoliation when sprinkled with chloride waters (Ehlig and Bernstein 1959). In fact, concentrations of as low as 2 mol/m^3 or 3 mol/m^3

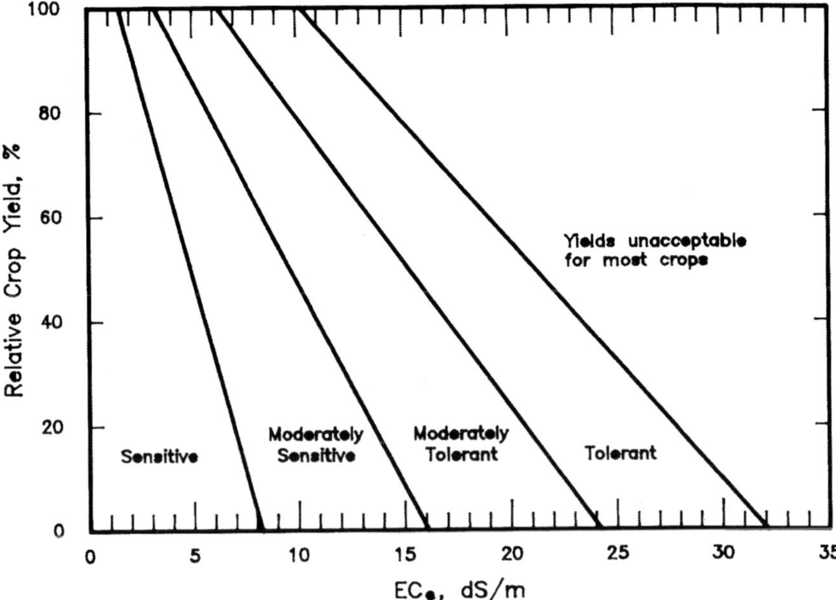

Fig. 13.3 Divisions for Classifying Crop Tolerance to Salinity

have caused severe leaf burn on citrus (Harding, Miller, and Fireman 1958). Avocado, in contrast, absorbs salt so slowly that foliar absorption is a negligible factor in sprinkling (Ehlig and Bernstein 1959). Similar results were obtained for strawberry, which is sensitive to soil salinity (Ehlig 1961). Among herbaceous crops, species belonging to the Solonaceae family, including pepper, potato, and tomato, seem to be the most sensitive to foliar injury. Greenhouse tests showed that the susceptibility to leaf injury for the following species of crops increased in the indicated order: sugarbeet < cotton, sunflower, cauliflower < safflower, barley, sorghum < alfalfa < sesame < < tomato, potato (Maas et al. 1982b).

Rates of salt absorption increase when concentrations of the ions and the frequency of sprinkling increase. In greenhouse experiments with bell peppers, increasing the frequency of sprinkling from one to eight times per day without changing the total daily duration of sprinkling doubled the rate of salt absorption (Maas et al. 1982a). In contrast, threefold increases in the duration of once-per-day sprinkling did not measurably affect the amount of salt uptake. Since salt continues to be absorbed by wetted leaves after sprinkling has stopped, frequent and short irrigations provide more of a chance for absorption than the total duration of sprinkling indicates. The rate of absorption, a function of salt concentration, increases rapidly as the solution evaporates from the leaf and the salt is concentrated. The more the wetting-drying cycle is repeated, the more salt is accumulated. Once the salts become dry, absorption stops (Grattan et al. 1981). If the salts are redissolved, e.g.,

TABLE 13.3 Salt Tolerance of Ornamental Shrubs, Trees and Ground Cover[a]

Common name (1)	Botanical name (2)	Maximum[b] permissible EC_e (dS/m) (3)
(a) Very sensitive		
Star jasmine	*Trachelospermum jasminoides*	1–2
Pyrenees cotoneaster	*Cotoneaster congestus*	1–2
Oregon grape	*Mahonia Aquifolium*	1–2
Photinia	*Photinia x fraseri*	1–2
(b) Sensitive		
Pineapple guava	*Feijoa Sellowiana*	2–3
Chinese holly, cv. Burford	*Ilex cornuta*	2–3
Rose, cv. Grenoble	*Rosa* sp.	2–3
Glossy abelia	*Abelia x grandiflora*	2–3
Southern yew	*Podocarpus macrophyllus*	2–3
Tulip tree	*Liriodendron Tulipifera*	2–3
Algerian ivy	*Hedera canariensis*	3–4
Japanese pittosporum	*Pittosporum Tobira*	3–4
Heavenly bamboo	*Nandina domestica*	3–4
Chinese hibiscus	*Hibiscus rosa-sinensis*	3–4
Laurustinus, cv. Robustum	*Viburnum Tinusm*	3–4
Strawberry tree, cv. Compact	*Arbutus Unedo*	3–4
Crape Myrtle	*Lagerstroemia indica*	3–4
(c) Moderately sensitive		
Glossy privet	*Ligustrum lucidum*	4–6
Yellow sage	*Lantana Camara*	4–6
Orchid tree	*Bauhinia purpurea*	4–6
Southern Magnolia	*Magnolia grandiflora*	4–6
Japanese boxwood	*Buxus microphylla var. japonica*	4–6
Xylosma	*Xylosma congestum*	4–6
Japanese black pine	*Pinus Thunbergiana*	4–6
Indian hawthorn	*Raphiolepis indica*	4–6
Dodonaea, cv. atropurpurea	*Dodonaea viscosa*	4–6
Oriental arborvitae	*Platycladus orientalis*	4–6

(continued)

TABLE 13.3 Salt Tolerance of Ornamental Shrubs, Trees and Ground Cover[a] (Continued)

Common name (1)	Botanical name (2)	Maximum[b] permissible EC_e (dS/m) (3)
Thorny elaeagnus	*Elaeagnus pungens*	4–6
Spreading juniper	*Juniperus chinensis*	4–6
Pyracantha, cv. Graberi	*Pyracantha Fortuneana*	4–6
Cherry plum	*Prunus cerasifera*	4–6
(d) Moderately Tolerant		
Weeping bottlebrush	*Callistemon viminalis*	6–8
Oleander	*Nerium oleander*	6–8
European fan palm	*Chamaerops humilis*	6–8
Blue dracaena	*Cordyline indivisa*	6–8
Spindle tree, cv. Grandiflora	*Euonymus japonica*	6–8
Rosemary	*Rosmarinus officinalis*	6–8
Aleppo pine	*Pinus halepensis*	6–8
Sweet gum	*Liquidamabar styraciflua*	6–8
(e) Tolerant		
Brush cherry	*Syzygium paniculatum*	>8[c]
Ceniza	*Leucophyllum frutescens*	>8[c]
Natal plum	*Carissa grandiflora*	>8[c]
Evergreen Pear	*Pyrus Kawakamii*	>8[c]
Bougainvillea	*Bougainvillea spectabilis*	>8[c]
Italian stone pine	*Pinus pinea*	>8[c]
(f) Very tolerant		
White iceplant	*Delosperma alba*	>10[c]
Rosea iceplant	*Drosanthemum hispidum*	>10[c]
Purple iceplant	*Lampranthus productus*	>10[c]
Croceum iceplant	*Hymenocyclus croceus*	>10[c]

[a]Species are listed in order of increasing tolerance based on appearance as well as growth reduction. Data compiled from Francois (1982), Francois and Clark (1978), and Bernstein et al. (1972).

[b]Salinities exceeding the maximum permissible EC_e may cause leaf burn, loss of leaves, excessive stunting, or all three.

[c]Maximum permissible EC_e is unknown. No injury symptoms or growth reduction was apparent at 7 dS/m. The growth of all iceplant species was increased by soil salinity of 7 dS/m.

when there is high humidity or dew, absorption resumes. Nighttime sprinkling reduces foliar absorption and injury (Eaton and Harding 1959; Busch and Turner 1967). Since rates of absorption increase with increasing temperature, salt accumulation is likely to be greatest during hot weather.

Salt Tolerance of Sprinkled Crops

Crops sprinkled with saline water are also exposed to soil salinity. Reductions in the yield caused by saline irrigation waters will be at least as great as they are for surface-irrigated crops. Additional reductions in the yield can be expected for those crops that are susceptible to foliar damage caused by salts absorbed directly through the leaves. Bernstein and Francois (1973) found that pepper yields decreased by 16% when furrow-irrigated with 4.5 dS/m water instead of 0.6 dS/m water and decreased by 54% when sprinkler-irrigated. A later study indicated that yields of sprinkler-irrigated peppers dropped by about 16% for each increase of 1 dS/m in the electrical conductivity of the water (Bernstein and Francois 1975). Sprinkling daily with 3.1 dS/m water similarly reduced the yield of peppers in Israel (Goldberg and Shmueli 1971). Tomatoes sprinkler-irrigated with 3.6 dS/m water produced only 38% as much fruit as plants that were drip-irrigated with the same water (Gornat et al. 1973). Even cotton, one of the most salt-tolerant crops, produced 15% less seed cotton when sprinkled during the day than when surface-irrigated with 4.4 dS/m water (Busch and Turner 1967). Sprinkling at night did not decrease yields of cotton.

Except for the data on peppers given above, no data yet exists to predict losses in yield as a function of the concentration of salt in water from a sprinkler irrigation system. The limited data for some crops can be used to estimate the threshold EC of the irrigation water (EC_i) based on foliar injury induced by sprinkling (Table 13.4). These thresholds are compared with EC_i thresholds based on reductions in yield caused by soil salinity. Those crops with thresholds of foliar injury below the soil salinity threshold are more likely to experience foliar injury when sprinkled with waters that have concentrations of salt equal to or above the threshold for soil salinity. At concentrations above both thresholds, foliar injury and a reduced yield can be expected. If the effects add up, losses in yield greater than those indicated in Table 13.1 will probably occur.

Saline irrigation water is obviously best distributed through surface distribution systems. If sprinkler irrigating with marginally saline water cannot be avoided, several precautions should be taken. When possible, irrigate susceptible crops below the plant canopy to avoid, or at least reduce, wetting the foliage. Since injury is related more to the number of sprinklings than to their duration, irrigations should be infrequent and heavy rather than frequent and light. Intermittent wetting by slowly rotating sprinklers that allow drying between cycles should be avoided. Propel lateral-move sprinkler systems downwind, when possible, so that salts accumulated on the leaves from salt drift are washed off as the sprinkler moves past. Irrigate at night, when evaporation and absorption of salt are reduced. If daytime sprinkling is necessary, avoid hot, dry, and windy days.

TABLE 13.4 Relative Susceptibility of Crops to Foliar Injury from Saline Sprinkling Waters[a]

Na or Cl concentration (mol/m^3) causing foliar injury[b]			
<5 (1)	5–10 (2)	10–20 (3)	>20 (4)
Almond	Grape	Alfalfa	Cauliflower
Apricot	Pepper	Barley	Cotton
Citrus	Potato	Corn	Sugar beet
Plum	Tomato	Cucumber	Sunflower
		Safflower	
		Sesame	
		Sorghum	

[a]Susceptibility based on direct accumulation of salts through the leaves. Data compiled from Ehlig and Bernstein (1959), Francois and Clark (1979b), Gornat et al. (1973), Maas et al. (1982a), and Maas et al. (1982b).
[b]Foliar injury is influenced by cultural and environmental conditions. These data are presented only as general guidelines for daytime sprinkling.

PLANT TOLERANCE TO SPECIFIC IONS AND ELEMENTS

As previously mentioned, the relative concentrations of Na^+, Ca^{2+}, Mg^{2+}, Cl^-, SO_4^{2-}, and HCO_3^- vary widely in saline soils and waters. Unless the ratios between these ions become extreme, though, most crops respond to the solutes in the soil solution as a function of its osmotic potential. Nevertheless, some herbaceous crops and many woody species are susceptible to direct ion toxicities or nutritional imbalances caused by these ions. These specific effects will be considered individually.

Sodium

Na^+ is not considered an essential element for most plants, but it beneficially affects the growth of some at concentrations below the threshold of salt tolerance. At concentrations above the threshold, Na^+ can directly and indirectly affect plants to their detriment. The accumulation of toxic levels of Na^+ cause direct effects, which are generally limited to woody species. Their ability to tolerate Na^+ varies considerably, for uptake of Na^+ varies widely among species and rootstocks. Injury by Na^+ in avocado, citrus, and stone-fruit trees is widespread and can take place at concentrations as low as 5 mol/m^3 in soil water. The symptoms may not appear immediately after exposure to saline water. Initially, Na^+ is retained in the roots and lower trunk, but after three or four years, the conversion of sapwood to heartwood appears to release the accumulated Na^+. It is then transported to the leaves, where it causes leaf burn (Bernstein et al. 1956).

Indirect effects include nutritional imbalance and impairment of the physical conditions of the soil. The nutritional effects of Na^+ are not simply related to the exchangeable Na^+ percentage of soils, but depend on the concentrations of Na^+, Ca^{2+}, and Mg^{2+} in the soil solution. In sodic, nonsaline soils, total soluble salt concentrations are low and, consequently, concentrations of Ca^{2+}, Mg^{2+}, or both, are often nutritionally inadequate. These deficiencies and not Na^+ toxicity usually and primarily cause poor growth among nonwoody species. Since uptake of Na^+ by plants is strongly regulated by Ca^{2+} in the soil solution, sufficient Ca^{2+} is needed to prevent the accumulation of toxic levels of Na^+. This is particularly important with woody crops that are sensitive to Na^+. In general, concentrations of Ca^{2+} and Mg^{2+} of above 1 mol/m^3 each in the soil solution, with Ca^{2+} greater than or equal to Mg^{2+}, are nutritionally adequate in nonsaline, sodic soils.

As the total concentration of salt increases into the saline-sodic range, concentrations of Ca^{2+} become adequate for most plants and osmotic effects begin to predominate. However, some species are susceptible to deficiencies of Ca^{2+} induced by salinity. Blackheart in celery and blossom-end rot in tomato and bell pepper occur when excess salts inhibit the uptake of Ca^{2+} (Geraldson 1957). Some cultivars of corn, sorghum, rice, wheat, and barley exhibit severe deficiencies of Ca^{2+} at high Na^+/Ca^{2+} (Maas and Grieve 1987; Grieve and Maas 1988; Grieve and Fujiyama 1987).

Sodic soil conditions affect virtually all crops due to the deterioration of the soil's physical conditions. Dispersion of soil aggregates in sodic soils decreases the permeability of the soil by water and air, reducing plant growth. Therefore, the reduced yield in crops not specifically sensitive to Na^+ reflect the combined effects of nutritional problems and the impaired physical condition of the soil.

Chloride

Chloride is an essential micronutrient for plants, but, unlike most micronutrients, it is relatively nontoxic (Maas 1986). In fact, most nonwoody crops are not specifically sensitive to Cl^-. There are exceptions: certain cultivars of soybeans that tend to accumulate excessive and toxic amounts of Cl^- (Abel and McKenzie 1964; Parker et al. 1983). Tolerant cultivars restrict the transport of Cl^- to the shoots. Many woody species are also susceptible to Cl^- toxicity, which varies among varieties and rootstocks within species. As in soybeans, these differences usually reflect the plant's ability to prevent or inhibit the translocation of Cl^- to the tops. Cooper (1951, 1961) found that the salt tolerance of avocado, grapefruit, and orange closely relates to the ability of the rootstocks to accumulate Cl^-. Similar effects have been reported for stone-fruit trees (Bernstein et al. 1956). Large differences in the salt tolerance of grape varieties have been linked with the chloride-accumulating characteristics of different rootstocks (Bernstein et al. 1969; Ehlig 1960; Groot, Obink, and Alexander 1973; Sauer 1968). To avoid this problem, select rootstocks that exclude chloride from the scions.

Table 13.5 lists the maximum concentrations of Cl^- permissible in the soil water that do not cause leaf injury for some cultivars of fruit and rootstocks. In some cases, the osmotic threshold may be exceeded and the yield is decreased without obvious injury. The list is incomplete, since quantitative data for many cultivars and rootstocks are unavailable.

Chloride contributes to osmotic stress. Comprehensive testing to determine the tolerances of crops to chloride salinity has yet to be conducted. However, most of the data on salt tolerance were obtained in fields salinized with chloride salts of Na^+ and Ca^{2+}, so they can be converted to express tolerances in terms of chloride concentration. If Cl^- is the predominant anion in the soil solution, then the concentration of Cl^-, denoted by $[Cl^-]$, can be estimated from the electrical conductivity by Eq. 13.6,

$$[Cl^-] = 10 \times EC \tag{13.6}$$

where $[Cl^-]$ is expressed in mol/m^3 and EC in dS/m (USSL 1954). Converting the data in Table 13.1 gives the values of the threshold and the slope for Cl^- tolerances of 54 different crops listed in Table 13.6.

Boron

Boron is an essential element but can become toxic to some plants when concentrations of soil water even slightly exceed the amount required for optimal growth. Generally, toxic concentrations in the soil are found only in arid regions of the world. Most surface-irrigation waters contain acceptable levels of boron, but well waters in some areas contain toxic levels. Plants vary in their tolerance to boron, so a water that is marginal for sensitive crops may be used for the more tolerant plants.

Most of the existing data on tolerance to boron were obtained from experiments conducted over 50 years ago by Eaton (1944). These data provided limits of threshold tolerance for more than 40 different crops. Although useful, they cannot be fitted to any reliable growth response function for most crops. Nevertheless, Eaton's data provided most of the limits of threshold tolerance presented in Table 13.7. Some recent data have shown that the two-piece linear response model can describe the response of plants to excess boron (Bingham et al. 1985; Francois 1984b, 1986, 1988a, 1989). These data provided the thresholds and the slopes for a limited number of crops and are included in Table 13.7. With few exceptions, the data on boron tolerance are based on the responses of crops to different levels of boron in sand cultures. The thresholds indicate the maximum concentrations in the soil water that do not reduce the yield. Some crops, however, may exhibit leaf injury at lower concentrations without experiencing a decrease in yield. The crops have been classified in six groups, ranging from very sensitive to very tolerant. Like salt tolerance, boron tolerance varies with the climate, the soil, and the crop. Therefore, the data will not invariably apply.

Different rootstocks of citrus and stone fruits absorb boron at different rates, so using a rootstock that restricts the uptake of boron may improve tolerance. Table 13.8 lists a number of these rootstocks.

Table 13.9 shows the tolerance limits of ornamental plants to boron. Since appearance is often more important than the growth rate for

TABLE 13.5 Chloride-Tolerance Limits of Some Fruit-Crop Cultivars and Rootstocks

Crop (1)	Rootstock or cultivar (2)	Maximum permissible Cl^- in soil water without leaf injury[a] (mol/m^3) (3)
Rootstocks		
Avocado	West Indian	15
(*Persea*	Guatemalan	12
americana)	Mexican	10
Citrus		
(*Citrus* sp.)	Sunki mandarin, grapefruit,	50
	Cleopatra mandarin, Rangpur lime	50
	Sampson tangelo, rough lemon[b],	30
	sour orange, Ponkan mandarin	30
	Citrumelo 4475, trifoliate orange,	20
	Cuban shaddock, Calamondin,	20
	sweet orange, Savage citrange,	20
	Rusk citrange, Troyer citrange	20
Grape		
(*Vitis* sp.)	Salt Creek, 1613-3	80
	Dog ridge	60
Stone fruit		
(*Prunus* sp.)	Marianna	50
	Lovell, Shalil	20
	Yunnan	15
Cultivars		
Berries[c]	Boysenberry	20
(*Rubus* sp.)	Olallie blackberry	20
	Indian Summer raspberry	10
Grape		
(*Vitis* sp.)	Thompson seedless, Perlette	40
	Cardinal, black rose	20
Strawberry		
(*Fragaria* sp.)	Lassen	15
	Shasta	10

[a]For some crops, these concentrations may exceed the osmotic threshold and cause some yield reduction. Data compiled from Bernstein (1965), Bernstein et al. (1969), Embleton et al. (1973), and Embleton et al. (1962).
[b]Data from Australia indicate that rough lemon is more sensitive to Cl^- than sweet orange (Grieve and Walker 1983).
[c]Data available for one variety of each species only.

TABLE 13.6 Chloride Tolerance of Agricultural Crops. Listed in Order of Increasing Tolerance[a]

Crop (1)	Maximum Cl^- concentration[b] without loss in yield (Threshold) (mol/m^3) (2)	Percent decrease in yield at Cl^- concentrations[b] above the threshold (Slope) (% per mol/m^3) (3)
Strawberry	10	3.3
Bean	10	1.9
Onion	10	1.6
Carrot	10	1.4
Radish	10	1.3
Lettuce	10	1.3
Turnip	10	0.9
Rice, paddy[c]	30[d]	1.2[d]
Pepper	15	1.4
Clover, strawberry	15	1.2
Clover, red	15	1.2
Clover, alsike	15	1.2
Clover, ladino	15	1.2
Corn	15	1.2
Flax	15	1.2
Potato	15	1.2
Sweet Potato	15	1.1
Broad bean	15	1.0
Cabbage	15	1.0
Foxtail, meadow	15	1.0
Celery	15	0.6
Clover, Berseem	15	0.6
Orchardgrass	15	0.6
Sugarcane	15	0.6
Trefoil, big	20	1.9
Lovegras	20	0.8
Spinach	20	0.8
Alfalfa	20	0.7
Sesbania[c]	20	0.7
Cucumber	25	1.3
Tomato	25	1.0
Broccoli	25	0.9
Squash, scallop	30	1.6
Vetch, common	30	1.1
Wild rye, beardless	30	0.6
Sudan grass	30	0.4
Wheat grass, standard crested	35	0.4
Beet, red[c]	40	0.9

(continued)

TABLE 13.6 Chloride Tolerance of Agricultural Crops. Listed in Order of Increasing Tolerance[a] (Continued)

Crop (1)	Maximum Cl^- concentration[b] without loss in yield (Threshold) (mol/m^3) (2)	Percent decrease in yield at Cl^- concentrations[b] above the threshold (Slope) (% per mol/m^3) (3)
Fescue, tall	40	0.5
Squash, zucchini	45	0.9
Hardinggrass	45	0.8
Cowpea	50	1.2
Trefoil, narrow-leaf bird's foot	50	1.0
Ryegrass, perennial	55	0.8
Wheat, Durum	55	0.5
Barley (forage)[c]	60	0.7
Wheat[c]	60	0.7
Sorghum	70	1.6
Bermuda grass	70	0.6
Sugar beet[c]	70	0.6
Wheat grass, fairway crested	75	0.7
Cotton	75	0.5
Wheat grass, tall	75	0.4
Barley[c]	80	0.5

[a]These data serve only as a guideline to relative tolerances among crops. Absolute tolerances vary, depending upon climate, soil conditions and cultural practices.
[b]Cl^- concentrations in saturated-soil extracts sampled in the rootzone. To convert Cl^- concentrations to ppm, multiply threshold values by 35. To convert % yield decreases to % per ppm, divide slope values by 35.
[c]Less tolerant during emergence and seedling stage.
[d]Values for paddy rice refer to the Cl^- concentration in the soil water during the flooded growing conditions.

ornamentals, injury to leaves and a reduction in growth were taken into account in setting maximum levels of boron. Symptoms of excess boron often include chlorotic and necrotic patterns of leaves, although some sensitive fruit crops, such as stone and pome fruits, may be damaged without showing injuries. Leaves normally contain about 40 mg to 100 mg of boron per kilogram of dry weight but often contain over 250 mg/kg when B in the soil approaches toxic levels. Leaf concentrations of boron may exceed 700 mg/kg to 1,000 mg/kg in extreme conditions of boron toxicity. Pome and stone fruits don't accumulate high levels of boron in their leaves but develop twig die-back and gummosis when exposed to toxic levels.

TABLE 13.7 Boron Tolerance Limits for Agricultural Crops

Common name (1)	Botanical name (2)	Threshold[a] (g/m^3) (3)	Slope (% per g/m^3) (4)	References (5)
(a) Very sensitive				
Lemon[b]	*Citrus limon*	<0.5		Eaton (1944), Haas (1929)
Blackberry[b]	*Rubus* sp.	<0.5		Eaton (1944)
(b) Sensitive				
Avocado[b]	*Persea american*	0.5–0.75		Haas (1929)
Grapefruit[b]	*C. x paradisi*	0.5–0.75		Haas (1929)
Orange[b]	*C. sinensis*	0.5–0.75		Haas (1929)
Apricot[b]	*Prunus armeniaca*	0.5–0.75		Woodbridge (1955)
Peach[b]	*P. persica*	0.5–0.75		Eaton (1944), Haas (1929)
Cherry[b]	*P. avium*	0.5–0.75		Eaton (1944)
Plum[b]	*P. domestica*	0.5–0.75		Woodbridge (1955)
Persimmon[b]	*Diospyros kaki*	0.5–0.75		Eaton (1944)
Fig, kadota[b]	*Ficus carica*	0.5–0.75		Eaton (1944)
Grape[b]	*Vitis vinifera*	0.5–0.75		Eaton (1944)
Walnut[b]	*Juglans regia*	0.5–0.75		Haas (1929)
Pecan[b]	*Carya illinoinensis*	0.5–0.75		Haas (1929)
Onion	*Allium cepa*	0.5–0.75		Eaton (1944)
Garlic	*A. sativum*	0.75–1.0		Singh and Singh (1974)
Sweet potato	*Ipomea batatas*	0.75–1.0		Eaton (1944)
Wheat	*Triticum aestivum*	0.75–1.0	3.3	Bingham et al. (1985), Khudairi (1961)
Sunflower	*Helianthus annuus*	0.75–1.0		Pathak et al. (1975)
Bean, mung[b]	*Vigna radiata*	0.75–1.0		Khudairi (1961)
Sesame[b]	*Sesamum indicum*	0.75–1.0		Khudairi (1961)
Lupine[b]	*Lupinus hartwegii*	0.75–1.0		Eaton (1944)
Strawberry[b]	*Fragaria* sp.	0.75–1.0		Eaton (1944)
Artichoke, Jerusalem[b]	*Helianthus tuberosus*	0.75–1.0		Eaton (1944)
Bean, kidney[b]	*Phaseolus vulgaris*	0.75–1.0		Eaton (1944)
Bean, snap	*P. vulgaris*	1.0	12	Francois (1989)
Bean, lima[b]	*P. lunatus*	0.75–1.0		Eaton (1944)
Peanut	*Arachis hypogaea*	0.75–1.0		Gopal (1971)
(c) Moderately sensitive				
Broccoli	*Brassica oleracea botrytis*	1.0	1.8	Francois (1986)
Pepper, red	*Capsicum annuum*	1.0–2.0		Eaton (1944)
Pea[b]	*Pisum sativa*	1.0–2.0		Eaton (1944)
Carrot	*Daucus carota*	1.0–2.0		Eaton (1944)
Radish	*Raphanus sativus*	1.0	1.4	Francois (1986)
Potato	*Solanum tuberosum*	1.0–2.0		Eaton (1944)
Cucumber	*Cucumis sativus*	1.0–2.0		El-Sheikh et al. (1971)
Lettuce[b]	*Lactuca sativa*	1.3	1.7	Francois (1988a)

(continued)

TABLE 13.7 Boron Tolerance Limits for Agricultural Crops (Continued)

Common name (1)	Botanical name (2)	Threshold[a] (g/m^3) (3)	Slope (% per g/m^3) (4)	References (5)
(d) Moderately tolerant				
Cabbage[b]	*Brassica oleracea capitata*	2.0–4.0		Eaton (1944)
Turnip	*B. rapa*	2.0–4.0		Eaton (1944)
Bluegrass, Kentucky[b]	*Poa pratensis*	2.0–4.0		Eaton (1944)
Barley	*Hordeum vulgare*	3.4	4.4	Bingham et al. (1985)
Cowpea	*Vigna unguiculata*	2.5	12	Francois (1989)
Oats	*Avena sativa*	2.0–4.0		Eaton (1944)
Corn	*Zea mays*	2.0–4.0		Eaton (1944), El-Sheikh et al. (1971)
Artichoke[b]	*Cynara scolymus*	2.0–4.0		Eaton (1944)
Tobacco[b]	*Nicotiana tabacum*	2.0–4.0		Eaton (1944)
Mustard[b]	*Brassica juncea*	2.0–4.0		Eaton (1944)
Clover, sweet[b]	*Melilotus indica*	2.0–4.0		Eaton (1944)
Squash	*Cucurbita pepo*	2.0–4.0		El-Sheikh et al. (1971)
Muskmelon[b]	*Cucumis melo*	2.0–4.0		Eaton (1944), El-Sheikh et al. (1971)
Cauliflower	*B. oleracea botrytis*	4.0	1.9	Francois (1986)
(e) Tolerant				
Alfalfa[b]	*Medicago sativa*	4.0–6.0		Eaton (1944)
Vetch, purple[b]	*Vicia benghalensis*	4.0–6.0		Eaton (1944)
Parsley[b]	*Petroselinum crispum*	4.0–6.0		Eaton (1944)
Beet, red	*Beta vulgaris*	4.0–6.0		Eaton (1944)
Sugar beet	*B. vulgaris*	4.9	4.1	Bingham et al. (1985), Vlamis and Ulrich (1973)
Tomato	*Lycopersicon lycopersicum*	5.7	3.4	Francois (1984b)
(f) Very tolerant				
Sorghum	*Sorghum bicolor*	7.4	4.7	Bingham et al. (1985)
Cotton	*Gossypium hirsutum*	6.0–10.0		Eaton (1944)
Celery[b]	*Apium graveolens*	9.8	3.2	Francois (1988a)
Asparagus[b]	*Asparagus officinalis*	10.0–15.0		Eaton (1944)

[a]Maximum permissible concentration in soil water without yield reduction. Boron tolerances may vary, depending upon climate, soil conditions, and crop varieties.
[b]Tolerance based on reductions in vegetative growth.

TABLE 13.8 Citrus and Stone-Fruit Rootstocks Ranked in Order of Increasing Boron Accumulation and Transport to Scions[a]

(1)	Common name (2)	Botanical name (3)
Citrus	Alemow	*Citrus macrophylla*
	Gajanimma	*C. pennivesiculata or C. moi*
	Chinese box orange	*Severina buxifolia*
	Sour orange	*C. aurantium*
	Calamondin	*x Citrofortunella mitis*
	Sweet orange	*C. sinensis*
	Yuzu	*C. junos*
	Rough lemon	*C. limon*
	Grapefruit	*C. x paradisi*
	Rangpur lime	*C. x limonia*
	Troyer citrange	*x Citronicirus webberi*
	Savage citrange	*x Citroncirus webberi*
	Cleopatra mandarin	*C. Areticulata*
	Rusk citrange	*x Citroncirus webberi*
	Sunki mandarin	*C. reticulata*
	Sweet lemon	*C. limon*
	Trifoliate orange	*Poncirus trifoliata*
	Citrumelo 4475	*Poncirus trifoliata x C. paradisi*
	Ponkan mandarin	*C. reticulata*
	Sampson tangelo	*C. x Tangelo*
	Cuban shaddock	*C. maxima*
	Sweet lime	*C. aurantiifolia*
Stone fruit	Almond	*Prunus duclis*
	Myrobalan plum	*P. cerasifera*
	Apricot	*P. armeniaca*
	Marianna plum	*P. domestica*
	Shalil peach	*P. persica*

[a]Data compiled from Embleton et al. (1973) and Hansen (1948 and 1955).

Bicarbonate

A few reports in the literature indicate that sensitivity to HCO_3^- differs among species. Gauch and Wadleigh (1951) found that 12 mol/m^3 of HCO_3^- was highly toxic to Dallisgrass (*Paspalum dilatatum*), but had no effect on Rhodes grass (*Chloris guayana*). Beans are much more sensitive than beets. Stolwijk and Thimann (1957) found that aerating the root medium with 6.5% CO_2 completely inhibited the root growth of beans, broad beans, peas, and sunflowers, but had no effect on that of oats and barley. Unfortunately, the data on the effects of HCO_3^- are so limited that the tolerance limits of plants to concentrations of HCO_3^- in soils or irrigation waters remain unknown.

TABLE 13.9 Boron Tolerance Limits for Ornamentals[a]

Tolerance (1)	Common name (2)	Botanical name (3)	Threshold[b] (g/m^3) (4)	References (5)
Very sensitive	Oregon grape	*Mahonia aquifolium*	<0.5	Francois and Clark (1979a)
	Photinia	*Photinia x fraseri*	<0.5	Francois and Clark (1979a)
	Xylosma	*Xylosma congestum*	<0.5	Francois and Clark (1979a)
	Thorny elaeagnus	*Elaeagnus pungens*	<0.5	Francois and Clark (1979a)
	Laurustinus	*Viburnum tinus*	<0.5	Francois and Clark (1979a)
	Wax-leaf privet	*Ligustrum japonicum*	<0.5	Francois and Clark (1979a)
	Pineapple guava	*Feijoa sellowiana*	<0.5	Francois and Clark (1979a)
	Spindle tree	*Euonymus japonica*	<0.5	Francois and Clark (1979a)
	Japanese pittosporum	*Pittosporum tobira*	<0.5	Francois and Clark (1979a)
	Chinese holly	*Ilex cornuta*	<0.5	Francois and Clark (1979a)
	Juniper	*Juniperus chinensis*	<0.5	Francois and Clark (1979a)
	Yellow sage	*Lantana camara*	<0.5	Francois and Clark (1979a)
	American elm	*Ulmus americana*	<0.5	Francois and Clark (1979a)
Sensitive	Zinnia	*Zinnia eleganus*	0.5–1.0	Eaton (1944)
	Pansy	*Viola tricolor*	0.5–1.0	Eaton (1944)
	Violet	*V. odorata*	0.5–1.0	Eaton (1944)
	Larkspur	*Delphinum* sp.	0.5–1.0	Eaton (1944)
	Glossy abelia	*Abelia x grandiflora*	0.5–1.0	Francois and Clark (1979a)
	Rosemary	*Rosmarinus officinalis*	0.5–1.0	Francois and Clark (1979a)
	Oriental arborvitae	*Platycladus orientalis*	0.5–1.0	Francois and Clark (1979a)
	Geranium	*Pelargonium x hortorum*	0.5–1.0	Kofranek et al. (1958)
Moderately sensitive	Gladiolus	*Gladiolus* sp.	1.0–2.0	Kofranek et al. (1957)

(continued)

TABLE 13.9 Boron Tolerance Limits for Ornamentals[a] (Continued)

Tolerance (1)	Common name (2)	Botanical name (3)	Threshold[b] (g/m^3) (4)	References (5)
	Marigold	*Calendula officinalis*	1.0–2.0	Francois and Clark (1979a)
	Poinsettia	*Euphorbia pulcherrima*	1.0–2.0	Kofranek et al. (1956)
	China aster	*Callistephus chinensis*	1.0–2.0	Kohl et al. (1957)
	Gardenia	*Gardenia* sp.	1.0–2.0	Lunt et al. (1956b)
	Southern yew	*Podocarpus marco- phyllus*	1.0–2.0	Francois and Clark (1979a)
	Brush cherry	*Syzygium panicu- latum*	1.0–2.0	Francois and Clark (1979a)
	Blue dracaena	*Cordyline indivisa*	1.0–2.0	Francois and Clark (1979a)
	Ceniza	*Leucophyllus frutescens*	1.0–2.0	Francois and Clark (1979a)
Moderately tolerant	Bottlebrush	*Callistemon citrinus*	2.0–4.0	Francois and Clark (1979a)
	California poppy	*Eschscholzia californica*	2.0–4.0	Eaton (1944)
	Japanese boxwood	*Buxus microphylla*	2.0–4.0	Francois and Clark (1979a)
	Oleander	*Nerium oleander*	2.0–4.0	Francois and Clark (1979a)
	Chinese hibiscus	*Hibiscus rosa-sen- ensis*	2.0–4.0	Francois and Clark (1979a)
	Sweet pea	*Lathyrus odoratus*	2.0–4.0	Eaton (1944)
	Carnation	*Dianthus caryo- phyllus*	2.0–4.0	Lunt et al. (1956a)
Tolerant	Indian hawthorn	*Raphiolepis indica*	6.0–8.0	Francois and Clark (1979a)
	Natal plum	*Carissa grandiflora*	6.0–8.0	Francois and Clark (1979a)
	Oxalis	*Oxalis bowiei*	6.0–8.0	Eaton (1944)

[a]Species listed in order of increasing tolerance based on appearance as well as growth reduction.
[b]Boron concentrations exceeding the threshold may cause leaf burn and loss of leaves.

SUMMARY

The response of plants to salinity and various constituent ions and solutes in the soil solution varies widely according to species. Considerable effort has been expended for over 50 years to quantify the relationship between the yield and the concentration of salt in the root medium for many crops. Although this relationship depends largely on many factors involving the plant, the soil, the water, and other parts of the environment, the expression of relative yield as a function of the solute concentration in the root zone has provided tolerance guidelines. This chapter summarizes the currently available data for the tolerances of plants to total soluble salts in the soil and several specific ions and solutes.

REFERENCES

Abel, G.H., and MacKenzie, A.J. (1964). "Salt tolerance of soybean varieties (*Glycine max* L. Merrill) during germination and later growth." *Crop Sci.* 4:157-161, 1964.

Aceves-N., E., Stolzy, L. H., and Mehuys, G. R. (1975). "Combined effects of low oxygen and salinity on germination of a semi-dwarf Mexican wheat." *Agron J.* 67(4):530-532.

Ahi, S. M., and Powers, W. L. (1938). "Salt tolerance of plants at various temperatures." *Plant Physiol.* 13:767-789.

Aubertin, G. M., Rickman, R. W., and Letey, J. (1968). "Differential salt-oxygen levels influence plant growth." *Agron. J.* 60:345-349.

Bernstein, L. (1962). "Salt-affected soils and plants." *Problems of the Arid Zones.* Proc. UNESCO Symp., (Paris, France), 18th, 139-174.

Bernstein, L. (1965). "Salt tolerance of fruit crops." *U. S. Dep. Ag. Inf. Bull.* 292. U. S. Government Printing Office, Washington D.C. 8 p.

Bernstein, L., Brown, J. W., and Hayward, H. E. (1956). "The influence of rootstock on growth and salt accumulation in stone-fruit trees and almonds." *Proc. Amer. Soc. Hort. Sci.* 68:86-95.

Bernstein, L., Ehlig, C. F., and Clark, R. A. (1969). "Effect of grape rootstocks on chloride accumulation in leaves." *J. of Amer. Soc. Hort. Sci.* Joseph, Mich. 94:584-590.

Bernstein, L., and Francois, L. E. (1973). "Comparisons of drip, furrow and sprinkler irrigation." *Soil Sci.* 115:73-86.

Bernstein, L., and Francois, L. E. (1975). "Effects of frequency of sprinkling with saline waters compared with daily drip irrigation." *Agron. J.* 67:185-190.

Bernstein, L., Francois, L. E., and Clark, R. A. (1972). "Salt tolerance of ornamental shrubs and ground covers." *J. Am. Soc. Hort. Sci.* 97: 550-566.

Bernstein, Leon, Francois, L. E., and Clark, R. A. (1974) "Interactive effects of salinity and fertility on yields of grains and vegetables." *Agron. J.* 66:412-421.

Bingham, F. T., Rhoades, J. D., and Keren, R. (1985). "An application of the Maas-Hoffman salinity response model for boron toxicity." *Soil Sci. Soc. Am. J.* 49:672-674.

Busch, C. D., and Turner, Jr., F. (1967) "Sprinkler irrigation with high salt-content water." *Trans. ASAE* 10:494-496.

Bytnerowicz, A., and Taylor, O. C. (1983). "Influence of ozone, sulfur dioxide, and salinity on leaf injury, stomatal resistance, growth, and chemical composition of bean plants." *J. Environ. Qual.* 12:397-405.

Cooper, W. C. (1951). "Salt tolerance of avocados on various rootstocks." *Texas Avocado Society Yearbook,* Weslaco, Tex., pp. 24-28.

Cooper, W. C. (1961). "Toxicity and accumulation of salts in citrus trees on various rootstocks in Texas." *Florida State Horticultural Society Proceedings,* Vol. 74, Lake Alfred, Fla., pp. 95-104.

Eaton, F. M. (1944). "Deficiency, toxicity, and accumulation of boron in plants." *Agric. Res.* 69:237-277.

Eaton, F. M., and Harding, R. B. (1959). "Foliar uptake of salt constituents of water by citrus plants during intermittent sprinkling and immersion." *Plant Physiol.* 34:22-26.

Ehlig, C. F. (1960). "Effects of salinity on four varieties of table grapes grown in sand culture." *Proc. Amer. Soc. Hort. Sci.* 76: Joseph, Mich., 323-331.

Ehlig, C. F. (1961). "Salt tolerance of strawberries under spinkler irrigation." *Proc. Amer. Soc. Hort. Sci.* 77:376-379.

Ehlig, C. F., and Bernstein, L. (1959). "Foliar absorption of sodium and chloride as a factor in sprinkler irrigation." *Proc. Amer. Soc. Hort. Sci.* 74:661-670.

El-Sheikh, A. M., Ulrich, A., Awad, S. K., and Mawardy, A. E. (1971). "Boron tolerance of squash, melon, cucumber, and corn." *J. Am. Soc. Hort. Sci.* 96:536-537.

Embleton, T. W., Jones, W. W., Labanauskas, C. K., and Reuther, W. (1973). "Leaf analysis as a diagnostic tool and guide to fertilization." *Chapt. 6. Citrus Ind.* 3:183-Appendix I.

Embleton, T. W., Matsumura, M., Storey, W. B., and Garber, M. J. (1962). "Chlorine and other elements in avocado leaves as influenced by root stock." *Proc. Am. Soc. Hort. Sci.* 80:230-236.

Francois, L. E. (1982). "Salt tolerance of eight ornamental tree species." *J. Am. Soc. Hort. Sci.* 107:66-68.

Francois. L. E. (1984a). "Salinity effects on germination, growth, and yield of turnips." *HortScience* 19(1):82-84.

Francois, L. E. (1984b). "Effect of excess boron on tomato yield, fruit size, and vegetative growth." *J. Am. Soc. Hort. Sci.* 109:322-324.

Francois, L. E. (1985). "Salinity effects on germination, growth, and yield of two squash cultivars." *Hort Sci.* 20(6):1102-1104.

Francois, L. E. (1986). "Effect of excess boron on broccoli, cauliflower and radish." *J. Am. Soc. Hort. Sci.* 111(4):494-498.

Francois, L. E. (1987). "Salinity effects on asparagus yield and vegetative growth." *J. Amer. Soc. Hort. Sci.* 112(3):432-436.

Francois, L. E. (1988a). "Yield and quality responses of celery and crisphead lettuce to excess boron. *J. Amer. Soc. Hort Sci.* 113(4):538-542.

Francois, L. E. (1988b). "U. S. Salinity Laboratory," Riverside, Calif. (Unpublished data).

Francois, L. E. (1988c). "Salt tolerance of kenaf." New Crops Symposium, Indianapolis, Ind. Oct. 23-26, 1988.

Francois, L. E. (1989). "Boron tolerance of snapbean and cowpea. *J. Am. Soc. Hort. Sci.* 114:615-619.

Francois, L. E., and Bernstein, L. (1964). "Salt tolerance of Sphaerophysa salsula." Report to Collaborators, U. S. Salinity Laboratory, Riverside, Calif. pp. 52-53.

Francois, L. E., and Clark, R. A. (1978). "Salt tolerance of ornamental shrubs, trees, and iceplant." *J. Am. Soc. Hort. Sci.* 103:280-283.

Francois, L. E., and Clark, R. A. (1979a). "Boron tolerance of twenty-five ornamental shrubs, trees, and iceplant." *J. Am. Soc. Hort. Sci.* 104: 319-322.

Francois, L. E., and Clark, R. A. (1979b). "Accumulation of sodium and chloride in leaves of sprinkler-irrigated grapes." *J. Am. Soc. Hort. Sci.* 104:11-13.

Francois, L. E., Donovan, T., and Maas, E. V. (1984). "Salinity effects on seed yield, growth and germination of grain sorghum." *Agron. J.* 76:741-744.

Francois, L. E., Donovan, T. J., Lorenz, K., and Maas, E. V. (1989a). "Salinity effects on rye grain yield, quality, vegetative growth, and emergence." *Agron. J.* 81:707-772.

Francois, L. E., Donovan, T. J., and Maas, E. V. (1990). "Salinity effects on emergence, vegetative growth, and seed yield of guar. *Agron. J.* 82:587–592.

Francois, L. E., and Donovan, T. J., Maas, E. V., and Rubenthaler, G. L. (1988). "Effect of salinity on grain yield and quality, vegetative growth, and germination of triticale." *Agron. J.* 80(4):642-647.

Francois, L. E., and Maas, E. V. (1978). "Plant responses to salinity: An indexed bibliography." USDA, ARM-W-6, October 1978. 291 pp.

Francois, L. E., and Maas, E. V. (1985). "Plant responses to salinity: A supplement to an indexed bibliography." USDA, ARS-24, 174 pp.

Francois, L. E., Maas, E. V., Donovan, T. J., and Youngs, V. L. (1986). "Effect of salinity on grain yield and quality, vegetative growth, and germination of semi-dwarf and durum wheat." *Agron. J.* 78(6):1053-1058.

Francois, L. E., and West, D. W. (1982). "Reduction in yield and market quality of celery caused by soil salinity." *J. Am. Soc. Hort. Sci.* 107: 952-954.

Gauch, H. G., and Wadleigh, C. H. (1951). "Salt tolerance and chemical composition of Rhodes and Dallis grasses grown in sand cultures." *Botanical Gazette* 112:259-271.

Geraldson, C. M. (1957). "Factors affecting calcium nutrition of celery, tomato, and pepper." *Soil Sci. Soc. Amer. Proc.* 21:621-625.

Goldberg, D., and Shumeli, M. (1971). "Sprinkle and trickle irrigation of green pepper in an arid zone." *Hort. Sci.* 6:559-562.

Gopal, N. H. (1971). "Influence of boron on growth and yield in groundnut." *Turrialba* 21:435-441.

Gornat, B., Goldberg, D., Rimon, R., and Ben-Asher, J. (1973). "The physiological effect of water quality and method of application on tomato, cucumber, and pepper." *J. Amer. Soc. Hort. Sci.* 98:202-205.

Grattan, S. R., Maas, E. V., and Ogata, G. (1981). "Foliar uptake and injury from saline aerosol." *J. Environ. Qual.* 10(3):406-409.

Grieve, A. M., and Walker, R. R. (1983). "Uptake and distribution of

chloride, sodium, and potassium ions in salt-treated citrus plants." *Aust. J. Agric. Res.* 34:133-143.
Grieve, C. M., and Fujiyama, H. (1987). "The response of two rice cultivars to external Na/Ca ratio." *Plant and Soil.* 103:245-250.
Grieve, C. M., and Maas, E. V. (1988). "Differential effects of sodium/calcium ratio on sorghum genotypes." *Crop. Sci.* 29(4):659-665.
Groot Obbink, J., and Alexander, D. McE., (1973). "Response of six grapevine cultivars to a range of chloride concentrations." *American Journal of Enological Viticulture* 24:65-68.
Haas, A. R. C. (1929). "Toxic effects of boron on fruit trees." *Bot. Gaz.* 88:113-131.
Hansen, C. J. (1948). "Influence of the rootstock on injury from excess boron in French (agen) prune and president plum." *Proc. Am. Soc. Hort. Sci.* 51:239-244.
Hansen, C. J. (1955). "Influence of the rootstock on injury from excess from excess boron in nonpareil almond and Elberta peach." *Proc. Am. Soc. Hort. Sci.* 65:128-132.
Harding, R. B., Miller, M. P., and Fireman, M. (1958). "Absorption of salts by citrus leaves during sprinkling with water salinity, leaching fraction and irrigation frequency." *Amer. Soc. Hort. Sci.* 71:248-256.
Heck, W. W., Taylor, O. C., Adams R., Bingham G., Miller, J., Preston, E., and Weinstein, L. (1982). "Assessment of crop loss from ozone." *J. Air Pollut. Contr. Assoc.* 32:353-361.
Heuer, B., Meiri, A., and Shalhevet, J. (1986). "Salt tolerance of egg-plant." *Plant and Soil* 95:9-13.
Hoffman, G. J., Catlin, P. B., Mead, R. M., Johnson, R. S., Francois, L. E. (1989). "Yield and foliar injury responses of mature plum trees to salinity." *Irrig. Sci.* 10:215-229.
Hoffman, G. J., Jobes, J. A., and Alves, W. J. (1983). "Response of tall fescue to irrigation water salinity, leaching fraction and irrigation frequency." *Agric. Water Manage.* 7:439-456.
Hoffman, G. J., and Rawlins, S. L. (1970). "Design and performance of sunlit climate chambers." *Trans. ASAE.* 13:656-660.
Hoffman, G. J., and Rawlins, S. L. (1971). "Growth and water potential of root crops as influenced by salinity and relative humidity." *Agron. J.* 63:877-880.
Hoffman, G. J., Rawlins, S. L., Garber, M. J., and Cullen, E. M. (1971). "Water relations and growth of cotton as influenced by salinity and relative humidity." *Agron.* J. 63:822-826.
Hoffman, G. J., Maas, E. V., Prichard, T., and Meyer, J. L. (1983). "Salt tolerance of corn in the Sacramento-San Joaquin Delta of California." *Irrig. Sci.* 4:31-44.
Hoffman, G. J., Maas, E. V., and Rawlins, S. L. (1973). "Salinity-ozone interactive effects on yield and water relations of pinto bean." *J. Environ. Qual.* 2:148-152.
Hoffman, G. J., Maas, E. V., and Rawlins, S. L. (1975). "Salinity-ozone interactive effects on alfalfa yield and water relations." *J. Environ. Qual.* 4:326-331.
Kaddah, M. T., Lehman, W. F., Meek, B. D., and Robinson, F. E. (1975). "Salinity effects on rice after the boot stage." *Agron. J.* 67:436-439.
Kaddah, M. T., Lehman, W. F., and Robinson, F. E. (1973). "Tolerance

of rice (*orzya sativa L.*) to salt during boot, flowering and grain-filling stages." *Agron. J.* 67:845-847.

Khudairi, A. K. (1961). "Boron toxicity and plant growth." In Salinity Problems in the Arid Zone, *Proc. Teheran Symp. Unesco.* 14:175-179.

Kofranek, A. M., Kohl, H. C., and Lunt, O. R. (1958). "Effects of excess salinity and boron on geraniums." *Proc. Am. Soc. Hort. Sci.* 71:516-521.

Kofranek, A. M., Lunt, O. R., and Kohl, H. C. (1956). "Tolerance of poinsettias to saline conditions and high boron concentrations." *Proc. Am. Soc. Hort. Sci.* 68:551-555.

Kofranek, A. M., Lunt, O. R., and Kohl, H. C. (1957). "Tolerance of gladioli to salinity and boron concentrations." *Proc. Am. Soc. Hort. Sci.* 69:556-560.

Kohl, H. C., Kofranek, A. M., and Lunt, O. R. (1957). "Response of china asters to high salt and boron concentrations." *Proc. Am. Soc. Hort. Sci.* 70:437-441.

Liberty Hyde Bailey Hortorium Staff, Hortus Third. (1976). "A concise dictionary of plants cultivated in the United States and Canada." *MacMillian Publishing Co.,* Inc., New York.

Lunt, O. R., Kohl, H. C., and Kofranek, A. M. (1956a). "Tolerance of carnations to saline conditions and boron." *Carnation Craft* 36:5-6.

Lunt, O. R., Kohl, H. C., and Kofranek, A. M. (1956b). "Tolerance of azaleas and gardenias to salinity conditions and boron." *Proc. Am. Soc. Hort. Sci.* 69:543-548.

Maas, E. V. (1985). "Crop Tolerance to Saline Sprinkling Waters." *Plant and Soil* 89:273-284.

Maas, E. V. (1986). "Physiological response of plants to chloride. In T. L. Jackson, ed. Chloride and Crop Production. *Proc. Amer. Soc. of Agron. Symp. Spec. Bull.* 2:4-20.

Maas, E. V. (1987). "Salt tolerance of plants." *Handbook of Plant Science in Agriculture.* B. R. Christie, ed. CRC Press Inc. Vol 2., pp. 57-75.

Maas, E. V., Clark, R. A., and Francois, L. E. (1982a). "Sprinkling-induced foliar injury to pepper plants: Effects of irrigation frequency, duration and water composition." *Irrig. Sci.* 3:101-109.

Maas, E. V., Donovan, T. J., and Francois, L. E. (1988). "Salt tolerance of irrigated guayule." *Irrig. Sci.* 9:199-211.

Maas, E. V., and Fisher, M. J. (1988). U. S. Salinity Laboratory, Riverside, Calif. (Unpublished data).

Maas, E. V., Grattan, S. R., and Ogata, G. (1982b). "Foliar salt accumulation and injury in crops sprinkled with saline water." *Irrig. Sci.* 3:157-168.

Maas, E. V. and Grieve, C. M. (1987). "Sodium-induced calcium deficiency in salt-stressed corn." *Plant. Cell. and Environ.* 10:559-564.

Maas, E. V., and Hoffman, G. J. (1977). "Crop salt tolerance—current assessment." *J. Irrig. and Drainage Div., ASCE* 103(IR2):115-134.

Maas, E. V., Hoffman, G. J., Chaba, G. D., Poss, J. A., and Shannon, M. C. (1983). "Salt sensitivity of corn at various growth stages." *Irrig. Sci.* 4:45-57.

Maas, E. V., Hoffman, G. J., Rawlins, S. L., and Ogata, G. (1973). "Salinity-ozone interactions on pinto bean: Integrated response to ozone concentration and duration." *J. Environ. Qual.* 2:400-404.

Maas, E. V., and Poss, J. A. (1989a). "Salt sensitivity of wheat at various growth stages." *Irrig. Sci.* 10:29-40.
Maas, E. V., and Poss, J. A. (1989b). "Salt Sensitivity of cowpea at various three growth stages. *Irrig. Sci.* 10:313-320.
Maas, E. V., Poss, J. A., and Hoffman, Glenn J. (1986). "Salinity sensitivity of sorghum at three growth stages." *Irrig. Sci.* 7:1-11.
Magistad, O. C., Ayers, Alvin D., Wadleigh, C. H., and Gauch, H. F. (1943). "Effect of salt concentration, kind of salt, and climate on plant growth in sand cultures." *Plant Physiol.* 18:151-166.
Nieman, R. H., and Poulsen, L. L. (1967). "Interactive effects of salinity and atmospheric humidity on the growth of bean and cotton plants." *Bot. Gaz.* 128:69-73.
Ogata, Gen, and Maas, E. V. (1973). "Interactive effects of salinity and ozone on growth and yield of garden beet," *Irrig. J. Environ. Qual.* 2:518-520.
Parker, M. B., Gascho, G. J., and Gaines, T. P. (1983). "Chloride toxicity of soybeans grown on Atlantic coast flatwoods soils." *Agron J.* 75: 439-442.
Pathak, A. M., Singh, R. K., and Singh, R. S. (1975). "Effect of different concentrations of boron in irrigation water on sunflower." *J. Indian Soc. Soil Sci.* 23:388-390.
Pearson, G. A., and Bernstein, L. (1959). "Salinity effects at several growth stages of rice." *Agron. J.* 51:654-657.
Ravikovitch, S., and Yoles, D. (1971). "The influence of phosphorus and nitrogen on millet and clover growing in soils affected by salinity, I. Plant Development." *Plant and Soil* 35:555-567.
Sauer, M. R. (1968). "Effects of vine rootstocks on chloride concentration in Sultana scions." *Vitis* 7:223-226.
Shannon, M. C., and Francois, L. E. (1978). "Salt tolerance of three muskmelon cultivars." *J. Am. Soc. Hort. Sci.* 103:127-130.
Siegel, S. M. (1982). University of Hawaii, Honolulu, Hawaii. (Personal communication).
Singh, R. N., and Singh, J. R. (1974). "Studies on the influence of boron nutrition on the growth characteristics of garlic (*Allium sativum* L)." *Indian J. Hort.* 31:255-258.
Stolwijk, J. A. J., and Thimann, K. V. (1954). "On the uptake of carbon dioxide and bicarbonate by roots, and its influence on growth." *Plant Physiol.* 32:513-520.
United States Salinity Laboratory Staff. (1954). *Diagnosis and improvement of saline and alkali soils.* U. S. Dep. Agr. Handbook 60, 160 p.
van Genuchten, Martinus Th. (1983). "Analyzing crop salt tolerance data: Model description and user's manual." *USDA-ARS-USSL Research Report No. 120.*
van Genuchten, M.Th., and Hoffman, G. J. (1984). "Analysis of crop salt tolerance data." Soil Salinity under Irrigation—Process and Management (I. Shainberg and J. Shalhevet, eds.). *Ecological Studies 51,* Springer-Verlag, N.Y. pp. 258-271.
Vlamis, J., and Ulrich, A. (1973). "Boron tolerance of sugar beets in relation to the growth and boron content of tissues." *J. Am. Soc. Sugar Beet Technol.* 17:280-288.

West, D. W., and Francois, L. E. (1982). "Effects of salinity on germination, growth and yield of cowpea." *Irrig. Sci.* 3:169-175.
Woodbridge, C. G. (1955). "The boron requirements of stone fruit trees." *Can. J. Agric. Sci.* 35:282-286.
Yermanos, D. M., Francois, L. E., and Tammadoni, T. (1967). "Effects of soil salinity on the development of jojoba." *Econ. Bot.* 21:69-80.
Yousif, Y. H., Bingham, F. T., and Yermanos, D. M. (1972). "Growth, mineral composition, and seed oil of sesame (*Sesamum indicum* L.) as affected by NaCl." *Soil Sci. Soc. Am. Proc.* 36:450-453.

NOTATION

a = salinity threshold expressed in dS/m; A = salinity threshold expressed in bars; b = slope expressed in % per dS/m; B = slope expressed in % per bar; c = average salinity of root zone; c_{50} = average salinity of root zone that reduces yield by 50%; EC = electrical conductivity; EC_e = electrical conductivity of saturated-soil extract; OP_{fc} = osmotic potential of soil solution of field capacity; Y_m = yield under nonsaline conditions; Y_r = relative yield; and p = empirical constant.

CHAPTER 14

CROP-WATER PRODUCTION FUNCTIONS UNDER SALINE CONDITIONS[a]

INTRODUCTION

Crop-water production functions provide relations between the yield of a crop, the quantity and quality of water, and other variables. They are necessary for analyzing alternate irrigation management strategies in agriculture and can be used to determine optimal irrigation volumes and timing. These functions can also be used to evaluate the losses from increased salinity in soil and water, the potential for reuse of saline drainage waters, the cost of reducing agricultural drainage water production, the demand for water in irrigated agriculture, and the response of growers to changes in irrigation and drainage policies.

In this chapter, crop-water production functions under saline conditions are reviewed. Definitions and general concepts are discussed first, and studies on crop-water production functions are then reviewed. In general, two approaches to estimate crop-water production functions are apparent in the literature. One approach synthesizes production functions from theoretical and empirical models of individual components of the crop-water process. Parameter values are obtained, in principle, by direct measurement. The second approach estimates production functions by statistical inference from observations on alternate levels of crop yield, water applications, soil salinity, and other variables. Transient and seasonal synthetic production functions are reviewed. Then, econometric/statistical production functions are examined.

Most production functions are estimated based on the assumptions that water applications are uniform and soil conditions are relatively homogeneous. However, in most fields, depths of applied water and conditions of the soil vary in space considerably. Thus, field-level production functions may differ from those estimated from small agronomic plots or theoretical models that assume homogeneous conditions. The last section of the chapter covers the integration of crop-water production functions and spatial variability.

DEFINITION OF PRODUCTION FUNCTIONS

The literature includes different definitions of the term, production function. For example, some authors refer to the relation between crop

---

[a]Prepared by: John Letey and Keith Knapp, Dept. of Soil and Envir. Sci., Univ. of Calif., Riverside; and Kenneth Solomon, Center for Irrig. Tech., Calif. State Univ., Fresno.

yield and seasonal evapotranspiration (ET) as a production function. Others refer to the relation between crop yield and applied water. Here, we will follow standard economic usage and refer to a production function as a relation between inputs to a production process and the outputs of that production process. More precisely, a production function gives the maximum possible output from alternate levels of inputs (Varian 1978). For example, suppose that output is crop yield and input is seasonal water applications. Different yields may be obtained, depending on how water is allocated during the season. A unique relation between crop yield and seasonal water applications can be obtained by defining the production function as the maximum yield that can be obtained from a given level of seasonal water applications. For the economic analysis of a season, this relation is also the most useful one, since growers are assumed to maximize profits.

Mathematically, let Y denote output. Then the production function can be written

$$Y = f^1 (X_1, \ldots, X_n) \tag{14.1}$$

where X_i, $i=1, \ldots, n$ denotes the quantity of the ith input to the production process. There are numerous inputs to the production process, but only some are considered to be variable or controlled by the farmer in a given analysis. In this case, we write

$$Y = f^2(X_1, \ldots, X_m \mid X_{m+1}, \ldots, X_n) \tag{14.2}$$

or, more simply,

$$Y = f^3(X_1, \ldots, X_m) \tag{14.3}$$

where the first m inputs are considered to be variable, and the last n-m inputs are fixed. For example, an agronomic experiment might vary water and nitrogen, while keeping pesticides, labor, and machinery constant.

The above production functions are for a single output and multiple inputs. In many cases, more than one output of interest may exist. Letting Y_i, $i=1, \ldots, n_y$, denote the quantity of the ith output, the production function can be written as

$$Y_i = f_i(X_1, \ldots, X_m \mid X_{m+1}, \ldots, X_n) \qquad i = 1, \ldots, n_y \tag{14.4}$$

where the f_i are component functions of the production function F, which is a vector function. For example, an empirical production function in a following section relates crop yield and quantity and quality of drainage water to the quantity and quality of irrigation water.

Production functions might be explicit, analytic functions. An example is

$$Y = a_o + a_1X + a_2X^2 \tag{14.5}$$

where the a_i are known constants. Such explicit functions are often estimated in agronomic experiments. A production function can also be implicit in a more general model. For example, consider the dynamic model

$$\begin{aligned} M_{j+1} &= M_j - \mathrm{ET_j} + W_j \\ \mathrm{ET_j} &= f(M_j; K_j) \\ Y &= g(\sum_{j=1}^{J} \mathrm{ET_j}) \end{aligned} \tag{14.6}$$

where M_j = soil moisture on day j; ET_j = daily evapotranspiration; K_j = a daily climatic variable; W_j = net water applied; and Y = yield. This model can be solved for Y given the initial moisture level M_1, the irrigation inputs W_j, and the functions f and g. Implicit in this dynamical system is the production function:

$$Y = h(W_1, \ldots, W_j \mid M_1) \tag{14.7}$$

The transient crop-growth models to be discussed later consist of dynamical systems that are solved numerically on the computer. The dynamic model can be solved repeatedly for different values of W_j and the production function (14.7) estimated from the resulting solutions. However, this has not been done very often.

Production functions can also be stochastic. For example, suppose

$$Y = F(X_1, \ldots, X_m \mid X_{m+1}, \ldots, X_n; \theta) \tag{14.8}$$

where θ is a random variable. In this case, the output Y is also a random variable, with a probability distribution that is conditional on the input quantities X_i. The random variable could represent stochastic variations in weather, environmental conditions during the season, or uncertainty about the parameter values or functional form of the production function. In any case, output cannot be predicted exactly, even if the quantities of all other inputs are known exactly. Stochastic production functions will not be considered further in this chapter, since there appears to be little, if any, work in this area under saline conditions.

TRANSIENT PRODUCTION FUNCTION MODELS

Bresler and Hanks (1969) combined the numerical method for predicting water flow in unsaturated soils developed by Hanks et al. (1969) with a salt distribution model described by Bresler (1967) to compute the water and concentration of salt in the soil as a function of time and depth. The model required data on water retention curves and the relationships between hydraulic conductivity and water in the soil for wetting and drying conditions. The model assumed no precipitation or dissolution of salts. Nimah and Hanks (1973a) expanded the analysis of Bresler and Hanks (1969) by including a plant root extraction term. It required information on effective water potential in the root at the surface of the soil, H_{root}, a root resistance term, and the distribution of active roots with depth in the soil. The root extraction term, A(z,t), is

$$A(z,t) = \frac{[H_{root} + (1.05z) - h(z,t) - \pi(z,t)] \cdot RDF(z) \cdot K(\theta)}{\Delta x \, \Delta z} \qquad (14.9)$$

where H_{root} is an effective water potential in the root at the surface of the soil; 1.05 is a combined term to account for a frictional loss due to flow in the root (0.05) and for the gravitational head within the root at different depths (1.0); h(z,t) is the soil matric potential; π(z,t) is the osmotic potential; RDF(z) is the proportion of total active roots in the depth increment Δz; K(θ) is the hydraulic conductivity; and Δx is the distance between the plant roots and the point in the soil where h(z,t) and π(z,t) are measured (assumed to be 1 cm in the model). All potentials are expressed in head units. The root extraction term is an equation that describes flux of water from the soil into the root. The value of the H_{root} term depends on the plant, climate, and conditions of the soil. In the model, a value of H_{root} is solved within limits selected by the modeler so that the plant root extraction over the total profile equals the potential transpiration. If the value of H_{root} reaches the lower limit, root extraction may be less than potential transpiration.

In addition to the previously mentioned data, the following are required for the model: potential transpiration and potential evaporation rate, irrigation or rainfall as a function of time for the entire run, osmotic potential of the irrigation water, initial osmotic potential and soil depth, and presence or absence of a water table at the bottom of the profile.

Childs and Hanks (1975) extended the model of Nimah and Hanks (1973a) to consider the effects of salinity and to compute the yield. DeWit (1958) proposed that the ratio of transpiration to potential transpiration, T_p, be used as an estimate of relative dry matter yield. Hanks (1974) explored this relationship in greater detail. Based on these analyses, the cumulative transpiration output from the Nimah and Hanks (1973a) model was translated into relative yield by using the deWit (1958) relationship. Data from Lunin and Gallatin (1965), Shalhevet and Bernstein (1968), and Bingham and Garber (1970) were used to demonstrate a linear relationship between Y/Y_{max} and T/T_p, when the transpiration was affected by soil salinity. Later data reported by Stewart et al. (1977) and Hanks et al. (1977) demonstrated a linear relationship between relative yield and relative evapotranspiration, when water stress was caused by high salinity or limited water. Childs and Hanks (1975) assumed that salinity influences yield through its influence on transpiration. Furthermore, they assumed that a decrease in osmotic potential decreased the water potential gradient between the soil and the plant, thus decreasing transpiration. In effect, the osmotic potential was a part of the root extraction term (Eq. 14.9) of the Nimah and Hanks (1973a) model so that the Childs and Hanks (1975) model merely focused attention on that aspect of the model.

Childs and Hanks (1975) modified the original model to allow the ratio of T_p to potential evapotranspiration (ET_p) to vary with time for row crops and to allow diurnal variation in potential evaporation and

T_p. The model of Bresler (1973) was incorporated into the salt-flow part of the model.

Bresler et al. (1982) modified the root extraction term (Eq. 14.9):

$$A(z,t) = -b(z) \cdot K(\theta) \cdot [H_r - h(z,t) - \pi(z,t)] \qquad (14.10)$$

The term $[H_{root} + (1.05z)]$ has been replaced by the total water potential in the root at the root-soil interface, H_r, and the term $RDF(z)/\Delta x \Delta z$ has been replaced by a coefficient of proportionality, b, that represents the geometry of water flow to the roots. A critical potential in the root (H_{cr}) is defined to be the lowest possible potential. Water extraction by roots ceases when $h(z,t) + \pi(z,t)$ is less than or equal to H_{cr}. Hereafter, the root extraction term as expressed by Eqs. 14.9 and 14.10 will be referred to as the Hanks model and the Bresler model, respectively.

The root extraction term links the soil-water-salinity state to plant response. Hanks (1984) stated, "The function described here is quite crude, particularly regarding plant factors." He indicated that attempts to make the model more realistic by including such plant factors as axial and radial root resistance did not improve the predictability when applied to actual problems.

The measured and computed water content-depth profile from the Hanks model correlated well with a field test in which alfalfa was a crop (Nimah and Hanks 1973b). No plant growth comparisons were made in that study. Childs and Hanks (1975) reported a good correlation between computed and measured transpiration, using data from experiments by Bingham and Garber (1970) and Lunin and Gallatin (1965). They also reported predictions of the model for crops with different rooting depths, different initial soil salinities, and multi-year cropping. The authors emphasized that the predictions had not been verified by measured data. Wolf (1977) used the Hanks model to determine whether it would estimate the results of irrigation timing and salinity on yield as reported by Stewart et al. (1977). According to Hanks (1984), Wolf found that computed and measured data correlated excellently when levels of salinity were low but that the model overestimated yields when levels of salinity were high.

Bresler and Hoffman (1986) compared computed results from the Bresler model with measured results from experiments using several crops at the U.S. Salinity Laboratory. A value for H_{cr} must be chosen for the Bresler model. Values of H_{cr} equal to -0.34 and -0.22 MPa were selected for computations to compare to results from studies involving nine crops. Computed and measured yields for the nine crops correlated better for H_{cr} of -0.34 than for -0.22 MPa. Computed relative differences in yield between using H_{cr} of -0.34 and -0.22 MPa were about 0.1 to 0.2, depending on the amount of applied water. They chose a value of H_{cr} of -1.5 MPa to compute results for an experiment with a crop of tall fescue. Computed and measured results correlated well. They failed to explain why they selected the value of H_{cr} or why tall fescue had a much lower H_{cr} value than other crops. They presented no information on a choice of the value for b(z) in the root extraction term (Eq. 14.10).

Bresler (1987) used his model to simulate crop-water production functions, as well as other response functions related to irrigation with saline waters. A value of −0.34 MPa was selected for H_{cr} in the simulation. Computed relative yields (RY) increased as the amount of applied water increased until RY = 1 was achieved, after which yield remained constant. The applied water needed to achieve RY = 1 increased as the salinity of the irrigation water increased. However, if irrigation water had an EC = 8 dS/m or greater, zero transpiration and zero yield were computed for all levels of water application. This is a consequence of the root extraction term (Eq. 14.10). When the summation of the matric and osmotic potential is less than H_{cr}, the root extraction (transpiration) is zero. High rates of water application provide high matric potential and hydraulic conductivity [$K(\theta)$] values, so that the value for root extraction remains high until the osmotic potential of the irrigation water approaches the H_{cr} value, after which extraction drops to zero. The root extraction term is considerably more sensitive to changes in matric potential than to changes in osmotic potential. Decreasing matric potential results in decreasing values of $K(\theta)$. The value of $K(\theta)$ is unaffected by changes in osmotic potential.

Computed relationships between RY and seasonal average root zone salinity were presented by Bresler (1987) for H_{cr} values of −0.22 and −0.34 MPa. A RY value equal to 1 was computed for low salinities, and then the RY decreased sharply with increasing salinity above threshold values. The threshold salinity value was greater for H_{cr} = −0.34 than for H_{cr} = −0.22. Maas and Hoffman (1977) suggested that the results be characterized by a piecewise linear relationship, in which maximum yield takes place until the average root zone salinity reaches a threshold value and decreases linearly with increasing soil salinity above the threshold value. They summarized threshold values for several crops as well as the slope of the curve at values greater than the threshold value. The simulated relationships reported by Bresler were similar to those reported by Maas and Hoffman, except that the simulated yield decreased much more sharply beyond the threshold value.

Bresler (1987) wrote, "Plants that have roots with a higher potential to extract water will tolerate a higher concentration in the soil solution than plants having lower water extraction potential when all other environmental conditions are identical and osmotic regulation has a negligible effect." Although not stated by Bresler, this consideration would provide an approach to selecting the value of H_{cr} for the model. The Maas and Hoffman (1977) threshold coefficient for a given crop could be matched with the appropriate H_{cr} value that would simulate the same behavior. However, this approach does not appear to provide valid results. Bresler and Hoffman (1986) found that computed and measured yields of several crops correlated well, using a H_{cr} value of −0.34 MPa for all crops. The Maas and Hoffman threshold value ranged from 1.3 to 6.1 for these crops.

In summary, the Hanks and Bresler models are similar and differ only in detail on the root extraction term (Eqs. 14.9 and 14.10). Each developed the root extraction term, assuming that matric potential and osmotic potential equally affected the rate of water extraction on roots and, therefore, equally affected transpiration. However, shifts in matric potential affect transpiration more than shifts in osmotic potential, because K(θ) varies with changes in matric potential but not with changes in osmotic potential. Neither Hanks nor Bresler addressed how to incorporate crop sensitivity to salinity into the model, although Bresler (1987) alluded to crop sensitivity as being related to the value of H_{cr}. Limits placed on H_{root} or selected values for H_{cr} could be used to adjust the model for plant salt tolerance. However, this would be valid only if a crop sensitive to osmotic potential was equally sensitive to matric potential on a quantitative basis, a condition yet to be scientifically established.

van Genuchten (1987) developed a numerical model for water and solute movement in and below the root zone that included a root extraction term. The approach, although comparable to that of Hanks and Bresler, differs significantly in detail. Only the root extraction term will be reviewed. Differences in the model with respect to water and solute flow in the soil profile will not be discussed.

The root extraction term, S (Feddes et al. 1978; Belmans et al. 1983), used by van Genuchten is

$$S = \alpha(H)\, S_m\,(z) \tag{14.11}$$

where α(H) is a dimensionless water stress response function equivalent to the ratio between the actual (S) and potential (S_m) extraction rates, and

$$S_m = \frac{T_p}{L} \tag{14.12}$$

where T_p is potential transpiration and L is rooting depth. $S_m(z)$ can be adjusted at varying depths to account for root distribution. The root extraction rate equals the potential transpiration rate when α(H) = 1 and becomes less than potential transpiration when α(H) < 1. Water potential (H) can be separated into matric potential (h) and osmotic potential (π).

A linear relationship between yield and transpiration is assumed, and maximum yield is associated with potential transpiration. Empirical relationships between yield and h, or π, can be used to establish the relationship between T and h, or π. van Genuchten and Hoffman (1984) suggested that measured salt tolerance response functions of various crops could be expressed by a smooth S-shaped curve between relative yield, i.e., yield relative to the maximum yield under nonsaline conditions, and the average salt concentration of the root zone. The expression, after converting salt concentration to osmotic potential, is

$$Y = \frac{Y_m}{\left[1 + \left(\dfrac{\pi}{\pi_{50}}\right)^p\right]} \tag{14.13}$$

where Y is yield; Y_m is maximum yield; π_{50} is the osmotic potential at which the yield is reduced by 50%; and p is an empirical constant. This model described salt tolerance data as well or better than the threshold-slope model of Maas and Hoffman (1977). Using the assumption $S/S_m = T/T_m = Y/Y_m$ from Eqs. 14.11 and 14.13, then

$$\alpha(\pi) = \frac{1}{\left[1 + \left(\frac{\pi}{\pi_{50}}\right)^p\right]} \tag{14.14}$$

If matric potential and osmotic potential have similar, but not necessarily linearly additive, effects on yield and, thus, on transpiration, then

$$\alpha(h) = \frac{1}{\left[1 + \left(\frac{h}{h_{50}}\right)^p\right]} \tag{14.15}$$

where h_{50} = the matric potential at which yield is reduced by 50%. The expression for $\alpha(H)$ would combine relationships expressed in Eqs. 14.14 and 14.15 in the form of

$$\alpha(H) = \frac{1}{1 + \left[\frac{(a_1 h + a_2 \pi)}{h_{50}}\right]^p} \tag{14.16}$$

where h_{50}, π_{50}, a_1, and a_2 are presumed to be parameters specific to the crop, the soil, and the climate.

Considerable data are available on the relationship expressed by Eq. 14.13. The exponent p was found to be approximately 3 for several crops (van Genuchten and Hoffman 1984). Plant response functions to matric potential are unavailable, as they are difficult to determine experimentally. Yields computed with van Genuchten's model and results from experiments have not been compared at the time of this writing.

Hanks, Bresler, and van Genuchten developed their root extraction terms on the basis that yield and transpiration are linearly related. Hanks and Bresler formulate their extraction term as an equation for water flow rate from the soil into the root, using the water potential gradient and hydraulic conductivity. When the computed flow rate equals potential transpiration, maximum yield is assumed. When the flow rate is less than potential transpiration, yield is assumed to be less than maximum at the same proportion as the transpiration is less than potential transpiration. van Genuchten, in contrast, uses empirical relationships between the matric potential or the osmotic potential and the relative yield. Relative yield is then used to compute relative transpiration. In other words, Hanks and Bresler used relative transpiration to compute relative yield, and van Genuchten uses relative yield to compute relative transpiration. The most significant difference between these approaches is that van Genuchten takes advantage of empirical plant response data in his model.

SEASONAL PRODUCTION FUNCTION MODELS

The dynamic models of Bresler, van Genuchten, Hanks and colleagues described above can be used to simulate seasonal crop-water production functions for various irrigation schedules, if appropriate input data for the given model is available.

Solomon (1985a) and Letey et al. (1985) presented seasonal water-salinity-production functions based on our current understanding of the response of crops to water, the salt tolerance of crops, and the leaching process. Solomon (1985a) presented the general theory and Letey et al. (1985) developed a special case of the theory.

The dynamic models described above and the seasonal models of Solomon and Letey et al. assume a unique relationship between yield and ET for a given crop and climate that is independent, regardless of whether the water stress leading to the reduced ET is caused by deficit water supply, excess salinity, or both. Beginning with this premise, Solomon (1985a) expressed the following:

> Irrigating with saline water will cause some degree of salinization of the soil. This, in turn, will cause a decrease in crop yield relative to yield under nonsaline conditions. This reduced yield ought to be associated with a decrease in plant size and a decrease in seasonal ET. But, as ET goes down, effective leaching will increase, mitigating the initial effect of the saline irrigation water. For any given amount and salinity of irrigation water, there will be some point at which values for yield, ET, leaching, and soil salinity all are consistent with one another. The yield at this point is the yield to be associated with a given irrigation water quantity and salinity.

Letey et al. (1985) combined the relationships of yield versus ET, yield versus average root zone salinity, and average root zone salinity versus leaching fraction to develop an equation that related yield to the amount of seasonal applied water of a given salinity. The combination of these relationships led to the point that, as Solomon stated, "The value for yield, ET, leaching and soil salinity are all consistent with one another."

A linear relationship between yield and ET was used in the model. This, as has been previously stated, is a basic premise for dynamic and seasonal models. Letey et al. (1985), Doorenbos and Kassan (1979), Vaux and Pruitt (1983), and others have referred to studies reporting relationships between yield and ET. Various investigators reported the linear relationship between total dry matter production and ET. However, cotton yield was found to be curvilinearly related to ET (Grimes et al. 1969, Davis 1983). Cotton lint and other crops with a marketable yield component not linearly related to ET must receive special consideration in a manner to be detailed later. This special consideration is required for the dynamic and the seasonal model, even though authors of the dynamic models have not addressed it.

The piecewise linear relationship proposed by Maas and Hoffman (1977) was used to relate yield to average root zone salinity in the seasonal model. This relationship between yield and root zone salinity is comparable to the one used by van Genuchten in his dynamic model. The difference is that van Genuchten used a smooth sigmoidal curve (van Genuchten and Hoffman 1984) instead of a piecewise-linear relationship. Use of these relationships represents a significant difference between the dynamic models of Hanks and colleagues and Bresler and both the dynamic model of van Genuchten and the seasonal model.

Hoffman and van Genuchten (1983) presented relationships between average root zone salinity and leaching fraction for given salinities of irrigation water for three root functions on the uptake of water. The exponential uptake function correlated best with experimental data, so that relationship was used to develop the model. This relationship is based on steady-state assumptions, a significant departure from the dynamic models that compute salt and water distributions under transient conditions.

Combination of these three relationships provided a model that could be used to compute the yield, leaching volume, or EC of the water percolating below the root zone for given quantities of seasonal applied water (AW) of given salinities. The seasonal AW includes quantities of precipitation or irrigation applied before planting the crop, but it excludes runoff. The model is for uniform water application. Nonuniform irrigation is discussed later in this chapter.

Letey and Dinar (1986) used the model to compute production functions for several crops. Yields were recorded on a relative scale, with maximum yield represented by 1.0 and AW scaled to pan evaporation, E_p, to facilitate transfer of the production function to locales other than those from which the data were measured. The computed production functions for alfalfa in Fig. 14.1 exemplifies production functions reported by Letey and Dinar (1986) for several crops.

In providing production functions for cotton, Letey and Dinar (1986) described a procedure for crops with no linear relationship between marketable yield and ET. Basically, it involves computing total dry matter production, which is linearly related to ET, and then converting the computed total dry matter value to marketable yield using an empirical relationship between the two. This procedure, or another that accomplishes the same goal, must be used with dynamic and seasonal models.

Letey et al. (1985) and Letey and Dinar (1986) compared yields from experiments and those computed by the seasonal model for several crops. The only data used in making these comparisons were the seasonal AW, E_p, and salinity of irrigation water. When the salinity of the irrigation water varied, the weight-averaged electrical conductivity of the irrigation waters was used. Measured and computed yields correlated well. Letey and Dinar (1986) discussed the usefulness and limits of the model in detail.

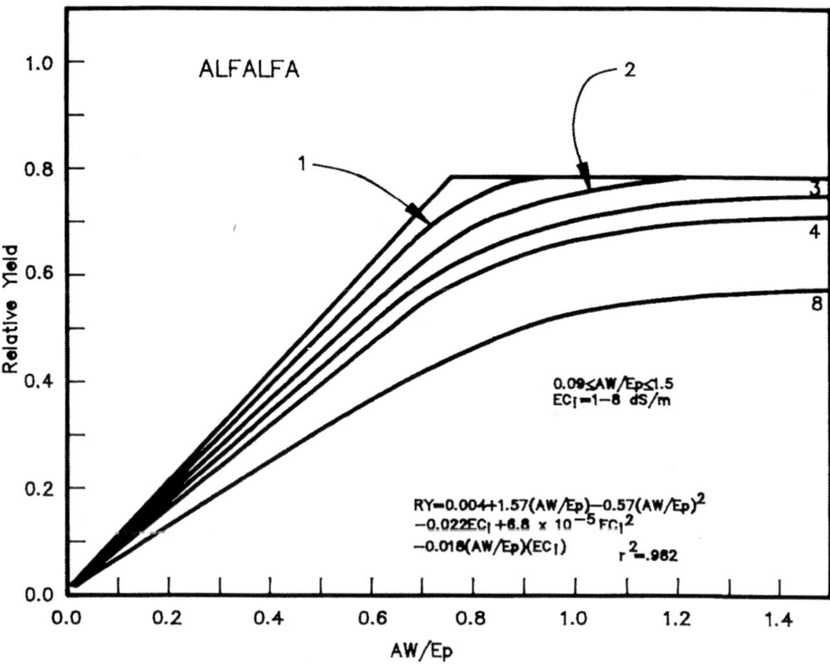

Fig. 14.1 Computed Relative Yield of Alfalfa for Various Quantities of Applied Water, Scaled to Pan Evaporation (Each curve is for a given EC of irrigation water.) (Letey and Dinar 1986)

ECONOMETRIC AND STATISTICAL ESTIMATION OF PRODUCTION FUNCTIONS

Suppose that the production function is defined by

$$Y = f(X_1, \ldots, X_m; \alpha) + e \qquad (14.17)$$

where Y is output; $X_1, \ldots, X_m$ are inputs; α is a vector of unknown parameter values; and e is a zero-mean random error term. The error term, e, captures the effects of factors that influence yield but are unobserved. Various combinations of inputs and the associated yields can be observed from experiments or other data. The unknown parameters can then be estimated with the use of econometric and statistical methods, primarily multiple regression, after specifying a functional form for the production function. Thus, the production function is inferred from observed data on inputs and outputs. This contrasts with the previously discussed production functions, in which outputs are predicted from given input levels using theoretically based models, and parameters are estimated based on direct observation.

A vast literature on crop-water production functions under nonsaline conditions exists. Vaux and Pruitt (1983), Solomon (1985b), and others provide extensive surveys. Statistical procedures have been used to provide parts of the synthetic production function models discussed in the previous two sections. Maas and Hoffman (1977) summarize the relations between relative yield and soil EC_e for a number of crops. Feinerman et al. (1982) and van Genuchten and Hoffman (1984) further discuss estimation procedures for relative yield-soil EC_e relations. Relations between relative yield and crop ET or transpiration are also typically estimated using statistical procedures. Shalhevet (1984) and Hoffman (1986) review the evidence and provide references to the literature. Yaron et al. (1972) estimate relations between crop yield and soil moisture and salinity. Relatively few studies estimate directly the relation between crop yield and variables under the grower's control. Existing studies are reviewed below.

Hanks et al. (1977) investigated the effects of irrigation volumes and initial soil salinities on crop yield and ET for corn and alfalfa. A line-sprinkler system provided irrigation and rain gauges installed at regular intervals measured the amount of applied water. Corn yield in 1974 declined as initial soil salinity increased, but showed little effect as water applications decreased. A high water table explained the latter effect. A production function relating corn yield to water applications and initial soil salinity levels was given in tabular form for 1975. Yields generally increased as applied water increased or as initial soil salinity decreased. Two treatments with saline irrigation water showed no measurable effect. Alfalfa production remained unaffected by irrigation volumes and soil salinity. This was also attributed to a high water table.

Selassie and Wagenet (1981) conducted field trials for corn. They related grain yield, dry matter yield, and nitrogen content to seasonal irrigation volumes, initial soil salinity, and nitrogen fertilizer. Grain yields increased significantly with increases in irrigation volume and decreases in initial soil salinity. Nitrogen did not appear to increase grain yields and, in some cases, decreased the yield. The authors hypothesized that the initial nitrogen level of the soil sufficiently met the crop's needs and that dissolution of the excess nitrogen contributed to decreased osmotic potential.

Hoffman et al. (1983) conducted rhizotron experiments for tall fescue to determine how yield, evapotranspiration, and root density are related to irrigation-water salinity, leaching fraction, and irrigation frequency. Irrigation frequency had no significant effect on yield. The increased salinity of irrigation water and a decreased leaching fraction significantly reduced the yield and evapotranspiration.

Campbell et al. (1986) conducted greenhouse experiments for snapbeans. They investigated the effect of irrigation-water salinity, irrigation frequency, and fertilizer applications on yield and nitrogen fixation. The yield significantly decreased as the salinity of the irrigation and the interval between applications of water increased. Applications of fertilizer had no effect.

Dinar and Knapp (1986) analyzed data from field experiments with cotton reported by Rhoades (1982) and data for alfalfa reported by the U.S. Salinity Laboratory staff (1981). For alfalfa, regressions for crop yield and soil salinity at the end of the growing season were estimated as functions of initial soil salinity and seasonal irrigation volumes. Similar relations were estimated for cotton, except that the quality of the irrigation water was also included as an explanatory variable.

The studies described above do not explicitly consider plant and soil dynamics during the season. Also, production functions, when estimated, are generally *ad hoc* functional forms, such as quadratic. Theoretical considerations are generally excluded, except for specifying relevant variables. Dinar et al. (1986) moves away from this with an econometric study where a daily crop-water production function model is estimated for cotton. In the model, soil moisture depends on evapotranspiration and irrigation volume. Drainage volumes, drainage salinity, and soil salinity after an irrigation depend on irrigation volume, irrigation salinity, and soil moisture and salinity before irrigation. Daily evapotranspiration is a function of soil moisture, soil salinity, date during the growing season, and pan evaporation. Crop yield depends on evapotranspiration throughout the season.

Dinar et al. (1986) estimated the production function model, using field experiment data reported by Rhoades and Rawlins (1980). Evapotranspiration is the product of maximum ET and relative ET. Maximum ET equals the product of an exogenous crop factor and pan evaporation. Relative ET is a function of matric plus osmotic potential. The unknown parameters are estimated econometrically from the field data. The soil salinity function was specified using a piston-flow model that explained the field data quite well. The yield-ET relation was estimated statistically.

The statistical/econometric approach to production function estimation differs from the approach taken in the previous two sections. The models in those sections tend to be formulated on conceptual and theoretical grounds. The statistical models often use *ad hoc* functional forms, although Dinar et al. (1986) indicates that this need not always be the case. A more significant difference is the method used to estimate unknown parameter values. The dynamic models discussed earlier presumably rely on actual measurements of the relevant parameters. In the statistical approach, parameter values are inferred from observations on alternate levels of yields and inputs. Statistical models can predict the conditions under which they are estimated reasonably well but will likely be less transferable to other areas than the more general models of the previous sections.

NONUNIFORMITY

The production functions presented above relate yield to applied water (AW), where AW is expressed as an equivalent depth of water. AW in the field can be ambiguous. It can represent water discharged to the field, including runoff, or the amount that infiltrated the soil, which is the difference between discharge and runoff. In this chapter, AW will denote infiltrated water because runoff does not contribute to crop production. Water discharge and runoff are usually measured by volume, so AW is computed by dividing the volume by the area of the land to which the water added. Applied water computed in this way represents an average value ($\overline{AW}$) for the field. The actual AW at any given location in the field may be considerably different from $\overline{AW}$ because irrigation water is never applied uniformly. Variable AW values at different locations in the field lead to variable yields at different locations. The production function for a field would have the average yield, $\overline{Y}$, related to $\overline{AW}$. As reported below, the relationship between $\overline{Y}$ and $\overline{AW}$ differs from the relationship between Y and AW.

A number of researchers recognize and have studied the effects of irrigation uniformity on crop production (Howell 1964, Zaslavsky and Buras 1967, Varlev 1976, Seginer 1978, Hart et al. 1980, Warrick and Gardner 1983, Solomon 1983, Peri et al. 1979, Feinerman et al. 1983, Letey et al. 1984, Letey 1985, Seginer 1987, Warrick and Yates 1987). The general effects of irrigation uniformity on yield are well documented. Methodologies for measuring and quantitatively incorporating irrigation uniformity in crop production functions have yet to be fully developed or tested in the field.

Characterizing Irrigation Uniformity

Irrigation uniformity can be characterized by measuring AW at several locations in the field and plotting the frequency distribution. It is assumed that the frequency distribution of AW at the measured sites represents the frequency distribution for the field, and that the distribution is static in time. The distribution has been characterized by various indices, such as the mean, the variance, the skewness, and the kurtosis. Christiansen (1942) proposed a uniformity coefficient (CUC) for irrigation based on the absolute deviation of individual values from the mean. This relationship is

$$CUC = 100\left(1 - \frac{\sum|X_n - \overline{X}|}{\overline{X}N}\right) \tag{14.18}$$

where x_n = the individual observation value; $\bar{x}$ = the average of the observations; and N = the total number of observations. The distribution uniformity, another common way to characterize irrigation uniformity, is the ratio of the average x_n for the lowest one-fourth x_n values to $\bar{x}$, expressed as a percentage. Irrigation engineers have used various relationships to characterize irrigation systems (Merriam and Keller 1978). Warrick (1983) reported interrelationships between various irrigation uniformity terms. However, irrigation uniformity distribution may not be of any particular analytic form. Indeed, Howell (1964) stated, "The distributions of application depth resulting from sprinkler irrigation observed by the writer have generally been unlike any standard statistical distributions, and they exhibit a wide variety of dispersion, skewness, and kurtosis." Thus, presenting the frequency or cumulative curve graphically is the most accurate way to record uniformities.

Measuring the infiltrated water distribution poses more of a problem than does recording the distribution. When pressurized irrigation systems, such as sprinkler or drip systems, are designed to avoid surface ponding, they shift control of the infiltration rate from the soil to the design and performance of the irrigation system. Techniques for evaluating the uniformity of sprinkler irrigation, such as measuring the volume of water in catch-cans distributed within the sprinkler pattern, have been developed and extensively used. Computed distributions, however, depend on the size and spacing between the catch-cans. The scale of uniformity observation affects the results. It must be matched with the scale of influence of individual plant root zones before the data can be used to develop crop production functions. Since plants tend to integrate the effects of water within the root zone, some nonuniform distribution of water within the root zone can be tolerated. Cogels (1983) and Seginer (1979) address this matter. Letey (1985) also suggests approaches to matching catch-can and root zone scales.

Properties of the soil control infiltration under surface irrigation systems, such as furrow or border. Therefore, the variability of these irrigation systems relates to the variability of these properties. The opportunity time—the time during which water remains on the soil surface—also affects the amount of infiltration. Opportunity time is a factor for furrow or border irrigation in which water applied at one end of the field eventually flows across the field. Most techniques for evaluating furrow systems measure the time that it takes for water to move down the furrow. The measurements reflect opportunity time. Wallender (1986) found that nonuniformity related to variable soil conditions that exceed those associated with differences in opportunity time. Yitayew et al. (1985) concluded that assessing soil variability was a major factor in determining the uniformity of a furrow irrigation system. Field measurements on infiltration uniformity are scarce. Indeed, some questions on how to make these measurements exist (Letey 1985). For example, Sisson and Wierenga (1981) used cyclinders with inside diameters of 5-cm, 25-cm, and 127-cm in a relatively small plot to measure steady-state rates of infiltration. The uniformity results depended on the size of infiltration cylinders. Consequently, the infiltration measurement should be scaled with the root zone scale. Jury et al. (1987) presented the results of field studies on the spatial variability of rates of infiltration. The smallest coefficient of variation for any field-size study was 40%. Variability of soil infiltration rates on a field basis is quite high, based on reported measurements. This variability significantly determines the crop production function.

Production Function

The Y versus AW relationship can be combined with distribution of water application to obtain $\overline{Y}$ versus $\overline{AW}$ (the field production function). Empirical crop water production functions obtained in the field where the irrigation was nonuniformly applied already have nonuniformity included in the production function.

Analytical ways to combine Y versus AW functions with irrigation uniformity have been proposed. Howell (1964) concluded that $\overline{Y}$ depends only on the mean value of application depth and is independent of irrigation uniformity for a linear Y versus AW relationship. However, this is valid only for a very limited range. Linear relationship between Y and AW exists only until maximum yield is achieved. Thereafter, yield remains constant or declines with increased application of water. The response curve is piecewise linear. Howell's conclusion is valid when

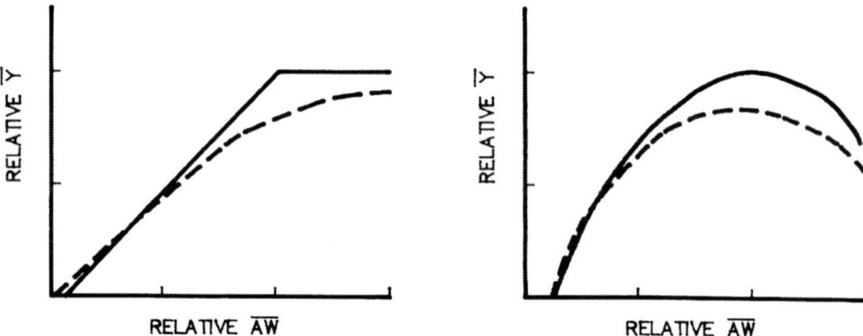

Fig. 14.2 Relationship Between Relative Yield and Relative Applied Water for Uniform Irrigation (Solid Line) and Irrigation with a Coefficient of Variability Equal to 0.5 (Warrick and Yates 1987)

the field receives water less than or equal to the quantity that provides maximum yield. Howell determined that $\overline{Y}$ for a parabolic relationship between Y and AW depends on the mean value of the depth of application and the second moment of distribution of application depth about the mean. Similarly, if the yield curve is cubic, then $\overline{Y}$ depends on the mean value application depth and the second and third moments about the mean of the distribution of application depth, etc.

Warrick and Yates (1987) presented tables for response functions and average yields for three water distributions. Fig. 14.2 shows the relationships between relative $\overline{Y}$ and relative $\overline{AW}$ for uniform irrigation and irrigation with a coefficient of variability equal to 0.5 for Y versus AW functions that are piecewise linear and quadratic. Except for very low applications of water, yields under nonuniform application are lower than yields under uniform application for a given $\overline{AW}$ value. For the piecewise linear production function, applying more water overcomes the effects of nonuniformity, i.e., maximum yield can be achieved by applying high amounts of water. It is assumed that the soil can transmit the higher applications of water without becoming waterlogged. If the production function is quadratic, nonuniformity cannot be overcome by applying more water; yields under these conditions will be lower than yields under uniform conditions. Feinerman et al. (1983) concluded that the economically optimal value of AW increases with decreasing uniformity for piecewise linear production functions, but the optimal level of AW is lower for the nonuniform application than for the uniform application if the production function is quadratic.

Analytical solutions are usually restricted to specific crop-water production function and applied water distribution (Solomon 1984). Letey et al. (1984) presented numerically computed production functions for nonuniform irrigation. With high-speed computers, this procedure can be used for any infiltrated water distribution or Y versus AW function. AW at any location on the field, a, can be related to $\overline{AW}$ by

$$\mathrm{AW}(a) = \beta(a) \cdot \overline{\mathrm{AW}} \tag{14.19}$$

where $\beta(a)$ is a parameter whose distribution over the field must be determined (the irrigation uniformity). The distribution function of β can be approximated by a discrete distribution in which β takes only a finite number (n) of distinct values, $\beta_1, \beta_2, \ldots, \beta_n$. The fractional area of the field (A_i), in which $\beta = \beta_i$, is equal to the probability that $\beta = \beta_i$. Under uniform conditions, the yield, Y_i, for a given crop-water production function, f, in that portion of the field, A_i, is given by

$$Y_i = f(\beta_i \overline{AW}) \tag{14.20}$$

and the yield for the field is given by summing the fractional areas of the field from n=1, . . . ,N:

$$\overline{Y} = \sum_{i=1}^{n} A_i Y_i \tag{14.21}$$

To compute a total production function, these computations are repeated for various values of $\overline{AW}$.

Knapp et al. (1986) used a calibration scheme to estimate irrigation uniformity distributions. They assumed that β is distributed normally over the field with an expected value of unity; they based production functions under uniform conditions on Letey and Dinar (1986); and they assumed that farmers maximize profits. Unknown parameters to be estimated were the standard deviation for β and the maximum yield. Profit-maximizing water applications and the associated yields were first computed over a range of values for standard deviations and maximum yields. The values of standard deviation and maximum yield that resulted in the closest match between calculated water applications and yields and observed water applications and yields were then used as the parameter estimates. They obtained CUC values of 73, 86, 93, 80, and 73 for alfalfa, wheat, sugar beets, cotton, and barley, respectively, under furrow irrigation.

In conclusion, crop water production functions for nonuniform irrigation depend greatly on the distribution of infiltrated water over a field and the shape of the crop-water production function under uniform irrigation. The scale of uniformity measurement should be related to the dimension of the root zone. Investigations in the field on irrigation uniformity for surface irrigation systems have yet to be conducted. Such application in the field is limited by uncertainty in measuring irrigation uniformity.

CONCLUSION

Substantial progress has been made in developing empirical models that can be used to relate crop yields and irrigation management under saline conditions. Further work is needed before these empirical models can be reliably applied under a wide variety of field conditions. Further work also is required on the relation of ET to soil and environmental conditions. In many instances, potential ET or transpiration is determined externally to the model. However, potential ET depends in part on the size of the plant, which depends on irrigation management during the previous part of the irrigation season. Hence, relative and maximum or potential ET should be endogenous variables.

Nonuniform applications of water and spatial variations in soil parameters significantly affect seasonal water production functions. To date, little or no work has been done to estimate transient production functions under nonuniform conditions. Procedures for estimating uniformity distributions on a scale relevant to the plant also are needed. Variations in the environment affect the growth of the plant, so random effects related to the weather need to be included in models of the growth of plants under saline conditions.

REFERENCES

Belmans, C., Wesseling, J.G., and Feddes, R.A. 1983. "Simulation model of the water balance of a cropped soil. SWATRE. *Journal Hydrology* 63: 271–286.

Bingham, F.T., and Garber, M.J. 1970. "Zonal salinization of the root system with NaCl and boron in relation to growth and water uptake of corn plants." *Soil Sci. Soc. Am. Proc.* 34: 122–126.

Bresler, E. 1973. "Simultaneous transport of solute and water under transient unsaturated flow conditions." *Water Resour. Res.* 9: 975–986.

Bresler, E. 1967. "A model for tracing salt distribution in the soil profile." *Soil Sci.* 104: 227–233.

Bresler, Eshel. 1987. "Application of a conceptual model to irrigation water requirement and salt tolerance of crops." *Soil Sci. Soc. Am. J.* 51: 788–793.

Bresler, E., and Hanks, R.J. 1969. "Numerical method of estimating simultaneous flow of water and salt in unsaturated soils." *Soil Sci. Soc. Am. J.* 33: 827–832.

Bresler, E., and Hoffman, G.J. 1986. "Irrigation management for soil salinity control: theories and field tests." *Soil Sci. Soc. Am. J.* 50: 1552–1559.

Bresler, E., McNeal, E.L., and Carter, D.L. 1982. *Saline and Sodic Soils.* Springer-Verlag, New York.

Campbell, W.F., Wagent, R.J., and Rodriguez, R.R. 1986. "Salinity, water management and fertility interactions on yield and nitrogen fixation in snap-beans." *Irrig. Sci.* 7:195–204.

Childs, S.W., and Hanks, R.J. 1975. "Model of soil salinity effects on crop growth." *Soil Sci. Soc. Am. J.* 39: 617–622.

Christiansen, J.E. 1942. "Irrigation by sprinkling." University of Calif., Berkeley. Agric. Expt. Sta. Bull. 670.

Cogels, O.G. 1983. "An irrigation system uniformity function relating effective uniformity of water application to the scale of influence of the plant root zones." *Irrig. Sci.* 4: 289.

Davis, K.D. 1983. "Trickle irrigation of cotton in California." Proc. Western Cotton Production Conf., 9-11, August 1983, Las Cruces, N.M. pp. 34–38.

deWit, C.T. 1958. *Transpiration and crop yield.* Verslad, van Landbouck, Onderzoek No. 64.6, 88 p.

Dinar, A., and Knapp, K.C. 1986. "A dynamic analysis of optimal water use under saline conditions." *West. J. Agric. Econ.* 11(1): 58–66.

Dinar, A., Knapp, K.C., and Rhoades, J.D. 1986. "Production function for cotton with dated irrigation quantities and qualities." *Water Resour. Res.* 22(11): 1519–1525.

Doorenbos, J., and Kassam, A.H. 1979. "Yield response to water." FAO Irrigation Drainage Paper NO. 33, FAO, Rome.

Feddes, R.A., Kowalik, E.J., and Caradny, H. 1978. *Simulation of Field Water Use and Crop Yield.* Halstead Press, John Wiley and Sons, New York, 188 p.

Feinerman, E., Yaron, D., and Bielorai, H. 1982. "Linear crop response functions to soil salinity with a threshold salinity level." *Water Resour. Res.* 18(1): 101–106.

Feinerman, E., Letey, J., and Vaux, H.J. 1983. "The economics of irrigation with non-uniform infiltration." *Water Resour. Res.* 19: 1410–1414.

Grimes, D.W., Yamada, H., and Dickens, W.L. 1969. "Function for cotton (*gossypium hirsutum, L*) production from irrigation and nitrogen fertilization variables. I. Yield and evapotranspiration." *Agron. J.* 61: 769–773.

Hanks, R J. 1974. "Model for predicting plant yield as influenced by water use." *Agron. J.* 66: 660–665.

Hanks, R.J. 1984. "Predictions of crop yield and water consumption under saline conditions." I. Shainberg and J. Shalhevet, eds., *Soil Salinity under Irrigation,* Ecological Studies 51. Springer-Verlag, New York. pp. 272–283.

Hanks, R.J., Sullivan, T.E., and Hunsaker, V.E. 1977. "Corn and alfalfa production as influenced by irrigation and salinity." *Soil Sci. Soc. of Am. J.* 41: 606–610.

Hanks, R.J., Klute, A., and Bresler, E. 1969. "A numerical method for estimating infiltration, redistribution, drainage and evapotranspiration of water from soil." *Water Resour. Res.* 5: 1064–1069.

Hart, W.E., Norum, D.I., and Peri, G. 1980. "Optimal seasonal application analysis." *J. Irrig. Drain. Div. ASCE* 106(IR3): 221.

Hoffman, G. 1986. "Yield of irrigated crops under saline conditions." G. J. Horner, ed., *Proceedings: Decision Criteria for Residuals Management in Agriculture,* Giannini Foundation of Agricultural Economics and Kearney Foundation of Soil Science, April 1986.

Hoffman, G.J., Jobes, J.A., and Alves, W.J. 1983. "Response of tall fescue to irrigation water salinity, leaching fraction, and irrigation frequency." *Agric. Water Management* 7: 439–456.

Hoffman, Glenn J., and van Genuchten, M.Th. 1983. "Soil properties and efficient water use: water management for salinity control." H.M. Taylor et al., eds. *Limitations to Efficient Water Use and Crop Production.* American Society of Agronomy, Madison, Wisc. 53711. pp. 73–85.

Howell, D.T. 1964. "Sprinkler uniformity characteristics and yield." *J. Irrig. Drain Div. ASCE* 90(IR): 55–67.

Jury, William A., Russo, David, Sposito, Garrison, and Elabd, Hesham.

1987. "The spatial variability of water and solute transport properties in unsaturated soil. I. Analysis of property variation and spatial structure with statistical models." *Hilgardia* 55(4): 1–32.

Knapp, K., Dinar, A., and Letey, J. 1986. "On-farm management of agricultural drainage water: an economic analysis." *Hilgardia* 54(4): 1–31.

Letey, J. 1985. "Irrigation uniformity as related to optimum crop production—additional research is needed." *Irrig. Sci.* 6: 253–263.

Letey, J., and Dinar, Ariel. 1986. "Simulated crop-water production functions for several crops when irrigated with saline waters." *Hilgardia* 54(1): 1–32.

Letey, J., Dinar, Ariel, and Knapp, Keith C. 1985. "Crop-water production function model for saline irrigation waters." *Soil Sci. Soc. Am. J.* 49: 1005–1009.

Letey, J., Vaux, H.J., and Feinerman, E. 1984. "Optimum crop water application as affected by uniformity of water infiltration." *Agron. J.* 76: 435–441.

Lunin, J., and Gallatin, M.H. 1965. "Zonal salinization of the root in relation to plant growth." *Soil Sci. Soc. Am. Proc.* 29: 608–612.

Maas, E.V., and Hoffman, G.J. 1977. "Crop salt tolerance—current assessment." *J. Irrig. Drain. Div. ASCE* 103(IR2): 115–134.

Merriam, John L., and Keller, Jack. 1978. "Farm irrigation system evaluation: a guide for management. Agricultural and Engineering Dept., Utah State University, Logan, Utah 84321. 271 pp.

Nimah, M.N., and Hanks, R.J. 1973a. "Model for estimating soil, water, plant and atmospheric interrelations: I. Description and sensitivity." *Soil Sci. Soc. Am. J.* 37: 522–527.

Nimah, M.N., and Hanks, R.J. 1973b. "Model for estimating soil, water, plant and atmospheric interrelations: II. Field test of model." *Soil Sci. Soc. Am. J.* 37: 528–532.

Peri, G., Hart, W.E., and Norum, D.I. 1979. "Optimal irrigation depths—a method of analysis." *J. Irrig. Drain. Div. ASCE* 105(IR4): 341.

Rhoades, J.D. 1982. "Use of Saline Drainage Water for Irrigation—A Field Test in the Westside Project, Kern County, California." 4th Annual Report, U.S. Salinity Laboratory, Riverside, Calif.

Rhoades, J.D., and Rawlins, S.L. 1980. "Use of Saline Drainage Water for Irrigation: A Field Test in the Westside Project, Kern County, California." 2nd Annual Report, U.S. Salinity Laboratory, Riverside, Calif.

Seginer, Ido. 1987. "Spatial water distribution in sprinkler irrigation." *Advan. Irrig.* 4: 119–168.

Seginer, I. 1978. "A note on the economic significance of uniform water application." *Irrig. Sci.* 1: 19.

Seginer, I. 1979. "Irrigation uniformity related to horizontal extent of the root zone." *Irrig. Sci.* 1: 89.

Selassie, T.G., and Wagenet, R.J. 1981. "Interactive effects of soil salinity, fertility, and irrigation on field corn." *Irrig. Sci.* 2: 67–78.

Shalhevet, J. 1984. "Management of Irrigation with Brackish Water." I. Shainberg and J. Shalhevet, eds. *Soil Salinity Under Irrigation*, Springer-Verlag, Berlin.

Shalhevet, J., and Bernstein, L. 1968. "Effects of vertically heterogeneous soil salinity on plant growth and water uptake." *Soil Sci.* 106: 85–93.

Sisson, J.V., and Wierenga, P.J. 1981. "Spatial variability of steady-state infiltration rates as a stochastic process." *Soil Sci. Soc. Am. Proc.* 45: 699–703.

Solomon, K.H. 1983. "Irrigation uniformity and yield theory," thesis presented to Utah State Univ. in Logan, in partial fulfillment of the requirements for the degree of doctor of philosophy.

Solomon, K.H. 1984. "Yield-related interpretations of irrigation uniformity and efficiency measures." *Irrig. Sci.* 5: 161–172.

Solomon, K.H. 1985a. "Water-salinity-production functions." *Trans. ASAE* 28(6): 1975–1980.

Solomon, K. 1985b. "Typical crop water production functions." Paper No. 85-2596, American Society of Agricultural Engineers. 19 pgs.

Stewart, J.I., Danielson, R.E., Hanks, R.J., Jackson, E.B., Hagan, R.M., Pruitt, W.O., Franklin, W.T., and Riley, J.P. 1977. "Optimizing crop production through control of water and salinity levels in the soil." Utah Water Lab. PRWG 151-1, Logan, Utah. 191 pp.

U.S. Salinity Laboratory Staff. 1981. "Minimizing salt in return flow through irrigation management. Report to the U.S. Environmental Protection Agency, EPA-IAGD6-037L.

van Genuchten, M.Th. 1987. "A numerical model for water and solute movement in and below the root zone." Research Rep. 121, USDA-ARS, U.S. Salinity Laboratory, Riverside, Calif.

van Genuchten, M.Th., and Hoffman, G.J. 1984. Analysis of crop salt tolerance data. I. Shainberg, I., and Shalhevet, J., eds. *Soil Salinity under Irrigation,* Ecological Studies 51. Springer-Verlag, New York. pp. 258–271.

Varian, H. 1978. *Microeconomic Analysis.* W. W. Norton and Co., New York.

Varlev, I. 1976. "Evaluation of non-uniformity and irrigation on yield." *J. Irrig. Drain Div. ASCE* 102(IR1): 149.

Vaux, H.J., and Pruitt, W.O. 1983. "Crop-water production functions." *Adv. Irrig.* 2: 61–97.

Wallender, W. 1986. Furrow model with spatially varying infiltration. *Trans. ASAE* 29: 1012–1016.

Warrick, A.W. 1983. "Interrelationships of irrigation uniformity terms." *J. Irrig. Drain. Div. ASCE* 109: 317–332.

Warrick, A.W., and Gardner, W. R. 1983. "Crop yield as affected by spatial variations of soil and irrigation." *Water Resour. Res.* 19: 181.

Warrick, A.W., and Yates, S. R. 1987. "Crop yield as influenced by irrigation uniformity." *Advan. Irrig.* 4: 169–180.

Wolf, J.K. 1977. "The evaluation of a computer model to predict the effects of salinity on crop growth," thesis presented to Utah State Univ., Logan.

Yaron, D., Bielorai, H., Shalhevet, J., and Gavish, Y. 1972. "Estimation Procedures for Response Functions of Crops to Soil Water Content and Salinity." *Water Resour. Res.* 8(2): 291–300.

Yitayew, M., Letey, J., Vaux, H.J., Jr., and Feinerman, E. 1985. "Factors

affecting uniformity and optimal water management in furrow irrigation." *Irrig. Sci.* 6: 201–210.
Zaslavsky, D., and Buras, M. 1967. "Crop yield response to non-uniform application of irrigation water." *Trans. ASAE* 10: 196–198, 200.

NOTATION

a_i = known constants; a_1, a_2 = parameters specific to the crop, the soil, and the climate; A_i = fractional area of the field; AW = applied water; CUC = uniformity coefficient; e = zero-mean random error; E_p = pan evaporation; ET_j = daily evapotranspiration; f, g = functions; f_i = component function of the production function; F = production function; h = matric potential; h (z,t) = soil matric potential; h_{50} = matric potential at which yield is reduced by 50%; H = water potential; H_{cr} = critical potential in the root; H_{root} = effective water potential in root at the surface of soil; K_j = daily climatic variable; $K(\theta)$ = hydraulic conductivity; L = rooting depth; M_j = soil moisture on day j; M_1 = initial moisture level; N = total number of observations; p = empirical constant; RDF(z) = proportion of total active roots in the depth increment; RY = computed relative yields; S = actual extraction rate; S_m = potential extraction rate; T_p = potential transpiration; W_j = net water applied; x_n = individual observation value; $\bar{x}$ = average of the observations; X_1, . . . ,X_m = inputs; X_i, i=1, . . . ,n = quantity of the *i*th input to the production process; Y = output; Y = yield; Y_m = maximum yield; Y_i, i=1, . . . ,n_y = quantity of the *i*th output to the production process; α = vector of unknown parameter values; $\alpha(H)$ = dimensionless water stress response function; π = osmotic potential; π_{50} = osmotic potential at which yield is reduced by 50%; π (z,t) = osmotic potential; $\beta(a)$ = parameter whose distribution over the field must be determined; θ = random variable; Δx = distance between the plant roots and the point in the soil where h(z,t) and π(z,t) are measured; and Δz = depth increment.

CHAPTER 15

SPATIAL AND TEMPORAL VARIABILITIES IN SALINITY[a]

INTRODUCTION

Agricultural irrigation and drainage are often essential for crop growth and salinity management (Bernstein 1974; Hoffman et al. 1980). Drainage systems have been used for many years to control rising water tables and remove salts in irrigation drain water. Today, a growing awareness exists of how irrigation practices can contaminate ground water (Bouwer 1987; Schmidt and Sherman 1987; Sonnen et al. 1987) and surface water (*The Sacramento Bee* 1985; *The Reno Gazette-Journal* 1987; Tanji 1981; Tanji et al. 1986). This awareness includes recognition of concerns regarding the variation of water quality over space and time. Statistical models can help to determine and evaluate these spatial and temporal variabilities.

Engineers consider spatial and temporal variabilities to be explicit parts of salinity analysis, yet they rarely apply the statistical techniques used to characterize variability as routinely as they do more traditional methods. Presumably, they are less familiar with techniques like time series and kriging than with regression and analysis of variance. The former rely on interdependence of data for modeling, whereas the latter require independence of observations. At first, unfamiliar terms and definitions may present obstacles, but computer software is available to enable a user to learn the methods and terminology and evaluate data output without time-intensive personal programming.

Statistical techniques used for characterizing and estimating temporal and spatial variability include time series analysis (Box and Jenkins 1976; Chatfield 1975; Hoff 1983, Jenkins and Watts 1968; Shumway 1988), kriging (Clark 1979; David 1982; Davis 1973; Journel and Huijbrechts 1976), and co-kriging, which is kriging with more than one simultaneous variable. These techniques characterize and estimate temporal and spatial data sets. They make use of specialized forms of standard statistical parameters, such as correlation, cross correlation, variance, estimation, and spectral analysis.

Time series analysis makes use of the autocorrelogram. This is a plot of the autocorrelation coefficient for incremental increases between serial data pairs for one variable of a data set. As examples will show, the autocorrelation indicates the degree of correlation between all data pairs separated by the same temporal or spatial sampling interval. A

---

[a]Prepared by: J.C. Guitjens, Dept. of Range Wildlife and Forestry, Univ. of Nevada-Reno, Reno, NV 89557; and B.R. Hanson, Cooperative Extension, Dept. of Land, Air, and Water Resour., Univ. of Calif., Davis, CA 95616.

high degree of correlation indicates a large interdependence and, frequently, uniformity. Once a trend has been discerned, a random error remains. The random error is involved in making forecasts based on probability. In kriging, the autocorrelation function (covariance function) is embedded in the variogram. The autocorrelation and random error can be used to make estimates, such as *b*est, *l*inear, *u*nbiased *e*stimate (BLUE) values for points intermediate to points of measurement.

Cross-correlation is the analysis of two sets of data for common patterns of variability. This technique helps to evaluate how well the data sets correlate to each other and can be used to find the cross-correlogram. It gives the cross-correlation coefficient as related to data interval.

Spectral analysis identifies periodicity in data by partitioning the variance into frequency intervals. Co-spectral analysis does the same for covariance between two data sets. Spectral and co-spectral analysis estimate the contribution of the periodicity to the variance of the data. The distribution of the statistical variance, or covariance, among frequency intervals indicates whether patterns exist in the data distribution. High and low frequencies may provide clues to the spatial character of soil salinity. Nielsen et al (1983) showed an example in which the covariance between two data sets equaled zero but was, according to the co-spectrum, positive for one frequency interval and negative for another.

Characterization of salinity and salinity components with descriptive statistics, such as mean and variance, comparisons among means, or correlation tests between variables, is inadequate for spatial and temporal characterization, as several case studies in this chapter will illustrate. Drainage water salt concentration and salt flux can be explained in spatial and temporal terms, using the time series technique developed by Box and Jenkins (1976) or using co-kriging.

An intervention analysis involves evaluating a bivariate model in which the second variable represents an important consideration or action. For example, after developing a time series model of the sodium-adsorption ratio (SAR) in drain water from an irrigated field, irrigation modifications can be added, and the effects of drainage SAR analyzed temporally and spatially.

Soil salinity data can exhibit divergent statistical behavior; however, individual parameters can still be modeled for the purpose of forecasting or for estimation of data at points between the actual data locations. Soil salinity data will be used below to illustrate the interpretation of data by autocorrelation, cross-correlation, and spectral and co-spectral analysis.

Time Series

The purpose of time series models is to forecast values beyond the exising data series. In ordinary least square regression methods of deterministic statistics, error terms, the difference between estimated and actual values, are assumed not to correlate for adjacent data points:

$$\mathrm{COV}(e_t, e_{t-1}) = 0 \qquad (15.1)$$

If they are correlated, the predicted values will be biased. However, time series analysis depends on the least-square regression of both the deterministic and error components, as well as a stochastic random error component

One common time series model is known as an ARIMA model. ARIMA stands for *a*utoregressive, *i*ntegrated, *m*oving *a*verage. ARIMA models are a function of three filters: an autoregressive filter (p), a differencing filter (d, which corresponds with the integrated aspect), and a moving average filter (q). Filters are used to create a random error component that has a mean of zero and a constant variance.

The purpose of an ARIMA model is to forecast values beyond the existing data series. To test a given model's accuracy, after-the-fact forecasting is used. Existing data are divided into two sets representing early and late intervals, with the later one representing an interval for estimation. Once a new model has been generated with the early data, data can be forecasted for the later interval. The estimated and actual data can be compared to calculate errors and evaluate the model's usefulness.

McDowall et al. (1987) list six software distributors of ARIMA time series software. These various programs do not always give identical technical output. For example, Minitab (Ryan et al. 1982) does not provide a 95% confidence interval of statistical significance when it plots the autocorrelation function.

Kriging and Co-Kriging

Kriging is one of several methods for data estimation collectively known as geostatistics. Co-kriging refers to kriging, i.e., estimating data, with more than one variable.

The statistical part of a kriging technique consists of a variogram. Variogram variance is given by:

$$\gamma = \frac{1}{2N} \sum_{i=1}^{N} [Z(X_i) - Z(X_i + h)]^2 \qquad (15.2)$$

where h = the separation distance; X_i a sampling location or sampling event for temporal modeling; and $Z(X_i)$ the value of an observation at location X_i. Fig. 15.1 explains the elements of a variogram: the covariance, C; the nugget, C_o, which is also known as the noise or micro spatial variability; the sill, which is C and C_o combined; and the range. Range is the maximum separation distance, also known as maximum lag, with non-zero covariance values. Covariance indicates the autocorrelation among the data. In Fig. 15.1, the error, μ, represents the difference between the sill and the statistical variance of the data set. Usually, spherical models are used to describe and fit variogram variance curves. Specific software may present other options, such as a linear and Gaussian methods of curve fitting (Carr et al. 1985).

Kriging is an estimation method that gives BLUE values. Its estimate of a variable at any point, X_o, is linear, described by

$$Z^*(X_o) = \sum_{i=1}^{M} \lambda_i Z(X_i) \qquad (15.3)$$

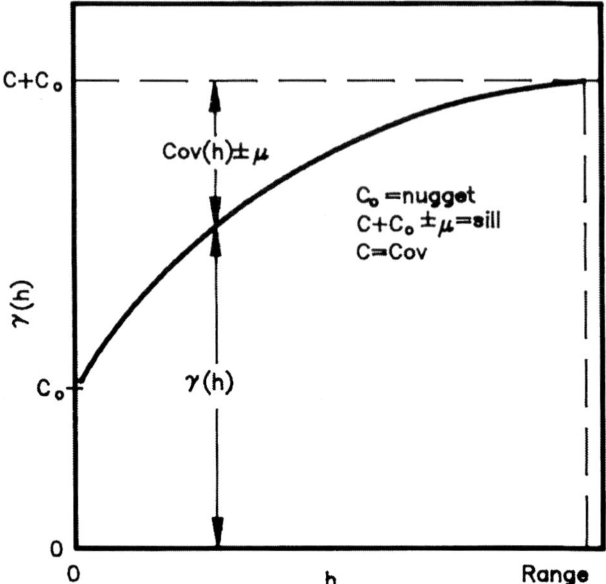

Fig. 15.1 Spherical Variogram Model

under the unbiased condition that

$$\sum_{i=1}^{M} \lambda_i = 1 \tag{15.4}$$

where λ_i = the weights that represent a computer-generated matrix calculation involving the variogram's covariance model and a Lagrangian multiplier to represent the error, μ. As the program moves from estimation location to estimation location, the window of M neighboring measured values, $Z(X_i)$, moves accordingly.

In co-kriging, since more than one variable is being estimated, values of C_o, C, range, and a model of the variogram variance are required for each variable and for the sums of all paired combinations (cross-variograms) of the variables. If the absolute values of the variables differ considerably in scale, they must first be normalized by dividing the individual values by the mean of the variable data set. Otherwise, the impact of the variable with smaller magnitude would be overshadowed in the cross-variogram, which represents the variogram of the sum of two variables.

Kriging requires normally distributed data. A histogram of the data can be used to determine whether the data distribution is normal. Excessively skewed data may need to be transformed to create a more normal distribution. The estimated results can be retransformed with a log transformation, i.e., by taking the antilog of the kriging estimates. Kurtosis indicates data variability.

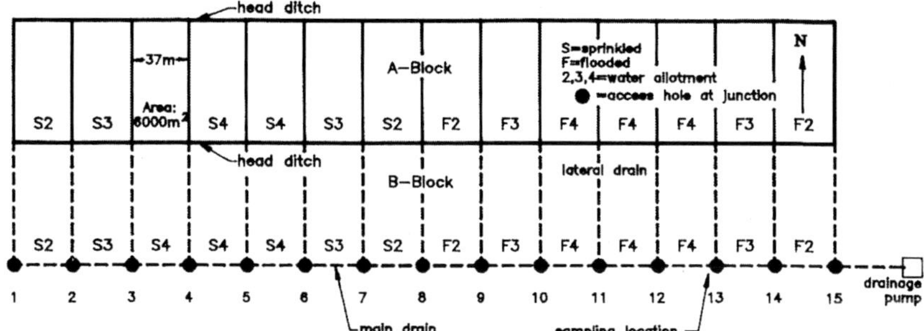

Fig. 15.2 Layout of 18-ha Study Site for Example

DRAIN WATER SALINITY

Drainage water salinity can be expressed in terms of salt concentration (mg/l) or mass flux (mg/min). Both measurements will vary over time and space because the drainage is a function of water application and the soil's salt conditions. Drain flow also depends on the soil's hydraulic properties. In highly variable soils, modeling a physical system of drainage flow, salt concentration, and mass flux in response to irrigation poses a complex temporal and spatial problem. As an alternative, the drain response can be monitored and evaluated statistically and then temporal and spatial forecasting models produced for individual variables with the use of time series techniques. With a co-kriging model, data can be estimated where gaps exist.

An example of how to apply the aforementioned methodologies is as follows: Alfalfa is grown on an 18-ha field, as shown in Fig. 15.2, where half the field has natural drainage (A Block) and half is equipped with a subsurface drainage system (B Block). A tile drainage system is used, consisting of 15 parallel drains placed about 2 m below the ground surface. Both irrigation and natural lateral ground water contribute to drainage outflow. Due to decreasing water supply during the growing season, irrigation water is applied under allotment so the full irrigation occurs for two, three, or all four growth cycles of a crop. Consequently, the physical characteristics of the field plot exhibit natural and imposed variabilities. The following sections will illustrate spatial and temporal series, as well as co-kriging methods of data manipulation and estimation for the drain's discharge rate (Q), salt concentration (TDS), and salt flux (Q × TDS).

Spatial ARIMA Modeling

Sets of 27 consecutive weekly data for each drain have been joined to form one data series, starting with drain #1 and ending with drain #15. A spatial ARIMA was developed. Fig. 15.3 shows autocorrelation functions (ACF) and partial autocorrelation functions (PACF) for the spatial Q × TDS distribution of the original data, differenced data, and ACF of the residual error after filtering. Table 15.1 presents a corresponding ARIMA model as produced by BMDP2T (Liu 1981).

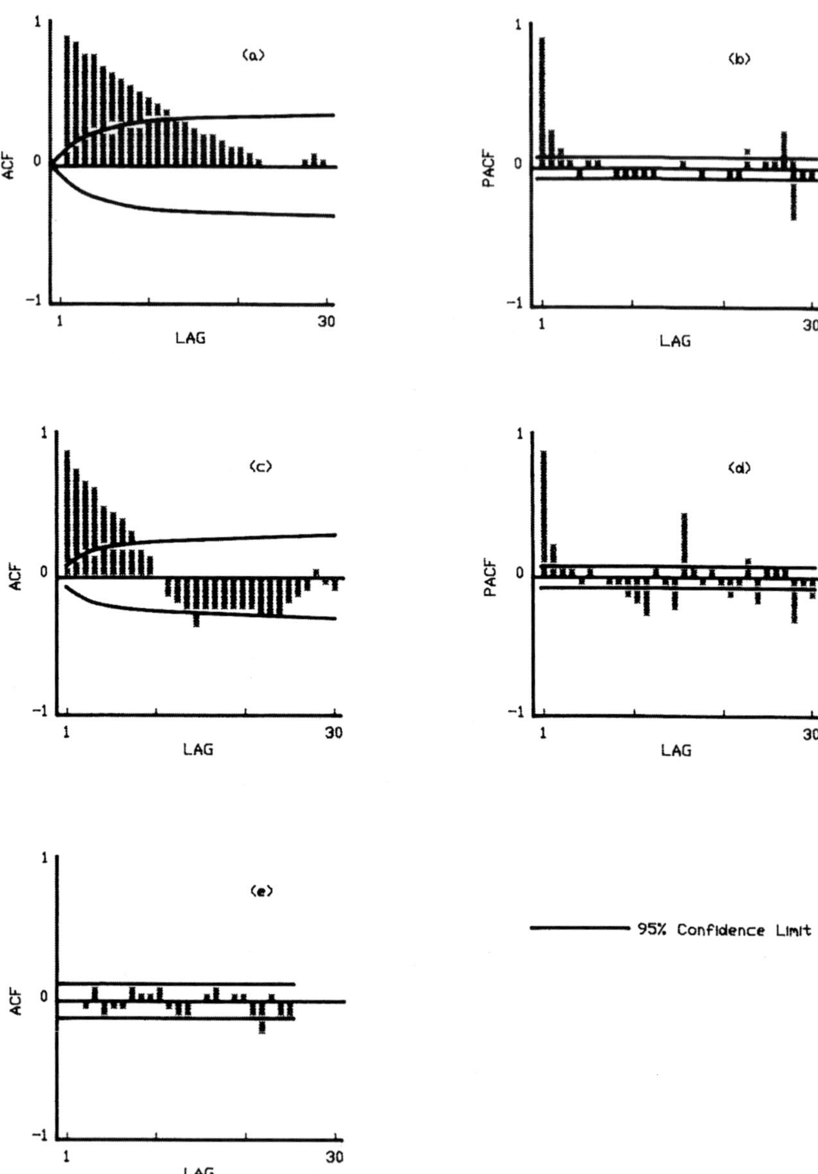

Fig. 15.3 Auto- and Partial Auto-Correlation Functions for Spatially Distributed Data for Mass Flux of Salts From Drains [Original data are represented in (a) and (b), differenced data in (c) and (d), and residual error in (e).]

Figs. 15.3a and 15.3b give the ACF and PACF of the original spatial series, respectively. They show a definite declining trend in the ACF and PACF but no cyclical, spatial seasonality because the shape of the ACF does not repeat itself after one cyclical lag (lag equals the separation distance of 27), represented by 27 weekly intervals. The largest significant autocorrelations indicate the differencing orders. In Fig. 15.3a, lag of 1 shows the largest spike. This again occurs in Fig. 15.3b. If a large autocorrelation had been repeated for lag equals 27 for the data in Fig. 15.3a, a second differencing order of 27 (d = 27) would have been tried due to the significant peak at 27 in Fig. 15.3b. Since the lag at 1 falls within a cycle of 27 weeks rather than at multiples of 27 weeks, the differencing order of 1 is called regular differencing. A regular differencing order of d = 1 produced the ACF (Fig. 15.3c) and PACF (Fig. 15.3d) of the differenced data series.

The ACF and PACF of the differenced data can be studied to select the value of p, the moving average filter, and q, the autoregressive filter. Fig. 15.3c suggests the moving average filters of p = 1, p = 15, and perhaps other choices. After trial runs, p = 1 provided a significant coefficient, as will be indicated by a significant t-value in the output shown in Table 15.1. After similarly testing several filter choices (Fig. 15.3d), a suitable autoregressive filtering order of q = 27 was selected.

Whether to accept a filtering order depends on significant t-ratios (Table 15.1) and the ACF of the residual error. Fig. 15.3e shows the ACF of the residual error, also known as white noise, after filter choices were included. The ACF should have no significant spikes at early lags, or the residual error might not be random. A test of the randomness can be made on software that prints the Q-statistic, e.g., BMDP2T. If the Q-statistic is smaller than the corresponding Chi-square value from statistical tables, this confirms the randomness. All t-ratios are significant and the ACF of the white noise series (Fig. 15.3e) appears stationary, as indicated by non-significant values at early lags.

Discharge rate and total dissolved solids concentrations were likewise modeled. They behaved similarly to the ACF of the original spatial series. Table 15.1 also tabulates their ARIMA model results.

Temporal ARIMA Modeling

The temporal Q × TDS series consists of combining sequential spatial data from drains 1 through 15 during 27 consecutive weeks, starting with week one. It involves using the same software as the spatial example. The series shows strong temporal seasonality, which repeats in lags containing sets of 15 drains (Figs. 15.4a and 15.4b). A differencing filter of 15 (d = 15) and further filtering with orders 1 and 15 for p and q (Figs. 15.4c and 15.4d) result in a stationary residual error series (Fig. 15.4e). Table 15.2 shows the ARIMA model details for Q, TDS, and Q × TDS. Seasonality in the original series, indicated by large, cyclical, weekly spikes at multiples of 15 regular lags that correspond to the field's 15 drains, e.g., Figs. 15.4a and 15.4b, persisted over 11 weekly lags for Q, 10 lags for TDS, and 12 lags for Q × TDS.

TABLE 15.1 Summary of ARIMA $(p,d,q)_1$ Model for Q(l/min), TDS (mg/l), and Q × TDS (mg/min) Spatial Series -1982 (Guitjens 1987)

Parameter (1)	Type[a] (2)	p (3)	d (4)	q (5)	Factor (6)	Order (7)	Estimate (8)	t-ratio[b] (9)
Q	MA	1	2	2	1	1	0.3640	8.07
Q	MA	1	2	2	2	27	0.5424	8.22
Q	AR	1	2	2	1	22	−0.2332	−4.83
Q	AR	1	2	2	2	27	0.8923	28.86
TDS	MA	1	1	1	1	5	0.1413	2.83
TDS	AR	1	1	1	1	17	0.1521	3.04
Q × TDS	MA	1	1	1	1	1	0.3415	7.72
Q × TDS	AR	1	1	1	1	27	0.5338	14.31

[a]MA = Moving Average; AR = Autoregressive.
[b]The estimator for all values given is significant (<0.05).

After-the-Fact Forecast Modeling.

The next analysis will make after-the-fact forecasts using ARIMA modeling of Q, TDS, and Q × TDS of the data previously discussed and shown in Figs. 15.3 and 15.4. Arbitrarily, data from 27 consecutive weeks for drain 15 for the spatial case and data for weeks 26 and 27 of all 15 drains for the temporal case were forecasted. Table 15.3 shows the comparisons of these after-the-fact forecasts. In a comparison between the means of forecasts and actual values, the temporal models are better than the spatial models. The same is true for mean percent error. However, these results are not consistently reflected by the other error tests. Whether these models are acceptable depends on the criteria incorporated in the standards of the user.

The diagnosis of salinity conditions with ARIMA models involves the use of features that are unavailable from deterministic regression equations or analysis of variance tests for comparing mean values. Regression does not consider temporal and spatial influences or the effect of systematic errors inherent in each equation and each calculation (Guitjens 1987).

In addition to helping to develop a forecasting model, the time series technique provides certain insights from the step-wise model building. For example, the (a) and (b) parts of Figs. 15.3 and 15.4 reveal a poor cyclical trend for the spatial case and a much stronger one for the temporal case. The estimates of equation coefficients in Tables 15.1 and 15.2 show larger t-ratios for the temporal series. This means that autocorrelation is much stronger in the temporal series, and perhaps sampling frequency could be reduced without losing an unacceptable degree of accuracy. Since temporal autocorrelation extends over 10 or more weeks, weekly sampling is probably much more intense than needed. On the other hand, the rapid decay of significant spikes in Fig. 15.3a for the spatial case occurs before the first cyclical lag of 27 weeks has been completed, and no significant spikes reappear in the second cycle. This implies a lack of autocorrelation between adjacent drains and beyond.

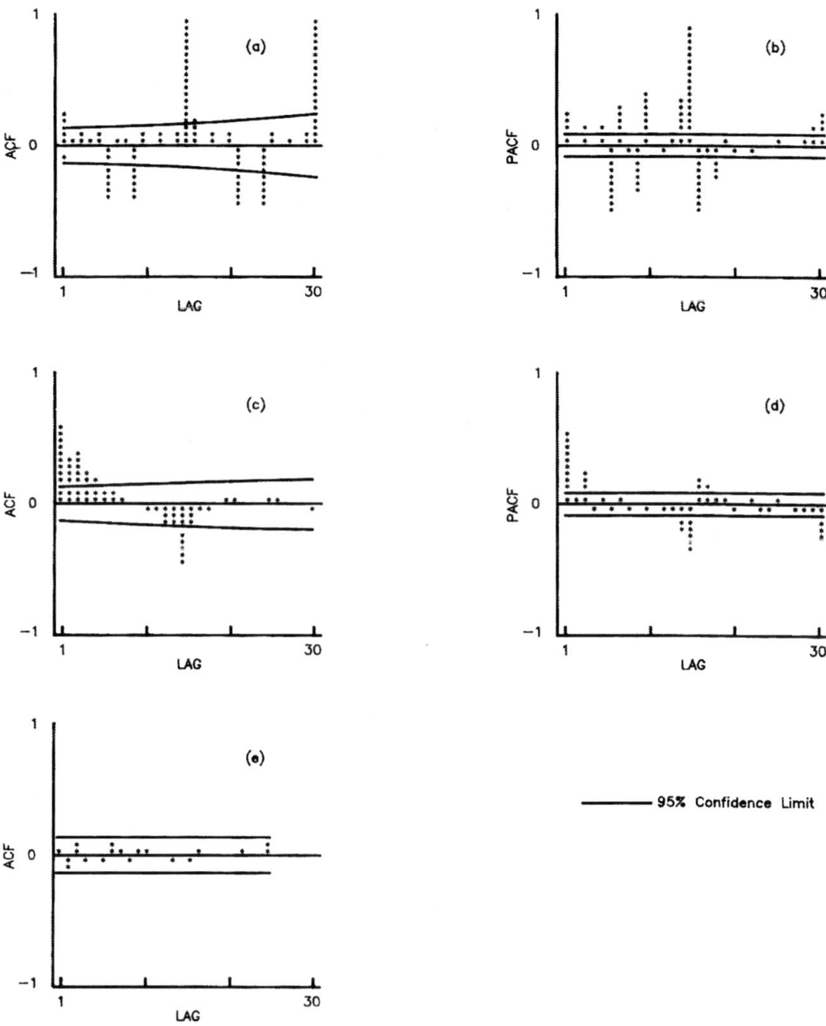

Fig. 15.4 Auto- and Partial Auto-Correlation Functions for Temporally Distributed Data for Mass Flux of Salts from Drains [Original data are represented in (a) and (b), differenced data in (c) and (d), and residual error in (e).]

Engineers can use data characterization as displayed by the ACF of the original data to determine the spatial and temporal behavior of drainage constituents. Poor ability to forecast indicates that sampling must be modified and increased before acceptable results are possible. As changes are tracked with future samples, the models can be updated so that subsequent forecasts will incorporate the new trends. Such information is critical in evaluating management practices for improving drainage quality.

TABLE 15.2 Summary of ARIMA $(p,d,q)_{15}$ model for Q(l/min), TDS (mg/l), and Q × TDS (mg/min) Temporal Series-1982 (Guitjens 1987)

Parameter (1)	Type[a] (2)	p (3)	d (4)	q (5)	Factor (6)	Order (7)	Estimate (8)	t-ratio[b] (9)
Q	MA	1	2	2	1	1	0.4312	4.88
Q	MA	1	2	2	2	15	0.9118	49.59
Q	AR	1	2	2	1	1	0.7748	12.46
Q	AR	1	2	2	2	15	0.3974	8.26
TDS	MA	1	1	2	1	1	−0.2194	−4.36
TDS	MA	1	1	2	2	15	0.9156	40.29
TDS	AR	1	1	2	1	15	0.6522	13.40
Q × TDS	MA	1	2	2	1	1	0.4539	5.51
Q × TDS	MA	1	2	2	2	15	0.7988	18.31
Q × TDS	AR	1	2	2	1	1	0.8058	14.60
Q × TDS	AR	1	2	2	2	15	0.2425	3.51

[a]MA = Moving Average; AR = Autoregressive.
[b]The estimator for all values given is significant (<0.05).

Spatial and Temporal Co-Kriging Models

The data points estimated by after-the-fact forecasts were removed from the data sets to make co-kriging estimates. The results are also presented in Table 15.3. Data from 27 consecutive weeks for drain #15 for the spatial estimates and data from weeks #26 and #27 of all 15 drains for the temporal estimates were chosen arbitrarily.

Descriptive statistics of the data are shown in Table 15.4. The Q, TDS, and Q × TDS were first normalized by dividing each value by the mean of its set. The statistics indicate the least amount of variability and skewness in variable TDS and the highest amount in variable Q. Positive values for skewness (g) and the corresponding histograms reflect a skewness to the right in the data. However, it was decided to conduct the analysis without a log transformation, even though the latter would give a more normal tendency in the data. Kurtosis (k) indicates that Q has the largest spread in the data.

Table 15.5 lists variogram parameters for the three variables and their cross-variograms. Nugget, sill, and range estimations were manipulated by trial and error until the co-kriging program estimated missing data with the accuracy measures listed in Table 15.6. The user's judgment dictated acceptance. One step in cross-validation was to compare mean square error (MSE) and mean variogram variance (MVAR). These values should be close. Methods of testing estimates were the same as those used in time series forecasts (Tables 15.3).

Generally, time series analysis implies a temporal investigation, but here it was also applied to a spatial data series. Likewise, kriging implies spatial data estimation, but here the data will be used for a spatial and a temporal analysis by co-kriging involving three simultaneous variables. An overall comparison of temporal and spatial statistics

TABLE 15.3 Statistics for After-the-Fact Forecasts and Co-Kriging Estimates for Q, TDS, and Q × TDS

Variable (1)	Units (2)	Statistical method (3)	Model[a] (4)	Actual mean (5)	Forecast mean (6)	Mean absolute error (7)	Residual standard error[b] (8)	Mean absolute % error (9)	Mean % error (10)
Q	l/min	Time series	Temporal	6.2	8.6	3.2	5.0	146	−30.1
			Spatial	40.9	19.3	21.6	25.7	49.5	−49.5
		Co-kriging	Temporal	6.2	6.2	3.2	4.8	194	−171
			Spatial	40.9	12.1	2.9	9.6	70	70
TDS	mg/l	Time series	Temporal	1,360	1,320	54	84	3.7	−2.3
			Spatial	480	780	300	314	63.5	63.5
		Co-kriging	Temporal	1,360	660	700	395	51	51
			Spatial	480	1,140	660	72	138	−138
Q × TDS	mg/min	Time series	Temporal	6,110	8,770	3,150	4,770	59	23
			Spatial	19,680	13,940	6,120	7,230	28	24
		Co-kriging	Temporal	6,110	5,310	3,410	4,800	107	−70
			Spatial	19,680	12,900	6,890	4,510	33	32

[a]Temporal values apply to all 15 drains for weeks 26 and 27, and spatial values apply to all 27 weeks for drain 15.
[b]Standard error of the estimate.

TABLE 15.4 Descriptive Statistics for Q, TDS, and Q × TDS of Co-Kriging Example[a]

(1)	Q (2)	TDS (3)	Q × TDS (4)
Mean	1.00	1.00	1.00
Maximum	5.25	2.74	5.19
Minimum	0.017	0.258	0.047
VAR	1.05	0.180	0.678
g	1.68	0.100	0.690
k	5.80	0.172	2.16

[a]All original data for each parameter were normalized through dividing by their means.

TABLE 15.5 Variogram Parameters for Co-Kriging Example

	Temporal			Spatial		
Variable (1)	**Nugget** (2)	**Sill** (3)	**Range** (4)	**Nugget** (5)	**Sill** (6)	**Range** (7)
Q	0.06	1.06	40	1.06	1.06	1
TDS	0.00	0.18	40	0.10	0.18	2
Q × TDS	0.50	0.68	30	0.68	0.68	1
Q + TDS	0.06	0.73	40	0.73	0.73	1
Q + Q × TDS	0.07	3.18	35	3.18	3.18	1
TDS+Q × TDS	0.05	0.57	35	0.57	0.57	1

between time series techniques and co-kriging (Tables 15.3) favors time series. For variable Q, the temporal estimate looks better than time series forecasts for the first four statistics but poorer for the last two. Inspection of temporal and spatial results for all three variables shows varying degrees of accuracy. The multivariate nature of co-kriging may be one reason.

With time-series analysis, model building is considered complete when the error (white noise) is random and the Q-statistic is less than the Chi-square-value. However, tests for acceptability of co-kriging models are not available in the referenced software. The results of MVAR and MSE, shown in Table 15.6, are one guide for model acceptability. A second testing procedure can be used when the variogram choices are made by making histograms of individual error distribution to check for randomness. For the example data set, histograms of individual error distributions showed that not all means equal zero and not all distributions are random as required.

If manipulation of the variogram parameters (Table 15.5) will not improve co-kriging estimates, the modeler may have to return to the original data and consider options for improving the randomness, such

TABLE 15.6 Mean Square Error[a] (MSE) and Average Co-Kriging Variance (MVAR) for Example Data Set

Variable	Temporal Case		Spatial Case	
	MSE	MVAR	MSE	MVAR
(1)	(2)	(3)	(4)	(5)
Q	0.15	0.12	0.92	1.18
TDS	0.01	0.00	0.13	0.17
Q × TDS	0.14	0.09	0.60	0.74

[a]Cross validation:

$$MSE = 1/405 \cdot \sum_{i=1}^{405}(Z_i - Z_i^*)^2 \text{ where } i$$

$= (1,2,3,. . .405)$, Z^* is estimated value, and Z is original value.

as data transformation, or remove outliers in the construction of variograms. If the outliers are removed for the variograms, they should be replaced for co-kriging. Sometimes, neither removal of outliers nor data transformation permits successful modeling.

Intervention Analysis Modeling

An intervention analysis involves the evaluation of an ARIMA model of a bivariate case in which the second variable represents an important consideration or action. This will be illustrated using sodium adsorption ratio (SAR) values from the drainage system in the aforementioned example. Here, the first variable is SAR of drainage water and the second variable will be irrigation. If irrigation occurs within the one week before sampling, the second variable is given the value of 1. If not, then the variable is given a value of 0. Significant influence of irrigation on the SAR level of drainage water would, in turn, give insight into water-quality management. Drainage water disposal could be implemented over the specified period when the SAR is high, while conversely allowing beneficial use while the drain water has a low SAR value.

Time series ARIMA techniques from BMDP2T software (Liu 1981) produced the output. Results for the univariate spatial case (the intervention variable has been excluded) and for the bivariate equivalent (when the intervention variable is present) are shown in Table 15.7. Similarly for the temporal case, both univariate and bivariate model results appear in Table 15.8.

Analysis is made through statistical significance tests of the models. The t-tests of the irrigation variable indicate a non-significant intervention in both the temporal and spatial cases; a higher drainage flow after irrigation had no significant influence on SAR relative to the stated times and spacing of sampling. Moreover, little has changed between corresponding ARIMA model information of univariate and bivariate cases, as demonstrated by a comparison of the data within Table 15.7 and Table 15.8. The Q-statistic and the Chi-square-values are footnotes

TABLE 15.7 Summary of ARIMA $(2,2,3)_{1,27}$ Model for SAR_{dw} Spatial Series—Before and After Applying Intervention Function (Ahsan and Guitjens 1988)

(1)				Before Intervention Function		After Intervention Function	
Parameter (1)	Type (2)	Factor (3)	Order (4)	Estimate (5)	t-ratio[a] (6)	Estimate (7)	t-ratio[a] (8)
SAR_{dw}	MA	1	1	0.6493	7.22	0.6554	7.35
SAR_{dw}	MA	2	21	0.2577	5.06	0.2589	5.07
SAR_{dw}	MA	3	27	0.9125	61.06	0.9117	61.03
SAR_{dw}	AR	1	1	0.3041	2.68	0.3117	2.76
SAR_{dw}	AR	2	11	−0.1393	−2.69	−0.1389	−2.67
Irrigation	Upper polynomial	1	0	—	—	−0.8145	−0.95[b]

(2)	Before Intervention Function	After Intervention Function
Number of observations	405	405
Degrees of freedom	360	359
Mean of the residuals for differenced series	−0.1155	−0.1064
Q-statistics for residual ACF (lags 1 to 81)	60.36	61.49
Chi-square	97.3	96.20
Degrees of freedom at 0.05-level	76	75

[a]All parameter estimates significant at 0.05-level of t-test.
[b]Parameter estimate not significant at 0.05-level of t-test.

to Tables 15.7 and 15.8. The conditions stipulated in Eq. 15.1 were satisfied. They support the requirement that the ACF of the residual series be random and non-significant.

SPATIAL VARIABILITY OF SOILS AND SOIL SALINITY

The following case studies illustrate several statistical methods of analyzing spatial variability of soil salinity. The first study uses soil chemical characteristics from soil samples collected along transects in Blocks A and B of the field in Fig. 15.2. The second uses transect data from a field in California. The added features in the second study are the use of cross-correlation and spectral and co-spectral analysis.

Statistical Variability and Spatial Modeling

Two fields were sampled for soil chemistry characteristics every 10 m along a 500-m transect. Table 15.9 shows the characteristics, their means and coefficients of variation for four sampling depths, and the

TABLE 15.8 Summary of ARIMA $(2,1,2)_{15}$ Model for SAR_{dw}, Temporal Series—Before and After Applying Intervention Function (Ahsan and Guitjens 1988)

(1)				Before Intervention Function		After Intervention Function	
Parameter (1)	Type (2)	Factor (3)	Order (4)	Estimate (5)	t-ratio[a] (6)	Estimate (7)	t-ratio[a] (8)
SAR_{dw}	MA	1	1	0.5825	8.61	0.5858	8.80
SAR_{dw}	MA	2	15	0.5019	11.01	0.5045	11.03
SAR_{dw}	AR	1	1	0.8970	22.93	0.9020	23.69
SAR_{dw}	AR	2	8	−0.1550	−3.29	−0.1578	−3.35
Irrigation	Upper polynomial	1	0	—	—	−0.5246	−0.77[b]

(2)	Before Intervention Funotion	After Intervention Funotion
Number of observations	405	405
Degrees of freedom	377	376
Mean of the residuals for differenced series	−0.2289	−0.2308
Q-statistics for residual ACF (lags 1 to 60)	70.30	70.78
Chi-square	74.45	73.29
Degrees of freedom at 0.05-level	56	55

[a]All parameter estimates, except where noted, significant at 0.05-level of t-test.
[b]Parameter estimate not significant at 0.05-level of t-test.

maximum separation distances (lag) with significant autocorrelation. Each characteristic can be modeled with time series techniques and kriging as before.

Variability, as expressed by the coefficient of variation, is small for pH and much more inconsistent and larger for the other characteristics. Autocorrelation, as indicated by lags with significant autocorrelation, is inconsistent and frequently less than one separation distance of 10 m. Smaller sampling distances could reveal whether a stronger autocorrelation exists over shorter distances, but such sampling intensity may be impractical or uneconomical.

Table 15.10 shows the results of a Box-Jenkins time series analysis of the electrical conductivity of the soil extract in field A for the sampling increment of 0 mm to 150 mm (Table 15.9). The differencing filter is I with order $d = 3$ and the autoregressive filter is AR with order $p = 2$. There was one significant spike in the ACF and PACF of the original series and one significant spike in the ACF of the differenced series. A significant spike in the PACF of the differenced series yielded a non-significant coefficient, hence, $q = 0$. Standard errors of all 18 after-the-fact forecasts equalled zero. No tests were performed for random-

TABLE 15.9 Variability of Soil Chemical Characteristics in Two Adjacent Fields (Guitjens et al. 1986)

	Lag (meters × 10)				Mean				Coefficient of Variation (%)			
Depth[a] (mm) (1)	0–150 (2)	150–300 (3)	300–450 (4)	0–450 (5)	0–150 (6)	150–300 (7)	300–450 (8)	0–450 (9)	0–150 (10)	150–300 (11)	300–450 (12)	0–450 (13)
FIELD A												
EC_e (dS/m)	0	1	1	4	0.42	0.34	0.30	0.35	26	32	27	31
pH	2	1	2	4	7.4	7.4	7.5	7.5	5	4	5	5
Ca (meq/l)	0	0	1	4	2.3	1.6	1.5	1.8	39	55	45	50
Mg (meq/l)	2	0	1	2	0.7	0.6	0.5	0.6	53	107	63	80
Na (meq/l)	0	4	1	4	0.8	0.7	0.8	0.7	80	42	59	65
SAR	0	4	1	1	0.8	0.7	0.8	0.7	128	48	68	90
FIELD B												
Ec_e (dS/m)	1	1	0	1	0.66	0.55	0.56	0.60	42	35	54	45
pH	1	3	0	3	7.7	7.7	7.9	7.7	2	2	3	3
Ca (meq/l)	1	1	1	4	4.7	3.4	2.8	3.6	65	37	58	62
Mg (meq/l)	1	1	0	4	1.2	0.8	0.7	0.9	61	39	64	62
Na (meq/l)	1	3	0	0	1.6	1.3	2.3	1.7	111	55	167	146
SAR	1	0	0	1	0.9	1.1	1.6	1.2	92	151	153	148

[a]Data for depths 0 mm–150 mm, 150 mm–300 mm, and 300 mm–450 mm consist of 51 samples; data for 0 mm–450 mm consist of 153 samples.

TABLE 15.10 Summary of ARIMA $(1,1,0)_3$ Model for EC_e (dS/m), Spatial Series—1984

Parameter (1)	Type (2)	Factor (3)	Order (4)	Estimate (5)	t-ratio (6)
EC_e	AR	1	1	−0.32	−2.25[a]

[a]The estimator in the model is significant (<0.05).

ness of forecast errors because the white noise series had no significant spikes.

The corresponding variogram model had characteristics as follows: nugget = 0; sill = 0.012; and range = 80 m. The modeled data have a mean square error of 0.012 and a mean variogram variance of 0.00.

A kriging model could be used to estimate values at intermediate points between observation locations. Other options for modeling and estimation could include the use of any of the other three depth increments (Table 15.9) or estimation at any spatial distance for each depth increment.

Correlation and Spectral Analysis Modeling

Data were collected from an experiment aimed at evaluating the effect of a pre-plant irrigation on salt accumulation from the previous summer's irrigations with saline water and determining the depth of salt accumulation. In March, 1985, soil salinity measurements were made along a transect crossing three blocks, each having six treatments of varying irrigation water salinity. The sequence of salinity treatments in each block was 9,000 mg/l, 4,500 mg/l, 1,500 mg/l, 400 mg/l, 3,000 mg/l, and 6,000 mg/l. Each block was 100 m wide and of 3 ha in area. Measurements were made with a four-electrode salinity probe at 30-cm, 60-cm, and 90-cm depths. The sampling interval was 5 m. Software by Shumway (1988) was used for the statistical analysis.

Salinity at 30 cm ranged from 1.5 dS/m to 2.0 dS/m (Fig. 15.5). Valleys in the data at about 30 m, 130 m, 230 m, and 330 m along the transect suggests a periodic behavior at that depth. The salinity at 90 cm showed a higher mean value and considerably more variance. Patterns of variability at 90 cm were not readily apparent.

The correlograms for three depths are shown in Fig. 15.6. Fig. 15.6a shows a large peak at lag = 95 m in the correlation for the 30-cm depth. This indicates a cyclic behavior every 95 m (about equal to the width of one block). Fig. 15.6b for the 60-cm depth shows a somewhat similar result. The plot, Fig. 15.6c, for the 90-cm depth is different, with slight peaks at 35 m and 70 m.

Power spectra plots, shown in Fig. 15.7, show that the highest cross-correlation (r = 0.81) occurred at lag = 0 between depths 30 cm and 60 cm. A much smaller value of r = 0.25 occurred for the co-spectral analysis between depths at 30 m and 90 m.

A plot of the spectral analysis, the power spectrum, for the 30-cm depth (Fig. 15.8a) shows a very large peak for frequencies corresponding

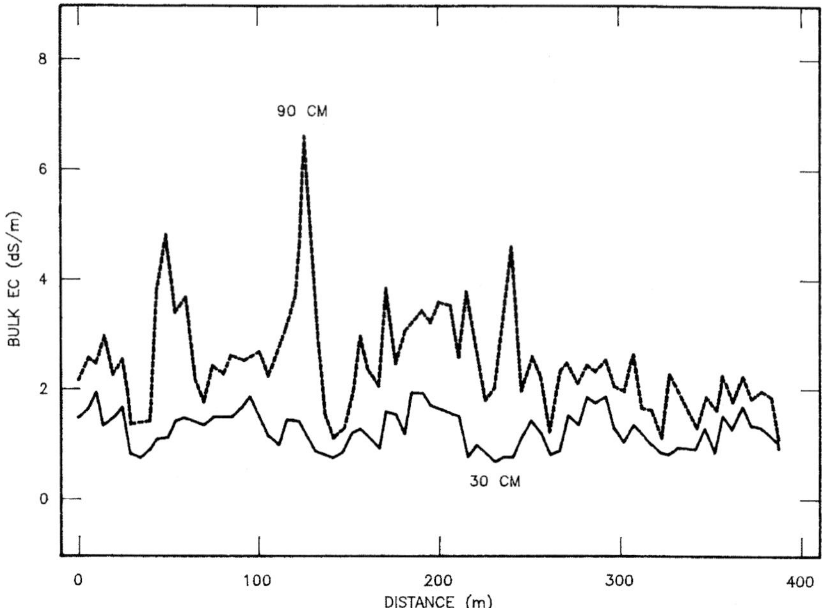

Fig. 15.5 Salinity Along a Transect Crossing Four Blocks

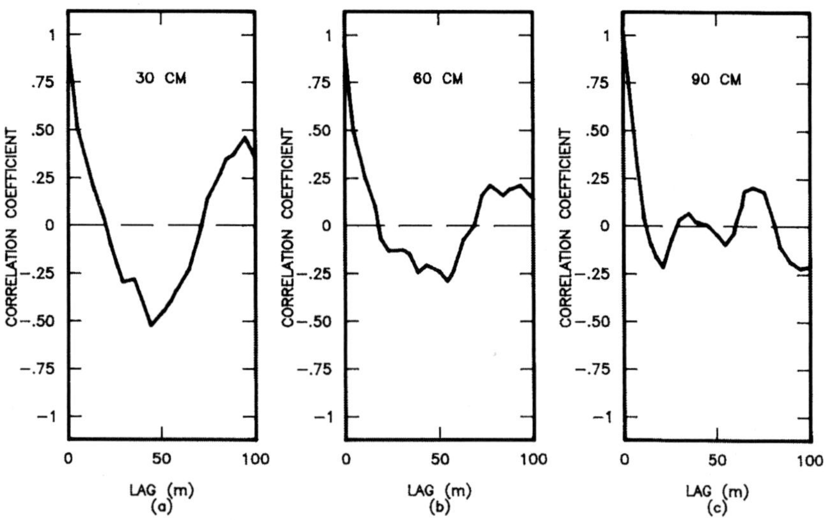

Fig. 15.6 Correlograms of Salinity Data at 30-cm, 60-cm, and 90-cm Depths

to periods of 80 m to 106 m. Calculations showed that 54% of the data variance is contributed by that frequency interval. The remaining variance appeared to be random. The spectrum of the 60-cm depth (Fig. 15.8b) also shows a large peak over the same frequency interval but with

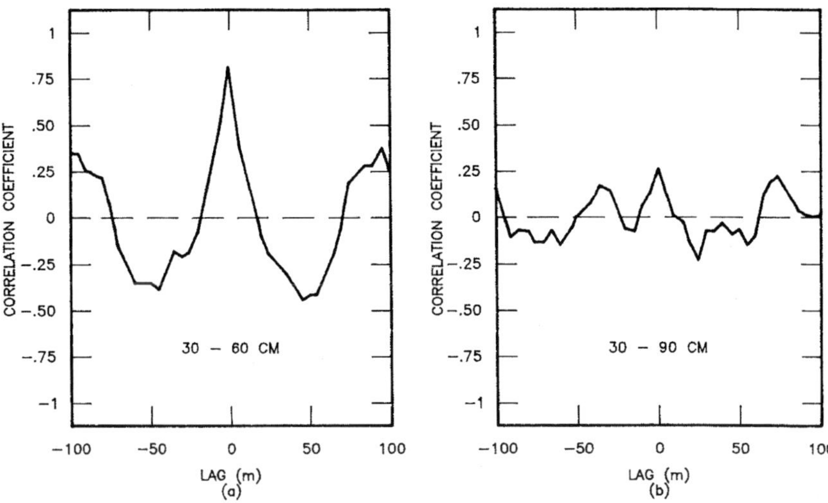

Fig. 15.7 Power Spectra of Salinity Data at 30-cm, 60-cm, and 90-cm Depths

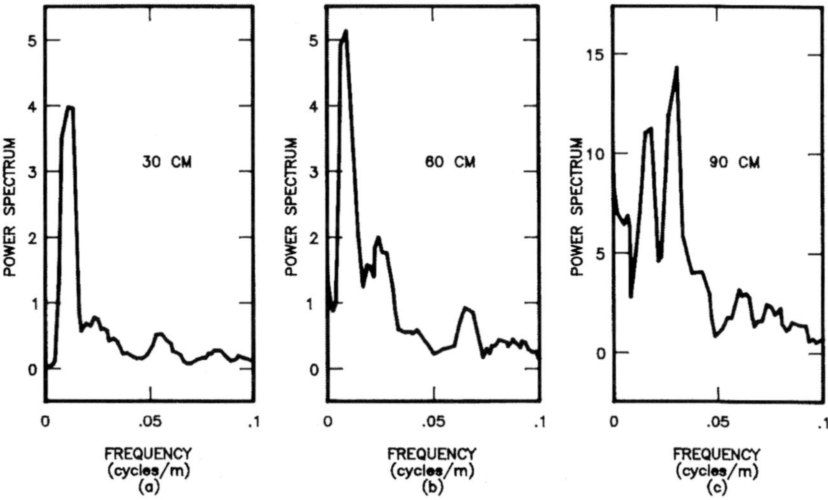

Fig. 15.8 Cross-Correlograms Comparing 30-cm to 60-cm Depths and 30-cm to 90-cm Depths

additional peaks corresponding to periods of 36 m to 43 m and a small peak at period 16 m. The power spectrum of the 90-cm depth (Fig. 15.8c) shows very different patterns of variance. Several large peaks, unrelated to those in the other spectrums, occur at frequencies corresponding to periods of about 36 m and 64 m to 71 m.

The co-spectrum of the 30-cm to 60-cm depths (Fig. 15.9a) shows that most of the covariance occurs at frequencies corresponding to 80 m

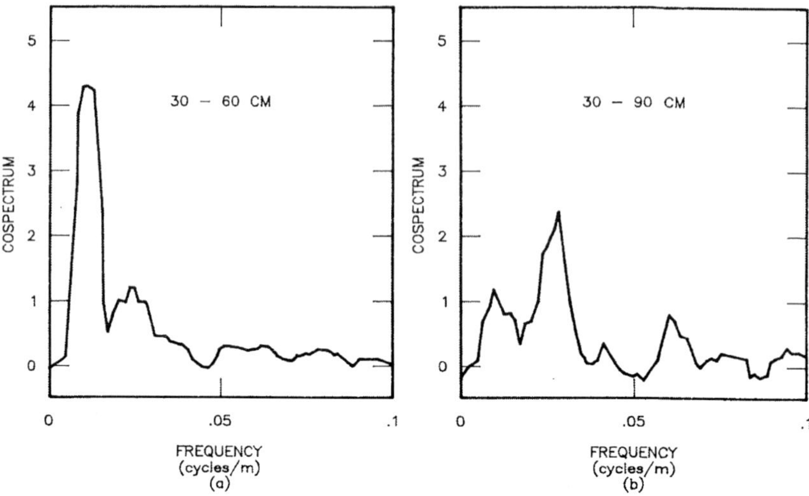

Fig. 15.9 Cospectra Describing Covariance Contributions with Frequency for 30-cm to 60-cm Depths and 30-cm to 90-cm Depths

to 106 m. The co-spectrum of the 30-cm to 90-cm depths (Fig. 15.9b), however, revealed little covariance between the two depths at that frequency.

Spectral analysis reveals a cyclic recurrence in soil salinity over intervals of about 80 m to 106 m at the 30-cm and 60-cm depths. This behavior corresponds to block width and a constant replicate structure in each block. The 90-cm depth was not strongly affected.

SUMMARY

The material presented in this chapter introduces engineers to statistical techniques that are not commonly used today. With time-series models, the quality characteristics of water or soil over time or space can be forecasted. Similarly, kriging can be used to provide estimation models. Intermediate information in model development includes the autocorrelation model, which can be used to determine the maximum lag with significant correlation in the data series, and the variogram model, which can be used to determine the range. These and other statistical characterizations are becoming increasingly important in agriculture because they provide information for evaluating the effectiveness of water quality management practices.

REFERENCES

Ahsan, M.A., and Guitjens, J.C. (1988). "Irrigation intervention analysis of drainage SAR." *ASCE Proceedings of Irrigation and Drainage Division Conference,* p 331–340.

Bernstein, L. (1974). "Crop growth and salinity." *Drainage for Agriculture.* van Schilfgaarde, J., ed. ASA Monograph No. 17:39–54.
Bouwer, H. (1987). "Effect of irrigated agriculture on ground water." *J. Irrig. and Drainage Engrg., ASCE,* 113(1): 4–15.
Box, G.E.P., and Jenkins G.M. (1976). *Time Series Analysis: Forecasting and Control,* 2nd ed. Holden-Day, Inc., San Francisco, Calif.
Carr, J.R., Myers, D.E. and Glass, C.E. (1985). "Co-kriging—A computer program." *Computers and Geosciences* 11(2): 112–127.
Chatfield, C. (1975). *The Analysis of Time Series: Theory and Practice.* John Wiley & Sons, Inc., New York, N.Y.
Clark, I. (1979). *Practical Geostatistics.* Applied Science Publishers, Ltd, London, England.
David, M. (1982). *Geostatistical Ore Reserve Estimation.* Elsevier, New York.
Davis, J.C. (1973). *Statistics and Data Analysis in Geology.* John Wiley and Sons, Inc., New York.
Guitjens, J.C. (1987). "Monitoring drainage for water quality evaluation." ASAE Proc. Fifth National Drainage Symposium, pp. 178–188.
Guitjens, J.C., Leedy, C.D. and Thran, D.F. (1986). "Grain yield variability under scheduled irrigation." ASAE 1986 Summer Mtg., San Luis Obispo, Calif. 12 p.
Hoff, J.C. (1983). *A Practical Guide to Box-Jenkins Forecasting.* Wadsworth, Inc., Calif.
Hoffman, G.J., Ayers, R.S., Doering, E.J. and McNeal, B.L. (1980). "Salinity in irrigated agriculture." *Design and operation of farm irrigation systems.* Jensen, M.E., ed. ASAE Monograph No. 3: 145–188.
Jenkins, G.M., and Watts, D.G. (1968). *Spectral Analysis and its Application.* Holden-Day, Inc., San Francisco, Calif.
Journel, A.G. and Huijbregts, C.J. (1976). *Mining Geostatistics.* Academic Press, New York.
Liu, L.M. (1981). "Box-Jenkins time series analysis." *BMDP Statistical Software.* Dixon, E.J., ed. University of California Press, Berkeley.
McDowall, D., McClearly, R., Meidinger, E.E., and Hay, R.A. (1987). "Interrupted time series analysis." Sage University Paper, 21, Sage Publications, Inc., Beverly Hills, Calif.
Nielsen, D.R., Tillotson, P.M., and Vieira, S.A. (1983). "Analyzing field-measured soil-water properties." *Agricultural Water Management,* 6:93–109.
Reno Gazette-Journal. (1987). "Bird carcasses torched in effort to kill bacteria." February 19.
Ryan, T.A., Joiner, B.L. and Ryan, B.F. (1982). *Minitab.* The Pennsylvania State Univ. State College, Pa.
Schmidt, K.D., and Sherman, I. (1987). "Effect of irrigation on ground water quality in California." *J. Irrig. and Drainage Engrg., ASCE,* 113(1): 16–29.
Shumway, R.H. (1988). *Applied Statistical Time Series Analysis.* Prentice-Hall, N.J.
Sonnen, M.B., Thomas, J.L. and Guitjens, J.C. (1987). "Irrigation effects in six Western States." *J. Irrig. and Drainage Engrg., ASCE,* 113(1): 57–68.

Tanji, K.K. (1981). "California return flow case studies." *J. Irrigation and Drainage Div., ASCE,* 107(IR2): 209–220.

Tanji, K., Lauchli, A and Meyer, J. (1986). "Selenium in the San Joaquin Valley: a challenge to western irrigation." *Environment* 28(6): 6–39.

The Sacramento Bee. (1985). Selenium. Sept. 8–10.

NOTATION

C = covariance; C_o = nugget, also known as noise or micro-spatial variability; d = differencing filter; h = separation distance; p = auto-regressive filter; q = moving average filter; Q = drain's discharge rate; Q × TDS = salt flux of drainage water; TDS = salt concentration of drainage water; X_i = sampling location or sampling event; $Z(X_i)$ = value of an observation at location X_i; λ_i = weights that represent a computer-generated matrix calculation involving the variogram's covariance model and a Lagrangian multiplier; and μ = error.

VI. SALINITY MANAGEMENT OPTIONS

CHAPTER 16

ON-FARM IRRIGATION AND DRAINAGE PRACTICES[a]

INTRODUCTION

Irrigators have many options for managing their irrigation and drainage systems. The options that they choose significantly affect the salinization of the soil and the depth of the water table, which, in turn, influence the productivity and profitability of their farms. On-farm water management decisions can also change the quantity and quality of surface and subsurface return flows, affecting both neighboring farms and the general environment.

This chapter contains brief descriptions of the most common types of irrigation and drainage systems. It outlines management options that help to prevent or correct salinity problems and minimize the build-up of water tables, demonstrates the importance of drainage for salinity management, and describes methods for management of shallow water tables.

IRRIGATION AND SALINITY CONTROL

Irrigation must be adequate, over the long term, to prevent harmful accumulations of salt in the root zone. To prevent high water tables, which often contribute to salt accumulations at the soil surface, irrigation applications must not be excessive. Infiltrated depths of water must be relatively uniform to meet the crop's needs and leach salts adequately, without excessive surface runoff or deep percolation. To meet such depth and uniformity requirements, irrigation systems must be suited to the site, well-designed, and well-managed.

Surface Irrigation

In surface irrigation systems, such as the furrow, border dike, and basin systems described below, water flows over the field's surface and part of the water infiltrates as the stream advances. When the water has advanced to the end of the field, runoff may occur or the water may pond on the soil surface. This ponding will continue until inflow is

---

[a]Prepared by: E. Gordon Kruse, USDA-ARS, Agric. Engrg. Res. Center, Foothills Campus, Fort Collins, CO 80523; Lyman Willardson, Dept. of Agric. and Irrig. Engrg., Utah State Univ., Logan, UT 84332; and James Ayars, USDA-ARS Water Management Lab., 2021 S. Peach Avenue, Fresno, CA 93727.

stopped and all surface water has infiltrated. For uniform infiltration, the variability of soil across the field must be small and the infiltration opportunity times must be uniform in all parts of the field.

Furrow and Corrugation Irrigation. Furrow irrigation is commonly used for trees, vines, and annual row crops. Corrugations, similar to furrows, but with smaller cross sections, are used for alfalfa, small grains, and other close-growing crops. Furrow and corrugation irrigations can be uniform if advance times to the end of the field are small, perhaps one-fourth to one-fifth of the time needed to infiltrate the necessary water. Advance times can be decreased by using the maximum flow rate of the nonerosive irrigation stream and by shortening field lengths. Excessive runoff can sometimes be prevented by reducing the size of each furrow stream after advance is complete.

Mathematical techniques for simulating the advance and recession of the stream and depth of infiltration have been under development since the mid 1970's. Such techniques can be used to determine the optimal stream size and water application time for irrigation of furrows. However, the infiltration conditions must be known accurately and expressed mathematically. Infiltration varies from irrigation to irrigation, year to year, and furrow to furrow. Choosing measurement sites that can represent an entire field is difficult, and direct measurements of infiltration are time-consuming. Therefore, supplying representative intake information for simulation techniques can be difficult. Walker and Skogerboe (1987) summarize the simulation procedures and list primary references.

On some soils, especially those with coarser textures, surge flow (cycling the furrow inflow stream on and off) causes advance to be completed with less water than with continuous furrow inflow (Stringham and Keller 1979). More uniform infiltration results. Cablegation delivers a continuously decreasing inflow rate to each furrow (Kemper et al. 1981). The high initial furrow inflow rate contributes to rapid stream advance; the low final rate minimizes runoff.

Border Irrigation. Border dike irrigation is used primarily for close growing crops. The land is prepared by constructing parallel earthen dikes to contain the irrigation streams. The dikes are generally parallel to the greatest land slope. When this slope exceeds 4% for sod-forming crops, and 2% for other crops, the dikes may be angled to reduce slope in the direction of water flow. Slight cross slopes (up to 0.1%) are allowable if streams are sufficiently large and slopes in the direction of flow are sufficiently gradual. The irrigation stream is usually introduced from a single turnout at the upslope end of each strip. Streams need to be relatively large to cover the width of the border and to advance to the end in reasonable time. Stream size and application time are set so that recession rates will parallel advance rates to permit uniform depth of infiltrated water. For detailed information on border irrigation design, refer to the Soil Conservation Service National Engineering Handbook (U.S. Department of Agriculture 1974). Strelkoff (1985) outlines programs for mathematical simulation of border irrigation, similar to those described above for furrow irrigation and with similar limitations.

Basin Irrigation. Basins are used to irrigate close-growing crops, orchards, and vineyards. Each basin surface is leveled to eliminate cross slope and leave little or no slope in the direction of irrigation. Each basin is surrounded with dikes and may be any shape. To irrigate, a predetermined volume of water is introduced at a high flow rate from one or more inlets along one end of the basin. When water has covered the basin, it ponds and remains ponded until infiltration is complete. A variation, basin-furrow irrigation, can be used to irrigate row crops. The furrows are similar to those for sloping furrows, but they have no slope, and are larger to accommodate the larger stream sizes. For detailed information on level basin design, see the aforementioned Soil Conservation Service National Engineering Handbook.

Land Leveling. Earth moving on irrigated fields to improve surface slope and smoothness is often referred to as land leveling. It establishes uniform field topography, even though all slope may not be removed. Land planing, including laser leveling, removes minor irregularities in the surface just before planting. Land smoothing removes high and low areas but leaves the field with variable slope. Land leveling will produce uniform slopes in the direction of irrigation-stream advance and may totally eliminate slope for level basins. Land planing and land smoothing improve conditions for salinity control by improving uniformity of surface irrigation. Thus, water for salt leaching can be applied more effectively, and deep percolation from non-leaching irrigations can be minimized, restricting the water table rise.

When soil is moved during land leveling to create a field of constant or zero slope, the depth of soil removed from some portions of the field may be a significant fraction of the root zone depth. Subsoils that are highly saline or low in plant nutrients may be exposed. Earth-moving cuts may also lower the soil surface close to slowly permeable soil lenses or perched water tables. If the lowered soil surface is within the capillary fringe of a water table, preventing salt accumulation in the soil surface will be difficult. Thus, soil profiles and ground-water conditions should be carefully investigated before significant leveling is begun. Surface soils may be removed and stockpiled before reconfiguring and replaced on the field afterward.

Distribution of Applied Irrigation Water in Soil. With surface irrigation, intake opportunity time and, thus, the infiltrated depth of water is normally greatest at the upstream end of each furrow, border, or basin. If applications merely replace the soil-water deficit at the upstream end of a furrow, the downstream end will be under-irrigated. Refilling the entire profile at the downstream end causes deep percolation upstream. Furrow irrigation uniformity can be increased by using larger furrow streams for faster stream advance or by reducing furrow length. If needed, tailwater recovery systems can be used to allow surface runoff to be reused.

Infiltrated water from furrows is not distributed uniformly in a direction perpendicular to the furrows because of nonuniform soil conditions and variable furrow geometry. On some soils, movement of infiltrated water from the furrow to the adjacent ridges is limited. This poses no problem for plants with well-developed root systems but may hamper seed germination or cause stress in young plants. In such cases, shallower furrows or longer water application times can be used to increase lateral water movement to the centers of the ridges.

If irrigation water is applied to all furrows equally, less infiltration will occur in furrows compacted by tractor traffic. The lateral distribution of infiltrated water after irrigation will vary from site to site, being less uniform for coarser textured soils and every-other-furrow irrigation.

Distribution of infiltrated water is most uniform for border dike systems if the borders have low cross-slope and evenly sloping or slightly convex profiles in the direction of flow. Irrigation stream size and duration may need to be adjusted for the infiltration and flow resistance conditions that exist at the time of each water application.

Water application can be highly efficient and uniform for level-basin and basin-furrow systems if leveling is precise and the application rate and volume are appropriate for the basin's size and soil conditions. Irrigation streams must be large enough to inundate the basin's surface in a fraction of the time needed for the desired net irrigation depth to infiltrate. The irrigation stream must be turned off when the applied volume of water is equal to the design irrigation application. Since no runoff can take place, the net infiltration will nearly equal the gross water application.

Salt Accumulation Patterns. Soluble salts in the soil are dissolved and carried in solution by infiltrated water. When the water is adsorbed by roots or evaporated from the soil surface, most of the salts are left behind, precipitated or concentrated in the remaining soil water. Where deep percolation occurs, the concentrated saline water leaches below the root zone.

Salinity can vary widely from the bottom of furrows to the tops of ridges. Capillary flow carries infiltrated water into the ridges, where water use by plants increases the salt concentration and evaporation leaves salt deposits at the soil surface. This accumulation of salt is often great enough by the end of an irrigation season to hamper seedling development in the following year. Fig. 16.1 shows typical patterns of salt accumulation in ridges between furrows. It illustrates seed locations where young plants can avoid the highest salt concentrations. Plowing or other tillage can redistribute this salt to allow continued crop production. Variation of salinity in the direction of furrows is generally unimportant if the infiltration is uniform enough to meet crop water needs.

For border and basin irrigation, in the absence of high water tables, water and salt flows downward, and significant surface salt accumulations are unlikely. Salt distribution in basin furrows is much like that with graded furrows.

Sprinkler Irrigation

Types of Systems. Sprinkler systems can be categorized as set or mobile. With set systems, the sprinkler heads are stationary while applying water. As the name implies, mobile systems move continuously, in

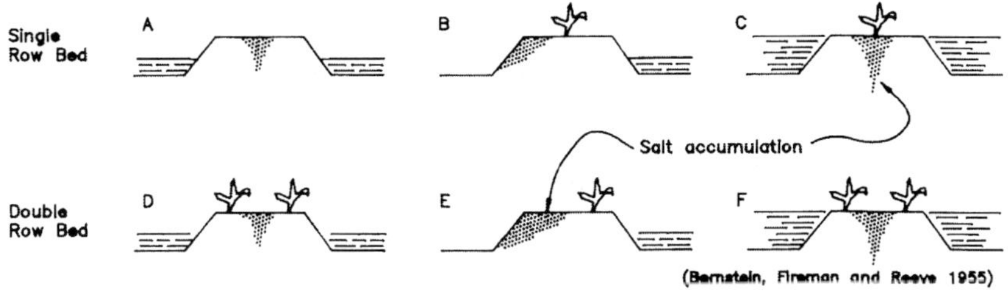

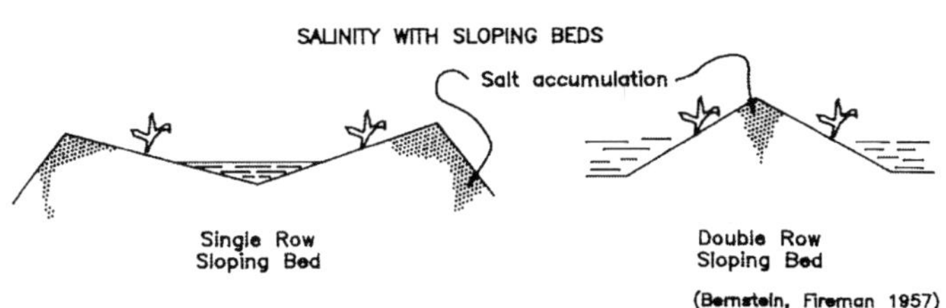

Fig. 16.1 Typical Salt Accumulation Patterns in Ridge and Bed Cross Sections Under Furrow Irrigation (Bernstein et al. 1955; Bernstein and Fireman 1957)

straight lines or circular patterns, while irrigating. They generally cost more than set systems but require less labor.

Distribution of Applied Water. Pressure losses in lateral lines can cause different operating pressures at individual sprinklers and, thus, different discharge rates. Sprinkler heads at different elevations on fields of irregular topography may operate at different pressures. Pressure regulators or appropriately sized nozzles for individual sprinkler heads can minimize differences in discharge rates and provide uniform irrigation.

Uniformity of sprinkler application, especially for set systems, is affected by both the water application pattern of each individual sprinkler head and the degree of overlap. Computer simulation programs are available to determine optimal sprinkler spacings for any given sprinkler distribution pattern.

Windy conditions during sprinkler application distort the water distribution pattern. Set systems can be designed to maximize uniformity for any predominant wind condition, principally by selecting the proper type of sprinkler head and by correct spacing of sprinkler heads and laterals (U.S. Department of Agriculture 1983).

Water applied to the soil surface uniformly may not infiltrate uniformly. If application rates exceed the soil intake capacity, some water will pond or flow across the soil surface to finally accumulate in low areas or run off the field. If application rates everywhere are less than intake capacity and application amounts are less than the soil water

TABLE 16.1 Factors Affecting Selection of Irrigation Method Under Saline Conditions

Water application method (1)	Application (2)	Pattern of salt accumulation (3)	Leaching effectiveness (4)	Special considerations (5)
Furrow	Row crops, low to medium infiltration-rate soils.	High in ridges between furrows, may increase in direction of slope if irrigations are nonuniform.	Effective leaching beneath furrow channels, salt left in ridges. Leaching requires more water than for methods with lighter, intermittent applications.	None
Corrugation	Close-growing crops.	Leaves saltier strips between corrugation channels unless entire field surface inundated.	Similar to furrows, above.	None
Border dike	Close-growing crops.	Leaves salt in dikes that separate borders.	Areas between dikes leached uniformly, but more water required than for light, intermittent applications.	None
Sprinkler: set	Most crops, all but very fine-textured soils.	No salt concentrations in root zone, if system designed and managed properly.	Uniform leaching. Can be used to leach salt accumulations left by other irrigation methods.	May encourage disease in sensitive crops, e.g., beans. Salty irrigation water may leave harmful deposits on leaves.

(*continued*)

TABLE 16.1 Factors Affecting Selection of Irrigation Method Under Saline Conditions (*Continued*)

Water application method (1)	Application (2)	Pattern of salt accumulation (3)	Leaching effectiveness (4)	Special considerations (5)
Sprinkler: Mobile	Most crops, except trees, vines. Can be used to irrigate fields on rolling topography.	No salt concentrations in root zone, if system designed and managed well.	Uniform leaching. Same as for set sprinklers.	None
Micro irrigation (drip, trickle, sub-irrigation)	Because of high initial costs, used mostly for high value crops or crops with high irrigation labor costs.	Salt concentrates at outer fringes of the soil mass wetted by each emitter.	Soil mass wetted by each emitter is well-leached. Difficult to leach all soil to depth of root zone.	When automated for light, frequent irrigations, saline water can be used, because low matric stress compensates for osmotic stress.

holding capacity, uniformity of stored soil water will be good, even for nonuniform soils. Tillage or residue management practices may be varied to maintain adequate infiltration conditions.

Salt Accumulation Patterns. Since well-designed, well-operated sprinkler systems apply water uniformly over the complete soil surface, they tend to leach salts evenly. Surface salt accumulations are unlikely to occur. For descriptions of typical salt patterns in a soil profile during and after leaching with sprinklers or other irrigation methods, see Bresler (1981) or Sammis et al. (1979).

Micro-Irrigation

System Description. Micro-irrigation refers to irrigation methods such as drip/trickle, subsurface, bubbler, and spray irrigation. A common feature of micro-irrigation systems is water delivery to the soil near plants through a network of tubing with closely spaced, low-flow-rate emitters. Water typically infiltrates where it is applied. The soil volume wetted is, therefore, controlled by the number of application points and the lateral movement of water in the soil. Systems of emitters are easily automated, making frequent, light water applications possible without high labor costs.

A high level of management is needed because a large soil water reservoir is typically not maintained, and plants can be stressed if irrigations are improperly timed.

Distribution of Applied Water. Micro-irrigation systems do not cover the soil surface uniformly with applied water. Emitters apply water to small areas within the rooting pattern of the plants. On many fields, the limited wetting of the surface soil is an advantage, as it inhibits weed growth and evaporation from the soil surface.

In micro-irrigation with porous or perforated tubing or closely spaced in-line emitters, the wetted soil pattern is constant in the direction of the irrigation line. Such systems are especially well adapted for use with row crops, where the roots occupy the wetted areas and dry zones occur between rows. When such systems are used with densely spaced crops, tubes should be placed closely enough to minimize dry soil zones between them.

One or more emitters may be needed to apply enough water to the root zone of each tree or other large plant. Each emitter produces a volume of wetted soil that may or may not intersect adjacent wetted volumes.

Salt Accumulation Patterns. The movement of water in the soil wetted by a micro-irrigation system dictates the movement and relocation of salts. Water and salt flow from each emitter to the boundary of the wetted soil volume. Salts remain when the water evaporates or is absorbed by roots. Thus, salt concentrations become highest on portions of the soil surface and at the boundaries of the wetted soil volumes. West and Merrigan (1979) concluded that when tomatoes are drip-irrigated with saline water, irrigation must be managed to provide enough wetted soil volume with low salt concentration to minimize contact between roots and zones of high salinity.

Irrigation Management Practices for Salinity Control

Table 16.1 summarizes the most common irrigation methods, their applications, and their effects on salinity control. The following paragraphs cover additional aspects of salt management.

Crops. Different plant cultivars have different salt tolerances. The profitability of crop production can be increased by selecting the proper crop for the site's salinity conditions (Chapter 13). Meiri and Plaut (1985) review practices that maximize yield under saline conditions.

Management of different irrigation methods often depends on the crop's characteristics. Temporary sprinklers are sometimes used to germinate and establish salt-sensitive crops. Sprinklers can apply small depths of water uniformly, keeping the seed bed adequately moist and salt-free (Robinson and Mayberry 1976). Surface irrigation is then used to grow the established crop.

If the sprinkler irrigation water is saline (Chapters 11 and 13), some crops may be adversely affected by salt deposits on leaves (Maas 1985; Collier 1984). Deciduous fruit trees are especially susceptible (Hoffman et al. 1980). The salts that accumulate on leaves are largely those that remain in solution in the intercepted water at the end of an irrigation. As this water evaporates, the salts are concentrated and deposited on the foliage. Larger, less frequent irrigations may reduce the salt deposits relative to the volume of irrigation applied. Irrigation at night or another low evaporation period also minimizes salt concentration and absorption.

Planting Practices. If micro-irrigation emitters are located near individual plants of perennial crops, salts tend to move away from the roots and concentrate in intermediate soil areas. To avoid problems with germination or salt stress on seedlings of annual crops, it is important to plant precisely where previous micro-irrigation has left low concentrations of salt. Planting seeds of furrow-irrigated crops on the sides of furrow ridges may keep seedlings out of the most saline soil zone (Fig. 16.1).

Tillage. Deep plowing can redistribute salt in the soil profile. The practice should be evaluated on a small land area before plowing entire fields. Slowly permeable soil layers that prevent leaching water from flowing downward can be chiseled or ripped.

Minimum Tillage. Some minimum tillage techniques allow furrow ridges formed for one crop to remain undisturbed for the next. Planting on such ridges may place seed in the most highly saline zone of the soil profile. Seeds should be planted on sides of ridges to avoid the most saline zones. Also, ridge splitting, a minimum tillage method of planting deep in the ridge, may be used to provide a suitable seed bed for row crops. Organic residue left on the soil surface by minimum tillage practices presents a problem for furrow irrigation. It slows furrow stream advance, increasing longitudinal nonuniformity. Surge flow may be useful in furrows with large amounts of residue. Surge-caused reductions in intake rates tend to speed stream advance, compensating for the slowing caused by the residue (Evans et al., 1987).

Timing of Irrigation. Irrigation scheduling is important when management includes salinity considerations. Proper timing can help to avoid low levels of soil water that cause salts in the soil solution to become highly concentrated. Frequent water applications maintain low matric water stress, which can compensate for the osmotic stresses caused by saline water. Frequent irrigations also keep the salts moving through and away from the root mass. If irrigations are applied frequently, each irrigation must be small. Small irrigations can seldom be applied as uniformly with surface methods as with sprinkler or micro-irrigation. Shainberg and Shalhevet (1984) reviewed the effect of the frequency of saline water irrigation on yield, using saline water and concluded that higher frequencies result in higher yields.

Scheduling each irrigation application amount allows the depth of leaching water to be accurately determined and prevents excessive deep percolation. Jensen et al. (1970), Burman et al. (1980), and the American Society of Agricultural Engineers (1981) presented methods for estimating rates of ET by irrigated crops and scheduling individual irrigations. Limiting irrigation applications to amounts necessary for replacement of root zone water and leaching also helps to minimize deep percolation and buildup of the water table.

Micro-irrigation systems have helped to maintain suitable water potentials in the root zones of plants, even with saline water. Irrigation intervals of several days to allow for internal drainage are unnecessary because large soil volumes are not saturated. While frequent irrigations may allow the use of saline irrigation water, irrigators should be aware of the hazards of omitting irrigations. Matric stress may occur quickly and, added to the osmotic stress, cause immediate crop damage. If micro-irrigation causes high salt concentrations to accumulate near the soil surface, unexpected rainfall can move the salt down into the root mass. Irrigations should be scheduled during or after rains to leach the salts before they damage the crop.

Water Measurement

To follow irrigation scheduling recommendations and irrigate efficiently, irrigation-water applications should be measured. Sprinkler, level basin, and micro-irrigation systems can usually be designed so that none of the applied water runs off the irrigated area. This simplifies water measurement, because the total water volume applied per set, divided by the surface area, gives the average net depth of application. For surface irrigation on sloping fields, the net depth of application can be calculated by measuring the volumes applied to and running off each set. A less accurate alternative is to measure the water applied and multiply this by a water application efficiency determined from past measurements or experience.

Numerous types of weirs, flumes, and meters exist for measuring open channel or pipeline flows with sufficient accuracy for irrigation management. [See Robinson and Humphreys (1967) and Replogle et al. (1980).]

Leaching

A wide range of waters containing dissolved salts can be used for leaching. With waters of higher salinity, larger applications are needed to maintain a salt balance. The use of sodic waters (SAR $>$ 15) is not recommended, except on very sandy soils, since such waters may damage soil structure and reduce permeability, making continued leaching difficult. Chapter 12 discusses in detail the control of salinity in the root zone. Chapter 19 reviews chemical amendments that aid in correcting salinity and sodicity problems.

The volume of water needed for a given degree of leaching may be greater for furrow and other surface irrigation methods than for sprinkler irrigation. The minimum depth of water that can be applied uniformly by most surface methods is several times greater than the minimum for sprinkler applications. Hoffman (1980) suggests amounts for different degrees of leaching (Fig. 16.2). Small, intermittent applications

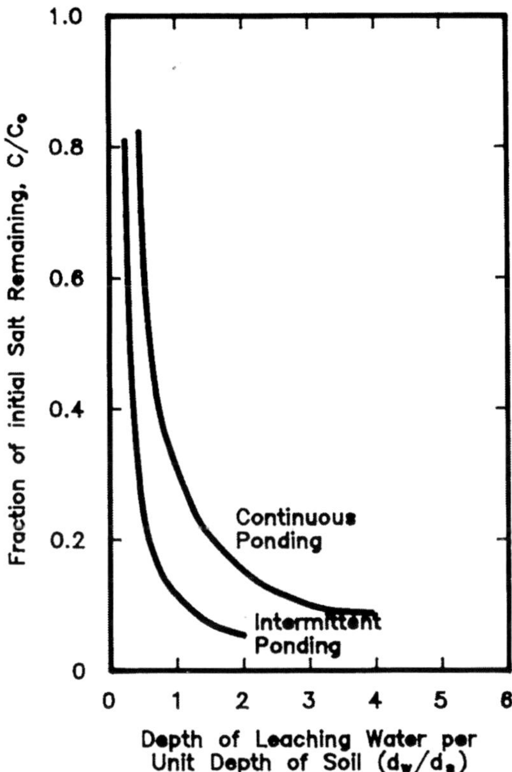

Fig. 16.2 Depth of Water per Unit Depth of Soil Required to Leach Saline Soil by Continuous or Intermittent Ponding (Hoffman 1980)

are more effective for leaching than continuous flooding applications. The unsaturated flow of soil water achieved by the former more effectively removes salt from the smaller soil pores. Much of the flow achieved by the latter occurs in large pores and root channels.

With level basins, water can be applied uniformly over the entire surface of the soil. If the soil water deficit in the root zone at the time of irrigation is known accurately and is uniform, the portion of the application that will cause leaching can be precisely controlled.

Border irrigations require greater care to adjust flows to varying infiltration rates and application depths.

Sprinkler irrigation also applies water uniformly over the surface of the soil. Application depths are easily controlled. Sprinklers can be operated just long enough to replace the moisture deficit of the root zone and apply the desired additional depth for leaching. Uniformity of application must be considered. Application rates must not exceed the soil intake capacity.

Some types of micro-irrigation systems, especially those using mini-sprinklers, wet most of the soil surface and move salts downward. For most types of micro-irrigation, however, salt moves and accumulates in a more complex manner. The type and spacing of emitters, soils, and crop rooting patterns vary, making general recommendations for leaching difficult. Salts should be carefully monitored by direct soil measurements or by frequent, careful observations of crop conditions. Under-irrigation should be avoided and additional amounts of water applied for leaching when salt accumulations are observed.

Long-term use of micro-irrigation may create large, irregularly distributed salt accumulations. If a crop with different row spacing or rooting pattern is to be produced on such an area, special leaching methods may be needed. Similarly, nonuniform surface irrigation applications may cause zones of salt concentrations that are inadequately leached. In such cases, portable sprinkler systems can be used every few years.

DRAINAGE AND SALINITY CONTROL

Drainage of Irrigated Fields

Applying irrigation water to soil disturbs the natural hydrologic balance of the soil profile. Since irrigation water cannot be applied with total uniformity, some water will percolate below the root zone. If the amount of deep percolation is less than the natural drainage capacity of the soil, the water table will remain low and the net movement of salt in the profile will be downward. If deep percolation exceeds the natural drainage capacity of the soil, the water table will rise. Soil layers of low permeability may restrict percolating water from flowing downward and cause perched water tables. When the water table is too close to the soil surface in an arid region, water and salt will be carried upward by capillary action and the upper soil profile and surface may become salinized as the water evaporates. Annually, if enough irrigation water is applied for net downward movement of water through the profile, a favorable salt balance can exist, even in the presence of a high water table. The shallower the water table, the more care needs to be taken with water applications to assure a net downward movement.

If the natural drainage capacity is so limited that normal deep percolation of irrigation water causes the water table to rise close to the soil surface, drains must be installed. Poor irrigation uniformity and excessive deep percolation increase the amount of water that drainage systems must remove.

Interception of Lateral Flow. If the subsurface geology contains shallow impermeable layers, underground water may be forced to the surface. Interceptor drains are installed perpendicular to the subsurface flow, upstream of the land to be protected, to collect excess lateral seepage water before the water table gets too close to the surface. Intercepted water must be discharged to a suitable outlet. By intercepting subsurface soil water that exceeds the soil's natural drainage capacity, the soil may more easily carry the rest of the water, helping to maintain the water table at an acceptable depth. The design and location of interceptor drains requires local topographic and ground-water surveys.

Relief Drains. The most common type of subsurface drains, relief drains usually consist of perforated pipes installed in a regularly spaced grid pattern. They are placed at a depth of one to two times the desired depth of the water table and spaced to prevent the water table from rising too close to the surface. Relief drains remove water that comes from irrigation or rainfall. Their spacing is a function of the soil permeability, drain depth, and the amount of water to be removed. The amount of water to be removed is the difference between the total infiltrated and the amount stored in the root zone for use by the crop. Leaching water, nonuniform applications, or over-irrigation will increase the drainage requirement.

Steady or non-steady state (transient) methods can be used to calculate the spacing of relief drains. Two techniques of designing relief drain depth and spacing, the Hooghoudt steady-state method and the U.S. Bureau of Reclamation transient method (Luthin 1978), must be used with corrections for equivalent depth to barrier. The calculated water table depth is adjusted according to the design method used (Food and Agriculture Organization 1980). Generally, the transient method better represents irrigated conditions.

Pumped drainage wells, closely spaced so that their areas of influence overlap, can be used as relief drains to lower water tables if subsurface geologic conditions are suitable (Luthin 1978).

Water Table Depths. Deep-rooted crops require a deeper water table than shallow-rooted crops. The water table should be deep enough to allow adequate aeration in the active plant root zone. The water table should be deeper in clay soils than in sandy soils, as the capillary rise of water is greater in the clay soils. Salt deposits on the soil surface or crops that appear to be suffering from poor soil aeration may indicate a water table that is too shallow. The desired water table depth varies with crop, soil, and spacing of drains (Food and Agriculture Organization 1980). In arid regions, drains that are placed approximately 1.8 m deep and designed to remove 2.0 mm to 5.0 mm of water per day generally maintain the water table at a depth that prevents salinization of the root zone. Such drains also provide an adequately aerated root zone for all crops.

Drainage of Nonirrigated, Saline Fields

Nonirrigated cropped areas may need subsurface drains, depending on the annual amount and distribution of rainfall, soils, and plant water use. Seeps or saline areas may develop when the amount of water present exceeds the natural drainage capacity of the soil. For example, if an area of natural vegetation is converted to cropland and then fallowed, excess water from precipitation that is not used by the crops may cause drainage and salinity problems.

Saline seeps sometimes appear on nonirrigated, arid lands when precipitation exceeds evapotranspiration during the process of soil fallowing. Excess water percolates down to the water table and then moves laterally to lower soil surfaces, dissolving salts in the material through which it flows. Subsurface conditions eventually force the flow to the surface, where it evaporates and results in a saline seep. Chapter 17 covers saline seep management.

Control of Upward Flow from Confined Aquifers

In stratified alluvial soils, confined aquifers may contain water under artesian pressure. When a well punctures the confining layer, the water may flow without pumping. Unbroken confining layers may allow slow, vertically upward seepage over large areas, creating high water tables and saline areas. Some form of relief drainage must be used to reverse the net annual flow direction in the root zone from upward to downward.

In confined aquifers, it may be possible to drill wells into the aquifer and pump them to relieve artesian pressure. Such action will reverse the direction of water flow in the overlying soil if the pressure head is lowered well below the soil surface. Gravity relief wells have been proposed to reduce artesian pressure, but they must be closely spaced and discharge into the bottom of deep drains. If the confining layer is very impermeable, the artesian pressure may have no effect other than to prevent water from moving downward, and pumped or gravity relief wells will be ineffective. In this case, relief drains above the confining layer may solve high water table problems. Luthin and Robinson (1969) discuss capillary rise from shallow water tables, including how drain depth and soil stratification affect salt movement in the soil.

Drainage Wells

Wells in unconfined aquifers can be pumped to lower the water table. Pumped wells have continuing operation and maintenance costs. If the aquifer water is of good quality and useful for irrigation or domestic purposes, the costs of lowering the water table can be offset by other benefits. The deep water table that results from pumping will eliminate salinity and drainage problems if the overlying soil has no layers of low permeability.

Water Extraction by Plants

Healthy plants rapidly remove excess water from the soil. Normal rates of evapotranspiration are two to three times greater than subsurface drainage rates in arid areas. Plants are not normally used to remove excess water from soil because crops with high economic value will suffer physiological root damage before they can remove enough water to aerate the root zone adequately.

Disposal of Drainage Effluent

Concerns and regulations about water quality may prevent the discharge and disposal of drainage effluents in natural or man-made channels. Tanji et al. (1985) describe ways to design and manage on-farm evaporation ponds. Hanson (1984) discusses general solutions to drainage water quality and disposal, especially in those areas where off-farm disposal is impossible. Drainage water must be discharged to a gravity outlet, such as an open drain, a river, or an evaporation basin. A suitable outlet must be located before undertaking any drainage work.

MANAGEMENT OF SHALLOW GROUND WATERS

Subirrigation is practiced as an active form of water table management in humid areas (Skaggs 1978). In arid and semi-arid areas, ground-water management historically has been passive and almost an incidental part of irrigation schemes. Recently, however, ground-water management has been considered as a means to use water more effectively in semi-arid regions because new knowledge indicates that crop production can be maintained with water of higher salinity than previously thought possible (van Schilfgaarde et al. 1974; Rhoades 1984), problems with water disposal have intensified the need to minimize the volume of drainage water, and other demands for water have increased the need for agriculture to maximize its use of available water. The salinities of the soil, ground water, and irrigation water influence the proper management of shallow ground water.

Operation of Shallow Ground-Water Systems

Shallow ground-water management requires a systems approach that includes the design and operation of irrigation systems and consideration of irrigation and ground-water qualities, soil salinity limits, crop tolerances, and cropping patterns.

In arid areas, the objectives of drainage are to maintain favorable oxygen availability and salinity levels in the crop root zone. Drains in arid and semi-arid areas are generally placed as deep as possible, often in excess of 2 m, and no attempt is made to control the rate of decline or position of the water table, except to maintain some minimum depth to the water table at the midpoint between drains.

Doering et al. (1982) proposed a shallow water table control concept for drainage design that allows plants to use some water from the water table. The water table control is passive, a function of drain placement and irrigation management. Based on studies of the use of water by corn and sugar beets, Benz et al. (1981) found that drain depths of 1.5 m adequately protect the crop from high water tables and provide a substantial part of the crop water requirement. Maintaining the water table at a depth of 1 m resulted in maximum yields of corn, sugar beets, and alfalfa. A design proposed by Hanson (1987) uses drain placement and outlet control on individual drain lines to regulate water tables caused by deep percolation from irrigation. Drain lines are laid perpendicularly to the direction of the field's maximum surface slope. Each line has a valve at the outlet to control drain discharge and water table depth. The drain orientation prevents excessive buildup of the water table at the field's low end. Flow-restricting devices may be installed in existing drain lines to reduce the rate of water table decline, and, thus, improve the crop's opportunity to use the water.

Salt Accumulation and Distribution

The irrigation-induced movement of salt into the profile and the transport of salt by water moving up from a shallow water table result in the pattern of salt distribution observed by Grimes and Henderson (1984). Salinity increases with depth to a value equal to the ground-water

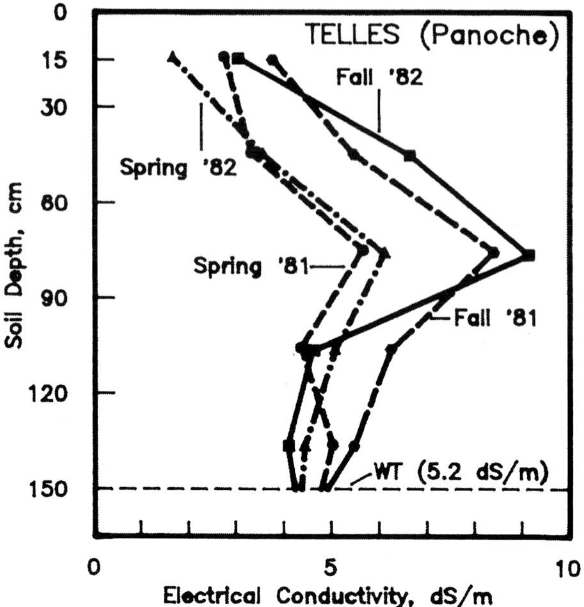

Fig. 16.3 Salinity Changes in Soil Profile over a Two-Year Period (Grimes and Henderson 1984)

salinity before irrigation (Fig. 16.3) and increases uniformly throughout the profile during the growing season. The total salt increase is a function of the water extracted from ground water, effective leaching during irrigation, and salt added by the irrigation water. A pre-season salinity profile can be restored by leaching during a fallow period. If not leached, salt will accumulate in the profile until plants can no longer grow. Rainfall and pre-plant irrigation can leach areas where drains are installed.

Management Practices

Crop Use of Saline Subsurface Water. The depth of the water table, the quality of the ground water, and the salt tolerance of the plants determine the amount of controlled, shallow ground water available to a crop. For instance, a salt-sensitive crop grown in the presence of ground water at a depth of 1 m and with an electrical conductivity (σ) of 8 dS/m probably would extract little ground water and might suffer from the salt carried up through the profile by water that evaporates from the soil surface. The more closely the crop's rooting depth and salt tolerance match the ground water's depth and salinity, the more likely the plants are to extract ground water. Maas and Hoffman (1977) tabulated salt tolerances for a wide range of crops (Chapter 13). Grimes and Henderson (1984) found that the ground water's maximum direct contribution to crop water requirement was a function of both water quality and water table depth. They attributed reduced yields with shallow water-table depths to limited aeration and restricted root volumes, while reductions under deeper water table depths were due to limited capillary movement of water to the roots.

Management of Soil Salinity. The change in profile salinity caused by the crop's water use from ground water can be approximated for a particular situation if the initial average soil salinity, irrigation water salinity, and ground-water salinity are known. Knowing the crop water requirement and proportioning water use between irrigation and ground water, the final average soil profile salinity may be estimated by

$$\sigma_F = \frac{(\sigma_I D_I + \sigma_i D_i + \sigma_{gw} D_{gw})}{D_F} \tag{16.1}$$

where σ_F and σ_I represent the final and initial average electrical conductivities (EC) of the soil solution in the profile; D_I and D_F are the initial and final depths of water stored in the profile at saturation; σ_i and σ_{gw} represent the conductivities of the irrigation and ground water; and D_i and D_{gw} are the depths of irrigation and ground water used. It is assumed that no leaching occurs, no salt precipitation or dissolution takes place, and the relation between salt concentration and electrical conductivity is approximately linear (Hoffman and van Genuchten 1983). A sample calculation for a 1.2-m profile with a 55% volumetric water content at saturation, σ_I = 7.00 dS/m, σ_i = 0.2 dS/m, σ_{gw} = 8 dS/m, D_i = 397 mm, and D_{gw} = 174 mm yields a final salinity, σ_F = 9.6 dS/m (although the average over the growing season would be 8.3 dS/m). Using this type of calculation, a manager can select crops and develop a strategy to maximize use of ground water without a loss in productivity due to salinity. The average soil profile salinity should remain below the salt tolerance level of the crop.

If ground water supplies part of the crop water requirement, the concentration of the applied water is determined on a volume-weighted basis. It is assumed that the leaching requirement, Lr, can be estimated by Lr = σ_a/σ_D^*, where σ_a = the weighted EC of the water available to the crop, including irrigation water supplied and ground water used, and σ_D^* is the desired soil salinity value (Hoffman 1985). In the example above, σ_a would equal $(\sigma_i D_i + \sigma_{gw} D_{gw})/(D_i + D_{gw})$ = 2.6 dS/m, not simply the 0.2 dS/m in the irrigation water. If the required σ_D^* was 7.0, Lr = 2.6/7 = 0.37. As the ground water becomes more saline, the potential for soil salinization appears to increase rapidly if significant quantities of ground water are used. If rainfall and pre-plant irrigation water are included, then the volume weighted concentration of applied water decreases and the Lr decreases.

To avoid raising the water table to the extent that the crop is damaged, leaching should take place during a fallow period or early in the growing season, when the crop's root system is shallow and the water demand is small (Rhoades 1984). Chapter 12 discusses the required depth of leaching.

Irrigation Scheduling. The depth of the water table in uncontrolled areas is a function of lateral inflows and deep percolation from irrigation and rainfall. Good management of irrigations will allow the crop to use some shallow ground water.

Scheduling irrigation by use of traditional water balance methods is difficult if shallow ground water exists. Quantifying the ground water contribution to crop evapotranspiration (ET) poses a problem. Most scheduling programs assume no contribution from the ground water and assume irrigation and precipitation to be the only sources of water in calculating a soil-water depletion. When the ground water does contribute, the root zone's water content decreases more slowly than estimated by such scheduling programs. Fig. 16.4 shows approximate upward flux rates from water tables at different depths in soils of various textures.

Scheduling programs that specify the allowable depth of water to be depleted before irrigation often do not consider the osmotic stress that takes place in a saline environment. Excessive stress levels to the crop can result as the soil dries and salts in the soil water become more concentrated. Howell et al. (1984a) developed ways to use plant stress, either matric, osmotic, or the sum of both, to indicate the time to irrigate. Leaf-water potential and crop-water stress index (CWSI) provide the bases for two such methods (Idso et al. 1982). The CWSI method uses plant canopy temperature along with the vapor pressure deficit to indicate stress. Howell et al. (1984b) developed values of CWSI to use in scheduling irrigations for cotton. The needed depth of application at the time of irrigation can be estimated from soil water content as determined by neutron probe, gravimetric analysis, or another method.

Irrigation scheduling and water tables can be managed jointly for either the maximum contribution from the shallow ground water, little or no ground water contribution, or some intermediate value of groundwater use as defined by the user.

In the case of maximum contribution from the ground water, the salt tolerance of the crop and the availability of shallow ground water limit total water use. To maximize availability, the water table position is controlled. Its elevation is carefully monitored to ensure an adequately aerated zone. Water table depths of 1 m to 1.5 m have provided the maximum contribution from ground water for a wide range of soil types. The shallower water table depth applies to sandy, nonsaline soils.

With maximum ground-water contributions, irrigations can best be scheduled using plant-based measures. The optimum time to irrigate is at the highest stress level that does not reduce yield. Irrigation at lower stress levels would result in more frequent irrigations, more deep percolation, and less contribution to ET from the water table. The depth of irrigation water to apply is estimated from soil-water measurements. The time of the first irrigation is critical for unrestricted plant growth and root development (Ayars and Schoneman 1986). Soil salinity measurements at the end of the irrigation season can be used to calculate the leaching required for re-establishing a favorable soil salinity profile for the next growing season. Rainfall plus pre-plant irrigation can provide necessary leaching. Most types of irrigation systems allow this management.

To obtain little or no ground-water contribution, management is somewhat simpler. Only irrigation and rainfall supply the crop water

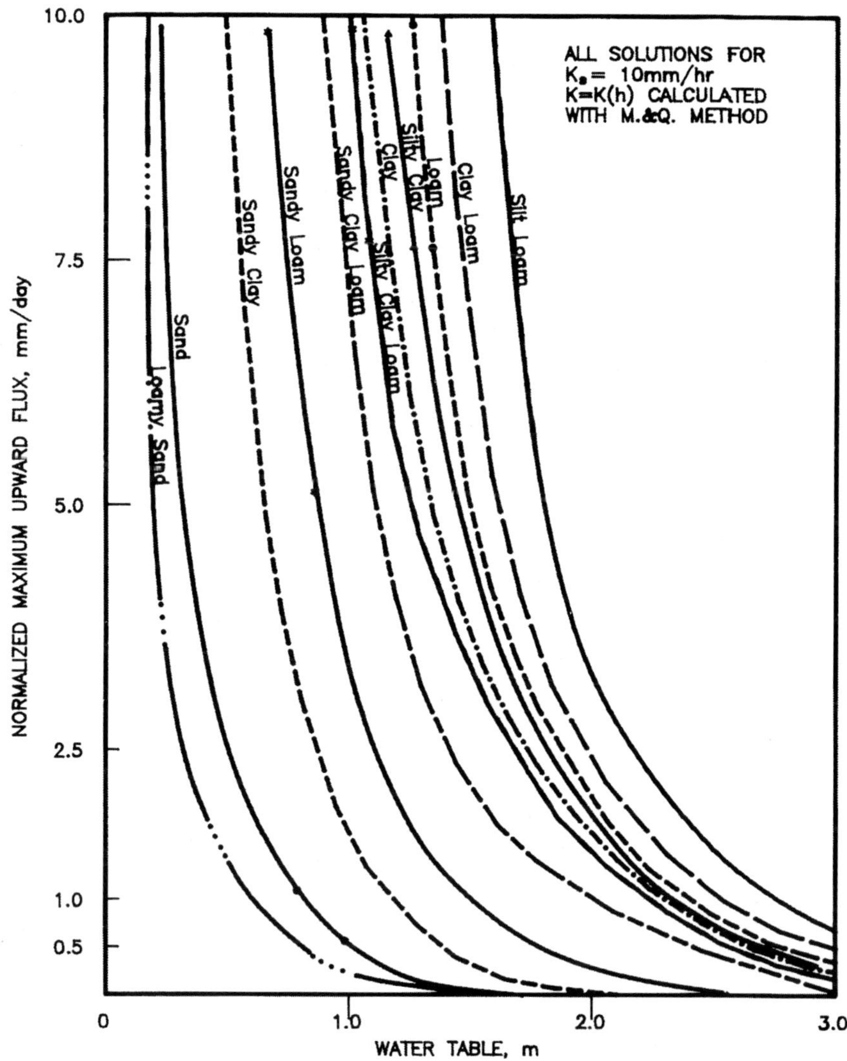

Fig. 16.4 Approximate Upward Flux Relationships for a Range of Soil Textures (Skaggs 1980)

requirement. Deep percolation is minimized. This can be done most easily by using daily irrigations to resupply water used by the crop, thus preventing significant contribution from the water table. Periodic leaching during the season will prevent the buildup of salts. An automated irrigation system that can provide highly uniform applications will be required. Solid set sprinklers or drip irrigation systems are preferred. In some situations, surge irrigation is acceptable.

Lack of data on the plant's temporal extraction of ground water presents an obstacle to obtaining an intermediate amount of ground water use, which can achieved on a seasonal basis by eliminating the final irrigation of the season. Hutmacher et al. (1986) found that cotton and wheat grown in the presence of a shallow saline water table did not suffer a reduction in yield when the last irrigation of the season was eliminated.

REFERENCES

American Society of Agricultural Engineers (ASAE). 1981. *Irrigation scheduling for water and energy conservation in the 80's.* Proceedings of the ASAE Irrigation Scheduling Conference. ASAE, St. Joseph, Michigan. 231 pp.

Ayars, J. E. and R. A. Schoneman. 1986. "Use of saline water from a shallow water table by cotton." *Trans. ASAE* 29: (6): 1674–1678.

Benz, L. C., E. J. Doering, and G. A. Reichman. 1981. Water table management saves water and energy. *Trans. ASAE* 24(4): 995–1001.

Bernstein, Leon, and Milton Fireman. 1957. "Laboratory studies on salt distribution in furrow-irrigated soil with special reference to the pre-emergence period." *Soil Science* 83: 249–263.

Bernstein, Leon, Milton Fireman, and R. C. Reeve. 1955. Control of salinity in the Imperial Valley, California. *USDA, ARS*-41-4. 16 pp.

Bresler, E. 1981. Chapter 4 in *Salinity in Irrigation and Water Resources.* D. Yaron, ed. Marcel Dekker, New York.

Burman, R. D., P. R. Nixon, J. L. Wright, and W. O. Pruitt. 1980. "Water requirements." *Design and Operation of Farm Irrigation Systems,* M. E. Jensen, ed. ASAE, St. Joseph, Mich. pp. 189–234.

Collier, F. W. 1984. "The use of saline water in irrigation." *Irrigation, Drainage and Flood Control.* State-of-the-Art Publication #3. International Commission on Irrigation and Drainage. K. K. Framji, ed. New Delhi, India. pp. 203–229.

Doering, E. J., L. C. Benz, and G. A. Reichman. 1982. "Shallow water-table concept for drainage design in semiarid and sub-humid regions." *Advances in Drainage.* Proceedings of the ASAE Fourth National Drainage Symposium. ASAE, St. Joseph, Michigan. ASAE 12-82, pp. 34–41.

Evans, R. G., J. S. Aarstad, D. E. Miller and M. W. Kroeger. 1987. "Crop residue effects on surge furrow irrigation hydraulics." *Trans. ASAE* 30(2): 424–429.

Food and Agriculture Organization. 1980. Drainage design factors. FAO Irrigation and Drainage Paper No. 38. Rome, Italy.

Grimes, D. W., and D. W. Henderson. 1984. "Developing the resource potential of a shallow water table." Completion Report on OWRT Project B-216-Cal, to the California Water Resources Center, Univ. Calif. at Davis. 63 pp.

Hanson, B. R. 1984. "Effects of increasing drainage in the San Joaquin Valley." *California Agriculture* 38(10): 40–41.

Hanson, B. R. 1987. "A systems approach to drainage reduction." *California Agriculture* 41(9/10): 19–24.

Hoffman, G. J. 1980. "Irrigation management—salinity control." *Irriga-*

tion Challenges of the 80's, Proceedings of the ASAE Second National Irrigation Symposium. ASAE, St. Joseph, Mich. pp. 166–175.

Hoffman, G. J. 1985. "Drainage required to manage salinity." *J. Irrig and Drainage Div., ASCE* 111(IR3): 199–206.

Hoffman, G. J., R. S. Ayers, E. J. Doering and B. L. McNeal. 1980. "Salinity in irrigated agriculture." *Design and Operation of Farm Irrigation Systems,* M. E. Jensen, ed. ASAE, St. Joseph, Mich. pp. 145–188.

Hoffman, G. J. and M. Th. van Genuchten. 1983. "Soil properties and efficient water use: water management for salinity control." *Limitations to Efficient Water Use in Crop Production.* H. M. Taylor, W. R. Jordan and T. R. Sinclair, eds. Am. Soc. of Agronomy. Madison, Wisc., pp. 73–85.

Howell, T. A., J. L. Hatfield, J. D. Rhoades, and M. Meron. 1984a. "Response of cotton water stress indicators to soil salinity." *Irrig. Sci.* 5: 25–36.

Howell, T. A., J. L. Hatfield, H. Yamada and K. R. Davis. 1984b. "Evaluation of cotton canopy temperature to detect crop water stress." *Trans.* ASAE 27: 84–88

Hutmacher, R. B., J. E. Ayars, R. A. Schoneman and S. S. Vail. 1986. "Furrow irrigation management of cotton, wheat and sugar beets in the presence of a shallow, saline watertable." Water Management Research Laboratory, Annual Report. Fresno, Calif. pp. 3–4.

Idso, S. B., R. J. Reginato, S. M. Farah. 1982. "Soil and atmosphere-induced plant water stress in cotton as inferred from foliage temperature." *Water Resources Res.* 18: 1143.

Jensen, M. E., D. C. N. Robb and C. E. Franzoy. 1970. "Scheduling irrigations using climate-crop-soil data." *J. Irrig. and Drainage Div., ASCE* 96(IR1): 25–38.

Kemper, W. D., W. H. Heinemann, D. C. Kincaid, and R. V. Worstell. 1981. "Cablegation: I. Cable controlled plugs in perforated supply pipes for automatic furrow irrigation." *Trans. ASAE* 24(6): 1526–1532.

Luthin, J. N. 1978. *Drainage Engineering.* Robert E. Krieger Publishing Co. Huntington, N.Y.

Luthin, J. N. and Frank Robinson. 1969. "Depth of drainage in irrigated areas." *Trans. ASAE* 12(1): 27–28,31.

Maas, E. V. 1985. "Crop tolerance to saline sprinkling water." *Plant and Soil* 89: 273–284.

Maas, E. V., and G. J. Hoffman. 1977. Crop salt tolerance current assessment. *J. Irrig. and Drainage Div., ASCE* 103(IR2): 115–134.

Meiri, A., and Z. Plaut. 1985. Crop production and management under saline conditions. *Plant and Soil* 89(1/3): 253–271.

Replogle, J. A., J. L. Merriam, L. R. Swarner and J. T. Phelan. 1980. "Farm water delivery systems." *Design and Operation of Farm Irrigation Systems,* M. E. Jensen, ed. Monograph No. 3, ASAE, St. Joseph, Mich. pp. 317–346.

Robinson, A. R., and A. S. Humphreys. 1967. "Water control and measurement on the farm." Chapter 42. *Irrigation of Agricultural Lands,* R. M. Hagan, H. R. Haise, and T. W. Edminster, eds. Agronomy Monograph No. 11, American Society of Agronomy, Madison, Wisc. pp. 828–884.

Robinson, F. E., and K. S. Mayberry. 1976. "Seed coating, precision planting, and sprinkler irrigation for optimum stand establishment." *Agronomy Journal* 68: 674–675.

Rhoades, J. D. 1984. "New strategy for using saline waters for irrigation." *Water Today and Tomorrow.* Proceedings of the Specialty Conference of Irrigation and Drainage Division, ASCE. Flagstaff, Ariz. July 24–26, pp. 231–236.

Sammis, T. W., D. L. Weeks and E. G. Hanson. 1979. "Influence of irrigation methods on salt accumulation in row crops." *Trans. ASAE* 22(4): 791–796.

Shainberg, I., and J. Shalhevet, eds. 1984. *Soil Salinity Under Irrigation Processes and Management.* Springer-Verlag, New York. 349 pp.

Shalhevet, J. A., A. Meiri and Bruria Heuer. 1981. "Irrigation timing as a factor in crop tolerance in salinity." *The Problem of Salinity in Agriculture.* A Conference on Biosalinity. Davis, Calif. September, 1981. pp. 54--56.

Skaggs, R. W. 1978. "A water management model for shallow water table soils." Report No. 134. Water Resources Research Institute of the University of North Carolina.

Skaggs, R. W. 1980. DRAINMOD—reference report; Methods and design and evaluation of drainage-water management systems for soils with high water tables. USDA-SCS, South National Technical Center, Fort Worth, Tex. 330pp.

Strelkoff, Theodor. 1985. BRDRFLW: A mathematical model of border irrigation." ARS-29. Agric. Res. Serv. USDA. 100 pp.

Stringham, Glen E. and Jack Keller, 1979. Surge flow for automatic irrigation. *Irrigation and Drainage in the Nineteen-eighties.* Proceedings of the Specialty Conference of Irrigation and Drainage Division, ASCE. Albuquerque, NM. pp. 132–142.

Tanji, K. K., M. E. Grismer and B. R. Hanson. 1985. "Subsurface drainage evaporation ponds." *California Agriculture* 39(9/10): 10–12.

USDA. 1974. Border irrigation. Chapter 4, Section 15 (Irrigation) SCS Nat'l. Engr. Handbook. 50 pp.

USDA. 1983. "Sprinkle irrigation." Chapter 11, Section 15 (Irrigation) SCS Nat'l. Engr. Handbook, 2nd ed. 121 pp.

van Schilfgaarde, J., L. Bernstein, J. D. Rhoades, and S. L. Rawlins. 1974. Irrigation Management for Salt Control. *J. Irrig. and Drainage Div., ASCE* 100(IR3): 321–338.

West, D. W., and I. F. Merrigan. 1979. "Soil salinity gradients and growth of tomato plants under drip irrigation in saline soils." *Soil Science* 127(5): 281–291.

Walker, Wynn R. and Gaylord V. Skogerboe. 1987. *Surface Irrigation Theory and Practice.* Prentice-Hall, Inc., Englewood Cliffs, N.J. 386 pp.

NOTATION

D_i = depth of irrigation water; D_I = initial depth of water stored in profile at saturation; D_F = final depth of water stored in profile at saturation; D_{gw} = depth of ground water; Lr = leaching requirement; σ_a = weighted electrical conductivity of water available to crop; σ_{D*} =

desired soil salinity value, expressed as electrical conductivity; σ_F = final average electrical conductivity of soil solution in profile; σ_{gw} = electrical conductivity of ground water; σ_i = electrical conductivity of irrigation water; and σ_I = initial average electrical conductivity of soil solution in profile.

CHAPTER 17

MANAGEMENT OF DRYLAND SALINE SEEPS[a]

INTRODUCTION

Discussed in this chapter are the diagnosis, control, and reclamation of dryland saline seep problems.

A saline seep results from a salinization process, often accelerated by dryland farming, which allows water to move through salt-laden substrata below the root zone. It refers to intermittent or continuous saline water discharge at or near the surface of the soil, downslope from recharge areas under dryland conditions. It reduces or eliminates the growth of crops in the discharge area due to increased soluble concentrations of salt in the root zone. Saline seeps can be differentiated from other saline soil conditions by their recent and local origin, saturated root zone profile, shallow water table, and sensitivity (short-term response) to precipitation and cropping systems (Brown et al. 1983).

Saline seeps occur frequently in dryland farming areas throughout the Great Plains (Ballantyne 1963; Berg et al. 1986; Brown et al. 1987; Colburn 1983; Doering and Sandoval 1976b; Halvorson and Black 1974; Neffendorf 1978; Vander Pluym 1978). Miller et al. (1981) estimated that nearly 1 million ha of productive cropland has been salinized in the northern Great Plains. Saline seep problems also exist in Australia (Malcolm 1982; Matheson 1968), India, Iran, Turkey, and Latin America (Olson 1978). Saline seeps result from a combination of geologic, climatic, hydrologic, and cultural (land-use) conditions. The primary cause is a change from grassland or forest to a cropping system, such as crop-summerfallow rotation, that allows rainfall in watershed recharge areas to move below the root zone and provide seepage water.

Factors Contributing to Saline Seep Development

The characteristics and causes of saline seeps in the Great Plains are similar (Berg et al. 1986; Brown et al. 1983; Doering and Sandoval 1976a; Doering and Sandoval 1976b, Halvorson and Black 1974; Vander Pluym 1978). Typically, native grasses or naturally occurring vegetation have been replaced with agricultural fields and cropping systems with lower potential evapotranspiration requirements. Precipitation that exceeds the root zone's storage capacity, which takes place primarily during summerfallow periods, is the source of water.

The crop-summerfallow system of dryland farming has contributed significantly to the development of saline seep problems in the

---

[a]Prepared by: Ardell D. Halvorson, U.S. Dept. of Agriculture-Agricultural Res. Service, Central Great Plains Research Station, P.O. Box 400, Akron, CO.

northern Great Plains but is not the sole cause (Brown et al. 1983; Christie et al. 1985; Halvorson and Black 1974). Also contributing to the development of seeps are periods of above-normal precipitation, restricted surface and subsurface drainage due to the building of roads and pipelines, large snowdrifts, gravelly and sandy soils, obstructions across natural drainageways (e.g., roads), uncapped or poorly cased artesian water wells, leaky ponds and dugouts, and crop failures. These factors, combined with the right geologic conditions, can result in saline seeps years after vegetation has changed. Water conservation practices in the southern Great Plains, such as level bench terraces, have contributed to the development of saline seeps (Berg et al. 1986; Naney et al. 1986).

Types of Saline Seeps

Miller et al. (1981) discussed geologic formations of the northern Great Plains. Materials in the soil profile vary from glacial till deposited over shales to highly stratified, water-deposited geologic materials. Brown et al. (1983) diagrammed several geologic conditions that can result in seeps in the northern Great Plains (Fig. 17.1). Seeps generally develop on sidehills or toe slopes of rolling to undulating topography, where permeable material lies above less permeable strata that are conducive to the development of perched water tables. Characteristics of different types of seep are as follows:

1. *Geologic outcrop seep.* The recharge area lies above material of low hydraulic conductivity (HC) such as shale, dense till, or clay. Soil above the low HC layer varies in texture and depth. Most of the seeps expand laterally and downslope. Only limited expansions occur upslope.
2. *Coal seam seep.* The recharge area lies above coal, which, in turn, lies above clay of low HC. Soil above the coal seam varies in texture. Water moves laterally through the coal-related material at a rapid rate. Seepage occurs where outcropping and truncated coal beds exist. Coal seeps typically expand laterally and downslope.
3. *Glaciated Fort Union seep.* The recharge area of glacial till lies above sandstone, siltstone, lignite, and clay strata of the Fort Union Formation. Water in the recharge area enters permeable strata to form a water table above a zone of low HC. Seepage water moves downslope to glacial till. The till is of lower HC and truncates the permeable zone, causing the water table to rise. Seep expansion is generally upslope, with some downslope and lateral expansion.
4. *Textural change seep.* The recharge area lies above material with a low HC. The soil above the zone of low HC has a coarse or medium texture. Water moves through the root zone to a zone of low HC, then laterally downslope until it encounters a zone of lower HC, which slows movement and causes the water table to rise to the surface of the soil. Seep expansion is lateral and downslope.
5. *Slope change seep.* The recharge area lies above geologic material of low HC. Soil above the zone of low HC varies in texture. Water moves through the root zone to the zone of low HC, then laterally downslope to where the slope decreases. The reduced gradient at that location

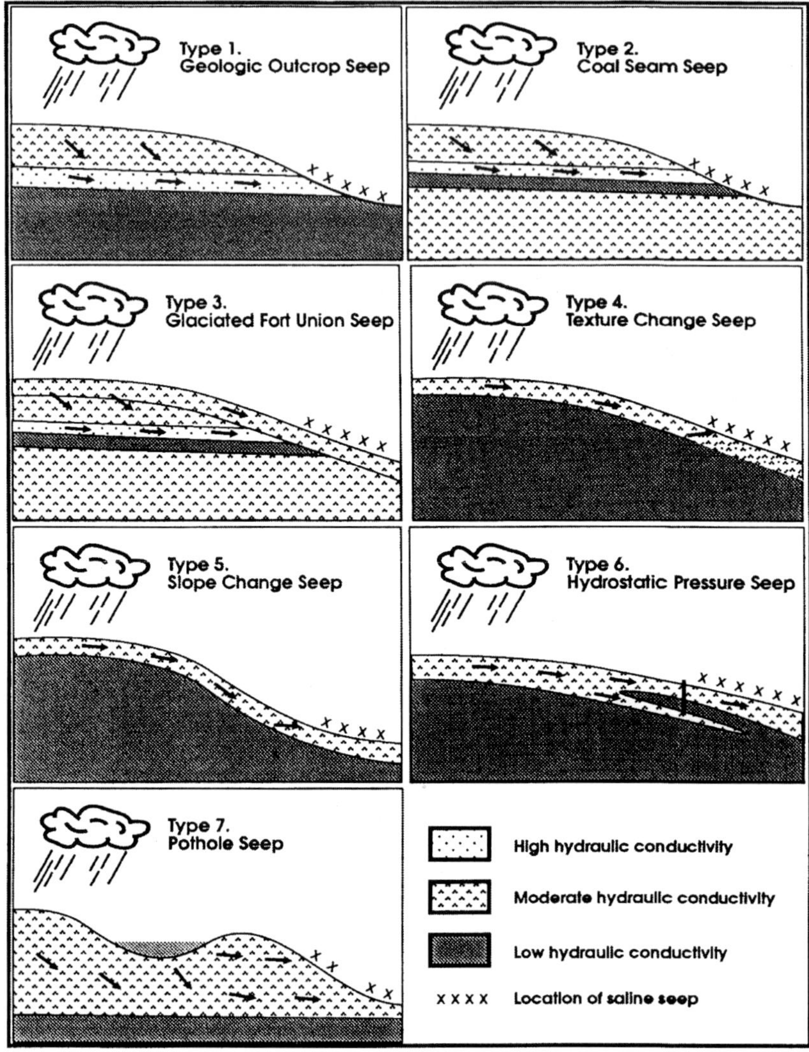

Fig. 17.1 Schematic Diagrams Illustrating Seven Geologic Conditions for Saline-Seep Development (Brown et al. 1983)

causes water to move slower and the water table rises, forming a seep. Seep expansion is mostly lateral and downslope with some upslope expansion.

6. *Hydrostatic pressure seep*. The recharge area lies above geologic material of low HC. Soil above the dense layer varies in texture. Water moves through the root zone to the zone of low HC and then laterally downslope to a zone of low HC located above the saturated zone. The confined water is under hydrostatic pressure. This often forces water through fractures to the surface of the soil and causes a saline seep. The

seep expands mostly laterally and downslope. The recharge area may be located at a greater distance and at a higher elevation than for other types of seeps.

7. *Pothole seep.* The recharge area has potholes or poorly drained areas that lie above material of low HC. Water moves through slowly permeable material in a pothole to a zone of low HC, then downslope, where it may encounter a zone of higher HC that outcrops at or near the surface of the soil to form a saline seep. The seep primarily expands laterally and downslope. The rate of seepage increases rapidly during periods of high precipitation, when ponded water volumes are greatest, and contracts during dry periods.

The above types of seeps vary in the field. Understanding the geology and circumstances that cause a saline seep to form will help in designing ways to control or prevent them. Agronomic practices generally work well for some types of seeps: geologic outcrop, coal seam, glaciated Fort Union, texture change, and slope change. They need to be combined with drainage and land leveling for hydrostatic pressure seeps and pothole seeps.

WATER QUALITY ASSOCIATED WITH SALINE SEEPS

As water passes through the soil profile toward the perched or permanent water table, salts dissolve and move downward. Hydrologic studies show that seeps are generally sustained by local recharge areas (Doering and Sandoval 1976a; Halvorson and Black 1974; Halvorson and Reule 1980; Hendry and Schwartz 1982; Naney et al. 1986).

Numerous studies document the movement of soluble salts and NO_3-N toward and into shallow water tables. Ferguson and Batteridge (1982) have estimated that as much as 90 Mg/ha of salt has migrated toward the ground-water table in glacial till soils of north-central Montana after several decades of grain production. Christie et al. (1985) reported that the soil-profile salinity of cultivated land decreased more than that of an adjacent native non-cultivated area, indicating movement of salt to lower depths. Doering and Sandoval (1981) reported that a drained seep area had lost salt and 50 kg NO_3-N/ha.

The data in Table 17.1 show the chemical composition of waters associated with several saline seeps in the northern and southern Great Plains. The shallow ground water often associated with saline seeps is unsuitable for consumption by humans and livestock due to high levels of salt and NO_3 (>0.7 mmol/l) and unsuitable for irrigation due to total salt concentration. Calcium, magnesium, and sodium are the dominant cations and sulfate is the dominant anion in most of the shallow ground water associated with saline seeps. Compared to sulfates, chlorides exist in water and soil at relatively low concentrations in the northern Great Plains. They occur at slightly higher concentrations in the southern Great Plains. Soils in seep areas are generally in equilibrium with gypsum, lime, and other types of Ca-Mg sulfate minerals (Brun and Deutch 1979; Doering and Sandoval 1981; Oster and Halvorson 1978; Timpson et al. 1986).

TABLE 17.1 Chemical Composition of Waters Associated with Saline Seeps in the Great Plains

Location (1)	pH (2)	EC (dS/m) (3)	Ca[a] (4)	Mg[a] (5)	N[a] (6)	HCO_3[a] (7)	NO_3[a] (8)	Cl[a] (9)	SO_4[a] (10)	Reference (11)
MT recharge	8.4	5	7	11	18	3.8	4.3	0.7	21	Halvorson and Black (1974)
MT seep	8.2	9	8	21	66	9.8	0.4	0.8	52	Halvorson and Black (1974)
MT seep	7.9	14	10	37	109	8.1	29.5	2.6	80	Halvorson and Black (1974)
MT seep	8.4	26	1	108	211	4.0	5.4	7.6	225	Miller (1971)
MT recharge	8.2	7	3	21	39	2.4	6.2	11.2	44	Miller (1971)
ND seep	3.7	10	9	36	59	—	5.7	2.1	70	Doering and Sandoval (1981)
ND seep	4.6	8	9	30	40	—	4.7	2.5	55	Doering and Sandoval (1981)
OK seep	8.1	5	15	16	26	—	0.6	12.3	27	Berg et al. (1986)
OK seep	8.2	3	3	17	13	—	—	16.0	15	Naney et al. (1986)

[a]Chemical elements in mmol/l.

Researchers in the northern Great Plains have concluded that the NO_3 in ground water originates mainly from exchangeable NH_4 of geologic origin located deep in the profile and from NO_3 that comes from mineralized organic matter and is leached from the root zone during periods of summerfallow (Doering and Sandoval 1981; Hendry et al. 1984; Power et al. 1974). The NH_4 is oxidized to NO_3. Little, if any, of the NO_3 had its origin as fertilizer N because little fertilizer N was used by dryland farmers in the northern Great Plains before the early 1970's, when saline seeps became a problem.

IDENTIFICATION OF RECHARGE AND DISCHARGE (SEEP) AREAS

Early detection and diagnosis of a saline-seep problem may allow a farmer to change current cropping systems to minimize the damage. Postponing the use of control practices obviously leads to a problem that is more difficult to control.

Visual Assessment

Brown (1976) described several visual symptoms of the impending development of saline seeps: vigorous growth of kochia (*Kochia scoparia* L.) or other weeds after the harvest of grain in areas where the soil normally would be too dry to support the growth of weeds; the presence of salt crystals on the surface of the soil; prolonged wetness of the surface in localized areas after rainfall; the slipping of tractor wheels or

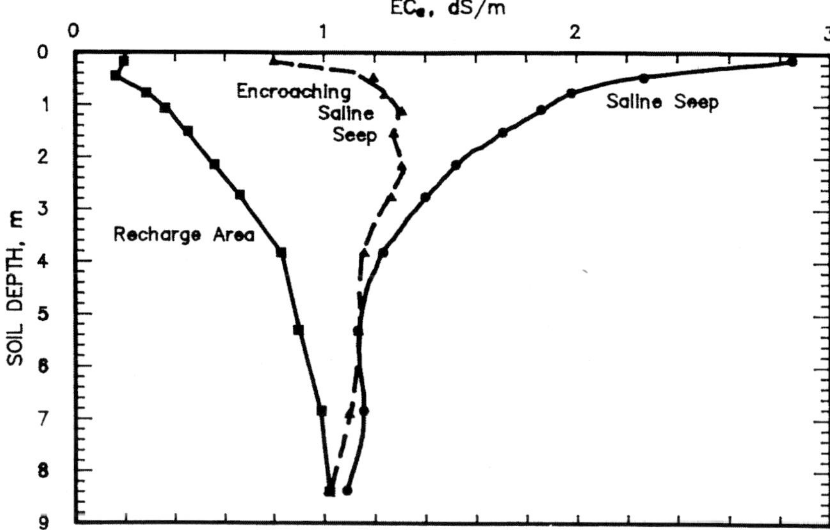

Fig. 17.2 Typical Four-Probe Electrical Conductivity (EC_a) Readings as a Function of Soil Depth in Saline-Seep Recharge Area, Encroaching Saline-Seep Area, and Saline Seep Itself (Halvorson and Rhoades 1974)

the bogging down of equipment in areas of a field, or the seepage of water into the tracks of wheels, with salt crystals becoming visible as the soil dries; crop growth accompanied by lodging in areas of the field that previously produced normal growth; increased infestations of foxtail barley (*Hordeum jubatum* L.); stunted or dying trees in a shelterbelt or windbreak; and poor seed germination.

Field Assessment of Soil Salinity

Electrical conductivity (EC) methods for measuring soil salinity have been developed to identify areas where saline seeps may develop. Halvorson and Rhoades (1974, 1976) and Halvorson et al. (1977) used a four-electrode resistivity technique to characterize the salinity levels of the soil profile and to identify recharge areas and incipient and existing saline seep areas (Fig. 17.2). Rhoades and Halvorson (1977) also used electrical conductivity methods to delineate saline seep areas and map soil salinity.

Saline seeps generally have high levels of salinity at the surface of the soil, decreasing with depth. Developing seep areas generally have low to medium levels of salinity at the surface of the soil, with higher salinity at a depth of 30 cm to 90 cm and lower salinity at deeper depths. Salinity generally increases gradually with increasing depth in recharge area.

Cameron et al. (1981) described the use of electromagnetic inductive techniques for measuring and mapping soil salinity in the field. This method can also be used to verify areas of high and low salinity in the field without laboratory analyses.

Locating Recharge Areas

The size of recharge areas must be delineated and estimated before treatments for controlling saline seeps can be designed. Generally, recharge areas are located a short distance (180 m–600 m) upslope from the discharge or seep area. If gravel beds and sandy soils are involved, recharge areas may be within 30 m. The recharge area usually is located directly upslope or at an angle across the slope from the discharge area (Brown et al. 1983). The following methods may be useful for determining the location and size of recharge areas.

Soil Maps and Geologic Information. Soil survey maps can be used to locate sandy or gravelly soils located upslope from the discharge area, as well as poorly drained areas, such as potholes. Geologic maps can provide information on subsurface stratification, the type of saline seep, and depths to permanent ground-water tables.

Soil Moisture Probes and Test Holes. If a seep is surrounded by elevated topography on several sides, a soil moisture probe (Brown 1958) can be used to identify the location of recharge area relative to the seep by locating abnormally wet soil in one general direction. Augering or coring machines can be used to examine and sample soil profiles to greater depths. Each drilled hole should be carefully logged during drilling. The depths at which dense materials, such as clay and shale, or highly permeable materials, such as sand, gravel, silt, and lignite, are encountered should be recorded. The depth to the water table should be noted and the hole cased with perforated pipe so that depths of the water table can be periodically monitored and water samples can be collected. Information collected from the test holes, including well log data, water depth, and salinity measurements, can be combined with visual observations and topography to delineate the recharge area. Often, soil moisture probes and a few well-placed test holes will provide the most economical way to locate a recharge area.

Visual Inspection. When soil survey maps, drill rigs, and equipment for measuring soil salinity in the field are unavailable, visually locate the upslope area, direction of seep expansion, and upslope factors that may contribute water to the discharge area. Bear in mind that: 1) The recharge area is higher in elevation than the seep area; 2) the recharge area is generally within 600 m of the seep area; 3) saline seeps in glacial till areas generally expand laterally and upslope toward the recharge area; 4) saline seeps in nonglaciated areas tend to expand laterally and downslope away from the recharge area; and 5) if the seep does not begin to dry up within two to three years of implementing control measures, such as planting alfalfa or grasses or annually cropping the suspected recharge area, the boundary of the recharge area was incorrectly identified, or the recharge area was larger than anticipated, or the seepage water may be coming from an artesian source.

METHODS FOR CONTROLLING SALINE SEEPS

Since seeps are caused by water moving below the root zone in the recharge area, the saline seep problem will not be permanently solved unless control measures are applied to the recharge area. Two procedures for managing seeps are: 1) Mechanically drain ponded surface water before it infiltrates, and intercept lateral flow of subsurface water with drains before the water reaches the discharge area; and 2) agronomically use the water before it percolates below the root zone.

Drainage

Undulating, near-level land with poor surface drainage (potholes) can create recharge areas for saline seeps. Runoff takes place after rainfall and snowmelt, causing these areas to fill with water temporarily. Where possible, surface drains are installed to prevent the temporary ponding of surface water. Drainageways under roadbeds should be cleared of debris and sediment so that they do not temporarily pond surface water. In the central Great Plains, level bench terraces serve as temporary water impoundments that may be contributing water to saline seeps (Berg et al. 1986; Naney et al. 1986). Their use may need to be evaluated if saline seepage is a problem.

Drainage studies have shown that hydraulic control can be accomplished quickly with subsurface interceptor drains located on the upslope side of the seep area (Doering and Sandoval 1976a; Sommerfeldt et al. 1978). However, a suitable outlet for disposal of the saline water must be available. Outlet considerations include easement for transport of drainage water across intervening lands and the effect of drainage waters on the quality of receiving streams or reservoirs. Because seep waters are saline and typically high in nitrate, disposal into downstream surface waters or ground waters is difficult due to physical and legal constraints and costs. Therefore, subsurface drainage is generally not a satisfactory solution to the problem. The best approach is to use the soil water for growing crops when the soil water is in the root zone of the recharge area and relatively nonsaline.

Mole drains have been used in Alberta, Canada, to maintain water tables at a sufficient depth to prevent the accumulation of salts on the surface of the soil (Sommerfeldt et al. 1978). Procedures for using mole-type drains are specific to the site. With moist, cohesive, fine-textured soils and shallow water tables (<100 cm), drains installed on proper grade work well. Such drains may not work in noncohesive soils and with a water table that is >100 cm below the surface of the soil (Sommerfeldt 1976).

Oosterveld (1978) used seep discharge water to irrigate the recharge area, thus, recycling the salts. Limited water supplies for irrigation, the cost of an irrigation system to deliver the water, and the buildup of soil salinity in the recharge area may reduce the usefulness of this technique.

Agronomic Practices

Hydraulic control of saline seep areas can be achieved by planting crops that use available soil water supplies in the root zone of the re-

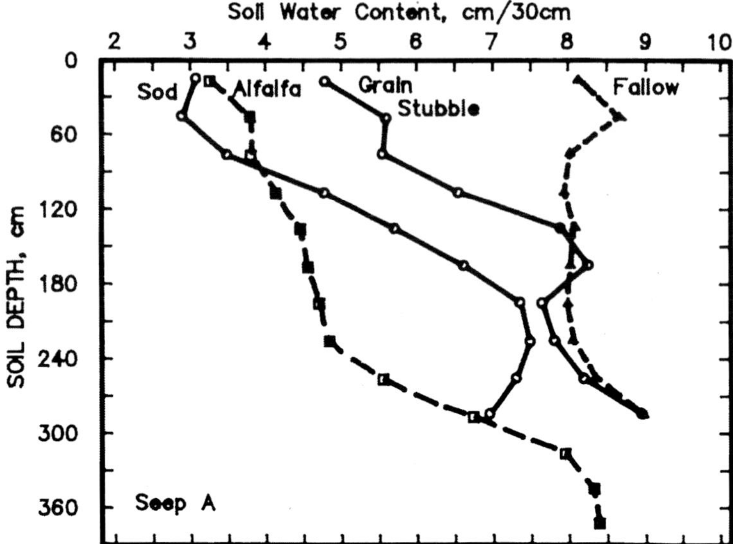

Fig. 17.3. September 1976 Soil Water Profiles of a Saline-Seep Recharge Area in Native Range Sod, Alfalfa (Seeded in 1973), Spring Wheat Stubble, and Summerfallow (Halvorson and Reule 1980)

charge area. To do so, the recharge area must be delineated; and practices that maximize the use of soil-water and minimize deep percolation must be adopted.

Alfalfa (*Medicago sativa* L.), seeded in recharge areas, is one crop that helps to control seep discharge areas hydraulically (Brown et al. 1983; Brun and Worcester 1975; Halvorson and Reule 1980). In Montana, Halvorson and Reule (1980) found that alfalfa extracted more water from the soil profile than native grass sod or small grain crops (Fig. 17.3). Alfalfa depleted more soil water to a depth of 3.0 m than did other crops. This created a larger reservoir in the soil to store precipitation more effectively, thus, reducing the potential loss of water via deep percolation. With the upslope recharge area in alfalfa, the saline seep area dried sufficiently to once again obtain normal yields (Halvorson 1984). In Colorado, alfalfa established in a terraced recharge area in 1984 hydrologically controlled an active saline seep (Halvorson 1988). By the autumn of 1985, the seep had dried sufficiently to allow the seep area, once again, to be worked with farm machinery. In 1987, three cuttings of alfalfa were harvested from the discharge area, where only salt-tolerant weeds had grown in 1984.

Brown and Miller (1978) and Miller et al. (1981) showed that alfalfa controlled saline seeps effectively, while Brown (1983) further showed that it took seven to eight years to recharge the dried soil profile to field-capacity water content, when a summerfallow-winter wheat-barley rotation followed three years of alfalfa (Table 17.2). Halvorson and Reule (1980) reported a rise in the level of the water table where a farmer

TABLE 17.2 Total Soil-Water Content (0.0 m–4.6 m) at the End of Each Growing Season Following Three Years of Alfalfa (Brown 1983)

Year (1)	Crop/ summerfallow (2)	Fall Soil Water (mm H_2O/4.6 m) (3)	Annual Precipitation (mm) (4)
1973	Alfalfa (3rd yr)	217	—
1974	Summerfallow (no crop)	342	278
1975	Winter wheat	330	563
1976	Barley	393	371
1977	Summerfallow (no crop)	448	363
1978	Winter wheat	461	418
1979	Barley	461	208
1980	Summerfallow (no crop)	524	380
1980	Estimated field capacity	573	—

reverted to a crop-summerfallow system of farming in the recharge area after several years of alfalfa production, during which hydraulic control of the seep area had been achieved and the seep area supported near normal crop production. These studies indicate that once a saline seep area has been controlled, reclaimed, and returned to normal crop production, a farmer cannot permanently return to a conventional crop-summerfallow system of farming in the recharge area. The soil water needs to be managed continually to prevent the recurrence of saline seep.

Other work has shown that small grain crops can be used to control saline seep areas (Alberta Agric. 1986; Bramlette 1971; Halvorson and Reule 1976; Holm 1983; Steppuhn and Jenson 1984). Using annual small-grain cropping systems to control seep discharge areas hydraulically is slower than using alfalfa because less soil water is used, and rooting depths are shallower. Oil-seed crops that are deeper-rooted than small grains, such as safflower and sunflower (Table 17.3), can help to deplete the stored soil water to greater depths, thereby increasing the capacity of the soil to store precipitation between crops or during periods of summerfallow.

Black et al. (1981) describe several dryland cropping strategies for controlling saline seeps in the northern Great Plains. They suggest using intensive, flexible cropping systems with adapted crops and soil, water, and crop management practices to improve the crop-production-water-use relationship enough to eliminate or reduce the need for summerfallow.

Flexible cropping involves planting a crop only in years when soil-water and precipitation are expected to be sufficient to produce an economic crop yield. Each year, based on data regarding soil water and expected precipitation during the growing season, the farmer decides to crop or summerfallow the area where the seep occurs (Alberta Agric. 1986; Brown et al. 1981; Naney et al. 1986). Re-cropping or annual crop

TABLE 17.3 Rooting Depth and Soil-Water Use by 11 Dryland-Grown Crops (Black et al. 1981)

Crop	Fort Benton, Montana		Culbertson, Montana	
	Rooting Depth (m)	Soil Water Use (mm)	Rooting Depth (m)	Soil Water Use (mm)
(1)	(2)	(3)	(4)	(5)
Alfalfa (1st yr)	2.1	178	—	—
Alfalfa (4th yr)	5.5	666	—	—
Sanfoin (1st yr)	1.5	150	—	—
Sanfoin (4th yr)	4.0	561	—	—
Russian wild rye (1st yr)	2.1	318	—	—
Russian wild rye (4th yr)	3.0	475	—	—
Sweet clover (1st yr)	1.8	276	—	—
Sweet clover (2nd yr)	2.7	403	—	—
Safflower	2.2	249	2.1	229
Sunflower	2.0	206	—	—
Winter wheat	1.8	200	1.6	190
Rapeseed	1.5	170	—	—
Spring wheat	—	—	1.2	152
Barley	1.4	190	1.1	135
Corn	1.2	94	—	—

ping is ill-advised when less than about 76 mm of soil water is available at planting time (Alberta Agric. 1986; Black and Ford 1976).

Farmers can use a moisture probe to determine soil moisture profiles (Brown 1958), or they may determine soil water content in another way. Halvorson and Kresge (1982) have developed a computer model, FLEXCROP, to help farmers select the best cropping and soil management strategies for wheat (*Triticum aestivum* L.), barley (*hordeum sativum*, Jess.), oats (*Avena sativa* L.), and safflower based on stored soil water and expected precipitation (the program is available from the author). Weed control and soil fertility are also critical factors in developing flexible dryland cropping systems.

Black et al. (1981) reported that crops grown under annual cropping systems used an average of 75% to 81% of the precipitation received between crop harvests within a grass barrier system. Conventional spring wheat-summerfallow systems used only 40% (Table 17.4). The amount of unused available water between crops, a portion of which may contribute to the development of saline seeps, averaged 473 mm for spring wheat-summerfallow systems and only 72 mm to 98 mm for annual cropping systems. These data show that more water, nitrates, and dissolved salts can be moved below the root zone with a spring wheat-summerfallow system than with an annual cropping system. Adequate fertility is essential for optimizing yields with annual cropping systems (Black et al. 1982, deJong and Halstead 1986, Halvorson et al. 1976, Schneider et al. 1980).

If intensive, flexible cropping systems are to succeed, more effi-

TABLE 17.4 Average Precipitation-Use Efficiency (PUE) per Cropping Sequence, as Influenced by Cropping System Within a Tall Wheat Grass Barrier System over a 12-Year Period (Black et al. 1981)

Cropping system (1)	Number of crops per year (2)	Total precip-itation per crop (mm) (3)	Total[a] water use per crop (mm) (4)	PUE[b] (%) (5)	Annual grain yield[c]		WUE[c,d]	
					Without Nitrogen (kg/ha) (6)	With Nitrogen (kg/ha) (7)	Without Nitrogen (kg/ha-mm) (8)	With Nitrogen (kg/ha-mm) (9)
Annual cropping								
1. 6WW-B-S-B-WW-S-B	1.00	396	322	81	1328	1794	3.4	4.5
2. 5SW-S-B-WW-B-WW-B-WW	1.00	394	296	75	993	1822	2.5	4.6
3. 4SW-S-B-WW-S-SW-B-WW-B	1.00	390	318	82	969	1590	2.5	41
Three-year rotation								
1. SW-WW-F	0.66	569	333	59	997	1416	2.6	3.7
Crop-summerfallow								
1. WW-F	0.50	788	404	51	1019	1247	2.6	3.1
2. SW-F	0.50	786	313	40	853	1065	2.2	2.7

WW = winter wheat; SW = spring wheat; B = spring barley; S = safflower; F = Summerfallow.

[a]Water-use per crop is based on soil water use to 120-cm depth plus precipitation received from seeding to harvest.

[b]PUE = [(total water use/crop)/(total precipitation received/crop)] × 100.

[c]Applied nitrogen of 34 kg N per ha each crop year.

[d]WUE = water use efficiency = [(grain yield/ha)/(total precipitation/crop rotation)] × 100.

cient methods for storing soil water during fallow periods must be found. In the northern and central Great Plains, supplies of soil water can be increased by controlling the growth of weeds and volunteer grain after harvest, leaving standing stubble to trap snow, using annual or perennial barriers or windbreaks for snow trapping and using reduced- or no-tillage cropping systems (Black and Siddoway 1976; Nicholaichuk and Gray 1986; Smika and Whitfield 1966). All of these practices enhance the efficiency of soil water storage. However, more intensive cropping systems than the conventional crop-summerfallow system must be used. Otherwise, the development of saline seep will intensify.

Cropping Strategies

Crops need different amounts of water to produce an economical yield because they have different rooting depths and water extraction patterns. Black et al. (1981) reported that safflower in Montana used more soil water and withdrew water from greater depths in one year than any other annual dryland crop (Table 17.3). Alfalfa used only slightly less water the first year than safflower and sweet clover (*Melilotus officinalis* L.), but its ability to use precipitation plus soil water from progressively deeper depths in successive years makes it the best crop to use first to hydrologically control seep recharge areas. Crops listed in order of decreasing rooting depth and soil water use are alfalfa, sweet clover, safflower, sunflower, winter wheat, rapeseed (*Brassica napus* L.), spring wheat, barley, and corn (*Zea mays* L.).

To decide which crop should succeed another, knowledge of the amount and depth of soil water depleted by the previous crop is needed. Crops should be grown in sequential order with increasing rooting depths, until the depth and amount of soil water removed exceeds soil water recharge during fallow periods (Black et al. 1982). Summerfallow should be used only when needed, e.g., after planting alfalfa or safflower, or when less than 76 mm of soil water exists at planting.

Crops must be rotated in a sequence that avoids weeds, diseases, and insect infestations. Rotating oilseed crops and small grain crops allows grass herbicides to be used, helping to control the buildup of grassy weeds in the small grain crops (Berg et al. 1979; Naney et al. 1986).

Soil fertility is almost as important as water in an annual cropping system. As cropping frequency increases, the need for N increases and responses to P fertilizer depend on the level of soil P and crop N's needs (Halvorson and Black 1985). Nitrogen needs should be balanced carefully with expected water supplies and the potential yield of the crop.

Strict adherence to a crop-summerfallow rotation restricts farmers to a fixed cropping system with limited flexibility to adjust cropping patterns to fit available water supplies. Selection of alternate cropping strategies to use available water supplies effectively requires a knowledge of the amount of water available at any given time, potential evapotranspiration requirements and rooting depths of adapted crops, and expected growing-season precipitation. A knowledge of the depth to some restricting or impermeable geologic strata and water table is

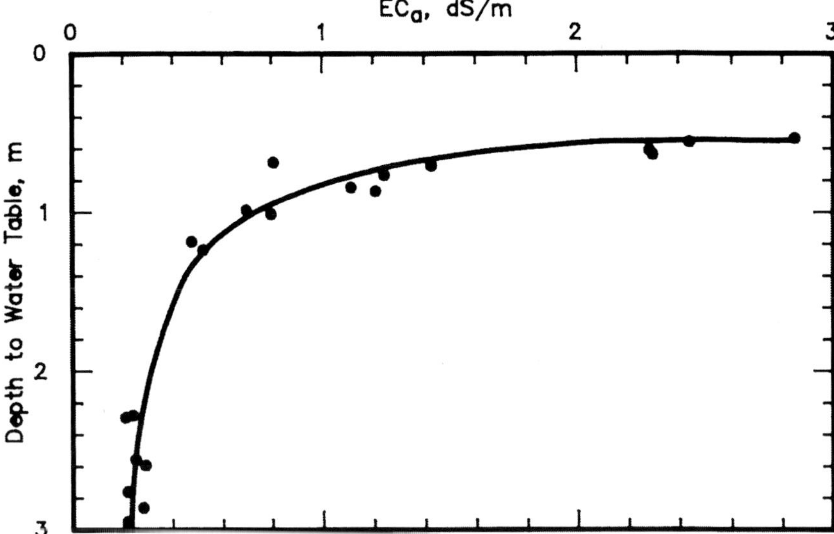

Fig. 17.4 Four-Probe Soil Electrical Conductivity (EC_a) Readings as a Function of Water Table Depth in a Saline Seep Area (Halvorson and Rhoades 1974)

essential if a cropping strategy to control or prevent saline seeps is to be developed.

RECLAMATION OF CONTROLLED SALINE SEEP AREAS

Before reclaiming a saline seep area, the flow of water from the recharge area must be reduced so that the depth of the water table in the area of the seep is low enough to prevent salts from moving up by capillary action into the root zone. If a saline water table is less than 90 cm below the surface, salts can move to the surface by capillary action. The depth of the water table often varies during the year and are shallower in spring and early summer than during the rest of the year. Fig. 17.4 illustrates the relationship between the depth of the water table and soil salinity in the upper 30 cm of soil.

Observation wells should be installed at strategic locations in recharge and seep areas to monitor water tables. A drill rig is needed to install deep wells in recharge areas, but a tractor-mounted post-hole auger or bucket auger can be used to install wells that are less than 180 cm deep. The level of the water table should be monitored monthly. A rising water table that persists into the summer months indicates that cropping practices should be intensified to increase the use of soil water.

The results of research and the experiences of farmers indicate that reclamation occurs quite rapidly (Brown and Miller 1978; Halvorson 1988; Halvorson 1984; Halvorson and Reule 1976 and 1980). If the depth of the water table in the seep area exceeds 150 cm, reclamation procedures to remove salts from the root zone can proceed. The rate of rec-

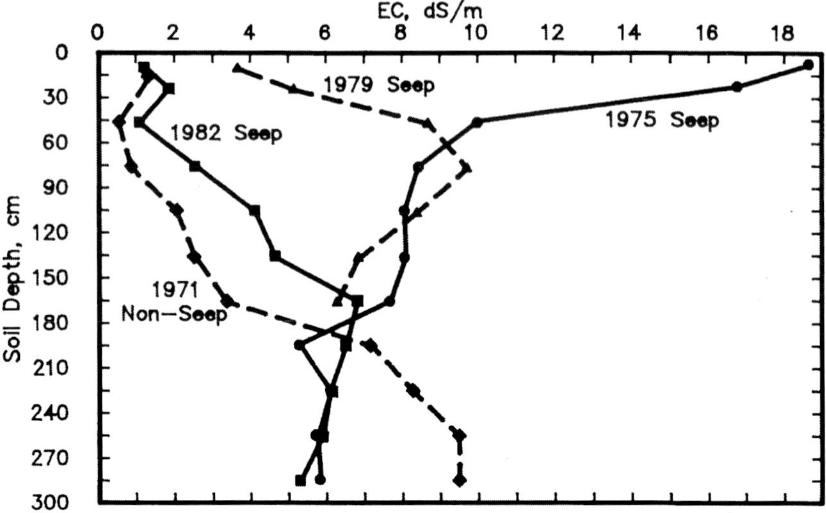

Fig. 17.5 Electrical Conductivity (EC) of Saturated Soil Extracts as a Function of Soil Depth and Time, for a Saline Seep Area that Was Brought Under Hydrologic Control and an Adjacent Non-Seep-Affected Soil (Halvorson 1984)

lamation depends on the amount of precipitation available to leach salts. Therefore, practices that enhance the downward movement of water in the salt-affected area, such as snow trapping or summerfallowing, accelerate reclamation. Summerfallowing can be used during reclamation to help increase downward leaching of salts from the soil profile. None of these practices will be effective, however, until hydrologic control is achieved in the recharge area and the water table is significantly lowered in the seepage area.

Halvorson (1984) reported that soil salinity at the 0-cm to 30-cm depth was markedly reduced two years after a seep was arrested, and various crops could be grown. Adding straw mulch to reduce loss through evaporation from the fallow area helped to accelerate the removal of salt from the 0-cm to 90-cm depth. The application of gypsum did not accelerate reclamation, probably because sufficient naturally occurring gypsum had been precipitated in the soil profile during the formation of saline seep. Because the salts present were Ca, Mg, and Na sulfates, neither the permeability nor the structure of the soil deteriorated during reclamation. Seven years after hydrologic control was achieved, soil salinity was still higher in the area of arrested saline seep than in adjacent areas (Fig. 17.5).

Miller et al. (1981) reported the control of a serious saline seep problem (4 ha in size) located in an 32-ha field near Fort Benton, Mont. "Ladak 65" alfalfa was seeded over the entire field in 1971, when the water table was 0.3 m below the surface in the seep area and 5.8 m below the surface in the recharge area. Six years later, the water table had dropped to 3 m below the surface in the seep area and to 8.5 m below the surface in the recharge area. Alfalfa roots had penetrated to a depth

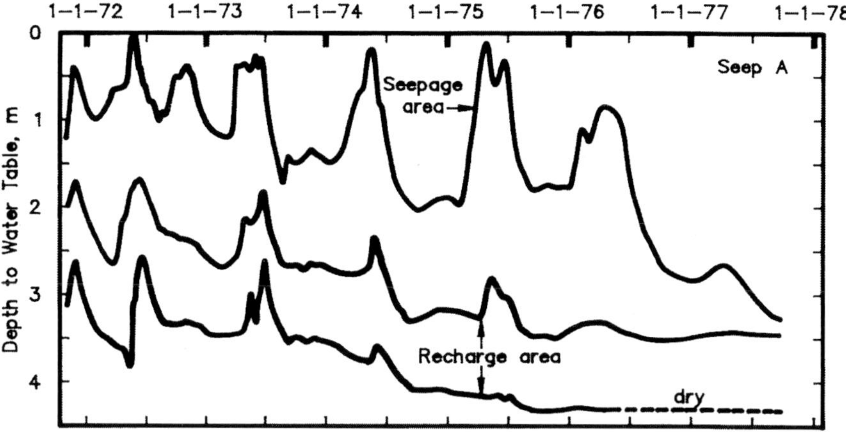

Fig. 17.6 Effect of Alfalfa, Seeded in Summer of 1973, on Water Table Depths in Recharge and Seep Areas on a Farm in Northeast Montana (Halvorson and Reule 1980)

of 4.6 m and had depleted 48 cm of water from the soil profile in the recharge area. The receding water table in the seep area was caused by the reduced flow from the recharge area. Soil salinity in the seep area was 21.3 dS/m and 13.9 dS/m for the 0-cm to 30-cm and 30-cm to 60-cm depths, respectively, in 1971, and 4.3 dS/m and 6.3 dS/m in 1977. With the drop in the level of the water table in the seep, salts had been leached below 60 cm, and the land once again supported economical crop production. In 1977, winter wheat yield in the area of the arrested saline seep was 70% of the surrounding area; in 1978, it was 100%.

Halvorson and Reule (1980) controlled a saline seep that developed in 1971 near Sidney, Mont., where the land was farmed in a crop-summerfallow system. "Ladak 65" alfalfa was seeded on about 80% of the recharge area in 1973. Fig. 17.6 shows changes in the level of the water table before, during, and after the alfalfa was established. The water table in the recharge and seepage areas began to recede shortly after the alfalfa was seeded. By 1975, the surface of the seepage area was dry enough to cross with farm machinery. By 1977, the water table had receded to about 2.4 m below the surface of the soil in the seepage area. In the recharge area, one of two observation wells was dry by 1977. The level of water in the other well had receded from 1.8 m to 3.3 m. Salinity in the top 30 cm of soil in the seep area decreased from 20 dS/m in 1972 to about 5 dS/m in 1978 (Fig. 17.5). Crop yields in the area of the arrested saline seep equalled average county yields after three to four years of hydrologic control (Table 17.5).

Saline seeps would reappear if the crop-summerfallow system were resumed. Halvorson (1984) reports that when a saline seep recharge area that had been hydraulically controlled was converted from alfalfa production back to a crop-summerfallow cropping system in 1979, soil salinity had within three years begun to increase at the 30-cm to 60-cm depths in the saline seep area. Data from the site of the Fort

TABLE 17.5 Yields of Several Crops Grown in Two Reclaimed Saline Seeps in 1978 and 1979, Compared to County Yields in Northeastern Montana (Halvorson 1984)

Crop	Yield (kg/ha)			Average County Yield (kg/ha)	
	1978	1979	Average	1978	1979
(1)	(2)	(3)	(4)	(5)	(6)
		Seep A		Richland County	
Spring wheat	2,462	1,586	2,024	2,184	1,398
Barley	4,547	2,135	3,341	2,382	1,333
Oats	3,385	1,577	2,481	1,971	1,247
Alfalfa	5,708	9,834	7,771	4,346	3,360
		Seep B		Roosevelt County	
Spring wheat	2,426	1,781	2,104	1,848	1,270
Barley	3,861	3,279	3,570	2,091	1,409
Oats	5,273	2,175	3,724	1,756	1,247
Corn (silage)	16,948	3,474	10,211	17,920	11,200

Benton seep indicated that six years of crop-summerfallow rotation had recharged the 4.6-m soil profile to field capacity and that continued crop-summerfallowing in the recharge area would reactivate the former saline seep area.

SOCIOECONOMIC CONCERNS

Saline seeps do not respect property lines. A recharge area on one farmer's property can supply water to a discharge area on a neighbor's farm, or the seep discharge can contaminate a stream, natural drainageway, or farm pond. Except for small, uncomplicated seeps, such as geologic outcrop and coal seam seeps, most farmers need help in diagnosing their saline seep problem and in developing cropping systems or other control measures. When a recharge area is on an adjacent farm, landowners need to cooperate. Knowledgeable individuals or agencies can help by characterizing the problem and recommending control measures. Legislation could provide ways for farmers to form salinity control districts and achieve collectively what cannot be done individually.

A saline seep is not just one farmer's problem. Any loss of farmland decreases the nation's food and tax base. Unless saline seeps are controlled, salty water from seeps can pollute fresh surface waters and add to the salinity of ground water. The problem of saline seep has political implications, involving such questions as subsidies, crop-acreage allotments, and landowner rights. Federal farm programs sometimes have inadvertently adversely affected the progress of saline seep control programs by restricting the acreage that can be planted with small grains or other crops to provide economic control of a saline seep problem. Hectares of summerfallow are often increased, magnifying the saline seep problem.

REFERENCES

Alberta Agriculture. 1986. "Dryland saline seep control." *Alberta Agriculture* AGDEX 518-11, Edmonton, Alberta, Canada. 12p.

Ballantyne, A.K. 1963. "Recent accumulation of salts in the soils of southeastern Saskatchewan." *Can. J. Soil Sci.* 43: 52-58.

Benz, L.C., F.M. Sandoval, E.J. Doering, and W.O. Willis. 1976. *Managing saline soils in the Red River of the North.* U.S. Dept. Agric. ARS-NC-42, 54p.

Berg, W.A., C.R. Cail, D.M. Hungerford, J.W. Naney, and G.A. Sample. 1986. "Saline seep on wheatland in northwest Oklahoma." *Proc. National Conference on Ground Water Quality and Agricultural Practices.* Lewis Publ. Co., Chelsea, Mich. pp. 265-271.

Black, A.L., P.L. Brown, A.D. Halvorson, and F.H. Siddoway. 1981. "Dryland cropping strategies for efficient water use to control saline seeps in the northern Great Plains, U.S.A." *Agricultural Water Management* 4: 295-311.

Black, A.L., and Ford, R.H. 1976. "Available water and soil fertility relationships for annual cropping systems." *Proc. Regional Saline Seep Control Symposium.* Montana State Univ., Bozeman, Coop. Ext. Serv. Bull. 1132.

Black, A.L., and F.H. Siddoway. 1976. "Dryland cropping sequences within a tall wheatgrass barrier system." *J. Soil Water Conserv.* 31: 101-105.

Black, A.L., F.H. Siddoway, and J.K. Aase. 1982. "Soil moisture use and crop management-(DRYLAND)." *Proc. Soil Salinity Conference.* Lethbridge, Alberta, Canada, Nov. 29-Dec. 2, 1982.

Bramlette, G. 1971. "Control of saline seeps by continuous cropping." *Proc. Saline Seep-Fallow Workshop,* Feb. 22-23, 1971, Great Falls, Mont. Highwood Alkali Control Association, Highwood, Mont.

Brown, P.L. 1958. Soil moisture probe. U.S. Patent No. 2,860,515.

Brown, P.L. 1976. *Saline seep detection by visual observation.* Montana Coop. Ext. Serv. Bull. No. 1132: 59-61.

Brown, P.L. 1983. "Saline seep control and soil water recharge under three rotations following alfalfa." Paper presented at Montana Chapter Soil Conservation Society of America meeting, Feb. 4-5, 1983, Bozeman, Mont.

Brown, P.L., A.L. Black, C.M. Smith, J.W. Enz, and J.M. Caprio. 1981. *Soil-water guidelines and precipitation probabilities for growing barley, spring wheat, and winter wheat in flexible cropping systems in Montana and North Dakota.* Montana Coop. Ext. Serv. Bull. No. 356.

Brown, P.L., H. Ferguson, and J. Holzer. 1987. "Saline seep development and control in Montana." *A Century of Action, Natural Resource Development and Conservation in Montana,* J.W. Bauder, ed. Montana Chapter of Soil Conservation Soc. of Am., Bozeman.

Brown, P.L., A.D. Halvorson, F.H. Siddoway, H.F. Mayland, and M.R.

Miller. 1983. *Saline-seep diagnosis, control and reclamation*. U.S. Dept. Agric. Cons. Res. Rpt. No. 30.

Brown, P.L. and M.R. Miller. 1978. "Soil and crop management practices to control saline seeps in the U.S. Northern Plains." Proc. of Meeting of Sub-commission on Salt-Affected Soils, 11th International Soil Sci. Soc. Congress, Edmonton, Alberta, Canada, June 21-24.

Brun, L.J., and R.L. Deutch. 1979. *Chemical composition of salts associated with saline seeps in Stark and Hettinger Counties, North Dakota*. North Dakota Agric. Exp. Sta. Farm Res. 37(1): 3-6.

Brun, L.J., and B.K. Worcester. 1975. "Soil-water extraction by alfalfa." *Agron. J.* 67: 586-589.

Cairns, R.R., and W.E. Bowser. 1977. *Solonetzic soils and their management*. Agric. Canada Public. No. 1391.

Cameron, D.R., E. DeJong, D.W.L. Read, and M. Oosterveld. 1981. "Mapping salinity using resistivity and electromagnetic inductive techniques." *Can. J. Soil Sci.* 61: 67-78.

Christie, H.W., D.N. Graveland, and C.J. Palmer. 1985. "Soil and subsoil moisture accumulation due to dryland agriculture in southern Alberta." *Can. J. Soil Sci.* 65: 805-810.

Colburn, E. 1983. "Salt buildup in soil, slicing Texas yields." *Crops and Soil Magazine* 35(4): 26.

de Jong, E., and E.H. Halstead. 1986. "Field crop management and innovative acres." Proc. Moisture Management in Crop Production Conference, Nov. 18-20, 1986, Calgary, Alberta. Alberta Agriculture, Edmonton, Alberta.

Doering, E.J., and F.M. Sandoval. 1976a. "Hydrology of saline seeps in the northern Great Plains." *Trans. ASAE* 19: 856-861, 865.

Doering, E.J., and F.M. Sandoval. 1976b. "Saline-seep development on upland sites in the northern Great Plains." U.S. Dept. Agric. ARS-NC-32; 26p.

Doering, E.J. and F.M. Sandoval. 1981. "Chemistry of seep drainage in southwestern North Dakota." *Soil Science* 132:142-149.

Ferguson, H., and T. Batteridge. 1982. "Salt status of glacial till soils of north-central Montana as affected by the crop-fallow system of dryland farming." *Soil Sci. Soc. Am. J.* 46: 807-810.

Halvorson, A.D. 1984. "Saline-seep reclamation in the northern Great Plains." *Trans. ASAE* 27: 773-778.

Halvorson, A.D. 1988. "Role of cropping systems in environmental quality: Saline seep control." *Cropping Strategies for Efficient Use of Water and Nitrogen*. Spec. Pub. no. 51, ASA-CSSA-SSSA, Madison, Wisc.

Halvorson, A.D., and A.L. Black. 1974. "Saline seep development in dryland soils of northeastern Montana." *J. Soil and Water Conservation* 29: 77-81.

Halvorson, A.D., and A.L. Black. 1985. "Long-term dryland crop responses to residual phosphorus fertilizer." *Soil Sci. Soc. Am. J.* 49: 928-933.

Halvorson, A.D., A.L. Black, F. Sobolik, and N. Riveland. 1976. "Proper management: Key to successful winter wheat recropping in Northern Great Plains." *North Dakota Agric. Exp. Sta. Farm Res.* 33 (4): 3-9.

Halvorson, A.D., and P.O. Kresge. 1982. "FLEXCROP: A dryland cropping systems model." U.S. Dept. Agric. Production Research Report No. 180; 49p.

Halvorson, A.D., and C.A. Reule. 1976. "Controlling saline seeps by intensive cropping of recharge areas." Proc. Regional Saline Seep Control Symposium. Montana State Univ., Coop. Ext. Serv. Bull. 1132.

Halvorson, A.D., and C.A. Reule. 1980. "Alfalfa for hydrologic control of saline seep." *Soil Sci. Soc. Am. J.* 44: 370-373.

Halvorson, A.D., and J.D. Rhoades. 1974. "Assessing soil salinity in identifying potential saline-seep areas with field soil resistance measurements." *Soil Sci. Soc. Am. Proc.* 38: 576-581.

Halvorson, A.D., and J.D. Rhoades. 1976. "Field mapping soil conductivity to delineate dryland saline seeps with four electrode technique." *Soil Sci. Soc. Am. J.* 40: 571-575.

Halvorson, A.D., J.D. Rhoades, and C.A. Reule. 1977. "Soil salinity—four-electrode conductivity relationships for soils of the northern Great Plains." *Soil Sci. Soc. Am. J.* 41: 966-971.

Hendry, M.J., R.G.L. McCready, and W.D. Gould. 1984. "Distribution, source and evolution of nitrate in a glacial till of southern Alberta, Canada." *J. Hydrology* 70: 177-198.

Hendry, M.J., and F. Schwartz. 1982. "Hydrogeology of saline seeps." *Proc. Soil Salinity Conference.* Lethbridge, Alberta, Canada, Nov. 29-Dec. 2, 1982.

Henry, J.L., P.R. Bullock, T.J. Hogg, and L.D. Luba. 1985. "Groundwater discharge from glacial and bedrock aquifers as a soil salinization factor in Saskatchewan." *Can. J. Soil Sci.* 65: 749-768.

Holm, H.M. 1983. "Soil salinity: A study in crop tolerances and cropping practices." Saskatchewan Agriculture, Plant Industry Branch, Regina, Saskatchewan. 17p.

Malcolm, C.V. 1982. "Wheat Belt salinity: a review of the salt land problem in southwestern Australia." Western Australian Dept. of Agric. Technical Bull. No. 52, 65p.

Matheson, W.E. 1968. "When salt takes over." *J. of Agric., South Australia,* 71: 266-272.

Miller, M.R. 1971. "Hydrology of saline-seep spots in dryland farm areas, a preliminary evaluation." Proc. Saline Seep-Fallow Workshop. Feb. 22-23, 1971, Great Falls, Mont. Highwood Alkali Control Association, Highwood, Mont.

Miller, M.R., P.L. Brown, J.J. Donovan, R.N. Bergatino, J.L. Sonderegger, and F.A. Schmidt. 1981. "Saline seep development and control in the North American Great Plains—hydrologic aspects." *Agric. Water Management* 4: 115-141.

Naney, J.W., W.A. Berg, S.J. Smith, and G.A. Sample. 1986. "Assessment of ground-water quality in saline seeps." *Proc. Agricultural Impacts on Ground water,* Aug. 11-13, 1986, Omaha, Neb. National Water Well Association, Dublin, Ohio.

Neffendorf, D.W. 1978. "Statewide saline seep survey of Texas." M.S. Thesis, Texas A&M University, College Station, Texas.

Nicholaichuk, W., and D.M. Gray. 1986. "Snow trapping and moisture infiltration enhancement." Proc. Moisture Management In Crop Production Conference, Nov. 18-20, 1986, Calgary, Alberta. Alberta Agriculture, Edmonton, Alberta.

Olson, G.W. 1978. "Some observations on dryland saline seepage in several countries and some comments on causes, effects, and cures." *Proc. Sub-Commission on Salt-Affected Soils,* 11th International Soil Sci. Soc., Congress, Edmonton, Alberta, Canada, Sec. 2: 7-29.

Oosterveld, M. 1978. "Disposal of saline drain water by crop irrigation." *Proc. of Meeting of Sub-commission of Salt-Affected Soils,* 11th International Soil Sci. Soc. Congress, Edmonton, Alberta, Canada, June 21-24.

Oster, J.D., and A.D. Halvorson. 1978. "Saline seep chemistry." *Proc. of Meeting of Sub-commission of Salt-Affected Soils,* 11th International Soil Sci. Soc. Congress, Edmonton, Alberta, Canada, June 21-24.

Power, J.F., J.J. Bond, F.M. Sandoval, and W.O. Willis. 1974. "Nitrification in paleocene shale." *Science* 183: 1077-1079.

Rasmussen, W.W., G.C. Lewis, and M.A. Fosberg. 1964. "Improvement of the Chilcott-Sebree (solodized-solonetz) slick spot spoils in southwestern Idaho." U.S. Dept. Agric., Agric. Res. Serv., ARS 41-91, 39p.

Rhoades, J.D. and A.D. Halvorson. 1977. "Electrical conductivity methods for detecting and delineating saline seeps and measuring salinity in northern Great Plains soils." U.S. Dept. of Agric., Agric. Res. Serv., Western Region, ARS W-42. 45p.

Schneider, R.P., B.E. Johnson, and F. Sobolik. 1980. "Saline seep management: Is continuous cropping an alternative? North Dakota Agric. Exp. Sta. Farm Res. 37(5): 29-31.

Smika, D.E., and C.J. Whitfield. 1966. "Effect of standing wheat stubble on storage of winter precipitation." *J. Soil Water Conserv.* 21: 138-141.

Sommerfeldt, T.G. 1976. "Mole drains for saline seep control." Montana Coop. Ext. Serv. Bull. 1132, p. 296-302.

Sommerfeldt, T.G., H. Vander Pluym, H. Christie. 1978. "Drainage of dryland saline seeps in Alberta." *Proc. of Meeting of Subcommission of Salt-Affected Soils,* 11th International Soil Sci. Soc. Congress, Edmonton, Alberta, Canada, June 21-24.

Steppuhn, H. and D. Jenson. 1984. "Barley can help control dryland salinity." *Crops and Soils Magazine* 36(8): 22-23.

Timpson, M.E., J.L. Richardson, L.P. Keller, and G.J. McCarthy. 1986. "Evaporite mineralogy associated with saline seeps in southwestern North Dakota." *Soil Sci. Soc. Am. J.* 50: 490-493.

Vander Pluym, H.S.A. 1978. "Extent, causes and control of dryland saline seepage in the Northern Great Plains of North America." *Proc., Sub-commission on Salt-Affected Soils,* 11th International Soil Sci. Soc. Congress, Edmonton, Alberta, Canada, June 21-24.

CHAPTER 18

PROJECT-LEVEL SALINITY MANAGEMENT OPTIONS[a]

INTRODUCTION

A regional, or project-level, program to manage water and soil salinity covers thousands of acres of land and involves a comprehensive plan to control salinity via numerous local-scale projects. Such projects might include augmentation of water supplies, restructuring of water use, enactment of water-related laws, development of interstate and international agreements on water use, improvement of the technology used to apply irrigation water, and water management techniques (Fig. 18.1).

The U.S. National Principles and Guidelines (Water Resources Council 1983) should be consulted for concepts that can be used to identify, analyze, plan, and evaluate soil and water resources projects. Economic conditions and current political objectives and policies are key factors in getting funding for water resource salinity control projects.

A regional management program must be built around objectives related to environmental quality or national and regional economic development, such as the following:

1. Improve surface and ground-water quality by reducing irrigation water salinity.
2. Reduce the need to develop additional water-supply structures by conserving water, as the construction of such structures often adversely affects fish and wildlife.
3. Reduce downstream water salinity damages.
4. Increase the overall economic efficiency of agricultural production.
5. Reduce salinity damages to ground water.
6. Increase crop production by reducing soil salinity.

For a project-level salinity management program to succeed, federal, state, and local agencies, farmers, and the public all must participate.

---

[a]Prepared by: John Hedlund, William Evans, Soil Conservation Service, Federal Building, Room 248, 511 N.W. Broadway, Portland, OR 97209; and Jim Thomas, U.S. Bureau of Reclamation, P.O. Box 25007, Denver Federal Center, Denver, CO 80225.

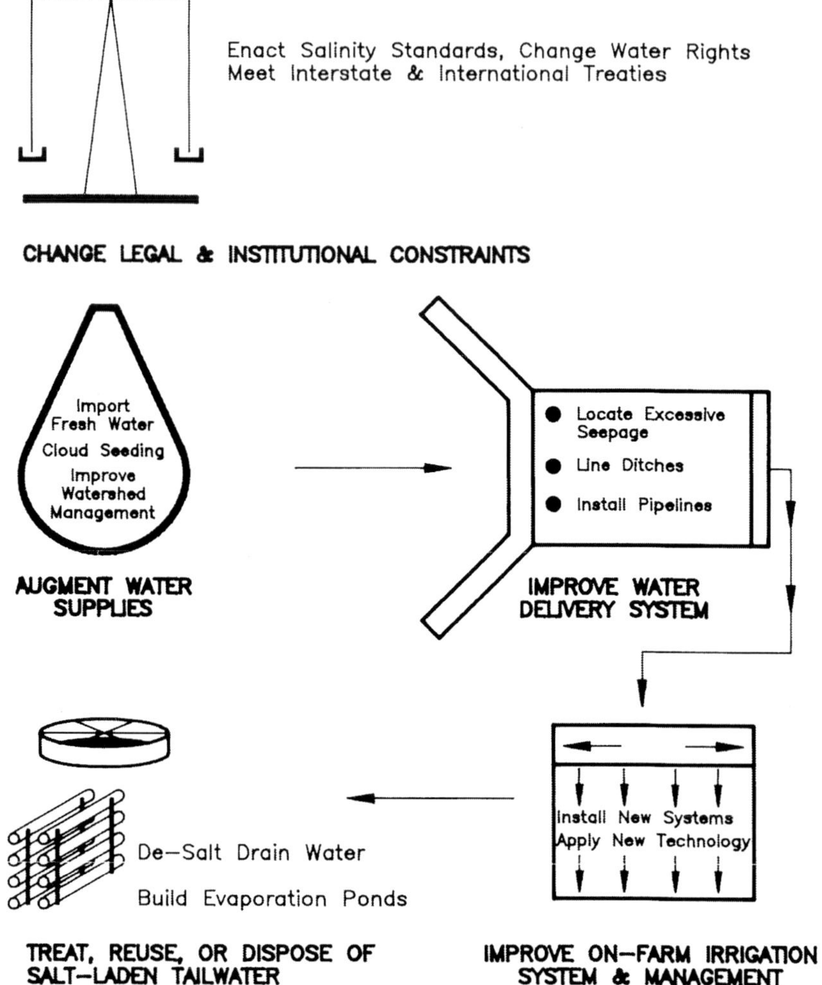

Fig. 18.1 Project-Level Salinity Management Options

PLANNING REGIONAL MANAGEMENT PROGRAMS

To plan and implement a regional soil and water salinity management program, the following steps must be taken:

1. Define the problem.
2. Analyze irrigation practices.
3. Develop alternative plans.
4. Compare alternative plans.
5. Select a plan for implementation.

Defining the Problem

One must go beyond saying that the main problem is downstream salinity or soil salinization and then quantify the problem.

Physical Factors. To analyze how salt loading affects streams, the seepage and deep percolation reductions that will result from improved delivery and on-farm irrigation systems must be determined. For example, many saline formations underlying soils have an unlimited salt supply, and saline subsurface return flows will continue to the river. If irrigation systems and management are improved so that seepage from canals and ditches and percolation from fields into the saline aquifer are reduced, then salt loading will be reduced proportionately. In other cases, deep percolation or seepage may displace residual saline ground water. Projects that control the saline water's rate of displacement could be used. The analysis is based on the technical ability to estimate accurately the reduction in seepage and deep percolation occurring with irrigation improvement. This translates into salinity changes. Basin-wide models are maintained to balance water and salt contributions from each study unit. Salt and water budgets form the nucleus for evaluating plan alternatives.

Physical budget analysis is never straightforward. Concurrent flow and water quality records for all desired locations in the basin over a representative time period are seldom available. Further, more than one mechanism could be at work to increase salinity between measuring points. Such mechanisms include irrigation, natural runoff, mining activity, and point discharges. Moreover, the number of these mechanisms varies over the years in response to hydrologic and economic activity.

In summary, data-intensive, interdisciplinary effort needs to be made to identify and quantify the salinity-producing mechanisms in an irrigated river basin. Many diverse pieces of data must be integrated to understand overall the movement of water and salt through the system. Sufficient evidence should be obtained from water and salt budgets to conclude whether salinity will be reduced by implementating the proposed salinity control program.

Economic Factors. A salinity control program is evaluated similarly to other kinds of water-related land resource developments (Hedlund et al. 1978). Since a program affects on-farm water users and downstream water users, the extent to which both groups are benefited should be analyzed. Also, several levels of resource development should be analyzed, so that the optimum level can be determined. This permits the selection of the best combination of control measures within an area and the best combination of areas to treat within a river system (Evans et al. 1981).

In formulating a program, the physical effects of each alternative are measured and translated into economic terms so that each alternative's benefits can be compared with its cost (Fig. 18.2).

To identify the optimal level of economic development, the point of maximum net benefit (point 2), which is when the incremental benefit equals the incremental cost, is determined.

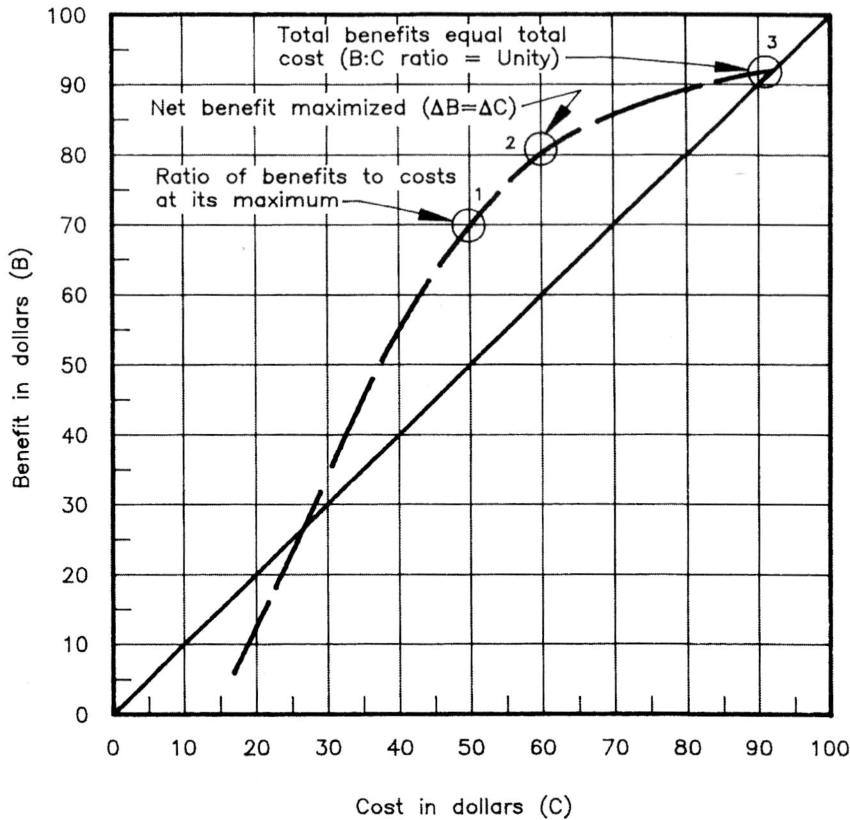

Fig. 18.2 Benefits and Costs for Alternate Plans

Environmental Factors. Improving the use and management of water for conservation makes full use of limited water supplies, increases instream flows for fish and wildlife, and reduces the salinity of downstream water. Irrigation improvements may adversely affect wildlife by eliminating artificially created wetlands along canals or within irrigated farms.

The processes that link the area's physical, chemical, and biological elements, including fish and wildlife, threatened and endangered species, cultural resources, and the value of prime farmland are analyzed. The effects of proposed actions, as well as the options and tradeoffs, are identified and measured. Water users, the concerned public, and state and federal agencies can then evaluate the choices for that area between water quality (salinity) and fish and wildlife habitat (Skogerboe et al. 1979).

Analyzing Irrigation Practices

Alternative irrigation systems used within the project area are listed, and how those systems and associated irrigation practices are adapted to specific site conditions are described.

Developing Alternative Plans

Available saline water and soil treatment opportunities (Fig. 18.1) are analyzed. Several promising technologies are selected and then the contribution that each alternative makes toward solving the salinity problem is evaluated. Acceptable alternative plans are evaluated in terms of physical, environmental, and economic factors.

Comparing Alternative Plans

A display that summarizes significant physical, economic, environmental, and social effects helps to identify readily the tradeoffs between alternatives. Alternative plans are analyzed to identify the plan that maximizes net on-farm monetery benefits, net off-farm or downstream benefits, total (on-farm and downstream) net monetary benefits, water conservation, salinity reduction, and environmental protection for fish and wildlife.

Table 18.1 summarizes the results of four typical alternatives for improving irrigation management and systems. Salt load and salinity concentration reduction and off-farm monetary benefits are used to illustrate salinity improvement. Summaries, such as this, can fully reveal the benefits and costs of each and provide a basis by which all interested groups can decide on salinity control. Plan A achieves a nominal amount of salinity reduction at the least cost. Plan B maximizes net benefits to the farmer. Plan C maximizes total net benefits. Plan D maximizes the reduction of salinity downstream.

Selecting a Recommended Plan

Local farmers select an acceptable plan that meets national and international water conservation and interstate water quality goals. It may include canal and lateral improvements, structural on-farm improvement measures, improved irrigation water management, reuse, selective land retirement, technical assistance, and cost-share support. For example, Plan C in Table 18.1, which maximizes net benefits, would be recommended for federal implementation under 1988 executive policies in the United States. It also provides nearly maximum salinity reduction and gives farmers some on-farm benefits (Hedlund et al., 1978).

SALINITY CONTROL OPTIONS

A good salinity control program involves a strategy for total management of a river system. A broad systems viewpoint and long-term perspective are essential to the national planning of solutions to salinity problems. To illustrate project-level actions, four case studies will be examined.

San Joaquin Valley Drainage Dilemma

In California, 1.6 million ha, or nearly 50% of all irrigated land, is threatened by salinity and high water table problems (Backlund and Hoppes 1984). This is in a state which, in 1987, produced nearly one-

TABLE 18.1 Summary of Alternate Plans for Improving Irrigation Systems and Irrigation Management

Item (1)	Alternatives: Plan A: Least cost salinity reduction (2)	Plan B: Maximize farmer's net benefits (3)	Plan C: Maximize total net benefits (4)	Plan D: Minimize downstream salinity (5)
Annual benefits				
On-farm	$115,000	$160,000	$175,000	$180,000
Off-farm	100,000	150,000	180,000	190,000
Total	$215,000	$310,000	$355,000	$370,000
Annual cost	$100,000	$140,000	$160,000	$180,000
Total net benefits	$115,000	$170,000	$195,000	$190,000
Net Benefits				
On-farm[a]	$ 15,000	$ 20,000	$ 15,000	$ 0
Off-farm[b]	$ 0	$ 10,000	$ 20,000	$ 10,000
Salt load reduction (tons)	4,000	6,000	7,200	7,600
Sediment load reduction (tons)	12,000	18,000	21,600	22,800
Loss of wetland (acres)	100	140	160	180
Conversion of wetland habitat to upland habitat (acres)	300	420	480	540
Decrease in electrical energy use (MWh)	11	14	16	18
On-farm irrigation efficiency (%)	58	71	78	81

[a]Net benefits accruing on-farm when all costs are borne by on-farm interests.
[b]Net benefits off-farm when all costs are borne by off-farm interests.

fourth of all agricultural goods in the United States, an output worth some $14 billion.

Installing subsurface drainage systems and improving the management of irrigation water can reduce production losses caused by high water tables and soil salinity. However, insufficient disposal sites for the saline drainage water, especially in the San Joaquin Valley, have prevented landowners and the district from implementing needed measures.

Solving the disposal of saline drainage water, which includes other toxic materials, is California's biggest challenge to sustaining a viable agricultural economy. Evaporation ponds provide an interim solution. However, evaporation may concentrate salts and toxic materials to hazardous levels.

One proposed solution involves building a pipeline over the mountains for ocean disposal, but the cost is prohibitive. Another less expensive option is to route return flows to the ocean via the San Francisco Bay. This has met with continued and vehement political opposition on environmental grounds.

Beside this search for a disposal solution, there exists an intensive on-farm water management effort to reduce the volumes of subsurface drainage and the loading of pollutants. Improved irrigation methods, significantly more efficient than existing furrow irrigation, could reduce drainage volumes by about 50%.

Colorado River Basin Salinity Control Program

The U.S. Congress recognized the severity of the Colorado River's salinity problems and passed Public Law 93-320, the Colorado River Basin Salinity Control Act (88 CFR 266, June 24, 1974). It provides for the construction, operation, and maintenance of certain works in the seven Colorado Basin states to control the salinity of water for users in the United States and Mexico.

Public Law 98-569, an amendment to PL 93-320 that was signed into law on October 30, 1984, authorizes construction of additional salinity control units and directs the secretary of agriculture to establish a major voluntary on-farm cooperative salinity control program (U.S. Bureau of Reclamation and USDA 1986). The Salinity Control Act requires full coordination and cooperation between the U.S. Department of the Interior, the U.S. Department of Agriculture (USDA), and the U.S. Environmental Protection Agency.

Salinity control planning has been oriented toward reducing the salt load to the river (Hedlund 1984). Options include preventing saline water from entering the river, and reducing salt pickup by improving irrigation systems.

The salinity control problem has many technically complex and institutionally challenging aspects. The cooperative state and federal program is based on maintenance of 1972 salinity concentrations in the lower mainstem while the upper basin develops its compact-apportioned waters.

Salt reduction at minimal cost is what determines which salinity control units or practices should be initiated (Walker et al. 1977). However, each on-farm alternative undergoes a complete evaluation to make sure on-farm benefits offset the farmers' costs.

Upper Arkansas River—Irrigation with Saline Water

The Arkansas River in southeastern Colorado is one of the most saline rivers in the United States. The total dissolved solids concentration of irrigation water used by farmers there is at least 2,000 mg/l and frequently exceeds 4,000 mg/l in the lower reaches of the valley. Water containing more than 2,000 mg/l salt would normally be assumed to be unsuitable for irrigation, but such water is used successfully on more than 80,000 ha of land in the Arkansas Valley, Colo.

The salinity in the Arkansas River Basin is more a problem of reuse and full development of a limited water resource than of salt loading.

About 85% of the total surface water supply is consumed before the river leaves Colorado.

Although controlling salt loading as much as practically possible is desirable, it is expected that salinity levels will remain high as long as water is used. Therefore, the greatest potential for reducing salinity damages is apparently in learning how to use and reuse highly saline water.

Great Plains—Saline Seep Management Projects

Saline seeps reduce production on over 200,000 ha of dry cropland throughout the Great Plains from Texas to Montana. They also pose a serious problem in Canada's prairie provinces, India, Iran, Turkey, and western and southern Australia. Examples of current project-level action that illustrate two different solutions to saline seep problems follow.

The Triangle Conservation District saline seep project in Montana involves a joint effort between 10 conservation districts in nine counties, the USDA Soil Conservation Service, the USDA Agriculture Research Service, the Montana Cooperative Extension Service, the USDA Agricultural Stabilization and Conservation Service, and the Montana Bureau of Mines and Geology (Dodd et al. 1984). A planning team works with individual farm operators using a three-stage approach. First, the team performs a field review to establish discharge and recharge areas. Next, test holes, or wells, or both, are installed to provide quantitative hydrologic data. Finally, the planning team develops a plan for reclaiming and managing the seep. In most cases, a change from growing shallow to deep-rooted crops or a change in land use management on the recharge areas alleviates the problem.

The saline seep program for the Red Plains Area of central Texas focuses on "salt spots" that hamper crop production in cultivated fields in all or parts of 55 counties (USDA 1983). The solution involves project-level planning and community action. The saline-seep problems in this region typically are solved or alleviated by installing subsurface drains. Deep-rooted vegetation are planted to make greater use of available soil moisture and prevent accumulation in shallow water tables. Tests of the quality of water discharged from existing subsurface drainage systems have shown that the overall quality is often better than the quality of the surface water in the adjoining local stream.

LEGAL AND INSTITUTIONAL FACTORS

Legal and institutional constraints often make implementing salinity control projects more difficult. Some progress has been made in modifying these constraints to make implementation of projects more feasible.

Water Quality Standards for Salinity

In 1965, the U.S. Congress passed the Water Quality Act, which requires each state to establish water quality standards for its portions of any interstate waters or waterways. In 1972, Congress passed the Federal Water Pollution Control Act Amendments, Public Law 92-500. The U.S. Environmental Protection Agency, which was designated as the agency that will enforce these amendments, interpreted them as a mandate to establish numeric salinity standards for problem river systems, such as the Colorado River (Colorado River Basin Salinity Control Forum 1984). This created specific standards that facilitate the implementation of salinity control projects.

Basin-Wide Jurisdiction

For an effective salinity control program, a single entity with legislative authority to plan and implement basin-wide projects is necessary. That same entity also would be responsible for pushing for appropriate legislation and funding.

For example, the seven Colorado River Basin states introduced legislation in Congress in 1972 and 1973 to authorize a basin-wide salinity control program. These states, to ensure cooperation, formed the Colorado River Basin Salinity Control Forum, comprised of up to three governor-appointed representatives from each state. With this basin-wide approach to controlling salinity, the Basin states have been able to adopt numeric water quality standards for salinity, develop a plan for salinity control, get federal legislation passed, and obtain funding.

Effect of Water Rights

State water rights constrain, to some degree, virtually all salinity control options. In some states, water rights prevent the retirement of high-salt-yielding farmland, the transferral of "saved" water to nonpolluting use, the collection and reuse of saline water, the collection and disposal of saline flows, or reducing the exportation of good-quality water from a river system.

Lack of Urgency

Much of the control problem arises from how the salinity issue is perceived locally and nationally. Irrigation return flow is an off-site or downstream problem; the upstream farmer may be unaffected and, therefore, unaware of it. At the national level, the salinity problem is perceived as a regional problem to be solved by the states. The lack of urgency that sometimes prevails may be due to confusions or disbelief about hydrologic predictions for the future. Federal agencies, regional interest, and environmental groups often perceive predictions as self-serving.

International Treaties

Treaties between the United States and the Republic of Mexico on the problem of salinity have fostered the development of project-level salinity management programs in the Colorado River Basin (Hobart 1984).

Approximately 90% of the water in the Colorado River is consumed within the United States before the river flows across the border into Mexico. In 1964, the Colorado River was completely dry from the

Mexicali Valley, Mexico, to the Gulf of California. The river was, and still is, fully consumed before reaching the ocean.

On August 30, 1973, the United States reached an agreement with Mexico in Minute No. 242 of the International Boundary and Water Commission, titled "Permanent and Definitive Solution to the International Problem of the Salinity of the Colorado River." That agreement states that the average annual total dissolved solids (TDS) content of all waters delivered to Mexico upstream from Morelos Dam may be no higher than 115 mg/l above the TDS content of Colorado River water arriving at Imperial Dam. The United States has been in compliance since the agreement was made.

Project planning and implementation, under Public Law 93-320, has involved cooperative actions by federal, state, and local entities, and individuals. Title I of the project involves several water management measures below the Imperial Dam, which include lining the Coachella Canal, construction of a regulatory pumping station, and construction of a desalting complex on the Wellton-Mohawk drain.

The USDA's involvement specifically relates to on-farm treatment and irrigation water management improvement in the Wellton-Mohawk Irrigation and Drainage Districts near Yuma, Ariz. (Advisory Committee on Irrigation Efficiency 1974). Since 1975, farmers have voluntarily and individually managed their water sufficiently to reduce diversions to the valley by 20% and irrigation return flow by 50%. About 1.23×10^8 m^3 less water per year is now being pumped into the project, delivered to the farm, applied to the crops, and pumped out to the desalting plant.

SALINITY REDUCTION BY IMPROVING WATER SUPPLIES AND IRRIGATION MANAGEMENT

Augmentation of Water Supplies

The expansion of water supplies to achieve greater dilution may in some cases be an important salinity control option. For example, in southeastern Australia, the River Murray Commission has implemented a plan to contain and reduce salinity in the River Murray by augmenting water supplies and increasing water storage capacity. The plan also calls for establishing numeric water quality standards for various points along the river, monitoring and modeling of spatial and temporal water quality trends, and diverting saline drainage waters away from the river. The salinity control options for the Colorado and Rio Grande Rivers in the United States and Mexico are similar to those for the River Murray. As water is used and then discharged back to the river, it becomes more saline to downstream users.

Earlier in this century, the United States vigorously augmented its water supplies in the Colorado River Basin and throughout the rest of the country by construction of numerous dams, reservoirs, canals, and pipelines. However, the potential for expansion of the nation's water supply system has become very limited. Few rivers remain undammed, and the construction of additional canals, reservoirs, and other water resources system infrastructures typically involve considerable environmental and legal problems and extremely high construction costs.

Imports of Fresh Water, Interbasin Transfers. In the arid and semi-arid portions of the West with serious salinity problems, water is scarce and nearly all existing water supplies are committed to beneficial uses. The opportunities for importation and transfers for augmenting water supplies are very limited. Little public support exists at either the local or national level for water resource development and interregional water transfers. In addition, during the last 20 years, construction costs have escalated and a strong, broadly-based, largely anti-development environmental movement has emerged.

Weather Modification. Current weather modification research may develop a reliable system for increasing precipitation in some local areas. The net benefits, the flexibility of use, and the opportunity to obtain even greater water yields with new techniques point to cloud seeding as a desirable tool for increasing water supplies to dilute salinity.

The technical, legal, and institutional problems of measuring and allocating increased flows will present major obstacles to a basin-wide weather modification program. Even if additional high-quality water could be produced, no assurances that it would be used for salinity control exist. On the contrary, junior appropriators could claim the water and use it to irrigate more land, thereby worsening the salinity problem.

Desalting Processes. Current technology has made several commercial desalting processes, such as reverse osmosis, electrodialysis, and ion exchange, available for use in the treatment of saline water. Process selection depends upon the volume and quality of the saline water and major economic factors, such as energy costs and the costs for disposing concentrated salt brine (U.S. Bureau of Reclamation 1972).

Saline water may be desalted at the point of diversion or prior to use. For irrigation purposes, this would involve large desalting plants. An alternative involves desalting the irrigation return flows before they are discharged to a stream. In some areas, desalting sea water and ground water to augment fresh water supplies may prove economical. Desalting is technically feasible, but the costs are high and the disposal of brines presents a problem.

Vegetation and Watershed Management. In most of the arid and semiarid West, the opportunity for water salvage by watershed management or phreatophyte control remain very limited. Environmentalists and preservationists want to retain wetland and wildlife habitat associated with irrigation development.

According to the U.S. Forest Service, removal of riparian vegetation on perennial or intermittent streams seldom saves water. Shade removal increases evaporation and transmission losses may be high. Ecologically sound watershed-management techniques could reduce the amount of sediment and associated salt from some areas.

Reuse of Saline Drainage Water for Irrigation. One strategy for controlling the salinity of river systems involves intercepting drainage returns before they are mixed back into the river and using them to irrigate certain crops in the rotation (Rhoades 1984). When the drainage water quality is such that its potential for reuse is exhausted, then this drainage is discharged into evaporation ponds, or injected into deep wells, or desalted.

Improvement of Water Delivery Systems

An important part of salinity management is the reduction of canal and lateral seepage into underlying saline aquifers. If a water delivery system is unlined and soils are permeable, considerable water may be lost through seepage. Not only is this seepage unavailable to the farmer, it may pick up other pollutants and salt before returning to a stream or entering a ground-water aquifer.

Locating and Quantifying Canal Leakage. In the United States, about one-quarter of all the water withdrawn from streams and reservoirs for irrigation does not reach the farm. On a delivery system, determining the rate at which seepage occurs and locating sections of high seepage is difficult.

Leaky reaches of canals are typically identified through geologic maps, construction records, aerial photographs, soil surveys, ground reconnaissance, interviews with local farmers, canal company and cooperative extension service personnel, hydrologic measurements, and ground-water data. A walking survey of the canal is recommended. The presence of wetland vegetation below a canal can help to identify seepage.

The rate of seepage from a given canal depends on such factors as the physical characteristics of the material in the channel and subsoil, depth of water in the canal, depth to water table, sediment concentration in the water, temperature of canal water, ground water, biological growth in the water and along the canal, and the time of year. To obtain reliable estimates of seepage, field measurements are needed.

With hundreds of miles of canals and laterals, seepage measurements on all canals and laterals is impossibly expensive and time-consuming. To extrapolate seepage measurements, the canal system must be characterized adequately, and reaches should be classified by soil type, conveyance properties, and geohydrologic setting.

Several methods for estimating seepage exist, but none is entirely satisfactory for all situations. They include ponding tests, inflow-outflow measurements, seepage meters, barrel tests, and mathematical determinations (Dhillon 1967).

Reducing Salt Loading. Efforts to reduce seepage alone do not necessarily lead to reduced salinity. Unless the seepage water dissolves minerals or displaces saline ground water on its journey to the underlying ground-water system or receiving stream, reducing seepage may negligibly affect salinity. Thus, a basin-wide or irrigation-project-wide investigation is needed to identify salinity-producing mechanisms and develop a control strategy.

Reduction of Canal Losses. Canals and laterals with large seepage losses used to be lined to conserve water, with no thought given to salinity control. It is now recognized that reducing seepage in areas where soluble salts are present saves water and reduces salt loading to streams. Seepage has been successfully controlled by lining canals with compacted earth, bentonite, concrete, and various geomembranes, or combinations of these materials. For smaller canals and laterals, closed conduits of concrete or polyvinyl chloride are practical alternatives. Pipelines reduce evaporation, produce a pressure head, and can be buried below the land surface. They are often useful for linking or otherwise combining canals.

Water Control Structures. Water measurement devices and control structures facilitate better management and reduce salt pickup independent of a seepage reduction program. In open channels, various culverts, checks, drops, and dividers are commonly required to regulate water levels, control velocities, prevent erosion, and direct water deliveries.

Improved Management. A modern irrigation delivery system, with appurtenant water control structures and flow measurement devices, must be operated efficiently to achieve crop production and salinity control goals. Lining canals or installing pipelines will control seepage and the associated salinity contributions. It is equally important to manage water deliveries to match crop requirements (Clemmens 1987). Good district records of water deliveries are needed. Accurate measurement and accounting for most of the water is essential for good irrigation-water management.

It is vitally important to match physical improvements to the delivery system with on-farm water conservation programs.

Improving On-Farm Irrigation Systems and Management

Surface and subsurface irrigation return flows carry agricultural pollutants. These include dissolved salts, sediment, nitrogen, phosphorus compounds, pesticides, and oxygen-consuming organic compounds. The most persistent pollution problem in the West is salt loading of streams and ground-water aquifers by ditch seepage and deep percolation of irrigation water.

Improving on-farm irrigation with practices that can reduce ditch seepage, deep percolation, and the associated salt load picked up in irrigation return flow is an important project-level strategy for salinity control (Table 18.2).

Soil salinization poses a critical problem on millions of acres of irrigated land, especially where on-farm drainage systems have not been installed and where no permanent disposal site for highly saline drainage effluent exists. Proper leaching of salts, management of irrigation water, and the disposal of saline drainage flow all must be considered in planning a program to prevent salinization of irrigated soils (Willardson and Hanks 1976). Water and salt budgets are needed to evaluate the project's effects on soil salinity and downstream water quality.

On-farm Irrigation Water Management. One part of a salinity control

TABLE 18.2 Sources of On-Farm Salt Loading and Practices for Attaining Salt-Load Reduction

Salt Sources (1)	Application Management (2)	Structural Practices (3)
Seepage:		
On-farm ditches	Utilize water measuring devices. Practice good maintenance.	Installing ditch lining or pipelines and appurtenant water structures.
Tailwater ditches	Follow irrigation water management (IWM) plan to reduce tailwater.	Install lined collection ditches or lined ponds.
Deep percolation:		
excessive and non-uniform application of irrigation water.	Adjust the number and frequency of irrigations, time of set, and flow rate. Follow IWM plan. Change method of irrigation.	Install measuring devices. Land level to improve length of run. Automate the system.
Inefficient layout of fields and irrigation systems.	Combine fields and reorganize or change to more efficient method of irrigation.	Convert existing open drains to buried drains. Relocate ditches and pipelines. Re-level and select proper length of run.

plan is the improvement of on-farm salinity water management. This may only involve providing a farmer with an irrigation water management plan that includes the use of devices to measure the rate and volume of water delivered to the field, the monitoring of soil moisture, the scheduling of irrigations, training the irrigator on how and when to apply the water and otherwise follow an irrigation-water-management plan, maintenance of the irrigation system, and following good agronomic practices to maintain water infiltration and promote full root development.

Ditch Lining and Pipelines. Another part of a salinity control plan is to line ditches or install pipelines to reduce seepage. Seepage from ditches can contribute significantly to salt load pickup. On-farm improvements should include concrete ditches with headgates and ports or notches, or pipelines and gated pipes with control valves. Delivery systems can be improved by including water control structures with built-in measuring devices. Gated pipe may be semi-automated by including timing devices on control valves, surge controls, and cablegation. Improving a significant reach of the off-farm water delivery system may be needed to operate the on-farm improvements properly. A pipeline could be installed to develop gravity pressure head to operate a sprinkler system.

Installing New or Improved Systems. Yet another part of a salinity-control plan, sometimes inseparable from the other two, is an improvement of the on-farm irrigation system to reduce deep percolation and associated salt loading. Obtaining precise land leveling and proper length of run on surface-irrigated fields will help to achieve uniform application. Combining fields or changing to a more efficient method of irrigation may also help reduce deep percolation.

On-Farm Drainage, Tailwater Ditches, Collector Ponds, Reuse Systems. To combine fields and reorganize irrigation systems, existing open drains must be converted to buried drains. Installing on-farm drainage systems to handle the leaching fraction of irrigation water will prevent soil salinization. Lining collector tailwater ditches and collection ponds can help to prevent seepage into underlying saline aquifers and reduce salt pickup. Irrigation drain water can sometimes be reused. Pump-back systems permit such reuse on the same field.

Selective Land Retirement and Irrigation Water Acquisition. For optimal salinity-control strategies in project-level planning, selective land retirement must be considered. In Arizona's Wellton-Mohawk Irrigation and Drainage District, the salinity control program involved the purchase of 2,520 ha of land with poor irrigation efficiency. This land was then retired from irrigation to reduce the volume and cost of desalting return flows for use in Mexico.

Monitoring and Evaluation. Another part of a salinity control plan is to coordinate with the farmer and adjust his irrigation water management plan to fit the system and changing farm practices. Monitoring and evaluation provides valuable economic, environmental, and irrigation information (USDA, SCS 1982).

Managing Wastewater

To alleviate drainage water disposal problems, point sources of drainage water can be intercepted and diverted to other outlets. Several alternatives for reusing or disposing of drain water are described.

Export to the Ocean or Another Basin. Irrigation drainage water may be disposed of in the ocean or in terminal lakes. The Imperial Valley, Calif., situated in the once-desolate Colorado Desert, has become one of the world's most productive farming regions. However, the Salton Sea, a terminal lake in the basin, has been the receptor of saline wastewater collected from 32,000 km of drains in the 186,000 ha irrigation project. The recent increases in surface area and salinity of the Salton Sea caused millions of dollars in damages to shoreline properties. Improved irrigation water management and installation of level basin surface irrigation would reduce the amount of irrigation drainage water that flows into the Salton Sea.

Evaporation Ponds. The situation in California's San Joaquin Valley illustrates some of the conflicting issues of using evaporation ponds. Irrigation-drainage water there is disposed of in on-farm and district-wide evaporation ponds. Although the ponds provide temporary relief, a permanent solution is necessary because evaporation results in the deposition of salts that contain hazardous levels of trace elements (Hall 1986).

Desalting. The Wellton-Mohawk Salinity Control Project in Arizona was created to help treat irrigation drainage water from the Colorado River basin and improve the quality of water sent downstream to Mexico. The keystone of the Wellton-Mohawk Salinity Control Project is the Yuma Desalting Plant, which makes use of reverse osmosis (RO) desalting equipment. When complete, this plant will treat 273,000,000 l of drainage water daily (Trompter 1987). The 8.3×10^7 m^3 per year of product water, at 295 mg/l, will be mixed with the flow of the Colorado River and delivered to Mexico. The RO reject flows through a lined bypass drain to the Santa Clara Slough above the Gulf of California.

Deep Well Injection. Injection of saline water into an isolated deep aquifer may be a feasible disposal method in many areas of the West. Deep well injection currently is a key part of the Paradox Valley Unit of the Colorado River Basin Salinity Control Program (U.S. Bureau of Reclamation 1978).

REFERENCES

The Advisory Committee on Irrigation Efficiency. (1974). "Special report on measures for reducing return flows from the Wellton-Mohawk Irrigation and Drainage District," U.S. Bureau of Reclamation, Yuma, Ariz. 109 pp.

Backlund, V. L., and Hoppes, R. R. (1984). "Status of soil salinity in California." *Calif. Agric.* 38(10): 8–9.

Clemmens, Albert J. (1987). "Delivery System Schedules and Required Capacities, in Planning Operation, Rehabilitation and Automation of Irrigation Water Delivery Systems." Proc. of ASCE Symposium in Portland, Oregon, July 1987.

Colorado River Basin Salinity Control Forum. (1984). "Water Quality Standards for Salinity, Colorado River System," U.S. EPA, Denver, Colo.

Dhillon, G. S. (January 1967). "Measurement of seepage losses from irrigation canals." *Ind. J. of Power and River Valley Dev.* 17(1).

Dodd, T., Holzer, J., Adams, W., Farkell, J., and Miller, M. R. (1984). "A Case Study on Dryland Saline Seep Control, Triangle Conservation District." Proc. International Symposium on State-of-the-Art Control of Salinity, ASCE, pp. 561–570.

Evans, R. G., Walker, W. R., and Skogerboe, G. V. (1981). "Optimizing Salinity Control Strategies for the Upper Colorado River Basin." Colorado State Univ., Ft. Collins, Colo., AER80-81RGE-WRW-GVSI.

Hall, S. K. (1986). "Evaporation Basins—Environmental Disaster, or

Economic Necessity?" Proc. of 1986 Regional Meeting on Toxic Substances in Agric. Water Supply and Drainage, U.S. Committee on Irrigation and Drainage, Denver, Colo., pp. 27–35.

Hedlund, J. D., Russell, C., and Collins, H. (1978). "Techniques for Planning and Evaluating Water Conservation and Water Quality Measures in Irrigated Agriculture," 33rd Annual SCSA Meeting, Soil Conservation Society of America, Ankeny, Iowa.

Hedlund, J. D. (1984). "USDA Planning Process for Colorado River Basin Salinity Control." Proc. of 1983 International Symposium on State-of-the-art Control of Salinity, ASCE, pp. 63–77.

Hobart, M. S. (1984). "The Lower Colorado, a Salty River." *California Agriculture*. 38(10): 6–8.

Rhoades, J. D. (1984). "Reusing Saline Drainage Waters for Irrigation: A strategy to reduce salt-loading of Rivers." R. H. French, ed., *Salinity in Water Courses and Canals*, Chap. 43, Ann Arbor/Butterworth Publishing Co., Boston, pp. 455–464.

Skogerboe, G. V., Walker, W. R., and Evans, R. G. (1979) "Enviromental Planning Maual for Salinity Management in Irrigated Agriculture." Report EPA-600/2-79-062, Ada OK.

Trompoter, K. M. (1987). "The Yuma Desalting Plant—A Water Quality Solution." Draft paper, U.S. Bureau of Reclamation, Yuma, Ariz.

U.S. Bureau of Reclamation and USDA. (1986). "Joint Evaluation of Salinity Control Programs in the Colorado River Basin," Denver.

U.S. Bureau of Reclamation. (1978). "Paradox Valley Unit Definite Plan Report," Denver.

U.S. Bureau of Reclamation. (1972). *Desalting Handbook for Planners*, 1st ed. Denver.

U.S. Department of Agriculture. (1983). "Dryland Salinity Study." Texas Cooperative River Basin Survey, Soil Conservation Service, Temple, Tex.

U.S. Department of Agriculture, SCS (1982). "Monitoring and Evaluation Plan for Grand Valley Unit, Colorado and Uintah Basin Unit, Utah." Soil Conservation Service, Denver, 95 pp.

Walker, W.R., Skogerboe, G.V., and Evans, R.G. (1977). "Development of Best Management Practices for Salinity Control in Grand Valley." Proc. National Conference on Irrigation Return Flow Quality Management. Colorado State Univ., Fort Collins, Colo. pp. 385–395.

Willardson, L. S., and Hanks, R. J. (1976). "Irrigation Management Affecting Quality and Quantity Return Flow." Report EPA-600/2-76-226. Ada, Okla., 206 pp.

Water Resource Council. (1983). "Economic and Environmental Principles and Guidelines for Water and Related Land Resource Implementation Studies." U.S. Printing Office, Washington, D.C.

VII. LAND RECLAMATION, TREATMENT AND DISPOSAL OF DRAINAGE WATERS

CHAPTER 19

RECLAMATION OF SALINE, SODIC, AND BORON-AFFECTED SOILS[a]

INTRODUCTION

For irrigated agriculture to succeed, salinity, sodicity, and boron (B) levels in soils and irrigation water must be controlled.

Saline soils exist in arid and semi-arid climates and in regions with poor drainage. The composition and concentration of salt in the soil solution influence the growth of plants by osmotic and specific ion toxicity effects and by changing the physical properties of soils. The accumulation of dispersive cations, such as Na, on the exchange phase affects the physical properties of the soil. This, in turn, affects the production of crops.

A low infiltration rate (IR) is associated with sodic soils. Irrigating with water of low electrolyte concentration (such as rainwater) results in ponding and runoff of water. Plants, which respond only to water in the root zone, consequently have less available water. The reclamation of sodic soils ameliorates adverse conditions, such as surface crusting, and increases the hydraulic conductivity (HC), which affects the infiltration and storage of water, the emergence of seedlings, and the development of roots.

Whereas adsorbed Na and Ca ions affect plant growth through changes in soil physical properties, B may cause injury to plants even when present at very low concentrations in the soil solution. Because plants obtain B from the soil solution, and since there is a relatively small range between B levels in the soil solution that cause deficiency and toxicity symptoms in plants, B distribution between solid and liquid phases of soil under various conditions is important. The suitability of irrigation waters is evaluated on the basis of B concentration, which determines the potential of the irrigation water to injure plants and reduce yield. It is important, therefore, to clarify the conditions that allow the soil to act as a buffer with respect to B concentration in the soil solution. This knowledge can improve the efficiency of using waters of varying B concentrations.

A successful irrigated agriculture requires permanent control of salinity, sodicity, and B levels in soils and irrigation water. Reclamation

[a]Prepared by: R. Keren, Inst. of Soils and Water, The Volcani Center, ARO, Bet Dagan, Israel; and S. Miyamoto, Agric. Res. and Extension Center, Texas A&M Univ., El Paso, TX 79927.

of affected soils by means of improved drainage, chemical treatments and leaching of salts, can reduce natural hazards and make a stable background for intensive cropping. Such reclamation will be the focus of this chapter.

EFFECTS OF SALINITY AND ADSORBED IONS ON SOIL PROPERTIES

Excess adsorbed Na most adversely affects the permeability of the soil (Alperovitch et al. 1981; Frenkel et al. 1978; McNeal et al. 1968), which decreases with the square of the pore radius. Swelling, or the movement of clay, or both, significantly affects permeability. The extent of swelling, or dispersion of clays, or both, depends on the mineralogy of the clays, the composition of the adsorbed ions, and the concentration of salt in the solution (Goldberg and Glaubig 1987; Oster et al. 1980; Shainberg et al. 1971). The swelling of clay increases as the electrolyte concentration of the solution decreases. The dispersion of clay occurs only at electrolyte concentrations below the flocculation value (FV) of the clay. The degree to which the clay swells before the solution is replaced with one that has an electrolyte concentration below the clay's FV determines whether clay particles will move into the conducting pores and leave the root zone, or remain trapped in the narrow pores and decrease the HC of the soil (Keren and Singer 1988).

When leached with water of low salt concentrations, the soil's potential for mineral weathering, i.e., for releasing electrolytes, influences the HC. Calcareous soils and soils with feldspar minerals release electrolytes into the solution in the presence of exchangeable Na. If the electrolyte concentration of the soil solution is maintained above the FV, the clay does not disperse and the HC changes minimally. The electrolyte concentration of the applied water controls the HC of soils without weathered minerals. When these soils are leached with low salt waters (below FV), clay disperses and the HC decreases.

RECLAMATION OF SALT-AFFECTED SOILS

Concepts and Principles

Leaching is the usual way to reclaim salt-affected soils, since plant uptake removes insignificant amounts of salt. It is suited to areas where leaching water is available. Nonirrigated areas must rely on natural precipitation for leaching.

Salt leaching involves the dissolution of soluble salts in the soil, the passage of water through soil profiles, and the removal of salt from the root zone. Thus, soils to be reclaimed must be permeable and have outlets for drainage. Drainage systems may need to be established before soils with high water tables are leached.

The extent of leaching required depends largely on initial soil salinity, the salt tolerance of the crops, and the depth of the water table. The prevailing idea in earlier reclamation trials was to leach all excess salts from the entire depth of the root zone. However, under good drainage conditions, the aim is to reduce salinity in the top 45 cm to 60 cm of the soil to below the threshold values of the crop (Chapter 13).

Land-use objectives dictate the amount of time to allot for leaching. To reclaim salt-affected virgin land, leaching may be conducted for several months or more. To reclaim cropped fields, e.g., permanent pastures and orchards, leaching must take place in a short time period to avoid lack of aeration.

Leaching generates highly saline drainage water, which can contaminate water and contribute to rising water tables. Evaluating the environmental and legal implications must be a part of any reclamation program.

Efficiency of Salt Leaching and Water Requirements

The efficiency of salt leaching can be defined as the quantity of soluble salts leached per unit volume of water applied. When leaching must take place within a short time or under high evaporative conditions, also consider the time factor. Factors that determine efficiency as defined above include the initial salinity of the soil, the amount of leaching water applied, the content of the soil water during leaching, hydrodynamic dispersion, spatial variation, and, when applicable, layout of drainage systems.

One-Dimensional Leaching. The efficiency of salt transport—the salt leaching efficiency normalized by the initial concentration of salt—increases as the contents of the soil water decrease during leaching. The range of soil pore size distribution also affects the efficiency of salt transport. Salt-transport efficiency is especially low in deeply plowed, wet, clay-rich fields, because the leaching water percolates between large clods that are too wet to slake. Drying and disking the soil reduces this problem, yet pore sizes may vary enough to cause substantial hydrodynamic dispersion, especially during ponded leaching. Such hydrodynamic dispersion decreases under unsaturated flow, resulting in the more efficient transport of salt (Nielsen et al. 1966).

Hoffman (1980) proposed the following empirical formula for the salt transport efficiency under one-dimensional leaching based on field data from various parts of the world:

$$\frac{C}{C_o} = \frac{k}{\left(\frac{D}{D_s}\right)} \tag{19.1}$$

where C = salt concentration in the soil; C_o = initial salt concentration in the soil; D = depth of leaching water applied; D_s = depth of the soil to be leached; and k an empirical coefficient that ranges from 0.1 (for sandy loam) to 0.3 (for clay) under ponded leaching as shown in Fig. 19.1a. Hoffman found that the coefficient of 0.3 also applies to silty clay loam, silty clay, and clay loam. He also found that, under intermittent ponding with 5 cm to 15 cm per application, the empirical coefficient is about 0.1, irrespective of the type of soil. Oster et al. (1972) found that sprinkler irrigation of silty clay produced a similar coefficient. The above

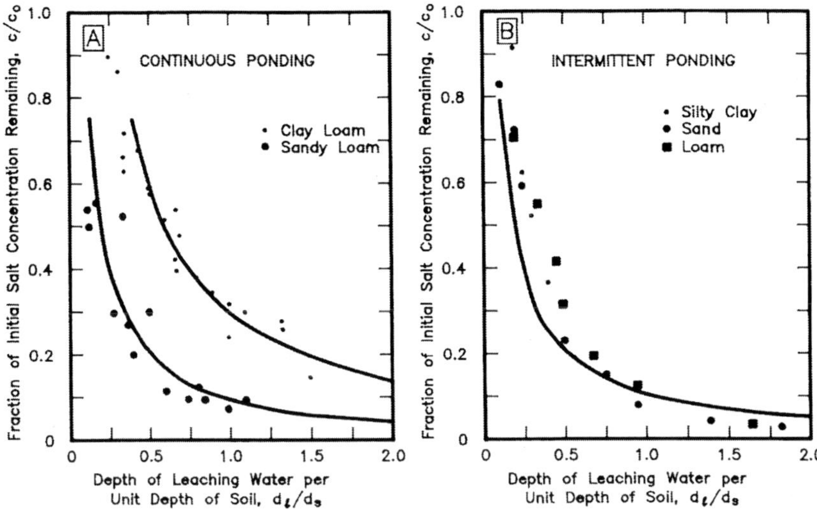

Fig. 19.1 Relationship Between Fraction of Initial Salt Concentration Remaining in Soil, C/C_o, and Depth of Leaching Water Applied per Unit Depth of Soil, D/D_s, (Hoffman 1980): (a) for Continuous Ponding; (b) for Intermittent Ponding

equation applies after the actual drainage has begun or in the range of $D/D_s > k$.

Eq. 19.1 can readily be used to calculate the amount of water needed to leach salts from a desired depth of soil for a specified initial and desired salinity. The salt transport efficiency decreases sharply when D to D_s ratios exceed 0.5 for sandy loam and 0.75 for clay loam to clay.

Even after adjusting for lesser water requirements, leaching by sprinkling can take more time than continuous ponding. This is because leaching water is applied at the rate below the saturated hydraulic conductivity. It increases the leaching time and water evaporation, which, in turn, increases the salinity of leaching water. Carter and Fanning (1964) and Minhas and Khosla (1986) demonstrated that leaching by sprinkling or intermittent ponding works better at low evaporation rates.

Drainage conditions affect redistribution of soil water and salts. In well-drained uniform soils, water after infiltration penetrates deeper into the profile than in poorly drained soils with high water tables or abrupt soil stratification. Consequently, salts will be leached deeper (Terkeltoub and Babcock 1971).

Two-Dimensional Leaching. Leaching into tile or open-ditch drains rapidly removes salt directly above and near the drains and removes little midway between the drains. Leaching efficiency decreases as soon as the rapid flow reaches the drains. This uneven leaching is more evident in water-saturated fields (Luthin et al. 1969) and less evident in unsaturated fields, where leaching above the water table is a one-dimensional process (Talsma 1967). To leach more efficiently, avoid ponding water near the drains (Miyamoto and Warrick 1974) or intermittently apply water to keep the water table as low as possible (Talsma 1967).

Leaching of Spatially Variable Fields. When fields consist of several types of soils, leaching takes place—first, in sandy portions, then, in clay-rich portions. If leaching is continued until the clay-rich portion is sufficiently leached, an excessive amount of water percolates into the sandy area. This problem is pronounced under ponded leaching, during which ponded water percolates predominantly through the sandy portion. In fact, the distribution of the salinity of the soil in basin, border, or furrow irrigated fields coincides closely with the distribution of soil type (Miyamoto and Cruz 1986, 1987). To leach much more efficiently, leach based on the distribution of type of soil (Miyamoto and Cruz 1986) or use the stochastic analysis of solute transport in spatially variable fields (Russo 1984) to divide a field into small sections that need different amounts of water.

Leaching Methods

Continuous Ponding. This method is used extensively in surface-irrigated areas, and usually accomplishes leaching in the shortest period of time with the least cost in slowly permeable fields. The leaching efficiency is less than with other methods, especially when used in water-saturated fields with tile or open drains (Talsma 1967). The land must be leveled and, when permeability is low, must also consist of subsoils or be deeply plowed (Longenecker and Lyerly 1974). Ideally, the soils should be dry before leaching. The permeability and initial salinity largely control the rate of leaching. For more details, refer to Reeve et al. (1948), Reeve et al. (1955), Rasmussen et al. (1972), and Hulsbos and Boumans (1960). In clay-rich soils with very low permeability, ponded leaching may be conducted along with the temporary cultivation of rice or stock fish.

Intermittent Ponding. This method is especially suitable in fields with tile drains, as it allows the water tables to draw down, which greatly increases leaching efficiency (Talsma 1967). When soils develop a surface seal, intermittent ponding may also help water to infiltrate by forming cracks. However, the longer period of wet soil exposure increases evaporation. Carter and Fanning (1964) demonstrated that combining intermittent ponding with mulching greatly improved the performance of intermittent ponding over a five-month period. Intermittent ponding with mulch is especially suitable when low rates of drainage and high water tables exist. Most soils in the middle Rio Grande project have been reclaimed by deep plowing and then intermittent ponding.

Sprinkling. This method can be used advantageously when the field is unprepared for ponded leaching. It has high leaching efficiency (e.g., Nielsen et al. 1966), except under high evaporative conditions, such as windy days, but is energy- and capital-intensive. It is compatible with unleveled sandy soils, especially when water for leaching is scarce and costly. It can also used advantageously when a large drainage load is prohibited.

Alternate Row or Border Leaching. Alternate row irrigation is used in saline areas to minimize the accumulation of salt in crop beds. Stewart et al.(1976) found that a similar system removed salts in areas with a high water table. High ridge rows are made. Every other furrow is then flooded. This moves salts laterally into the unirrigated furrows. The seeped water, because of high water table, can then be surface-drained away.

Another method used successfully in salt-affected orchards with high water tables is alternate border leaching (Miyamoto 1988). Two borders are placed along each row of trees. Leaching water is applied to the border strip, but not to the strip between the rows. This provides rapid leaching of salt near the trees and pushes part of the salts into the dry strip between the rows. Salts accumulated between the rows would be leached after lowering the water table. This method kept trees alive until drainage was completed.

Surface Flushing. High concentrations of soluble salts are often found at and near the surface of the soil, especially in soils with shallow water tables. It is desirable to remove surface-accumulated salts instead of leaching into the soil profile. However, Reeve et al. (1955) found that passing water over the surface by sheet flow failed to remove the salts. When plowed fields with a clay substratum are bedded and the aforementioned alternate row leaching technique used, the efficiency may improve. The physical removal of salt crusts by mechanical means is an alternative.

Post-Leaching Operations

The last phase of reclamation is drying the field and, when necessary, taking steps to enhance soil structure, such as applying organic matter or gypsum and cropping with grains or legumes. Prolonged ponding destroys the structure of the soil and depletes nitrogen in the soil. Sprinklers or intermittent ponding generally causes less destruction and may not need elaborate post-leaching measures.

RECLAMATION OF SODIUM-AFFECTED SOILS

Concepts and Principles

Reclamation of sodium-affected soils usually involves replacing exchangeable Na^{+} with Ca^{2+}. This Ca may originate from the dissolution of Ca-containing minerals in the soil, such amendments as gypsum and calcium chloride, or irrigation water with Ca ions. A significant factor in reclaiming sodic soils is the maintenance of HC by providing a sufficiently high electrolyte concentration in the soil solution to counter the influence of exchangeable Na. Generally, the higher the electrolyte concentration, the higher the exchangeable Na fraction (E_{Na}) at which a relatively high permeability can be maintained (Quirk and Schofield 1955). McNeal and Coleman (1966) showed that soils responded differently to the same combination of electrolyte concentration and E_{Na}. Each soil has a unique threshold of salinity concentration. The electrolyte concentration affects the HC less when the contents of the soil water is low (Russo and Bresler 1977).

If the electrolyte concentration of the percolating solution is adequate to reduce clay swelling, the permeability of the soil remains high. When low-salinity water (or rainwater) follows the saline water, permeability can be maintained by applying an electrolyte source on the surface of the sodium-affected soil. Applying a slow-dissolving salt adds sufficient electrolytes to the rainwater to prevent clay dispersion (Keren and Shainberg 1981), even if the Na-Ca exchange process in the adsorbed phase is limited (Keren et al. 1983). The effectiveness of amendments in reclaiming sodic soils depends on their dissolution properties (Kemper et al. 1975; Keren and O'Connor 1982b).

In spite of the documented effect of the electrolyte concentration in the soil solution on the permeability of the soil, the main criteria for determining the sodic hazard are the soil E_{Na} and the sodium adsorption ratio (SAR) of the saturation extract or irrigation water (Chapter 3). However, soil sodicity cannot be assessed without information on the total electrolyte concentration of the soil solution.

Reclamation of the Soil Profile by Adding Soil Amendments

Amendments commonly used to provide soluble Ca include gypsum ($CaSO_4 \cdot 2H_2O$) and calcium chloride dihydrate ($CaCl_2 \cdot 2H_2O$). Amendments that produce Ca in calcareous soils by enhancing the conversion of soil $CaCO_3$ to gypsum include sulfuric acid, sulfur, and iron and aluminum sulfates. Gypsum, sulfuric acid, and sulfur are the amendments most commonly used for reclamation due to their relatively low cost and effectiveness. Chemical amendments may improve water penetration caused by excessive sodium, if the texture of the soil, compaction, or water-restricting layers is not the cause of low permeability.

Gypsum. Due to its solubility, low cost, and availability, gypsum is the most commonly used amendment for reclaiming sodium-affected soil and reducing the harmful effects of high-sodium irrigation waters. Gypsum comes from both mines and as a byproduct of the phospate fertilizer industry. Under comparable conditions, the rate of dissolution of industrial gypsum is much higher than that of mined gypsum (Fig. 19.2).

Gypsum added to a sodic soil increases permeability by increasing electrolyte concentration and by reducing exchangeable Na (Keren and Shainberg 1981; Loveday 1976). Shainberg et al. (1982) evaluated the relative significance of the electrolyte effect and the cation exchange on the HC by comparing the effects of gypsum and $CaCl_2$ in chemically equivalent amounts. Infiltration of distilled water into a noncalcerous soil initially treated with $CaCl_2$ decreased and eventually stopped. The

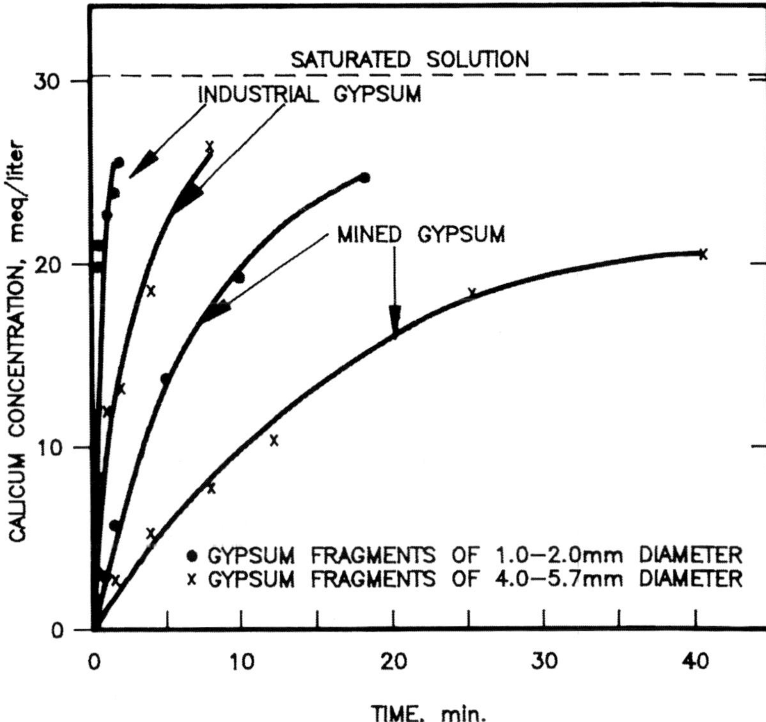

Fig. 19.2 Rate of Dissolution of Industrial and Mined Gypsum Particles of Two Fragment Sizes (Keren and Shainberg 1981)

gypsum treatment maintained high hydraulic conductivity. They observed no difference between the two amendments in a calcareous soil. Since both amendments had similar Na replacement, they attributed the differences to an electrolyte effect. Thus, the release of electrolytes by gypsum particles must be sustained to maintain the high HC of soils irrigated by waters with a low electrolyte concentration.

The relative significance of the two effects should be noted. If the electrolyte effect is sufficient to prevent soil clays from dispersing and swelling, the surface application of gypsum may be worthwhile. In this case, the amount of gypsum required depends on the amount of low salt water applied and the rate of gypsum dissolution.

In soils where the electrolyte concentration effect is less significant and the main effect results from cation exchange, the amount of gypsum required depends on the amount of exchangeable Na in the depth of soil. The amount of exchangeable Na to be replaced during reclamation depends on the initial exchangeable Na fraction (E_{Nai}), the soil cation exchange capacity (CEC, mmol/M_g), soil bulk density (M_g m^{-3}), the desired final exchangeable sodium fraction (E_{Naf}), and the depth of soil to be reclaimed (D_s, m). After the parameters are determined, the amount of exchangeable Na to be replaced per unit of land area (ΔNa, mol_c ha^{-1}) can be calculated from:

$$\Delta Na = 10^4 \, (D_s) \, (\rho_b) \, (CEC) \, (E_{Nai} - E_{Naf}) \quad (19.2)$$

The value of E_{Naf} depends on the response of the physical conditions of the soil and the sodium tolerance of the crop. The gypsum requirement, GR, the amount of gypsum needed to reclaim a sodium-affected soil, (metric ton ha^{-1}) can be calculated from

$$GR = 86.1 \times 10^{-6} \, \Delta Na \quad (19.3)$$

As the soil depth increases, the concentration of Ca and Na in the soil solution decreases and increases, respectively, as Ca-Na exchange occurs. Eventually, the SAR of the soil solution reaches a value in equilibrium with the existing E_{Na}. The remaining soluble Ca will not react with exchangeable Na and is removed by drainage. To correct for this loss, Doering and Willis (1975) suggest including a correction factor in Eq. 19.3.

The efficiency of applied Ca to remove adsorbed Na varies with E_{Na}. It is much greater at high ESP values (Chaudhry and Warkentin 1968). The efficiency of Na exchange at E_{Na} levels below about 10 is low (about 30%), because a greater fraction of applied Ca displaces exchangeable Mg (Greene and Ford 1985; Loveday 1976). Efficiency may also be low (20% to 40%) in fine-textured soils due to the slow kinetics of Na/Ca exchange inside the structural elements (Manin et al. 1982). A factor seldom considered is that applying large amounts of gypsum may temporarily reduce HC, since small gypsum particles lodge in the soil pores (Keren et al. 1980).

Gypsum must dissolve to effect reclamation. Gypsum contents and the velocity of the soil water influence gypsum-dissolution rates (Kemper et al. 1975; Keren and O'Connor 1982b). Increasing velocity increases the dissolution rate coefficient but decreases the time during which gypsum and the flowing solution are in contact. The net result is a decreasing dissolution rate with increasing velocity. Thus, the chemical equilibrium in reclamation models probably can be assumed only when the velocity of the soil water is lower than the rate of diffusion or when sufficient surface area of gypsum is available to reach equilibrium during leaching. Chemical equilibrium for gypsum dissolution may not be reached for soils with comparatively large HC, soils with low gypsum content, or soils with large particles of gypsum (Keren and Shainberg 1981) or particles of gypsum coated by $CaCO_3$ (Keren and Kauschansky 1981). Thus, lowering the rate of water application whenever possible, for example, with sprinkler irrigation, is preferable. It increases the rate of gypsum dissolution and enhances the efficiency of exchange (Fig. 19.3).

Generally, soil water penetrates too slowly to reclaim sodium-affected soils in a single leaching. For example, a 50-cm depth of water applied for leaching can only dissolve about 12 Mg/ha of gypsum. Larger gypsum treatments should not be made unless sufficient soil water penetrates to allow larger applications of water. Thus, sodium-affected soil normally can only be reclaimed to a limited depth in the first year. This will often permit a shallow-rooted crop to be established after leaching.

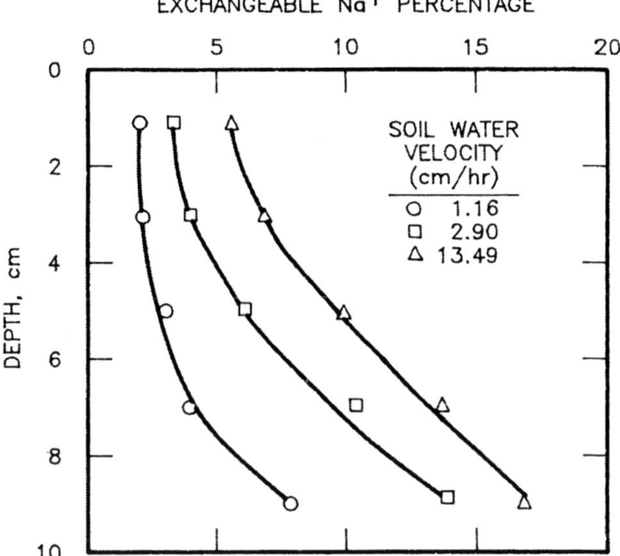

Fig. 19.3 Distribution of Exchangeable Na Fraction with Depth as a Function of Soil Water Velocity for a Given Amount of Dissolved Calcium (Keren and O'Connor 1982)

Subsequently, amendments and leaching volumes can be applied annually to reclaim the entire profile.

Acids and Sulfur. Sulfuric acid is an amendment that treats sodium-affected soils. In some areas, acid waste products from mining and industrial activities are available. Using these waste products as amendments may provide a safe way to dispose of them (Miyamoto et al. 1975b). The acid reacts with soil calcium carbonate to form gypsum (using H_2SO_4) or calcium chloride (using HCl). Sulfur requires an initial phase of microbiological oxidation to produce H_2SO_4.

In soil column studies, Yahia et al. (1975) and Prather et al. (1978) found that H_2SO_4 increases the penetration of water into calcareous sodium-affected soils more effectively than gypsum. Field experiments on yield response have generally shown results in favor of H_2SO_4 (Overstreet et al. 1951). Equivalent amounts of gypsum reduced soluble and exchangeable Na in the surface soil, similarly to H_2SO_4, but produced smaller yield responses.

The concentrated acid, applied directly on the surface of the soil, results in better distribution, less destruction of soil aggregates, and more efficient leaching of salts (Miyamoto et al. 1975a). Another technique used commercially (Tisdale 1970) is direct application by chiseling into the soil in bands about 45 cm apart. H_2SO_4 and SO_2, being highly corrosive, should not be added to water that will pass through metal or concrete irrigation systems. The amount of acid required can be calculated from equations using CEC and concentrations of Na, Ca, and Mg in the saturation extract made with the leaching water (Miyamoto et al. 1975a).

Acidulants that first must be oxidized, such as sulfur, pyrites, and polysulfides, act more slowly than H_2SO_4. Their effectiveness in field experiments has varied. It may relate to the presence or absence of microbial populations. For elemental S application, dust poses a problem. This can be overcome by using conventional fluid fertilizer equipment to apply water suspensions containing 55% to 60% S (Thorup 1972) or granulated S. Acids applied at high rates lower the soil's pH in a portion of the root zone and consequently increase the availability of P, Zn, Mn, and Fe. This is important in obtaining improved crop responses (Miyamoto et al. 1975b).

Calcium Chloride. Although $CaCl_2 \cdot 2H_2O$ generally is too expensive to compete with other amendments, it is available as an industrial waste product and could, therefore, be considered for reclamation. Similarly to GR (Eq. 19.3), the $CaCl_2 \cdot 2H_2O$ requirement, CCR, (metric ton ha^{-1}) can be calculated from:

$$CCR = 75.5 \times 10^{-6} \Delta Na \qquad (19.4)$$

Being highly soluble, calcium chloride initially yields high electrolyte levels and high rates of water intake. This makes it a more efficient amendment than gypsum for soils with high E_{Na} (Alperovitch and Shainberg 1973; Prather et al. 1978). Since $CaCl_2$ is highly soluble, it is quickly leached from the soil profile. Despite the lower E_{Na} value after treatment, noncalcareous soils may seal when the electrolyte concentration of the percolating solution falls below the FV (Shainberg et al. 1982). Therefore, a combination of $CaCl_2$ and gypsum may more quickly and effectively reclaim soil with high E_{Na} than gypsum alone (Prather et al. 1978).

Reclamation Without Adding Soil Amendments

Calcareous Soils. $CaCO_3$ particles isolated from soils have ion activity product (IAP) values expected for calcite (Suarez and Rhoades 1982), but a calcite supersaturation is usually found in soils (Levy 1981). The supersaturation appears to be due to the presence of Ca-silicates in the soil more soluble than calcite and is the result of unstable $CaCO_3$ phases (Suarez and Rhoades 1982).

The rate at which $CaCO_3$ dissolution in water approaches equilibrium depends on a number of factors, including the surface area solution volume ratio, the ionic composition of the solution, the ion composition of the adsorbed phase, the affinity of the clay minerals to cations, the temperature, and the local partial pressure of CO_2. Amrhein et al. (1985) concluded that the kinetics of $CaCO_3$ dissolution in soils is not a simple diffusion-controlled or first-order reaction. Soil $CaCO_3$ may be dissolved to contribute Ca, especially in reclaiming saline and sodium-affected soils in which its solubility is enhanced (Oster 1982). Adding $CaCO_3$ to nonsaline sodium-affected soils is of doubtful value, though. Its dissolution is not rapid enough, unless an acid is applied concurrently. However, the dissolution of $CaCO_3$ may maintain the concentration levels of the soil solution above the FV of the soil clays and prevent HC reduction in the profile of soils at low E_{Na} levels exposed to rainwater (Alperovitch et al. 1981).

Successive Dilutions of High-Salt Waters Containing Divalent Cations. The high salt dilution method is particularly effective for soils with expanding clay-type minerals that have an extremely low HC at low electrolyte concentration in the soil solution and need an excessive amount of reclamation time or amendment. It involves successively diluting a high salt water containing divalent cations (Reeve and Bower 1960). It tested successfully in experiments. A monogram was developed to predict the reclamation at various stages of leaching (Reeve and Doering 1966). In the early phase, high salinity of the water flocculates soil colloids and provides a source of Ca for exchange with Na. When diluted with water with a low concentration of salt, the SAR of the irrigation water is reduced by the square root of the dilution factor. Equations were derived for calculating the depth of water need to reclaim a sodic soil by leaching with such water. Misopolinos (1985) theoretically analyzed the use of high-salt water to add a constant quantity of Ca^{2+} and thus reclaim sodic soils.

Jury et al. (1979) demonstrated that leaching reclamation without amendments can take place. Good drainage through the soil profile, adequate leaching water, and a soil source of Ca that can be mobilized are all needed. Saline sodic soils frequently contain precipitated $CaCO_3$ and gypsum that, upon leaching, dissolve to provide sufficient Ca to exchange with adsorbed Na. Intermittent ponding requires less water than continuous ponding to achieve the same degree of leaching (Miller et al. 1965). Sprinkling irrigation is more efficient than other methods at removing salt from small pores in the soil profile (Nielsen et al. 1966). This method has also reclaimed a slowly permeable, sodium-affected soil in a humid environment, where HC and IR increased from 30 to over 100% (Rahman et al. 1974).

Soil Profile Modification. Deep tillage may benefit stratified, sodium-affected soils with impermeable layers between permeable layers, which are often found along the Snake River and its tributaries in the Northwestern United States, in much of Arizona's Salt River Valley, in parts of the Rio Grande Valley in New Mexico and Texas, and in California's San Joaquin Valley. In locations where soil with significant quantities of gypsum or $CaCO_3$ lies beneath sodium-affected soils or subsoils, deep plowing has effectively broken up and mixed the layers while supplying soluble Ca to aid reclamation. The depth of plowing required varies from 0.5 m to more than 1.0 m, depending on the concentration and depth of the sodic- and Ca-rich layers. A procedure to predict the optimum depth of plowing to maintain adequate permeability during reclamation is available (Rasmussen and McNeal 1973).

Soil Reclamation Models

Two currently available types of models that describe the reclamation of sodic soils are chromatographic (plate) models and miscible dis-

TABLE 19.1 Effect of Industrial Gypsum Treatments on Surface Runoff in Rainstorms from Loess Soil at Two Levels of E_{Na} (Keren et al. 1983)

			SURFACE RUNOFF (PERCENT OF RAINFALL)				
			E_{Na} 0.046		E_{Na} 0.193		
Storm no. (1)	Amount of rainfall (mm) (2)	Time between storms (3)	Control (4)	Gypsum[a] spread over (5)	Control (6)	Gypsum[a] mixed in (7)	Gypsum[a] spread over (8)
1	16	—	0	0	23.1	2.6	0
2	20	40	6.5	0	13.3	3.0	0.7
3	19	20	1.3	0.5	27.3	10.5	3.8
4	59	6	21.5	3.8	41.0	21.2	17.2
5	44	9	13.4	2.1	45.0	22.5	12.4
6	12	1	12.5	2.5	40.0	25.0	12.5
7	12	15	10.1	2.5	31.7	13.8	6.7
Total rainfall (mm)			22.8	3.8	64.2	30.1	18.7
Percent of annual rainfall (182 mm)			12.5	2.1	35.3	16.5	10.3

[a]The rates of industrial gypsum application were 5 Mg/ha and 10 Mg/ha for the soils having E_{Na} values of 0.046 and 0.193, respectively.

placement models. The former are generally simpler to use. The latter demand extensive input data. For both types, spatial variability needs to be taken into account when applied in the field.

Chromatographic models have been used to describe sodic soil reclamation with gypsum (Dutt et al. 1972; Tanji et al. 1972). The models, which include such reactions as the dissolution of $CaCO_3$ and gypsum and cation exchange, are based on the assumption that these reactions reach equilibrium in each soil segment of plates. When these models were field-tested, differences between predicted and measured values were found to be less than horizontal variations typically found in sodic soils. More recently, Tanji and Deverel (1984) expanded this approach to include the kinetic rate of several chemical reactions, including sulfur oxidation. Though they can apparently predict the chemistry of reclamation, chromatographic models are less able to predict soil water status or the rate of reclamation. Determining the appropriate number of soil segments for simulation also poses a problem.

The solution of solute transport equations about convective transfer and hydrodynamic dispersion is the core of miscible displacement models (Chapter 23). It rigorously describes solute transfer processes. Miscible displacement models, when coupled with chemical reaction models (Robbins et al. 1980), provide insight into the processes of reclamation. However, they have not been used extensively in the field, probably because solving them requires extensive data specific to the site. Also, existing models neither fully account for solute effects on hydraulic conductivity nor consider spatial variability, both of which are among the most important aspects of sodic soil reclamation.

Preventing Crust Formation

Gradual deterioration of soil structure and the formation of a surface crust can decrease an initially high rate of IR. The IR is also sensitive to the electrolyte concentration in the applied solution (Agassi et al. 1981), which is important in controlling infiltration of sodium-affected soils.

Applying gypsum reduced surface runoff (Table 19.1). Although the gypsum minimally affected the E_{Na} of the surface soil, it significantly affected the IR (Keren et al. 1983). The IR of a sodium-affected soil exposed to rain depends on the source, amount, and size of the gypsum fragments (Fig. 19.4). The efficiency of gypsum in maintaining a high IR is a function of its rate of dissolution. Industrial gypsum, because of its finely sized particles, is more effective than mined gypsum in maintaining a high IR (Keren and Shainberg 1981). The dissolution rate of gypsum is important because rainwater and gypsum particles at the surface of the soil are in contact for a short time.

Surface spreading of gypsum was more beneficial than mixing it in the upper layer because the former concentrates the amendment in the zone of crusting (Agassi et al. 1982, Keren et al. 1983). The amount of gypsum needed depends on the amount of low salt concentration water applied (or rain) and the rate of gypsum dissolution. Typical application rates in semi-arid regions range from 5 Mg $ha^{-1}$to 10 Mg ha^{-1}.

RECLAMATION OF BORON-AFFECTED SOILS

Boron Hazard

Boron (B) is one of the essential micronutrients required for the growth of plants. However, a relatively narrow margin exists between levels of B that cause deficiency and levels that cause toxicity. Boron toxicity occurs most frequently in soils of arid and semi-arid regions, where soils are irrigated with water high in B, while B deficiency is found primarily in humid regions.

Concepts and Principles

Boron can be specifically adsorbed by different clay minerals, hydroxy oxides of Al, Fe, and Mg, and organic matter. Keren and Bingham (1985) review the factors that affect the adsorption and desorption of B by soil constituents and the mechanisms of adsorption.

The Langmuir equation describes the adsorption of B by soils (Hatcher and Bower 1958) and clays (Hingston 1964). Deviations occur at higher solution concentrations at high pH (Hingston 1964). Hingston (1964) related the variation in B adsorption with pH to changes in the

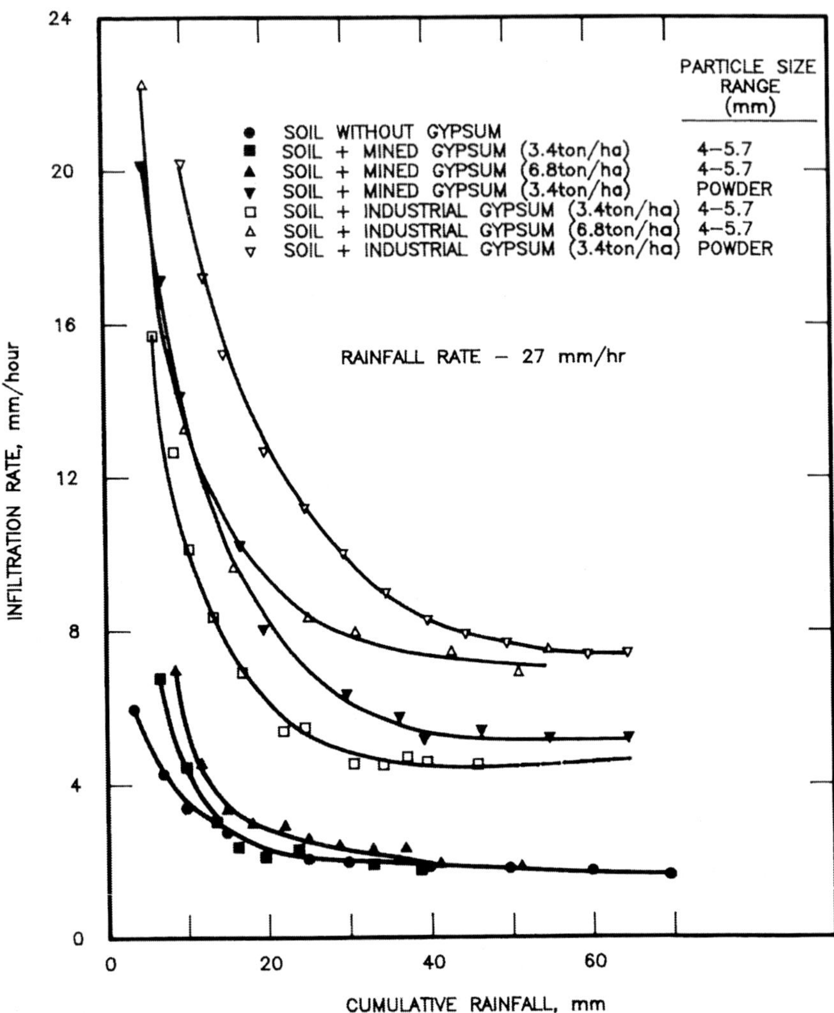

Fig. 19.4 Effect of Industrial and Mined Gypsum on Infiltration Rate of Loess Soil as a Function of Cumulative Rainfall (Keren and Shainberg 1981)

surface of the clays, which induced more adsorption sites to form at high pH but decreased adsorption due to the decrease of the concentration of boric acid relative to borate ions (Hingston 1964). This adsorption model has two disadvantages: It does not consider that two B species are involved and that their relative presence in solution varies with pH; and different values of the Langmuir coefficients, namely, maximum adsorption and binding energy constant, are needed to predict adsorption at any given pH. Another approach for describing B adsorption by clays, hydroxy-Al, and soils has been derived by Keren et al. (1981). It is assumed that $B(OH)_3$, $B(OH)_4^-$, and OH^- compete for the same adsorp-

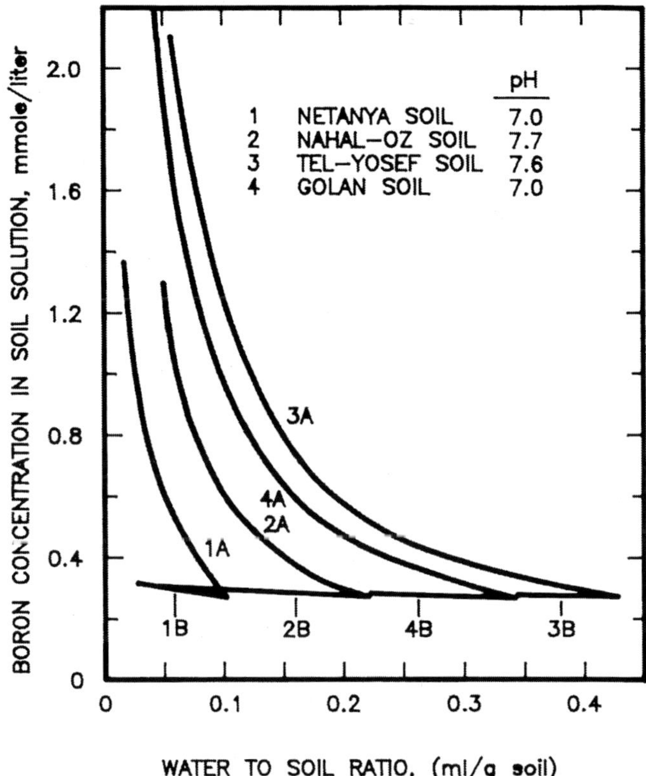

Fig. 19.5 Boron Concentration in Soil Solution as a Function of Solution-to-Soil Ratio for a Given Total Amount of Boron (Mezuman and Keren 1981): (a) No Interaction Between Boron and Soil, (b) Adsorption Accounted For

tion sites. Calculated and experimental results correlated well (Keren and Gast 1983; Keren and Mezuman 1981; Keren and O'Connor 1982a; Mezuman and Keren 1981).

Boron concentrations in the soil solution after drying or wetting tend to remain constant due to buffering by sorption and desorption reactions (Fig. 19.5). This buffering capacity depends on the soil affinity for B and the maximum B adsorption value. Thus, plants grown in the coarse-textured soil had the greatest uptake of B and plants grown in the fine-textured soil had the least (Keren et al. 1984, 1985).

Reclamation Methods

Before high-boron soils can be successfully used for agriculture, their soluble B contents must be reduced to nonphytotoxic levels. Such reclamation is typically accomplished by leaching the soil. The relative decrease of soluble B in field soils during reclamation is described as

$$\left(\frac{C}{C_o}\right)\left(\frac{D}{D_s}\right) = 0.6 \tag{19.5}$$

where C and C_0 are the final and initial soluble B concentrations and D/D_s is the depth of leaching water per unit depth of soil (Hoffman 1980). Eq. 19.5 is independent of the method of water application—be it sprinkling or ponding. Native soil B appears to be more difficult to leach than B accumulated from previous irrigations.

Hatcher and Bower (1958) and Tanji (1970) have used mathematical models to predict the movement of B in stratified soil profiles, considering adsorption isotherms and a chromatographic displacement theory. They found that theoretical and experimental results correlated well. The computer model (Tanji 1970) predicts profile distribution of *in-situ* soil water contents (field capacity and saturation percentage). These models may be used to estimate the amount of water required to reclaim B-affected soils only when boron is associated with the soil adsorption sites, however. The reduced B concentrations that follow the leaching of native high B soils may be temporary. Longer storage times of water in the soil profile resulted in larger increases in effluent B concentrations (Rhoades et al. 1970; Bingham et al. 1972). Peryea et al. (1985) inversely related this phenomenon, termed B regeneration, to the amount of water used for the initial leaching. Boron regeneration is therefore of primary concern during the early stages of soil reclamation, when appreciable residual sources of regenerable B are present. The potential for re-establishing phytotoxic concentrations in a soil is keyed to the relative completeness of the reclamation processes and the rate of B dissolution after reclamation.

REFERENCES

Agassi, M., Morin, J., and Shainberg, I. (1982). "Infiltration and runoff control in the semi-arid region of Israel." *Geoderma* 28: 345-356.

Agassi, M., Shainberg, I., and Morin, I. (1981). "Effect of electrolyte concentration and soil sodicity on the infiltration rate and crust formation." *Soil Sci. Soc. Am. J.* 45: 848-851.

Alperovitch, N., and Shainberg, I. (1973). "Reclamation of alkali soils with $CaCl_2$ solutions." Hadas, A. et al. (eds), *Physical Aspects of Soil, Water and Salts in Ecosystems*. Springer, Berlin, pp. 431-440.

Alperovitch, N., Shainberg, I. and Keren, R. (1981). "Specific effect of magnesium and hydraulic conductivity of sodic soils." *Clays and Clay Minerals* 33: 443-450.

Amrhein, C. Jurinak, J.J., and Moore, W.M. (1985). "Kinetics of calcite dissolution as affected by carbon dioxide partial pressure." *Soil Sci. Soc. Am. J.* 49: 1393-1398.

Bingham, F.T., Marsh, A.W., Branson, R., Mahler, R., and Ferry, G. (1972). "Reclamation of salt-affected high boron soils in Western Kern County." *Hilgardia* 41: 195-211.

Carter, D.L., and Fanning, C.D. (1964). "Combining surface mulches and periodic water applications for reclaiming saline soils." *Soil Sci. Soc. Am. Proc.* 28: 564-567.

Chaudhry, G.H., and Warkentin, B.P. (1968). "Studies on exchange of sodium from soils by leaching with calcium sulfate." *Soil Sci.* 105: 190-197.

Doering, E.J., and Willis, W.O. (1975). "Chemical reclamation for sodic strip-mine spoils." U.S. Department of Agriculture, Agricultural Research Service, ARS-NC-20.

Dutt, G.R., Terkeltoub, R.W., and Rauschkolb, R.S. (1972). "Prediction of gypsum and leaching requirements for sodium-affected soils." *Soil Sci.* 114: 93-103.

Frenkel, H., Goertzen, J.O., and Rhoades, J.D. (1978). "Effect of clay type and content, exchangeable sodium percentage, and electrolyte concentration on clay dispersion and soil hydraulic conductivity." *Soil Sci. Soc. Am. J.* 42: 32-39.

Goldberg, S., and Glaubig, R.A. (1987). "Effect of saturating cation, pH and aluminum and iron oxide on the flocculation of kaolinite and montmorillonite." *Clays and Clay Minerals* 35: 220-227.

Greene, R.S.B., and Ford, G.W. (1985). "The effect of gypsum on cation exchange and leaching in two red duplex wheat soils." *Aust. J. Soil Res.* 23: 61-74.

Hatcher, J.T., and Bower, C.A. (1958). "Equilibria and dynamics of boron adsorption by soils." *Soil Sci.* 85: 319-328.

Hingston, F.J. (1964). "Reaction between boron and clays." *Aust. J. Soil Res.* 2: 83-95.

Hoffman, G.J. (1980). "Guidelines for the reclamation of salt-affected soils." G. A. O'Connor, ed. 2nd Inter-American Conf. on Salinity and Water Management Technology, Juarez, Mexico. 11-12 Dec., 1980; pp. 49-64.

Hulsbos, W.C., and Boumans, J.H. (1960). "Leaching of saline soils in Iraq." *Netherlands J. Agr. Sci.* 8: 1-10.

Jury, W.A., Jarrell, W.M., and Devitt, D. (1979). "Reclamation of saline-sodic soils by leaching." *Soil Sci. Soc. Am. J.* 43: 1100-1106. Errata (1987), *Soil Sci. Soc. Am. J.* 51: 1092.

Kemper, W.D., Olsen, J., and DeMooy, C.J. (1975). "Dissolution rate of gypsum in flowing water." *Soil Sci. Soc. Am. J.* 39: 458-463.

Keren, R., and Bingham, F.T. (1985). "Boron in water, soils, and plants." *Advances in Soil Science, Vol. 1,* B. A. Stewart, ed., Springer-Verlag, New York. pp. 229-276.

Keren, R., Bingham, F.T., and Rhoades, J.D. (1984). "Plant uptake of boron as affected by boron distribution between liquid and solid phases in soils." *Soil Sci. Soc. Am. J.* 48: 297-302.

Keren, R., Bingham. F.T., and Rhoades, J.D. (1985). "Effect of clay content in soil on boron uptake and yield of wheat." *Soil Sci. Soc. Am. J.* 49: 1466-1470.

Keren, R., and Gast, R.G. (1983). "pH-dependent boron adsorption by montmorillonite hydroxy-aluminum complexes." *Soil Sci. Soc. Am. J.* 47: 1116-1121.

Keren, R., Gast, R.G., and Bar-Yosef, B. (1981). "pH-dependent boron adsorption by Na-montmorillonite." *Soil Sci. Soc. Am. J.* 45: 45-48.

Keren, R., and Kauschansky, P. (1981). "Coating of calcium carbonate on gypsum particle surfaces." *Soil Sci. Soc. Am. J.* 45: 1242-1244.

Keren, R., Kreite, J.F., and Shainberg, I. (1980). "Influence of size of gypsum particles on the hydraulic conductivity of soils." *Soil Sci.* 130: 113-117.

Keren, R., and Mezuman, U. (1981). "Boron adsorption by clay minerals using a phenomenological equation." *Clays and Clay Minerals* 29: 198-204.

Keren, R., and O'Connor, G.A. (1982a). "Effect of exchangeable ions and ionic strength on boron adsorption by montmorillonite and illite." *Clays and Clay Minerals* 30: 341-346.

Keren, R., and O'Connor, G.A. (1982b). "Gypsum dissolution and sodic soil reclamation as affected by water flow velocity." *Soil Sci. Soc. Am. J.* 46: 726-732.

Keren, R., and Shainberg. I. (1981). "The efficiency of industrial and mined gypsum in reclamation of a sodic soil-rate of dissolution." *Soil Sci. Soc. Am. J.* 45: 103-107.

Keren, R., Shainberg, I., Frenkel, H., and Kalo, Y. (1983). "The effect of exchangeable sodium and gypsum on surface runoff from loess soil." *Soil Sci. Soc. Am. J.* 47: 1001-1004.

Keren, R., and Singer, M.J. (1988). "Effect of low electrolyte concentration on hydraulic conductivity of Na/Ca-montmorillonite sand systems." *Soil Sci. Soc. Am. J.* 52: 368-373.

Levy, Rachel (1981). "Effect of dissolution of alumosilicates and carbonates on ionic activity products of calcium carbonate in soil extracts." *Soil Sci. Soc. Am. J.* 45: 250-255.

Longenecker, D.E., and Lyerly, P.J. (1974). "Control of soluble salts in farming and gardening." Texas Agr. Exp. Sta. Bulletin 876, 36 pp.

Loveday, J. (1976). "Relative significance of electrolyte and cation exchange effects when gypsum is applied to a sodic clay soil." *Aust. J. Soil Res.* 14: 361-371.

Luthin, J.N., Fernandez, P., Woerner, J., and Robinson, F. (1969). "Displacement front under ponded leaching." *J. Irrig. Drain. Div. ASCE* 95 (IR1): 117-125.

Manin, M., Pissarra, A., and Van Hoorn, J.W. (1982). "Drainage and desalinization of heavy clay soil in Portugal." *Agric. Water Manag.* 5: 227-240.

McNeal, B.L., and Coleman, N.T. (1966). "Effect of solution composition on soil hydraulic conductivity." *Soil Sci. Soc. Am. Proc.* 30: 308-312.

McNeal, B.L., Layfield, D.A., Norvell, W.A., and Rhoades, J.D. (1968). "Factors influencing hydraulic conductivity of soils in the presence of mixed salt solutions." *Soil Sci. Soc. Am. J.* 32: 187-190.

Mezuman, U., and Keren, R. (1981). "Boron adsorption by soils using a phenomenological adsorption equation." *Soil Sci. Soc. Am. J.* 45: 722-726.

Miller, R.J., Biggar, J.W., and Nielsen, D.R. (1965). "Chloride displacement in Panoche clay loam in relation to water movement and distribution." *Water Resour. Res.* 1: 63-73.

Minhas, P.S., and Khosla, B.K. (1986). "Solute displacement in a silt loam soil as affected by dry method of water application under different evaporation rates." *Agr. Water Manag.* 12: 64-74.

Misopolinos, N.D. (1985). "A new concept for reclaiming sodic soils with high-salt water." *Soil Sci.* 140: 69-74.

Miyamoto, S. (1988) "Reclamation of salt-affected orchards." *Proc. Texas Pecan Conf.* College Station, Tex. pp.25-37.

Miyamoto, S., and Cruz, I. (1986). "Spatial variability and soil sampling for salinity and sodicity appraisal in surface-irrigated orchards." *Soil Sci.Soc. Am. J.* 50: 1020-1025.

Miyamoto, S., and Cruz, I. (1987). "Spatial variability of soil salinity in furrow-irrigated torrifluvents." *Soil Sci. Soc. Am. J.* 51: 1019-1025.

Miyamoto, S., Prather, R.J., and Stroehlein, J.L. (1975a). "Sulfuric acid and leaching requirements for reclaiming sodium-affected calcareous soils." *Plant Soil* 43: 573-585.

Miyamoto, S., Ryan, J., and Stroehlein, J.L. (1975b). "Potentially beneficial uses of sulfuric acid in south-western agriculture." *J. Environ. Qual.* 4: 431-437.

Miyamoto, S., and Warrick, A.W. (1974). "Salt displacement into drain tiles under ponded leaching." *Water Resour. Res.* 10: 275-278.

Nielsen, D.R., Biggar, J.W., and Luthin, J.N. (1966). "Desalinization of soils under controlled unsaturated flow conditions." 6th Congr. Int. Comm. on Irrig. and Drainage, New Delhi, 19.15-19.24.

Oster, J.D. (1982). "Gypsum usage in irrigated agriculture: a review." *Fertil. Res.* 3: 73-89.

Oster, J.D., Shainberg, I., and Wood, J.D. (1980). "Flocculation value and gel structure of Na/Ca montmorillonite and illite suspension." *Soil Sci. Soc. Am. J.* 44: 955-959.

Oster, J.D., Willardson, L.S., and Hoffman, G.J. (1972). "Sprinkling and ponding techniques for reclaiming saline soils." *Transactions of the Amer. Soc. Agric. Eng.* 15: 115-117.

Overstreet, R., Martin, J.C., and King, H.M. (1951). "Gypsum, sulfur and sulfuric acid for reclaiming an alkali soil of the Fresno series." *Hilgardia* 21: 113-127.

Peryea, F.J., Bingham, F.T., and Rhoades, J.D. (1985). "Regeneration of soluble boron by reclaiming high boron soils." *Soil Sci. Soc. Am. J.* 49: 313-316.

Prather, R.J., Goertzen, J.O., Rhoades, J.D., and Frenkel, H. (1978). "Efficient amendment use in sodic soil reclamation." *Soil Sci. Soc. Am. J.* 42: 782-786.

Quirk, J.P., and Schofield, R.K. (1955). "The effect of electrolyte concentration on soil permeability." *J. Soil Sci.* 6: 163-178.

Rahman, M.A., Hiler, E.A., and Runkles, J.R. (1974). "High electrolyte water for reclaiming slowly permeable soils." *Transactions of the Amer. Soc. Agric. Eng.* 17: 129-133.

Rasmussen, W.W., and McNeal, B.L. (1973). "Predicting optimum depth of profile modification by deep plowing for improving saline-sodic soils." *Soil Sci. Soc. Am. Proc.* 37: 432-437.

Rasmussen, W.W., Moore, D.P., and Alban, L.A. (1972). "Improvement of a Solonetizic (slick spot) soil by deep plowing, subsoiling, and amendments." *Soil Sci. Soc. Am. Proc.* 36: 137-142.

Reeve, R.C., Allison, L.E., and Peterson, D.F. (1948). "Reclamation of saline-alkali soils by leaching." Utah Agr. Expt. Sta. Bulletin 335, 52 pp.

Reeve, R.C., and Bower, C.A. (1960). "Use of high-salt waters as a flocculant source of divalent cations for reclaiming sodic soils." *Soil Sci.* 90: 139-144.

Reeve, R.C., and Doering, E.J. (1966). "The high salt-water dilution method for reclaiming sodic soils." *Soil Sci. Soc. Am. Proc.* 30: 498-504.

Reeve, R.C., Pillsbury, A.F., and Wilcox, L.V. (1955). "Reclamation of a saline and high boron soil in the Coachella Valley of California." *Hilgardia* 24(4), 69-91.

Rhoades, J.D., Ingualson, R.D., and Hatcher, J.T. (1970). "Laboratory determination of leachable soil boron." *Soil Sci. Soc. Am. J.* 34: 938-941.

Robbins, C.W., Wagenet, R.J., and Jurinak, J.J. (1980). "A combined salt transport-chemical equilibrium model for calcareous and gypsiferous soils." *Soil Sci. Soc. Am. J.* 44: 1191-1200.

Russo, D. (1984). "Spatial variability considerations in salinity management." *Soil Salinity Under Irrigation*. Shainberg, I., and Shalhevet, J., Eds. Springer Verlag. pp. 198-219.

Russo, D., and Bresler, E. (1977). "Effect of mixed Na/Ca solutions on the hydraulic properties of unsaturated soils." *Soil Sci Soc. Am. J.* 41: 713-717.

Shainberg, I., Bresler, E., and Klausner, Y. (1971). "Studies on Na/Ca montmorillonite systems. I. The swelling pressure." *Soil Sci.* 111: 214-219.

Shainberg, I., Keren, R., and Frenkel, H. (1982). "Response of sodic soils to gypsum and calcium chloride application." *Soil Sci. Soc. Am. J.* 46: 113-117.

Stewart, E.H., Alberts, R.R., and Orth, P.G. (1976). "Water and salinity relationships in Perrine marl soils of south Florida." *Soil and Crop Soc. Sci. Florida* 36: 89-93.

Suarez, D.L., and Rhoades, J.D. (1982). "The apparent solubility of calcium carbonate in soils." *Soil Sci. Soc. Am. J.* 46: 716-722.

Talsma, T. (1967). "Leaching of tile-drained saline soils." *Aust. J. Soil Res.* 5: 37-46.

Tanji, K.K. (1970). "A computer analysis on the leaching of boron from stratified soil columns." *Soil Sci.* 110: 44-51.

Tanji, K.K., and Deverel, S.J. (1984). "Simulation modeling for reclamation of sodic soils." Shainberg and Shalhevet, eds., *Soil Salinity Under Irrigation*. Springer-Verlag, N.Y. pp. 238-251.

Tanji, K.K., Doneen, L.D., Ferry, G.V., and Ayers, R.S. (1972). "Computer simulation analysis on reclamation of salt-affected soil in San Joaquin Valley, California." *Soil Sci. Soc. Am. J.* 36: 127-133.

Terkeltoub, R.W., and Babcock, K.L. (1971). "Calculation of the leaching required to reduce the salinity of a particular soil depth beneath a specific value." *Soil Sci. Soc. Am. Proc.* 35: 411-413.

Thorup, J.T. (1972). "Soil sulphur application". *Sulphur Inst. J.* 8: 16-17.

Tisdale, S.L. (1970). "The use of sulphur compounds in irrigated arid-land agriculture." *Sulphur Inst. J.* 6: 2-7.

Yahia, T.A., Miyamoto, S., and Stroehlein, J.L. (1975). "Effect of surface applied sulfuric acid on water penetration into dry calcareous and sodic soils." *Soil Sci. Soc. Am. J.* 39: 1201-1204.

NOTATION

C = salt concentration in soil; C = final soluble B concentrations; C_o = initial salt concentration in soil; C_0 = initial soluble B concentrations; CCR = $CaCl_2 \cdot 2\ H_2O$ requirement; CEC = soil cation exchange capacity; D = depth of leaching water; D/D_s = depth of leaching water per unit depth of soil; D_s = depth of soil to be leached; D_s = depth of soil to be reclaimed; E_{Naf} = final exchangeable sodium fraction; E_{Nai} = initial exchangeable sodium fraction; GR = gypsum requirement; and k = empirical constant

CHAPTER 20

IRRIGATION WITH SALINE GROUND WATER AND DRAINAGE WATER[a]

INTRODUCTION

In many arid regions of Australia, Egypt, India, Israel, Pakistan, the United States, and the Soviet Union, large quantities of drainage water and saline ground water exist. Since they contain more dissolved salts and other undesirable residuals than most irrigation waters, they often are considered to have little value. However, several potential uses of saline ground water and saline drainage water might be feasible, depending on the quality and availability of the water. This chapter will focus on the use of shallow ground water and agricultural drainage water for irrigation.

PRACTICAL REUSE SITUATIONS

Many situations exist where drainage water or saline ground water could be used for irrigation if the water is readily accessible and available. Possible sources are waters that are discharged from irrigated land in drainage pipes or canals and waters that are present in shallow ground water systems. Reuse is facilitated if good-quality, i.e., nonsaline, water is available, or if sufficient rainfall occurs at certain times of the year to periodically meet some of the crop's water needs and leaching requirements. The reuse strategies described in this chapter presuppose the availability of two water sources: one of good (nonsaline) quality and the other of poorer (saline) quality water. These sources need not be available simultaneously during the entire season or year.

Saline water reuse might be most practical in areas where nonsaline water is available during the early growing season but too costly or limited in supply to meet the crop water requirements for the entire irrigation season. Where the costs of nonsaline water are prohibitive, saline drainage water could be used to irrigate crops of moderate-to-high salt tolerance at later growth stages with economical advantages—even if the yield is somewhat reduced. However, such reuse should be undertaken only if long-term deleterious effects on soil properties can be avoided. Various scientists (Rhoades et al. 1988; Rhoades 1977, 1984, 1987; Meiri et al. 1986; Grattan et al. 1987; Shennan et al. 1987; Ayars et al. 1986a, b; Rains et al. 1987) have concluded that the use of saline

---

[a]Prepared by: Stephen R. Grattan, Cooperative Extension, Dept. of Land, Air, and Water Resour., Univ. of Calif., Davis, CA 95616; and James D. Rhoades, USDA/ARS Salinity Lab., 4500 Glenwood Dr., Riverside, CA 92501.

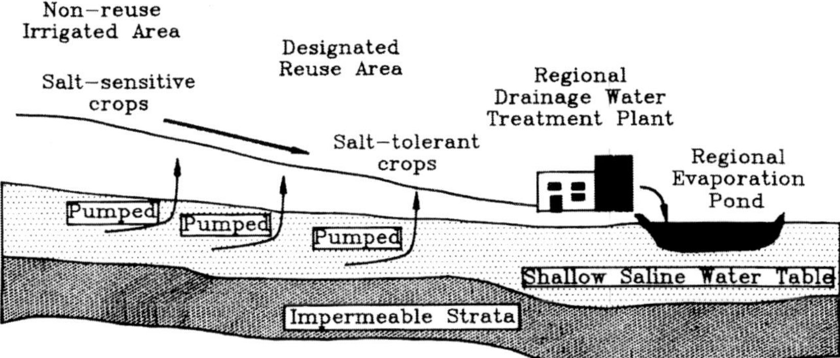

Fig. 20.1 Drainage Water Reuse Plan for Implementation on a Regional Scale

waters for irrigation is feasible, especially when waters are alternated or combined with good-quality water supplies. This option is being seriously considered in parts of the India (Gupta, personal communication) and Israel (Shalhevet 1984).

Another situation for reuse is in regions where water tables are shallow and disposal of drainage water is impossible due to physical, environmental, social, or political factors. Since many crops are sensitive to shallow or fluctuating water tables, the water table must be lowered and stabilized to sustain crop production. Reuse of the water for irrigation would reduce the amount of nonsaline water required by crops and would decrease the volume of drainage water to be disposed of, thereby reducing the costs of disposal and treatment. Furthermore, reducing the drainage return reduces salt loading of receiving waters.

Many growers in California's San Joaquin Valley are considering this reuse option at least as a temporary solution to reducing drainage volume. Its long-term feasibility would probably be increased if the reuse were implemented on a regional scale instead of on a farm scale. Regional reuse, as shown in Figure 20.1, avoids the successive increase in concentration that takes place when reuse repeatedly utilizes the same water supply. Certain areas in the region could be devoted to reuse. Other areas, such as upslope areas, could continue to be irrigated solely with nonsaline water. The secondary drainage water from the reuse area could be discharged to regional evaporation ponds or treatment plants. Growers in a region could coordinate their efforts and share costs.

Another, more extreme, use of saline waste waters is to apply them to plants that can accumulate large quantities of undesirable constituents, such as Se, Mo, and NO_3. This would be done before the secondary drainage subsequently discharges into a saline sink, such as an evaporation pond or the ocean. The accumulator plants might be used to supplement the feed of grazing animals, depending on the type of plant and the concentration of elements , such Se or Mo, that accumulate within them. However, caution is advised, as the danger of toxicosis exists. This process, known as biofiltration, has been studied by Cervinka et al. (1987) and Wu et al. (1987). They found that certain grasses and native species in California accumulate substantial amounts of Se in their shoots. This alternative management practice is best suited where problems with drainage disposal occur, the bioaccumulator has economic value, and other treatment processes (Chapter 21) are unavailable or prohibitively expensive.

EVALUATION OF THE REUSE POTENTIAL

Historically, waters that are suitable for many specialized irrigation situations have gone unused. The yield-threshold levels of salinity often used as standards for irrigation water quality reflect the response of crops to the average root zone salinity after the establishment of seedlings. However, salt tolerance of many crops increases as the plant matures (Lunin et al. 1963; Maas et al. 1986; Pasternak et al. 1986; Maas and Poss 1988). For this reason, irrigation water of relatively high salinity could be used late in the growing season on certain crops without a loss in yield. Furthermore, applying saline water later in the season reduces crop exposure to salinity, allowing waters of higher salinity to be used.

Studies conducted in many arid regions of the world have shown that water conventionally classified as too saline for agricultural use can be used to irrigate certain crops without a significant loss in yield or quality. Rhoades et al. (1988) provided references of such studies in a recent publication. Alfalfa, sorghum, and wheat have been grown in Colorado's Arkansas Valley with water containing 1,500 mg/l to 5000 mg/l TDS (Miles 1977). In the Pecos Valley, Texas, water averaging 2,500 mg/l TDS has been used as irrigation water for decades (Moore and Hefner 1976). In Iraq, pear trees have been grown with water ranging up to 4,000 mg/l without a loss in yield (Hardan 1976). In Israel, cotton is grown commercially with 4.6 dS/m EC water (about 3,000 mg/l TDS), according to Frenkel and Shainberg (1975) and Keren and Shainberg (1978). Evidently, this practice has been successful for nearly 15 years (Meiri, personal communication). However, an average of 400 mm of annual rainfall occurs in this region and large amounts of low-salinity water are applied before planting. These conditions make supplemental irrigation with saline water more feasible. In India, irrigation water in the 4 dS/m to 5 dS/m EC range has been used successfully over the long-term (Paliwal 1972). In Russia's Uzbekistan region, irrigation of cotton with drainage water between 5,000 to 6,000 mg/L TDS has been successful for many years (Bressler 1979).

After a worldwide survey of irrigation and salinity, Shalhevet and Kamburov (1976) concluded that water with salinity as high as 6,000 mg/l TDS was acceptable for irrigation. Pillsbury and Blaney (1966) suggested that the upper limit placed on irrigation water should be an EC of 7.5 dS/m (about 4,800 mg/l TDS). Dhir (1976) reported that water with twice this concentration was used to produce wheat in India. However, these areas are subject to annual monsoons. The most extreme case is reported by Epstein and Norlyn (1977), where several selections of barley were able to grow and reproduce when irrigated with Pacific Ocean water.

This study may not have much agronomic importance, but it does suggest the potential of certain crop species to tolerate excessive levels of salinity. Although several of the aforementioned studies describe multi-year success with saline-water irrigation, a crop-water management strategy to sustain such a practice year after year in areas without sufficient rainfall for leaching and crop establishment was not proposed.

REUSE STRATEGIES

Three strategies exist for using saline ground water or drainage water as a supplemental source of irrigation water. One is to manage the shallow saline ground water depth so that certain crops , such as cotton, can use this water directly. Chapter 16 discusses this concept in more detail. Here, the focus will be on the remaining two strategies, both of which use drainage water for irrigation (Rhoades 1977, 1984, 1987; Rhoades et al. 1988; Ayers et al. 1986 a and b; Grattan et al. 1987; Rains et al. 1987; Shalhevet 1984; Pasternak et al. 1986; and Meiri et al. 1986). One method is the cyclic strategy first proposed and tested by Rhoades (1984, 1987). The other is the traditional strategy, which involves blending surface water supplies before irrigation to obtain a water of acceptable quality. Either strategy can be used when two sources of water of different qualities are available. At least one source must be low in salts, while others can be saline within limits (less than about 6,000 mg/l TDS). Each method will be described and relevant work will be cited. The advantages and disadvantages of each will also be discussed.

Cyclic Strategy

The cyclic strategy is as follows:

1. Develop a crop rotation plan that makes use of salt-tolerant and moderately salt-sensitive crops. Moderately salt-sensitive crops can be included in the rotation because of the transient salinity of the soil profile.

2. Use at least two sources of irrigation water. One must be of high-quality water, i.e., low in salts. Others can be higher in TDS. Possible sources include saline well water or drainage water. Saline waters would never be blended with the higher-quality water before irrigation.

3. Develop an irrigation plan for the entire crop rotation based on crop tolerance to salinity, salt sensitivity of specific growth stages, and the salinity of the irrigation water. Use the nonsaline water for preplant and early irrigations of the salt-tolerant crop and all irrigations of the moderately salt-sensitive crop. Irrigate the salt-tolerant crops with saline water after they have reached a salt-tolerant stage of growth. After a salt-tolerant crop is grown, irrigate with low-salinity water to reclaim the soil profile's upper portion enough to establish the salt-sensitive crop. Subsequent irrigations during the rest of the season move the previously

accumulated salts farther down in the soil profile ahead of the advancing roots.

This cyclic use of high quality and saline water can be repeated again and again, if certain physical properties of the soil related to tilth and permeability can be sustained over the long-term.

Field studies conducted by Rhoades et al. (1984, 1987) have shown that the cyclic strategy is viable. In one seven-year study, cotton was grown for five consecutive years in California's San Joaquin Valley using irrigation waters of two different salinities. California aqueduct water (300 mg/l TDS) was used to irrigate the cotton until the seedlings were established. All other irrigations were made with drainage water containing 6,000 mg/l TDS and 6 mg/l boron. Wheat was subsequently grown with use of aqueduct water for reclamation. The cyclic strategy was then used on sugar beets during the following two years in the presence of a shallow (less than 1.2 m), saline water table (Rhoades, unpublished results).

Rhoades et al. (1988) conducted a second more extensive study on a 20 ha commercial field located in the Imperial Valley, Calif. The objective was to test the cyclic strategy on two cropping patterns. The first was a rotation of wheat, sugar beets, and melons. Colorado River water (900 mg/l TDS) was used to irrigate the melons and for the preplant and early irrigations of wheat and sugar beets. Alamo River drainage water (3,500 mg/l TDS) was used for all of the other irrigations. The two-year rotation was repeated a second time. Sugar beet and wheat yields were not reduced, even when drainage water supplied up to 75% of the irrigation water. Subsequent melon yields also were not reduced. The second cropping pattern tested was a four-year rotation of two years of dual cropping of cotton/wheat, followed by two years of alfalfa. Drainage water was used only on cotton after seedlings were established. Colorado River water was used for all other irrigations. In this crop pattern, no significant yield losses occurred for any crop grown with the cyclic strategy. However, crop qualities were often improved from the use of saline drainage water.

Another study was conducted in the San Joaquin Valley by Ayars et al. (1986a and b). Drip irrigation was used for three consecutive years to apply drainage water (EC = 8.0 dS/m) to cotton after the crops were established. A wheat crop then followed. It was irrigated only with water of good quality. Sugar beets were planted after the wheat and irrigated with drainage water after stand establishment. Yields under these conditions were the same as those that were irrigated continuously with the water of good quality.

Shennan et al. (1987) are currently testing a modified cyclic irrigation strategy on a crop rotation that consists of two years of cotton followed by processing tomatoes. Saline drainage water (EC = 7.9 dS/m) is applied to the moderately salt-sensitive processing tomato crop at a relatively tolerant growth stage to improve fruit quality, for example, soluble solids. High-quality water is used for the reclamation of subsequent cotton crops. Grattan et al. (1987) observed sustained yield and quality improvement in short-term tomato studies, in which drainage water, applied after first flower, supplied over 65% of the irrigation water requirements. Pasternak et al. (1986) found that soluble solids in the tomato fruit increased from use of brackish water (EC = 7.5 dS/m), but that yields were reduced by 30% when saline water was applied after the fourth or eleventh leaf stage. Rhoades et al. (1988) observed that the use of drainage water for irrigation improved the quality of wheat, cantaloupes, and alfalfa. More long-term research on irrigation of crops having intermediate salt tolerance must be conducted before the feasibility of the cyclic strategy can be assessed.

Using the cyclic strategy may be difficult on a farm of greater than 100 ha, where the amount of drainage water produced does not coincide with crop water demand. In California's San Joaquin Valley, most farms are large, and the flow of drain water from their fields peaks early in the irrigation season, when most crops require good-quality water. Later in the season, the water needed to irrigate a field could exceed that supplied by the drains. To use the surplus of early season saline waters and to avoid later shortages, an option is to build a surface storage reservoir for drainage water. Another is to plug the drains and to allow the soil to act as the reservoir. Regardless, drainage water must be stored and readily accessible to use the cyclic strategy. In concept, this can be more readily achieved on a regional than a farm-by-farm basis (Rhoades et al. 1988).

An advantage of the cyclic strategy is that steady-state salinity conditions in the soil profile are never reached. This behavior occurs because the quality of irrigation water changes over time. The intermittent leaching that takes place under this strategy can be more effective at leaching salts than continuous leaching strategies, i.e., imposing a leaching fraction with each irrigation (Shalhevet 1984). Another advantage of this strategy is that a facility for blending waters of different qualities is unnecessary.

Blending Strategy

The second reuse strategy is to blend water supplies before or during irrigation. Blending involves mixing two waters of different qualities to obtain water that is suitable for irrigation.

Blending was studied on an 8-ha site in the Tulare Lake Basin, California. A rotation of two years of cotton was followed by one year of safflower (Rains et al. 1987; Rolston et al. 1988). Crops were either irrigated continuously with California Aqueduct water (400 mg/l TDS) or with a blended mixture of drainage water and aqueduct water at different salinities (1,500, 3,000, 4,500, 6,000, and 9,000 mg/l TDS). All plots received a pre-plant irrigation with aqueduct water. Safflower was not irrigated after the preplant irrigation. The salinity treatments did not significantly reduce cotton yields for the first two years. However, the yield of the safflower, a more salt-sensitive crop than cotton, was reduced by almost 40% at the highest salinity level of 9,000 mg/l TDS. This was partly due to poor stand establishment. When cotton was planted the following year, yields were reduced by the 9,000 mg/l TDS treatment, again due in part to poor stand establishment.

To use a blending strategy, a controlled way of mixing the water supplies must exist. Shalhevet (1984) discussed two blending processes:

network dilution and soil dilution. With network dilution, water supplies are blended in the irrigation conveyance system. A facility for blending must be built into the system. With soil dilution, the soil acts as the media for mixing water of different qualities. Different water qualities are alternated, according to availability, between or within an irrigation event. Dinar et al. (1986) described a method to calculate optimal ratios of saline and nonsaline irrigation water for crop production.

Meiri et al. (1986) conducted a three-year study in Israel to compare crop performance under network and soil blending systems. They tested a rotation of potatoes and peanuts under drip irrigation. They concluded that crops responded to the weighted mean water salinity, regardless of the blending method.

Blending is often proposed as a way to extend the water supply. The above studies describe experiments where water of different salinities were blended, but the objective of these studies was not to evaluate how such blending affected the usefulness of the total water supply. The water supply is expanded beneficially only by blending with poor-quality water that is suitable for irrigation. For example, if the concentration of the poor-quality water is less than 1,500 mg/l TDS, blending would be a valuable means to expand the water supply. Blending good-quality irrigation water with water that is too saline, on the other hand, can negate the effect of expanding the water supply. It is difficult to conclude from the studies described above whether the saline fraction of the blended water was ever consumed by the crop, or whether blending resulted in an increased volume of water supply.

Blending Strategy Versus Cyclic Strategy

Intentional blending of water supplies of different quality in most situations does not provide additional benefits to the plant over the cyclic strategy. Furthermore, blending may deprive plants of the opportunity to use the good quality water fully. (Rhoades et al. 1988). A plant extracts water from the soil through transpiration, leaving most of the salts concentrated in the remaining soil water. The plant cannot use the remaining soil water for transpiration after a maximum concentration is reached without suffering a loss in yield. After adding pure water to the remaining soil water, the plant could, at best, extract the pure water fraction up to the same maximum concentration. Therefore, blending in this case would not increase the crops' usable water supply. If the following rotational crop grown with such a blended water is less salt-tolerant than the first crop, then the blending would result in a net loss of usable water.

The following is an extreme example of how blending can decrease usable water for crop transpiration. Suppose a vegetable grower has access to two water supplies. One is nonsaline and the other is drainage water as saline as sea water. If one liter of nonsaline water is blended with one liter of drainage water, the result will be zero liters of usable water. Normal vegetable crops cannot tolerate water with salinity values half that of sea water. This example is exaggerated, but the principle involved is valid. Good-quality water can become unusable when it is blended with drainage waters that are too highly saline (Rhoades et al. 1988).

The cyclic strategy is preferred over the blending strategy in most cases, as more highly salt-sensitive crops can be included in the rotation, and a blending facility is not required. Also, less danger of losing water usable by the crop exists, and drainage water of higher salinity can be used. However, larger quantities of drainage water may be needed during some cyclic irrigations, because the water is not blended and easy access to stored drainage water for irrigation is consequently needed. If the supply of drainage water is insufficient, the size of the reuse area can be reduced accordingly.

The authors believe the cyclic strategy has more potential and flexibility than the blending strategy. However, the latter may be more practical for situations where the drainage water or shallow ground water is not too saline for the crop and insufficient quantities of drainage water are available to meet crop water requirements.

POTENTIAL LONG-TERM ADVERSE EFFECTS

Several studies have suggested that irrigation water containing salt concentrations exceeding conventional suitability standards can be used successfully on many crops for at least seven years without a loss in yield. However, uncertainty still exists about the long-term effects of these practices.

Long-term effects on the soil could include soil crusting, reduced water infiltration capacity, and accumulation of toxic elements. Crusting and reduced water infiltration are a concern where soils are poorly structured and the drainage water's SAR exceeds the permeability threshold corresponding to the electrical conductivity of the irrigation water. As discussed in Chapter 5, water infiltration decreases as the exchangeable sodium percentage of the soil increases and as the electrical conductivity of the infiltrating water decreases. Adverse effects are more likely to occur during the preplant irrigation period, when low salinity water irrigates a field previously irrigated with saline-sodic water. Rolston et al. (1988) described such problems at their experimental site in California, where blended water [9,000 mg/l TDS; SAR = 30] had been applied for four years. The result was poor stand establishment. Whether the problem will arise depends on whether the EC of the good-quality water is less than the threshold value, given the SAR of the drainage water. Some combinations of the two waters are not permissible. See Chapter 3 for assessments of the compatibility of irrigation waters.

Long-term effects on soil salinity should also be considered. Chapter 9 describes techniques for monitoring soil salinity. The levels of soil salinity within the root zone and the crop response are more easily predicted under the steady-state conditions that occur with blending. The soil salinity fluctuates more with cyclic strategy than with blending strategy both spatially and temporally. The cyclic strategy fosters seedling establishment, as less salinity accumulates in the upper profile. Rhoades et al. (1988) found that cyclic changes in soil salinity were kept within acceptable levels for all crops and found no losses in soil permeability. With both strategies, management practices that keep the average root zone salinity within acceptable limits and soil salinity measurement techniques that monitor changing conditions help to achieve the desired control.

Many drainage waters contain elements, such as boron and chloride, that accumulate in plants to levels that cause foliar injury and reduce yield (Chapters 7 and 13). Toxic effects may not become evident for several years. In many cases, the accumulation of such elements may produce a longer-term detrimental effect than does salinity alone. In the San Joaquin Valley, much of the drainage water contains boron in excess of 5 mg/l. More leaching is required to remove large quantities of soil boron than is required to remove excess salts. Irrigation water that contains greater than 5 mg/l boron can be detrimental to many crops. Water of this quality is not advised for the long-term irrigation of perennial crops, since perennials are often more sensitive to specific-ion effects. Also, perennials represent a long-term investment, and they have a long time to accumulate these elements to toxic levels. Toxic effects may not become evident for several years. This also applies to planting perennial crops grown in the presence of a shallow ground water that contains elements toxic to the plant, such as chloride and boron.

Another long-term effect to consider is the accumulation in plants and soils of elements that are toxic to consumers. Many trace elements, e.g., selenium (Se), and molybdenum (Mo), are toxic to humans and animals. Toxic elements are of particular concern since they can concentrate as they move up the food chain, a process called biomagnification. In the San Joaquin Valley, drainage water in several locations contains unusually high levels of Se ($\geq$50 μg Se/l). Although Se is essential to humans and animals in trace amounts, excessive amounts can cause Se toxicosis. Drainage water that contained 250 μg Se/l to 350 μg Se/l was used to irrigate melons and processing tomatoes in the San Joaquin Valley (Grattan et al. 1987).

Although the concentrations detected in the fruit (250 μg Se/kg to 750 μg Se/kg, dry wt.) were not considered to be hazardous (Fan and Jackson 1987), they were significantly higher than the concentrations in those fruits irrigated with high-quality water ($<$5 μg Se/l). Selenium concentrations in fruit tissue could escalate if drainage-water reuse is practiced over a longer term, especially when waters are low in sulfate because sulfate competes with selenate during plant uptake.

Many forages and native species can accumulate excessive amounts of Se in their tissue (Wu et al. 1987). Since grazing animals consume larger quantities of plant mass than do humans, they are more prone to exceed their recommended intake.

Although Se and Mo are of major concern in the drainage water of the San Joaquin Valley, other elements in drainage water could dictate the reuse potential of drainage water in not only this area, but in other regions of the world.

MANAGEMENT CONSIDERATIONS AND GUIDELINES

Numerous management considerations must be addressed in the reuse of drainage water. An approach with rigid criteria is impractical, since most management decisions are subjective and innumerable combinations exist. Therefore, we will suggest parameters and provide approximate guidelines.

Drainage Water Availability

The availability and accessibility of drainage water must be determined. As indicated earlier, the cyclic strategy can have more stringent demands on availability than does the blending strategy, so a system for collection and redistribution of the drainage water must be developed to avoid the need for storage reservoirs. This is important where large areas will be irrigated by furrow or flood. The size of area that can be irrigated by drainage water will vary, depending on the capacity of the drainage system. Generally, at least 1 m^3/min-ha is required to surface-irrigate (Hanson, personal communication).

In areas with a high water table, drainage water can be collected by installing a network of interceptor drains. A submersible pump is placed in collector sumps to lift the drainage water to the field level. Another way to collect drainage water in areas with a shallow water table involves installing a network of shallow wells . All wells are connected to a common drainage-water manifold. Irrigation and drainage engineers might be consulted before installing any drainage water collection system.

A network blending system must be installed if the blending strategy is adopted. Sinai et al. (1988) provide the theory, design criteria for a network blending system, and methods of automatically diluting saline water sources for irrigation.

Water Analysis

To evaluate reuse potential, standard water sampling techniques and analysis are used to determine the quality of shallow ground water or drainage water (Chapman and Pratt, 1982). The most important water quality parameters are EC, SAR, and boron concentration. The water should also be analyzed for the presence of toxic elements. If available, water-quality data for different times of the year should be examined to assess temporal variations in water quality. See Chapter 11 to assess the suitability of the water for irrigation.

Crop Selection Criteria

Perhaps the most important management decision before implementing the reuse of drainage water is crop selection. A list of crop selection criteria is presented in Table 20.1.

Economics. Economics is an important consideration. Growing a high-yielding crop without a marketable product or without a positive cash flow would be ill-advised. For example, salt and boron tolerance often limit the crops that can be grown in a given location. In the San Joaquin Valley, a negative correlation exists between crop tolerance to salinity and economic value per area (Fig. 20.2). Few crops are both tolerant to salinity and high in economic value. Asparagus is tolerant to salinity and comparatively valuable, but harvesting it is labor-intensive and the production per area is relatively low.

Crop Salinity Tolerance. The abilities of various crops to withstand salinity and high concentrations of trace elements, discussed in Chapter

TABLE 20.1 Criteria for Selecting Crops for Water Reuse

Selection criterion (1)	Desirable (2)	Undesirable (3)
Economic value/marketability	High/marketable	Low/unmarketable
Crop salt tolerance	Tolerant	Sensitive
Crop boron/chloride tolerance	Tolerant	Sensitive
Crop potential to accumulate toxic constituent	Excludes toxic element	Toxic element accumulation
Crop quality	Improved by saline water	Unaffected or adversely affected by saline water
Crop rotation	Compatible	Incompatible
Managemental/ environmental conditions and requirements	Easy management; able to grow under diverse conditions	Requires intensive management; able to grow only under specific conditions

13, should be considered before selecting a crop. Field crops are generally more tolerant to salinity than annual vegetable crops (Fig. 20.2). Many trees and vines are prone to damage by ions, such as chlorine and boron. These sensitive perennials should, whenever possible, be excluded from reuse consideration.

Toxic Elements. Plants vary in their ability to absorb and translocate toxic elements (Chapter 17). Crops or cultivars that accumulate large quantities of toxic elements should not be grown in areas where drainage water leaves potentially hazardous residuals. If the wastewater in question contains high levels of a toxic element, the literature should be consulted for information on the different accumulation potential of crops under field conditions.

Drainage water underlying the San Joaquin Valley has been found to contain abnormal amounts of Se. Selenium is chemically similar to sulfur, and the Se accumulation potential of a crop is related to its sulfur requirement (Hurd-Karrer 1935). Fleming (1962) studied crops grown in seleniferous soils in Ireland and found that Se concentrations were higher in Cruciferae (cabbage, cress, radish, rape, and turnips), liliacea (onion), and leguminosae (clover and peas) than in compositae (artichoke and lettuce), gramineae (barley, oats, rye grass, and wheat), and umbeliferae (parsnip and carrots). It is also important to understand how toxic constituents are partitioned within the plant. For example, in most annual fruit and vegetable crops, selenium accumulates more in the leaves than in the fruit (Mikkelson et al. 1988). High levels of molybdenum, uranium, and arsenic also have been detected in the drainage water underlying the San Joaquin Valley.

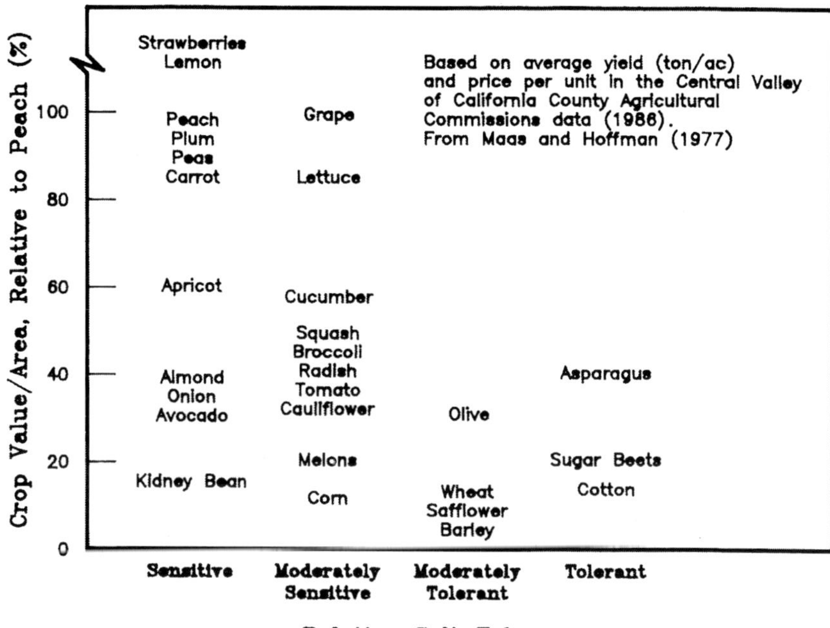

Fig. 20.2 Relationship Between Crop Salt Tolerance and Relative Economic Value per Area

Management Practices and Climate. Both crop management practices and the climate dictate the types of crops that can be grown in a region. The most desirable crops are those that can be grown under a wide range of environmental conditions without intensive management. Most cooperative-extension agents will have information on the management practices that are most successful for particular crops and locations. These include approximate planting dates, crop rotations, preferred management practices, and soil types. Extension agents can also provide tips on crop rotation compatibility and suggest practices for avoiding other problems, such as disease, weeds, and insect infestations.

When saline drainage water is to be used for irrigation, highly salt-tolerant crops should be selected. A crop of intermediate tolerance, such as alfalfa, melons, tomatoes, or wheat, can be used, especially if the salinity enhances the quality of the crop. For example, saline drainage water with EC of 4 dS/m-8 dS/m improved the soluble solids content of melons and tomatoes (Grattan et al. 1987), raised the protein content of wheat and alfalfa, increased total digestible nutrients in alfalfa, improved the color and netting of cantaloupe (Rhoades et al. 1988), and increased the peelability of processing tomato (May, personal communication). While higher plant quality should not be a major factor in selecting the reuse of drainage water, it can be an important factor in selecting a crop. Using drainage water to irrigate crops of intermediate salinity tolerance is feasible only after plants have been established with high quality water. The crop-water management strategy used must maintain soil salinity, boron, and other undesirable residuals within acceptable limits.

Irrigation Water Management

Proper management of irrigation water, regardless of its quality, is essential for good crop production. It is even more important when saline waters are used. Management decisions must be made based on stand establishment, irrigation method, and water suitability.

Stand Establishment. Precautions should be taken to ensure an adequate crop stand. Hard crusts limit the emergence of young seedlings, so management practices that promote aggregate stability should be used. Excessive crusting can take place after the soil's surface dries if water low in electrolytes is applied to a field that contains a high percentage of exchangeable sodium from previous applications of saline-sodic drainage water (Chapter 5). Aggregate stability is facilitated by maintaining in the upper portion of the soil profile a relatively high ratio of divalent cations to sodium ratio, i.e., by keeping the ESP low. The most common way to achieve this condition is by applying and incorporating gypsum into the topsoil.

Irrigation Method. The irrigation method influences crop performance greatly, especially when the irrigation water is saline. Shalhevet (1984) described three factors that should be considered before selecting an irrigation method: (1) The salt distribution in the soil; (2) the sensitivity of the crop to foliar wetting; and (3) the ease with which high osmotic and matric potentials can be achieved.

The distribution of salts in the root zone is influenced by the leaching fraction, the crop's water extraction patterns, the shape of the seed bed, and the method of water application (Oster et al. 1984). Bernstein and Fireman (1957) showed that, with furrow irrigation, salts accumulate in certain regions of the seed bed. Information from this study helped to improve the design of the seed bed and furrow to minimize this problem. Oster et al. (1984) described such seed placement and surface irrigation strategies as the irrigation of alternative furrows and the control of depth of water in furrows to move salts out of the seed area. Under drip irrigation, the salt content is lowest in the soil immediately below and next to the emitters and highest around the wetted zone. The salt that has accumulated in this front of the wetting zone must be removed. Sprinkler irrigation can be effective in leaching salts from the topsoil and producing a favorable low-salinity environment in the upper soil profile, a necessary condition for the establishment of salt-sensitive seedlings. Chapter 16 discusses salt distribution patterns in more detail.

Many crops are sensitive to saline water under sprinkler irrigation, since the foliage absorbs salts upon wetting. Salts can accumulate in leaves by foliar absorption until they reach lethal concentrations. However, the crop's sensitivity to saline sprinkling water relates more to the rate of foliar salt accumulation than to the crop's tolerance to soil salinity. Bernstein and Francois (1973) found that the application of 4.4 dS/m water through a sprinkler system reduced the yield of bell peppers by 59%, compared to a drip system. Meiri et al. (1982) found similar results for potatoes. Sprinkling at night may reduce this problem.

Theoretically, plant performance should be optimized by maintaining the average root zone at the highest possible water potential. Therefore, irrigation intervals with saline water should be brief compared to those with nonsaline water. Frequent irrigations reduce the time-averaged salt concentration in the soil solution and maintain a higher potential. Based on the ease with which it maintains a high soil water potential, drip irrigation would seem to be the method to use under saline conditions. Little evidence exists in the literature, however, to demonstrate that increasing the frequency of saline water irrigations improves crop performance in the field (Shalhevet 1984). This may be because problems induced by increased irrigation frequency, such as poor aeration and increased root disease, negate the benefits of increased water potential.

Water Suitability. The suitability of water for irrigation depends on many factors, including quality, type of crop, soil, climate, cropping patterns, and irrigation methods.

Rhoades and Merrill (1976) developed a computer model that can be used to predict the salinity, sodicity, and toxic-solute concentration of the soil water resulting after irrigation with water of known composition and at a specified leaching fraction. They developed the model to assess the effects of salinity on crop yield and sodicity on soil permeability. The program, called Watsuit, calculates the soil water concentrations of Ca, Mg, Na, CO_3, HCO_3, Cl, and SO_4; the sum of cations and anions; the Ca/Mg ratio; and the pH, EC, SAR, and osmotic potential (OP) at five depths in the root zone (surface, 25%, 50%, 75%, and 100% depth fractions) for each of five leaching fractions (5%, 10%, 20%, 30%, and 40%). It also calculates average and upper profile EC, OP, and SAR, and water-uptake-weighted salinity after corrections for the precipitation and dissolution of $CaCO_3$ and gypsum. The water-uptake-weighted or upper profile levels of salinity can be used to assess the suitability of the water for conditions of high-frequency irrigation. The average profile levels are used for conditions of infrequent irrigation management. Toxicity can be assessed by substituting specific solute concentrations for EC and OP. With the cyclic strategy, the model can be used to predict crop response.

Soil permeability and crusting problems can be determined by comparing the predicted ESP and EC of the infiltrating water with appropriate EC/SAR threshold values (Oster et al. 1984).

Maas and Hoffman (1977) provided crop salt tolerance values. A soil salinity problem may occur if the predicted salinity exceeds the threshold tolerance level of the crop to be grown; a permeability problem is likely if the predicted EC/SAR threshold levels are exceeded at the soil surface.

CONCLUSIONS

The reuse of drainage water might be practiced to advantage in several situations. It can be practiced where nonsaline water is too costly or unavailable during portions of the growing season. Reuse is possible if disposal problems must be reduced and water supplies conserved. Lastly, drainage water can be reused when toxic constituents must be removed from the drainage water before it is discharged to other water supplies.

Numerous studies indicate that water normally deemed too saline for irrigation can be used for irrigation successfully. Many of them, however, were conducted in regions with sufficient rainfall to leach a substantial amount of salts from the root zone. In low-rainfall areas, e.g., India, Israel, and California, reuse of drainage water may be feasible if several sources of irrigation water are available within the season. Such reuse can expand the existing water supply and decrease drainage volume.

Two reuse strategies have been presented. One, the cyclic strategy, involves using fresh, nonsaline water to irrigate moderately salt-sensitive crops and establish salt-tolerant crops. Saline water is then used to irrigate the salt-tolerant crops during later (tolerant) growth stages. The other strategy, blending, involves combining the saline and nonsaline waters before or during an irrigation.

Advantages of the cyclic strategy are that more salt-sensitive crops can be included in the rotation, a blending facility is not required, soil salinity can be lowered at critical times, the use of the entire water supply is maximized, and water of higher salinities can be used. Methods of collecting sufficient drainage water for irrigation must be developed, however. Blending may be advantageous if the cost of drainage water disposal exceeds the consequences of reduced yield and loss of usable water, or if the salinity of the poor-quality water is, by itself, suitable for irrigation. In most cases, the cyclic strategy is preferred over blending.

The numerous factors that facilitate reuse must be considered. Drainage water must be available and accessible. Water supplies should be analyzed and managed, taking account for EC, SAR, HCO_3, Cl, and potentially toxic elements. Criteria for selecting crops includes economic considerations; tolerances to salinity, specific elements, and toxic elements; and crop quality. Many decisions about water management beyond irrigation scheduling must be made, e.g., irrigation method, crop stand establishment, and assessing the suitability of water for irrigation. If blending is used, the manager must determine the optimal mix of saline and nonsaline water.

Although most studies conducted to date support the feasibility of irrigating crops with saline drainage water, uncertainty still exists about the long-term effects of this practice on the physical characteristics of the soil. We caution those considering the reuse of drainage water, first, to evaluate their specific conditions carefully.

REFERENCES

Ayars, J.E., Hutmacher, R.B., Schoneman, R.A., Vail, S.S., and Felleke, D. 1986a. "Drip irrigation of cotton with saline drainage water." *Transactions of the ASAE* 29(6): 1668-1673.

Ayars, J.E., Hutmacher, R.B., Schoneman, R.A., and Vail, S.S. 1986b. "Trickle irrigation of sugar beets with saline drainage water." Annual Report of the Water Management Research Laboratory USDA-ARS, Fresno, Calif., pp. 5-6.

Bernstein, L. and Fireman, M. 1957. "Laboratory studies on salt distribution in furrow irrigation soil with special reference to pre-emergence period." *Soil Sci.* 83: 249-263.

Bernstein, L. and Francois, L.E. 1973. "Comparison of drip, furrow, and sprinkler irrigation." *Soil Sci.* 115: 73-86.

Bressler, M.B. 1979. "The use of saline water for irrigation in the U.S.S.R." Joint Commission on Scientific and Technical Cooperation, Water Resources, Bureau of Reclamation, Engineering Research Center, Denver.

Cervinka, V., Finch, C., Beyer, J., Menezens, F., and Ramirez, R. 1987. "The agroforestry demonstration program in the San Joaquin Valley." Progress report. California Dept. of Food and Agriculture, Agricultural Resources Branch, Sacramento, Calif. 95814.

Chapman, H.D. and P.F. Pratt. 1982. "Methods of analysis for soils, plants, and waters." University of California Publication 4034. 309 pp.

Dhir, R.D. 1976. "Investigations into use of highly saline waters in an arid environment. I. Salinity and alkali hazard conditions in soil under a cyclic management system." Proc. International Management Conference on *Managing Saline Water on Irrigation*, Texas Tech. Univ., Lubbock, Texas, Aug. 1976, pp. 608-609.

Dinar, A., Letey, J., and Vaux, H.J. 1986. "Optimal ratios of saline and non-saline irrigation waters for crop production." *Soil Sci. Soc. Amer. J.* 50: 440-443.

Epstein, E. and Norlyn, J. D. 1977. "Seawater-based crop production: A feasibility study." *Science* 197: 247-251.

Fan, A.M. and Jackson, R.J. 1987. "Health assessment of human dietary selenium intake from agricultural crops." A report of the U.C. Salinity/Drainage Task Force. Div. of Ag. and Natural Resources. Univ. of Calif., 22 pp.

Fleming, G.A. 1962. "Selenium in Irish soils and plants." *Soil Sci.* 94: 28-35.

Frenkel, H. and Shainberg, I. 1975. "Irrigation with brackish water: Chemical and hydraulic changes in soils irrigated with brackish water under cotton production." *Irrigation with Brackish Water*, International Symposium, Beer-Shevos, Israel, The Negev University Press, pp. 175-183.

Grattan, S.R., Shennan, C., May, D.M., Mitchell, J.P., and Burau, R.G. 1987. "Use of drainage water for irrigation of melons and tomatoes." *California Agriculture* 41(9/10): 27-28.

Hardan, A. 1976. "Irrigation with saline water under desert conditions." Proc. International Salinity Conference, *Managing Saline Water for Irrigation*, Texas Tech. Univ., Lubbock, Texas, Aug. 1976, pp. 165-169.

Hurd-Karrer, A.M. 1935. "Factors affecting the absorption of selenium from soils by plants." *J. Agr. Res.* 50: 413-427.

Keren, R. and Shainberg, I. 1978. "Irrigation with sodic and brackish water and its effect on the soil and on cotton fields." *Harrade* 58: 963-976.

Lunin, J., Gallatin, M.H., and Batchelder, A.R. 1963. "Saline irrigation of several vegetable crops at various growth stages. I. Effect on Yields." *Agron. J.* 55: 107-110.

Maas, E.V., and Hoffman, G.J. 1977. "Crop salt tolerance-Current assessment." *J. Irrig. and Drain. Div., ASCE* 193(IR2): 115-134.

Maas, E.V. and Poss, J.A. 1988. "Salt sensitivity of wheat at various growth stages." *Irrigation Sci.* (In press)

Maas, E.V., Poss, J.A., and Hoffman, G.J. 1986. "Salinity sensitivity of sorghum at three growth stages." *Irrigation Sci.* 7: 1-11.

Meiri, A., Shalhevet, J., Shinshi, D., and Tibor, M. 1982. "Irrigation of spring potatoes with saline water." Annual Report, Agric. Res. Org., Inst. Soil and Water, Volcani Center, Bet Dagan, Israel.

Meiri, A., Shalhevet, J., Stolzy, L. H., Sinai, G., and Steinhardt, R. 1986. "Managing multi-source irrigation water of different salinities for optimum crop production." BARD technical report #1-402-81. Volcani Center, Bet Dagan, Israel, 172 pp.

Mikkelson, R.L., Page, A.L., and Bingham, F.T. 1988. "Factors affecting selenium accumulation by agricultural crops." *Selenium in Agriculture and the Environment*, L. W. Jacobs, ed., Special Publication, Amer. Soc. Agronomy, Madison, Wisc.

Miles, D.L. 1977. Salinity in the Arkansas Valley of Colorado. Interagency Agreement Report EPA-IAG-D4-0544; Environmental Protection Agency, Denver, CO, 80 pp.

Moore, J. and Hefner, J.V. 1976. "Irrigation with saline water in the Pecos Valley of West Texas." Proc. International Salinity Conference "Managing Saline Water for Irrigation," Texas Tech. Univ., Lubbock, Aug. 1976, pp. 339-344.

Oster, J.D., Hoffman, G.J., and Robinson, F.E. 1984. "Management alternatives: Crop, water, and soil." *California Agriculture* 38(10):29-32.

Paliwal, K.V. 1972. "Irrigation with saline water." Water Technology Centre, Indian Agricultural Research Institute, New Delhi, 198 pp.

Pasternak, D., DeMalach, Y., and Borovic, J. 1986. "Irrigation with brackish water under desert conditions. VII. Effect of time of application of brackish water on production of processing tomatoes (*lycopersion esculentum* Mill)." *Agricultural Water Management* 12: 149-158.

Pillsbury, A.F. and Blaney, H.F. 1966. "Salinity problems and management in river systems." *Proc. ASCE* (IRI) 92: 77-90.

Rains, D.W., Goyal, S., Weyranch, R., and Lauchli, A. 1987. "Saline drainage water reuse in a cotton rotation system." *California Agriculture* 41(9/10): 24-26.

Rhoades, J.D. 1977. "Potential for using saline agricultural drainage waters for irrigation." *Proc. Water Management for Irrigation and Drainage, ASCE.* Reno, Nevada, July 1977, pp. 85-116.

Rhoades, J.D. 1984. "Use of saline water for irrigation." *California Agriculture* 38(10): 42-43.

Rhoades, J.D. 1987. "Use of saline water for irrigation." *Water Quality Bulletin* 12: 14-20.

Rhoades, J.D. and Merrill, S.D. 1976. "Assessing the suitability of water for irrigation: Theoretical and empirical approaches in prognosis of salinity and alkalinity." *FAO Soils Bull.* 31: 69-109.

Rhoades, J.D., Bingham, F.T., Letey, J., Hoffman, G.I., Dedrick, A.R.,

Pinter, P.J., and Replogle, J.A. 1988. "Use of saline drainage water for irrigation: Imperial Valley study." *Agricultural Water Management* (In press).

Rolston, D.E., Rains, D.W., Biggar, J.W., and Luchli, A. 1988. "Reuse of saline drain water for irrigation." Paper presented at UCD/INIFAP Conference. Guadalajara, Mexico, March 1988.

Shalhevet, J. 1984. "Management of irrigation with brackish water." *Soil Salinity Under Irrigation, Processes and Management*. Shainberg, I. and Shalhevet, J., eds. Springer-Verlag, pp. 298-318.

Shalhevet, J. and Kamburov, J. 1976. *Irrigation and salinity: A world-wide survey*. International Commission on Irrigation and Drainage. 106 pp.

Shennan, C., Grattan, S., May, D., Burau, R., and Hanson, B. 1987. "Potential for the long-term cyclic use of saline drainage water for the production of vegetable crops." Technical Progress Report, U.C. Salinity/Drainage Task Force, Div. of Ag. and Natural Resour., Univ. of Calif., Davis, pp. 142-146.

Sinai, G., Jury, W.A., and Stolzy, L.A. 1988. "Methods for automated dilution of saline water sources for irrigation." *J. Irrig. Drain. Div. ASCE* (In press)

Wu, L., Burau, R.G., and Epstein. E. 1987. "Study of selenium accumulation and selenium-salinity cotolerance in turf, forage, and salt marsh species for crop production in deteriorated soils and water quality improvement in the San Joaquin Valley." Technical progress report. U.C.Salinity/Drainage Task Force. Div. of Agric. and Natural Resour., Univ. of Calif.

NOTATION

Term	Abbreviation	SI Unit	Other Unit
electrical conductivity	EC	dS/m	mmho/cm
exchangeable sodium percentage	ESP	—	—
sodium adsorption ratio	SAR	mmol/l	meq/l
total disolved solids	TDS	—	mg/l

CHAPTER 21

DRAINAGE WATER TREATMENT AND DISPOSAL OPTIONS[a]

INTRODUCTION

Treating and disposing of subsurface drainage water from irrigated agricultural lands presents unique technical challenges. A review of the literature reveals limited experiences in the management of such waters. The challenge is made more difficult by the complex chemical characteristics of most drainage water. Drainage usually contains a heavy salt load—a perennial cause for concern—and residual pesticides, herbicides, fungicides, fertilizers, and toxic trace elements—a more recent cause for concern. While crop management practices and conventional treatment processes can control salts and residuals to some degree, a new treatment technology needs to be developed to control toxic trace elements. Conventional methods appear to be ineffective in meeting the requirements set by many regulatory agencies.

In this chapter, the treatment and disposal of subsurface drainage from irrigated lands will be covered, disposal options will be presented, the technology of drainage water treatment and disposal options will be reviewed, and current research on treatment technology will be discussed.

TECHNOLOGY DEVELOPMENT

Treatment Technology Needs

The management of drainage from irrigated lands is an important part of any agricultural development plan. History has recorded many instances of fertile lands subsequently made barren by salt (FAO 1973). More recently, traditional approaches to sustaining and optimizing agricultural productivity by salinity control have been complicated by the need to protect public health and the environment from being affected by the residual fertilizers, pesticides, herbicides, fungicides, and trace toxic substances in drainage from irrigated lands. The alarm was raised when dead and deformed birds—most likely the result of selenium, were found in the Kesterson Reservoir, the San Joaquin River Basin, California (Ohlendorf 1984). All of this points to the need to develop a technology that will control and manage drainage from irrigated lands.

---

[a]Prepared by: Edwin W. Lee, U.S. Bureau of Reclamation, Mid-Pacific Regional Office, 2800 Cottage Way, Sacramento, CA 95825.

TABLE 21.1 Water Quality Objectives for San Joaquin River, California

Constituent (1)	Objective (2)
Selenium (wetland use)	2 μg/l
Selenium (in river)	5 μg/l
Electroconductivity	1.0 dS/m
Boron	700 μg/l
Molybdenum	10 μg/l

Source: California State Water Resources Control Board Technical Committee Report for Regulation of Agricultural Drainage to the San Joaquin River (1987).

Disposal, Reclamation, and Reuse Objectives

In any management plan for agricultural drainage, the objective is to dispose of or reuse wastewater in a way that will optimize water resources. Treatment processes that can affordably reduce levels of contaminants to meet the standards for receiving waters or reuse are needed.

Discharge limits set by regulatory agencies form the objectives for developing the technology to treat and dispose of discharge. These limits vary according to the beneficial uses of receiving waters and its reuse, the geographic area, and the regulatory agency.

In the past, disposal limits on agricultural drainage have not been established, since most regulatory authorities considered agricultural drainage as nonpoint sources. Until the 1987 amendments to the Federal Water Quality Act, the U.S. Environmental Protection Agency (EPA) did not regulate nonpoint agricultural discharges. Under the new amendments (Section 319 of the Federal Water Quality Act), states now need to establish plans for the control of nonpoint sources, including agricultural drainage.

The California State Water Resources Control Board (SWRCB 1987) in the aftermath of the ecological problems at Kesterson Reservoir directed the development of plans to regulate agricultural drainage in the San Joaquin River Basin. The experiences in this drainage control program can provide guidelines for others since basic issues are similar.

This program developed water quality objectives for the San Joaquin River (Table 21.1). It focuses on selenium, boron, molybdenum, and total dissolved solids, but other trace elements may be considered later. The receiving-water objectives for selenium and molybdenum are based on maintaining the aquatic environment, while boron and total dissolved solids are for agricultural uses. Reviewing the chemical characteristics of the local drainage water reveals the wide disparity between water-quality objectives for trace elements and drainage water quality (Table 21.2). The chemical characteristics of drainage water at other geographical areas will probably differ from the San Joaquin Valley, as will objectives for the quality of the receiving water, but many common problems in the management of drainage exist.

TABLE 21.2 Drainage Water Analysis/San Luis Drain, Mendota, California

Constituent (1)	Units (2)	Average (3)	Maximum (4)
Sodium	mg/l	2,230	2,820
Potassium	mg/l	6	12
Calcium	mg/l	554	714
Magnesium	mg/l	270	326
Alkalinity (as $CaCO_3$)	mg/l	196	213
Sulfate	mg/l	4,730	6500
Chloride	mg/l	1,480	2,000
Nitrate/nitrite (as N)	mg/l	48	60
Silica	mg/l	37	48
TDS	mg/l	9,820	11,600
Suspended solids	mg/l	11	20
Total organic carbon	mg/l	10.2	16
COD	mg/l	32	80
BOD	mg/l	3.2	5.8
Temperature[a]	°C	19	29
pH	—	8.2	8.7
Boron	μg/L	14,400	18,000
Selenium	μg/l	325	420
Strontium	μg/l	6,400	7,200
Iron	μg/l	110	210
Aluminum	μg/l	<1	<1
Arsenic	μg/l	1	1
Cadmium	μg/l	<1	20
Chromium (total)	μg/l	19	30
Copper	μg/l	4	5
Lead	μg/l	3	6
Manganese	μg/l	25	50
Mercury	μg/l	<0.1	<0.2
Molybdenum	μg/l	88	120
Nickel	μg/l	14	26
Silver	μg/l	<1	<1
Zinc	μg/L	33	240

Source: U.S. Bureau of Reclamation (1985).
[a]Temperature varied from 23–25°C (summer) to 12-15°C (winter).

AVAILABLE TREATMENT TECHNOLOGIES

The literature includes numerous wastewater treatment processes that can be applied to agricultural drainage water (Jenkins 1987, Lee et al. 1988). Many processes have the potential to treat drainage, but relatively few have been studied or developed. Desalinization technology is the most advanced and frequently applied for the production of drinking water. Many other processes that reduce trace elements have been screened out due to high costs or reported ineffectiveness. Conventional treatment technology is not easily applied to the complex chemical characteristics of agricultural drainage water.

Lee (1988) reviewed studies on treatment technologies for application to drainage water in California's San Joaquin Valley. The following discusses selected treatment methods.

Desalination

Considered one of the more promising technologies for treating drainage water, desalination has been developed over several decades. It is used mainly to convert seawater and brackish water for drinking and other purposes in many places throughout the world.

Of the many desalting processes, reverse osmosis (RO) is considered to have the most promise for agricultural drainage. This technology has been well advanced (Culp/Werner/Culp 1979; PRC Toups 1982) and applied worldwide. A semipermeable membrane separates water from dissolved salts and suspended materials. Pressure is applied to the saline feedwater and water is forced through the membrane, leaving behind the salts in a brine stream. CH2M Hill (1986) evaluated off-the-shelf RO technologies for drainage water in their study of a 38×10^3 m^3/day (10 MGD) reverse osmosis desalting plant for treating water from the San Luis Drain in the San Joaquin Valley. The U.S. Bureau of Reclamation desalting test facility at Yuma, Ariz. (USBR 1985), and the California Department of Water Resources at Los Banos, California (DWR 1986), reported other experiences with desalting drainage water.

Pretreatment is critical to RO operation. In its study, CH2M Hill selected a proven softening process in the water treatment industry that uses lime and soda ash to reduce carbonate and noncarbonate (calcium and magnesium) hardness. Due to the low carbonate hardness of the San Joaquin Valley drainage water, a large amount of soda ash would be needed. The RO process studied by CH2M Hill would have a TDS concentration of 550 mg/l to 650 mg/l in the product water, while the brine would have a TDS concentration of 55,000 mg/l to 67,000 mg/l. The concentration of selenium would be about 10 μg/l to 20 μg/l, while other trace metals would also be reduced to very low levels. However, seven to eight mg/l of boron would remain in the water (San Luis Drain water averaged about 14 mg/l).

Brine disposal is a major consideration in desalting systems. Evaporation ponds must dispose of the entire reject brine stream, if disposal to surface water is impossible. Acreages for ponds are based on the ability to hold mixtures of salt and water at depths of 6 ft in each cell for 20 years. After 20 years, ponds are almost filled with accumulated salts

TABLE 21.3 Reverse Osmosis Process for Desalting Drainage Water, San Joaquin Valley, California

Factor (1)	Value (2)
Raw Water Feed	10.7×10^6 (m^3/Yr)
Treatment Water Recovery[a]	9.4×10^6 (m^3/Yr)
Treated Water Quality[b]	
TDS (mg/l)	600
Selenium (μg/l)	10 to 20
Boron (μg/l)	7,600
Capital Investment[c]	$21,200,000
Capital Charge ($/$m^3$)[d]	0.17
Operating Cost ($/$m^3$)[e,f]	0.71
Desalting Cost ($/$m^3$ raw water)	0.88

[a]All costs shown are based on 87.5 % recovery in the reverse osmosis unit. Plant size 38×10^3 m^3/day (10 MGD), San Joaquin Valley, California.
[b]Influent drainage water quality: TDS = 9,820 mg/l, Se = 325 μg/l, B = 14,400 μg/l. Pretreatment: Lime-soda with acid scale control.
[c]Capital investment does not include reverse osmosis brine reject disposal.
[d]Capital charge is based on annual interest rate of 6 %, amortized over 20 years.
[e]Power costs are based on 0.075 $/kWh.
[f]Product water credit from resale is not included.

and must be abandoned or cleaned. The costs of ponds are a significant part of RO treatment cost. This chapter discusses details of evaporation ponds in a later section.

Solving the drainage problem by desalting provides the best option for the reuse of drainage water, but desalting is expensive. CH2M Hill estimated that treatment would cost $0.88/$m^3$ ($1,090/AF) of drainage water, with the lime and soda ash pretreatment process (Table 21.3).

Trace Element Treatment

The reverse osmosis process removes dissolved solids and trace elements, but is costly. Less costly methods are being developed. They fall into three categories: biological, chemical, and physical methods. The more promising of these processes are described here.

Biological Processes. Several biological processes are being developed, but only two have advanced to a stage where basic performance can be reported at this time (1989). The first process involves anaerobic bacteria, and the second involves algae and bacteria.

The anaerobic bacteria process has been under investigation since 1985 in California's San Joaquin Valley (EPOC Agriculture 1987; Binnie California 1988). The process consists of biological reactors for bacteria-

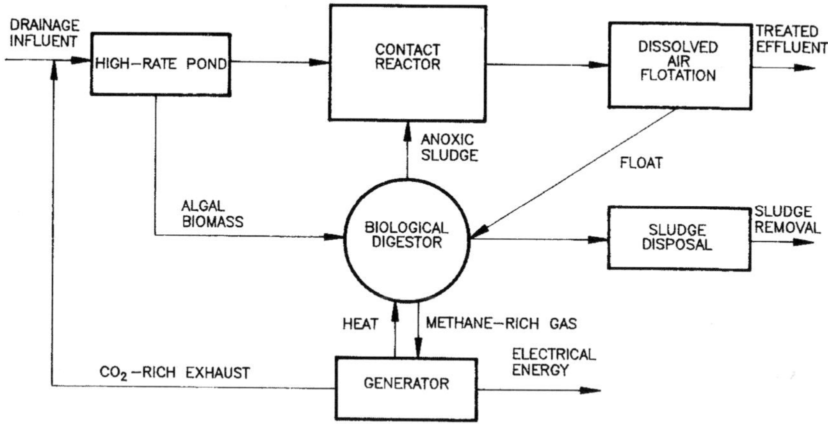

Fig. 21.1 Microalgal Bacterial Process.

mediated reduction of selenium, microfiltration for removing fine solids, and ion exchangers for polishing the effluent stream. A carbon source is injected into the drainage water and enters the biological reactors. It provides the energy for the bacteria-cellular buildup to drive the process. Methanol is usually fed to the reactors, but other carbon sources can be used. The reaction time in the biological reactors is less than an hour.

The studies evolved through three configurations of biological reactors, starting with upflow fixed films (gravel), followed by fluidized beds (sand), and, finally, sludge blanket reactors (SBR). SBR provided the best operational features and form the basis of a prototype plant design.

Biological reactors can reduce the selenium in drainage from about 330 μg/l to 550 μg/l to about 16 μg/l to 50 μg/l. After microfiltration, it is reduced to 10 μg/l to 40 μg/l (EPOC Agriculture 1987; Binnie California 1988). After polishing in the ion exchange resin or soil column, it is reduced to less than 10 μg/l. In the ion exchange process, boron was reduced from about 10 mg/l to less than 1 mg/l. Treatment costs are estimated to range from \$118 to \$132 per thousand m^3 for a 38×10^3 m^3/day (10 MGD) plant to \$182 per thousand m^3 for a 3.8×10^3 m^3/day (1 MGD) plant. These include capital, maintenance, and operation costs, but exclude drainage collection or disposal costs.

Oswald (1985, 1987, 1988a, 1988b) has been developing a microalgal bacteria process (Fig. 21.1) to reduce selenium in agricultural drainage in the San Joaquin Valley, Calif. Microalgae and bacteria are cultured in open ponds under conditions that induce a high rate of growth. Then they are harvested by an inexpensive bioflocculation process and fermented in digestors to generate methane gas. The digested algal material is recovered and used to reduce the oxidation state of selenium in drainage water. In a reactor, the digested algal material reduces selenium in drainage water to an insoluble state. The selenium is then separated from the water column. Methane gas powers the plant operation, and combustion gases are fed back as a supplemental carbon source to the high-rate algal pond. The effluent from the reactor is clarified in a dissolved air flotation chamber or a gravity separator and then discharged.

In laboratory experiments, the digested algal material reduced the concentration of soluble selenium in drainage water from 367 μg/l to 20 μg/l, which is a removal of about 95%. Part of the remaining selenium may be colloidal and can be removed by a solids separation system to achieve a higher degree of reduction.

Preliminary studies indicate that treatment costs by the microalgal bacterial process range from $55 per thousand m^3 in a 380×10^3 m^3/ day (100 MGD) plant to $83 per thousand m^3 in a 38×10^3 m^3/ day (10 MGD) plant. These include capital, maintenance, and operation costs, but exclude drainage collection and disposal costs.

Chemical Processes. Chemicals are commonly used to remove trace metals for industrial pollution control, but are not easily applied to drainage waters due to their complex chemical characteristics. Selenium in the selenate form can be reduced to elemental selenium and removed from the aqueous phase with ferrous hydroxide (Murphy et al. 1988). The process can be conducted in two major components, a chemical reactor and a solids gravity separator.

In preliminary laboratory tests, ferrous hydroxide reduced and precipitated selenium by 99% within 30 minutes of contact time (Moody 1988). However, field tests indicated a slower reaction. It could be reduced to 1 μg/l, but only with a longer contact time. The process depends on the pH, but dissolved oxygen and nitrates appear to inhibit the rate of reaction. The U.S. Bureau of Reclamation (1987 and 1988) issued progress reports on the development of the process. Preliminary costs for treatment are estimated to be $57 per thousand cubic meter for a 190×10^3 m^3/day (50 MGD) plant, but can rise to over $122 if drainage contains an excessively high amount of nitrates and oxygen. These estimates include capital, maintenance, and operational costs, but exclude collection and disposal costs.

Physical Processes. The treatment of wastewater involves adsorption of ions on natural and synthetic surfaces of active materials, including ion exchange resins. Many natural materials and synthetic substances can adsorb ions, but few can adsorb the trace elements needed to treat wastewater. Iron filings were used as an adsorption material for heavy metals in a minipilot plant at the Panoche Drainage District near Firebaugh, California (Harza 1986) using a process patented by Mayenkar (1986), who claimed that oxygen could activate the surface of the iron filings, which could then adsorb selenium.

The cementation of iron filings clogged the columns used in the studies. The problem, which was left unexplained, might be due to precipitation, coagulation, or occlusion of drainwater constituents in the iron filing beds. This can affect the life of the bed and, hence, operating costs. Preliminary studies indicated that costs would range from $57 to $231 per thousand cubic meter, depending on the life of the bed and the removal expectancies.

In-situ Biological Processes. Frankenberger (1988a) studied an in-situ biological process, in which microbial action volatilized selenium. The microbial methylation of selenium can significantly accelerate volatilization with metal and organic additives. Studies support using the process to detoxify seleniferous soils, sediments, and water.

Frankenberger showed in the laboratory that fungi could convert selenium salts into methylated species, primarily dimethylselenide. Methylated gas forms are naturally volatilized into the atmosphere and dispersed. Carbon sources, moisture, aeration, and activators stimulate the process. Zinc, nickel, and cobalt appear to be good activators, but high applications of nitrogen inhibit the reaction. Adequate aeration and moisture are needed.

Studies in a soil system with isolates of selenium-methylating fungi, including *acremonium folciforme, penicillium citrinum,* and *ulocladium tuberclatum,* showed enhanced gas evolution. Field studies began in 1987 on soils at Kesterson Reservoir, where contamination by selenium ranges from 1 mg/kg to over 200 mg/kg, with the higher levels at the surface. Citrus pulp, cattle manure, straw, and protein substances were used as carbon sources. Ammonium nitrate, ammonium sulfate, and zinc sulfate were added to test plots. The upper layer of the soil was kept moist.

A dome capture device sampled the volatilization of selenium. It recorded the highest rate of emission, more than 800 $\mu g/m^2$/hour, during the summer (Frankenberger 1988b). Citrus pulp enhanced volatilization more than other organic sources. Temperature and moisture levels also greatly influenced the process and emission rate.

Studies indicate no minimal threshold limit. Therefore, the process can potentially reduce selenium to zero level. Estimates indicate that reducing the selenium in the Kesterson Reservoir to an acceptable level by this process would take over two years.

Frankenburger (1988c) reported on the microbial methylation of selenium from water. Studies in the laboratory and the field indicate that selenium can be volatilized from open water. Thus, microbial methylation is one way to detoxify contaminated ponds and aquatic systems.

DISPOSAL, RECLAMATION, AND REUSE

The treatment of drainage water to remove salt and reduce toxic constituents leave residues of brines and salt that must be disposed of, reclaimed, or reused. Regulations may constrain all of these options. Nishimura et al. (1988) reviewed these options for the drainage problems in the San Joaquin Valley, Calif.

Disposal Options

Three ways to dispose of drainage water exist: discharge to surface waters, evaporation in ponds, and injection into deep wells.

Discharge to Surface Waters. This is the most widely used practice today. Drainage water is discharged to oceans, bays, deltas, rivers, and streams, mainly from nonpoint sources. Water quality standards may exist for all receiving waters to protect beneficial uses. These standards determine the engineering design, dilution factors, and installation of conveyance and disposal facilities. Special consideration must be given to protecting the aquatic environment and evaluating the ecological impact.

Evaporation Ponds. Drainage water is disposed of in evaporation ponds, when physical factors preclude discharge to surface waters or regulatory agencies foreclose other options. However, evaporation ponds pose a risk to ground-water supplies and the environment.

Regulatory requirements may affect the siting, construction, and operation of evaporation ponds. Since drainage water with high levels of trace elements can be considered hazardous or attain toxic levels in evaporation ponds, they must be designed and operated with care.

In California, as in many other locations, wastes in ponds are classified according to toxic constituents. Any waste exceeding predetermined toxic limits is classified as hazardous and requires special construction (Class I) for containment, including double linings, leachate collection and removal systems, and precipitation and drainage-control facilities (Table 21.4). Also, toxic ponds must be monitored in a way approved by regulatory agencies. The requirements for other liquid waste containment (Class II) are less rigid. The ecological impact of evaporation ponds may determine their design, construction, and operation. Selenium in impounded water can bioaccumulate in the aquatic food chain by more than 2000 times and threaten the wildlife that inhabit these ponds (USFWS 1987).

Drainage water may be treated to remove toxic constituents before ponding,but this is costly. Moreover, the increasing concentration of toxic substances during evaporation stages may nullify such treatment. Consideration for evaporation ponds in the San Joaquin Valley, California, involves the following actions:

1. Maintain water at a depth of at least 0.6 m.
2. Make the slopes of waterside levees as steep as possible.
3. Remove all major vegetation from the pond before filling it, and then keep the ponds free of vegetation.
4. Provide year-round access to all ponds by truck and airboat. An airboat launch ramp may be constructed from local soil materials if compacted, and preferably have a slope of about 6:1 in all ponds.

Special requirements incur extra costs. For example, in California, the cost estimate for toxic ponds (Class I) can exceed $500,000 per hectare, as compared to about $50,000 per hectare for an ordinary evaporating pond. Also, the monitoring of ponds can incur added annual costs.

In evaluating evaporation ponds for the management of drainage water, land requirements, the site, and the disposal of accumulated salts must be considered. The land area required for an evaporation pond is largely determined by the local climate and geology. A pond must be

TABLE 21.4 Construction Standards for Waste Management Ponds*

Type of waste management unit[a] (1)	Clay liner and permeability[b] (2)	Synthetic liner (3)	Leachate collection and removal system (4)	Permeability of Subsurface Barriers	
				Cutoff walls (5)	Grout curtaiins (6)
Class I					
Surface impoundment	Double liner[c] $\leq 1 \times 10^{-6}$ mm/sec	Required	Required, blanket type	$\leq 1 \times 10^{-6}$ mm/sec[e]	$\leq 1 \times 10^{-6}$ mm/sec
Class II					
Surface impoundment	Double or single liner required[d], $\leq 1 \times 10^{-5}$ mm/sec	Not required	Required with double liner, blanket type	$\leq 1 \times 10^{-5}$ mm/sec[f]	$\leq 1 \times 10^{-5}$ mm/sec

[a]Class I for hazardous waste, Class II for designated waste.
[b]All permeabilities specified in this table are maximum allowable permeabilities.
[c]Outer liner shall be a clay liner; inner liner may be a synthetic liner instead of a clay liner if inspected according to Subsection 2548(f).
[d]Single liner shall be a clay liner and removed or replaced as described in Section 2542. Double liner systems shall have either an outer clay liner or shall be underlain by a substantial thickness of natural geologic materials with a permeability of 1×10^{-5} mm/sec or less to act as an outer liner.
[e]Cutoff walls required where there is potential for lateral movement of fluid, including waste and leachate.
[f]Cutoff walls required where there is potential for lateral movement of fluid, including waste or leachate, and the permeability of natural geologic materials is used for waste containment.
*From the California Water Code, Article 4, Title 23.

designed to suit the site under consideration. U.S. Soil Conservation Service (SCS) provided design data for a sequential four-cell pond in the semi-arid San Joaquin Valley (Brown and Caldwell 1987.) They suggested that a pond surface of 0.86 ha would be necessary to contain each hectare-meter of drainage water (0.265 acre/acre-foot) produced per year. Allowing for levees and buffer zones, the SCS estimated each hectare-meter of drainage water to be disposed of annually requires about one hectare of land.

While evaporation ponds are viable disposal options, particularly for inland basins, using them to contain drainage water with toxic trace elements poses problems. They should be used only after carefully evaluating their environmental consequences.

Deep-Well Injection. The oil and gas industry has used this technology to enhance production and dispose of waste brine. It could be applied to the disposal of agricultural drainage (URS 1986).

Institutional constraints, such as the Underground Injection Control (UIC) Program of the U.S. Environmental Protection Agency, rigidly control deep-well injection systems. Injection of hazardous wastes recently received considerable attention from federal, state, and local regulators when some injection efforts failed. Also, the public may perceive that underground aquifers do not adequately control hazardous materials.

If waters to be injected are deemed hazardous under federal and state definitions, several provisions of the U.S. Resources Conservation and Recovery Act (RCRA) apply. The UIC program specifies the RCRA regulations that apply to deep-well injection of hazardous wastes.

Deep-well injections require numerous approvals. Potential liabilities should be considered. Leakage from pipes and confining zones bear liability risks. The EPA requires that certain financial responsibilities be maintained. Also, other complex, uncertain liability issues, such as adjoining property rights, exist.

Geohydrological conditions must be considered when selecting injection sites. Existing well drilling data can be a useful source of information. Fresh ground water must be protected, and geologic formations with low permeability must isolate the injection zones from fresh water formations.

The potential of seismicity induced by the injection process must be considered. In addition, the compatibility of the drainage water's chemistry with the geochemistry of the geologic formation should be evaluated. The biochemistry of the injecting fluids must not cause the formation to clog biologically and prematurely fail. Facilities are specially built to inject hazardous or nonhazardous wastes. A hole is drilled into a formation that can receive the wastes to a depth that meets UIC requirements and below the lowest formation that contains (within 400 m or 0.25 mile) a well used for drinking water. Porous, saline, water-bearing sands confined above and below by impermeable rock strata are the most desirable receiving formations.

Fig. 21.2 depicts a prototype injection well. An outside casing is placed throughout the length of the well hole, where it passes through zones of fresh water. A second steel casing is installed within the outer casing down through the confining layer. The space between the casings is filled with a cement grout. These barriers prevent waste from migrating into any zones of fresh water. A third pipe, the injection tubing, is installed within the second casing and sealed to the second casing at the top and bottom.

The annular cavity between the second and third casing should be filled with a noncorrosive fluid that is kept under constant pressure, which is continually monitored for leaks when wastewater is pumped under pressure through the injection tubing.

The size of the well depends largely on cost, which, in turn, is a

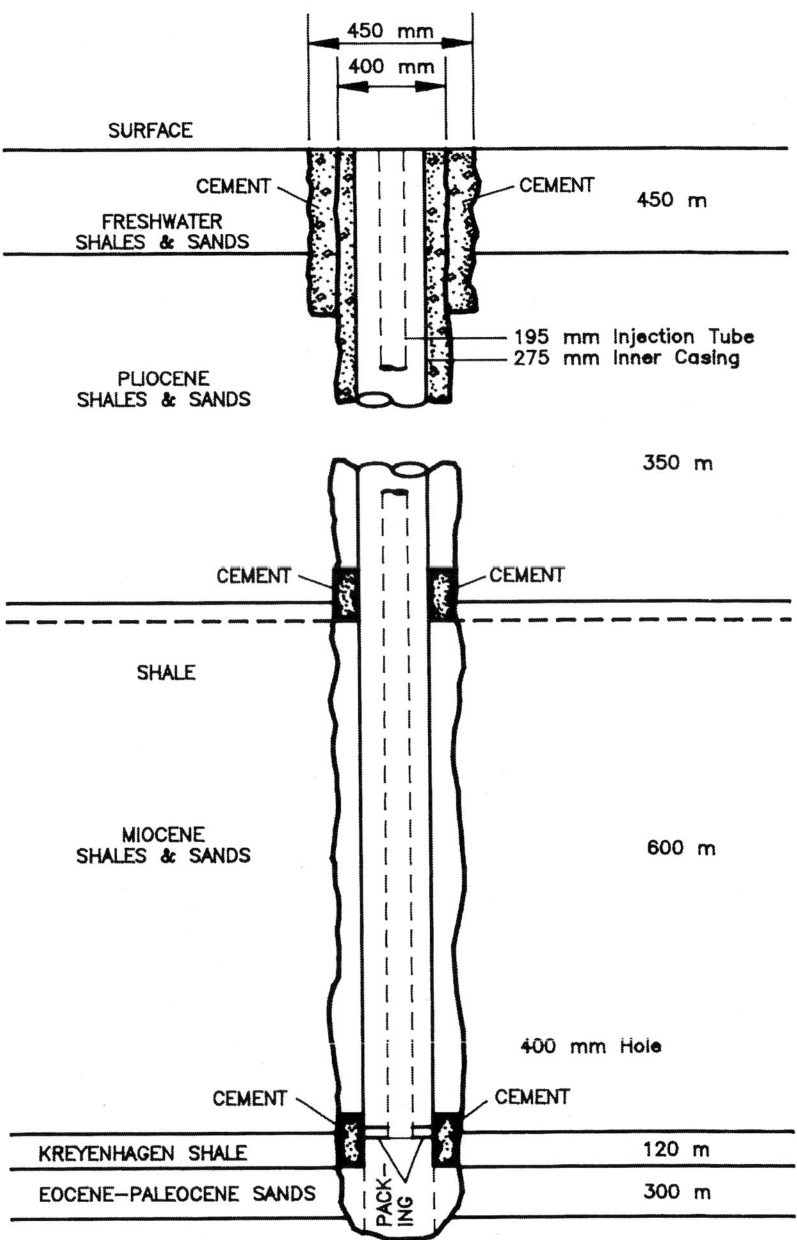

Fig. 21.2 Diagrammatic Section of a Prototype Disposal Well, San Joaquin Valley, Calif.

TABLE 21.5 Disposal Costs of Deep-Well Injection with Different Pretreatments/San Joaquin Valley, Calif.

Pretreatment option[a] (1)	Annual Cost per 1000 m^3, in Dollars		
	Capital (2)	O&M (3)	Total (4)
No solids separation	63	70	133
Sedimentation	92	104	195
Rapid Filtration	78	75	154
Microscreens	85	88	173

[a]For 38 × 10^3 m^3/day (10 MGD), 6 % interest; \$0.075/kWh power cost; and 25-year service life; at San Luis Drain, San Joaquin Valley, Calif. (1986).
Source: URS (1986).

function of capital construction and operating costs. The smaller the diameter of the well, the lower the capital costs and the higher the operating costs will be. The higher operating cost of wells with a smaller diameter is due mainly to higher head losses in pressurizing the wells and the accompanying higher annual energy cost.

The build-up of aquifer pressure largely governs the service life of the injection well. Such buildup in the receiving formation is a function of the volume injected and the rate of injection. Injection wells in the oil industry have had a service life of about 35 years. A conservative estimate of service life would be about 25 years. Drainage water may need to be treated before injection to minimize clogging or fouling of the well. The well or receiving formation could clog from one or more of the following:

1. Accumulated particulate matter or chemical precipitation, e.g., calcium sulfate (gypsum), may clog the injection face or formation.
2. Slime or nitrogen gas bubbles from the biological denitrification of nutrient-rich drainage water may clog the well or formation.

Ways to prevent physical clogging by particulate matter include injection after chlorination with no solids separation, pretreatment by flocculation and sedimentation in a clarifier system, pretreatment by rapid filtration through a single media bed, and microscreens.

Table 21.5 shows the unit costs for deep-well injection for various pretreatments of drainage effluents in the San Joaquin Valley. Biological activity and resultant fouling of the injection horizon appear to be serious concerns. Therefore, all injectate should be chlorinated to reduce any biological activity, which may still prevail due to the high nutrient content of agricultural drainage water. Biological fouling can take place because microbial activity causes slime to accumulate across the pipe, across the injection face, and in the receiving formation, or because it causes nitrogen bubbles to form.

Reclamation and Reuse Options

Many ways to reclaim and reuse agricultural drainage water, a good management practice, exist. A few of the more promising ones are described below.

Salt-Tolerant Crops. Chapter 20 covers in detail the use of saline drainage water for irrigating salt -tolerant crops. A special area of agroforestry exists, in which drainage water is used for irrigation. Salt-tolerant trees have the potential to lower water tables and reduce the volume of drainage through evapotranspiration and consumptive use.

An agroforestry program for the management of saline drainage water was demonstrated in the San Joaquin Valley (CDFA 1987 and 1988). In 1985, eucalyptus, casuarina, poplar, mesquite and Elderica pine trees were planted in about 75 ha. The seedlings were irrigated by fresh water during the first year to establish the trees. They were then irrigated by saline drainage water from the ground water system. Findings to date indicate that drainage water can be used to irrigate trees. Once established, trees can draw water from the ground water system for growth and evapotranspiration. An added incentive is the uptake and disposal of trace elements. Harvesting trees provides biomass to fuel power plants that generate electricity and produce pulp.

Power Plant Cooling. Drainage water can be used to cool power plants. Due to its chemical characteristics, it needs to be treated beforehand for corrosion and scale control. A test program in California used water from a tiled drainage system for power plant cooling (California Department of Water Resources and the University of California 1978). The study was based on a plant designed to treat 43.5×10^3 cubic meter/day (11.5 million gallons per day) for a 1,000-MW powerplant. Treatment processes softened water by ion exchange resins and regenerated the resin, using concentrated cooling tower blowdown without adding new chemicals. The estimated cost of treating drainage water was about $132 per thousand cubic meters ($163 per acre-ft), adjusted to 1984 cost, and not including the disposal of any wastewater. Additional costs for brine disposal in evaporation ponds added $88 per thousand cubic meter ($108/acre-ft). Therefore, the sum of treatment and evaporation pond disposal costs totaled about $220 per thousand cubic meter ($271 per acre-ft). If selenium or other contaminants in the waste brine were to reach hazardous levels, the disposal costs could be substantially higher because special holding ponds might be required.

The use of drainage water from the Palo Verde Irrigation Outfall Drain in the Colorado River basin for a hypothetical power generating station near Las Vegas, Nev., was studied. Laughlin (1986) prepared a report as part of the Colorado River Water Quality Improvement Program of the U.S. Bureau of Reclamation. He studied several ways to treat drainage water and concluded that drainage was a viable source of cooling water, which would help lower salt loads in the basin.

Reclamation for Reuse. Treated water can be used for municipal, industrial and agricultural purposes if it can meet reuse standards. Limi

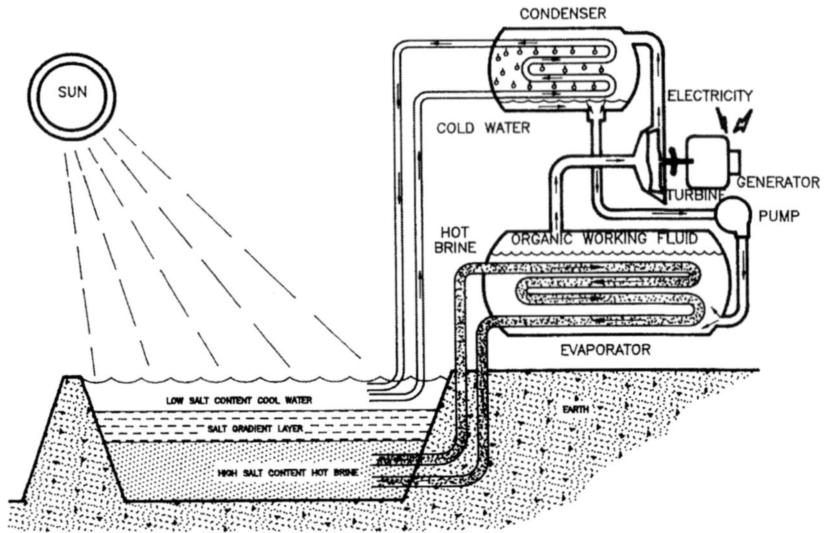

Fig. 21.3 Schematic of a Solar Pond Generator

tations on reuse depend on the desalination process and the drainage water's chemical composition. Not only must the total dissolved solids be reduced, but substances, such as boron, and toxic trace elements must meet standards for beneficial uses. For example, in reuse for agricultural irrigation, levels of TDS and boron are particularly important. For natural habitat and human consumption, toxic trace elements are important to evaluate. Desalination is expensive for agricultural reuse, as indicated in the analysis for reverse osmosis. Desalination for municipal or industrial purposes may be affordable, if other supplies are too expensive.

To date, the desalination of drainage water for reuse has not been widely practiced. Although studies have been made on the potentials for reuse, the large volume of drainage and related high costs have not favored the installation of large facilities.

Solar Gradient Ponds. The use of evaporation ponds and desalination technology results in end products that must somehow be disposed. Solar-gradient-pond technology offers the possibility of using the waste products of brine or salt to produce electrical energy.

The literature describes several types of solar ponds for converting solar energy to electrical energy, but the discussion is limited to the nonconvective, salt-gradient solar ponds (U.S. Bureau of Reclamation 1982). These are about 2 m to 5 m deep, with three distinct zones (Fig. 21.3). The top layer, or the upper convecting zone, is about 0.2 m to 0.5 m deep and consists of uniform, relatively low-salinity water of up to about 60,000 mg/l (MIT/EPRI 1985). The intermediate zone is the nonconvective gradient zone, in which both temperature and salinity increase with depth. This zone, about 0.75 m to 1.5 m thick, helps to insulate the underlying zone of heat storage. The third and lowest zone, usually about 1.0 m to 3 m thick, consists of very saline brine of about 150,000 mg/l to 250,000 mg/l.

Ordinarily, heated bottom water in water bodies expands to become lighter and rises to the surface to lose its absorbed heat by convection and radiation. However, in salinity gradient ponds, the salt content of the lower zones causes them to remain denser than the zones above, even when heated. Short-wave solar radiation penetrates the upper zones into the heat storage zone and raises its temperature. The stored heat can be used as a low-temperature energy source.

The study of solar ponds in Israel indicates that about 20% to 25% of the incident irradiation can be collected and extracted at about 82° C to 93° C (Ormat Turbines, Ltd. 1981). Using specially designed turbines and generators, Ormat showed that the low-temperature energy can be effectively converted into electrical energy. The U.S. Bureau of Reclamation demonstrated the Ormat technology at El Paso, Texas (Hightower 1987).

The energy budget for solar ponds depends on the following factors: net incident solar radiation, penetration of solar radiation to the zone of heat storage (as affected by solar angle and water clarity), diffusion of heat from the zone of heat storage to the intermediate and nonconvecting gradient zone, loss of ground heat from the zone of heat storage, and thermal energy extraction (MTI/EPRI 1985).

Energy extraction is accomplished by one of two methods. One is to place the heat exchangers in the zone of heat storage. The other way is to pass it through an external heat exchange and return the cooled brine to the pond.

Solar ponds may need to meet stringent construction standards due to the concentration of trace element contaminants to hazardous levels. If toxic containments were required, the cost would increase substantially. The ponds would cost at least $50 per square meter. Solar ponds are not now a part of the solution to drainage water treatment, but technology may advance to make them feasible. Such advancement will be governed largely by the supply and demand for energy.

Salt Recovery. The recovery of salts or minerals from agricultural drainage water, after concentration in evaporation ponds, may present the ultimate solution to the problem of salt disposal. The products could be sold to cover some, if not all, of the disposal costs. The U.S. Bureau of Reclamation (1979) studied the possibility for salt recovery. However, respondents to a survey of chemical and mineral extraction companies expressed no interest in extracting minerals from agricultural drainage water. The trends in worldwide markets and proximity to markets are important factors in the recovery of salt. Unless these favor a specific location, the commercial marketing of recovered salt would be limited.

Other Reuse Options. The reuse of reclaimed drainage water has been suggested for many other purposes, including aquaculture and the creation of wetlands. These methods remain in the conceptual phase. Trace elements in drainage water pose a major problem. However, drainage water can be treated to reduce accumulated toxic substances. A novel way to integrate several reclamation and reuse options is cogeneration-desalination. Explored for the Westlands Water District, California (URS Corporation 1987, Westlands Water District 1988), it involves using gas turbines to generate electrical power with the bypass heat used to desalinate drainage water. Salt from the brine stream crystallizes, and the desalinated water is made available for reuse. Surplus electrical power from plant operation is sold. Although the technology of each process is well developed, the processes have yet to be integrated into an economical system. Economic considerations include whether the sale of surplus electrical energy and salt products can cover the costs of processing.

REFERENCES

Binnie California and California Dept. of Water Resources (1988). "Performance Evaluation of Research Pilot Plant for Selenium Removal," Fresno, Calif.

Brown and Caldwell Consulting Engineers (1987). "San Joaquin Valley Drainage Program, Screening Alternative Geographic Disposal Areas," prepared for the San Joaquin Valley Drainage Program, Sacramento, Calif.

California Department of Food and Agriculture (1987). "The Agroforestry Demonstration Program in the San Joaquin Valley—Progress Report," Sacramento, Calif.

Calif. Dept. of Food and Agriculture (1988). "The Agroforestry Demonstration Program in the San Joaquin Valley—Progress Report," Sacramento, Calif.

California Department of Water Resources and University of California (1978). "Agricultural Wastewater for Powerplant Cooling Development and Testing of Treatment Processes," Volume II, Sacramento, Calif.

California Department of Water Resources (1986). "Los Banos Demonstration Desalting Facility, Status of Operations," Sacramento, Calif.

California State Water Resources Control Board (1987). "Regulation of Agricultural Drainage to the San Joaquin River," Report of Technical Committee, Sacramento, Calif.

CH2M Hill Inc. (1986). "Reverse Osmosis Desalting of the San Luis Drain Conceptual Level Study." U.S. Department of the Interior, Bureau of Reclamation.

Culp/Wesner/Culp (1979). "Water Reuse and Recycling Evaluation of Technology." U.S. Department of the Interior, Office of Water Research and Technology.

EPOC Agriculture (1987). "Removal of Selenium from Subsurface Agricultural Drainage by an Anaerobic Bacterial Process," Fresno, Calif.

Food and Agriculture Organization, United Nations (1973). "Irrigation Drainage and Salinity," Rome, Italy.

Frankenberger, W. (1988a). "Microbial Volatilization of Selenium at Kesterson Reservoir," Interim Report for the U.S. Bureau of Reclamation, Sacramento, Calif.

Frankenberger, W. (1988b). "Dissipation of Soil Selenium by Microbial

Volatilization at Kesterson Reservoir," Final Report for U.S. Bureau of Reclamation, Sacramento, Calif.

Frankenberger, W. (1988c). "In-Situ Volatilization of Selenium, Evaporation Ponds," Report for San Joaquin Valley Drainage Program, Sacramento, Calif.

Harza Engineering Co. (1986). "Selenium Removal Study," prepared for Panoche Drainage District, Firebaugh, Calif.

Hightower, S., and Bronicki, L. (1987). "Installation and Operation of the First 100 kW Solar Pond Power Plant in the United States," U.S. Bureau of Reclamation, Denver, Colo.

Jenkins, D. (1987). "Treatment Technology Relative to San Joaquin River River Water Quality Objectives," California State Water Resources Control Board, Sacramento, Calif.

Laughlin, J.K. (1986). "Study of Saline Water Use at the Etiwanda Generating Station." Prepared for U.S. Bureau of Reclamation, Denver, Colo.

Lee, E.W., Nishimura, G.H., and Hansen, H.L. (1988). "Agricultural Drainage Water Treatment, Reuse and Disposal in the San Joaquin Valley of California—Part I: Treatment." San Joaquin Valley Drainage Program, Sacramento, Calif.

Massachusetts Institute of Technology (MIT) (1985). "A State-of-the- Art Study of Nonconvective Solar Ponds for Power Generation." Report prepared for the Electrical Power Research Institute (EPRI), Palo Alto, Calif.

Mayenkar, A. (1986). "Removal of Dissolved Heavy Metals from Aqueous Waste Effluents," U.S. Patent 4,565,633, Chicago, Ill.

Moody, C.D., et al. (1988). "Experimental Results for Removing Selenate from Agricultural Drainage Waters." U.S. Bureau of Reclamation, Denver, Colo.

Murphy, A.P. (1988). "Removal of Selenate from Water by Chemical Reduction," *Ind. Eng. Chem. Res.* 27: 181-191.

Nishimura, G.H., Lee, E.W., and Hansen, H.L. (1988). "Agricultural Drainage Water Treatment, Reuse and Disposal in the San Joaquin Valley of California—Part II: Reuse and Disposal." San Joaquin Valley Drainage Program, Sacramento, Calif.

Ohlendorf, H.M. (1984). "The Biological System." Conference on Toxicity Problems at Kesterson Reservoir, California, Proceeding of a Research Meeting, Sacramento, December 5-7, 1983.

Ormat Turbines, Ltd. (1981). "A Study of the Feasibility of a Solar Pond Generating Facility in the State of California, USA", Final Report, Volume I of a report prepared for the Southern California Edison Company, Calif.

Oswald, W.J.. (1985). "Potential for Treatment of Agricultural Drain Water with Microalgal-Bacterial Systems," University of Calif., Berkeley.

Oswald, W.J. (1987). "Microalgal-Bacterial Treatment for Selenium Removal from San Joaquin Valley Drainage Waters." University of Calif, Berkeley.

Oswald, W.J. (1988a). Status Report, Microalgal-Bacterial Treatment for Selenium Removal from San Joaquin Valley Drainage Waters, University of Calif, Berkeley.

Oswald, W.J. (1988b). Annual Report, "Microalgal-Bacterial Treatment for Selenium Removal from San Joaquin Valley Drainage Waters." University of Calif, Berkeley.

PRC Toups (1982). "Evaluation of Desalination Technology for Wastewater Reuse," prepared for U.S. Department of the Interior, Office of Water Research and Technology, Washington, DC.

U.S. Bureau of Reclamation (1979). "Agricultural Drainage and Salt Management in the San Joaquin Valley," Final Report Including Recommended Plan and First-Stage Environmental Impact Report, Interagency Drainage Program (IDP), Sacramento, Calif.

U.S. Bureau of Reclamation (1982). "Preliminary Study of Solar Ponds for Salinity Control in the Colorado River Basin," Appendix A, Solar Pond Brine Properties and Pretreatment Options, Denver, Colo.

U.S. Bureau of Reclamation (1985). Colorado River Basin Salinity Control Project, Background, Plan, and Status Report, Denver, Colo.

U.S. Bureau of Reclamation (1987). "Experimental Results for Removing Selenate from Agricultural Drainage Waters," Denver, Colo.

U.S. Bureau of Reclamation (1988). "Selenium Removal from Agricultural Drainage Waters by Chemical Reduction: Preliminary Process Design," Denver, Colo.

U.S. Fish and Wildlife Service (1987). Aquatic Cycling of Selenium: Implication for Fish and Wildlife, Washington, D.C.

URS Corporation (October 1986). "Deep-Well Injection of Agricultural Drain Waters." Prepared for the San Joaquin Valley Drainage Program, Sacramento, Calif.

URS Corp. (1987). "Agricultural Drainage Salt Disposal," prepared for Westlands Water District, Sacramento, Calif.

Westlands Water District (May 1988). "Selenium Removal/Cogeneration Strategy." Drainage Update, Fresno, Calif.

VIII. THE USE OF MODELS IN SALINITY ASSESSMENT

CHAPTER 22

STEADY- STATE ROOT ZONE SALT BALANCE[a]

INTRODUCTION

Chemical reactions in the root zone of an irrigated soil affect soil salinity and sodicity and the salt load of the drainage water. Significant reactions include the dissolution and precipitation of sulfate, carbonate, and silicate minerals; adsorption and desorption of carbon dioxide gas and boron; cation exchange between the solution and adsorbed (solid) phases; and the formation of sulfate and carbonate ion pairs in the solution phase.

Dutt and Tanji (1962) and Dutt (1962) were the first to demonstrate the use of computer programs to predict the inorganic compositions of soil solutions. Oster and McNeal (1971) evaluated their usefulness in calculating changes in the soil-solution composition with changing soil-water content. Others have used them to evaluate salinity, sodicity, and environmental hazards of drainage waters that result from irrigation with different waters (Oster and Rhoades 1975; Rhoades and Suarez 1977) and the predictable minimum solution concentrations that result from constraints imposed by the presence of soluble minerals (Oster 1982, Oster and Tanji 1985). Rhoades and Merrill (1976) and Rhoades (1977) incorporated the soil chemistry computer programs into another computer program, called Watsuit, to assess the suitability of irrigation water. Also, computer programs have been used to calculate the influence of chemical reactions on the composition of the soil solution for transient conditions within the root zone (Jury et al. 1978; Robbins et al. 1980) and for sodic-soil reclamation (Dutt et al. 1972; Tanji et al. 1972; Oster and Frenkel 1980).

In this chapter, the interactions among the irrigation-water composition, chemical reactions, and leaching fraction will be reviewed. The use of Watsuit to evaluate how these interactions affect the suitability of irrigation water and the selection of management options and amendments will be described.

CALCULATING SOIL SOLUTION COMPOSITION

The soil solution begins as irrigation water or rainwater at the surface of the soil. As it moves downward, plant water uptake concen-

[a]Prepared by: J.D. Oster, Univ. of Calif. Cooperative Extension, Soil and Water, Univ. of Calif., Riverside, CA 92521; and J.D. Rhoades, U.S. Salinity Lab., 4500 Glenwood Dr., Riverside, CA 92501.

trates it, and adsorption, dissolution, and precipitation modify it. Varying the leaching fraction modifies these effects. With its reduction, total dissolved salts decrease as chemical reactions shift from dissolution of soil lime ($CaCO_3$) and gypsum to precipitation of the same minerals.

The assumption of steady-state conditions makes it unnecessary to consider exchange and adsorption reactions when calculating the soil solution composition of mixed salt solutions containing Ca, Mg, Na, K, $CO_3 + HCO_3$, Cl, and SO_4. Further simplification occurs with the use of annual or longer averages of the following variables: the composition of irrigation water corrected for rainfall, the leaching fraction, the depth distribution of water uptake by plants, and the partial presssure of CO_2, pCO_2. In a soil chemistry program, RB6, developed by Oster and Rhoades (1975), saturation with soil lime may be assumed, thereby accounting for the potential dissolution of soil lime, or soil silicates, or both (Rhoades et al. 1968; Suarez and Rhoades 1982). The program adjusts the concentration of Ca, CO_3, and HCO_3 for cases where this assumption is made so that the activity product, $a_{Ca}a_{CO_3}$, equals the solubility product of soil lime. Other assumptions include: soil lime equilibria governs the solution's pH; ion activities are a function of ionic strength, expressed by the extended form of the Debye-Huckel equation if ion pair formation is taken into account; and the solubility products of gypsum and soil lime are 2.45×10^{-5} and 6.47×10^{-9}. The equilibrium solution composition is calculated by successive iteration, taking into account ionic strength and ion-pair formation constants for sulfate, carbonate, and bicarbonate ion pairs with Na, Ca, and Mg. The final composition meets simultaneously the following conditions: ion-pair equilibrium constants; charge balance, including hydrogen and hydroxyl concentrations; equilibrium among carbonate, bicarbonate, hydrogen, and hydroxyl ion species with the assigned pCO_2; the solubility products of soil lime if this option is selected; and the solubility products of soil lime and gypsum if the appropriate activity products are exceeded. In RB6, if the input charge concentrations of the cations and anions are not equal, the charge balance constraint results in adjustments in the concentrations of carbonate, bicarbonate, hydrogen, and hydroxyl ion species. Once the equilibrium composition has been calculated, the electrical conductivity, EC, of the resulting solution composition is calculated by method 3 of McNeal et al. (1970), and the sodium adsorption ratio, SAR, is calculated from the total concentrations of Na, Ca, and Mg.

CALCULATING SOIL SOLUTION COMPOSITION WITHIN ROOT ZONE

Rhoades and Merrill (1976) and Rhoades (1987) combined RB6 with an assumed distribution of plant water uptake to develop Watsuit, a computer program for assessment of irrigation water. The basic approach is to predict the salinity, sodicity, and toxicity within the root zone that result from irrigation with a water of known composition, with and without amendments.

TABLE 22.1 Terminal Display during Watsuit Start-Up.

```
Wish to send output to (D)isk or (S)creen
To print results in screen mode, hit: control P
SATURATE WITH CACO3? Y
CASE ID
____________
PORT
ENTER DELIMITED BY COMMAS: CA, MG, NA, K, CL, ALK,
  and SO4
    .1, .05, 4.9, .05, 1.7, 2.14, 1.25
WHICH AMENDMENTS?
    (B) H2SO4? Y
    (C) 1 CASO4? Y
    (D) 20 CASO4? Y
WHICH LEACHING FRACTIONS TO ACCEPT
    .05? Y
    .10? Y
    .20? Y
    .30? Y
    .40? N
```

The composition of the soil water (Ca, Mg, Na, CO_3, HCO_3, Cl, SO_4), electrical conductivity, sodium adsorption ratio, and osmotic potential are calculated for five soil depths—the surface of the soil and 1/4, 1/2, 3/4, and full depths of the root zone. The relative water uptake for the four successive intervals from the surface downward is assumed to be 40%, 30%, 20%, and 10% of the total, respectively. The corresponding depth distribution of pCO_2 is 0.07 kPa at the soil's surface, and .5 kPa, 1.5 kPa, 2.3 kPa, and 3 kPa for each successive root zone interval downward. Leaching fraction choices include 0.05, 0.1, 0.2, 0.3, and 0.4. Amendment choices include gypsum and sulfuric acid. The model runs on IBM-compatible personal computers. With 16-bit technology, the calculation time for one leaching fraction is about five minutes. With 32-bit technology, it is about 30 seconds.

Table 22.1 shows the computer display during data entry. The following choices and entries need to be made: Are the results to be printed, stored on disk, or displayed on screen? Is the assumption of soil lime saturation to be made? How is the case to be identified? What is the ionic composition of the water in units of $mmol_c/l$ (=meq/l)? Which amendments and leaching fractions should be included? Amendment choices include the following: (1) Sulfuric acid to replace 90% of the alkalinity with sulfate; (2) gypsum to add 1 $mmol_c/l$ of $CaSO_4$ simulating water or top dress soil treatment; and (3) gypsum to add 20 $mmol_c/l$ to simulate incorporated soil treatment. Both gypsum amendments can be chosen for the same computer run. No amendment is the default condition: It is always run.

The sum of cation and anion charge concentrations entered in response to the fourth request should be about equal. If they are not, Watsuit adjusts the concentrations to make them equal as follows: if the sum of the cation concentrations exceeds that for the anions, i.e., the difference in charge concentrations are positive, the difference is added to the anion that has not been initialized in the order Cl, SO_4, ALK (HCO_3 + CO_3). If all were initialized, the difference is added to Cl. If the difference is negative, the program will subtract the difference from the anion in the order Cl, SO_4, ALK. If the difference is greater than any single anion concentration, the program stops and a message questioning the analysis is displayed. This modification of RB6, where the sum of H, OH, HCO_3, and CO_3 was adjusted to achieve charge balance under appropriate equilibria constraints, facilitates use of chemical analyses where one ion, typically SO_4, has not been determined.

For each root-zone interval, calculation begins by multiplying the ion concentations by the value of 1/L appropriate to the leaching fraction, L, at the bottom boundary of the interval. Average EC and SAR are calculated for the top half and the entire root zone. The EC, weighted by the water uptake, is also calculated. Rhoades and Merrill (1976) discuss the underlying rationale for the use of these different indices and how they relate to the suitability of different methods of irrigation.

MIXED SALT SOLUTION CHARACTERISTICS

Electrical Conductivity, EC

For steady-state conditions, the ratio of the electrical conductivity of the drain water to that of the applied water (EC_d/EC_i) is proportional to 1/L. For predominately chloride salt solutions, the proportionality constant is about 1.0. The relationship for lime-saturated NaCl and $CaCl_2$ is

$$\frac{EC_d}{EC_i} = \frac{0.98}{L} \tag{22.1}$$

The ratio is less than 1.0 because the ability of charged ions to conduct current decreases with increasing concentration. Ion pairing further reduces the proportionality constant for sulfate and bicarbonate salt solutions; the number of ions conducting current are reduced, and neutrally charged ion pairs conduct little current (Tanji 1969). The proportionality constants in Eq. 22.1 for lime-saturated solutions of calcium sulfate and sodium sulfate are 0.61 and 0.77.

For irrigation waters, the proportionality constant is generally less than 1.0 (Fig. 22.1). In addition to the concentration and ion pair effects on ionic conductivity, $CaCO_3$ and $CaSO_4 \cdot 2H_2O$ (gypsum) often precipitate as L decreases. When 1/L is less than six for Colorado River water and less than ten for Gila River water (Table 22.2), the slopes are less than 1.0 due to ion pairing and lime precipitation. For larger values of 1/L, gypsum precipitation further reduces the slope for both waters.

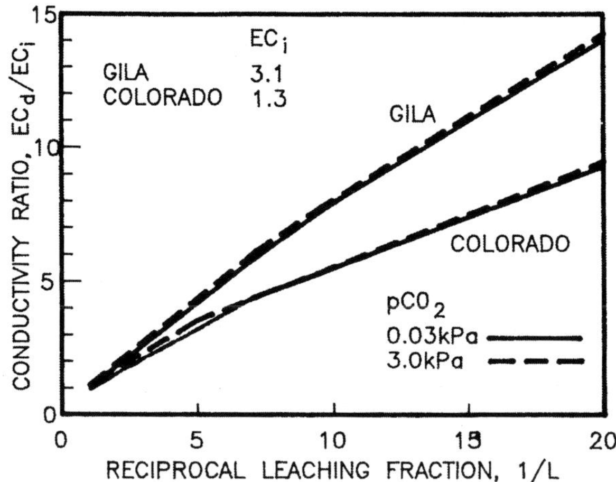

Fig. 22.1 Conductivity Ratio, EC_d/EC_i, as a Function of the Reciprocal Leaching Fraction ($1/L$), and pCO_2 of Gila and Colorado River Water. (Calculations by RB6. The subscripts *d* and *i* represent the drain and irrigation water.)

TABLE 22.2 Chemical Composition of Irrigation Waters for Five Rivers

River (1)	Na[a] (2)	K[a] (3)	Ca[a] (4)	Mg[a] (5)	$HCO3_3$[a] (6)	Cl[a] (7)	SO_4[a] (8)	EC[b] (9)	SAR[c] (10)
Feather	0.20	0.04	0.45	0.36	0.86	0.08	0.16	0.1	0.3
Porterville	4.90	0.05	0.10	0.05	2.10	1.70	1.25	0.6	12.8
Colorado	3.40	0.22	6.95	3.63	3.73	1.03	9.31	1.3	1.5
Gila	18.55	0.09	7.22	5.88	3.17	20.17	8.48	3.1	7.3
Pecos	11.38	0.08	16.98	9.07	3.11	12.13	22.39	3.3	3.2

[a]in $mmol_c/l$
[b]in dS/m
[c]in $(mmol_c/l)^{0.5}$

Partial Pressure of CO_2, pCO_2

Since pCO_2 affects the solubility of soil lime, it also affects the relationships between EC_d/EC_i and 1/L. For Gila River and Colorado River water, the effects are small and only significant for the Colorado River when $1/L < 6$ (Fig. 22.1). For the dilute Feather River water (EC_i = 0.1), the effects of pCO_2 are far greater (Fig. 22.2). At a 1/L of 10, EC_d/EC_i ranges from 6.3 to 12.1 for pCO_2's of .03 and 10 kPa. The pH of the soil is also affected: at a 1/L of 10, the predicted pH's are 8.8 and 6.9 for pCO_2's of 0.03 and 10 kPa.

Sodium Adsorption Ratio, SAR

Dissolution and precipitation reactions also affect SAR. Without reactions, the SAR_d/SAR_i ratio would be proportional to $(1/L)^{0.5.}$ Chem-

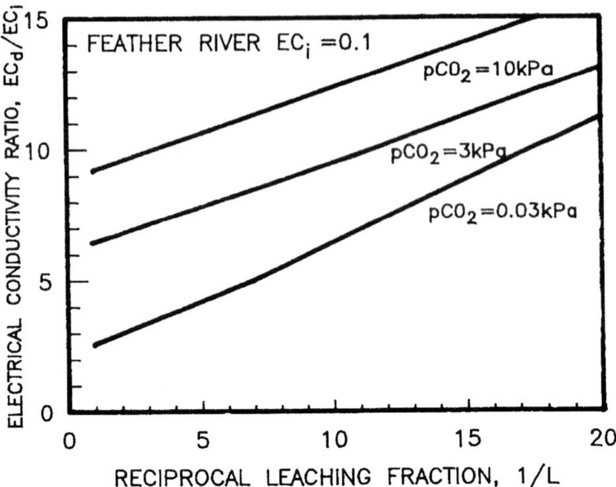

Fig. 22.2 Conductivity Ratio, EC_d/EC_i, as a Function of the Reciprocal Leaching Fraction, (1/*L*), and pCO_2 of Feather River Water. (Calculations by RB6. The subscripts *d* and *i* represent the drain and irrigation water.)

ical reactions cause slopes to increase with increasing $(1/L)^{0.5}$ (Fig. 22.3). Slopes less than 1.0 result from lime dissolution. Slopes greater than 1.0 result from lime and gypsum precipitation.

Salt Balance

Salt concentrations in the root zone must be controlled to sustain irrigation over the long-term. Soluble salts added with the irrigation water must be removed by leaching to prevent toxic conditions. Because of precipitation and dissolution reactions, the amount of salt leached may be greater than, equal to, or less than that applied by irrigation. Under steady state conditions, the salt leached decreases with decreasing *L*. Lysimeter studies (Rhoades et al. 1973, 1974) demonstrated that minimizing *L* reduces salt application to the soil and salt discharge in the drain water from the root zone of irrigated crops. Eight river waters, differing in total salinity and ionic composition, were used and managed to achieve *L* values of 0.1, 0.2, and 0.3. Salt budgets were determined for soil-filled lysimeters during three years of alfalfa production. The salt loads in the drain waters were reduced by 2 Mg ha^{-1} y^{-1} to 12 Mg ha^{-1} y-1 by reducing *L* from 0.3 to 0.1. Based on the salt balance for individual ionic species, most of the reduction was due to a shift from dissolution of soil lime, at an *L* of 0.3, to precipitation of applied Ca, HCO_3, and SO_4 salts as soil lime and gypsum at an *L* of 0.1. The data agreed with those calculated using RD6, thereby demonstrating its usefulness in calculating compositions of drain water for different irrigation waters and leaching fractions.

The data in Table 22.3 illustrate the use of Watsuit to predict how leaching fraction affects the loss or gain of Ca in the root zone of a crop irrigated with Pecos River water. This water is gypsiferous (Table 22.2).

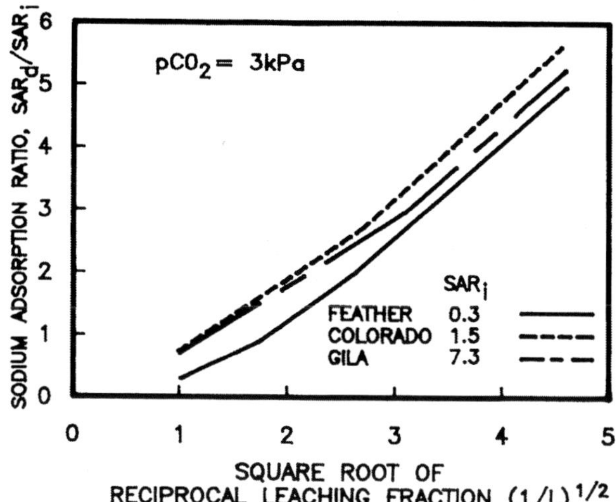

Fig. 22.3 Sodium Adsorption Ratio, SAR_d/SAR_i, as a Function of the Square Root of the Reciprocal Leaching Fraction $(1/L)^{0.5}$. (Calculations by RB6. The subscripts d and i represent the drain and irrigation water.)

TABLE 22.3 Water and Calcium Balance of Root Zone After Irrigation with Pecos River Water, as Calculated by Watsuit

Leaching fraction (1)	Depth interval[a] (2)	Infiltrated calcium mass ($mmol_c$) (3)	Leachate calcium concentration ($mmol_c$/l) (4)	Leachate calcium mass ($mmol_c$) (5)	Change in calcium for depth interval[b] ($mmol_c$) (6)
0.1	1	1,700	24.5	1,740	−40
	2	1,700	33.0	1,354	386
	3	1,700	33.1	698	656
	4	1,700	32.8	364	334
0.3	1	2,186	22.1	2,273	−90
	2	2,186	31.5	2,298	−25
	3	2,186	33.4	1,764	535
	4	2,186	33.7	1,440	324

[a]Root zone depth is divided into four quarter intervals, with 1 representing the top quarter and 4 the bottom.
[b]The differences in Ca mass entering and leaving the root zone depth intervals.

The Ca concentration, 16.98 mmol/l, is 45% of the total millimolar charge concentration of cations, and the sulfate concentration, 22.39 mmol/l, is 60% of the total millimolar charge concentration of anions. The leachate volume leaving each quarter depth of the root zone (Table 22.3) was calculated based on the following assumptions: 100 units of water uptake by plants; a leaching fraction of 0.3 and 0.1 and corresponding total units of applied water of 142.9 and 111.1; and the uptake of water and pCO_2 depth distributions described above for Watsuit. The concentrating effects due to the decreasing leachate volume with depth, i.e., water uptake by plants, and, to a smaller extent, the dissolution of soil lime, results in an increased Ca concentration (Table 22.3) in the leachate from the second depth as compared to that from the first. The Ca concentrations of the second, third, and fourth depths are about the same because gypsum and lime precipitation counteract the concentrating effects of water uptake by plants. Consequently, the mass of Ca (Table 22.3) in the leachate ($V \times C_{ca}$) decreases with depth in all cases, but one. A small increase occurs from depth one to two for a leaching fraction 0.3. The loss of Ca (Table 22.3) within the upper root-zone intervals results from soil-lime dissolution. The gain of Ca within the lower root-zone intervals results from precipitation of soil lime and gypsum. The gains in Ca show that the amount of solids precipitating in the soil is appreciable for such gypsiferous waters, even at high ($L=0.3$) leaching fractions.

The preceding data clearly demonstrate that where drainage waters can be intercepted before returning to surface water or ground water, the volume and salt load of drainage can be reduced substantially with reduced leaching. The Wellton-Mohawk Project in Arizona is an example. The drainage water is collected by pumps and conveyed in discharge canals to a desalinization plant. Water diversions from the Colorado River flowing into and drain water flowing out of the Wellton-Mohawk Project could be reduced by 227×10^6 m^3, salt return could be reduced by 324,000 Mg, and the drainage water could be concentrated to the point that it would have virtually no remaining value for irrigation, according to results (Table 22.4) obtained from Watsuit (Rhoades and Suarez 1977).

While reduced leaching will always reduce the salt discharged from the root zone, it may not always reduce degradation of the receiving water. Rhoades and Suarez (1977) analyzed a hypothetical river system at steady state, where the drainage waters from ten sequentially irrigated areas returned directly to the river. Three different types of river water were assumed: Type 1, undersaturated with lime; Type 2, saturated with lime; and Type 3, saturated with lime and nearing saturation with gypsum. Two leaching fractions were assumed: 0.4 for high leaching and 0.1 for low leaching. For Type 1 and 3 waters, the salinity levels at equivalent distances downstream increased with increasing leaching fraction. However, the composition of Type 2 water was unchanged by leaching differences because it was initially saturated with lime.

Under nonsteady state conditions, cation exchange reactions reduce or increase the salt load of drainage water (Jury et al. 1978, Rhoades et al. 1973). Irrigation of new saline lands, or changing the irrigation water used for irrigation, can affect the salt loads of drain water for many years. Exchange of sodium and magnesium for calcium can double the predicted mineral precipitation in a soil profile initially high in exchangeable calcium. Exchangeable calcium precipitation will be larger for reduced leaching when bicarbonate > calcium or sulfate > calcium.

TABLE 22.4 Terminal Display of Soil-Water Composition Resulting From Irrigation with Porterville Well Water at a Leaching Fraction of 0.10

**** CASE: PORT *** (A) UNTREATED ***
**** LF TREATMENT: .10

DEPTH	LF	1/LF	CA	MG	NA+K	CL	CO_3	HCO_3	SO_4
			mmol$_c$/l						
0	1.00	1.00	1.14	.05	4.95	1.75	.47	2.67	1.25
1	.64	1.56	2.26	.08	7.73	2.73	.43	4.95	1.95
2	.37	2.70	3.03	.14	13.38	4.73	.43	8.01	3.38
3	.19	5.26	2.53	.26	26.05	9.21	.45	12.60	6.58
4	.10	10.00	1.78	.50	49.50	17.50	.54	21.23	12.50

DEPTH	PH	CA/MG	SUM CAT.	LIME	GYP	EC	SAR
			mmol$_c$/L			(dS/m)	(mmol$_c$/L)$^{0.5}$
0	8.37	22.87	6.14	−1.03	.00	.61	6.34
1	7.77	28.92	10.07	−2.10	.00	.94	7.08
2	7.48	22.41	16.54	−2.76	.00	1.5	10.53
3	7.48	9.62	28.85	−2.01	.00	2.69	21.81
4	7.56	3.55	51.78	−.78	.00	4.90	45.94

When the conditions are reversed, less of the exchangeable calcium precipitates. The reverse effect—increased dissolution of salt due to cation exchange reactions—can also occur. If calcium from the irrigation water replaces exchangeable sodium or magnesium, the dissolution of lime and gypsum is increased, resulting in a corresponding increase in the salt load of the drainage water.

Drainage water can encounter a variety of chemical conditions as it travels to receiving water. If more saline waters or readily soluble salts, such as gypsum, are present in the flow path, reducing the leaching fraction results in reduced salt emissions. This is the case in the Grand Valley of Colorado (Inman and King 1983). The annual salt pickup in the Colorado River due to displacement and dissolution is estimated to be 528,000 Mg. The corresponding increase in dissolved salts is 55 g/m^3. The west side of the San Joaquin Valley in California is another example of how salt displacement and dissolution by drainage water from irrigation results in increased salt load in the receiving waters, including the San Joaquin River, underlying vadose zones, and ground waters. In these cases, reducing the drainage water may be expected to reduce salt emissions proportionately.

WATER QUALITY PROGNOSIS AND DIAGNOSIS USING WATSUIT

The chemical data generated by Watsuit simplify the evaluation of irrigation-water suitability when used with and without chemical amendments. Soil-salinity problems are indicated if the predicted salinity of the root zone exceeds the tolerance of the crops to be grown.

Similar assessments are made for boron, chloride, and sodium. In Chapter 13, the salt and specific ion tolerances of different plant species are related. If some reduction in yield is tolerable, a higher level of salinity can used in place of the threshold levels. Soil permeability is judged to be a problem if the combination of predicted near-surface SAR and irrigation water EC are expected to result in significant aggregate slaking, dispersion, and clay swelling. See Chapters 3 and 11 for the diagnostic criteria for soil permeability. The benefits of soil and water amendments on water suitability are evaluated based on their effects on soil-solution composition. The following example illustrates these techniques.

Porterville Well Water

This water is used to irrigate citrus and avocado groves in Tulare County in the southeastern San Joaquin Valley, Calif. The climate during the summer is hot and dry. Rainfall between December and March varies, ranging from 50 mm to 500 mm.

The Porterville well water (Table 22.2) is relatively nonsaline, EC = 0.5 dS/m, but it is sodic, SAR = 12.8 $(mol/m^3)^{0.5}$. The predominant salt is sodium bicarbonate. Table 22.4 provides soil-solution compositions at the soil surface and through the root zone for untreated Porterville water at an *L* of 0.1. Also given are the calculated Ca/Mg and SAR ratios, the EC, and the gain in salt concentration due to soil-lime dissolution. The latter reduces the SAR to levels less than that of the irrigation water for the soil surface and depths 1 and 2. Note the increasing depth, total ion concentrations, EC, and SAR with depth due to increasing 1/L. The decrease in pH with depth reflects the assumed increase in pCO_2 with depth.

The summary data for several leaching fraction and amendment combinations for the Porterville water (Table 22.5) include average-profile EC, SAR, and chloride; upper-profile EC, SAR, and chloride; and salinity weighted by water uptake, C′, in $mmol_c/l$, and osmotic potential, PI′, in bar. These data indicate several potential problems. Excessive surface SAR relative to the surface EC, and high SAR's relative to EC's in the root zone may result in poor water infiltration and reduced hydraulic conductivities. Rainfall would increase this problem. The consequent reduction in soil-solution EC would increase the likelihood of poor soil physical properties due to slaking, dispersion, and swelling of soil clays (Shainberg and Letey 1983). Applying gypsum to the surface of the soil or injecting it into the irrigation water (Wildman et al. 1989; Rivers 1989, personal communication) would reduce these hazards.

Salinity is not indicated as a problem for citrus because the average root zone salinities range from 1.1 dS/m to 2.8 dS/m for the untreated case (Table 22.5). These values are less than twice the EC_e threshold salinity, 1.7 dS/m for citrus. EC_e is assumed to be 1/2 the average root zone salinities (see Chapter 3). Consequently, only root zone salinities greater than 3.4 dS/m would indicate potential yield reduction due to excessive soil salinity. However, salinity may be a problem for avocados, which are more sensitive than citrus based on the California experience with both crops.

TABLE 22.5 Terminal Display of Summary Data for Porterville Well Water as Calculated by Watsuit

**** CASE: PORT *** (A) UNTREATED ***

LF TR.	AVG.EC	UP.EC	AVG.SAR	UP.SAR	AVG.CL	UP.CL	C′	PI′
.05	2.79	1.05	26.00	8.30	9.75	3.17	17.77	.64
.10	1.97	1.00	16.39	7.76	6.57	2.99	14.98	.54
.20	1.44	.94	10.25	7.01	4.51	2.72	12.53	.45
.30	1.22	.90	7.88	6.44	3.60	2.51	11.29	.41

SUR.EC = .62 SUR.SAR = 6.34

****CASE: PORT *** (B) AMENDED WITH SULFURIC ACID ***

LF TR.	AVG.EC	UP.EC	AVG.SAR	UP.SAR	AVG.CL	UP.CL	C′	PI′
.05	3.12	1.21	13.69	6.00	9.75	3.17	20.61	.74
.10	2.27	1.16	9.88	5.70	6.57	2.99	17.58	.63
.20	1.67	1.08	7.20	5.27	4.51	2.72	14.80	.53
.30	1.40	1.02	5.92	4.93	3.60	2.51	13.31	.48

SUR.EC = .71 SUR.SAR = 4.65

****CASE: PORT **** (C) AMENDED WITH 1 mmolc CASO4 ***

LF TR.	AVG.EC	UP.EC	AVG.SAR	UP.SAR	AVG.CL	UP.CL	C′	PI′
.05	2.95	1.13	19.64	6.94	9.75	3.17	18.96	.68
.10	2.11	1.08	12.51	6.55	6.57	2.99	16.14	.58
.20	1.55	1.01	8.42	5.99	4.51	2.72	13.61	.49
.30	1.31	.96	6.73	5.56	3.60	2.51	12.27	.44

SUR.EC = .66 SUR.SAR = 5.37

****CASE: PORT **** (D) AMENDED WITH 20 mmolc CASO4 ***

LF TR.	AVG.EC	UP.EC	AVG.SAR	UP.SAR	AVG.CL	UP.CL	C′	PI′
.05	4.57	2.78	7.33	2.36	9.75	3.17	42.59	1.53
.10	3.78	2.73	4.84	2.22	6.57	2.99	39.92	1.43
.20	3.20	2.60	3.29	2.05	4.51	2.72	37.25	1.34
.30	2.94	2.50	2.62	1.92	3.60	2.51	35.78	1.29

SUR.EC = 1.86 SUR.SAR = 1.58

Calcium levels may become a problem if dissolution of soil lime and silicates is inadequate. Assuming the soil solution is in equilibrium with soil lime, the average calcium levels for leaching fractions of 0.1 $mmol_c/l$ and 0.2 $mmol_c/l$ are 2.6 $mmol_c/l$ and 3.2 $mmol_c/l$, which are adequate for nutritional needs (> 1—2 $mmol_c/l$). However, in the absence of soil lime, calcium levels would be about 0.5 $mmol_c/l$, which are inadequate. Adding lime or gypsum amendments would be recommended if the native soil does not contain soil lime, or soil lime is depleted by prolonged irrigation with this water.

The average EC and SAR values for soil samples obtained beneath citrus trees exceeded the upper-profile EC and SAR for an L of 0.05 (Table 22.5), indicating inadequate irrigation, poor infiltration, or both. The recommendation given to the farmer (Pehrson et al. 1985) was to apply 7 Mg/ha of gypsum to a lightly tilled soil surface every two to three years to increase water infiltration and continue irrigation into the start of the rainy season to assure adequate leaching. The injection of sulfuric acid or sulfur dioxide gas, using a sulfur burner to neutralize 90%

of the bicarbonate, would also be a possibility (Table 22.5). This could result in lower soil pH. Prolonged use of this water without gypsum or lime amendments will reduce the amount of soil lime, and the use of acid amendments would increase the rate of the dissolution of soil lime. Thus, soil-lime contents should be determined before recommending the addition of acid amendments to the Porterville irrigation water.

CONCLUSION

More efficient water use or reduced leaching increases the salt concentration in the root zone and reduces the volume and salt load of the drainage. Increased salt concentration induces decreased salt dissolution and increased salt precipitation of sparingly soluble salts, such as lime and gypsum. Computer programs exist that can help to evaluate the potential effects of dissolution and precipitation reactions on soil-water compositions at different leaching fractions. With the rapid advance in computer technology, greater use of computer-based chemistry software can be expected to assess the suitability of water for irrigation and to evaluate the potential environmental effects of irrigation.

REFERENCES

Dutt, G.R., and K.K. Tanji. 1962. "Predicting concentrations of solutes in effluents." *J. of Geophysical Research* 67: 3437-3439.

Dutt, G.R., R.W. Terkeltoub and R.S. Rauschkolb. 1972. "Prediction of gypsum and leaching requirements for sodium-affected soils." *Soil Science* 114: 93-103.

Dutt, G.R. 1962. "Prediction of the concentration of solutes in soil solutions for soil systems containing gypsum and exchangeable Ca and Mg." *Soil Science Society of America Proceedings* 26: 41-343.

Inman, R., D. Olson and D. King. 1983. "Grand Valley salt pickup calculation." U.S. Dept. of the Interior, Bureau of Reclamation, Grand Junction, Colo.

Jury, W.A., H. Frenkel and L.H. Stolzy. 1978. "Transient changes in the soil-water system from irrigation with saline water: I. Theory." *Soil Science Society of America Proceedings* 42: 579-585.

McNeal, B.L., J.D. Oster and J. Hatcher. 1970. "Calculation of electrical conductivity from solution composition data as an aid to in-situ estimation of soil salinity." *Soil Science* 110: 405-414.

Oster, J.D. 1982. "Gypsum usage in irrigated agriculture: A review." *Fertilizer Research* 3: 73-89.

Oster, J.D., and B.L. McNeal. 1971. "Computation of soil solution composition variation with water content for desaturated soils." *Soil Science Society of America Proceedings* 35: 436-442.

Oster, J.D., and J.D. Rhoades. 1975. "Calculated drainage water compositions and salt burdens resulting from irrigation with river waters in the Western United States." *J. Environmental Quality* 4: 73-79.

Oster, J.D., and H. Frenkel. 1980. "The chemistry of the reclamation of sodic soils with gypsum and lime." *Soil Science Society of America Proceedings* 44: 41-45.

Oster, J.D., and K.K. Tanji. 1985. "Chemical reactions within the root zone of arid zone soils." *J. Irrig. & Drainage Engrg.* 111: 207-217.

Pehrson, J., N. O'Connell, and J.D. Oster. 1985. "Salinity management for San Joaquin Valley citrus." *Citrograph,* November. pp. 15-17.

Rhoades, J.D. 1977. "Potential for using saline agricultural drainage waters for irrigation." Proceedings of the specialty conference on water management for irrigation and drainage, ASCE/Reno, Nev. 1987, pp. 85-116. July.

Rhoades, J.D. 1987. "Use of saline water for irrigation." *Water Quality Bulletin* 12: 14-20.

Rhoades, J.D., D.B. Krueger and M.J. Reed. 1968. "The effect of soil mineral weathering on the sodium hazard of irrigation waters." *Soil Science Society of America Proceedings* 32: 643-647.

Rhoades, J.D., R.D. Ingvalson, J.M. Tucker, and M. Clark. 1973. "Salts in irrigation drainage waters. I. Effects of irrigation water composition, leaching fraction, and time of year on the salt composition of irrigation drainage waters." *Soil Science Society of America Proceedings* 37: 770-774.

Rhoades, J.D., J.D. Oster, R.D. Ingvalson, J.M. Tucker and M. Clark. 1974. "Minimizing the salt burdens of irrigation drainage waters in the Western United States." *J. Environ. Qual.* 4: 311-316.

Rhoades, J.D., and S.D. Merrill. 1976. "Assessing the auitability of water for irrigation; theoretical and empirical approaches." *Prognosis of salinity and alkalinity, FAO Soils Bull.* 31: 69-109

Rhoades, J.D., and D.L. Suarez. 1977. "Reducing water quality degradation through minimized leaching management." *Agricultural Water Management,* 1(2): 123-197.

Rivers, Jerry B. Sales representative of Domtar Gypsum Inc., 2424 E. Valley Oaks Dr., Suite D., Visalia, CA 93291.

Robbins, C.W., R.J. Wagenet, and J.J. Jurinak. 1980. "A combined salt transport-chemical equilibrium model for calcareous and gypsiferous soils." *Soil Science Society of America Proceedings* 44: 1191-1194.

Shainberg, I. and J. Letey. 1983. "Response of soils to sodic and saline conditions." *Hilgardia* 52(2): 1-57.

Suarez, D.L., and J.D. Rhoades. 1982. "The apparent solubility of calcium carbonate in soils." *Soil Science Society of America Journal* 46: 716-722.

Tanji, K.K. 1969. "Predicting specific conductance from electrolytic properties and ion association in some aqueous solutions." *Soil Science Society of America Proceedings* 33: 656-661.

Tanji, K.K., L.D. Doneen, G.V. Ferry and R.S. Ayers. 1972. Computer simulation analysis on reclamation of salt-affected soils in San Joaquin Valley, California. *Soil Science Society of America Proceedings* 36: 127-133.

Wildman, W.E., W.H. Krueger, and R.E. Pelton. 1989. "Calcium amendments for water penetration in flooding systems." *California Agriculture* 43: 14-15 and 37.

CHAPTER 23

MODELING TRANSIENT ROOT ZONE SALINITY[a]

INTRODUCTION

Soil models necessarily are simplifications of real soils. The science of modeling consists of formulating mathematical equations to describe chemical, biological, and physical processes and developing algorithms to solve those equations. The art of modeling lies in judging, based on experience and intuition, which of the processes to include and the degree to which a given process may be approximated mathematically.

The management of saline or potentially saline soils can involve a compromise between several, often conflicting, objectives, such as minimizing salinity in the root zone and salt leaching. A model enables a number of alternative management practices to be compared and may indicate the need for modified management or further study of processes not included in the model.

This chapter describes some of the major processes typically included in transient salinity models. The focus is on modeling approaches, rather than specific computational techniques. This chapter lists some of the most widely used equations and describes some of the factors that must be considered when selecting a model.

EARLY SALINITY MODELS

Early salinity models focused on irrigation return flows and the mineralization of rivers. Walker (1978) summarized the main features of many of the early irrigation return flow models. They ranged from simple infiltration models to those describing the chemistry of watersheds.

A problem of incompatibility of scale existed in many early modeling efforts because root zone models that can include only a few combinations of crop, soil, and management practices were used to simulate leachate quality and quantity from large areas of irrigated soil. The models often were calibrated so that simulated solute discharge matched measured salt loads in rivers.

Several recent reviews summarize much of the available knowledge about the modeling of leaching in saline soils. Bresler et al. (1982) reviewed the physics and chemistry of saline and sodic soils and described modeling techniques in detail. Tanji and Deverel (1984) reviewed

---

[a]Prepared by: J.L. Hutson, Dept. of Agronomy, Cornell Univ., Ithaca, NY 14853; L.M. Dudley, Utah State Univ., Logan UT 84322; and R.J. Wagenet, Dept. of Agronomy, Cornell Univ., Ithaca, NY 14853.

the chemical processes involved in modeling the reclamation of sodic soils. Nielsen et al. (1986) reviewed processes of water and chemical transport in the unsaturated zone and the various models used to simulate them. Grove and Stollenwerk (1987) reviewed the chemical reactions simulated by ground-water quality models but listed very few multi-species models.

FUNDAMENTAL MODELING CONCEPTS

Addiscott and Wagenet (1985) have delineated two main categories of models. In deterministic models, individual processes and the interactions between those processes are defined mathematically, with each set of input data leading to a unique and reproducible prediction. This type of model typically is poor at accounting for the spatially variable nature of soil. Stochastic models, on the other hand, place less emphasis on process and more on predicting the statistical distribution or probability of a given characteristic. Models formulated to describe the complexities and interactions of the saline soil environment must emphasize process and are primarily deterministic rather than stochastic.

A deterministic simulation model typically comprises several modules, each of which describes a single process or aspect of soil behavior that operates within a given interval of time and space. The output from one module serves as the input for another. As an example of a typical salinity model, we have used the salinity version of the Leaching Estimation and Chemistry Model, LEACHM (Wagenet and Hutson 1987). Other versions of this model describe nitrogen and pesticide fate in soil.

Fig. 23.1 shows a schematic outline of the structure of LEACHM. Each box represents a group of processes. The lines represent data transfer. The core processes in a salinity model describe water flow, solute transport, and chemical reactions. Peripheral to, but strongly influencing the core processes, are modules describing plant growth, plant behavior, and boundary conditions. Core routines usually are developed in greater detail than peripheral routines.

Under certain circumstances, a peripheral module may come to dominate and determine the outcome of a simulation. The ability to recognize situations in which output values are largely determined by a peripheral approximation is an important part of the evaluation of a modeling exercise. A common example is the simulation of drainage fluxes, which may be calculated in a detailed water flow routine, but which are often determined by coarse approximations of evapotranspiration and plant root development.

MODELING FLUID FLOW

Water Content, Potential, and Flux

Three possible approaches to modeling water flow in soils exist, none of which is universally applicable.

The simplest is the capacity-type model. With this, field capacities, wilting coefficients, and evaporation limits are defined for each of sev-

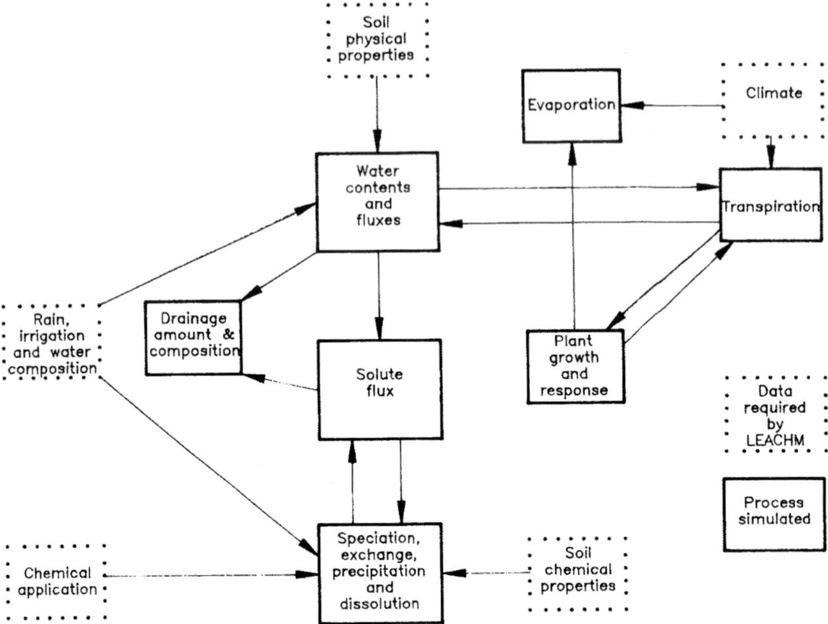

Fig. 23.1 Schematic Outline of Structure of LEACHM

eral layers of soil in a profile and conceptually link the layers. For example, infiltrating rain water or irrigation water increases the water content of the uppermost soil layer. Water in excess of field capacity then drains to the next layer, where further downward drainage may occur. At the end of this process, all infiltrating water is allocated to soil layers, none of which exceed field capacity. Excess water draining from the bottom of the profile represents net drainage. Evapotranspiration demand is satisfied by removing water from each layer, weighted according to water content, root distribution, and depth.

Advantages of the capacity approach are relatively few data are required, the data typically are easily obtained or estimated, and minimal computer time is required because the mathematics involved is relatively simple and daily time steps are usually employed. The main disadvantage of this method is that the dynamic nature of soil water is oversimplified, leading to the inaccurate estimation of water fluxes.

A second approach for simulating soil water flow assumes that water flux density is proportional to the hydraulic potential gradient, with the proportionality constant being the hydraulic conductivity (K). This is known as the mechanistic, or Darcian, approach. Relationships between water content (θ), matric potential (h), and hydraulic conductivity are required. These usually, but not necessarily, are assumed to be single-valued, or nonhysteretic. The water-content distribution in time and space is obtained by solving the Richards equation,

$$\frac{\partial \theta}{\partial t} = \frac{\partial h C_w}{\partial t} = \frac{-\partial}{\partial z}\left[K(\theta)\frac{\partial H}{\partial z}\right] + U \quad (23.1)$$

where C_w = the differential water capacity ($\partial\theta/\partial h$); H = the hydraulic head (the sum of the pressure and gravitational components of the soil water potential), z = depth; and U = absorption of water by plants.

Hanks and Bowers (1962) described a numerical solution that has subsequently been used as the basis for many finite-difference and finite-element solution techniques. The finite-difference solution technique often is used to solve one-dimensional models because it is relatively simple to implement on the computer and boundary conditions can be incorporated easily. The merits of various solution techniques have been compared by Haverkamp et al (1977). LEACHM uses an implicit, central-difference method to solve Eq. 23.1.

When infiltration fluxes are high and the soil is strongly structured or contains large cracks or pores open to the surface, Darcian water flow models may be inappropriate. Water can follow preferential pathways that constitute a small proportion of the potentially water-filled pore space. Thus, associated solutes move downwards more rapidly than would otherwise be expected. Similarly, in soils with abruptly discontinuous fine-to-coarse textures, water can flow through macropores (large voids) in the soil. Quantitative prediction of macropore flow and the conditions that cause it are presently the subject of much research.

Most models that include macropore flow define two distinct regions of the soil volume— a mobile or macropore region in which water is retained only poorly and solute can move rapidly and an immobile or micropore region in which flow is slow and solute is not subject to rapid leaching. Darcian flow may operate in the micropore region. Beven and Germann (1982) have found that gravity is the dominant force moving water in macropores. They developed a model in which water moves in the macropores under gravity and is sorbed into the matrix via a diffusive sink term (Germann and Beven, 1985). Van Stiphout et al. (1987) modeled a cracked clay soil in a much simpler way by establishing an infiltration threshold, above which water and solute were immediately allocated to the soil at the lower depth of the cracked horizon.

Agricultural soils usually have extensive areas with vertical macroscopic water fluxes. For this reason, most transient root zone models are one-dimensional. Surface micro-relief, drip, or furrow irrigation systems, and plant root and leaf distribution patterns can cause non-uniform water application and leaching of water and solutes. In these situations, two- or three-dimensional models would be more appropriate. However, the increased complexity, data demands, and computer processing times required when running a three-dimensional model necessarily limit the degree of process detail. This illustrates a familiar root zone modeling dilemma: a more detailed model requires more extensive data, but detailed data is difficult to obtain and introduces more uncertainty.

Boundary Conditions, Sources and Sinks

Water applied to the soil surface can be characterized by time, rate, or amount of application. These parameters may be measured or, as in the case of rainfall, generated to meet some statistical pattern. Fixed-time-period data, e.g., total daily infiltration data, can be used to execute capacity models. For mechanistic models, especially those predicting runoff or macropore flow, exact times and rates of water application must be used.

The spatial variability of infiltration results in spatially variable root distribution and water extraction patterns. These inhomogeneities make prediction of chemical solute movement and distribution more difficult. English et al. (1986) described a model that allows calculating the water balance components for a spatially variable, sprinkler-irrigated field. Combining their approach with a salinity model could help to simulate inhomogeneous systems, which cannot be simulated accurately by applying a simple one-dimensional model.

Water is lost from soils by surface evaporation, transpiration, and drainage. After relatively short periods of infiltration and redistribution, root extraction determines the distribution of water in the soil profile. Small errors in estimating actual ET can greatly affect estimates of drainage. Over the long-term, drainage should equal the difference between infiltration and evapotranspiration. Many methods exist for calculating potential evapotranspiration. LEACHM uses a relatively simple routine that is based on pan evaporation data. Root distribution, matric potential, osmotic potential, and hydraulic conductivity then determine extraction of water from the soil profile.

Soil Hydrological Properties

Capacity-type models require as input an estimate of the drained upper limit of soil water content (the field capacity), the wilting coefficient, and evaporation limits. Salter and Williams (1965), Salter et al. (1966), Shaykewich and Zwarich (1968), Rivers and Shipp (1978), Cassel et al. (1983) and Ratliff et al. (1983) have derived regression equations relating field capacity and wilting coefficients to soil textural criteria. When measured data are unavailable, such relationships are extremely useful.

Measured data relating $\theta(h)$ and $K(h)$ are rarely available. These relationships, which are necessary for mechanistic models, can be measured by standard techniques (Klute 1986) or they can be estimated from other soil properties.

Hall et al. (1977), Gupta and Larson (1979), Rawls and Brakensiek (1985), and Hutson (1986) present regression equations relating water retained at various matric potential values to soil particle size, bulk density, and organic carbon content. The modeler may use an empirical retentivity function (Campbell 1974; Clapp and Hornberger 1978; van Genuchten 1980; Hutson and Cass 1987), or interpolate tabulated data to incorporate $\theta(h)$ data into models. Ideally, the retentivity function should be continuous and sigmoidal. If it is to be related to soil properties, it should contain as few parameters as possible. The effects of hysteresis are often ignored because they typically are small relative to the effects of spatial variability and data uncertainty.

Defining the relationship between K and soil properties is more difficult. The simplest approach uses an empirical retentivity function like that of Campbell (1974) in which K(θ) incorporates the saturated hydraulic conductivity K_s and two retentivity parameters and scales the relationship according to the conductivity measured or desired at a particular matric potential.

MODELING SOLUTE TRANSPORT

Water flow in soil convectively transports dissolved chemical constituents. Velocity distributions between the large and small pores disperse the flow. The distribution of solute concentration due to convection is further modified by diffusion and chemical reactions. The type of transport model selected to describe convection and dispersion depends on the approach used to model water flow.

If a capacity model is used, then the solute transport model is usually based on a mixing and displacement process. Mechanistic water flow models based on Darcy's law are used to calculate water flux densities. These flux densities can, in turn, be used to solve the convection-dispersion equation (CDE). Neither approach has proven to provide reliable simulation of solute transport when macropore or preferential flow is important. The quantitative description of solute transport in these soils is presently the subject of much research.

In layer mixing models or chromatographic models, incoming solution from an overlying layer mixes with water already present in the layer. Any water in excess of field capacity then moves downward to the next layer, along with its share of dissolved salts. In this way, the "smearing" of a concentration front is simulated. The layer-mixing approach cannot simulate diffusion of solute or redistribution at water contents below field capacity unless additional assumptions are made. A model by Tanji et al. (1972) used this approach but retained an immobile fraction of soil water within each layer. Dutt et al. (1972) calculated water fluxes by using Eq. 23.1 and used long-term cumulative fluxes as input to a chromatographic solute transport model. Kapoor and Pal (1986) described a simple model for predicting salinization, sodification, and salt distribution based on a modification of Burn's (1974) layer-mixing model.

The CDE is formulated on the premise that net solute flux (J_t) is the sum of convection (J_q) and diffusion (J_d) fluxes. Thus

$$J_q = -D_m(q)\frac{\partial c}{\partial z} + qc \tag{23.2}$$

$$J_d = D(\theta)\frac{\partial c}{\partial z} \tag{23.3}$$

where q = water flux density; c = solute concentration; D(θ) = the molecular diffusion coefficient in unsaturated soil; D_m = flux density-dependent hydrodynamic dispersion coefficient; and $\partial c/\partial z$ = the vertical concentration gradient (Wagenet 1983). The total of any chemical species (T) consists of both solution and sorbed components

$$T = \theta C + pS \tag{23.4}$$

where p = bulk density (M L^{-3}) and S = sorbed chemical (M M^{-1}). For a multi-species transient salinity model, the relationship between C and S for any one chemical species is complex, since it involves exchange, mineral equilibria, and possibly sorption equilibria. Solute concentrations are usually high relative to the amount taken up by plants; however, plant uptake can be included as a sink term, if desired. Ignoring sources and sinks, the chemical species balance equation is

$$\frac{\partial T}{\partial t} = \frac{-\partial J_t}{\partial z} \tag{23.5}$$

Combining Eqs. 23.2 through 23.5 gives the CDE:

$$\frac{\partial(\theta C + \rho S)}{\partial t} = \frac{\partial}{\partial z}\left\{\theta\left[D_m(\theta,q) + D(\theta)\right]\frac{\partial c}{\partial z} - qc\right\} \tag{23.6}$$

Examples of transient root zone models that use the CDE are SALTFLOW (Robbins et al. 1980b), models by Melamed et al. (1977) and Tillotson et al. (1980), and LEACHM (Wagenet and Hutson 1987). Some models apply the Richards equation in calculating water fluxes and a plate-mixing model to simulate salt transport (Dutt et al. 1972).

Some equations for calculating diffusion coefficients have a theoretical basis, such as that used by Jury (1983), which is based on the Millington and Quirk tortuosity model

$$D(\theta) = D_o\left(\frac{\theta^{3.33}}{\theta_s^2}\right) \tag{23.7}$$

where D_o = the ionic diffusion coefficient in water and θ_s = saturated water content.

Other equations are empirical, for example (Bresler 1973):

$$D(\theta) = D_o a e^{b\theta} \tag{23.8}$$

where a and b are constants.

Jurinak et al. (1987) showed how ionic diffusion coefficients for Ca, Na, and Cl may be calculated from electrical conductivity measurements.

Hydrodynamic dispersion coefficients are usually calculated from

$$D_m = |q|\,\frac{\lambda}{\theta} \tag{23.9}$$

where λ = the dispersivity, which depends on scale and water flux density. Unless a correction is applied, finite-difference solutions may include numerically-induced dispersion (Bresler 1973). Campbell (1985) and others have substituted numerical dispersion for dispersivity because the latter is usually unknown. Dispersivity is one of the most uncertain parameters in a diffusion-convection model, being virtually impossible to measure in the field for all water flux densities.

Solute transport in aggregated and structured soils is difficult to model because the assumption that equilibrium exists between solution and solid phases is invalid in these soils during periods of rapid water flux. As for water-flow simulations, the large range of macropore sizes complicates accurate modeling of macropore effects. Addiscott (1977) used a plate-mixing model that divides the pore volume into mobile and immobile fractions. The model allows infiltrating water and solute to move only in the mobile phase, and slow equilibration between the two phases may occur only after the infiltration ceases. A model described by van Genuchten and Dalton (1986) uses a convection-dispersion approach to describe flow in cracks and macropores, with intra-aggregate diffusion lumped into a dispersion coefficient. The model includes equations derived for well-defined macropore geometries; these can be extended to more complex systems by using shape factors.

Solutes in field soils do not always move only in the vertical direction. Surface application patterns, root distribution patterns, soil layering, and regional drainage patterns can induce two- or three-dimensional flow. Bresler (1975) described a two-dimensional finite difference model for predicting water and salt distributions under trickle irrigation. Increases in computer power undoubtedly will prompt the development of new multidimensional transient models, but one-dimensional field-scale models presently predominate.

NUMERICAL SOLUTIONS TO RICHARDS EQUATION AND SOLUTE TRANSPORT EQUATION

It is impossible to obtain general analytical (exact) solutions to the partial differential equations used to describe transient water movement and solute transport through soils; therefore, solutions to the equations typically are approximated by finite-difference or finite-element methods.

The system of finite-difference equations that must be solved when approximating the second-order partial differential equations for both water movement and solute transport is nonlinear, and its solution requires iterative methods. Small increments in time and space may be necessary to cause the iterative numerical routine to converge toward a single solution. A stable numerical solution is one for which the difference between the theoretical solution and the approximate solution is well-bounded for a large number of iterations. An accurate numerical solution agrees well with the theoretical exact solution. To be useful, a finite-difference scheme must be convergent, stable, and accurate under the boundary conditions and initial conditions imposed by the physical situation to be represented.

Usually, smaller time and space increments increase accuracy, promoting stability and convergence. They also increase computation time.

Thus, a compromise exists between accuracy and computational efficiency.

There are many ways that finite-difference schemes may be formulated; each has its own convergence, stability, and accuracy criteria. The Richards equation and the one-dimensional convection-dispersion equation are examples of parabolic equations with specified initial and boundary conditions. The implicit Crank-Nicholson (central difference) finite-difference scheme often is applied to this problem. LEACHM uses a Crank-Nicholson formulation, which permits increasing the time increment as water content and flux decrease with no loss of accuracy.

MODELING CHEMICAL INTERACTIONS

Some problems do not require a precise understanding of the chemical composition of the soil solution; for these, a model that simulates the movement and distribution of a non-interacting salt is adequate. Other problems require modeling the interactions between dissolved ions and between the aqueous ions and solid phases. Research-level salinity models typically model solid-phase precipitation or dissolution, sorption, and ion-exchange processes.

The simplest way to calculate chemical equilibria in a model is to assume that reactions are complete at a specified point in space and time. This often is referred to as the assumption of local equilibrium. LEACHM and many other models rely on this assumption. The assumption is weakest when water fluxes are high.

Changes in solution composition and concentration caused by movement of solutes, changes in water content, and uptake by plants can upset the equilibrium between the solution and the exchangeable and precipitated phases. Equilibrium is re-established by adjusting the levels of each ion in the solution and solid phases in accord with chemical equilibria, while preserving chemical mass balance. Since the changes that occur within each time step of a finite-difference model are small, it is usually unnecessary to call the chemical equilibrium routines during each time step. In LEACHM, chemical equilibrium is re-established at a user-specified frequency, usually every four to ten time steps.

Concentrations, Activities, and Ion Speciation

Two approaches may be used to calculate the equilibrium composition of a soil water system: mass balance and minimization of Gibbs free energy. The mass-balance formulation, which is the method of choice for multi-species transport models, provides direct values for species concentrations and may be readily adapted to problems in which only conditional constants are available. (Mass-action equations that incorporate conditional constants have the same form as those with thermodynamic equilibrium constants. However, the value of the conditional constant depends on the ionic strength, pH, or solution composition for which it was determined.)

The first step in the mass-balance procedure is to compute single-ion activities from ion-association and ionic strength relationships and the total solution concentrations. The single-ion activities are, in turn, used when simulating exchange processes and dissolution-precipitation processes. Ionic strength may be computed directly as the weighted sum of each ionic species in solution or approximated from the electrical conductivity of the solution (Griffin and Jurinak 1973). As described in Chapter 3, the Davies equation is used to compute single-ion activity coefficients.

The Davies equation becomes inaccurate at ionic strengths greater than about 0.5 M, which roughly corresponds to the ionic strength of sea water (Sposito 1984). Thus, computations of chemical equilibrium in concentrated soil solutions, such as salt-laden irrigation return flows, are of questionable accuracy. This limitation also exists at low-water content, but it probably does not greatly affect model leaching predictions because water fluxes are also low. Other functional relationships exist for computing ionic strength and activity coefficients in systems of 1 molal to 6 molal ionic strengths (Sposito 1984). However, those equations have not yet been used in salt leaching models.

Ions in solution form associations by electrical attraction between unlike charges (ion pairs) or by chemical bonding (complexes). The total or analytical concentration of an ion is the sum of the concentrations of a number of ion-pairs, complexes and the free ion. For example, the mass balance equation for total Ca concentration (Ca_T) is the sum of the solution species:

$$Ca_T = [Ca^{2+}] + [CaCO_3^0] + [CaOH^+] + [CaCl^+] + [CaHCO_3^+] + [CaSO_4^0] \tag{23.10}$$

One mass balance equation is required for each ion that has a free concentration not fixed by a model input parameter. For example, LEACHM contains mass balance equations for Ca, Mg, Na, K, Cl, SO_4, CO_3, and HCO_3.

Mass-action laws for formation of associations between a cation and an anion result in the definition of a stability constant

$$K_d = \frac{(Cat^{m+})^a(An^{n-})^b}{(Cat_a\ An_b^{m-n})} \tag{23.11}$$

where Cat^{m+} = a cation of positive charge m and An^{n-} = an anion of negative charge n. Mass-action laws based on thermodynamic (activity-based) equilibrium constants are preferred over those using conditional (concentration-based) constants because the thermodynamic constants can be used to calculate equilibrium compositions over a broader range of solution conditions. In LEACHM, conditional constants are used in the chemical equilibration routines because it is easier to preserve mass balance using concentrations, but the values of these constants are recalculated whenever ionic strength and activity coefficients change.

These equations are used to rewrite the mass balance equations in terms of free ion concentrations. The system of equations is then solved for the free concentrations of each of the ions. For example, by combining Eq. 23.10 and the conditional form of Eq. 23.11 for each of the five ion pairs, the total analytical concentration of Ca may be calculated from $[Ca^{++}]$, the conditional stability constants (K_d'), and free anion concentrations:

$$Ca_T = [Ca^{2+}]\left(1 + \frac{[CO_3^{2-}]}{K'_{d1}} + \frac{[OH^-]}{K'_{d2}} + \frac{[Cl^-]}{K'_{d2}} + \frac{[HCO_3^-]}{K'_{d4}} + \frac{[SO_4^{2-}]}{K'_{d5}}\right) \tag{23.12}$$

A similar equation can be written for each cation. The equation for a given cation can be rearranged in terms of the free-ion concentration, the analytical concentration, or the ion-species concentration.

Soil Atmosphere

Carbon dioxide dissolves in water to form carbonic acid. Carbonic acid in the soil solution influences the soil pH and, in turn, is a principal driving agent of weathering reactions. Saline soils often are calcareous. Thus, PCO_2 (the partial pressure of CO_2) and the CO_3^{2-}-HCO_3^-pH-$CaCO_3$ equilibria are an integral part of the description of the reactions that affect salt mobility. Robbins (1985) summarized the thermodynamic relationships between P_{CO_2} and the Ca^{2+} activity, (Ca^{2+}), and between the pH and CO_3^{2-} activity, (CO_3^{2-}), in a calcareous soil:

$$(CO_3^{2-}) = \frac{K_{a1}K_{a2}K_H p_{CO_2}}{(H^+)^2} \tag{23.13}$$

$$(H^+)^2 = \frac{p_{CO_2}K_H K_{a1}K_{a2}(Ca^{2+})}{K_{sp2}} \tag{23.14}$$

where K_{a1} and K_{a2} are the first and second dissociation constants of H_2CO_3, and K_H, the Henry's Law constant for CO_2, is defined as $K_H = P_{CO_2}/(H_2CO_3)$.

The level of CO_2 in a soil depends on the balance between sources and sinks. Diffusion of CO_2 into water is a slow process. Amrhein et al. (1985) identified the diffusion of CO_2 as the rate-limiting step in the dissolution of calcite. Considering the wide variation in CO_2 levels that occurs in biologically active agricultural soils and the relatively slow diffusion rate of CO_2 through solution, the assumption of equilibrium with atmospheric CO_2 is doubtful.

Temporal variations in soil CO_2 are difficult to predict; salinity models generally use P_{CO_2} values that are constant over time, but which change with depth.

Precipitation and Dissolution

Gypsum and calcite are the precipitates commonly considered in salinity models. While other weatherable minerals may contribute to the soluble salt load in soils, their dissolution rates are considered to be slow enough to be negligible. Gypsum and lime precipitation and dissolution are usually described by means of solubility products (K_{sp}):

$$K_{sp1} = (Ca^{2+})(SO_4^{2-}) \quad (23.15)$$

$$K_{sp2} = (Ca^{2+})(CO_3^{2-}) \quad (23.16)$$

The activity of CO_3^{2-} is also influenced by pH and P_{CO_2} levels.

Again, local equilibrium is usually assumed for mineral precipitation and dissolution. This may be a reasonable approximation for gypsum (Glas et al. 1979) but certainly is not true for calcite. Chemical kinetics or factors that can inhibit dissolution precipitation, such as organic coatings on existing crystals (Inskeep and Bloom 1986, Amrhein and Suarez 1987), may prevent calcite from reaching equilibrium with a soil solution.

Finally, the assumption that Ca-carbonates, when present, determine the Ca^{2+} activity in soils is questionable. Suarez and Rhoades (1982) found that the presence of silicate minerals more soluble than calcite leads to ion activity products higher than those for calcite. Modeling solid-solution phase transition is further complicated by dependency of K_{sp} on the crystal structure, age, and composition of the precipitate.

Chemical kinetic descriptions of calcite behavior could be included in a model, but determining the parameters for a specific soil would be difficult. Melamed et al. (1977), in their salt movement model, assumed the rates of dissolution and precipitation to be a function of the difference between the ambient and equilibrium concentrations and defined an empirical coefficient, which they then evaluated in column studies. To achieve accurate predictions under field conditions, the transfer coefficient had to be defined separately for each soil layer.

Dissolution and diffusion of salts from saline sediments can contribute significantly to soil salinity. Jurinak et al. (1977) found salt release to be a diffusion-controlled process. They identified three distinct stages of dissolution: first to dissolve are the very soluble salts, then lime and gypsum, then silicates.

The expression of chemical input data must be carefully considered, because analytical data rarely satisfy the lime-gypsum-pH-P_{CO_2}-(Ca^{2+}) equilibria and exchange equilibria. This applies particularly to CO_3^{2-}, HCO_3^-, Ca^{2+} and SO_4^{2-}. Certain data must be given priority during the initial execution of the chemical equilibrium routine, and many models have a means of absorbing any ionic charge imbalance that is generated. In LEACHM, a fictitious balancing ion is used for this purpose. It is typically assumed that laboratory-determined alkalinity values do not represent the HCO_3^- and CO_3^{2-} concentrations in the soil solution. Therefore, those values are calculated by the model using equations that relate pH, P_{CO_2}, and the K_{sp} of calcite.

Cation Exchange

Cation exchange processes can be modeled using exchange equations. In general, the less sophisticated the equation and the fewer the cation species, the smaller the concentration and composition range over which it will be useful for field soils. Evangelou and Coale (1987) found that for some soils a concentration-based Gapon constant holds for all ratios of exchangeable Na to Ca, but for others, generally those having high CEC, the Gapon constant changes when the exchangeable Na percentage exceeded 0.4.

In LEACHM, an activity-based Gapon equation is used (Robbins et al. 1980a). The selectivity coefficients (K_G) for the equilibrium between a pair of cations M^{m+} and N^{n+} is defined as

$$K_G = \frac{(M^{m+})^{1/m} XN_{1/n}}{(N^{n+})^{1/n} XM_{1/m}} \tag{23.17}$$

where X = the exchangeable cation. If four cations are modeled, then there are six possible pairs and six selectivity coefficients.

In LEACHM, the cation exchange capacity and selectivity coefficients are assumed constant, anion exchange is ignored, and the CEC is assumed to equal the sum of exchangeable Ca, Mg, Na, and K.

The model could be refined for acid (non-calcareous) soils by including pH-dependent CEC, aluminum chemistry, ionic strength or pH-dependent selectivity coefficients, and anion exchange. Forster (1986) described a model that included aluminum chemistry and mobile and immobile water fractions. It was validated with data from soil columns subjected to steady-state, unsaturated flow.

Most salinity models assume that the exchange reaction rate is fast relative to solute velocities, i.e., all phases reach equilibrium. Although the exchange reaction itself probably is fast, the overall exchange rate often is retarded because ions are slow to diffuse through particles or water films that may exist at the water-solid interface (Sparks 1987). Jardine and Sparks (1984) examined the equilibration rates of K/Ca adsorption and desorption reactions and distinguished between readily accessible sites and those having difficult access. Ogwada and Sparks (1986) presented diffusion and rate equations to describe the equilibration process. Dahiya and Abrol (1973) found in soil columns that the apparent CEC for soils subjected to steady-state, unsaturated flow decreases as saturation decreases. From this, they concluded that some exchange sites become inaccessible as water content decreases.

When water fluxes are high, lack of equilibrium limits model accuracy. The inaccuracy will be greatest in structured and aggregated soils. Slow equilibration causes ions already in solution to move further than expected, and fewer will enter the solution from the solid phase. Thus, in real soils, ions present in applied water will move further than expected, but dissolution from solid soil amendments, such as gypsum, will be lower and less effective in replacing exchangeable Na than simulated.

Research to date has elucidated some of the rate-limiting steps in soil chemical reactions, but there is no general mechanism, theory, or equation that can be used in models of flowing systems. It is perhaps sufficient at present to be aware of the limitations of the local equilibration assumption and to interpret simulations accordingly. Detailed research models that include kinetic processes may be helpful in this regard, but only if sensitivity studies show that kinetics are indeed important and if reliable data and equations are available.

Calculating Equilibrium Chemical Composition

The system of mass balance equations, Eqs. 23.10 through 23.17, usually is solved by successive approximation (Öster and Rhoades 1975, Oster and McNeal 1971, Robbins et al. 1980a). In LEACHM, the system of mass balance equations is solved using the following iterative steps:

1. Calculate ionic strength and activity coefficients (Davies equation).
2. Convert thermodynamic (activity-based) stability, solubility, and exchange constants and conditional (concentration-based) constants using activity coefficients from the most recent iteration.
3. Equilibrate with lime if it is present.
4. Equilibrate with gypsum if it is present.
5. Adjust H^+ and OH^- to satisfy the dissociation constant of water.
6. Adjust H^+, HCO_3^- and $CO_3^=$ to be in equilibrium with P_{CO_2}.
7. Bring soil solution into equilibrium with the exchange phase.
8. Calculate new ionic strength and activity coefficients.
9. Repeat the cycle if the absolute change in any chemical species or ion pair is greater than a previously defined threshold value.
10. Between each of the above steps, use Eqs. 23.10, 23.11, and 23.12 to update the free ion concentrations and, if necessary, ion speciation or pairing.

Generally, this type of technique has been adequate for transport models that contain only a few species and where irrigation waters are similar to the equilibrium composition of the soil solution. However, the equilibrium chemistry subroutines sometimes fail to converge, particularly when irrigation water and soil solutions are dissimilar. As salt transport models are applied to problems of increased chemical complexity and problems are addressed with increased chemical rigor, the successive approximation approach will be replaced with the more efficient and more accurate numerical methods that are currently used in equilibrium soil solution chemistry models (Sposito and Mattigod 1979).

PLANT RESPONSES

Most salinity models are created for better understanding of the dynamic nature of the chemical and physical environment in which plants grow. Therefore, the response of plants to salinity is an important component of the model's structure and output. Plants suffer water stress sooner when osmotic potentials are high, and salinity can upset plant nutrition when an imbalance of certain anions and cations occurs. These factors, in turn, can influence plant development, root distribution patterns, and water extraction patterns. The net effect of salinity is to reduce plant growth and transpiration.

The major problem in modeling plant response to salinity is the lack of quantitative knowledge of the short-term response of plants to fluctuating salinity. Salinity levels fluctuate in response to wetting and drying cycles. Alternating low- and high-salinity infiltration may be superimposed on the wetting and drying cycles, such as when saline irrigation water is interspersed with rainfall. Traditionally, plant response to salinity is expressed in terms of a salinity threshold, above which the plant's yield decreases as a function of increasing salinity. In many models, simulated transpiration is linked directly to crop yield (Hanks 1974; Childs and Hanks 1975):

$$Y_r = T_r = \frac{Y}{Y_p} = \frac{T}{T_p} \tag{23.18}$$

where T_p = potential or maximum transpiration; Y_p = the corresponding potential or maximum yield; Y = actual yield; T = actual transpiration; Y_r = relative yield; and T_r = relative transpiration.

Water uptake by plant roots often is modeled according to concepts described by Nimah and Hanks (1973). Total uptake is the sum of extraction from each soil segment, calculated as:

$$Q = \frac{\sum \Delta t[RF_i \cdot K_i(H_{root} + R_{root} \cdot z - h_i - s_i)]}{\Delta x \Delta z} \tag{23.19}$$

where subscript i = the soil layer; RF = the fraction of active roots; H_{root} = an effective water potential in the root at the soil surface (or crown potential); and R_{res} = root resistance. Q is the desired potential transpiration flux during a time step Δt. The matric potential, h; bulk soil osmotic potential, s; and hydraulic conductivity, K, are assumed to represent measured values at a distance Δx from a plant root. The thickness of the soil layer (Δz) in Eq. 23.19 ties Q to a change of θ. Using an iterative procedure to determine H_{root} allows the water extraction pattern through the profile to be simulated. Imposing lower limits on H_{root} and s causes actual transpiration to fall below potential as the soil dries.

The success of this approach depends on accurate estimates of root distribution and potential transpiration. The former can be predicted using empirical equations relating root depth and density to time (Feddes et al. 1974; Davidson et al. 1978), but such equations are specific for a particular plant and soil and do not reflect the effects of an abnormal physical or chemical environment. Bresler (1987) and Bresler and Hoffman (1986) suggested that differences in crop salt tolerance to salinity could be explained by variations in depth and distribution of plant roots, but currently no method exists for predicting root distributions and effects of transient changes in the root environment on plant growth and development. The interactions between plant growth, salinity, evapotranspiration, and drainage are often ignored when calculating leaching fractions (du Plessis 1986).

STABILITY OF SOIL STRUCTURE

The hydrological and mechanical properties of a soil are intrinsically related to the geometry and nature of the solid matrix. The solid matrix is usually assumed to remain constant throughout the simulation period so that constant $\theta(h)$ and $K(h)$ relationships can be used, but this is rarely true for real soils. Cultivation, compaction, surface sealing, root development, wetting and drying cycles, swelling and shrinking, faunal activity, and dispersion all can cause changes in structure that influence retentivity, conductivity, and solute flow. Some of these changes are gradual, leaving soil properties mostly unaltered. Others, such as dispersion, are rapid and drastic.

Repulsive forces between clay particles increase as levels of exchangeable sodium increase and salinity decreases. In soils containing smectite or other mixed-layer clays, electrical repulsion leads to swelling and, eventually, as repulsive forces overcome attractive forces, aggregate dispersion. The pore size distribution shifts to smaller diameters, K_s values decrease, and dry strength and crusting become more pronounced.

The transition from a flocculated to a dispersed state can cause almost complete impermeability. Surface crusting and sealing can cause infiltration to decrease drastically. Drop impact can accentuate chemical dispersion at the surface, allowing further dispersion to then occur at much lower concentrations of Na (Agassi et al. 1981). The responses of a soil to sodicity and salinity—and the mechanisms by which these reponses occur—are described in detail in Chapter 5.

The onset of dispersion can be predicted from measured relationships between ESP, SAR, EC, and structural stability, but its effect on soil hydrological parameters is difficult to quantify. Russo (1988a) described a transient Ca and Na transport model in which soil structural changes were calculated from physicochemical principles. This was extended to layered soils (Russo 1988b). Although it was not validated, the model provides a theoretical approach that may prove useful for interpreting and quantifying empirical relationships used for predicting the effects of sodium.

VERIFYING AND VALIDATING MODEL RESULTS

Presently, much controversy and uncertainty exists surrounding the terms "model verification," "model validation," and "model calibration." To address this issue, the American Society for Testing and Materials (ASTM, 1984) has drawn up a standard practice for evaluating environmental chemical fate models. ASTM defines verification as "the examination of the numerical technique in the computer code to ascertain that it truly represents the conceptual model and that there are no inherent numerical problems with obtaining a solution."

Validation is "the comparison of model results with numerical data independently derived from experiments or observations of the environment," the test group maintains. Calibration is related to validation;

according to ASTM, it involves the adjustment of uncertain parameter values to obtain a match between prediction and measurement.

The scarcity of reliable field data often limits calibration and model validation. A user unfamiliar with the algorithms and mathematical intricacies used in a model usually can partially verify and validate a model by obtaining answers to the following questions:

1. Is mass balance preserved without continual correction?
2. Do simulated results for certain well-defined situations, such as steady-state water flow, correspond to those calculated using analytical solutions of the water and solute transport equations?
3. Do the various chemical equilibria appear to be satisfied?
4. Does the model simulate all of the processes considered to be important in the field? (The fact that data required for a certain process are contained in the input file does not guarantee that the process is modeled appropriately.)
5. Do conditions that can cause incorrect results exist? For example, can a model designed for unsaturated flow work properly under saturated conditions?

SUMMARY

Salinity problems may exist over a spatial dimension as small as millimeters, as in a soil column study of the root zone—or as large as kilometers, as in a watershed study. They may occur over a temporal scale that ranges from minutes to years. Numerous kinds of salinity models have been written to help evaluate these problems.

The structure and content of models are continually adapted to reflect our increasing understanding of the soil-plant-atmosphere system. Models can be used to organize, integrate, and evaluate our concepts and improve our understanding of how the soil-plant-atmosphere system functions, but they cannot generate new basic knowledge. Model development must accompany or follow observation, experimentation, and the development of theoretical concepts.

Increasingly fewer restrictions are imposed on modelers, as computers become more accessible, less expensive, and more powerful. Future modeling research and development must focus on: (1) Evaluating the sensitivities of models under a wide variety of conditions; (2) incorporating into models routines that handle data uncertainty and data variability; and (3) clarifying criteria that will allow constructing and validating models at different spatial and time scales.

How useful are salinity models for management purposes? Since any model is a simplification of the real world and there are many possible model structures and levels of sophistication, much of the art in successful modeling lies in the user's ability to select a model suitable for a particular application and to understand its limitations.

REFERENCES

Addiscott, T.M. 1977. "A simple computer model for leaching in structured soils." *J. Soil Sci.* 28: 54-563.

Addiscott, T.M., and R.J. Wagenet. 1985. "Concepts of solute leaching in soils: a review of modeling approaches." *J. Soil Sci.* 85: 411-424.

Agassi, M., I. Shainberg, and J. Morin. 1981. "Effect of electrolyte concentration and soil sodicity on infiltration rate and crust formation." *Soil Sci. Soc. Am. J.* 45: 848-851.

Amrhein, C., J.J. Jurinak, and W.M. Moore. 1985. "Kinetics of calcite dissolution as affected by carbon dioxide partial pressure." *Soil Sci. Soc. Am. J.* 49: 1393-1397.

Amrhein, C. and D.L. Suarez. 1987. "Calcite supersaturation in soils as a result of organic matter mineralization." *Soil Sci. Soc. Am. J.* 51: 932-937.

American Society for Testing and Materials 1984. *Standard Practice for evaluating environmental fate models of chemicals.* New York.

Beven, K., and P. Germann. 1982. "Macropores and water flow in soils." *Water Resour. Res.* 18: 1311-1325.

Bresler, E. 1973. "Simultaneous transport of solutes and water under transient unsaturated flow conditions." *Water Resour. Res.* 9: 975-986.

Bresler, E. 1975. "Two-dimensional transport of solute during nonsteady infiltration from a trickle source." *Soil Sci. Soc. Am. Proc.* 39: 604-613.

Bresler, E. 1987. "Application of a conceptual model to irrigation water requirement and salt tolerance of crops." *Soil Sci. Soc. Am. J* 51: 788-793.

Bresler, E., and G.J. Hoffman. 1986. "Irrigation management for soil salinity control: Theories and tests." *Soil Sci.Soc. Am. J.* 50: 1552-1559.

Bresler, E., B.L. McNeal, and D.L. Carter. 1982. *Saline and Sodic Soils: Principles, Dynamics, Modeling.* Advanced Series in Agricultural Sciences 10: Springer-Verlag, Berlin, W. Germany.

Burns, I.G. 1974. "A model for predicting the redistribution of salts applied to fallow soils after excess rainfall or evaporation." *J. of Soil Sci.* pp. 165-178.

Campbell, G.S. 1974. "A simple method for determining unsaturated conductivity from moisture retention data." *Soil Sci.* 117: 311-314.

Campbell, G.S. 1985. "Soil physics with BASIC-transport models for soil-plant systems." *Developments in Soil Science 14.* Elsevier Science Publishers, Amsterdam, The Netherlands. 150 pp.

Cassel, D.K., L.F. Ratliff, and J.T. Ritchie. 1983. "Models for estimating in-situ potential extractable water using soil physical and chemical properties." *Soil Sci. Soc. Am. J* 47: 764-769.

Childs, S.W., and R.J. Hanks. 1975. "Model of soil salinity effects on crop growth." *Soil Sci. Soc. Am. Proc.* 39: 617-622.

Clapp, R.B., and G.M. Hornberger. 1978. "Empirical equations for some soil hydraulic properties." *Water Resour. Res.* 14: 601-604.

Dahiya, I.S., and I.P. Abrol. 1973. "Dynamics of calcium-sodium exchange under unsaturated flow conditions." *Comm. in Soil Science and Plant Analysis* 4: 443-453.

Davidson, J. M., D.A. Graetz, P.S.C. Rao, and H.M. Selim. 1978. "Simulations of nitrogen movement, transformation, and uptake in the plant root zone." EPA-600/3-78-029, Washington, D.C.

Du Plessis, H.M. 1986. "On the concept of leaching requirement for salinity control." *S. Afr. J. Plant and Soil* 4: 181-184.

Dutt, G.R., M.J. Shaffer, and W.J. Moore. 1972. "Computer simulation model of dynamic bio-physicochemical processes in soils." Univ. of Ariz. Agric. Expt. Stn. Tech. Bull. 196.

English, M., A. Taylor and P. John. 1986. "Evaluating sprinkler system performance." *New Zealand Agric. Sci.* 20: 32-38.

Evangelou, V.P., and F.J. Coale. 1987. "Dependence of the Gapon coefficient on exchangeable sodium for mineralogically different soils." *Soil Sci. Soc. Am. J* 51: 68-71.

Feddes, R.A., E. Bresler, and S.P. Neumann. 1974. "Field test of a modified numerical model for water uptake by root systems." *Water Resour. Res.* 10: 1199-1206.

Forster, R. 1986. "A multicomponent transport model." *Geoderma* 38: 261-278.

Germann, P., and K. Beven. 1985. "Kinematic wave approximation to infiltration into soils with sorbing macropores." *Water Resour. Res.* 21: 990-996.

Glas, T. K., A. Klute, and D. B. McWhorter. 1979. "Dissolution and transport of gypsum in soils. II. Experimental." *Soil Sci. Soc. Am. J* 42: 268-273.

Griffin, R.A., and J.J. Jurinak. 1973. "Estimation of activity coefficients from the electrical conductivity of natural aquatic systems and soil extracts." *Soil Sci.* 116: 26-30.

Grove, D.B., and K.G. Stollenwerk. 1987. "Chemical reactions simulated by ground-water-quality models." *Water Resources Bull.* 23: 601-615.

Gupta, S.C., and W.E. Larson. 1979. "Estimating soil water retention characteristics from particle size distribution, organic matter percent and bulk density." *Water Resour. Res.* 15: 1633-1635.

Hall, D.G.M., M.J. Reeve, A.J. Thomasson, and V.F. Wright. 1977. "Water retention, porosity and density of field soils." Tech. Monograph No. 9, Soil Survey of England and Wales.

Hanks, R.J. 1974. "Model for predicting plant yield as influenced by water use." *Agron. J.* 66: 660-664.

Hanks, R.J., and S.A. Bowers. 1962. "Numerical solution of the moisture flow equation for infiltration into layered soils." *Soil Sci. Soc. Am. Proc.* 26: 530-535.

Haverkamp, R., Vauclin, M., Touma, J., Wierenga, P.J., and G. Vachaud. 1977. "A comparison of numerical simulation models for one-dimensional infiltration." *Soil Sci. Soc. Am. J* 41: 285-294

Hutson, J.L. 1986. "Water availability of some South African soils in relation to particle size criteria and bulk density." *S. Afr. J. Plant Soil* 3: 151-155.

Hutson, J.L., and A. Cass. 1987. "A retentivity function for use in soil-water simulation models." *J. Soil Sci.* 38: 105-113.

Inskeep, W.P., and P.R. Bloom. 1986. "Kinetics of calcite precipitation in the presence of water-soluble organic ligands." *Soil Sci. Soc. Am. J* 50: 1167-1172.

Jardine, P.M., and D.L. Sparks. 1984. "Potassium/calcium exchange in a multireactive soil system: 1. Kinetics." *Soil Sci. Soc. Am. J* 48: 39-45.

Jurinak, J.J., S.S. Sandhu, and L.M. Dudley. 1987. "Ionic diffusion coefficients as predicted by conductometric techniques." *Soil Sci. Soc. Am. J* 51: 625-629.

Jurinak, J.J., J.C. Whitmore and R.J. Wagenet. 1977. "Kinetics of salt release from a saline soil." *Soil Sci. Soc. Am. J* 41: 721-724.

Jury, W.A., W.F. Spencer, and W.J. Farmer. 1983. "Behavior Assessment model for trace organics in soil: I. Model development." *J. Environ. Qual.* 12: 558-564.

Kapoor, A.K. and R. Pal. 1986. "Predicting salinization and sodification of a bare sandy loam soil after irrigation with poor-quality water interspersed with rain." *Soil Sci.* 141: 281-288.

Klute, A. (ed). 1986. "Methods of Soil Analysis, Part 1: Physical and Mineralogical Methods." Agronomy Monograph No. 9, 2nd Ed. American Society of Agronomy, Inc., Madison, Wisc.

Melamed, D., R.J. Hanks, and L.S. Willardson. 1977. "Models of salt flow in soil with a source-sink term." *Soil Sci. Soc. Am. J* 41: 29-33.

Nielsen, D.R., M.Th. van Genuchten, J.W.Biggar. 1986. "Water flow and solute transport processes in the unsaturated zone." *Water Resour. Res.* 22: 895S-108S

Nimah, M.N., and R.J. Hanks. 1973. "Model for estimating soil water, plant and atmospheric interralations 1. Description and sensitivity." *Soil Sci. Soc. Am. Proc.* 37: 522-527.

Ogwada, R.A., and D.L. Sparks. 1986. "Kinetics of ion-exchange on clay minerals and soil: II. Elucidation of rate-limiting steps. " *Soil Sci. Soc. Am. J* 50: 1162-1166.

Oster, J.D., and B.L. McNeal. 1971. "Comutation of soil solution composition variation with water content for desaturated soils." *Soil Sci. Soc. Am. Proc.* 35: 436-442.

Oster, J.D., and J.D. Rhoades. 1975. "Calculated drainage water compositions and salt burdens resulting from irrigation with river waters in the western United States." *J. Environ. Qual.* 4: 73-78.

Ratliff, L.F., J.T. Ritchie, and D.K. Cassel. 1983. "Field-measured limits of water availability as related to laboratory-measured properties." *Soil Sci. Soc. Am. J* 47: 770-775.

Rawls, W.J. and D.L. Brakensiek. 1985. "Agricultural management effects on soil water retention." DeCoursey, D.G., ed. *Proceedings of the 1983 Natural Resources Modeling Symposium.* U.S. Dept of Agriculture, Agricultural Research Service, ARS-30, pp. 115-117.

Rivers, E.D., and R.F. Shipp. 1978. "Soil water retention as related to particle size in selected sands and loamy sands." *Soil Sci.* 126: 94-100.

Robbins, C.W. 1985. "The $CaCO_3$-CO_2-H_2O system in soils." *J. Agron. Educ.* 14: 3 -7.

Robbins, C.W., J.J. Jurinak, and R.J. Wagenet. 1980a. "Calculating ion-exchange in a salt transport model." *Soil Sci. Soc. Am. J* 44: 1195-1192.

Robbins, C.W., R.J. Wagenet, and J.J. Jurinak. 1980b. "A combined salt transport-chemical equilibrium model for calcareous and gypsiferous soils." *Soil Sci. Soc. Am. J* 44: 1191-1194.

Russo, D. 1988a. "Numerical analysis of the non-steady transport of interacting solutes through unsaturated soil 1. Homogeneous systems." *Water Resour. Res.* 24: 271-284.

Russo, D. 1988b. "Numerical analysis of the non-steady transport of interacting solutes through unsaturated soil 2. Layered systems." *Water Resour. Res.* 24: 285-290.

Salter, P.J., G. Berry, and J.B. Williams. 1966. "The influence of texture on the moisture characteristics of soils, III: Quantitative relationships between particle size, composition and available water capacity." *J. Soil Sci.* 17: 93-98.

Salter, P.J., and J.B. Williams. 1965. "The influence of texture on the moisture characteristics of soils. II. Available-water capacity and moisture release characteristics." *J. Soil Sci.* 16: 310-317.

Shaykewich, C.F., and M.A. Zwarich. 1968. "Relationships between soil physical constants and soil physical components of some Manitoba soils." *Can. J. Soil Sci.* 48: 199-204.

Sparks, D.L. 1987. "Kinetics of soil chemical processes: past progress and future needs." Future developments in Soil Science Research. Soil Science Society of America, Madison, Wisc. pp. 61-73.

Sposito, G. 1984. "The future of an illusion: ion activities in soil solutions." *Soil Sci. Soc. Am. J* 48: 531-536.

Sposito, G., and S. V. Mattigod. 1979. "A computer program for the calculation of chemical equilibria in soil solutions and other natural waters." Kearney Foundation of Soil Science, Univ. of Calif., Riverside.

Suarez, D.L., and J.D. Rhoades. 1982. "The apparent solubility of calcium carbonate in soils." *Soil Sci. Soc. Am. Proc.* 46: 716-722.

Tanji, K.K., and S.J. Deverel. 1984. "Simulation modeling for reclamation of sodic soils." Shainberg, I. and J. Shalhevet, eds. *Soil Salinity Under Irrigation.* Ecological Studies, Vol. 51. Springer-Verlag, Berlin, W. Germany.

Tanji, K.D., L.D. Doneen, G.V. Ferry, and R.S. Ayers. 1972. "Computer simulation analysis on reclamation of salt-affected soils in San Joaquin Valley, California." *Soil Sci. Soc. Am. Proc.* 36: 127 -133.

Tillotson, W.R., C.W. Robbins, R.J. Wagenet and R.J. Hanks. 1980. "Soil water, solute and plant growth simulation." Bull. Utah Agr. Exp. Stn., 502. 53pp.

Van Genuchten, M.Th. 1980. "A closed-form equation for predicting the hydraulic conductivity of unsaturated soils." *Soil Sci. Soc. Am. J* 44: 892-898.

Van Genuchten, M.Th., and F. N. Dalton. 1986. "Models for simulating salt movement in aggregated soils." *Geoderma* 38: 165-183.

Van Stiphout, T.P.J., H.A.J. van Lanen, O.H. Boersma and J. Bouma. 1987. "The effect of bypass flow and internal catchment of rain on the water regime in a clay loam grassland soil." *J. Hydrol.* 95: 1-11.

Wagenet, R.J. 1983. "Principles of salt movement in soils." Chemical Movement and Reactivity in Soil Systems. American Society of Agronomy, Madison, Wisc.

Wagenet, R.J., and J.L. Hutson. 1987. LEACHM —Leaching Estimation and Chemistry Model. Continuum 2, Water Resources Institute, Cornell Univ., Ithaca, N.Y.

Walker, W.R. 1978. "Identification and initial evaluation of irrigation return flow models." Report EPA-600/2-78-144, Office of Research and Development, U.S. Environmental Protection Agency, Ada, Oklahoma 74820, 123 pp.

NOTATION

a, b = constants; An^{n-} = anion of negative charge n; c = solute concentration; $\partial c/\partial z$ = vertical concentration gradient; Cat^{m+} = cation of positive charge m; C_w = differential water capacity; D_m = flux density-dependent hydrodynamic dispersion coefficient; D_o = ionic diffusion coefficient in water; $D\theta$ = molecular diffusion coefficient in unsaturated soil; h = matric potential; H = hydraulic head, or sum of pressure and gravitational components of soil water potential; H_{root} = effective water potential in root at soil surface, or crown potential; i = soil layer; J_d = diffusion flux; J_q = convection flux; J_t = net solute flux; k = hydraulic conductivity; K = hydraulic conductivity; K_{a1} = first dissociation constant of H_2CO_3; K_{a2} = second dissociation constant of H_2CO_3; K_d^1 = conditional stability constants; K_H = Henry's Law constant; p = bulk density; q = water flux density; Q = desired potential transpiration flux; R_{res} = root resistance; RF = fraction of active roots; s = bulk soil osmotic potential; S = sorbed chemical; T = total of any chemical species; T = actual transpiration; T_p = potential or maximum transpiration; T_r = relative transpiration; u = adsorption of water by plants; X = exchangeable cation; Y = actual yield; Y_p = corresponding potential or maximum yield; Y_r = relative yield; z = depth; θ = water content; θ_s = saturated water content; λ = dispersivity; Δt = time step; Δx = distance from plant root; and Δz = thickness of soil layer.

CHAPTER 24

CONCEPTUAL IRRIGATION RETURN FLOW HYDROSALINITY MODEL[a]

INTRODUCTION

Agricultural irrigation return flows (IRF) are receiving much attention in the United States and elsewhere due to increasing water-quality regulations and the public's mounting concern about water conservation. Of particular interest are the concentration and mass of salt in IRF. In regions where salinity from IRF degrades surface water and ground water, programs to control and manage salinity are increasingly needed.

One of the most effective ways to reduce the salt received by surface and ground-water systems is to limit subsurface drainage from irrigated lands. This can be achieved by improving irrigation application systems so that water is more evenly applied, scheduling irrigation on the basis of evapotranspiration (ET) estimates to maintain low leaching fractions, managing shallow ground water so that water extracted from the water table by roots is included if its quality is suitable, and redesigning under-drainage systems so that they capture more root-zone percolation. Salinity control and management also require a knowledge of the sources and sinks of salt and the principal flow paths of salt. A conceptual hydrosalinity model allows the concentration and mass of salt in IRF to be more fully ascertained and the effectiveness of various salt control alternatives to be evaluated.

This chapter summarizes hydrosalinity modeling efforts over the last two decades, describes a conceptual model developed by Tanji (1977) and modified by Aragües et al. (1985) and presents results of the calibration, verification, and application of this model.

HYDROSALINITY MODEL PHILOSOPHY AND THEORY

Hydrosalinity models range from a simple, mass balance approach to much more complicated routines, which consider the individual processes that may contribute to the salinity of IRF.

Fig. 24.1 shows the IRF system. Water is diverted from a river for

---

[a]Prepared by: R. Aragües, Diputacion General de Aragon, Departamento de Agricultura, Ganaderia Montes, Servicio de Investigacion Agraria, Apartado 727, 50080 Zaragoza, Spain; K. K. Tanji and D. Quilez, Department of Land, Air, and Water Resourc., Univ. of Calif., Davis, CA 95616; and J. Faci, Diputacion General de Aragon, Departmento de Agricultura, Ganaderia Montes, Servicio de Investigacion Agraria, Apartado 727, 50080 Zaragoza, Spain.

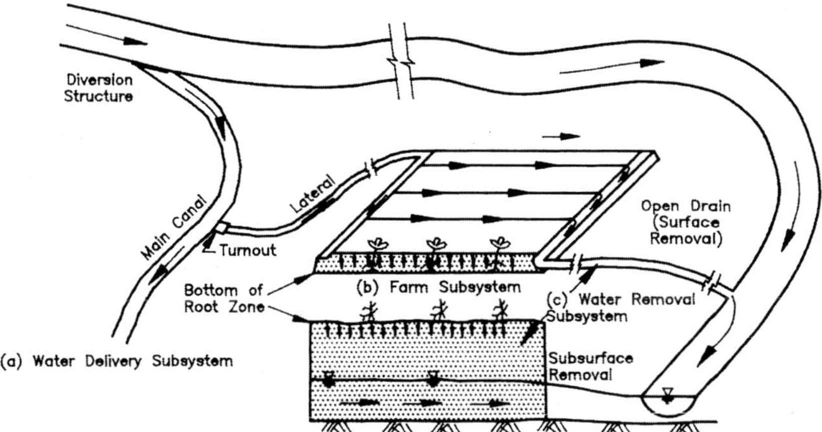

Fig. 24.1 Idealized Drawing of Irrigation Return Flow System (Skogerboe and Walker 1981)

irrigation, applied to agricultural fields, and returned as drainage water to the river. IRF includes the water delivery subsystem's canal seepage and operational spills and the farm subsystem's surface and subsurface drainage.

Hornsby (1973) classified IRF into two categories: surface return flows and subsurface return flows. Surface return flows, if irrigation is properly managed, retain approximately the same salinity level as the applied irrigation water. In contrast, subsurface return flows, which percolate through the root zone, typically contain levels of salinity that have increased due to evapotranspiration and soil chemical weathering processes. The latter include cation exchange reactions, mineral precipitation-dissolution reactions, dissolution and leaching of fertilizers and soil amendments, ion association, temperature changes that affect the reaction rates, oxidation-reduction reactions, and interaction with ground water.

Chapter 23 analyzed the dynamics of salt movement (accumulation and leaching) in the soil profile. The steady state hydrosalinity models described in this chapter are based on the principles of continuity of mass, momentum, and energy. They typically model salinity as a reactive parameter through chemical equilibrium, which considers only the difference between initial and final states and does not consider time as variable. Most chemical reactions occur rapidly enough that attaining chemical equilibrium is relatively easy, especially in systems with small water-flow rates. Chemical kinetic equations are used as necessary, e.g., to describe the precipitation of lime and dissolution of gypsum.

A general salt balance in an irrigated area is

$$C_iV_i - C_dV_d - M_{hc} + (M_{sp} - M_{sd}) + M_{is} + M_{pre} + M_{afm} + M_{pr} + M_{iwr} - M_{dp} = 0 \tag{24.1}$$

where V_i and V_d are the volumes of the irrigation and drainage water, and

C_i and C_d their concentrations. M_{hc} = salt removal with the harvest of crops; ($M_{sp} - M_{sd}$) = precipitation-dissolution of salts; M_{is} = initial amount of salts present in the soil solution dissolved from natural mineral deposits; M_{pre} = salt mass in precipitation or snow; M_{afm} = salts contributed by soil amendments and fertilizers; M_{pr} and M_{iwr} = salts picked up by irrigation or precipitation runoff; and M_{dp} = salt mass leaching to ground water (deep percolation). Ionic exchange is not considered explicitly. It produces a change in the ionic composition rather than a change in concentration. When natural mineral deposits are present or chemical amendments are applied, the most important sinks and sources of salt generally are M_{dp}, M_{is}, and $M_{sp} - M_{sd}$. Scofield (1940) presented one of the first salt balance equations for irrigated lands. His equation considered the volumes and concentrations of irrigation and drainage waters and the amount of salts dissolved from natural geologic minerals, removed by the crops, and precipitated from irrigation water.

Salt balance equations have since been extensively used. They have been applied not only to total dissolved solids (TDS), but to other parameters. Woods and Orlob (1963) applied mass balance techniques to study pesticides in Oregon's Klamath River. Woods (1967) applied mass balance to study EC, hardness, bicarbonate, sodium, phosphorus, and boron in the Sacramento River basin. The California Department of Water Resources (1969) studied salinity in the lower San Joaquin River basin. Water Resources Engineers, Inc. (1969) analyzed salinity (TDS) as a conservative parameter, including interaction with ground water, in the upper Santa Ana River Basin. Walker (1970) developed a hydrosalinity model of the Grand Valley that considered cation exchange, nitrogen transformations, redistribution of solutes in the soil, and absorption of nitrogen by plants. Hyatt et al. (1970) applied a model also based on mass balance to study salinity (TDS) in the upper Colorado River basin. Thomas et al. (1971) and Hill et al. (1973) developed a hydrosalinity model of the Bear River in Utah that considered cation exchange and dissolution-precipitation of gypsum and lime. Fapohunda et al. (1981) incorporated a hydrosalinity model with an economic model to study the economic effects of salinity in the Hadejia River of Nigeria.

Other models have been developed or applied by Dutt et al. (1972), the U.S.Bureau of Reclamation (1975), Ribbens and Shaffer (1976), Skogerboe et al. (1979), Moolman et al. (1983), Bay-Valley Consultants (1974), Tanji (1977), Labadie and Khan (1979 a and b), and Khan (1982).

Hornsby (1973) reviewed individual process- and system-level IRF models, especially salinity control models. He concluded that simulation modeling will play an increasingly important role in water resources management, especially for examining alternative approaches to irrigation management and salinity control. Walker (1978) evaluated system-, subsystem-, and process-level IRF models. His review of over 40 models includes the scope, input requirements, spatial and time scales, structure of the code, and mathematical or analytical approach for each model.

Tanji (1982) evaluated hydrologic-salinity modeling at the river sub-basin and basin levels. He reviewed 10 representative large-scale hydrosalinity models applied in the western United States and concluded that much more work is needed in the field validation of models to improve their usefulness for management and policy-making decisions. He pointed out that, if modeling objectives involve scientific questions and refined simulations, the requirements of the data base are very demanding; therefore, one of the first considerations that must be addressed is the availability of hydrologic data.

Many models, formulated for and applied to site-specific problems, cannot be applied to more generalized situations. To apply a model from one situation to another, the mathematical model, the computer model, or both, may need to be modified.

The testing of a computer model should incorporate sensitivity analysis, which determines the relative importance of the various input parameters on the model's output; calibration of the model, which involves fine-tuning the model's coefficients and parameters so that the computed data fit more closely to the observed data; and verification, which involves testing the calibrated model with observed data other than the set previously used for calibration. Without a complete verification for the area to which the model is to be applied, the model may give unsatisfactory results.

DESCRIPTION OF CONCEPTUAL HYDROSALINITY MODEL

The simple Conceptual Irrigation Return Flow (CIRF) hydrosalinity model by Tanji (1977) focuses on the crop's root zone and considers only the main flow pathways of water and salts. The model assumes that the masses of water and salt (except gypsum) are conservative and that steady state conditions can model long-term transient conditions approximately. All of the model's spatial and temporal variables and parameters are taken as average values. Therefore, the model can be applied to large systems, including irrigation districts, and for long periods of time, such as an irrigation season, a hydrologic year, or a series of consecutive years. The CIRF model consists of a hydrologic submodel coupled to a salinity submodel.

Hydrologic Submodel

Fig. 24.2 shows an idealized representation (freebody diagram) of the hydrology and flow pathways in a typical irrigation project, with particular focus on the root zone. With the continuity equation, the rate of change of root-zone water storage can be described

$$\frac{dQ_s}{dt} = \sum^{n} Q_i - \sum^{m} Q_o \tag{24.2}$$

where Q = the quantity of water in m^3; and t = time (hydrological year, irrigation, or non-irrigation season); subscripts s, i, and o, respectively, denote storage, inputs, and outputs from the crop's root zone; and n and m, respectively, are the number of inputs and outputs.

The hydrologic inputs are defined by

$$\sum^{n} Q_i = Q_{diw} + Q_p + Q_{rim} \tag{24.3}$$

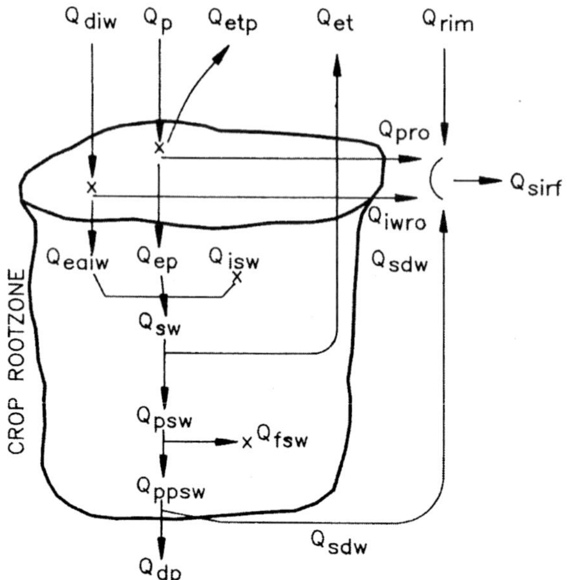

Fig. 24.2 Freebody Diagram of Hydrologic Submodel, Focusing on Crop Root Zone (The symbol Q refers to quantity of water in m^2.)

where the subscripts diw, p, and rim, are diverted irrigation water, precipitation, and rim inflows from adjacent systems, respectively.

The hydrologic outputs are defined by:

$$\sum^{m} Q_o = Q_{et} + Q_{etp} + Q_{dp} + Q_{pro} + Q_{iwro} + Q_{sdw} + Q_{rim} \quad (24.4)$$

where Q_{et}, Q_{etp}, Q_{dp}, Q_{pro}, Q_{iwro}, and Q_{sdw} are evapotranspiration of soil water to the atmosphere, evaporation of precipitation, deep percolation from root zone to ground-water system, precipitation runoff, irrigation runoff, and collected subsurface water, including tile drainage effluent and seepage into natural and man-made channels, respectively. If tile drainage effluents and surface runoff are collected in a single drain, the last four terms in the equation constitute the surface irrigation return flow system.

The net change in water storage in the root zone is also given by

$$\frac{dQ_s}{dt} = Q_{fsw} - Q_{isw} \quad (24.5)$$

where isw and fsw denote initial and final stored soil water, respectively.

Table 24.1 presents the detailed mass balance formulation of this hydrological submodel. Table 24.2 includes definitions of symbols. From a management standpoint, several variables linked to the hydrologic subsystem may be defined:

TABLE 24.1 Mathematical Equations Describing Hydrologic Submodel

Hydrological inputs and outputs

(1) $Q_i = Q_{diw} + Q_p + Q_{rim}$

(2) $Q_o = Q_{iwro} + Q_{pro} + Q_{et} + Q_{etp} + Q_{dp} + Q_{sdw} + Q_{rim}$

Diverted irrigation water

(3) $Q_{diw} = Q_{eaiw} + Q_{iwro}$

(4) $Q_{eaiw} = Q_{diw} \cdot E_{iae}$

(5) $Q_{iwro} = Q_{diw} \cdot (1 - E_{iae})$

(6) $Q_{sweaiw} = Q_{eaiw} / ETCF$

Precipitation

(7) $Q_p = Q_{pro} + Q_{etp} + Q_{ep}$

(8) $Q_{pro} = a \cdot Q_p$

(9) $Q_{etp} = b \cdot Q_p$

(10) $Q_{ep} = c \cdot Q_p$

(11) $a + b + c = 1$

(12) $Q_{swep} = Q_{ep} / ETCF$

Initial stored soil water

(13) $Q_{swisw} = Q_{isw} / ETCF$

Water inputs to root zone

(14) $Q_{sw} = Q_{eaiw} + Q_{ep} + Q_{isw}$

(15) $Q_{psw} = Q_{sw} - Q_{et} = Q_{sweaiw} + Q_{swep} + Q_{swisw}$

(16) $ETCF = Q_{sw} / Q_{psw}$

(17) $Q_{ppsw} = Q_{psw} - Q_{fsw}$

Water outputs from root zone

(18) $Q_{sw} = Q_{et} + Q_{fsw} + Q_{dp} + Q_{sdw}$

(19) $Q_{et} = e \cdot D_{pet}$

(20) $Q_{dp} = f \cdot Q_{ppsw}$

(21) $Q_{sdw} = (1 - f) \cdot Q_{ppsw}$

Change in water storage

(22) $Q_{fc} = D_{whc} \cdot D_r \cdot A$

(23) $Q_{isw} = D_i \cdot Q_{fc}$

(24) $Q_{fsw} = D_f \cdot Q_{fc}$

(25) $Q_{csw} = (D_f - D_i)\, Q_{fc}$

Surface irrigation return flows

(26) $Q_{sirf} = Q_{iwro} + Q_{pro} + Q_{sdw} + Q_{rim}$

Irrigation efficiency, water application efficiency, leaching fraction

(27) $EIE = (Q_{eaiw} + Q_{ep} + Q_{isw} - Q_{fsw}) / (Q_{diw} + Q_{ep} + Q_{isw})$

(28) $WAE = Q_{et} / (Q_{eaiw} + Q_{ep})$

(29) $LF = Q_{ppsw} / (Q_{eaiw} + Q_{ep})$

TABLE 24.2 Symbols used in the hydrological submodel

Symbol (1)	Definition (2)
A	Surface (ha)
a	Precipitation runoff coefficient
b	Precipitation evaporation coefficient
c	Effective precipitation coefficient
D_f	Final soil water as a fraction of Q_{fc}
D_i	Initial soil water as a fraction of Q_{fc}
D_{pet}	Potential evapotranspiration (ha-m)
D_r	Average rooting depth (m)
D_{whc}	Unit water holding capacity (m/m)
Eiae	Irrigation application efficiency
e	Evaporation coefficient or crop coefficient
EIE	Irrigation efficiency
ETCF	Evapotranspiration concentration factor
f	Deep percolation coefficient
LF	Leaching fraction
Q_{csw}	Change in soil water storage (ha-m)
Q_{diw}	Diverted irrigation water (ha-m)
Q_{dp}	Deep percolation (ha-m)
Q_{eaiw}	Effective applied irrigation water (ha-m)
Q_{ep}	Effective precipitation (ha-m)
Q_{et}	Evapotranspiration (ha-m)
Q_{etp}	Evaporation of precipitation (ha-m)
Q_{fc}	Field capacity (ha-m)
Q_{fsw}	Final stored soil water (ha-m)
Q_i	Hydrologic inputs (ha-m)
Q_{isw}	Initial stored soil water (ha-m)
Q_{iwro}	Irrigation water runoff (ha-m)
Q_o	Hydrologic outputs (ha-m)
Q_p	Precipitation (ha-m)
Q_{ppsw}	Water available for subsurface drainage and deep percolation (ha-m)
Q_{pro}	Precipitation runoff (ha-m)
Q_{psw}	Soil water after ET (ha-m)
Q_{rim}	Rim inflow/outflow (ha-m)
Q_{sdw}	Subsurface drainage water (ha-m)
Q_{sirf}	Surface irrigation return flow (ha-m)
Q_{sw}	Soil-water before ET (ha-m)
Q_{sweaiw}	Effective applied irrigation water after ET (ha-m)
Q_{swep}	Effective precipitation after ET (ha-m)
Q_{swisw}	Initial soil water after ET (ha-m)
WAE	Water application efficiency

• Evapotranspiration concentration factor (ETCF), which gives the proportion of soil water consumed in the ET process:

$$\text{ECTF} = \frac{Q_{sw}}{Q_{psw}} \tag{24.6}$$

• Leaching fraction (LF), the portion of the infiltrated water that percolates through the root zone:

$$\text{LF} = \frac{Q_{ppsw}}{(Q_{eaiw} + Q_{ep})} \tag{24.7}$$

• Water application efficiency (WAE), the amount of effective applied irrigation water, irrigation plus precipitation, that undergoes evapotranspiration:

$$\text{WAE} = \frac{Q_{et}}{(Q_{eaiw} + Q_{ep})} \tag{24.8}$$

• Irrigation application efficiency (E_{iae}), the proportion of applied irrigation water that effectively infiltrates the soil:

$$E_{iae} = \frac{Q_{eaiw}}{Q_{diw}} \tag{24.9}$$

Salinity Submodel

Fig. 24.3 is a freebody diagram that describes salinity and its flow pathways. The symbol C represents total dissolved solids (TDS, mg/l), and M the mass of salts (kg) obtained from the product of salt concentration (TDS) and water volume.

Based on conservation of salt mass, salinity is described with

$$\frac{dM_s}{dt} = \sum^{1} M_i - \sum^{k} M_o \tag{24.10}$$

where dM_s/dt = the rate of change in mass of salts stored in the crop's root zone; and l and k are the number of inputs and outputs, respectively.

The salt inputs are defined by

$$\sum^{1} M_i = M_{diw} + M_p + M_{rim} \tag{24.11}$$

where all subscripts have been defined in the hydrologic submodel section.

Salt outputs are defined by

$$\sum^{k} M_o = M_{dp} + M_{pro} + M_{iwro} + M_{sdw} + M_{rim} \tag{24.12}$$

where all subscripts have been previously defined.

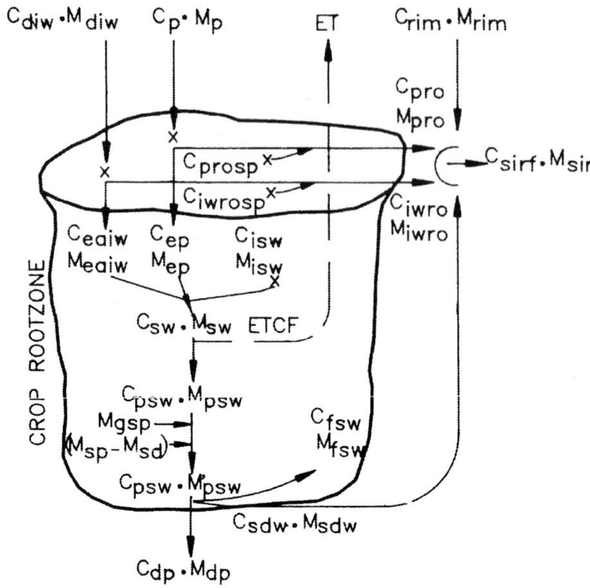

Fig. 24.3 Freebody Diagram of Salinity Submodel, Focusing on Crop Root Zone. (The symbols C and M denote dissolved solids (in mg/l) and mass of salts (in kg), respectively.)

The rate of change in mass of salt stored in the crop's root zone, including sources and sinks, is also defined by

$$\frac{dM_s}{dt} = M_{fsw} - M_{isw} + M_{sd} - M_{sp} - M_{gsp} \tag{24.13}$$

where M_{sd} denotes calcium carbonate ($CaCO_3$) and gypsum ($CaSO_4 \cdot 2H_2O$) precipitation as the soil solution is concentrated by ET, and M_{sp} and M_{gsp}, respectively, denote lime and gypsum dissolution from the soil. The terms M_{isw} and M_{fsw}, respectively, represent the initial and final concentrations of soluble salts other than gypsum and calcium carbonate that are present in the soil solution.

It is assumed that waters flowing over the soil's surface may pick up salts, as is reflected by adding C_{iwrosp} and C_{prosp} to C_{diw} and C_p, respectively, in calculating M_{iwro} and M_{pro}. Eqs. 52 and 53 from Table 24.3 may be used to calculate the TDS of initial stored soil water, corrected by Gypsum C_{isw}, from EC_e (electrical conductivity of soil saturation extract). The volume-weighted average of effective applied irrigation water (C_{eaiw}), effective precipitation (C_{ep}), and C_{isw}, gives the TDS of soil water, C_{sw}, which is then concentrated by an evapotranspiration factor (ETCF) to account for evapotranspiration losses.

The term $M_{sp} - M_{sd}$ is computed with a soil chemistry program (Oster and Dell'Osso 1975) that calculates the extent to which applied irrigation water, as it becomes concentrated in the soil solution due to

TABLE 24.3 Mathematical Equations Describing Salinity Submodel

Salt inputs and outputs and change in storage

(30) $M_i = M_{diw} + M_p + M_{rim}$

(31) $M_o = M_{iwro} + M_{pro} + M_{dp} + M_{sdw} + M_{rim}$

(32) $M_s = M_{fsw} - M_{isw} + M_{sd} - M_{sp} - M_{gsp}$

Lateral contributions

(33) $M_{rim} = C_{rim} \cdot Q_{rim} \cdot SMCF$

Diverted irrigation water

(34) $M_{diw} = C_{diw} \cdot Q_{diw} \cdot SMCF$

(35) $C_{iwro} = C_{diw} + C_{iwrosp}$

(36) $M_{iwro} = C_{iwro} \cdot Q_{iwro} \cdot SMCF$

(37) $C_{eaiw} = C_{diw}$

(38) $M_{eaiw} = C_{eaiw} \cdot Q_{eaiw} \cdot SMCF$

(39) $C_{sweaiw} = C_{diw} \cdot ETCF$

(40) $M_{sweaiw} = M_{eaiw}$

Precipitation

(41) $M_p = C_p \cdot Q_p \cdot SMCF$

(42) $C_{pro} = C_p + C_{prosp}$

(43) $M_{pro} = C_{pro} \cdot Q_{pro} \cdot SMCF$

(44) $C_{ep} = C_p$

(45) $M_{ep} = C_{ep} \cdot Q_{ep} \cdot SMCF$

(46) $C_{swep} = C_{ep} \cdot ETCF$

(47) $M_{swep} = M_{ep}$

Initial soil water corrected by gypsum solubility

(48) $M_{gyp} = 2630 \cdot (Q_{sw} - Q_{et}) \cdot SMCF$

(49) $Gypsum = A \cdot D_r \cdot R_b \cdot (\%G)$

(50) If $M_{gyp} \leq Gypsum$: $EC_{gyp} = EC_{gype} = 2.2$, $TDS_{gyp} = 2630$ (Saturation)

(51) If $M_{gyp} > Gypsum$: $TDS_{gyp} = (Gypsum/Q_{psw}) / SMCF$
$EC_{gyp} = 0.003 \cdot (TDS_{gyp})^{0.849}$
$EC_{gype} = EC_{gyp} \cdot Q_{psw}/(SP \cdot D_r \cdot A \cdot R_b)$ or
$EC_{gype} = EC_{gyp} \cdot Q_{psw}/(2 \cdot Q_{fc})$

(52) $C_{se} = (EC_e - EC_{gype}) \cdot 640$

(53) $C_{isw} = C_{se} \cdot (SP \cdot D_r \cdot A \cdot R_b / Q_{isw})$ or $C_{isw} = C_{se} \cdot (2 \cdot Q_{fc} / Q_{isw})$

(54) $M_{isw} = C_{isw} \cdot Q_{isw} \cdot SMCF$

(55) $C_{swisw} = C_{isw} \cdot ETCF$

(56) $M_{swisw} = M_{isw}$

Gypsum, salt pickup-salt deposition

(57) If $Gypsum \leq M_{gyp}$: $M_{gsp} = Gypsum$

(58) If $Gypsum > M_{gyp}$: $M_{gsp} = M_{gyp}$

(59) $C_{sp} - C_{sd} = -102.04 + 122.87 \cdot LF$

(60) $M_{sp} - M_{sd} = (C_{sp} - C_{sd}) \cdot Q_{eaiw} \cdot SMCF$

(continued)

TABLE 24.3 Mathematical Equations Describing Salinity Submodel (continued)

Salt input to root zone

(61) $M_{sw} = M_{eaiw} + M_{ep} + M_{isw}$

(62) $C_{sw} = (M_{sw} / Q_{sw}) / SMCF$

(63) $M_{psw} = M_{sweaiw} + M_{swep} + M_{swisw} = M_{sw}$

(64) $C_{psw} = C_{sw} \cdot ETCF = (M_{psw} / Q_{psw}) / SMCF$

(65) $M'_{psw} = M_{psw} + M_{gsp} + (M_{sp} - M_{sd})$

(66) $C'_{psw} = (M'_{psw} / Q_{psw})/SMCF$

Salt output from root zone

(67) $M'_{psw} = M_{fsw} + M_{dp} + M_{sdw}$

(68) $C_{fsw} = C'_{psw}$

(69) $M_{fsw} = C_{fsw} \cdot Q_{fsw} \cdot SMCF$

(70) $C_{dp} = C'_{psw}$

(71) $M_{dp} = C_{dp} \cdot Q_{dp} \cdot SMCF$

(72) $C_{sdw} = C'_{psw}$

(73) $M_{sdw} = C_{sdw} \cdot Q_{sdw} \cdot SMCF$

Salt load in irrigation return Flows

(74) $M_{sirf} = M_{rim} + M_{iwro} + M_{pro} + M_{sdw}$

(75) $C_{sirf} = (M_{sirf} / Q_{sirf})/ SMCF$

Note: Equation (59) is valid only for the Violada Irrigation District.

evapotranspiration, effectively dissolves $CaCO_3$ from the soil or precipitates out $CaSO_4 \cdot 2H_2O$ and $CaCO_3$. This phenomenon is known as salt pickup-salt deposition, or mineral dissolution-mineral precipitation. Required input data include the chemical composition of the water (Ca^{2+}, Mg^{2+}, Na^+, HCO_3^-, Cl^-, SO_4^{2-}), the leaching fraction (LF), and the partial pressure of carbon dioxide (P_{CO_2}), which significantly influences the solubility of $CaCO_3$.

Because Oster's model does not consider the possibility that gypsum may be initially present in the soil as a source of salinity, salt pickup due to gypsum dissolution (M_{gsp}) must be incorporated separately. It is assumed, for this purpose, that the solubility of gypsum is 2.63 kg/m^3 (or 30.7 meq/l as in distilled water) and that gypsum will dissolve until saturation or until its supply is exhausted. The salt concentration of the initially-stored soil water has to be corrected for the solubility of gypsum (Table 24.3, Eq. 52). The TDS due to gypsum cannot be significantly concentrated during evapotranspiration because of gypsum's limited solubility.

The concentration and mass of soil water after evapotranspiration losses (C_{psw} and M_{psw}), adjusted by gypsum dissolution (M_{gsp}) and salt pickup minus salt deposition (M_{sp}-M_{sd}), gives C'_{psw} and M'_{psw}, the final salt concentration and mass in the soil solution. At this point, the soil solution subdivides into three components of equal salt concentration C'_{psw}: final stored soil water (fsw), deep percolation (dp), and collected subsurface drainage (sdw). The concentration of surface irrigation re-

TABLE 24.4 Definition of Symbols for the Salinity Submodel

Symbol (1)	Definition (2)
C_{diw}	TDS of diverted irrigation water (mg/l)
C_{eaiw}	TDS of effective applied irrigation water ($= C_{diw}$) (mg/l)
C_{ep}	TDS of effective precipitation ($= C_p$) (mg/l)
C_{dp}	TDS of deep percolation (mg/l)
C_{fsw}	TDS of final stored soil water (mg/l)
C_{isw}	TDS of initial stored soil water (mg/l)
C_{iwro}	TDS of tailwater (mg/l)
C_{iwrosp}	Salt pickup by tailwater (mg/l)
C_p	TDS of precipitation (mg/l)
C_{pro}	TDS of precipitation runoff (mg/l)
C_{prosp}	Salt pickup by precipitation runoff (mg/l)
C_{psw}	TDS of soil water after ET (mg/l)
C'_{psw}	TDS of soil water after ET, M_{gsp} and $M_{sp} - M_{sd}$ (mg/l)
C_{rim}	TDS of rim inflow/outflow (mg/l)
C_{sdw}	TDS of subsurface drainage (mg/l)
C_{se}	TDS of soil saturation extract corrected by gypsum solubility (mg/l)
C_{sirf}	TDS of surface irrigation return flow (mg/l)
$C_{sp} - C_{sd}$	Salt pickup − salt deposition (mg/l)
C_{sw}	TDS of soil water after ET (mg/l)
C_{sweaiw}	TDS of applied irrigation water after ET (mg/l)
C_{swep}	TDS of precipitation after ET (mg/l)
C_{swisw}	TDS of initial stored soil water after ET (mg/l)
EC_e	EC of soil saturation extract at 25°C (dS/m)
EC_{gyp}	EC of initial soil water due to gypsum at 25°C (dS/m)
EC_{gype}	EC of soil saturation extract due to gypsum at 25°C (dS/m)
ETCF	Evapotranspiration concentration factor
Gypsum	Amount of gypsum in the soil (t)
M_{diw}	Salt mass in diverted irrigation water (t)
M_{dp}	Salt mass in deep percolation (t)
M_{eaiw}	Salt mass in effective applied irrigation water (t)
M_{ep}	Salt mass in effective precipitation (t)
M_{fsw}	Salt mass in final stored soil water (t)
M_{gsp}	Amount of gypsum dissolved in soil water
M_{gyp}	Amount of gypsum potentially soluble in soil water (t)
M_i	Salt input (t)
M_{isw}	Salt mass in initial stored soil water (t)
M_{iwro}	Salt mass in tailwater (t)

(continued)

TABLE 24.4 Definition of Symbols for the Salinity Submodel (continued)

Symbol (1)	Definition (2)
M_o	Salt outputs (t)
M_p	Salt mass in precipitation (t)
M_{pro}	Salt mass in precipitation runoff (t)
M_{psw}	Salt mass in soil water after ET (t)
M'_{psw}	Salt mass in soil water after ET, M_{gsp} and $M_{sp}-M_{sd}$ (t)
M_{rim}	Salt in rim inflow/outflow (t)
M_s	Salt mass stored in the crop's root zone
M_{sdw}	Salt mass in subsurface drainage water (t)
M_{sirf}	Salt mass in surface irrigation return flow (t)
$M_{sp}-M_{sd}$ =	Salt pickup − salt deposition (t)
M_{sw}	Salt mass in soil water before ET (t)
M_{sweaiw}	Salt mass in Q_{sweaiw} (t)
M_{swep}	Salt mass in Q_{swep} (t)
M_{swisw}	Salt mass in Q_{swisw} (t)
SMCF	Salt mass concentration Factor
R_b	Soil bulk density (g/cm^3)
%G	Percentage of gypsum in the soil
TDS_{gyp}	TDS of initial soil water due to gypsum (mg/l)

turn flow (C_{sirf}) is the volume-weighted average of the concentrations of subsurface drainage (C_{sdw}), runoff components (C_{pro} and C_{iwro}), and lateral contributions (C_{rim}). Table 24.3 presents the complete mass balance formulation of the salinity submodel. Table 24.4 defines symbols.

The next two sections describe in detail how to calibrate, verify, and evaluate the model. These are the steps to follow to demonstrate the credibility of any model.

MODEL CALIBRATION AND VERIFICATION

The original model by Tanji (1977) was first calibrated with data from the Glenn-Colusa Irrigation District for the 1973 irrigation season. The water and soils in this district are low in dissolved mineral salts. The calibration required three iterations to fit the calculated flow and TDS of surface irrigation return flows to measured values. Adjusted in the process were two coefficients, the deep percolation coefficient (f) and the irrigation application efficiency (E_{iae}). After this calibration, subsequent outputs for flow, TDS, and salt mass were within ±2% of the observed data (Table 24.5). The calibrated model was then verified with data from the 1970 irrigation season. The model's predictions were within ±33% of the observed values (Table 24.5), acceptable results in light of the simplifying assumptions made in the model and the fact that 1970 was a normal year (0.461-m precipitation) compared to the relatively wet calibration year of 1973 (0.677-m precipitation).

TABLE 24.5 Results of Original Model Calibration and Validation Using Data from Glenn-Colusa Irrigation District for the 1973 and 1970 Irrigation Seasons (April–October)

Parameter (1)	Measured (2)	Calculated (3)	Difference (4)
(a) *Calibration* (1973 data)			
return flow (ha-m)	26,583	27,112	+2.0%
return flow TDS (mg/l)	220	215	−2.3%
salt mass in return flow (t)	58,522	56,328	−0.3%
(b) *Validation* (1970 data)			
return flow (ha-m)	34,512	23,127	−33.0%
return flow TDS (mg/l)	192	236	+22.9%
salt mass in return flow (t)	66,305	54,614	−17.0%

Note: 1 ha-m = $10^4 m^3$; and 1 t = 10^3kg.

TABLE 24.6 Results of Original Model Application on Panoche Drain District for Calendar Year 1975

Flow (1)	Measured (2)	Gypsum (3)	No Gypsum (4)	Difference[a] (5)
Return flow (ha-m)	5,585	6,552	6,552	+17.3%
Return flow TDS (mg/l)	1,955	458	1,818	−7.0%
Salt mass in return flow (t)	109,267	30,027	119,191	+9.1%

[a]Difference is calculated as the percentage difference between the measured and the calculated gypsum values.
Note: 1 ha-m = $10^4 m^3$; and 1 t = 10^3kg.

Tanji's model was then applied to a second irrigation project, the Panoche Drainage District, which contains salt-affected soils. Model-predicted volume of surface irrigation return flow was within +17% of the measured value (Table 24.6). However, TDS and salt load were greatly underestimated, as the original model did not consider the presence of native soil salts, particularly gypsum. When salt contributions from gypsum were considered, and assuming a solubility of 21.49 t/ha-m (25 meq/l), the calculated results for TDS and mass of salts in the surface irrigation return flows were within −7% and +9% of measured values, respectively (Table 24.6).

Aragues et al. (1985) modified Tanji's 1977 model and applied it to an irrigation project in Spain. They added routines that describe chemical weathering of soil gypsum, lateral contributions from adjacent areas, and initial and final stored soil water. The modified model was calibrated and verified for the Violada Irrigation District (VID) in the middle Ebro River basin (0°35′W, 42°05′N) of Spain. Fig. 24.4 shows a map of this district.

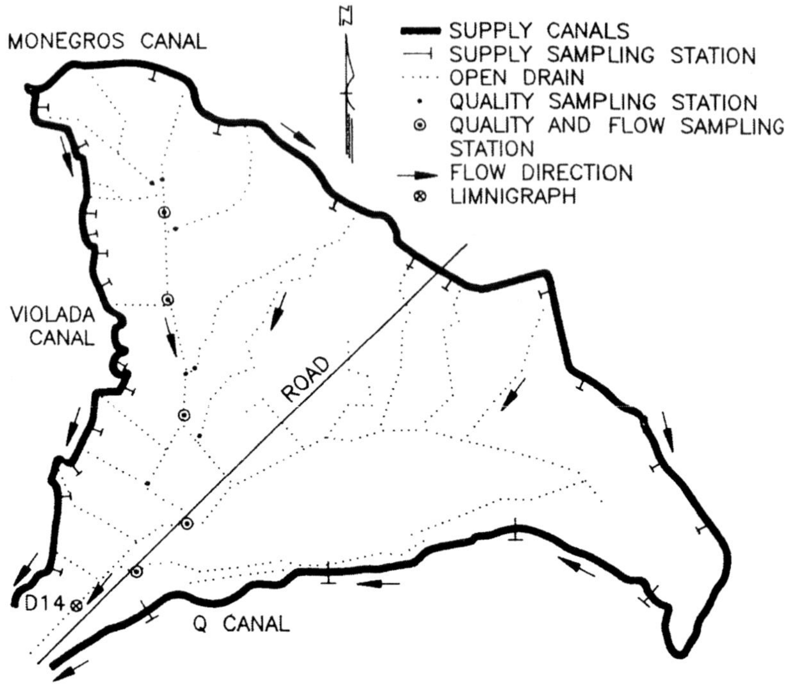

Fig. 24.4 Sampling Station Locations in Supply Canals and Main Drains of Violada Irrigation District

Table 24.7 summarizes some relevant measurements. Faci et al. (1985) describes in detail the study area and the methods used to collect data and estimate parameters. The model was calibrated for the hydrological year of 1982. Only one parameter, the irrigation application efficiency (E_{iae}), was adjusted (from an initial value of 0.935 to a final value of 0.916) in the process of fitting predicted and measured values of TDS in the surface irrigation return flows (C_{sirf}). After calibration, calculated values for volume, concentration, and salt load of irrigation return flows all were within 2% of measured values (Table 24.8).

The calibrated model was verified with data from hydrological years 1983 and 1984. Table 24.7 shows the input data and model coefficients for those years. The results of model verification show that model predictions were within 13% of the observed values (Table 24.9).

The modeling results demonstrate that the modified CIRF model could accurately simulate the average water and salinity flows in the aforementioned irrigated areas.

MODEL VALIDATION

Model validation involves a sensitivity analysis of model coefficients and variables, and a review of the model's inherent limitations. This section discusses both aspects.

TABLE 24.7 Violada Irrigation District: Required Input Data and Model Coefficients for 1982, Calibration Year, and 1983 and 1984, Validation Years

Input Parameter (1)	Hydrologic Year 1982 (2)	1983 (3)	1984 (4)
Total land area (ha)	3,913	3,913	3,913
Diverted irrigation water (ha-m)	3,880.5	4,419.5	3,970.1
Irrigation application efficiency	0.916	0.916	0.916
Precipitation (ha-m)	1,808.9	1,580.5	1,913.5
Precipitation runoff coefficient	0.013	0.013	0.013
Precipitation evaporation coefficient	0.038	0.038	0.038
Effective precipitation coefficient	0.949	0.949	0.949
Initial stored soil water (ha-m)	841	841	841
Final stored soil water (ha-m)	841	841	841
Unit water holding capacity (m/m)	0.36	0.36	0.36
Average rooting depth (m)	0.72	0.72	0.72
Evapotranspiration (ha-m)	2,821.8	2,677.4	2,373.1
Potential evapotranspiration (ha-m/ha)	1.17	1.07	1.01
Average crop coefficient	0.62	0.64	0.58
Deep percolation coefficient	0	0	0
Rim inflow (ha-m)	308	289.7	824
TDS of rim inflow (mg/l)	283	283	219.2
TDS diverted irrigation water (mg/l)	190	190	190
Salt pickup by tailwater (mg/l)	20	20	20
TDS of precipitation (mg/l)	51	67	33
Salt pickup by precipitation runoff (mg/l)	40	40	40
EC of soil saturation extract (dS/m)	2.21	2.21	2.21
Soil saturation percentage (g/100g)	49	49	49
Percent of gypsum in soil (g/100g)	3.6	3.6	3.6
Soil bulk density (g/cm^3)	1.5	1.5	1.5

Note: 1 ha-m = 10^4m^3; and 1 t = 10^3kg.

TABLE 24.8 Results of CIRF Model Calibration for Violada Irrigation District Using Data from Hydrological Year 1982

Parameter (1)	Measured (2)	Calculated (3)	Difference (4)
Return flow (ha-m)	3,164	3,107	−1.8%
Return flow TDS (mg/l)	2,273	2,273	0.0%
Salt mass in return flow (t)	71,910	70,615	−1.8%

Note: 1 ha-m = $10^4 m^3$; and 1 t = 10^3kg.

TABLE 24.9 Results of CIRF Validation for Violada Irrigation District from Hydrological Years 1983 and 1984

	Hydrologic Year 1983			Hydrologic Year 1984		
Parameters (1)	Measured (2)	Calculated (3)	Difference (4)	Measured (5)	Calculated (6)	Difference (7)
Return flow (ha-m)	3,643	3,552	−2.5%	4,307	4,282	−0.6%
Return flow TDS (mg/l)	2,054	2,330	13.4%	1,843	2,071	12.4%
Return flow (t)	74,850	82,782	10.6%	79,391	88,656	11.7%

Note: 1 ha-m = $10^4 m^3$; and 1 t = 10^3kg.

Sensitivity Analysis

For sensitivity analysis, one of the variables is varied while simultaneously keeping all other variables constant. Model results are compared. If outputs differ largely, the variable is said to be sensitive.

Table 24.10 summarizes results of the sensitivity analysis for the calibrated, VID version of the CIRF model using data for hydrological year 1982. Figure 24.5 shows that the order of sensitivity is $Q_{diw} > Q_{et} > Q_p >> Q_{rim}$, and that all variables behave linearly. The effect of other input parameters is nil. Analyzing for C_{sirf} (concentration of surface irrigation return flow, Fig. 24.6), demonstrates that the sensitivity behavior of some variables, especially Q_{et} and Q_{diw}, is not linear. An extreme example is gypsum content. It has no effect on C_{sirf} when it is present in amounts excessive enough to result in a soil solution saturated with gypsum. However, it exerts a sharp influence when it is present in amounts small enough to cause the solution to become undersaturated. The effect of gypsum on C_{sdw} is qualitatively similar (Fig. 24.7). In fact, C_{sdw} is only strongly affected by EC_e and gypsum (Fig. 24.7), due to the excess gypsum content and the absence of other soluble salts in the soils at the Violada Irrigation District.

Fig. 24.6 shows that the input parameter known as irrigation application efficiency (Eiae) strongly affects C_{sirf}, as Eiae establishes the

TABLE 24.10 Sensitivity Analysis with Classification of Input Parameters According to their Influence on Model Outputs

Model Output (1)	Input Parameter											
	Q_{diw} (2)	E_{iae} (3)	Q_p (4)	Q_{isw} (5)	Q_{et} (6)	Q_{rim} (7)	C_{rim} (8)	C_{diw} (9)	C_p (10)	EC_e (11)	SP (12)	%Gyp (13)
Q_{sirf}	+3	0	+3	0	−3	+1	0	0	0	0	0	0
C_{sirf}	+2	+3	+2	0	−2	+1	0	+1	0	+3	0	+2
C_{sdw}	0	0	0	0	0	0	0	+1	0	+3	0	+2

Note: Type 0. Input variation of 100% gives output variation less than 5%. (straight line)
Type 1. Input variation of 100% gives 5–20% output variation. (straight line)
Type 2. For small variations, type 1, for large, type 3. (curved line)
Type 3. Input variation of 100% gives output variation of greater than 20%. (straight line)

Plus signs indicate that the output variable increases as input variable value increases. Negative signs indicate that the output variable decreases as input increases.

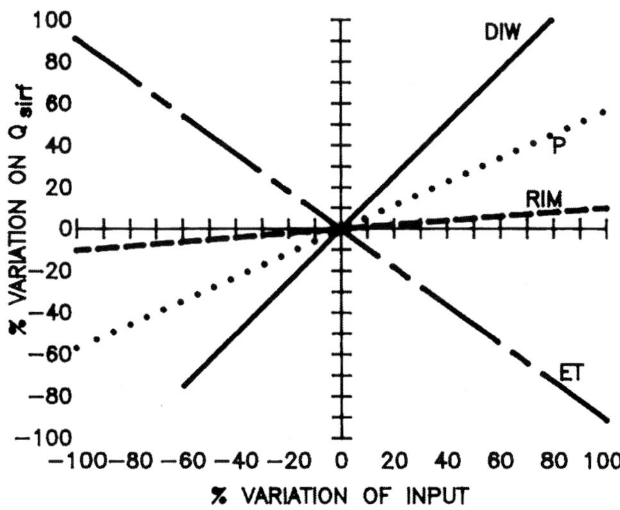

Fig. 24.5 Sensitivity Analysis: Effect of Different Variations of Input Variables (%) on Volume of Surface Irrigation Return Flow (Q_{sirf})

partitioning between runoff water of low salt concentration and subsurface drainage water of high and constant salt concentration. As the relative contribution of subsurface water decreases due to the decrease in diverted irrigation water (Q_{diw}), the TDS of return flow (C_{sirf}) also

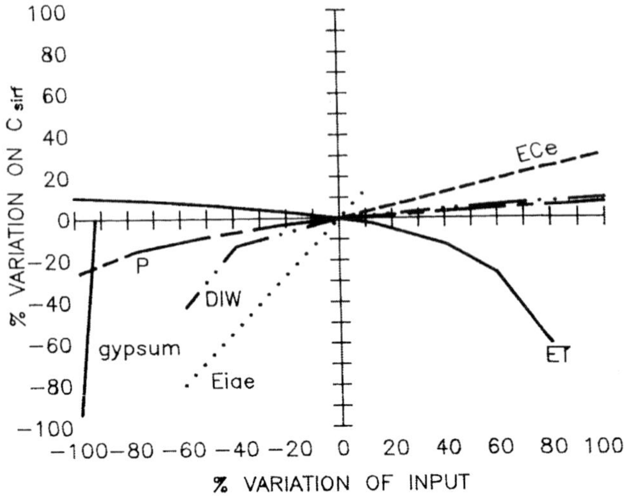

Fig. 24.6 Sensitivity Analysis: Effect of Different Variations of Input Variables (%) on Salinity of Surface Irrigation Return Flows (C_{sirf})

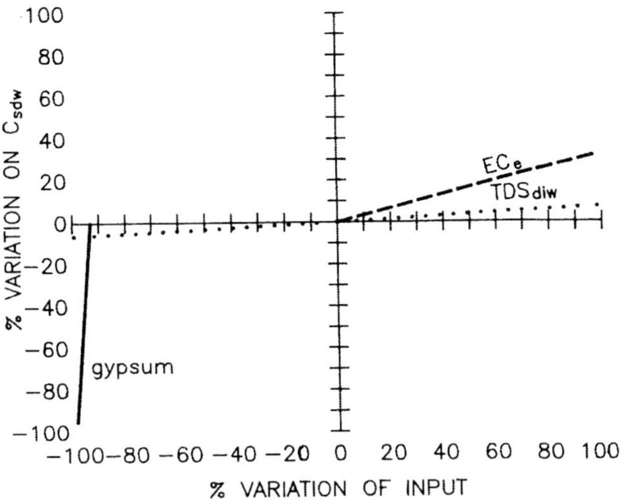

Fig. 24.7 Sensitivity Analysis: Effect of Different Variations of Input Variables (%) on Salinity of Subsurface Drainage Water (C_{sdw})

decreases. In other cases without gypsum or with other types of soluble salts, the concentration of subsurface drainage and return flow obviously tends to increase as Q_{diw} decreases. This is due to the ET effect, which is not significant in this case because of soil characteristics in VID.

The above demonstrates and emphasizes the importance of performing a sensitivity analysis for each particular case of interest, since it provides a way to rank the more important measurements and estimates and to better understand the system.

Limitations of Model

The CIRF, a management-oriented model, is inherently limited by its simplifying assumptions. Even with simplifications, many of the input parameters are difficult to measure or estimate.

One of the model's limitations is its assumption of steady-state conditions. Whether or not this holds true for the case to be simulated needs to be ascertained.

Another limitation is that the model's stationary characteristics are coupled with some nonstationary variables, such as volume and concentration of initial soil water and gypsum content. For example, gypsum content changes with time due to its dissolution, or its precipitation, or both. However, if gypsum is present in excess, the soil solution remains saturated at all times and this variable can be considered stationary. Although the alternative of a limited amount of gypsum in the soil profile has been considered in the model's mathematical formulation, this part of the model has yet to be substantiated.

Another nonstationary variable is the electrical conductivity of the soil saturation extract (EC_e). As mentioned before, the model has not been extensively verified in areas where soluble soil minerals other than gypsum are dominant. In its present configuration, the model predicts that all soluble salts and salts contributed from gypsum will be leached out of the soil profile during the time interval modeled.

Finally, problems could arise from the fact that input variables are averaged over time and space. This simplification could prove limiting in cases with highly heterogeneous soil properties or nonuniform application of irrigation water. In such cases, the studied area can be subdivided into multiple, homogeneous parts.

MODEL APPLICATION

An advantage of the CIRF model is that it only requires 25 input data and model coefficients and yet apparently has the characteristics needed to evaluate water management practices, irrigation water quality, soil properties, and evapotranspiration that affect salt loading of irrigation return flows. The sensitivity analysis of the model proved that the key to minimizing salt loading in IRF from the Violada Irrigation District in Spain was to decrease applied irrigation water and, thus, reduce subsurface drainage water. This option is clearly viable, as is shown by the irrigation district's low water application efficiency (WAE), an average of 0.45 for the three years studied. Concurrently, average calculated LF is excessively high (0.51). Considering the EC of the irrigation water and the crops grown in the area, the theoretical LF should be much less (LF= 0.02). The model was, therefore, used to evaluate water management alternatives for salinity control. Table 24.11 gives model projections for the main outputs of the Violada Irrigation District in 1983, assuming that salt control practices resulted in hypothetical reductions of 10% (case A) and 20% (case B) in diverted irrigation water. The final TDS of soil water, C'_{psw}, increases slightly due to an increase of ETCF, but the decrease in the amount of water beyond the

Table 24.11 Model Predictions for Hypothetical 10% (case A) and 20% (case B) Reductions in Diverted Irrigation Water During 1983 for Violada Irrigation District

	Hydrologic Year	Case A (−10%)		Case B (−20%)	
(1)	1983 (2)	Calculated (3)	Difference (4)	Calculated (5)	Difference (6)
Diverted water (ha-m/ha)	1.13	1.02	−10.0%	0.91	−20.0%
Leaching below root zone (ha-m/ha)	0.74	0.63	−14.0%	0.53	−28.0%
Final soil water TDS (mg/l)	2,824.0	2,829.0	0.2%	2,836.0	0.4%
Salt in soil water after ET (t/ha)	2.26	2.07	−8.0%	1.87	−18.0%
Salt pickup-salt deposition (t/ha)	−0.42	−0.37	−12.0%	−0.33	−21.0%
Salt mass due to gypsum (t/ha)	25.03	22.30	−11.0%	19.58	−22.0%
Surface irrig return flow (ha-m/ha)	0.91	0.80	−12.0%	0.69	−24.0%
Surface return flow TDS (mg/l)	2,326.0	2,291.0	−1.5%	2,244.0	−3.5%
Salt in return flow (t/ha)	21.23	18.31	−14.0%	15.39	−28.0%
LF	0.45	0.41	−9.0%	0.37	−18.0%
ETCF	1.13	1.81	5.0%	1.92	12.0%
WAE	0.45	0.49	9.0%	0.53	18.0%

root zone (Q_{ppsw}) and the corresponding decrease in the mass of dissolved gypsum (M_{gsp}) substantially reduces the mass of salts in surface irrigation return flow (M_{sirf}). This results in 14% and 28% less salt load for cases A and B, respectively.

Because gypsum is the main salt source in the Violada Irrigation District and other large and less saline flows contribute to the surface irrigation return flows, the TDS of the return flow is slightly decreased. If soil gypsum had been absent or more readily soluble salts had been present in the soil profile, the TDS of IRF would have increased. For the Violada Irrigation District, the salt concentration of the leachate is basically independent of the volume of the leachate. The concentration of salt below the root zone produces a gypsum- and lime-saturated condition and, hence, is relatively constant in salt concentra-

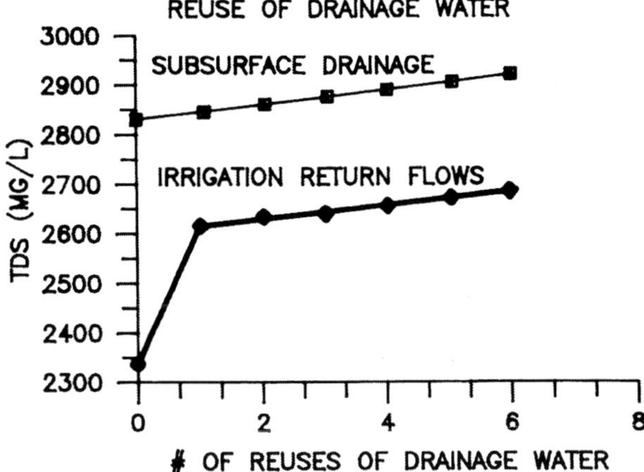

Fig. 24.8 Effect of Six Consecutive Reuses of Drainage Water for Irrigation on Salinity of Subsurface Drainage Water and Surface Irrigation Return Flow (Data from hydrologic year 1983.)

tion. The salt loading in the Violada Irrigation District due to irrigation return flows can, thus, be approximated from a knowledge of water balance only.

Table 24.11 shows that another consequence of reducing Q_{diw} is the precipitation of salts in the soil profile due to a reduced leaching fraction. Researchers at the U.S. Salinity Laboratory (Bernstein and Francois 1973) postulated the concept of minimum leaching fraction. Its applicability to waters, similar to those of the Violada Irrigation District, is expressed by the equation:

$$C_{sp} - C_{sd}\ (mg/l) = -102 + 122.9\ \text{LF (at } P_{CO_2} = 3.5 \times 10^{-4}\ \text{atm)} \tag{24.14}$$

which shows that the amount of salts precipitating at LF of 10% almost doubles that at LF of 40%. These LF reductions can be achieved with proper water management in the Violada District and do not affect crop yields.

Finally, another management alternative to consider is the reuse of return flows for irrigation purposes. The IRF in the Violada District are moderately saline (EC of 2–3 dS/m) and rich in gypsum. The model predicts that, after the first reuse, subsurface drainage waters and irrigation return flows become nearly saturated with gypsum. Subsequent reuses negligibly affect the concentration of these waters (Fig. 24.8). As an example of the model's usefulness for evaluating this alternative, Fig. 24.9a gives projections for hydrological year 1983 and Fig. 24.9b assumes that 25% of the irrigation water applied for that year is reused drainage water. It schematically shows the graphical contribution of each salinity source in the system and the model outputs. The results in Fig. 24.9 indicate that drainage water reuse reduces salts in return flows by 30% in the Violada District.

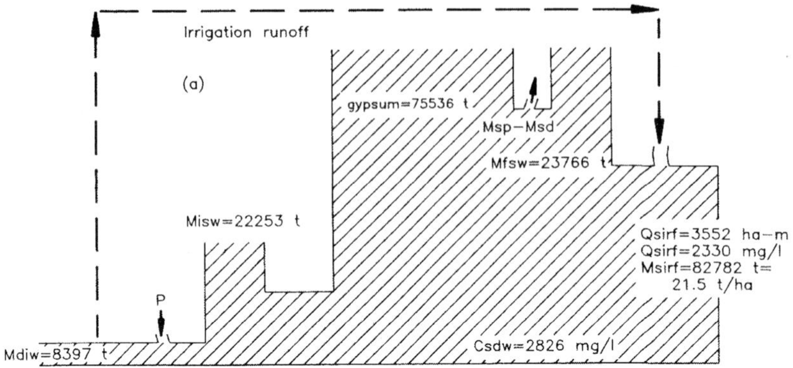

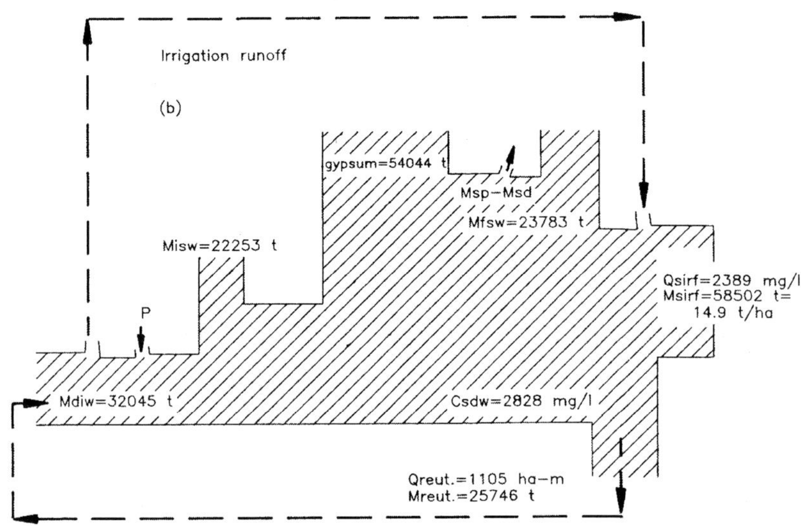

Fig. 24.9 Contribution of Each Salt Source to Irrigation Return Flow Salinity; (a) Hydrologic Year 1983; and (b) Assuming That 25% of Applied Water Comes From Reused Drainage Water

MODEL IMPROVEMENTS, FUTURE RESEARCH NEEDS

One limitation of the CIRF in its present configuration is that it cannot consider the inefficiency of the displacement of salts in the porous medium. This can be overcome by introducing an average leaching efficiency coefficient, LEFF, in the model and applying it to the term M'_{psw}. Salts not leached out $[(1 - \text{LEFF}) \times M'_{psw}]$ are assumed to remain in the soil profile and increase the concentration of the final stored water (fsw). After this change, the concentration of the fsw no longer equals the concentration of the subsurface drainage water. The value of this last variable is an average over the period of study, and the concentration of final stored water becomes an instantaneous value at the end of the studied period. The value of LEFF depends on the soil's physical characteristics, including macropore bypass in soils with large pore sizes and high macroscopic pore velocities, number of pore volumes displaced by the infiltrating solution, kind of salts to be leached out, and soil water content (Biggar and Nielsen 1967, Bresler et al. 1982). A value for this coefficient can be approximated by calibrating the model in the particular system of interest.

Future research must also be conducted on the nature of rising water tables and the nature of the salt-mixing process at the interface between percolating water and resident shallow ground water, an important consideration because many irrigation districts with soil salinity problems also have rising shallow ground-water problems. Assuming that no water is used from the water table, the situation could be represented in the model by the presence of a non-interacting area and a mixing area. Although the total volume of subsurface drainage water would remain unaffected by the mixing mechanism, its concentration will change according to the expression:

$$C_{sdw} = (1-f)\, C'_{ssw} + f\,[w\, C_{gw} + (1-w)\, C'_{ssw}] \qquad (24.15)$$

where C'_{ssw} = the subsurface drainage water concentration predicted by the model (assuming no mixing); C_{gw} = ground-water concentration; f = mixing coefficient for the interaction between subsurface water and ground water; and w = weighting coefficient that takes into account the relative magnitude of the interacting process. The last expression can be reduced to

$$C_{sdw} = g\, C_{gw} + (1-g)\, C'_{ssw} \qquad (24.16)$$

where g is a global interactive coefficient between ground water and subsurface drainage. Because extensively modifying a model to fit a particular salinity problem fosters the development of mathematical models that are too site-specific (Tanji 1982), further changes were rejected for the CIRF model.

SUMMARY AND CONCLUSIONS

Over the past two decades, significant advances have been made in the modeling of irrigation return flows. Models range from simple mass balance routines to state-of-the-art, modular digital codes. These models provide increased insight into water and salt flow in irrigated systems.

The CIRF model presented in this chapter is simple in concept and formulation and has reliably predicted the volume and salt concentration of IRF in areas with nonsaline to saline soils where gypsum is the main source of salinity. The model has also proven to be useful for evaluating alternative salinity control practices.

The CIRF model is written in BASIC and runs interactively, with numerous built-in comments and explanations. It also includes a subroutine for sensitivity analysis. Versions of the program that run on IBM-, PC-, and AT- compatible microcomputers and the Apple Macintosh microcomputer are available from the authors.

REFERENCES

Aragüés, R., Tanji, K.K., Quilez, D., Alberto, F., Faci, J., Machin, J., and Arrue, J.L. (1985). "Calibration and verification of an irrigation return flow hydrosalinity model." *Irrig. Sci.* 6: 85-94.

Bay-Valley Consultants (1974). "Recommended water quality management plan, Sacramento River basin, San Joaquin River basin, Sacramento-San Joaquin Delta." Report to the California State Water Resources Control Board, 1-1: 7-11.

Bernstein, L., and Francois, L.E. (1973). "Leaching requirement studies: sensitivity of alfalfa to salinity of irrigation and drainage waters." *Soil Sc. Soc. Am. Proc.* 37: 931-943.

Biggar, J.W., and Nielsen, D.R. (1967). "Miscible displacement and leaching phenomenon." Chapter 14, *Irrigation of Agricultural Lands.* Hagan et al., eds. *Agronomy Series No. 11,* American Society of Agronomy, Madison, Wisc.

Bresler, E., Mc Neal, B.L., and Carter, D.L. (1982). *Saline and sodic soils.* Springer-Verlag. Berlin, W. Germany, 236pp.

California Department of Water Resources (1969). "Lower San Joaquin River water quality investigation." *Bull. No. 143-5.* California Department of Water Resources, 207 pp.

Dutt, G.R., Shaffer, M.J., Moore, W.J. (1972). "Computer model of dynamic biophysiochemical processes in soils." *Tech. Bull. 196.* Univ. of Ariz., Tucson, 101pp.

Faci, J., Aragüés, R., Alberto, F., Quilez, D., Machin, J., and Arrue, J.L. (1985). "Water and salt balance in an irrigated area of the Ebro River Basin (Spain)." *Irrig. Sci.* 6: 29–37.

Fapohunda, H.O., and Hill, R.W. (1981). "River basin hydro-salinity-economic modeling." *J. of Irrig, and Drain. Div., ASCE* 197-IV: 53-69.

Hill, R.W., Israelsen, E.K., and Riley, J.P. (1973). "Computer simulation of the hydrologic and salinity flow systems in the Bear River basin." *Report No. PRWG 104-1.* Utah Water Research Laboratory. Utah State Univ., Logan.

Hornsby, A.G. (1973). "Predicting modeling for salinity control in irrigation return flows." *EPA-R2-73-168,* Washington, D.C. 56pp.

Hyatt, M.L., Riley J.P., McKee M.L., Israelsen E.L. (1970). "Computer simulation of the hydrological-salinity flow system within the Upper Colorado River Basin." *Report No. PRWG 104-1,* Utah Water Res. Lab. Utah State Univ., Logan.

Khan, I.A. (1982). Managing salinity in irrigated agriculture: *J. Irrig. and Drain. Div. ASCE* 108: 43-56.

Labadie, J.W., and Khan, I.A. (1979a). "River basin salinity management via the ASTRAN method, I: Model development." *J. Hydrol.* 42: 301-321.

Labadie, J.W., and Khan, I.A. (1979b). "River basin salinity management via the ASTRAN method, I: Case study." *J. Hydrol.* 42: 323-345.

Moolman, J.H., Van Rooyen P.C., and Weber, H.W. (1983). "The effect of irrigation practices in the Bree River Valley on the salt content of a small river." *Irrig. Sci.* 4: 103-116.

Oster, J.D., and Dell'Osso, G.P. (1975), "FORTRAN IV computer program for soil solution chemistry," U.S. Salinity Lab., Riverside, Calif.

Ribbens, R.W., and Shaffer M.J. (1976). "Irrigation return flow modeling for the Souris Loop environmental aspects of irrigation and drainage." Proc. of ASCE Specialty Conference. Ottawa. Canada: 545-557.

Scofield, C.S. (1940). "Salt balance in irrigated areas." *Agricultural Research* 61: 17-39.

Skogerboe, G.V., McWhorter, D.B., and Ayars, J.E. (1979). "Irrigation practices and return flow salinity in Grand Valley." R.S. Kerr Env. Res. Lab. U.S. EPA, Ada, Okla.

Skogerboe, G.V., and Walker, W.R. (1981). "Impact of irrigation on the quality of groundwater and river flows." *Salinity in Irrigation and Water Resources*. Dan Yaron, ed. 121-157.

Tanji, K.K. (1977). "A conceptual hydrosalinity model for predicting salt load in irrigation return flow." *Managing saline water for irrigation.* Texas Tech. Univ. pp.49-70.

Tanji, K.K. (1982). "River basin hydrosalinity modeling." *Agricultural Water Management* 4: 207225.

Thomas, J.O., Riley J.P., and Israelsen E.K. (1971). "A computer model of the quantity and chemical quality of return flow." *Report No. PRWG 77-1.* Utah Water Research Laboratory. Utah State Univ., Logan. 94pp.

U.S. Bureau of Reclamation (1975). *Predicting the Mineral Quality of Irrigation Return Flow.* Vol. I. Engineering and Research Center, Denver, Colorado. 58pp.

Walker, W.R. (1970). "Hydrosalinity model of the Grand Valley," thesis presented to the Colorado State Univ., at Fort Collins, Colorado, in partial fulfillment of the requirements for the degree of Master in Science.

Walker, W.R. (1978). "Identification and initial evaluation of irrigation return flow hydrosalinity models." U.S. Environ. Protection Agency, EPA-600/2-78-144, Washington, D.C.

Water Resources Engineers, Inc. (1969). "An investigation of salt balance in the upper Santa Ana River Basin." *Final Report to the California Dept. of Water Resour.* 198 pp.

Woods, P.C. (1967). "Management of hydrologic system for water quality control." *Water Resources Center Contribution No. 121.* Univ. of Calif., Davis, 121pp.

Woods, P.C., and Orlob G.T. (1963). "The lost river system: a water quality management investigation." *Water Resources Center. Contribution No. 68.* Univ. of Calif., Davis, 54 pp.

CHAPTER 25

DYNAMIC OPTIMIZATION MODELS FOR SALINITY AND DRAINAGE MANAGEMENT[a]

INTRODUCTION

Numerous management practices exist for handling salinity and drainage problems in irrigated agriculture. They include: modifying crop rotations, changing the volume and timing of irrigation water, investing in improved irrigation systems, installing subsurface drainage systems, reusing drainage water, and treating or disposing of water collected in subsurface drains.

One approach to selecting the best management practice is to simulate alternate management policies using crop-water production functions, such as those in Chapter 14, and then choose the best according to some criterion. Another approach is to formulate a dynamic optimization problem and then solve it with the appropriate algorithms. The simulation approach allows construction of a detailed physical, chemical, and biological processes model but does not optimize beyond simple enumeration or trial and error. Dynamic optimization finds the best management practice under specific conditions, but computational considerations usually limit model complexity. The two approaches may be combined for some applications. First, the various options are screened with an optimization model, and then one or more simulation models are used to evaluate the selected options.

This chapter focuses on the dynamic optimization approach. It describes the available dynamic optimization models and possible extensions of those models for applications, including irrigation scheduling, management of inter-seasonal irrigation, and whole-farm analyses.

DYNAMIC OPTIMIZATION CONCEPTS

Dynamic optimization involves the control of a dynamic system to maximize the value of an objective function. State variables summarize the current status of the dynamic system, while control variables are controlled directly by the decision-maker influence on evolution of the

[a]Prepared by: Keith C. Knapp, Dept. of Soil and Envir. Sci., Univ. of Calif., Riverside, CA 92521; and Dennis Wichelns, Dept. of Resource Economics, Univ. of Rhode Island, Kingston, RI 02881.

dynamic system. The objective function depends on the state and control variables at various points in time. In deterministic problems, the goal is to choose a time sequence for the control variables that makes the objective function as large as possible. In stochastic problems, the goal is to choose decision rules for the control variables that maximize the expected value of the objective function, where the decision rules give the optimal values of the control variables as a function of the state variables. Numerous references on dynamic optimization exist in the economic literature. Useful summaries are given by Kendrick (1981) for discrete-time problems and by Intriligator (1971) for continuous-time problems. In this chapter, discrete-time models will be considered exclusively, since they appear to be more amenable to empirical analysis.

Deterministic Control Problems

Let x_t and u_t represent the values of the state and control variables, respectively, at time period t. For example, x_t may represent salt concentration of the soil water at the beginning of the irrigation season, and u_t may represent seasonal water applications. Then

$$x_{t+1} = f(x_t, u_t) \tag{25.1}$$

is an equation of motion for the system that shows how the state variables change over time in response to alternate levels of the control variables u_t. Choice of the control variables is subject to various constraints that, in turn, depend on the level of the state variables. For example, seasonal water applications must be non-negative and may be restricted by system capacity or institutional considerations. Mathematically, we require

$$u_t \in U(x_t) \tag{25.2}$$

where U = the set of feasible values for the control variable u_t. This set depends on the state variable, x_t.

The objective function measures the system's performance as a function of the state and control variables. The objective function here is the present value of annual profits over the time horizon. The objective function for deterministic problems is written:

$$\sum_{t=1}^{T} \alpha^t g(x_t, u_t) \tag{25.3}$$

where T = the time horizon (possibly infinite); α = the discount factor, equal to one divided by one plus the interest rate; and g = annual net returns as a function of the state and control variables. The discount factor reflects the time preference for returns now received over returns received in the future and the fact that capital can earn positive rates of return in alternative uses. The problem is to choose a sequence of values for u_t, $t=1,\ldots,T$ that makes Expression 25.3 as large as possible, subject to the equation of motion (Eq. 25.1) and the constraints (Eq. 25.2).

This optimization problem is formulated as a nonlinear programming problem and solved with the use of gradient methods. Dennis and Schnaibel (1983) introduce basic principles. A popular optimization package is MINOS (Murtagh and Saunders 1982), which is also implemented in GAMS (Brooke et al.1988). The solution consists of a sequence of values for the control variables and state variables and a value for the objective function.

While gradient algorithms offer a useful approach, limitations exist. All functions must have the approximate convexity to guarantee a global optimum. The computational effort to solve mathematical programming problems increases geometrically as the number of constraints increases. The number of constraints here depends directly on the number of time periods, so practical problems may be limited to short time horizons if many state and control variables exist. Gradient-based approaches are also inappropriate where some variables are discrete, as is the case with many decisions at the field level (e.g., choice of crop, investment in irrigation, and installation of drainage systems).

An alternative solution method is dynamic programming. Dynamic programming algorithms proceed by calculating a series of optimal decision rules. These rules give the optimal values of the control variables in period t as a function of the state variables in period t. An optimal decision rule is calculated for each time period in a finite horizon problem. For time-autonomous infinite horizon problems, the optimal decision rule is the same in every time period. (Time-autonomous means that the functions f and g in Eqs. 25.1 and 25.3 and the set U in Eq. 25.2 remain the same in every time period.) Given initial values for the state variables, a time path of the optimal values for the control and state variables can be traced out after calculating the optimal decision rules. Bertsekas (1976) is a useful reference on dynamic programming.

Dynamic programming has several advantages over gradient-based methods. Global optima are found (subject to discretization error) regardless of convexity, and integer values for control variables, state variables, and nonlinear and non-analytic functions are readily handled. Computational effort increases linearly with the time horizon, in contrast to gradient methods. However, dynamic programming algorithms typically discretize the state and control variables, which can introduce error into the final solution. More seriously, computational effort increases geometrically as the number of state and control variables increases, so the method is generally limited to problems with relatively few state and control variables.

Stochastic Control Problems

Stochastic control problems occur when random variables enter into the equation of motion (Eq. 25.1) and, possibly, the objective function (Eq. 25.3). In every time period, the decision-maker observes the current value of the state variables and chooses the values for the control variables. The optimization problem, then, is to choose a sequence of decision rules that maximizes the expected value of the objective function, subject to the equations of motion and constraints on the control variables. This differs from the deterministic problem, where the optimal values for the control variables in each time period can be chosen at time zero. In the stochastic problem, information about previous realizations of the random variables can be used to improve decisions in each time period. This implies that optimization over sequences of decision rules is superior to optimization over sequences of actual values of the control variables.

Dynamic programming provides a general approach to solving stochastic control problems, but the computational effort may become prohibitive for more than a few state variables. Linear-quadratic control methods can be used when the objective function is quadratic, the equation of motion is linear, and the control variables are unconstrained. This method is very efficient and allows for large models with many state and control variables. However, the conditions are restrictive and unlikely for realistic salinity and drainage problems. Bertsekas (1976) discusses dynamic programming in the stochastic case. Bertsekas (1976) and Kendrick (1981) discuss linear-quadratic control. Bertsekas (1976) also discusses some suboptimal controllers that provide approximate solutions to the stochastic control problem.

A third class of dynamic optimization problems is adaptive control problems, sometimes referred to as stochastic control with active learning. In deterministic and stochastic control problems, the probability distributions for the random variables and the parameters defining the equations of motion and objective function are assumed to be known. However, in many instances, this is not the case. Substantial uncertainty over the actual parameters values may exist. For example, crop-water production functions or soil salinity dynamics are typically not known precisely. In many instances, the actual parameter values can be learned as the system evolves, thereby reducing uncertainty about their actual values. In such cases, values for the control variables should be chosen both to maximize expected net benefits, according to current beliefs about parameter values, and to promote learning about the true parameter values. Finding the best way is an adaptive control problem. Dynamic programming algorithms can be used to solve adaptive control problems, but actual applications that use this approach are very rare. Bellman (1961), Kendrick (1981), and Mangel (1985) provide relevant references.

OPTIMAL IRRIGATION SCHEDULING

Optimal irrigation scheduling refers to choosing irrigation timing, volumes, and qualities to maximize net benefits over the course of an irrigation season. Social net benefits equal revenue from selling the crop minus irrigation costs, production costs, and the cost of reusing or disposing of drainage water. Social net benefits will equal farm profits, if all inputs and outputs are correctly priced. This means, for example, that the farmer is charged for the off-farm environmental costs that result from disposal of drain water. The optimal choice of control variables requires consideration of crop revenues and all irrigation and drainage costs.

A conceptual model for determining optimal irrigation scheduling is formulated using a daily model of crop growth over the season. This model integrates over the root zone and applies where irrigation is relatively uniform. Some extensions are considered later. Major variables are defined by: W_j = volume of applied irrigation water on day j; C_j = electrical conductivity (EC) of the irrigation water; ET_j = evapotranspiration on day j; M_j = soil moisture in the root zone on day j; S_j = soil salinity measured as the EC of a saturated extract on day j; and Y = crop yield. In these definitions, j = {m,...,n}, m is the first day of the growing season, n is the last day in the growing season; M_j and S_j are state variables; and W_j and C_j are control variables. The grower does not directly control C_j. However, suppose two irrigation water sources of differing qualities and costs exist. The salt concentration of the applied irrigation water can then be changed by changing the mix of the two irrigation sources. The formulation here avoids extra control variables when more than two irrigation sources exist. In this case, a least-cost mix is calculated *a priori* for various combinations of W and C.

Daily evapotranspiration is a function of soil moisture, soil salinity, and climate:

$$\mathrm{ET}_j = f_{1j}(M_j, S_j; K_j) \tag{25.4}$$

where K_j includes climatic variables such as temperature, cloud cover, wind speed, and pan evaporation. The functional relation changes over time as the plant grows.

Soil moisture dynamics are given by

$$M_{j+1} = M_j + W_j + R_j - \mathrm{ET}_j + U_j - D_j \tag{25.5}$$

where R_j = rainfall on day j; U_j = upward flux of water from the water table; and D_j = deep percolation of water below the root zone. Deep percolation and upward flux are functions of the moisture content of the root zone and various parameters, such as hydraulic conductivity and depth to the water table.

Soil salinity dynamics are given by

$$S_{j+1} = f_2(M_j, S_j, W_j, C_j, R_j) \tag{25.6}$$

recognizing that ET_j, U_j, and D_j are functions of the other variables in Eq. 25.6. depends parametrically on such factors as the depth and salt concentration of the water table and the properties of the soil. Crop yield is a function of soil moisture, soil salinity, and climate during the growing season:

$$Y = f_3(M_m, \ldots, M_n; S_m, \ldots S_n; K_m, \ldots, K_n) \tag{25.7}$$

Social benefits net of water costs are defined by

$$PY - \sum_{j=m}^{n} c(W_j, C_j) - \sum_{j=m}^{n} d(D_j) \tag{25.8}$$

where P = the crop price; $c(W_j,C_j)$ are irrigation costs as a function of quantity and quality of irrigation water, and $d(D_j)$ are costs associated with deep percolation. The optimization problem is to choose W_j and C_j, j = m,...n, to maximize Eq. 25.8, subject to Eqs. 25.4 through 25.7, nonnegativity conditions and possibly other constraints, given the initial values M_m and S_m. This formulation implicitly assumes that rainfall and other climatic variables are deterministic. In the stochastic case, decision rules for the control variables are chosen to maximize the expected value of Eq. 25.8.

Several approaches can be taken to solve the optimization problem in Eqs. 25.4 through 25.8. Under deterministic conditions, nonlinear programming techniques could be used. However, most nonlinear programming algorithms are guaranteed to find global optima only for convex problems. Empirical specifications for the ET function (Eq. 25.4) are generally very nonlinear and may exhibit significant non-convexities. The evidence suggests that irrigation setup costs significantly affect the number of irrigations. These introduce non-convexities and discontinuities into the irrigation cost function that conventional optimization algorithms cannot handle.

Dynamic programming (DP) offers another approach. It handles arbitrary constraint sets and inclusion of irrigation setup costs is straightforward. The variables in the optimization problem must be discretized before solution. Subject to this limitation, DP provides a global optimum, regardless of convexity considerations. The major disadvantage is that the computational burden increases exponentially as the number of state variables increases.

Bras and Seo (1987) propose the use of extended linear quadratic control in the irrigation scheduling problem. The first step is to choose an initial path of control and state variables. The system is linearized around this path, and a new set of values for the control variables is chosen by using linear quadratic control techniques. Expectations are calculated where needed, and modifications are introduced to handle inequality constraints on the control and state variables. Iterations are repeated until convergence to an optimum is achieved. The method shows promise for handling problems with many state variables. The drawbacks include the use of a suboptimal controller for stochastic problems and possible convergence to a local, but not global, optimum for non-convex problems.

Several studies analyze optimal irrigation scheduling under saline conditions. Yaron et al. (1980) focus on sorghum. Crop ET is assumed to be a linear function of soil moisture. They do not consider rainfall and water movement between the root zone and the water table. Soil salinity is calculated using a model developed by Bresler (1967). In the Bresler (1967) model, soil salinity after an irrigation is calculated by mass balance from soil salinity before the irrigation, salts added in the irrigation water, and salts leached through deep percolation. The volume of deep percolation is the excess of irrigation over the moisture deficit (field capacity minus current moisture content), and the salt concentration is the average of soil salinity before and after irrigation. If the volume of deep percolation water exceeds two pore volumes, then soil salinity after the irrigation is assumed to equal the irrigation water's salt concentration.

Yaron et al. calculate yield as a function of the number of critical days in several sub-periods, where a critical day is any day in which soil water potential is less than a specified level. Irrigation can take place every six days, and irrigation costs include only the cost of water. They use a modified dynamic programming algorithm to solve the optimization problem.

Knapp and Dinar (1987) consider optimal irrigation scheduling for cotton. Crop ET is assumed to be a function of matric potential plus osmotic potential. They use a piston-flow model to compute soil salinity and assume yield to be a function of seasonal ET. They also do not consider rainfall and water movement between the root zone and the water table. Deep percolation is calculated as the excess, if any, of irrigation over the moisture deficit. Irrigation can occur on any day of the season and irrigation costs include a fixed setup cost plus the cost of water. Dynamic programming is used to solve the optimization problem.

Bras and Seo (1987) consider corn. Evaporation is a function of time, while transpiration depends on soil moisture and salinity. Rainfall is stochastic, upward flow from a water table is calculated exogenously, and deep percolation is a function of the soil moisture content. Crop yield equals a weighted average of transpiration in each stage. Irrigation expenses include water and fixed costs of irrigation, although the latter are not treated as an integer variable. They use the extended linear quadratic control algorithm discussed earlier to solve the problem.

These empirical studies provide a number of insights into irrigation scheduling under saline conditions. Yaron et al. and Knapp and Dinar find that initial soil salinity levels have a major effect on seasonal applied water quantities. In Knapp and Dinar, for example, seasonal water applications double when initial EC_e—the electrical conductivity of a saturated extract—increases from 2 dS/m to 12 dS/m. In practice, salt leaching typically occurs at the beginning of the season. However, Yaron et al. and Knapp and Dinar find that some leaching during the season is also optimal.

Yaron et al. find that an increase in the irrigation water's salinity significantly increases irrigation volumes and decreases net income. Changes in the price of irrigation water scarcely affect irrigation volumes for cotton (Knapp and Dinar 1987) but more significantly affect corn (Bras and Seo 1987). This is likely due to the crops' biological characteristics and their relative values. Drainage water disposal costs reduce irrigation volumes by up to 10%, depending on initial salinities (Knapp and Dinar 1987). The results in these studies are calculated assuming uniform irrigation water infiltration. The static analysis in Dinar, Knapp, and Letey (1989) indicates that the effects of irrigation and drainage costs will be more significant under nonuniform water applications and soil conditions.

Knapp and Dinar find that the setup costs of irrigation are a major determinant of the number of irrigation events during the season. Eliminating these costs increases irrigation events by a factor of four to six, depending on initial salinity levels. This corresponds to the observation that irrigations are more frequent under drip and sprinkler systems than under surface systems, where the costs of initiating irrigations are higher. To some extent, this result depends on the model. In the Knapp and Dinar model, crop ET is reduced continuously as moisture falls below field capacity. Yields will likely decline as a consequence, depending on total seasonal ET. However, if maximum ET can be maintained over a range of soil moistures and soil salinity is low, then setup costs will not affect the frequency and depths of irrigation as much.

Knapp and Dinar consider the use of low-quality water for irrigation as a way to augment fresh-water supplies and reduce the disposal and environmental costs of collected drain water. Depending on initial salinities, the low-quality water contributes from 33% to 100% of the total water applied for the particular situation considered.

Extensions must be made to the work described above for the optimal irrigation scheduling models under saline conditions to be fully operational. Current studies implicitly assume uniform irrigation conditions. However, irrigation water is generally applied nonuniformly over the field, and soil parameters are highly variable over space. These variabilities largely determine optimal water applications (see Chapter 14 for a discussion of spatial variabilities). Parameter variations by depth are also ignored here. Some evidence suggests that crops can minimize stress by extracting water from layers where water is relatively abundant or nonsaline. Depth variations also influence the quantity and quality of deep percolation.

As discussed in the next section, salinity and moisture levels at the end of the season influence production and profitability in future years. Hence, they need to be taken into account in the current year. This can be accomplished by including a terminal value function in the objective function, Eq. 25.8. The terminal value function gives discounted net benefits in future years as a function of the soil salinity and soil moisture at the end of the current year. Inclusion in Eq. 25.8 implies that both current and future profits will be taken into account when irrigation decisions can be made during the current year. The inter-seasonal models discussed in the next section can be used to specify terminal value functions.

High water tables, common in areas with saline soils, exacerbate the salinity problem. Decisions regarding irrigation influence and are influenced by the presence of an underlying water table. Fully-specified irrigation scheduling models should include the height of the water table, its salt concentrations, and upward flows from the water table to the root zone as endogenous variables.

Bras and Seo (1981) consider uncertain rainfall. However, crop prices, production costs, pest problems, and meteorological conditions during the growing season are also likely to be uncertain when irrigation decisions are made. Ideally, the probability distributions of these variables would be characterized and the problem solved as a stochastic control problem. However, this is likely to be computationally burdensome. A good alternative is Bras and Seo's (1987) strategy in which the optimization problem is solved repeatedly throughout the season as new observations on the state variables become available. Finally, many of the parameters underlying Eqs. 25.4 through 25.8 are not known with certainty. Accounting for parameter uncertainty with learning over time results in an adaptive control problem, as discussed in the previous section.

OPTIMAL INTER-SEASONAL IRRIGATION MANAGEMENT

Salt accumulation in the root zone changes with time, depending on factors such as irrigation practices, cropping patterns, rainfall, height of the water table, and salt concentrations. This section reviews research on the optimal management of soil salinity over a period of years. The analysis considers an individual field with the same crop grown in every year. Seasonal water quantities are assumed to be variable. Some extensions are considered in the following section.

The following variables are defined: S_t = soil salinity measured as the EC of a saturated extract; W_t = volume of applied irrigation water in year t; C_t = salt concentration (EC) of the irrigation water; and Y_t = crop yield. In these definitions, $t=1,\ldots,T$, and T is the time horizon (possibly infinite).

Soil salinity dynamics are given by

$$S_{t+1} = f(S_t, W_t, C_t) \tag{25.9}$$

The parameters in this equation depend on the type of crop, environmental conditions, soil properties, and other factors.

Crop yield is given by

$$Y_t = g(S_t, W_t, C_t) \tag{25.10}$$

The present value of returns to land and management is given by

$$\sum_{t=1}^{T} \alpha^t \left[PY_t - c(W_t, C_t) - F\right] \tag{25.11}$$

where P = crop price net of harvest costs; c = the cost of irrigation as a function of the quantity and quality of irrigation water; α = a discount factor equal to one divided by one plus the interest rate; and F = the non-water production costs, which are assumed to be independent of W_t and C_t. The optimization problem is to choose W_t and C_t, $t=1,\ldots T$, to maximize Eq. 25.11, subject to Eqs. 25.9 and 25.10 and non-negativity conditions.

Dinar and Knapp (1986) solved the above problem for alfalfa and cotton. They use data from field experiments to estimate the salt balance relation statistically (Eq. 25.9) and yield relation (Eq. 25.10). Dynamic programming is used to compute optimal decision rules for an infinite time horizon. These give the profit-maximizing volume of seasonal water that must be applied, expressed as a function of the salinity of the soil at the beginning of the irrigation season. In general, water applications increase as initial soil salinity increases. Optimal time paths for soil salinity can be calculated from the optimal decision rules for various initial conditions. For cotton, soil salinity (measured as the EC of a saturated extract) converges to a steady-state level of 6 dS/m, regardless of the initial conditions. More generally, annual fluctuations in salinity can be expected from crop rotations and changes in input costs and price received.

Dinar and Knapp (1986) also consider the optimum mix of a primary source of high-quality irrigation water for cotton and a second source of low-cost, low-quality irrigation water. The results show that no low-quality water is used until the price of the good-quality water exceeds $.80/ha-mm. A single-season decision rule, calculated by maximizing net returns in each year without consideration for future years, results in discounted net returns as much as 26% lower than those from an optimal interseasonal decision rule.

Yaron and Olian (1973) also examine the inter-seasonal dynamics of salinity management. They consider leaching and salt accumulation in the soil profile over time. The control variable in their analysis is the amount of leaching water applied during the winter. The amount of water applied in summer is fixed at levels determined by traditional farming practices. The quality of applied water is an exogenous variable that is varied parametrically, and the objective is to maximize the expected value of cumulative net returns over the time horizon.

The model is applied to citrus, a salt-sensitive perennial crop. Mass balance relations describe physical-system dynamics in a modified version of the leaching model developed by Bresler (1967). Coefficients describe yield losses due to salinity and are based on information from soil scientists and data from local sources. Results provide a decision rule for an optimal leaching policy over a range of observed levels of state variables. The decision rule suggests the optimal leaching fraction to use in the approaching summer as a function of soil salinity at the end of winter and the quality of available leaching and irrigation water. Economic losses from soil salinity for various water qualities and initial soil salinities are used to estimate the farm-level value of improvements in water quality. These results, as noted by the authors, depend heavily on assumptions about salt leaching, accumulation functions, and the relationship between yield and salinity.

Matanga and Mariño (1979) extend the Yaron and Olian model to include seasonal irrigation water as a decision variable. They use stochastic dynamic programming and simulation to choose the leaching depths and seasonal water applications that maximize discounted net benefits over finite and infinite planning horizons. The physical model is similar to the one used by Yaron and Olian (1973). Net additions to soil salinity are calculated as the amount of salt added to the soil profile by applied water minus the amount that plants leach out or absorb. The effects of soil salinity on crop yields are based on information from agricultural extension service specialists. Seasonal rainfall is treated as a stochastic variable, and probabilities conditional on rainfall during the previous period are used to compute its expected value in any season.

The model is applied to corn, grain sorghum, and pinto beans in a single-crop framework for each. The dynamic programming approach requires a finite number of levels of seasonal irrigation and leaching depths. The work of Stewart et al. (1974) is used to develop irrigation programs and production coefficients for this purpose. The dynamic programming model selects the optimal combination of these discrete decision possibilities. For each crop, results from the interseasonal model provide decision rules describing the optimal amount of seasonal irrigation water and leaching water to apply as a function of soil salinity and expected rainfall at the beginning of any stage. Empirical results are given for 30-year finite and infinite planning horizons.

The optimal irrigation and leaching policies are independent of the amount of rainfall received. In the case of a finite planning horizon, this results from the assumption that soil moisture is at field capacity when leaching takes place. This condition may not be true in drought years when additional water is needed for leaching. This extra water is not considered in the model.

OPTIMAL INTER-SEASONAL IRRIGATION MANAGEMENT: EXTENSIONS

The models in the previous section are based on the assumption that the same crop is grown every year. In most situations, this is unlikely, as crops are often rotated for agronomic and economic reasons. Suppose that the number of possible crops that can be grown is n and that there exists a time period, m, before a previous crop effects the current crop's yield. The salinity optimization model can be expanded to include endogenous crop rotations by defining Φ_{it} as the crop grown in the *i*th year before year t, $\Phi_{it} \in \{1,\ldots,n\}$, and where $i=0,\ldots,m$. Thus, Φ_{0t} is the crop grown in year t and

$$\Phi_{it} = \Phi_{i\text{-}1,t\text{-}1} \tag{25.12}$$

for $i=1,\ldots,m$. Eq. (25.12) is an accounting identity that records the previous crop history for every year. The Φ_{it} are ultimately determined by the choice of Φ_{0s} in previous years s ($s \leq t$).

Soil salinity dynamics are redefined as

$$S_{t+1} = f(S_t, W_t, C_t, \Phi_{0t}) \tag{25.13}$$

recognizing that the crop grown in year t influences evapotranspiration and, hence, deep percolation and soil salinity for a given S_t, W_t, and C_t. Crop yield is also redefined as

$$Y_t = g(S_t, W_t, C_t, \Phi_{0t}, \ldots, \Phi_{mt}) \tag{25.14}$$

This accounts for the type of crop grown in year t and the influence of previous crops on the current crop's yield. The formulation recognizes the need to rotate crops for pest management, soil fertility, and other such purposes.

The objective function for the crop rotation problem is

$$\sum_{t=1}^{T} \alpha^t [P(\Phi_{0t}) Y_t - c(W_t, C_t) - F(\Phi_{0t})] \tag{25.15}$$

where price received and non-water production costs are a function of the crop grown in year t. The optimization problem is to choose W_t, C_t, and Φ_{0t}, $t=1,\ldots,T$ so as to maximize Eq. 25.15 and subject to Eqs. 25.12 through 25.14 and non-negativity conditions. Apparently, no studies have been conducted for intertemporal salinity optimization with crop rotations. However, an optimal time path for soil salinities can be expected to fluctuate. Also, fixed crop rotations can be incorporated by imposing constraints on the choice of Φ_{0t} as a function of the crop history, or by appropriately defining Eq. 25.14. As an example of the latter, suppose the fixed rotation is one year of tomatoes followed by one year of cotton. Eq. 25.14 is then specified, so that tomato yield in year t is zero, unless Φ_{1t} = cotton. Likewise, cotton yield in year t is zero, unless Φ_{1t} = tomatoes. Another way to incorporate fixed rotations is to recognize that the problem is periodic and use the methods in Chapter 6 of Bertsekas (1976).

The previous section's models also ignore the disposal and environmental costs of drainage water. In many areas, agricultural drainage water degrades surface water and ground water and damages crops by raising perched water tables. It also may need to be disposed, imposing additional costs on farmers and society. Ways to reduce the volume of agricultural drainage water include reducing amounts of applied water, improving irrigation systems, changing crop rotations, and reusing drain water.

A dynamic optimization model for irrigation management under high-water table conditions can be formulated as follows. NIS types of irrigation systems are available. Let TIS_t be an integer variable that denotes the type of irrigation system in place at the beginning of year t, and let IIS_t be an integer variable denoting investment in a new irrigation system at the beginning of year t. Setting $IIS_t = 0$ means no new investment takes place and, hence, the irrigation system remains the same. Setting $IIS_t = i$, where $i \in \{1,\ldots,NIS\}$, implies investment in a new irrigation system of type i. Then

$$TIS_{t+1} = \begin{cases} TIS_t \text{ when } IIS_t = 0 \\ IIS_t \text{ when } IIS_t > 0 \end{cases} \tag{25.16}$$

is the equation of motion for the type of irrigation system. Each irrigation system has a maximum life, denoted by YRMAX(TIS). Letting YIS_t denote the age of the irrigation system during year t, then

$$YIS_{t+1} = \begin{cases} YIS_{t+1} \text{ when } IIS_t = 0 \\ 1 \quad \text{when } IIS_t > 0 \end{cases} \tag{25.7}$$

The constraint

$$IIS_t > 0 \text{ if } YIS_t = YRMAX(TIS_t) \tag{25.18}$$

forces the system to be replaced when it reaches the end of its physical life.

The height of the water table at the beginning of year t is denoted by HWT_t and evolves over time according to

$$HWT_{t+1} = f_1(HWT_t, QDP_t, QDW_t) \tag{25.19}$$

where QDP_t is the quantity of water that percolates below the root zone, QDW_t is the amount of water collected by the drainage system, and f_1 depends parametrically on specific yields, transmissivity, and lateral flows. HWT_t is constrained by

$$HMIN \leq HWT_t \leq HMAX \tag{25.20}$$

which specifies upper and lower bounds for HWT_t.

When the water table reaches a critical height, a drainage system is installed and the effluent is disposed off-farm. DS_t denotes the presence or absence of a drainage system:

$$\begin{aligned} DS_{t+1} = DS_t & \text{ when } IDS_t = 0 \\ 1 & \text{ when } IDS_t = 1 \end{aligned} \tag{25.21}$$

where $DS_t = 0$ is no drainage system; $DS_t = 1$ is a drainage system; and IDS_t is investment in a drainage system. $IDS_t = 0$ indicates no investment, while $IDS_t = 1$ indicates an investment. Investment in the drainage system is a function of DS_t and the height of the water table. Thus:

$$IDS_t = f_2(DS_t, HWT_t) \tag{25.22}$$

and f_2 is specified to guarantee that investment takes place whenever the water table reaches a critical height and no drainage system exists.

Crop yield (YLD_t) and quantity of deep percolation water (QDP_t) are given by:

$$[YLD_t, QDP_t] = f_3(CC_t, TIS_t, IIS_t, QAW_t) \tag{25.23}$$

where CC_t is the crop grown in year t and QAW_t is the quantity of applied water. Quantity of water removed from the drainage system is given by

$$QDW_t = f_4(HWT_t, DS_t, QDP_t) \tag{25.24}$$

and depends on the design of the drainage system and other parameters.

The state variables in this problem are TIS_t, YIS_t, DS_t, and HWT_t. The control variables are QAW_t and IIS_t. The optimization problem is to choose QAW_t and IIS_t, $t=1,\ldots,T$, to maximize the present value of profits subject to Eqs. 25.16 through 25.24 and non-negativity conditions. This formulation assumes exogenous crop rotations and steady state salinities. Extensions to include endogenous rotations and dynamic soil salinities can be made along the lines discussed earlier. Knapp et al. (1990) report some preliminary results, using this model.

As with the intra-seasonal irrigation scheduling problem, a major consideration is incorporating the field-level spatial variability in applied water or soil conditions. There are two ways to do this. With one, the field is considered as a collection of grid cells. Salinity, moisture, water table heights, and other state variables are specified for each cell, with applied water in each cell a function of the water applied to the field as a whole. Year-to-year dynamics can then be monitored for each cell. Conceptually, the algorithms discussed earlier could be used to solve this optimization problem. However, this approach would probably require impractical amounts of computer time.

A second approach is to assume that a probability distribution over the field with a finite number of moments characterizes the state variables of interest, e.g., soil salinity. The distribution moments are then treated as state variables in a new optimization problem and evolve over time as a function of the moments and control variables in the previous period. This problem is likely to be more computationally tractable if the number of moments characterizing the spatial distribution is small.

FARM-LEVEL ANALYSES

Previous sections have considered irrigation and drainage management at the field level. However, most farms consist of several fields and, in some instances, the operation of all fields must be considered simultaneously. This is especially the case if actions on one field affect either the profitability of other fields or the farmer's valuation of those profits. This can occur in one of the following ways: (1) Resource constraints for the farm imply that all fields must be simultaneously considered. For example, if the total quantity of water available to the farm is constrained, then the amount applied to one field affects the amount available for other fields. (2) Marketing quotas or contracts for specified crops affect the choice of crops that can be grown in a given year. (3) Technical externalities may exist between fields. For example, deep percolation from one field may influence water table heights under other fields, or pest management practices in one field may affect pest populations in other fields. (4) If the grower is risk-adverse or risk-loving and the problem is stochastic, then valuation of farm profits is the expected value of a nonlinear function of farm profits and not equal to the sum of the expected value of profits from individual fields. Each of these circumstances implies that decisions on all fields must be considered simultaneously.

One approach to farm-level modeling is to treat the farm as a collection of separate fields. For each field, a set of equations may be written that relates the field's state variables to the field's control variables and, possibly, to variables in other fields. The object is to choose the levels of the control variables in each time period that maximize the present value of profits subject to constraints on the controls and the equations of motion for the state variables in each of the fields.

Gradient methods may be used to solve this problem provided the problem is deterministic, the functions are well-behaved, no discrete variables exist, and the number of time periods and variables is limited. Dynamic programming offers an approach to more general problems.

However, computational considerations limit its usefulness to fairly small problems. Iterations can be made where a limited resource forces simultaneous consideration of all fields. For example, suppose that the farm's total water supply is limited. Decisions on each field can be optimized separately for a given water price. If total water used is greater than the supply, then the water price is increased. If the total is less than the supply, then the price is lowered. The procedure is continued until water use just equals water supply. Yaron and Dinar (1982) give an algorithm based on this approach.

Yaron and Voet (1982) give an example of this discrete area approach. They consider a farm with two orchards and a field that can be used for growing various crops. A fixed amount of water is available. The orchards require a fixed amount of water for irrigation and the remainder is allocated to leaching in each orchard and for growing field crops. Equations of motion describe soil salinity dynamics in each of the orchards as a function of the variable leaching quantities and the fixed irrigation quantities. The optimization problem is solved via dynamic programming.

A second approach to farm-level modeling is to consider the farm as one area with possibly different soil qualities. The areas devoted to various crops are assumed to be continuous variables. For example, suppose there are N possible crops, where X_i, $i=1,\ldots N$, denotes the area devoted to each crop, each crop uses a_i units of water per unit area, and the land area and water use are constrained. A simple problem would then be to choose the levels of X_i that maximize farm profits subject to

$$\sum X_i \leq L$$

$$\sum a_i x_i \leq W \qquad (25.25)$$

where L = the total land area on the farm; and W = the total quantity of water available. These problems are readily solved using linear and nonlinear programming methods. This approach is widely used in empirical studies of agricultural production economics. Hazell and Norton (1986) summarize the methods.

Lansford et al. (1986) give an example of this continuous area approach for the intra-seasonal salinity and drainage problem. They consider a farm with fresh and saline water sources, a fixed amount of land, and several crops that can be grown. An irrigation scheduling simulation model is used to determine the yield from a number of alternate irrigation scheduling policies. These policies vary according to irrigation timing and the respective amounts of fresh and saline water used during the season. A linear programming model is then formulated, where the variables include the amount of land devoted to a particular crop irrigated by a particular scheduling policy. The object is to maximize farm profit subject to constraints on land and the amount of fresh and saline water available to the farm.

For the inter-seasonal problem, each variable or activity in the linear/nonlinear programming model could represent the acreage devoted to a specified time path of decisions on crops and irrigation activities over the horizon. For example, X_1 could represent the acreage of the three-year rotation cotton/cotton/tomatoes over the horizon with a specified irrigation volume for each crop in each year. X_2 could represent the acreage of continuous alfalfa with a specified irrigation volume in each year. Implicit in each sequence, X_i is a time-path for soil salinity, drainage water volumes, crop yields, and so on. The object is to choose the values for X_i that maximize the present value of profits over the horizon subject to various constraints such as land, water supplies, and marketing quotas. We are not aware of any studies of farm-level salinity and drainage that use this approach for the interseasonal problem.

Feinerman and Yaron (1983) describe another variant of the continuous area approach. Suppose that the problem being considered is interseasonal soil salinity management; that a specified range of soil salinities is being considered, e.g., EC_e from 0 dS/m to 20 dS/m; and that this range of salinities is divided into n intervals. Let $A_i(t)$ be the area of the farm with soil salinity in the *i*th salinity interval in year t and $X_{jk}(t)$ be the area of the kth cropping activity planted on soil in the *j*th salinity interval in year t. Cropping activity here refers to a specific crop grown with a specific irrigation volume, specific fertilizer quantity, and so on. Evolution of the $A_i(t)$ is then governed by

$$A_i(t+1) = \sum_j \sum_k a_{ijk} X_{jk}(t) \tag{25.26}$$

where a_{ijk} = the fraction of land from cropping activity k planted on salinity interval j, which has soil salinity in interval i at the beginning of the next year. In a deterministic problem, a_{ijk} is a binary, zero-one variable with the summation of a_{ijk} over i equal to one. In a problem with stochastic rainfall, as considered by Feinerman and Yaron (1983), the a_{ijk} ranges between zero and one, but the summation over i still equals one. The X_{jk} must also satisfy

$$\sum_k X_{jk}(t) \leq A_j(t) \tag{25.27}$$

for all j and all t. This guarantees that the area of cropping activities planted on soil with salinity in interval j does not exceed the total area in that interval. The optimization problem is to choose values of X_{jk} in each year to maximize the present value of profits over the horizon, subject to Eqs. 25.26 and 25.27 and other constraints, such as water availability and marketing quotas. Feinerman and Yaron use linear programming to solve the optimization problem in their paper.

CONCLUSION

In an increasing number of studies, operations-research methods are being applied to problems in agricultural production economics. To date, however, relatively few studies have applied formal optimization methods to salinity and drainage problems in irrigated agriculture.

Some qualitative results have emerged from the existing studies.

The irrigation scheduling model studies suggest that initial salinity levels significantly affect optimal irrigation volumes during the season and that the setup costs of irrigation are a major determinant of the number of irrigation events during the season. The inter-seasonal model analysis described here finds convergence to an optimal steady state that is independent of initial conditions for deterministic problems when a single crop is grown continuously.

These results provide some evidence that reuse of drain water can exist at current prices when the only benefit is the replacement of fresh water. However, reuse will likely become a much more attractive option when the disposal costs and environmental costs of drain water are taken into account.

Insights into optimal salinity and drainage management can be gained with existing models. Still, much remains to be done before these models can be used in actual situations. Necessary extensions include improvement of crop-water production functions for saline conditions, incorporation of field-level spatial variability in irrigation scheduling and inter-temporal models, improved handling of crop rotations in inter-temporal models, inclusion of water table heights, and investment in irrigation systems. Some approaches to these and other problems have been outlined in this chapter. Logically, it would seem that realistic, verifiable field-level analyses need to be conducted before the more difficult farm-level problem is addressed.

REFERENCES

Bellman, R. E. 1961. *Adaptive Control Processes: A Guided Tour*. Princeton, NJ: Princeton University Press.

Bertsekas, D. P. 1976. *Dynamic Programming and Stochastic Control*. Academic Press, New York.

Bras, R. L. and D. J. Seo. 1987. "Irrigation control in the presence of salinity: extended linear quadratic approach." *Water Resour. Res.* 23(7): 1153-1161.

Bresler, E. 1967. "A model for tracing salt distribution in the soil profile and estimating the efficient combination of water quality and quantity under varying field conditions." *Soil Sci.* 104: 227-233.

Brooke, A., D. Kendrick, and A. Meeraus. 1988. GAMS: *A User's Guide*. The Scientific Press, Redwood City, Calif.

Dennis, J. E., Jr., and R. B. Schnabel. 1983. *Numerical Methods for Unconstrained Optimization and Nonlinear Equations*. Prentice-Hall, Englewood Cliffs, N.J.

Dinar, A., and K. Knapp. 1986. "A dynamic analysis of optimal water use under saline conditions." *Western J. of Agric. Economics* 11(1): 58-66.

Dinar, A., K. C. Knapp, and J. Letey. "Irrigation water pricing policies to reduce and finance subsurface drainage disposal." *Agric. Water Management* (in press).

Feinerman, E., and D. Yaron. 1983. "Economics of irrigation water mixing within a farm framework." *Water Resour. Res.* 19(2): 337-345.

Hazell, P. B. R., and R. D. Norton. 1986. *Mathematical Programming for Economic Analysis in Agriculture.* Macmillan Publishing Co., New York.
Intriligator, M. 1971. "Mathematical Optimization and Economic Theory." Prentice-Hall. Englewood Cliffs, N.J.
Kendrick, D. 1981. "Control theory with applications to economics." *Handbook of Mathematical Economics,* K. J. Arrow and M. D. Intriligator, eds. North-Holland. New York.
Knapp, K. C., and A. Dinar. 1987. "Optimum irrigation under saline conditions." *Irrigation Systems for the 21st Century,* L. G. James and M. J. English, eds. ASCE, New York.
Knapp, K. C., B. Stevens, J. Letey, and J. Oster. 1990. "A dynamics optimization model irrigation investment and management under limited drainage conditions." Water Resources Research, 26(7): 1335–1340.
Lansford, R. R., T. McGuckin, C. L. Mapel, and T. W. Sammis. 1986. "Optimization of irrigation scheduling with alternative saline water supplies." Presented at the Annual Meeting of the American Agricultural Economics Association, in Reno, Nev., July 1986.
Mangel, M. 1985. *Decision and Control in Uncertain Resource Systems.* Academic Press, New York.
Matanga, G. B., and M. A. Mariño. 1979. "Irrigation planning II: water allocation for leaching and irrigation purposes." *Water Resour. Res.* 15(3): 679-683.
Murtagh, B. A., and M. A. Saunders. 1982. "A projected Lagrangian algorithm and its implementation for sparse nonlinear constraints." *Mathematical Programming Study* 16: 84-117.
Stewart, J. I., R. M. Hagan, and W. O. Pruitt. 1974. "Functions to predict optimal irrigation programs." *J. of the Irrig. Div.,* ASCE 100(IR2): 179-199.
Yaron, D., and A. Dinar. 1982. "Optimal allocation of farm irrigation water during peak seasons." *American J. of Agric. Economics* 64(4).
Yaron, D., and A. Olian. 1973. "Application of dynamic programming in Markov chains to the evaluation of water quality in irrigation." *American J. of Agric. Economics* 55(3): 467-471.
Yaron, D., and H. Voet. 1982. "Application of an integrated dynamic and linear programming model to the analysis of optimal irrigation on a farm with dual quality (salinity) water supply." *Planning and Decisions in Agribusiness: Principles and Experience,* C. F. Hanf and G. W. Schierer, eds. Elsevier, Amsterdam.
Yaron, D., E. Bresler, H. Bielorai, and B. Harpinist. 1980. "A model for optimal irrigation scheduling with saline water." *Water Resour. Res.* 16(2): 257-262.

NOTATION

a_i = units of water per unit area; a_{ijk} = fraction of land from cropping activity, k, planted on salinity interval j; $A_i,^{(t)}$ = area of farm with soil salinity in the ith salinity interval in year, t; c = cost of irrigation as a function of quantity and quality of irrigation water; $c(W_j,C_j)$ = irrigation costs as function of quantity and quality of irrigation water; C_j = electrical conductivity (EC) of irrigation water; C_t = salt concentration (EC) of irrigation water; CC_t = crop grown in year, t; $d(D_j)$ = costs associated with deep percolation; D_j = deep percolation of water below root zone; DS_t = presence or absence of drainage system; ET_j = evapotranspiration on day, j; f_1 = function that depends parametrically on specific yield, transmissivity, and lateral flows; f_2 = function specified to guarantee that investment takes place whenever table reaches critical height and no drainage system exists; F = non-water production costs; g = annual net returns; IDS_t = investment in drainage system; j = {m,...,n}; K_j = climatic variables; L = total land area on farm; m = first day of growing season; M_j = soil moisture in root zone on day, j; n = last day of growing season; P = crop price; QAW_t = quantity of applied water; QDP_t = quantity of water that percolates below root zone; QDW_t = amount of water collected by drainage system; R_j = rainfall on day, j; S_j = soil salinity measured as EC of saturated extract on day, j; S_t = soil salinity measured as EC of saturated extract; T = time horizon; U = set of feasible values for control variable u_t; U_j = upward flux of water from water table; W = total quantity of water available; W_j = volume of applied irrigation water per day, j; W_t = volume of applied irrigation water in year, t; x_i, i=1,...N = area devoted to each crop; x_t = state variable; X_1 = acreage of three-year rotation of cotton/cotton/tomatoes; X_2 = acreage of continuous alfalfa; $X_{jk}^{(t)}$ = area of *k*th cropping activity planted on soil in the *j*th salinity interval in year, t; Y = crop yield; Y_t = crop yield; YIS_t = age of irrigation system during year, t; YLD_t = crop yield; YRMAX (TIS) = maximum life of irrigation system; α = discount factor; Φ_{it} = crop grown in *i*th year before year, t.

IX—ENVIRONMENTAL, LEGAL, AND ECONOMIC ISSUES

CHAPTER 26

INSTITUTIONAL AND LEGAL CONSTRAINTS[a]

ISSUES

The United States public is demanding clean water for all purposes. Continual contamination of the nation's water resources in some areas and the discovery of new water quality problems in others have aroused the public and various interest groups and caused the government to become deeply involved in the regulation of water quality. Many different agencies now have jurisdiction over water and its quality. Myriad federal, state, and local laws have been passed. These laws and regulations require individuals to change and restrict their activities. In some cases, they must change long-established practices.

The quality of water necessary and the responsibility for maintaining it are two of the most controversial resource issues facing our society. The cost of and ultimate economic responsibility for water quality control often become hot political topics, as the enforcement of regulation and control often requires public or private funds.

Farmers generally will not adopt procedures that decrease their net profit. They, therefore, might delay the installation of necessary on-farm drainage facilities if the increased production allowed by use of drains is not offset by the cost of the facilities.

Once on-farm drains are installed in areas with a saline water table, the production of saline drainage effluent is generally continuous (Pillsbury and Johnston 1965). The quantity and quality of the effluent depends on the origin of the irrigated soil, past farming practices, current on-farm water management, and, in particular, irrigation practices. Drainage effluent containing high levels of salts or elements toxic to some form of life is particularly troublesome. Regulatory agencies place requirements on the discharge of drainage water whenever the discharge of drainage effluent does or could cause problems for others or impairs downstream water quality. The farmers must act to meet the requirements. When the regulatory agencies do not act quickly enough to satisfy the public, concerned citizens and well-financed environmental interest groups act to force what is generally a confrontation between the farmers and the regulatory agencies.

---

[a]Prepared by: William R. Johnston, Search II, 1342 W. San Jose Ave., Fresno, CA 93711; and Stephen K. Hall, The Land Preservation Assoc., 770 East Shaw, Suite 205,Fresno,CA 93710.

INSTITUTIONS

Many governmental agencies have jurisdiction over the disposal of agricultural drainage effluent (Johnston and Heaton 1986). Therefore, numerous legal, political, and economic constraints exist for the disposal of such effluent.

Federal Agencies

Congress controls the purse strings of all federal agencies. It also establishes the laws of the land on water quality. Congress, therefore, plays an essential role in controlling and regulating water quality. The federal agencies that develop, regulate, and control agricultural drainage water are the U.S. Department of the Interior, the U.S. Department of Agriculture, and the U.S. Environmental Protection Agency (EPA). The U.S. Department of the Interior comprises the U. S. Bureau of Reclamation, the U.S. Fish and Wildlife Service, and the U. S. Geological Survey. The U. S. Soil Conservation Service is the primary agency of the U.S. Department of Agriculture.

The Bureau of Reclamation and others have built irrigation water supply systems throughout the western United States. Saline drainage effluent and other agricultural runoff from irrigated land appear in some areas to threaten fish, waterfowl, and their habitat. In addition, the public often finds agricultural runoff from irrigation projects aesthetically unacceptable (Deason 1987). The Fish and Wildlife Services operates many wildlife refuges and management areas which are recharged by surface or subsurface agricultural runoff water. In the past, the Service viewed such water as a resource. However, beginning in the early 1980's, the agency has increased its efforts to eliminate the discharge of potentially toxic agricultural drainage water and replace drainage water with other water supplies.

The Geological Survey conducts research and monitors the nation's water resources. It devotes considerable time, money, and energy to develop information on the source of toxic elements. This provides ideas for the best management practices to control such elements (U.S. Geological Survey 1984).

The EPA, an independent federal agency with broad jurisdiction, is responsible for protecting the quality of the nation's water supply. As Congress has directed, the EPA established national regulations on drinking water and wastewater disposal. Hence, the agency regulates the disposal of agricultural wastewater. Most of the nation's effort to clean up wastewater has so far been directed toward point source discharges; however, Congress more recently has directed the EPA to focus also on the cleanup of nonpoint source discharges.

Point source pollution discharges are generally those traceable to a single, discrete, readily identified source. Nonpoint pollution discharges are usually natural, uncontrolled runoff, seepage into water bodies, or other difficult-to-identify sources of pollution. The law defines agricultural runoff as a nonpoint source discharge. Water discharged from agricultural drains does not generally originate from a single source, but it may flow from a single pipe. For example, drainage water may originate from water that is applied in the area of the drain, on upslope land, seepage from water conveyance facilities, or in naturally occurring ground water.

The Clean Water Act and its amendments, passed by Congress to clean the nation's waterways of all pollution by 1985, provides that the EPA may transfer its authority to enforce all federal water quality regulations to any state which adopts regulations that are at least as strict as those adopted by the EPA (U.S. Federal Water Pollution Control Act 1972).

The Soil Conservation Service works with farmers to develop on-farm soil and water conservation plans, sometimes including advice on the design and construction of on-farm drainage and water control facilities.

State Agencies

State legislatures and state departments in charge of water resources, agriculture, environment, wildlife, and health usually take part in establishing water quality control policy. Such agencies operate under different names in different states. They are generally similar to their sister agencies in the federal government in their authority to control and protect water quality, though at the state level.

Special Interest Groups

Numerous special interest groups exist to provide national, state, and local leadership on behalf of the public to insure that government agencies adopt and enforce water quality control regulations. Environmental advocacy groups, including the Environmental Defense Fund, the Natural Resources Defense Council, the Sierra Club, and the Planning and Conservation League carefully watch the actions of the regulatory agencies. When they believe the agencies are not properly carrying out their duties to protect the public interest, they litigate to force the agencies to adopt a stronger environmental position. These groups are active in areas where agricultural wastewater is thought to threaten the beneficial uses of water.

REGULATION OF AGRICULTURAL DRAINAGE WATER

Most of the laws and regulations on water quality were written to control point sources. Many are now being expanded, piecemeal, to include agricultural drainage water and other nonpoint source discharges. Numerous other laws were written to control extremely toxic wastes from industrial discharges. They are, unless modified, inappropriate for the control of nonpoint discharges (Hall et al. 1987).

The following describes several national and state laws and local regulations on such discharges.

Federal Regulations

The Clean Water Act directs the EPA to set effluent standards and regulate all pollutants through the use of the following clauses: "Best

Practicable Control Technology," "Best Available Technology Economically Achievable," and, later, "Best Conventional Pollutant Control Technology." The act also requires that all beneficial uses of receiving water be protected from any discharge of pollutant. The National Pollutant Discharge Elimination System program, created by the act, provides for a permit system to regulate the discharge of pollutants from point sources to the waters of the nation. The act requires that all nonpoint discharges be identified and controlled but does not stringently regulate agricultural wastewater or other nonpoint discharges.

The little-known, seldom-invoked Migratory Bird Treaty Act was passed to protect migratory birds that fly between the United States and Canada (U.S. Migratory Bird Treaty Act 1918). It was enacted to prevent the "unlawful taking or possession" of a migratory bird. The law's terse language has led to some difficulty in its practical application. It came into prominence when the U. S. Secretary of the Interior used it to cut off the flow of agricultural drainage water into Kesterson National Wildlife Refuge after the deaths of migratory birds were attributed to high concentrations of selenium in the drainage water.

The Endangered Species Act is the most comprehensive law passed in the United States to protect wildlife species of all types (U.S. Endangered Species Act 1973). It has been increasingly invoked as resource agencies and public interest groups broaden the geographic areas where endangered species are protected. Since agricultural drainage water often changes the character of wildlife habitat, this act will likely be applied to drainage water management projects.

Environmental protection laws that require all federally funded projects to identify their potential effects on the environment have also been adopted (U.S. National Environmental Policy Act 1969).

State Regulations

Water Quality Control Acts. The Porter-Cologne Water Quality Control Act of California (1970) is an example of a state law under which a federal water-quality law is enforced. The California State Water Resources Control Board and nine regional water quality control boards administer the law. Each regional board develops basin plans, which are used as guidelines for regional water quality planning. Water quality standards are established and persons operating point source waste discharges that empty into state waterways are issued federal or state discharge permits. Nonpoint dischargers must use best management practices, under which they must meet the established standards. Even though agricultural runoff is generally considered a nonpoint discharge, when subsurface agricultural drainage is collected and discharged into artificial basins, the regional boards issue permits for those types of discharges.

Ground-Water Protection Laws. California adopted a strict law, the Toxic Pits Cleanup Act (1984), to protect ground water from the discharge of toxic wastes from industrial manufacturing plants. According to that act, impoundments containing specific contaminant levels must be made safe by designated deadlines or closed. Federal regulations, such as the 1976 Resource Conservation and Recovery Act (RCRA), contain the design and operating criteria for all hazardous waste management. The state water quality control board also developed waste classification regulations for liquid wastes disposed in surface impoundments. Categories range from inert to extremely hazardous. The Toxic Pits Cleanup Act and the criteria for hazardous waste impoundments apply to agricultural drainage water because trace elements that can concentrate to hazardous levels through evaporation are present in some drainage water.

Also, laws prohibit creating a general nuisance. In the case of the Kesterson Reservoir, the state board ruled that the discharge of the agricultural drainage water created a nuisance and, therefore, ordered the reservoir to be closed and cleaned up. The state board also declared the water and sediment in the reservoir to be hazardous waste because of the potential threat to ground water, public health, and the environment (WQ Order 85-1, 1985). The public health threat they perceived involves the possibility of humans eating wild birds that have accumulated selenium in their tissues by eating selenium-tainted seeds and other foods at the reservoir. No direct public health problems were identified.

Several states also have adopted environmental control acts that require any person or agency to evaluate the environmental impact of proposed projects before starting the project.

Local Regulations

Local land-use laws are often imposed on agricultural wastewater management projects. Frequently, they may affect a project as much as any federal or state law.

SAN JOAQUIN VALLEY, CALIF. — CASE STUDY

Subsurface agricultural drainage problems commonly exist in extensive areas of prime agricultural land on the west side of California's San Joaquin Valley. Drainage is needed for salt management and water table control (Letey et al. 1986).

The following discussion illustrates the institutional and legal constraints faced by San Joaquin Valley farmers, who must maintain agricultural production by drainage and salt management. The discussion offers an insight into most of the economical, regulatory, political, emotional, environmental, and water-quality problems that could face any area.

Early Irrigation and Drainage Development

Drainage and water quality problems developed soon after irrigation started on the west side of the valley in 1872, when a cattle rancher, Henry Miller, completed a diversion dam on the San Joaquin River near the city of Mendota. Large quantities of water were applied to the land served by the canals. It became apparent that drainage had to be provided to maintain productive irrigated agriculture.

Miller started to build deep, open drains around 1900 to reclaim some of the land that had become saline. The lowest areas of natural drainage, streams, and sloughs flowing into the San Joaquin River were enlarged to accommodate greater irrigation runoff.

The irrigated area expanded upslope as row crops were introduced. This expansion also caused the drainage problem to move upslope. Drainage water was collected in large open drains and reused as it flowed downslope through the low-lying riparian marshes known as the grasslands area. In many instances, the poor quality drainage water mixed with good quality river water. The reused water eventually ended up in the San Joaquin River, contributing to the deterioration of water quality in the river.

During the 1940's, the irrigated area continued to expand as ground-water pumping facilities and more efficient irrigation systems were developed. In the 1950s, buried subsurface drainage techniques were introduced to the area. As more subsurface drains were installed, the amount of poor quality subsurface drainage water increased to become a larger percentage of the total agricultural runoff. This made it increasingly difficult to reuse the drainage water (Miller and Anderson 1966).

Imported Federal Irrigation Water

During 1939, to prepare for the construction of the Friant and the Delta-Mendota Divisions of the federal Central Valley Project, the Bureau of Reclamation purchased the water rights of the grasslands area and planned a water exchange with other entities that had rights to the San Joaquin River water. This water was reserved for use in another area of the valley. The exchange caused the wetlands area to rely on drainage water exclusively or to purchase imported water. The upslope water users had to use the imported water for irrigation. Most of the wetlands area used available drainage water, since it was essentially free, and supplemented it with purchased imported water.

The imported water contained more salt than the San Joaquin River water that it replaced, i.e., 300 mg/l to 350 mg/l as compared to 40 mg/l to 50 mg/l total dissolved solids (TDS). Imported water was first delivered to the area about 1950, when the Delta-Mendota Canal was completed. The use of the lower-quality imported federal water supply resulted in an increasing rate of salt accumulation in the soil and the drains and a continued degradation of the San Joaquin River. The river's average salinity increased to about 160 mg/l TDS (U.S. Bureau of Reclamation and South Delta Water Agency 1980).

During the 1950s and 1960s, the Soil Conservation Service worked with farmers and conservation districts to help finance, design, and build a network of drains to route agricultural drainage water through the grasslands for use as a wetland-area water supply. After flowing through the wetlands, the drainage water was discharged into the San Joaquin River. This practice increased the San Joaquin River's water quality problem (Leach 1979).

Additional imported irrigation water came into the area through the federal San Luis Unit, which was authorized in 1960 (U.S. San Luis Unit Authorization Act 1960). Construction of the San Luis Unit took place between 1965 and 1974, with the first deliveries of water taking place in 1967.

A major controversy about the importation of water to the San Luis service area developed between local landowners, who were experiencing drainage problems, and the federal government. Landowners argued that drainage had to be provided if the land upslope from them was to receive imported water. Congress agreed and authorized a drainage system as part of the project. The plan was to convey saline drainage water from the irrigated land to the San Francisco Bay-Sacramento/San Joaquin Delta estuary through a 303-km (188 mi), concrete-lined drainage canal (U.S. Department of the Interior 1955).

The importation of water sparked the beginning of many years of confrontation between the federal government, the state of California, local agencies, land owners, environmental groups, and the water users. Environmental groups and individuals representing the Bay-Delta area expressed concern over potential damage to the estuary if agricultural drainage water discharged directly into the estuary. Congress, in 1965, prohibited the Bureau of Reclamation from determining the final point of discharge for the authorized drain. Congress required federal and state agencies to complete environmental studies and to agree on the actions needed to minimize the discharge's effect on estuarine resources before the drain outlet could be built (U.S. Appropriations Bill 1965).

State Service Area

California built a similar water project, the State Water Project, to import irrigation water from northern California into the San Joaquin Valley, south of the federal service area in the Tulare Lake Basin and the Valley's Kern County area .

This area, unlike the northern San Joaquin Valley, has no natural outlet to the Pacific Ocean, except during floods. The three major streams that flow into the Tulare Basin terminate in that closed basin. Before 1900, runoff from the Sierra Nevada created shallow, marshy areas and lakes in the low-lying areas of the basin. The streams were controlled and the runoff stored for irrigation, so the lakes dried up and the lake beds became excellent farmland. Although they were extremely fertile, they eventually developed a high water table and a difficult salt-balance problem.

Since the state serves imported irrigation water to both areas, it considered that it was partly responsible for ensuring adequate drainage.

Federal-State Plans

During 1960, at the time Congress was authorizing the federal San Luis Drain, California included a valley drain, a part of the voter-approved State Water Project included in the state's Burns-Porter Act (1960). By 1967, state officials concluded that the lack of willing customers within the state's service area and the state's unwillingness to finance its share of the construction and operating costs of a valley drain indicated that the Bureau of Reclamation should independently begin to build a drain for the federal service area. After many delays, the Bureau began to build the central part of the San Luis Drain in 1968. Only 137 km (85 mi) of the center portion was ever completed. In addition, a 486-ha (1,200-acre) regulating reservoir was built in the flood plain of the San Joaquin River, near the town of Gustine. However, the reservoir became the terminus for the constructed portion of the drain. The planned outlet portion of the drain was never completed. The reservoir consequently became an evaporation basin for the drainage water collected from approximately 3,200 ha (8,000 acres) of agricultural land in the federal San Luis Unit service area (WQ Order 85-1, 1985).

The Fish and Wildlife Service agreed with the Bureau of Reclamation to manage the reservoir area, knowing that the only reliable water supply for the area would be subsurface agricultural drainage water, and named it the Kesterson National Wildlife Refuge (U.S. Department of the Interior, Bureau of Reclamation, Special Task Force Report on San Luis Unit 1978). Previously, the area had no water supply. It was used for grazing, with portions sometimes used for waterfowl hunting.

Between 1967 and 1974, the Bureau conducted extensive monitoring programs and studies within and outside the Bay-Delta estuary. The studies included efforts to determine the technical feasibility of discharging treated or untreated agricultural drainage water into the estuary. Among the expressed concerns were the problems that might be caused by pesticides or biostimulents in the drainage water (Kaiser Engineers 1969).

A comprehensive federal-state study, the Interagency Drainage Program, was started in 1976. It evaluated 18 alternatives for the safe disposal of San Joaquin Valley agricultural drainage water. It concluded that, of the 18 locations, including several land and ocean disposal alternatives, a discharge to the estuary was the most economical and environmentally acceptable alternative (U. S. Bureau of Reclamation 1979). It also concluded that no pesticide or nitrate problems would likely take place in the estuary as a result of the discharge of agricultural drainage water from a San Joaquin Valley Drain. It further recommended that the Bureau of Reclamation complete a final Environmental Impact Statement for the San Luis Drain before the start of any additional construction. The Bureau began studies to get the information needed for a discharge permit for the drain.

In the state service area of the Tulare Basin, salt buildup in the soil required that over 11,000 ha (28,000 acres) be drained with subsurface drains by 1970. There was little interest in taking part in a valley drain, mainly because of the costs involved. Therefore, the Tulare Lake Drainage District was formed in the northern part of the Tulare Lake Basin to manage the water produced by draining this land. The water collector system and evaporation basins for handling the first drainage water were built in the mid-1970s (Summers 1968). Water from drained land outside the district is disposed of in on-farm evaporation ponds or in irrigation canals from which it is reused. Later, additional district evaporation basins were built and expanded to accommodate other land, as additional drainage problems developed.

In Kern County, the valley's southernmost area, drainage problems developed more slowly. However, as of 1979, over 56,000 ha (140,000 acres) were lay within the area of the valley floor where drainage problems potentially occur (Rector and Core 1979). By 1988, the area

TABLE 26.1 Drainage Water Concentrations of Trace Elements

	Concentration (mg/l)		
Constituent (1)	Minimum (2)	Maximum (3)	Median (4)
Arsenic	<1	50	<1
Boron	40	84,000	6,500
Cadmium	<1	4	<1
Chromium	<1	800	2
Copper	<1	510	6
Iron	<3	360,000	250
Lead	<1	73	<1
Manganese	<1	9,000	80
Mercury	<0.1	1.6	<0.1
Molybdenum	<1	1,500	13
Nickel	<1	900	
Selenium	<1	4,700	25
Silver	<1	10	<1
Zinc	<3	1,000	11

over which the water table depth is less than 1.25 m (5-ft) encompassed 33,600 ha (84,000 acres).

Drainage Water Quality Problems

Subsurface drainage water on the west side of the San Joaquin Valley is generally characterized as saline, with salt content ranging from 2,000 mg/l to 100,000 mg/l TDS. However, in addition to the common elements in TDS, there are several other naturally occurring elements, such as arsenic, boron, chromium, mercury, molybdenum, and selenium, that are potentially toxic to agriculture, wildlife, and human health. Table 26.1 presents the minimum, maximum, and median concentrations of drainage water constituents found in samples of drainage water collected by the U.S. Bureau of Reclamation and the U.S. Geological Survey between 1984 and 1986 from 40 observation wells, farm drain sumps, and collector drains in the federal service area and Kings County (U.S. Department of the Interior 1987).

Surface drainage water generally is similar in quality to the delivered irrigation supply. With use, it may pick up additional constituents applied to agricultural crops, such as fertilizer, herbicide residues, and pesticide residues. In general, the proper and safe use of new-generation, rapidly-degrading farm chemicals keep the surface drainage water from becoming hazardous to the health and welfare of humans or wildlife.

The discovery of high concentrations of selenium in wildlife and the correlation of the selenium to agricultural drainage water caused tremendous public concern over the way in which agricultural drainage water was being managed in the San Joaquin Valley (California Water Resources Control Board 1987) and elsewhere in the western United States (U. S. Committee on Irrigation and Drainage 1988).

The aforementioned closure of Kesterson Reservoir was the culmination of a long series of events that started when an individual landowner filed a petition with the state water quality control board, requesting that the board overturn a decision by the Central Valley Regional Water Quality Control Board on the operation of the reservoir by the Bureau of Reclamation.

CONCLUSION

The institutional and legal constraints facing entities with agricultural drainage water disposal problems, whether they be public agencies or private farmers, are becoming formidable. The laws, regulations, and controls placed on subsurface agricultural drainage water in the San Joaquin Valley of California exemplify the type of constraints that may soon face many farmers elsewhere.

The public demands good-quality water, and the agricultural industry supports this goal. However, public perception of water quality problems can frequently complicate the development of technical solutions. It is extremely important that agricultural representatives understand the political sensitivity, as well as the legal and institutional constraints, related to water-quality problems.

The agricultural drainage water storage sites necessary in California as a result of the closure of Kesterson Reservoir are viewed by many regulators as hazardous waste disposal sites. To avoid this type of dilemma elsewhere, agricultural interests must assist legislative and regulatory representatives in developing long-range plans for the management and disposal of agricultural waste water. Successful long-term plans will include a realistically timed strategy for action and an accurate assignment of costs.

REFERENCES

California Water Resources Control Board. 1987. "Regulation of Agricultural Drainage to the San Joaquin River". *Technical Committee Report*, Sacramento, Calif.

Deason, J.P. 1987. "Selenium: It's Not Just in California." *Proceedings Irrig. and Drain. Specialty Conf.*, ASCE., Portland, Or. pp. 475-482

Hall, S.K., Johnston, W.R., and Miller, W.J. 1987. "Agricultural Drainage Water—How Should it be Regulated?" *Proceedings Irrig. and Drain. Specialty Conf.*, ASCE., Portland, O. pp. 483-488.

Johnston, W.R., and Heaton, M.G. 1986. "Legal, Economic, and Political Constraints on Managing Agricultural Drainage Water in California," *Proceedings National Symposium on Water Resource Law*, ASCE., Chicago, Ill. pp. 188-195.

Kaiser Engineers. 1969. "San Francisco Bay-Delta Water Quality Control Program," *Final Report to the State of California*, Oakland, Calif.

Leach, R. 1979. "Water Conveyance Pollutants Into The Grassland Resource Conservation District". *Report of Grassland RCD and USDA Soil Conservation Service*, Los Banos, Calif.

Letey, J., Roberts, C., Penberth, M., and Vasek, C. 1986. "An Agricultural Dilemma: Drainage Water and Toxics Disposal in the San Joaquin Valley." *Special Publication 3319,* Div. of Agric. and Nat. Res., Univ. of Calif., Oakland, Calif.

Miller, R.S., and Anderson, C.F. 1966. "Progress Report—Factors Affecting Drainage on the West Side of the San Joaquin Valley". *USDA Soil Conservation Service,* Berkeley, CA.

Pillsbury, A.F., and Johnston, W.R. 1965. "Tile Drainage in the San Joaquin Valley of California." *Water Resources Center, Contribution No. 97,* University of California, Los Angeles, Calif.

Rector, M.R., and Core, F.R. 1979. "Monitoring and Evaluating Perched Water Table Conditions in Kern County, Calif. *Technical Report,* Kern County Water Agency, Bakersfield, Calif.

State of California, Burns-Porter Act of 1960. *California Water Code, Sections 12930-12944,* Sacramento, Calif.

State of California, Porter-Cologne Water Quality Control Act of 1970. *California Water Code, Division 7, Section 13000 et seq.,* Sacramento, Calif.

State of California, Toxics Pits Cleanup Act of 1984. *California Health and Safety Code, Section 25208.*

Summers, J.B. Aug. 1968, Drainage Needs, Collection and Disposal *Tulare Lake Drainage District,* Corcoran, Calif., 36 pp.

U.S. Appropriations Bill of 1965 *PL 89-299, USC 43 Section 377a,* Washington, D.C.

U.S. Bureau of Reclamation and South Delta Water Agency. 1980. "Effects of the CVP on the Southern Delta Water Supply, Sacramento San Joaquin River Delta, Calif." *Joint Report, U.S. Water and Power Resources,* Sacramento, Calif.

U. S. Bureau of Reclamation, Calif. Dept. of Water Resour. and Calif. Water Resour. Control Board, June 1979. "Agricultural Drainage and Salt Management in the San Joaquin Valley." *Final Report of the Interagency Drainage Program,* Sacramento, Calif.

U. S. Committee on Irrig. and Drainage, Report of the Panel of Experts, 1988. "How Can Irrigated Agriculture Exist With Toxic Waste Regulations?" Summary of an Expert Panel Meeting Oct. 1987, *Proceedings 1987 National Meeting on Toxic Substances in Agricultural Water Supply and Drainage—Searching for Solutions,* Denver, Co.

U. S. Dept. of Interior, 1987. "Developing Options: An Overview of Efforts to Solve Agricultural Drainage and Drainage-Related Problems in the San Joaquin Valley." *San Joaquin Valley Drainage Program,* Sacramento, Calif.

U. S. Dept. of Interior, Bureau of Reclamation, May 1955. *San Luis Unit Feasibility Report.* Washington D. C. 156 pp.

U. S. Dept. of Interior, Bureau of Reclamation, 1978. *Special Task Force Report on San Luis Unit.* Washington D.C. 331pp.

U. S. Endangered Species Act of 1973. *PL 93-205, 16 USC Section 1531, et seq.*

U. S. Federal Water Pollution Control Act of 1972, *PL 92-500, USC 33 Section 1251 et seq.*

U. S. Geological Survey. Nov. 1984. *Areal Distribution of Selenium and Other Inorganic Constituents in Shallow Ground Water of the San Luis Service Area.* Sacramento, Calif..

U. S. Migratory Bird Treaty Act of 1918. *16 USC Section 703 et seq.*

U. S. National Environmental Policy Act of 1969. *PL 91-190 42 USC Section 4321 et seq.*

U. S. Resource Conservation and Recovery Act of 1976. *PL 94-580 90 USC Section 2795 et seq.*

U. S. San Luis Unit Authorization Act of 1960. *PL 86-188, 74 Stat. 156*

WQ Order 85-1 Feb. 1985. *Calif. Water Resour. Control Board Decision,* Sacramento, Calif. 65 pp.

CHAPTER 27

ENVIRONMENTAL QUALITY AND AGRICULTURAL IRRIGATION—RECONCILIATION THROUGH ECONOMIC INCENTIVES[a]

INTRODUCTION

This chapter surveys the effects of agricultural irrigation and drainage on natural resources and the environment, discusses the economic incentives that determine management decisions, and suggests several improvements. Illustrations of the causes and effects described in this chapter are based in the western United States, but many exist elsewhere in the world.

EFFECTS ON NATURAL RESOURCES AND THE ENVIRONMENT

Irrigation and drainage affect a variety of natural resources and environments. Effects termed on-farm or internal are confined to the irrigator's property and operations. Those termed off-farm or external extend beyond those confines.

Soil Deterioration

Soil deterioration, an on-farm effect, imposes costs on property owners and farmers. The build-up of soil salinity and waterlogging has increasingly concerned irrigators in many agricultural regions. Perhaps no other deleterious effect of irrigated agriculture has so long a demonstrated history. It undermined early Middle Eastern civilizations. It continues to plague much of the world (Szabolcs 1975), particularly Africa (Antar 1981), the Middle East (El Gabaly 1976), South Asia (Tyagi 1986; Herath 1985) , Australia (Greig and Devonshire 1981), and the Western United States (U.S. Committee on Irrigation and Drainage 1986, El Ashry and Gibbons 1988).

Estimates of the total area of land in the Western United States affected by salinization and waterlogging vary greatly. Those estimates made during the decades of irrigation system expansion that ended in the 1970's were too high. The expansion ended, and some farmers and irrigators adopted increasingly efficient farming practices. Soil deterioration has come to be viewed as a problem to manage by changing

---

[a]Prepared by: Zach Willey, Envir. Defense Fund, Rockridge Market Hall, 5655 College Ave., Oakland, CA 94618.

cropping practices and technologies. A number of innovative programs have been developed in California's San Joaquin Valley, particularly since the contamination of Kesterson Reservoir in the early 1980's and other irrigated areas. For example, the San Joaquin Valley's Broadview Water District has used an inverted block rate structure to provide incentives to reduce drainage and recycle (Wichelns and Cone 1989), while the Westlands Water District is investigating drainage-water management and treatment technologies (Westlands Water District 1987). Salinity control measures for irrigation in the Colorado River Basin have been investigated and used for some time (U.S. Department of the Interior 1983).

Ground-Water Overdraft

Ground-water overdraft occurs when irrigators and other water users cumulatively pump water from aquifers faster than infiltration and inflows of ground water replenish it. Overdrafts are particularly serious in parts of the western United States (U.S. Government Accounting Office 1980, Water Efficiency Working Group 1987), where technical and economic resources enable long lifts of large amounts of water. In California, they provide up to 10% of the total average annual irrigation water supply in normal years and much more in dry years (California Department of Water Resources 1987).

Overdrafts reduce the contribution of subsurface water sources to surface habitats and bodies of water. They have been a key factor in the decline of Sonoran desert plant communities and riparian habitats in a number of areas in Arizona (Bryan 1928) and the reduction of instream flows—the volume and rate of stream and river flows—in a number of river basins in the Southwest (Groenvald and Griepentrog 1985; Zimmerman 1969; Sheridan 1981). In a related effect, the collapse of voids formerly filled by water in strata with a high clay content can irreversibly reduce the water storage capacity of some aquifer formations and simultaneously cause surface-land elevations to subside (McCauley and Gum 1972; Willey 1977).

When overdraft exhausts ground-water supplies, it undermines the agricultural economy of regions with no other supplies. Farmers in the lower High Plains region of Oklahoma and Texas pay increasingly more to pump ground water from greater depths, as the overdraft causes the water table to decline. These increased costs reduce the economic viability of irrigated agriculture in the region (U.S. Department of Commerce 1982). Overdrafts can also stimulate political demands to build new surface-water supply projects. The proposed Mid-Valley Canal in California's Tulare Basin, which would tap the water supplies that maintain environmental values in other parts of the state, is one such example (U.S. Department of the Interior 1981). The now-defunct High Plains recharge project is another (U.S. Department of Commerce 1982).

Surface- and Ground-Water Contamination

Surface- and ground-water contamination results when water containing salts, trace elements, or other contaminants runs off irrigated fields. This contaminated runoff either directly flows to surface waters

TABLE 27.1 Information About Primary Nonpoint Source Pollutants in Use-Impaired Waters[a] in the United States

	Total Waters Impaired (%)	
(1)	Rivers (2)	Lakes (3)
Pollutants		
Sediments	47	22
Nutrients	13	58
Pathogens	9	2
Physical habitat alteration	9	4
Acidity	7	4
Toxics	6	3
Oxygen demand	4	3
Pesticides	3	1
Salinity	2	3
Sources		
Agriculture	64	57
Hydro-modification	4	13
Urban runoff	5	12
Land disposal	1	5
Construction	2	4
Resource extraction	9	1
Silviculture	6	1
Other	9	7

Source: Association of State and Interstate Water Pollution Control Administrators (1985).
[a]Use-impaired waters are those in which uses are severely or moderately impaired by nonpoint sources or in which nonpoint sources threaten to impair the designated use.

or infiltrates through the soil layer into ground-water aquifers that connect to surface waters. It adversely affects rivers, lakes, wetlands, and estuaries. Salts and trace elements often exist in agricultural lands before irrigation or in irrigation-water supplies from the natural erosion of local rock and soil formations. In either case, irrigation concentrates these contaminants, as evaporation and crop transpiration occurs. Subsurface irrigation drainage mobilizes and transports salts and trace elements. The loading of these pollutants in surface waters often results.

Agricultural water pollution in general and total dissolved solids (various salt compounds) and trace elements in particular play a significant role in surface-water pollution problems. Table 27.1 (Association of State and Interstate Water Pollution Control Administrators 1985) provides information on nonpoint sources of pollutants by source and pollutant and the extent to which they impair uses of rivers and lakes in the

TABLE 27.2 Observed Concentration of Trace Elements in 22 Evaporation Ponds: Tulare Basin, San Joaquin Valley, California, 1985-1987

Trace element (1)	Maximum (2)	Minimum (3)	Mean (4)	Standard deviation (5)
Arsenic (μg/l)				
Inflow	900	3	199	277
Pond	13,000	2	1,040	2,710
Boron (mg/l)				
Inflow	64	2.8	19	15
Pond	840	11	115	172
Molybdenum (mg/l)				
Inflow	7,775	83	1,476	1,650
Pond	40,000	250	5,929	8,396
Selenium (mg/l)				
Inflow	943	1.2	109	233
Pond	1,900	1.6	174	431

Source: Westcot et al. (1988).

United States. While agriculture is, by far, the largest nonpoint source of water pollution, salinity impairs uses in a relatively minor way. However, total dissolved solids and trace elements can be major pollutants of surface water in some regions, such as California's Sacramento and San Joaquin River Delta (Robie 1988) and Central Valley wetlands (Ohlendorf et al. 1986, Schroeder et al. 1988, State of California 1987). Table 27.2 (Westcot et al. 1988) shows the presence of trace elements in drainage evaporation ponds in California's Tulare Basin. As the table indicates, trace elements vary extremely in soils; they are present in high concentrations in some cases and hardly at all in others. The impact of trace elements in the Delta and San Francisco Bay is still being studied. Limited tests show no immediate toxic effects of trace elements on larval striped bass (Doroshov and Wang 1984), but such effects have been observed on the prime food source of the bass (SRI International 1985). Selenium, the most studied trace element as to its effects on wildlife, may cause damage within a band of concentrations as low as under 1 ppb and as high as about 50 ppb (Eisler 1985)].

Irrigation-water drainage has deteriorated the habitat of wildlife in other regions of the western United States, including Nevada's Truckee and Carson River Basins (U.S. Bureau of Reclamation 1987, Yardas 1989), Utah's Green River Basin (Stephens 1988), and Montana's Sun River Basin (Knapton, Jones, and Sutphin 1987). The bioaccumulation of selenium and subsequent deaths and birth deformities among several species of waterfowl at California's Kesterson Wildlife Refuge has led to a growing recognition that trace elements carried in irrigation drainage waters have contaminated other waterfowl refuges (U.S. Fish and Wildlife Service 1986). The problem was extensively studied in California, Nevada, Utah, Wyoming, Arizona, Montana, and Texas in 1986 and

1987 (U.S. Department of the Interior 1988). Seventeen other refuges in the western United States are being studied (U.S. Fish and Wildlife Service 1986). These studies, which will likely continue for some time, resulted from a recognition during the early 1980's of irrigation's adverse effects on wetland habitats.

Irrigation water often mobilizes and transports pollutants other than total dissolved solids and trace elements. Without irrigation return flows, some of these pollutants would have less effect, and those effects would be more restricted to the on-farm environment. For example, significant amounts of pesticide residue in ground water poses potential health risks where wells deliver drinking water supplies (Pye et al. 1983; Smith and Vaughan 1988b; Holden 1986). Evaluating such risks, which include identifying the most hazardous substances, routes of exposure, and highly exposed population sectors, is a complicated and ongoing task (Society for Risk Analysis 1981). Also, irrigation and drainage have provided a habitat for disease vectors, causing human health problems. For example, schistosomiasis has been a serious public health problem for years in the Nile River Basin (Amin 1976).

Surface-Water Diversions and Habitat Degradation

Diversions of surface water and degradation of habitats result from building and operating irrigation-water-supply reservoirs and canals. For example, loss of wildlife and recreational values typically results, regardless of whether such facilities are designed to accommodate the migration of fish. Examples within California's Central Valley Project, a federal reclamation project designed to provide irrigation water supplies, include the adverse effect of the Red Bluff Diversion Dam on migrating salmon populations in the Sacramento River Basin, the total loss of the San Joaquin River salmon and steelhead fisheries due to the Friant Dam's removal of instream flows, and the decline of the white-water recreational industry on the Stanislaus River due to the New Melones Dam. Other examples exist throughout the western United States (Moore 1989, O'Keefe et al. 1986).

ECONOMIC INCENTIVES — EXISTING DEFICIENCIES

Irrigation accounts for approximately 80% of total net annual water use in much of the western United States. A similar share of water use is present in most of the world's irrigated regions (Repetto 1986). Agriculture is also a major source of water pollution in the United States, with irrigation making a signifant contribution through surface and subsurface runoff. To control irrigation's diverse and substantial effects on the environment, incentives and options facing farmers need to be changed. Deficiencies in existing incentives can be grouped into three classifications.

Market Externalities

The environmental effects of irrigation include the internal or on-farm, such as soil deterioration, that have been the subject of considerable economic analysis (Colacicco, Osborn, and Alt 1989; Ribaudo et al. 1989) and the external or off-farm, such as pollution of surface water or destruction of the river habitat, the economics of which are much less understood. Difficult methodological issues and expensive data collection must be overcome to quantify externalities economically (Randall 1981). Measurement of use values, option and existence values, and irreversibilities is the main problem.

Nevertheless, efforts to quantify external effects economically are underway. For example, recent studies in California have tried to determine the economic value of several activities affected by irrigation externalities, including waterfowl hunting in the Central Valley (Miller 1986) and whitewater rafting on rivers draining the western front of the Sierra Nevada Mountains (Environmental Defense Fund 1984). Economic valuations are becoming common for resources that often compete with irrigation for water supplies, such as fisheries (Vaughan and Russell 1982). Such studies apply various methods. All try to quantify their economic findings based on survey or expenditure data.

Why are these fledgling efforts at quantifying external effects important? Irrigators do not bear the burden of these external costs. External parties do. Irrigation systems were developed when external effects and costs were much less recognized. Due largely to increased monitoring and research in recent years, irrigation is now known to degrade aquatic ecosystems and other external environments. Quantifying many of the lost economic values is just beginning.

Estimating the economic value of external water uses destroyed or degraded by irrigation is new. However, estimating two components of overall external costs is relatively straightforward. Engineering project data is used to estimate the cost of cleaning up polluted water and habitats. Such costs can be significant. For example, the costs of some programs to clean up the Kesterson Wildlife Refuge exceed the value of the agricultural lands that contributed to the problem (Wahl 1986). Technologies to treat drainage water, which continue to be extensively researched, will probably have costs beyond the repayment capacity of some crops (Lee, Nishimura, and Hansen 1988). Nonagricultural income to support such treatment facilities, such as revenue from water marketing, will likely be needed for treatment to be viable for irrigators (Willey 1987).

A second cost is that of replacing water supplies displaced by irrigation. Ways to replace the water include storage of surface and ground water, water conservation, wastewater reclamation, and water marketing. The literature on these options is specific to the site and extensive (El Ashry and Gibbons 1988; Willey 1985; Blackwelder, Harding, and Colborn 1987).

Water Pricing and Allocation

The availability of inexpensive irrigation water was a key aspect of the economic development of the western United States. Water prices charged to irrigators in most western states have historically been one to two orders of magnitude less than the prices paid by municipal and industrial users (Fig. 27.1). State water laws based on the appropriation

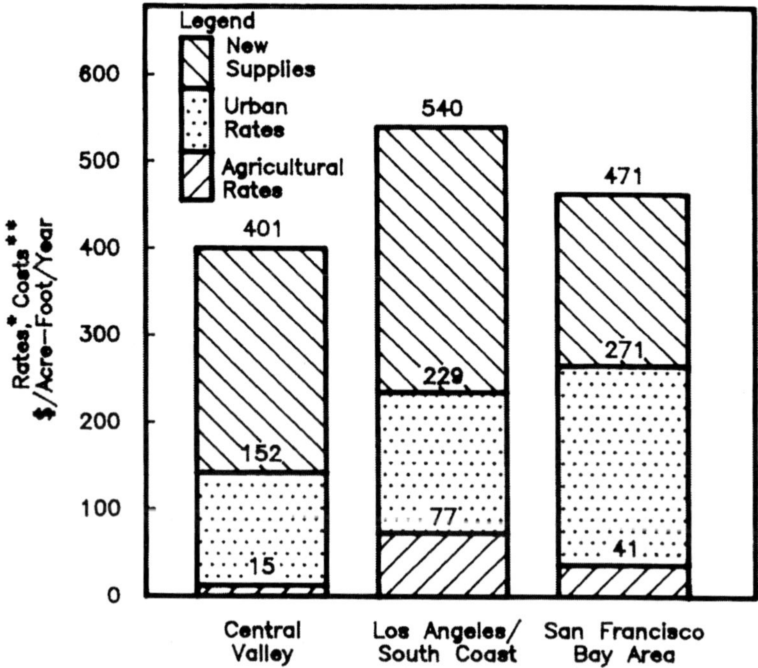

Fig. 27.1 Retail Rates and Costs of New Surface Supplies in California Water Systems in 1985 (Willey and Graff 1988)

doctrine of "first in time, first in right" have resulted in rights to irrigation water at virtually no cost other than conveyance to field headgates. Federal reclamation water contracts, based on the Reclamation Act of 1902 (amended in 1982), charge no interest to irrigators on the construction costs of water projects (Frederick 1982; El Ashry and Gibbon 1986).

The costs of developing structures that supply surface water for irrigation generally have been the same as those for municipal and industrial uses (Fig. 27.2). These costs are typically two orders of magnitude greater than prices paid for irrigation water. The cost of conserving water often exceeds the price paid by irrigators for the water. Hence, irrigators who pay exceptionally low water prices, as is the case in many areas in the western United States that have reclamation projects, have little economic incentive to reduce their use of water. Also, the fees paid by irrigators to discharge drainage and runoff are too low to create an incentive to reduce pollution loads by improving management of on-farm irrigation. As the problems associated with water scarcity and pollution become more acute, policies to create effective incentives, such as those discussed below, will become increasingly necessary.

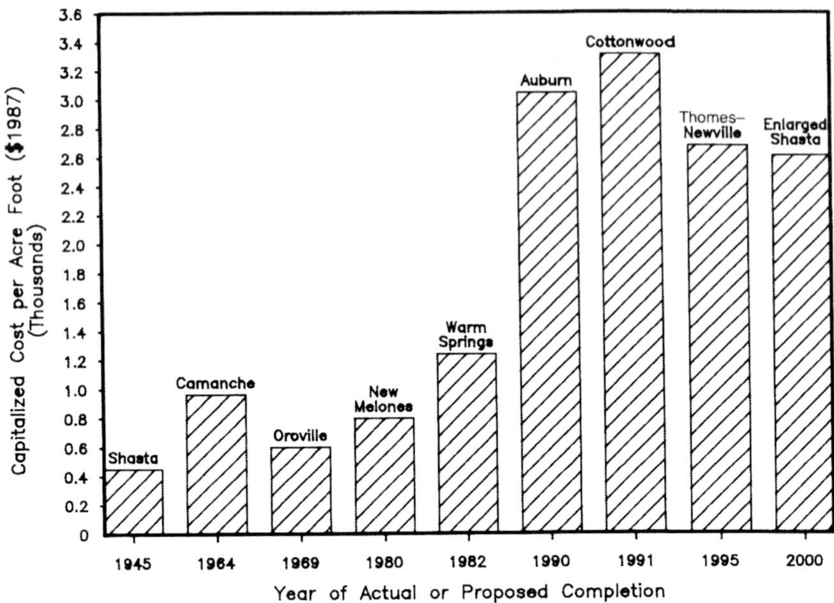

Fig. 27.2 Historical and Projected Costs of California's Surface Water Supply (Costs Capitalized over 50 Years at 8% per Annum) (California Department of Water Resources 1983)

Agricultural Commodity Supports

The agricultural commodity support programs in the United States have been extended in the federal farm bills that have been enacted every four years since the program began in 1933 (Glaser 1986). The programs create a major economic incentive to select crops and irrigate in ways that affect consumptive use and pollution of water. Fig. 27.3 shows that commodity support payments have grown more important in recent years as a source of farm income in the United States. Similar price subsidy programs also exist in other countries and with similar results (International Institute for Environment and Development and the World Resources Institute 1987).

In 1986, farmers in the 17 western states received a total of over $6 billion in commodity support payments (Table 27.3). These payments averaged nearly 15% of farm cash receipts in the Missouri Basin states and nearly 10% in the 17 western states. Nationally, cotton and feed grain payments accounted for approximately 75% of the total farm program payments in 1986.

These commodity price supports create surplus production (Paarlberg 1980), which results in larger irrigated acreages than would otherwise occur. Fig. 27.4 (U.S. Department of Agriculture 1987) shows that irrigated acreage in the western U.S. grew steadily until 1978 and then declined. These acreages would be less without the supports. Consumer prices would be somewhat higher, but costs to taxpayers would be lower. Reduced acreage in irrigated areas would reduce water use and

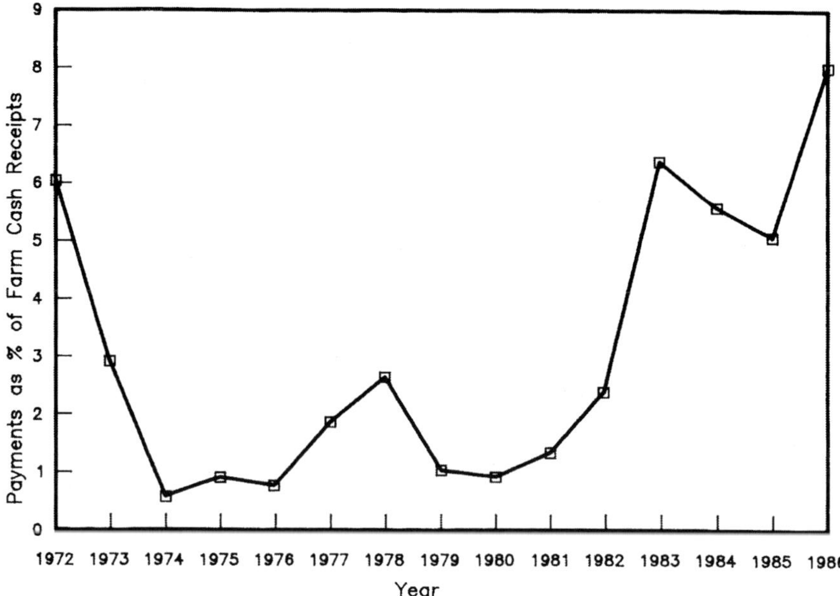

Fig 27.3 Government Cash Payments to Farms in The U.S. Commodity Support Program from 1972 to 1986 (U.S. Department of Agriculture, 1987)

pollution. In effect, costs would be redistributed from federal taxpayers and degraded environments to producers and consumers of these crops.

Without supports, more acreage would be available for other crops and other uses, such as irrigation. To estimate the amount of crop substitution and the changes in irrigation and drainage that would take place, economic and hydrologic models of agricultural production and markets can be used (Harris and Miller 1986). Substitute crops, such as fruits and vegetables, frequently have a more limited market than do the support crops, which often have large national and even international markets. Such specialized crops are often more profitable than support crops but also tend to be more risky in terms of produce quality, marketing, and price fluctuations. These risks would discourage a one-to-one reallocation of support-crop acreage to substitute-crop acreage.

Some support crops, such as cotton, rice, and alfalfa hay need more irrigation and drainage than many substitute crops. Other support crops, such as wheat and barley, need less. Thus, in places where cotton and rice are now grown, such as California and the lower Colorado River Basin, substitute crops would likely need less irrigation and drainage. In places where grains are now grown, such as the northern Great Plains, substitute crops might need more. As noted earlier, whether such substitution would, in fact, take place depends particularly on the marketing limits and opportunities facing producers of these substitute crops.

TABLE 27.3 Commodity Support Payments for 17 Western States of the United States for 1986

Region (1)	State (2)	Government payments ($$10^6$) (3)	Farm cash receipts ($$10^6$) (4)	Share (%) (5)
Missouri River Basin	Kansas	870.8	6,295.7	13.8
	Nebraska	867.8	7,796.3	11.1
	North Dakota	700.2	2,999.2	23.3
	South Dakota	382.9	2,845.5	13.5
	Montana	366.1	1,579.3	23.2
Subtotal		3,187.8	21,516.0	14.8
High Plains	Oklahoma	393.0	3,014.7	13.0
	Texas	978.4	9,422.3	10.4
Subtotal		1,371.4	12,437.0	11.0
Upper Colorado River Basin	Colorado	234.6	3,343.3	7.0
	Wyoming	29.3	595.2	4.9
	Utah	36.0	606.2	5.9
	New Mexico	62.4	1,072.4	5.8
Subtotal		362.3	5,617.1	6.4
Lower Colorado River Basin	Nevada	3.5	235.2	1.5
	Arizona	83.6	1,578.7	5.3
Subtotal		87.1	1,813.9	4.8
Pacific Northwest	California	387.9	14,436.6	2.7
	Washington	298.7	3,091.7	9.7
	Oregon	127.6	1,911.7	6.7
	Idaho	223.2	2,148.7	10.4
Subtotal		1,037.4	21,588.7	4.8
Total for the 17 states		6,046.0	62,972.7	9.6

Source: U.S. Department of Agriculture (1987).

POLICY REMEDIES — CREATING ECONOMIC INCENTIVES

Two types of public policy exist. One, often described as command-and-control, is based primarily on government regulation of production and other activities that affect society as a whole. The other, described as incentive-based, creates systems of incentives to induce businesses and people to act in ways that benefit society.

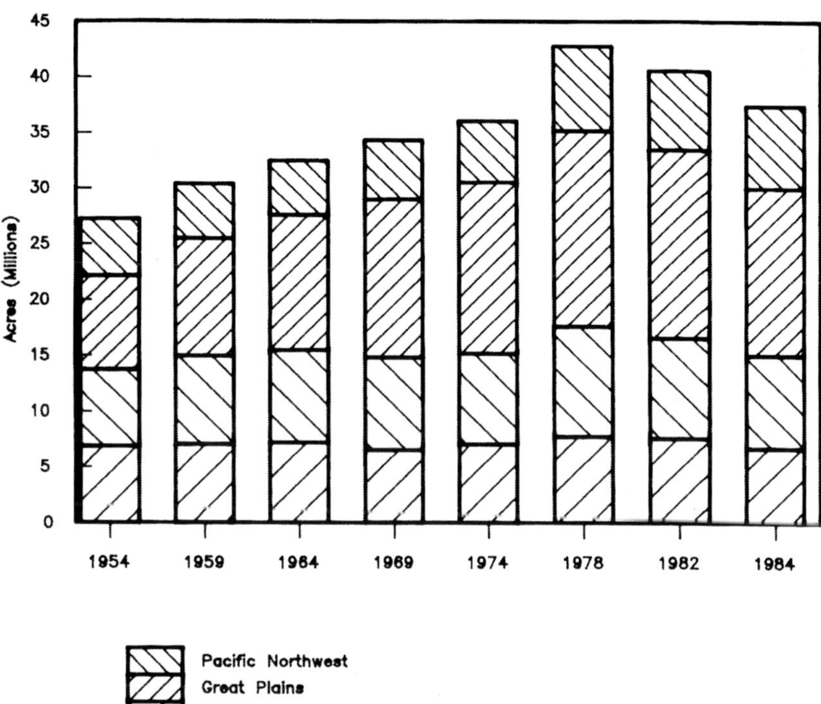

Fig. 27.4 Irrigated Acreage for Western Regions of the United States for Selected Years from 1954 to 1984 (Wiley 1989, U.S. Department of Agriculture 1987)

These competing philosophies have increasingly become the subject of discussion and debate (Columbia University School of Law 1988). Water-resource policies, especially those influencing irrigation, is part of that debate (Willey and Graff 1988). Incentive-based water policies have often been proposed, but partly implemented at best (National Water Commission 1973; Heinz and Wirth 1988). Proposals for irrigation range from invalidating irrigation-water rights to increasing contract water prices to full-scale implementation of water markets. Proposed policies about drainage and runoff pollution control have ranged from irrigation shut-downs to imposing disposal fees to trading wastewater discharge permits. The policies that appear to resolve environmental problems caused by irrigation practices with the least disruption to agriculture are described below.

Water Marketing and Transfers

All of irrigation's adverse effects on the environment stem from the scarcity or degradation of water supplies. Water supplies need to be shifted from existing uses to remedy these effects. Water marketing and transfers facilitate such shifts by providing incentives to voluntarily undertake them. Water marketing transactions can be of several types. Permanent transfers may involve outright selling of water rights or contracts in perpetuity. This can take place annually or intermittently, in dry years only, for example. Water rights can be leased as well, with the rights returning to the lessor after a specified time. The legal and technical arrangements for water marketing transactions can substantially vary. In any case, these transactions can help to mitigate and enhance environmental values.

Water marketing transactions benefit the environment in three ways. First, they can shift water from existing to new uses, thereby precluding the need for new supply and diversion projects that can adversely affect the environment. Irrigation, which accounts for an average of approximately 80% of all water uses in the Western United States, is the prime source of supply for water marketing transactions. Irrigators can use the income from water sales and leases to conserve water by improving irrigation systems and changing cropping patterns. Water transfers from agricultural to urban uses are increasingly common where municipal and industrial water uses are the primary source of new demand, as is the case in the Western United States.

Second, water marketing transactions can be used for the direct acquisition of water for environmental uses. A relatively new aspect of water marketing, such arrangements are being considered or used in several areas of the Western United States. Again, irrigation water supplies are the prime source for such acquisitions so that some of irrigation's adverse effects on the environment are mitigated and possibly reversed, while irrigators are compensated for adjusting their water use.

Finally, the availability of water marketing options provides irrigators with the economic incentive to decrease irrigation, which often reduces pollutant loadings. Also, to the degree that irrigators must pay for the right to discharge pollutants into the external environment, income from water marketing can help to build treatment or discharge facilities, buy discharge permits, or compensate liability claimants.

Water markets have developed in a number of regions in the Western United States. The feasibility of such markets in other irrigated regions of the world depends on each country's institutional, legal, and cultural setting. Case studies for the country, if not the region or province, would be required. Policies that foster water marketing in the Western United States have recently been supported in a number of ways. The Western States Governors' Association recommended that water transfers be facilitated as a way to encourge efficient use (Western Governors' Association 1987). The U.S. Bureau of Reclamation recommended water marketing of federal reclamation water contracts, as long as financial and third-party effects are considered (U.S. Department of the Interior 1987). Extensive state legislation has supported water marketing, subject to third-party and other external effects (Smith and Vaughan 1987). Congressional support is increasing for water marketing as a way to address environmental problems (Heinz and Wirth 1988) and unresolved Indian water claims in over 60 western river basins (U.S. Senate 1989).

Table 27.4 (Shupe 1987-89) provides some recent water market

TABLE 27.4 Sample Water-Market Transactions for Six Western States in 1987

State (1)	Source (2)	Region (3)	Amounts per transaction (acre-foot per year) (4)	Price (dollars per acre-foot) (5)	Price (dollars per acre-foot per year) (6)
Arizona	Ground water	Central	500–2,000	500–1,200	50–120
	Surface water	South	300–400	2,000	200
California	Ground water	South coast	100–1000	1,500	150
Colorado	Surface water	East	2,000–5,000	1,900–3,500	190–350
Nevada	Surface water	West central	10–1,000	2,000	200
New Mexico	Surface water	North central	50–300	1,000	100
Texas	Surface water	South-west	100–2,000	500–650	50–65

Source: Willey and Graff (1988); Shupe (1987, 1988, 1989).

prices in several western states. A comparison of these prices in southern California with the costs of new water-supply projects in northern California discussed earlier (Fig. 27.2) shows that buying existing water supplies tends to be much less expensive than building new projects. This difference would be considerably greater if the cost of transporting water from new project sites to the Los Angeles area were included. Further environmental damage that might have resulted from new projects is avoided. Also, these levels of water market prices enable irrigators to improve irrigation systems and practices and change crops and crop rotations. Irrigators can view water marketing as an attractive option, since the economic returns from water sales and leases exceed those from using water to produce some crops. Fig. 27.5 provides data on economic returns to water from several field crops produced in the Colorado River Basin.

Water marketing transfers have primarily been from irrigation to municipal and industrial uses. None of these transactions is without some degree of controversy, particularly those that involve poorly defined water rights. Such ownership conflicts continue to be the subject of considerable study and litigation (Tarlock 1988). Public rights to water have not been clearly defined, so attempts to define them continue to stir controversy in legislatures and the courts (Smith and Vaughan 1988a; Sax 1988). Ownership of water rights needs to be defined to implement water markets.

A number of ownership conflicts associated with water marketing

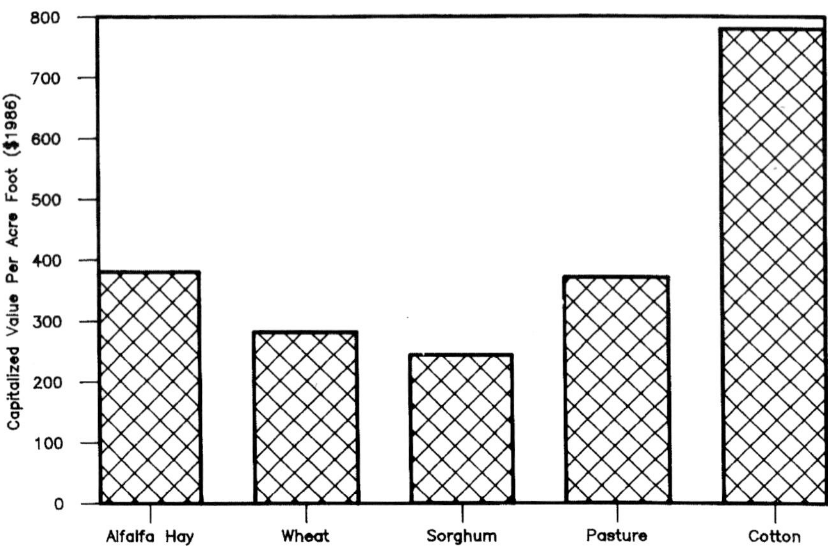

Fig. 27.5 Value of Water for Field-Crop Production in the Colorado River Basin for 1986 (Willey 1989)

transactions in the western United States have taken place. In California's Owens Valley and in some areas of Arizona, quantitative rights to ground water do not exist. Water transfers involve buying land, not water, to allow extensive pumping of the aquifer. Ground-water overdraft can lower other aquifer water levels and impose external economic and environmental damages. In addition, water transfers that retire irrigated land from production can adversely affect the economies of rural communities and result in extensive litigation (Smith and Vaughan 1987) and legislative activity (Smith and Vaughan 1989).

In spite of such problems, water marketing activity continues. The 1988 agreement between California's Imperial Irrigation District and Metropolitan Water District to transfer 100,000 acre-ft of water each year to Metropolitan's service area is based on improvements to the Imperial irrigation system financed by Metropolitan (Smith and Vaughan 1989). This transfer reduces Metropolitan's need for new water diversions in northern California and, consequently, avoids incremental environmental damages. With sufficient funds, an additional 300,000 acre-ft of water can be transferred from the Imperial system (Stavins and Willey 1983). This is enough water to avoid the environmental damages of, for example, the proposed Auburn Dam on California's American River or the proposed Animas-La Plata Dam on Colorado's Yampa River.

Water marketing for the direct benefit of specific environmental resources is also developing. In Colorado, where state law allows rights to instream flows, the Nature Conservancy is trying to acquire rights to some irrigation water (Harrison amd Wigington 1987). Proposals to acquire water rights for Lahontan Valley wetlands and settle the fisheries-based claims to water rights of the Pyramid Lake Paiute Tribe are aimed at mitigating the environmental effects of central Nevada's Newlands Reclamation Irrigation Project (Yardas 1989). In California, the purchase of replacement water supplies from irrigators in several regions (Mono Lake Committee and Los Angeles Department of Water and Power 1987) is part of an attempt to resolve long-standing legal disputes on the diversion of water from streams that flow into Mono Lake. Other environmental water marketing projects are being developed.

Tradable Water Pollution Discharge Permits

Water pollution control laws in the United States set performance standards for individual pollutant discharge points. The issuing of permits regulates the discharges of such point sources as paper processing plants, steel mills, and municipally owned wastewater treatment plants. Public regulatory agencies enforce the terms of these permits. Various forms of "best available technology" (BAT) for control determine allowable pollutant levels in discharges. This approach has been criticized in several ways. Construction of many of the legally required BAT treatment facilities has been heavily subsidized by federal taxpayers, and technological rigidities have impeded the use of the most economical control options (Russell 1981; Tietenberg 1985). Allowable discharge levels for individual points often fall short of the basinwide objectives of "swimmable and fishable" waters as originally targeted for 1982 in the Clean Water Act of 1972 (Smith, Alexander, and Wolman 1987; U.S. General Accounting Office 1989). Monitoring and enforcement by the U.S. Environmental Protection Agency and many state agencies is costly and often vulnerable to budget cuts.

Regardless of whether such criticisms are valid, the water pollution control benefits that have resulted from regulating the point sources will be much more difficult to achieve for nonpoint sources. Table 27.1 shows that such sources constitute most of the rest of this country's water pollution problem. Congress addressed these problems in its amendments to the Clean Water Act in 1987 but failed to formulate an effective incentive for nonpoint source controls (Ackerman and Stewart 1988). Being numerous and dispersed, nonpoint sources are difficult to monitor. Management-intensive production technologies and practices, particularly in irrigated agriculture, are unamenable to the BAT approach. Public subsidies, previously avail-able in significant amounts, are now much less available due to chronic budget deficits.

Tradable discharge permits (TDP) offer an incentive-based system for regulatory agencies to achieve maximum allowable pollutant loading targets. Existing authorities under the Clean Water Act establish these targets to avoid impairing other water uses. The use of TDP, unlike the BAT approach, allows polluters a degree of economic flexibility within the limits of the loading targets. According to the 1987 amendments to the Clean Water Act (Section 319, a, 41) , nonpoint source pollutant management should "to the maximum extent practicable develop programs on a watershed-by-watershed basis." The act does not specify the type of program. This is the states' responsibility. A watershed basis, which involves multiple sources of pollutants, makes possible the use of pollutant trading programs (Lence, Eheart, and Brill, Jr. 1988; O'Neill 1983;

DPRA, Inc. 1989). TDP's in the United States have been studied extensively (Tietenberg 1985; Hahn 1988; Dudek and Palmisano 1988), proposed by congressional leaders (Heinz and Wirth 1988). Similar approaches have been used in several European nations (Brown and Johnson 1984; Harrison and Sewell 1980). Extensive pilot studies and projects involving potential trading between point and nonpoint sources of water pollution have been implemented in the Great Lakes Basin, Tennessee, and Colorado (DPRA, Inc. 1989; Sessions and Fillmore 1989; Jaksch and Niedzialkowski 1985).

TDP's have the potential to control some of the most significant environmental effects of irrigation with minimal disruption of farming, but problems arise in implementing them. It is infeasible to issue and enforce a TDP for each of the millions of irrigated fields that exist, so TDP's would have to be issued to larger units, such as irrigation or drainage districts or large farming operations. Discharge rights within such units could be allocated internally in any number of ways, such as discharge assessments on irrigated lands, drainage fees imposed on applied water rates, and intra-unit water trading. To divide up the responsibilities and costs for discharge loads among farms and fields, uncertain hydrogeological information would have to used, and neighboring irrigators and landowners would have to reach agreements.

In California's Central Valley, the formation of a larger administrative unit that encompasses a number of irrigation districts to receive a trace element and salinity discharge permit to the San Joaquin River has been proposed (State of California 1987; Coontz 1989). The distribution of the initial permits, required before permit trades begin, is a difficult political decision due in part to the economic value of such permits. Each permit would specify a given permissible load for a pollutant at a given point of discharge into public waters. Regulatory agencies would have to estalish equivalencies with other discharge points and, possibly, other pollutants. Enforcing the terms of the permits would continue to be a problem.

Unlike water marketing, the use of TDP's is a policy in its infancy. Nevertheless, TDP's have the best potential to control pollution by irrigation with the minimal amount of regulations on irrigated agriculture. No other policy on the horizon has the potential to do both.

Liability and Compensation for External Environmental Effects

Water marketing and TDP's are farsighted policies. They seek to induce irrigators to make adjustments to avoid damaging the environment. Since such avoidance will sometimes fail, policies on environmental damages from irrigation are needed. The policies could assign liability and require compensation for external damages. This would provide additional economic incentives to prevent and mitigate irrigation's adverse effects on the environment.

Existing legal remedies involve property, nuisance, and tort laws that apply to a variety of liability and compensation disputes. These solutions place the burden of legal and technical costs on the damaged parties. Incorporating the concepts of liability and compensation into existing water use and pollution discharge rights could reduce this burden. Appropriative water rights and water pollution discharge permits could be structured to include these concepts in much the same way as many legal contracts include them. Procedures for assessing damage claims should be referenced. An example of such procedures already exists for environmental damages resulting from oil and chemical spills, i.e., regulations promulgated under the Comprehensive Environmental Response, Compensation, and Liability Act of 1980 by the U.S. Department of the Interior in 1987.

The dispute over paying for the cleanup of toxic materials at California's Kesterson Reservoir (U.S. Department of the Interior 1986) stems in large part from the underlying reclamation and water contracts' ambiguity about liability. A policy of incorporating liability provisions in water contracts, appropriative water rights, and discharge permits would help to avoid such disputes. Such a policy would remove public subsidies that cover liabilities of private parties, including irrigation districts. This would enhance the economic incentives that liabilities present to irrigators.

Liability provisions would apply to private and public parties. While irrigators can take actions that contribute to environmental damages, public agencies, such as the U.S. Interior and Agriculture departments, have adopted policies and otherwise taken actions that indirectly contribute to them. Clear liability provisions applied to public agencies would provide them with an added incentive to establish the rules of liability sharing with participants in their programs and projects.

CONCLUSION

The environmental effects of agricultural irrigation and drainage are widespread and significant. Most have become recognized only in recent times, long after irrigation development had peaked in the United States. Similar effects are occurring in other countries, with the response depending largely on each country's institutional capacities. The degradation of the habitat of wildlife, surface-water and ground-water supplies, and soil resources have become international problems closely linked with worldwide irrigated agricultural development.

Reconciling irrigated agriculture and environmental quality is only beginning in the United States and, in some countries, has not begun at all. The United States is attempting to adopt environmental and agricultural policies that will achieve agricultural and environmental goals simultaneously. A key issue is the degree to which government regulation of agricultural production practices are needed. The opportunity exists to strengthen current economic incentives and develop new ones that are suited to the industry and the agrarian ethic and that require less regulation by public agencies.

Implementing economic incentives that allow flexibility for irrigated agriculture and at the same time achieve environmental goals is difficult—technically and politically. Many agricultural and environmental interest groups favor the status quo and oppose changes with effects that they perceive to be uncertain. Nevertheless, the shortcomings of existing regulatory programs are increasingly recognized, incentive-based systems have demonstrated powerful potentials, and coalitions that support the incentive-based agenda are growing. Agriculturalists want flexibility in regulations. Environmentalists want protection and enhancement. Public officials want cost-effective programs. Rights advocates want a balance between the power of the state and the freedom of the individual. These interests, if combined, can lead the way to a reconciliation between irrigated agriculture and the environment.

REFERENCES

Ackerman, B., and Stewart, R. 1988. "Reforming environmental law: the democratic case for market incentives," *Columbia J. of Environ. Law*, Vol. 13, No. 2, School of Law, Columbia Univ., New York.

Amin, M. A. 1976. "Problems and effects of schistosomiasis in irrigation schemes in the Sudan," *Arid Land Irrigation in Developing Countries: Environmental Problems and Effects*, E. Barton Worthington, ed. Pergamon Press, Elmsford, N.Y.

Antar, I., 1981, "A study of salinization processes in some soils of Egypt," *A Conference on Biosalinity: The Problem of Salinity in Agriculture*, D.W. Rains, ed. Univ. of Calif., Davis.

Association of State and Interstate Water Pollution Control Administrators. 1985. *America's Clean Water, The States' Nonpoint Source Assessment*, Washington, D.C.

Blackwelder, B., Harding, B.L., and Colborn, T. 1987. "Alternatives to traditional water development in the United States," *Ambio*, Vol. 16, No. 1, Stockholm.

Brown, G., and Johnson, B. 1984. "Pollution control by effluent charges: it works in the Federal Republic of Germany, why not in the U.S.," *Natural Resources Journal*, Vol. 24, p. 929.

Bryan, K., 1928, "Changes in plants associated with changes in groundwater levels," *Ecology*, Vol. 9, No. 4.

California Dept. of Water Resour. 1983. *The California Water Plan: Projected Use and Available Water Supplies to 2010*, Bulletin 160-83, Sacramento.

California Dept. of Water Resour., 1987, *California Water: Looking to the Future*, Bulletin 160-87, Chapter 4, Sacramento.

Colacicco, D., Osborn, T., and Alt, K. 1989. "Economic damage from soil erosion," *Journal of Soil and Water Conservation*, Vol. 44, No. 1.

Columbia University School of Law. 1988. "Law and economics symposium: new directions in environmental policy," *Columbia Journal of Environmental Law*, Vol. 13, No. 2, Columbia Univ., New York.

Coontz, N. D., 1989. "Agricultural drainage management organizations in the drainage problem area of the grasslands area of the San Joaquin Valley," Report to the San Joaquin Valley Drainage Program, Ebasco Services, Inc., Sacramento, Calif.

Doroshov, S.I., and Wang, Y.L. 1984. "The effect of subsurface agricultural drainage water on larval striped bass, *Morone saxatilis* (Walbaum), Department of Animal Sci., Univ. of Calif., Davis.

Dudek, D.J. , and Palmisano, J. 1988. "Emissions trading: why is this thoroughbred hobbled?," *Columbia Journal of Environmental Law,* Vol. 13, No. 2.

Eisler, R. 1985. "Selenium hazards to fish, wildlife, and invertebrates: a synoptic review," *Biological Report 85(1.5),* U.S. Fish and Wildlife Service, U.S. Department of the Interior, Laurel, Md.

El Ashry, M., and Gibbon, D. 1986. *Troubled waters—new policies for managing water in the American West,* World Resources Institute, Washington, D.C.

El Ashry, M.T., and Gibbons, D.C. eds., 1988. *Water and arid lands of the Western United States,* Cambridge Univ. Press, New York.

El Gabaly, M.M., 1976, "Problems and effects of irrigation in the Near East," in E. B. Worthington ed. *Arid land irrigation in developing countries: environmental problems and effects,* Pergamon Press, New York.

Environmental Defense Fund. 1984. *The Tuolumne River: preservation or development?—an economic assessment,* Berkeley, Calif.

Frederick, K., 1982. *Water for Western agriculture,* Resources for the Future, Washington, D.C.

Glaser, L. 1986. "Provisions of the Food Security Act of 1985," Economic Research Service, U.S. Department of Agriculture, Washington, D.C.

Greig, P.J., and Devonshire, P.G. 1981. "Tree Removals and Saline Seepage in Victorian Catchments: Some Hydrologic and Economic Results," *Australian Journal of Agricultural Economics,* Vol. 25, No. 2.

Groenevald, D.P. and Griepentrog, T.E. 1985. "Interdependence of groundwater, Riparian vegetation, and streambank stability: a case study," *Riparian ecosystems and their Management: reconciling conflicting uses,* U.S. Department of Agriculture Forest Service General Technical Report RM-120.

Hahn, R. 1988. "Innovative approaches for revising the Clean Air Act," *Natural Resources J.,* Vol. 28, pp. 171-88.

Harris, T.R., and Miller, W.W. 1986. "Projected effects of product prices on producer response to agricultural water pollution control charges," *Applied Agric. Res.,* Vol. 1, No. 3.

Harrison, P., and Sewell, W.R.D. 1980. "Water pollution control by agreement: the French system of contracts," *Natural Resour. J.,* Vol. 20, p. 765.

Harrison, D.L., and Wigington, R. 1987. "Water rights: A protection tool for the West," Ecology Forum No. 64, *The Nature Conservancy Mag.,* August-October, Washington, D.C.

Herath, H.M.G.. 1985. "Economics of salinity control in Sri Lanka: some exploratory results," *Agric. Administration,* Vol. 18.

Holden, P.W. 1986. *Pesticides and Groundwater Quality,* National Academy of Science Press, Washington, D.C.

International Inst. for Envir. and Development; and the World Resour. Inst., 1987, *World Resources 1987,* Chapter 4, Basic Books, Inc., New York.

Jaksch, J., and Niedzialkowski, D., Oct. 1985. "Speeding water cleanup while saving money," *EPA J.,* Washington, D.C.

Knapton, J.R., Jones, W.E., and Sutphin, J.W. 1987. "Reconnaissance investigation of water quality, bottom sediment, and biota associated with irrigation drainage in the Sun River Area, West-Central Montana, 1986-87," *Water Resour. Investigations Report 87-4244,* U.S. Geological Survey, Denver, Colo.

Lee, E.W., Nishimura, G.H. and Hansen, H.L. 1988. "Agricultural drainage water treatment, reuse, and disposal in the San Joaquin Valley of California," Technical Report, San Joaquin Valley Drainage Program, Sacramento.

Lence, B.J., Eheart, J.W. and Brill, E.D. Jr. 1988. "Cost efficiency of transferable discharge permit markets for control of multiple pollutants," *Water Resour. Res.*, Vol. 24, No. 7.

McCauley, C.A., and Gum, R. 1972. "Subsidence damage in Southern Arizona," *Hydrol. and Water Resour. in Arizona and the Southwest,* Vol. 2, American Water Resour. Assoc., Arizona Section and Arizona Academy of Science, Proceedings, Tempe.

McGeorge School of Law. 1988. "Symposium: revisiting California water law," *Pacific Law Journal,* Vol. 19, No. 4, Univ. of the Pacific, Sacramento.

Miller, J.R., Jan. 1986, "An estimate of the value of a waterfowl hunting day in the Central Valley of California," Report to U.S. Fish and Wildlife Service, Salt Lake City.

Moore, D., June 1989, "Opportunities for riparian ecosystem preservation in the Verde River Basin, Arizona," *Proc. American Water Resour. Assoc. Symposium on Headwaters Hydro.*, Missoula, Mont.

National Water Commission. 1973. *Water policies for the future,* Final Report to the President and to the Congress, Washington, D.C.

Ohlendorf, H.M., Hothem, H.C., Bunck, C.M., Aldrich, T.W., and Morre, J.F. 1986. "Relationships between selenium concentrations and avian reproduction," *Trans. North American Wildlife and Natural Resour. Conf.*, pp. 330-42, Reno, Nev.

O'Keefe, M.D., Slocum, N.E., Snow, D.R., Thorson J.E., and Vandenberg, P. 1986. *Boundaries carved in water—an analysis of river and water management in the Upper Missouri River,* Northern Lights Institute, Missoula, Mont.

O'Neill, W. 1983, "Transferable discharge permit trading under varying stream conditions: a simulation of multiperiod permit market performance on the Fox River, Wisconsin," *Water Resour. Res.*, Vol. 19.

Paarlberg, D., 1980, *Farm and food policy—issues for the 1980's,* Chapter 4, Univ. of Nebraska Press, Lincoln.

Pye, V., Patrick, R., and Quarles, J. 1983. *Groundwater contamination in the United States,* University of Pennsylvania Press, Philadelphia.

Randall, A. 1981. "Problems in resource valuation," *Resource economics: an approach to natural resource and environmental policy,* Grid Publishing Inc., Columbus, Ohio.

Repetto, R., 1986, *Skimming the water: rent-seeking and the performance of public irrigation systems,* World Resources Instit. Washington, D.C.

Ribaudo, M.O., Colacicco, D., Barbarika, A., and Young, C.E. 1989. "The economic efficiency of voluntary soil conservation programs," *J. of Soil and Water Conservation,* Vol. 44, No. 1.

Robie, R.B. 1988. "The delta decision: the quiet revolution in California water rights," *Pacific Law J.*, Vol. 19, No. 4, McGeorge Law School, Univ. of Pacific, Sacramento.

Russell, C. 1981. "Controlled trading of pollution permits," *Envir. Sci. and Tech.*, Vol. 15, No. 1.

Sax, J.L. Nov. 1988. "The limits of private rights in public waters," *The public trust and the waters of the American West: yesterday, today, and tomorrow*, Lewis and Clark College, Boalt Hall Law School, Univ. of Calif., Berkeley.

Schroeder, R.A., Palawski, D.U., and Skorupa, J.P. 1988. "Reconnaissance investigation of water quality, bottom sediment, and biota associated with irrigation drainage in the Tulare Lake bed area, Southern San Joaquin Valley, California, 1986-87," U.S. Geological Survey, *Report 88-4001*, Denver, Colo.

Sessions, S., and Fillmore, L. 1989. "Evaluation of measures to control point and nonpoint sources of pollution to Boone Reservoir, Tennessee," Report to the U.S. Environmental Protection Agency, Sobotka & Co., Inc., Contract No. 68-01-7288.

Sheridan, D. 1981. *Desertification in the United States*, President's Council on Environmental Quality, Washington, D.C.

Shupe, S.J., ed. 1987-89. *Water market update*, Shupe and Associates, Vols. 1-3, Santa Fe., N.M.

Smith, B.A., Alexander, R.B., and Wolman, M.G. 1987. "Water Quality Trends in the Nation's Rivers," *Sci.*, Vol. 235, No. 1607.

Smith, R.T., and Vaughan, R., eds. 1987, "Irrigating the city: emerging markets for water transfers," *Water Strategist*, Vol. 1, No. 3; and "Just rewards: making water marketing work for local interests," *Water Strategist*, Vol. 1, No. 2, Stratecon, Inc., Claremont, Calif.

Smith, R.T., and Vaughan, R. eds. 1988a. "Water policy in the balance: water development and environmental interests in the era of the public trust doctrine," *Water Strategist*, Vol. 2, No. 1, Claremont, Calif.

Smith, R.T., and Vaughan, R., eds. 1988b. "Groundwater contamination: common ground for the common law," *Water Strategist*, Vol. 2, No. 4, Stratecon, Inc., Claremont, Calif.

Smith, R.T., and Vaughan, R., eds. 1989. "Let's make a deal: The IID/MWD water conservation agreement," *Water Strategist*, Vol. 3, No.1, Stratecon, Inc., Claremont, Calif.

SRI International. May 1985. "Chronic toxicity of San Luis drain effluent to *Neomysis Mercedis*," *Final Report to the U.S. Bureau of Reclamation*. Menlo Park, Calif.

State of California. 1987. *Regulation of agricultural drainage to the San Joaquin River*, State Water Resour. Control Board, Sacramento, Calif.

Stavins, R.N., and Willey, Z. 1983. "Trading conservation investments for water," R.J. Charbeneau, ed. *Regional and State Water Resour. Planning and Management*, American Water Resources Association, Bethesda, Md.

Stephens, D. 1988. "Reconnaissance investigation of water quality, bottom sediment, and biota associated with irrigation drainage in the Middle Green River Basin, Utah." Open-File Report, U.S. Geological Survey, Salt Lake City, Nev.

Szabolcs, I. 1975. "Present and potential salt affected soils: an introduction," *Prognosis of Salinity and Alkalinity*, FAO Soils Bulletin, No. 31, Rome.

Tarlock, A.D. 1988. *Law of water rights and resources*, Environmental Law Series, Clark Boardman Co., Ltd.

Tietenberg, T. 1985. *Emissions trading: an exercise in reforming pollution policy*, Resources for the Future, Washington, D.C.

Tyagi, N.K. 1986. "Optimal water management strategies for salinity control, (India)," *J. of Irrig. and Drainage Engrg*, ASCE Vol. 112, No. 2.

U.S. Bureau of Reclamation. 1987. "Final environmental impact statement for the Newlands Project proposed operating criteria and procedures," U.S. Department of the Interior, Sacramento, Calif.

U.S. Committee on Irrig. and Drainage. 1986. *Toxic substances in agricultural water supply and drainage*, Proc. of Regional Meetings, Denver.

U.S. Dept. of Agric. 1987. *Agric. Statistics*, Government Printing Office, Washington, D.C.

U.S. Department of Commerce. 1982. *Six-states high plains—Ogallala Aquifer Area regional study*, Economic Development Administration, Final Report to the U.S. Congress, Authorized Under Section 193 of PL 94-587, Washington, D.C.

U.S. Department of Interior. 1981. "A report on the Mid-Valley canal feasibility investigation," Central Valley Project, East Side Division, Water and Power Resources Service, Sacramento, Calif.

U.S. Department of Interior. 1983. "Colorado River water quality improvement program," Bureau of Reclamation, Colorado River Water Quality Office, Denver.

U.S. Department of the Interior. 1987. *Assessment '87 . . . A New Direction for the Bureau of Reclamation*, Washington, D.C.

U.S. Department of the Interior. 1988. "Interior begins detailed studies of irrigation drainage at four sites," Office of the Secretary, News Release, May 10, Washington, D.C.

U.S. Fish and Wildlife Service. 1986. *Contaminant issues of concern—national wildlife refuges*, U.S. Dept. of the Interior. Washington, D.C.

U.S. General Accounting Office. 1989. "Water pollution—more EPA action needed to improve the quality of heavily polluted waters," *GAO/RCED-89-38*. Washington, D.C.

U.S. Government Accounting Office. Sept. 1980. "Groundwater overdrafting must be controlled," Report to the U.S. Congress, *CED-80-96*. Washington, D.C.

U.S. Senate. April 6, 1989. Oversight hearings on Indian water policy, select committee on Indian Affairs. Washington, D.C.

Vaughan, W.J., and Russell, C. 1982. "Valuing a fishing day: an application of a systematic varying parameter model," *Land Economics*. Vol. 58, No. 4.

Wahl, R.W. Spring 1986. "Cleaning up Kesterson," *Resources Newsletter*, Resources for the Future. Washington, D.C.

Water Efficiency Working Group. 1987. *Water efficiency: opportunities for action*, Report to the Western Governors, Western Governors' Association. Denver, Colo.

Westcot, D.W., Rosenbaum, S, Grewell, B., and Belden, K. 1988. "Water and sediment quality in evaporation basins used for the disposal of agricultural subsurface drainage water in the San Joaquin Valley, California," Central Valley Regional Water Quality Control Board, State of California. Sacramento.

Western States Water Council. Dec. 1983. *Indian water rights in the West,* Salt Lake City, Utah.

Westlands Water District. 1987. "1985-86 water conservation and management program—review and evaluation," Fresno, Calif.; URS Corp., Oct. 1987, "Agricultural drainage and salt disposal—evaluation of options for brine concentrator salts and other systems," Report Prepared for the Westlands Water District, Fresno, Calif.

Wichelns, D. and Cone, D. 1989. "An increasing block-rate pricing program to motivate water conservation and drainage reduction," Presented at the Second Pan-American Regional Conference of the International Committee on Irrigation and Drainage, Ottawa, Canada.

Willey, Z. 1977. "Groundwater in the California water quandary," *Proc. of the Biennial Groundwater Conference,* Water Resources Center, Univ. of Calif., Davis.

Willey, Z. 1985. *Economic development and environmental quality in California's water system,* Inst. of Government Studies, Univ. of Calif., Berkeley.

Willey, Z. 1987. "Managing the Central Valley's agricultural salinity and toxic water pollution problems—is there a workable scenario during the next investment period?" *Applied Agric. Res.,* Vol. 2, No. 1.

Willey, Z. and Graff, T. 1988. "Federal water policy in the United States—an agenda for economic and environmental reform," *Columbia J. of Envir. Law.* Vol. 13, No. 2.

Willey, Z. April 6, 1989, "Water, economics, and native Americans," Testimony in Hearings before the Select Committee on Indian Affairs and the Subcommittee on Water and Power of the Committee on Energy and Natural Resources, U.S. Senate. Washington D.C.

Yardas, D. 1989. "Water transfers and paper rights in the Truckee and Carson basins," Indian Water Rights Section, *Proc. of the American Water Resources Assoc. Symposium on Headwaters Hydro.,* Missoula, Mont.

Zimmerman, R. C. 1969. "Plant ecology of an arid basin, Tres Alamo-Redington Area, Southeastern Arizona," Geological Survey Professional Paper 485-D, U.S. Geological Survey.

X—THE FUTURE OF IRRIGATED AGRICULTURE

CHAPTER 28

IRRIGATED AGRICULTURE: IS IT SUSTAINABLE?[a]

Irrigated agriculture has substantially contributed to world food production for thousands of years. In recent decades, many countries have made major investments to expand the irrigated base, which currently exceeds 24,000,000 ha in the United States. The unit production from irrigated land tends to be higher than from rainfed agriculture. In the United States, roughly 25% of the value of all crops produced comes from some 10% of the land farmed.

Irrigation in arid or semi-arid regions always degrades water quality. Without proper management, the land becomes waterlogged and salinized. With or without good management, drainage water from irrigated lands carries salt that requires disposal. The question arises as to whether salinization is inevitable. Does irrigation by nature have a finite life? Is it doomed to fail? Or is it possible to manage irrigation enterprises so that they can be sustained indefinitely?

A LOOK BACK

Irrigation has been practiced since time immemorial. Some civilizations that have depended on irrigation have come and gone. The fertility of Mesopotamia, the land between the Tigris and the Euphrates Rivers, is renowned. At the height of the Sumerian Empire, about 4,000 years ago, the region depended on a highly developed irrigation system. It is often argued that salinization of the farmland caused the collapse of the empire. That the two went together is beyond doubt. However, what was cause and what effect is uncertain. Erosion and sedimentation, as well as salinization, went hand in hand with the deterioration and eventual collapse of the social structure. Bilharziasis, spread by fresh-water snails, apparently was prevalent, indicating that irrigation affected public health (Encyclopedia Britanica).

Another ancient civilization, the Egyptian, depended on irrigation for three millennia. There have been ups and downs, and major changes. The Aswan Dam symbolizes recent dramatic changes. Tremendous efforts have gone into developing drainage systems over the last

---

[a]Prepared by: Jan van Schilfgaarde, Agric. Res. Service, 2625 Redwing Rd., Suite 350, Fort Collins, CO 80526

two decades to correct waterlogging and salinization problems aggravated by the post-Aswan-Dam increase in water supply. Still, irrigation in the Nile Delta has continued uninterrupted for thousands of years.

In the Indus Basin, British engineers developed an extensive water storage and distribution system in the nineteenth century. This system has enabled the irrigated base to greatly expand, but also has led to problems due to inadequate drainage. The verdict is still out on whether these problems will be solved.

More recently, after a calamity in 1905, a reliable supply of water for irrigation was brought to California's Imperial Valley. Since the area is a closed basin, its drainage water must be discharged into a sump, which became the Salton Sea. As irrigation developed here and in the Coachella Valley, which also drains to the Salton Sea, and drainage systems were improved, the salt concentration in the Salton Sea gradually increased (Meyer and van Schilfgaarde 1984).

Even more recent is the full development of the San Joaquin Valley in California. Here, waters that originate in the Sierra Nevada Mountains were harnessed and additional supplies transported from outside the valley to the north to develop one of the world's most productive agricultural areas. Yet salinity and sodicity have plagued this valley from the beginning. In the 1980s, there was also a serious concern with potentially toxic minor elements. Discovery of selenium in the biota depending on the Kesterson Reservoir, a set of ponds used for discharge of agricultural drainage water from part of the San Joaquin Valley, has led to a wholesale re-evaluation of the disposal of drainage water (Letey et al. 1986).

Other examples could be cited. The conclusion would be the same. All irrigation enterprises face problems of salinity management. All must consider the safe disposal of wastewater. Some have failed. Others have withstood the pressures for remarkably long periods of time. Does enough evidence exist to judge whether irrigation can indeed be sustained indefinitely?

SALINITY AND OTHER CONTAMINANTS

Salinity problems have been recognized for a long time. That they are often ignored when irrigation projects are planned has less to do with technical understanding than with political considerations. As E. W. Hilgard (1886) noted,

> "It is hardly necessary to go further into the details (of the problems occurring in India) to enforce the lesson and warning they convey to our irrigating communities... The evils now besetting (California's irrigation districts) are already becoming painfully apparent; and to expect them not to increase unless the proper remedies are applied is to hope that natural laws will be waived in favor of California. The natural conditions under which the irrigation canals of India have brought about the scourge, are exactly reproduced in the

great valley of California; and what has happened in India will assuredly happen there also."

In recognition of these facts, many scientists and engineers have devoted their careers to studying various aspects of salinity. They investigated the production of crops under irrigation, the reclamation of saline and sodic soils, and the maintenance of crop production over time. They focused on the carbonates, sulfates and chlorides of sodium/potassium, magnesium, calcium, and, to some extent, on boron. As earlier chapters in this book make evident, much has been learned about the management of salinity and sodicity. Though it has long been recognized that salinity management requires removing excess salts from the root zone and thus drainage, serious consideration of the problem of disposing of this saline drainage water is relatively recent. Though by no means a first, or even a turning point, passage of Public Law 93-320, the Colorado River Basin Salinity Control Act of 1974, illustrates the public's awakening to the signifigance of off-site effects of irrigation agriculture.

In 1982, U.S. Fish and Wildlife Service biologists first found high levels of selenium in the Kesterson Reservoir's mosquito fish. When plans were made for a Central Valley Project water delivery system to import water to the northern part of the San Joaquin Valley, they included a master drain, the San Luis Drain—to take the water to the Sacramento-San Joaquin Delta. For political reasons, only the upper part of this canal was built. It terminated at the Kesterson Wildlife Refuge and discharged into 490 ha of shallow ponds, called the Kesterson Reservoir. Shortly after drainage water began to be discharged, toxic levels of selenium were found in fish. Not long after, dead and deformed bird embryos were discovered. While the biological damage noted at Kesterson was attributed to selenium, high concentrations of other trace elements have been found in drainage waters since, including molybdenum, arsenic, and boron (Letey et al. 1986).

Rather than being introduced with the irrigation water, these trace minerals originate in the soils or the underlying geologic materials, where they are mobilized by irrigation and then displaced with irrigation drainage. Thus, the unexpected findings in the San Joaquin Valley suggest that similar conditions are likely to be found elsewhere. Indeed, preliminary investigations by the U.S. Department of the Interior indicate that several wildlife refuges across the West may have similar problems.

Concern for the disposal of irrigation drainage water, historically confined to major salts, now includes potentially toxic concentrations of several trace elements, potentially high levels of plant nutrients, especially nitrates, and low, but potentially dangerous, levels of various pesticides. Managing irrigation drainage waters has become substantially more difficult and controversial.

AN ASSESSMENT

Irrigation in semi-arid and arid regions must have drainage. A downward flux of water must move through the soil profile to prevent the concentration of solutes in the soil solution from rising to a level that cannot be tolerated by crop roots. In many instances, the natural drainage rate is sufficient to meet this need; in others, man-made drainage collector systems must be installed. In either case, this drainage water must go somewhere. Depending on the local geohydrology, it may be many decades, or just a season or two, before the need for drainage becomes evident.

Earlier chapters addressed in detail the concepts of leaching fraction, leaching requirement and salt balance; salt precipitation and dissolution; and salt displacement. It suffices to reiterate that, at equilibrium, enough salts must be removed from the system to maintain an acceptable concentration of solutes in the root zone.

In most natural systems, drainage from the uplands finds its way into rivers and then to the ocean. Ocean disposal is nature's way of moving dissolved salts out of the landscape. Irrigation tends to accelerate this phenomenon of displacement. In most instances, irrigation drainage water also finds its way to the ocean. For example, the United States Environmental Protection Agency (EPA) estimated that over one third of the salt carried by the Colorado River near its mouth could be attributed to irrigation (Jonez 1983).

In other instances, one encounters closed basins without an outlet to the ocean. The Dead Sea and Great Salt Lake are obvious examples, as are the Salton Sea in California, the Stillwater Reservoir in Nevada, and the Kesterson Reservoir. In these instances, drainage water collects at some terminus and evaporates, leaving its constituents behind. Early in their lives, these terminal lakes often have significant value for fish and wildlife and recreation. Over time, however, they must become ever more saline and, eventually, lose all biological value and become less attractive for recreation. It might be argued that an increase in salt concentration of whatever composition does not necessarily degrade the quality of the water. Degradation is a value concept and whether it occurs depends on the intended use. Also, the salt concentration in an already saturated river cannot be increased (Rhoades and Suarez 1977). These caveats notwithstanding, all irrigation ultimately degrades water quality off site for most uses. In the case of a river discharging to the ocean, it means an increase in the river's salt concentration. With a closed basin, it means increased salinity in the river and ever-increasing salinity in the terminal water body.

A permanent irrigation agriculture requires the sacrifice of some value elsewhere. A good example is the Colorado River, for which Lohman et al. (1988) recently estimated damages from salinity. From between 1976 and 1985, they estimated total annual damages from salinity to be $311 million when based on a salinity of 500 mg/l and $831 million when based on a salinity of 334 mg/l. The first is simply the Public Health Service standard for drinking water. The second is an EPA estimate of conditions on the Colorado River before human intervention. Damages affect agriculture, households, water utilities, and industry. Of the figures quoted, $113 million and $122 million, respectively, reflect damages to agriculture. One might ask rhetorically whether the damage to water heaters in Los Angeles is justified by the benefits of irrigation in Colorado.

There are ways to minimize the effects of irrigation on downstream salinity. As discussed in earlier chapters, more precise irrigation management to limit the leaching fraction to the amount needed to maintain full crop growth can substantially reduce the mass of salt discharged in the drainage water (van Schilfgaarde et al. 1974). Recent developments in irrigation technology make implementing that concept even more feasible. An example is the use of subsurface trickle irrigation, combined with fertilizer injection and automatic feedback, to control water application continuously (Phene et al. 1988). With such a system, Phene et al. were able to achieve exceptionally high yields of tomatoes while controlling the leaching fraction precisely.

It also is possible, and economically attractive in some instances, to reuse drainage water for irrigation (Chapter 20) or reduce the amount of irrigation water needed by making use of shallow, saline ground waters (Chapter 16). The use of these techniques has reduced fresh irrigation water requirements by as much as 50%.

These techniques can affect the amount and solute concentration of drainage water that needs disposal but cannot eliminate such a need. Ultimately, the drainage water must be transported out of the region, disposed of locally, or treated.

It is technically feasible to treat the water after the fact. Several studies have evaluated the use of reverse osmosis to remove salt from drainage water so that it can then be put to higher use. The world's largest desalination plant is under construction near Yuma, Arizona, for just such a purpose. However, it is difficult to justify such an approach economically (van Schilfgaarde 1982). Others have proposed the use of saline drainage water for cooling electric power plants, which may be successful if the cost of disposal can be passed on from the irrigator to the utility's customers. No instances have come to this author's attention in which drainage water has been treated for salt removal in an economically justified manner. An alternative to treating the water is to convey it to an evaporation pond. Experience in California indicates that this requires the use of between 10% and 14% of the land area for such ponds. This loss of land and the cost of construction makes ponds relatively unattractive. Care must be taken to avoid leakage and consequent salinization of the underlying ground water.

The introduction of trace elements adds an entirely new dimension. While damage due to salinity is substantial, the potential damage, real or perceived, from toxic concentrations of trace metals can be much more severe. Also, the energy required to remove a trace element, such as selenium in concentrations of 100 μg/l to 500 μg/l, is far less than that needed to remove total salts in concentrations of 2,000 mg/l to 8,000 mg/l. Spurred by concern for Kesterson Reservoir, a number of recent studies have evaluated ways to remove selenium. They include chemical filters that use iron filings or ferrous hydroxide, bacterial filters, and fungal digestive systems. To date, none has proven fully adequate. Except for the fungi, which dissipate the selenium to the atmosphere, all of them result in a waste product that again needs disposal (Frankenberger and Karlson 1988). Compared to the mass of salt in drainage water, however, the quantity of selenium or other trace element harvested would be extremely small.

The presence of trace elements in evaporation ponds also complicates their use. These trace elements could injure waterfowl and increase regulatory requirements.

With plant nutrients and pesticides, the situation is different. These substances are added in the farming operation. In most instances, the quantities involved are relatively small, and efficient management reduces losses into drainage water further. The recent surge in interest among the public and scientists in other forms of agriculture that depend less on chemical inputs might very well lessen the problems posed by these chemical inputs (National Research Council 1989a).

This chapter focuses on the permanence of irrigated agriculture, but a related problem is that of saline seeps. They form when water from precipitation or irrigation that exceeds plant use and soil water storage capacity percolates downward, picks up salts from underlying formations, moves laterally over a layer of lower hydraulic conductivity, and ultimately seeps out along a limited zone downslope from the area of recharge. Saline seeps characteristically consist of strips of saturated soils along a hillside. They prevail in many parts of the world, notably western Australia and the northern Great Plains of North America. In the late 1970s, they affected some 130,000 ha in Montana and North Dakota (Brown et al. 1982).

Saline seeps received much attention during the 1970's because the increase in wheat-fallow rotations in past years combined with a series of high-precipitation years to greatly increase their occurrence. Intended to make water available for the following crop, fallow clearly reduced consumptive use. Combined with above-normal precipitation, it led to increased percolation and subsequent lateral movement. To avoid or reclaim saline seeps, one uses agronomic practices in the recharge zone that increase water use. These include the use of a continuous cropping scheme instead of a rotation including fallow and the planting of deep-rooted, high-water-use crops, such as alfalfa, as much as possible. Such techniques as the use of vegetative barriers or standing stubble to trap snow can even increase available water to make a more intensive cropping pattern possible. Several investigators have worked on developing flexible cropping systems designed to condition the choice of crop on the available water in a particular season. Halvorson and Kresge (1982) developed the computer program FLEXCROP to aid in this decision-making process. Saline seeps also can be reclaimed by installing interceptor drains (Doering and Sandoval 1976). However, providing an outlet for these drains is often expensive and environmental considerations may preclude using such an outlet.

COMPARING THE OPTIONS

In writing a book about salinity and irrigation, one might assume that the use of land and water for irrigated agriculture is the preferred use. However, as pointed out earlier in this chapter, successful irrigation depends on adequate drainage, which implies some off-site disadvantages that offset irrigation's benefits. In a broader context, harnessing and diverting water for irrigation reduces the water available for other uses and carries with it a cost, termed the opportunity cost.

At one time, there were over 1,600,000 ha of wetlands in California. Today, more than 90% of these have been converted and only 120,000 ha of wetlands remain. As man has developed the water resources of the West, he has greatly reduced the abundance of streams, riparian vegetation, and wetlands suited for wildlife habitat. The flyways for migratory birds have been severely affected (National Research Council 1989b).

Water development affects not only the original ecosystem, but also the recreational and economic value of the resource. The loss of whitewater rivers and fishable streams reduces recreational opportunities. The loss of clean water for duck clubs affects the hunters. This, in turn, hurts the businesses that depend on recreation, such as the duck club owners, the raft operators, and the suppliers of fishing equipment.

As noted earlier, irrigation with water from the Colorado River increases salinity and adversely affects municipal and industrial water use. Had the San Luis Drain been completed as planned, there would have been additional discharge of saline waters into the Sacramento-San Joaquin Delta. Such a discharge, it was feared, would impair industries that depend on this source for their water.

Thus, the question of sustainability requires a two-part answer. Given the objective of maintaining irrigation agriculture indefinitely, this objective can be reached at a price. Whether society is prepared to pay the price may not be taken for granted.

At one time, the abundance of natural resources was taken for granted in the United States. The developmental ethic dominated the spirit of the land. Short-term economic gain was the primary goal. In this spirit, the Reclamation Act was born, stimulating development of the West's water resources and encouraging irrigation. Only about 25% of irrigation water in the West is supplied by the U. S. Bureau of Reclamation. Nonetheless, this agency has significantly influenced the development of the West, and its works symbolize the drive for development.

Such development was not without costs. One only has to recall the pollution of Lake Erie and the Chesapeake Bay, the acid rain in the Northeastern United States, or the air pollution in Denver and Los Angeles to realize that development has left its inadvertent scars. Just as we are not ready to do without electric power to reduce acid rain, we are not prepared to abandon irrigation to regain wild and scenic rivers.

But what of a balance? The question of sustainability for irrigation may now be restated: Are we prepared to maintain irrigation agriculture indefinitely?

Values are changing. Our increasing population is straining the limits of water resources. As we grow more prosperous, we place a higher value on recreation. Increased stress on natural resources has sharpened our awareness of potential damage to the environment. Injudicious use of chemicals by industry and agriculture has led to demands for reform. In this climate, it can no longer be taken for granted that irrigated agriculture is a protected, preferred practice and that its waste products will be accepted unconditionally.

Numerous institutions combine to establish the values by which society wishes to operate. Imperfect as they are, they and changes in them illustrate society's changing priorities (Chapter 27).

It is often claimed that the main goal of the Reclamation Act of 1902 was to develop the West by establishing family-sized irrigated farms. Though not originally intended, subsequent developments led to a system that included substantial subsidies to irrigators (Worster 1985). Our concern here is not whether such subsidies were desirable or whether the goal of family-sized farms was achieved (Goodall and Sullivan 1985). Of interest is the impact that these subsidies had continue to have on water allocation. The situation is compounded by the additional subsidies associated with the various commodity programs of the U.S. Department of Agriculture.

The Reclamation Act often limits the ability of districts or farmers to market their water to third parties or to use irrigation facilities to transport water not under contract. In other words, the institutions developed under the law lead to a distorted value for water, and restrict the possibility of finding its most valued use.

Such constraints are not restricted to the Reclamation Act. The opportunity has been demonstrated and its implementation mandated to conserve substantial amounts of irrigation water in the Imperial Valley through reduced transport losses and increased irrigation efficiency. However, attempts to reach an agreement with the Metropolitan Water District, which serves Los Angeles and surrounding communities, to transfer such saved water to the district have been stymied in part by the issue of whether California law gives the Imperial Irrigation District the right to water that can be conserved (Wahl and Davis 1986).

Numerous treatises on the dilemma of water allocation have been published in recent years (e.g., Huffman 1986). Society no longer automatically considers irrigation to be the preferred use. For example, instream use, long considered a stepchild in allocating water, is being increasingly recognized as a valued and protected right. States have successfully applied the public trust doctrine to protect instream uses (Lamb 1986).

The Migratory Bird Treaty Act and the Clean Water Act are two federal statutes that directly affect the disposal of drainage water. State laws can have an even greater impact. For example, California's Toxic Pits Cleanup Act severely limits the use of evaporation ponds for drainage disposal, and California's Water Resources Control Board has issued a series of orders on the management and clean-up of Kesterson Reservoir. Several bills on ground-water quality are pending in Congress. Laws, rules, and customs are clearly changing in the direction of greater protection of the environment and less preferential treatment of irrigation enterprises. Many changes reflect valid concern. Conversely, it is important to avoid overreaction and to recognize that irrigation has made a valuable and valid contribution to the economy of the West and the well being of the country. Its importance abroad is often far greater.

Political forces are driving the system since the issues are matters of policy. However, political decisions are influenced by facts and how they are perceived. Scientists and technicians who specialize in water resources should provide such facts in a comprehensible form. When

California and the U. S. Department of Interior faced the San Joaquin Valley drainage crisis, they turned to the National Research Council. A panel appointed by the Water Science and Technology Board of this council recommended, among other things, that a wide range of options be identified, evaluated and compared, and discussed in public forums and among policy makers (National Research Council 1989b).

Not all situations warrant or allow such a formal, scientific approach. However, the panel's recommendations should and could be implemented in other situations. The aim is to recognize the conflicting interests of various groups, quantify them as much as possible, and identify the costs associated with any one option. For example, if ocean disposal of drainage water is outlawed, as it is along parts of the California coast, what are the costs that society has to pay for that legislative decision? If irrigation becomes uneconomic under the constraints imposed by society, at what price does one maintain and enforce them?

Whether decisions are made in the political arena or based on scientific analysis, they must deal with uncertainty. Risk analysis and management try to deal with such uncertainty, but lack of an adequate data base is an inescapable fact of life. In the present instance, the problems of salinity are understood well enough to make reasonable recommendations. The potential problems from trace elements released by irrigation are far less well understood. The U.S. Department of Interior is surveying 21 sites across the West to determine the extent to which high levels of trace elements may endanger wildlife refuges. The results will provide a basis for speculating on future problems (Deason 1989).

CONCLUSIONS

Based on technical considerations, little question exists that irrigation in semi-arid climates can be sustained indefinitely. Salinization can be avoided by providing adequate drainage. However, drainage however, exacts a price. Drainage water needs disposal and degrades the quality of water along its disposal route. In closed basins, the disposal terminus, such as a lake or a sea, eventually becomes unfit for use as a wildlife refuge or any other biological use. The recent discovery of toxic levels of trace minerals in drainage waters does not change this conclusion, but implies a far greater price for disposal where such trace minerals are present.

Increased pressure on natural resources, especially on Western water resources, suggests that the benefits of irrigation and acceptance of the problems of irrigation drainage can no longer be taken for granted. Irrigation agriculture must adapt to changing physical and social conditions to survive. Greater concern with environmental values, increased demand for recreational opportunities and direct competition for water of good quality from industries and municipalities combine with a decrease in the farm block's political power to create a serious challenge to irrigated agriculture. The technology exists to boost irrigation efficiency, reducing the volume of drainage water that needs disposal. Still, certain concessions will be demanded from irrigating farmers.

Irrigation agriculture will not only survive, but flourish in the proper circumstances. Changes in institutional arrangements and reduced subsidies that more nearly reflect the costs of irrigation may force some economically inefficient operations out of business, but the bulk of irrigated agriculture will probably be able to adapt to the new circumstances.

This discussion drew heavily on the United States experience. The situation in other countries often differs drastically, although the parallels between New South Wales in Australia and the Colorado River Basin are striking. Notwithstanding such differences, most areas in the world where irrigation is practiced or planned will ultimately face the same problems that the United States now encounters. The issues differ in detail but not in concept. The time frame in which these issues arise can be very different, and others may put different values on alternate uses than does the American public. Just the same, the experience in California presages similar experiences in many other regions.

REFERENCES

Brown, P.L., Halvorson, A.D., Siddoway, F.H., Mayland, H.F., and Miller, M.R. 1982. *Saline-seep diagnosis, control and reclamation*. U.S. Dept. Agric. Conservation Res. Report, No. 30, 22 pp. illus.

Deason, J.P. 1989. "Irrigation-induced contamination: how real a problem?" *J. Irrig. & Drainage Division,* ASCE 115(IR1): 9-20.

Doering, E.J., and Sandoval, F.M. 1976. "Hydrology of saline seeps in the Northern Great Plains". *Trans.* ASAE 19(5):856-861, 865.

Encyclopedia Britannica, 15th Edition, Vol. 15, p. 875.

Frankenberger, W.T., and Karlson, U. Dec. 1988. "Dissipation of soil selenium by microbial volatilization at Kesterson Reservoir". Final report for U.S. Bureau of Reclamation, Sacramento, Calif. 380 pp.

Goodall, M.R., and Sullivan, J.D. 1985. "Water system entities in California: social and environmental effects." *Special Water Districts: Challenge for the Future,* James N. Carbridge, ed. Natural Resources Law Center, Univ. of Colo., Boulder. pp. 71-102.

Halvorson, A.D., and Kresge, P.O. 1982. *FLEXCROP—a dryland cropping systems model.* U. S. Department of Agriculture Production Research Report #180. 49 pp.

Hilgard, E.W. 1886. "Irrigation and alkali in India". College of Agriculture, Univ. of Calif., Berkeley, Bulletin #86, pp. 34-35.

Huffman, J.L. 1986. "Allocating water to instream uses: private alternatives". Proc. Nat. Symp., Water Resources Law. *ASAE Publication 10-86,* pp. 140-147.

Jonez, A.R. 1983. "Controlling salinity in the Colorado River basin, the arid west". *Salinity in Watercourses and Reservoirs*. R.H. French, ed. Butterworth Publishers. Boston, pp. 337-348.

Lamb, B.L. 1986. "In-stream flow litigation: a survey for managers". Proc. Nat. Symp. Water Resources Law, *ASAE Publ. 10-86,* pp. 148-159.

Letey, J., Roberts, C., Penberth, M., and Vasek, C. 1986. "An agricultural dilemma: drainage water and toxic disposal in the San Joaquin Valley". Univ. of Calif.

Lohman, L.C., Milliken, J.G., Dorn, W.S., and Tucky, K.E. 1988. "Estimating Economic Impacts of Salinity of the Colorado River". Report to the U.S. Bureau of Reclamation, Water Quality Office, Denver, Colo.

Meyer, J.L., and van Schilfgaarde, J. 1984. "Case History—Salton Sea Basin". *California Agric.* 38(10): 13-16.

National Research Council. 1989a. *Alternative Agriculture.* National Academy Press. (In Press) Washington, D.C.

National Research Council. 1989b. Irrigation-induced water quality problems: what can be learned From the San Joaquin valley experience? National Academy Press. (In Press) Washington D.C.

Phene, C.J., Davis, K.R., McCormick, R.L., and Heinrick, D. 1988. "Subsurface drip irrigation: management for maximizing yields and reducing drainage". *Proc. Drip Irrig. Symp.*, San Diego, Calif. pp. 34-54.

Rhoades, J.D., and Suarez, D.L. 1977. "Reducing water quality degradation through minimized leaching management". *Agric. Water Management*, 1(2):127-142.

van Schilfgaarde, J. 1982. "The Welton Mohawk dilemma." *Int'l. J. of Water Supply and Mgmt.*, 6:115-127.

van Schilfgaarde, J., Bernstein, L., Rhoades, J.D., and Rawlins, S.L. 1974. "Irrigation management for salt control." *J. Irrig. and Drainage Div., ASCE* 100(IR3): 321-338.

Wahl, R.W., and Davis, R.K. 1986. "Satisfying Southern California's thirst for water: Efficient alternatives." *Scarce Water and Institutional Change*, K.D. Frederick and D.C. Gibbons, eds. Resources for the Future. Washington, D.C. pp. 102-133.

Worster, Donald. 1985. *Rivers of Empire*. Pantheon Books, New York, xi +402 pp.

APPENDICES

Table A.1 Examples of Preferred (P) and Acceptable (A) Units for General Use[1]

Quantity/rate (1)	Application (2)	Unit (3)	Symbol (4)
Angle	X-ray diffraction pattern	radian (P)	θ
		degree (A)	°
Area	Land area	square meter (P)	m^2
		hectare (A)	ha
	Leaf area	square meter	m^2
	Specific surface area of soil	square meter per kilogram	$m^2\ kg^{-1}$
Interatomic spacing	Crystal structure	nanometer (P)	nm
		Angstrom (A)	Å
Bulk density (ρ_b)	Soil bulk density	megagram per cubic meter (P)	$Mg\ m^{-3}$
		gram per cubic centimeter (A)	$g\ cm^{-1}$
Electrical conductivity	Salt tolerance	decisiemen per meter	$dS\ m^{-1}$
Elongation rate	Plant	millimeter per second (P)	$mm\ s^{-1}$
		millimeter per day (A)	$mm\ d^{-1}$
Ethylene production	N_2_fixing activity	nanomole per plant per second	$nmol\ plant^{-1}\ s^{-1}$
Extractable ion	Soil, mass basis	centimole per kilogram (P)	$cmol\ kg^{-1}$
		milligram per kilogram (A)	$mg\ kg^{-1}$
Extractable ion	Soil, volume basis	mole per cubic meter (P)	$mol\ m^{-3}$
		gram per cubic meter (P)	$gm\ m^{-3}$
		centimole per liter (A)	$cmol\ l^{-1}$
		milligram per liter (A)	$mg\ l^{-1}$
Fertilizer rate	Soil	gram per square meter (P)	$gm\ m^{-1}$
		kilogram per hectare (A)	$kg\ ha^{-1}$

(continued)

Table A.1 Examples of Preferred (P) and Acceptable (A) Units for General Use[1] (continued)

Quantity/rate (1)	Application (2)	Unit (3)	Symbol (4)
Flux density	Gas diffusion	mole per square meter per second (P)	mol m^{-2} s^{-1}
		gram per square meter per second (A)	gm m^{-2} s^{-1}
	Water flow	kilogram per square meter per second (P)	kg m^{-2} s^{-1}
		cubic meter per square meter per second (A)	m^3 m^{-2} s^{-1}
		meter per second (A)	m $^{-1}$
Gas diffusivity	Gas diffusion	square meter per second	m2 s^{-1}
Growth rate	Plant growth	gram per square meter per second	gm m^{-2} s^{-1}
Hydraulic conductivity	Water flow	kilogram second per cubic meter (P)	kg s m^{-3}
		cubic meter second per kilogram (A)	m^3 s kg^{-1}
		meter per second (A)	m s^{-1}
Ion transport	Ion uptake	mole per kilogram (of dry plant tissue) per second	mol kg^{-1} s^{-1}
		mole of charge per kilogram (of dry plant tissue) per second	mol^c kg^{-1} s^{-1}
Leaf area ratio	Plant	square meter per kilogram	m^2 kg^{-1}
Length	Depth, width, and height	meter (P)	m
		centimeter (A)	cm
		millimeter (A)	mm

(continued)

Table A.1 Examples of Preferred (P) and Acceptable (A) Units for General Use[1] (continued)

Quantity/rate (1)	Application (2)	Unit (3)	Symbol (4)
Nutrient concentration	Plant	millimole per kilogram (P)	$mmol\ kg^{-1}$
		gram per kilogram (A)	$gm\ kg^{-1}$
Photosynthetic rate	CO_2 amount of substance flux density (P)	micromole per square meter per second (P)	$\mu mol\ m^{-2}\ s^{-1}$
	CO_2 mass flux density (A)	milligram per square meter per second (A)	$mg\ m^{-2}\ s^{-1}$
Precipitation	Rainfall	millimeter	mm
Soil texture composition	Soil	gram per kilogram (P)	$g\ kg^{-1}$
		percent (A)	%
Transpiration rate	H_2O flux density	gram per square meter per second (P)	$g\ m^{-2}\ s^{-1}$
		cubic meter per square meter per second (A)	$m^3\ m^{-2}\ s^{-1}$
		meter per second (A)	$m\ s^{-1}$
Volume	Field or laboratory	cubic meter (P)	m^3
		liter (A)	l
Water content	Plant	gram water per kilogram wet or dry tissue (P)	$g\ kg^{-1}$
	Soil (acceptable for plants)	kilogram water per kilogram dry soil (P)	$kg\ kg^{-1}$
		cubic meter water per cubic meter soil (A)	$m^3\ m^{-3}$
Yield	Grain or forage yield	gram per square meter (P)	$g\ m^{-2}$
		kilogram per hectare (A)	$kg\ ha^{-1}$
		megagram per hectare (A)	$Mg\ ha^{-1}$
		tonne per hectare (A)	$t\ ha^{-1}$

(continued)

Table A.1 Examples of Preferred (P) and Acceptable (A) Units for General Use[1] (continued)

Quantity/rate (1)	Application (2)	Unit (3)	Symbol (4)
	Mass of plant or plant part	gram (gram per plant or plant part)	g (g $plant^{-1}$ or $kernel^{-1}$)

[1]Reprinted with permission from *ASA-CSSA-SSSA Publications Handbook and Style Manual*, American Society of Agronomy, Crop Science Society of America, and Soil Science Society of America, Madison, Wisc., 1988.

Table A.2 Conversion Factors for SI and Non-SI Units[1]

To convert Column 1 into Column 2, multiply by (1)	Column 1 SI Unit (2)	Column 2 non-SI unit (3)	To convert Column 2 into Column 1, multiply by (4)
(a)	Length		
0.621	kilometer, km (10^3 m)	mile, mi	1.609
1.094	meter, m	yard, yd	0.914
3.28	meter, m	foot, ft	0.304
1.0	micrometer, μm (10^{-6} m)	micron, μ	1.0
3.94×10^{-2}	millimeter, mm (10^{-3} m)	inch, in	25.4
10	nanometer, nm (10^{-9} m)	Angstrom, Å	0.1
(b)	Area		
2.47	hectare, ha	acre	0.405
247	square kilometer, km^2 (10^3 m)^2	acre	4.05×10^{-3}
0.386	square kilometer, km^2 (10^3 m)^2	square mile, mi^2	2.590
2.47×10^{-4}	square meter, m^2	acre	4.05×10^3
10.76	square meter, m^2	square foot, ft^2	9.29×10^{-2}
1.55×10^{-3}	square millimeter, mm^2 (10^{-6} m)^2	square inch, in^2	645
(c)	Volume		
9.73×10^{-3}	cubic meter, m^3	acre-inch	102.8
35.3	cubic meter, m^3	cubic foot, ft^3	2.83×10^{-2}
6.10×10^4	cubic meter, m^3	cubic inch, in^3	1.64×10^{-5}
2.84×10^{-2}	liter, l (10^{-3} m^3)	bushel, bu	35.24
1.057	liter, l (10^{-3} m^3)	quart (liquid), qt	0.946
3.53×10^{-2}	liter, l (10^{-3} m^3)	cubic foot, ft^3	28.3
0.265	liter, l (10^{-3} m^3)	gallon	3.78
33.78	liter, l (10^{-3} m^3)	ounce (fluid), oz	2.96×10^{-2}
2.11	liter, l (10^{-3} m^3)	pint (fluid), pt	0.473

(continued)

Table A.2 Conversion Factors for SI and Non-SI Units[1] (continued)

To convert Column 1 into Column 2, multiply by (1)	Column 1 SI Unit (2)	Column 2 non-SI unit (3)	To convert Column 2 into Column 1, multiply by (4)
(d)	Mass		
2.20×10^{-3}	gram, g (10^{-3} kg)	pound, lb.	454
3.52×10^{-2}	gram, g (10^{-3} kg)	ounce (avoirdupois), oz	28.4
2.205	kilogram, kg	pound, lb	0.454
0.01	kilogram, kg	quintal (metric), q	100
1.10×10^{-3}	kilogram, kg	ton (2000 lb), ton	907
1.102	megagram, Mg (tonne)	ton (U.S.), ton	0.907
1.102	tonne, t	ton (U.S.), ton	0.907
(e)	Yield and Rate		
0.893	kilogram per hectare, kg ha^{-1}	pound per acre, lb $acre^{-1}$	1.12
7.77×10^{-2}	kilogram per cubic meter, kg m^{-3}	pound per bushel, lb bu^{-1}	12.87
1.49×10^{-2}	kilogram per hectare, kg ha^{-1}	bushel per acre, 60 lb	67.19
1.59×10^{-2}	kilogram per hectare, kg ha^{-1}	bushel per acre, 56 lb	62.71
1.86×10^{-2}	kilogram per hectare, kg ha^{-1}	bushel per acre, 48 lb	53.75
0.107	liter per hectare, l ha^{-1}	gallon per acre	9.35
893	tonnes per hectare, t ha^{-1}	pound per acre, lb $acre^{-1}$	1.12×10^{-3}
893	megagram per hectare, Mg ha^{-1}	pound per acre, lb $acre^{-1}$	1.12×10^{-3}
0.446	megagram per hectare, Mg ha^{-1}	ton (2000 lb) per acre, ton $acre^{-1}$	2.24
2.24	meter per second, m s^{-1}	mile per hour	0.447

(continued)

Table A.2 Conversion Factors for SI and Non-SI Units[1] (continued)

To convert Column 1 into Column 2, multiply by (1)	Column 1 SI Unit (2)	Column 2 non-SI unit (3)	To convert Column 2 into Column 1, multiply by (4)
(f)	Specific Surface		
10	square meter per kilogram, $m^2\ kg^{-1}$	square centimeter per gram, $cm^2\ g^{-1}$	0.1
1000	square meter per kilogram, $m^2\ kg^{-1}$	square millimeter per gram, $mm^2\ g^{-1}$	0.001
(g)	Pressure		
9.90	megapascal, MPa (10^6 Pa)	atmosphere	0.101
10	megapascal, MPa (10^6 Pa)	bar	0.1
1.00	megagram per cubic meter, $Mg\ m^{-3}$	gram per cubic centimeter, $g\ cm^{-3}$	1.00
2.09×10^{-2}	pascal, Pa	pound per square foot, $lb\ ft^{-2}$	47.9
1.45×10^{-4}	pascal, Pa	pound per square inch, $lb\ in^{-2}$	6.90×10^3
(h)	Temperature		
1.00 (K − 273)	Kelvin, K	Celsius, °C	1.00 (°C + 273)
(9/5 × °C) + 32	Celsius, °C	Fahrenheit, °F	5/9 × (°F − 32)
(i)	Energy, Work, Quantity of Heat		
9.52×10^{-4}	joule, J	British thermal unit, Btu	1.05×10^3
0.239	joule, J	calorie, cal	4.19
10^7	joule, J	erg	10^{-7}

(continued)

Table A.2 Conversion Factors for SI and Non-SI Units[1] (continued)

To convert Column 1 into Column 2, multiply by (1)	Column 1 SI Unit (2)	Column 2 non-SI unit (3)	To convert Column 2 into Column 1, multiply by (4)
0.735	joule, J	foot-pound	1.36
2.387×10^{-5}	joule per square meter, J m^{-2}	calorie per square centimeter (langley)	4.19×10^{4}
10^{5}	newton, N	dyne	10^{-5}
1.43×10^{-3}	watt per square meter, W m^{-2}	calorie per square centimeter minute (irradiance), cal cm^{-2} min^{-1}	698
(j)	Transpiration and Photosynthesis		
3.60×10^{-2}	milligram per square meter second, mg m^{-2} s^{-1}	gram per square decimeter hour, g dm^{-2} h^{-1}	27.8
5.56×10^{-3}	milligram (H_2O) per square meter second, mg m^{-2} s^{-1}	micromole (H_2O) per square centimeter second, μmol cm^{-2} s^{-1}	180
10^{-4}	milligram per square meter second, mg m^{-2} s^{-1}	milligram per square centimeter	10^{4}
35.97	milligram per square meter second, mg m^{-2} s^{-1}	milligram per square decimeter hour, mg dm^{-2} h^{-1}	2.78×10^{-2}
(k)	Plane Angle		
57.3	radian, rad	degrees (angle), °	1.75×10^{-2}

(continued)

Table A.2 Conversion Factors for SI and Non-SI Units[1] (continued)

To convert Column 1 into Column 2, multiply by (1)	Column 1 SI Unit (2)	Column 2 non-SI unit (3)	To convert Column 2 into Column 1, multiply by (4)
(l)	Electrical Conductivity, Electricity, and Magnetism		
10	siemen per meter, S m^{-1}	millimho per centimeter, mmho cm^{-1}	0.1
10^4	tesla, T	gauss, G	10^{-4}
(m)	Water measurement		
9.73×10^{-3}	cubic meter, m^3	acre-inches, acre-in	102.8
9.81×10^{-3}	cubic meter per hour, $m^3\ h^{-1}$	cubic feet per second, $ft^3\ s^{-1}$	101.9
4.40	cubic meter per hour, $m^3\ h^{-1}$	U.S. gallons per minute, gal min^{-1}	0.227
8.11	hectare-meters, ha-m	acre-feet, acre-ft	0.123
97.28	hectare-meters, ha-m	acre-inches, acre-in	1.03×10^{-2}
8.1×10^{-2}	hectare-centimeters, ha-cm	acre-feet, acre-ft	12.33
(n)	Concentrations		
1	centimole per kilogram, cmol kg^{-1} (ion exchange capacity)	millequivalents per 100 grams, meq 100 g^{-1}	1
0.1	gram per kilogram, g kg^{-1}	percent, %	10
1	milligram per kilogram, mg kg^{-1}	parts per million, ppm	1

(continued)

Table A.2 Conversion Factors for SI and Non-SI Units[1] (continued)

To convert Column 1 into Column 2, multiply by (1)	Column 1 SI Unit (2)	Column 2 non-SI unit (3)	To convert Column 2 into Column 1, multiply by (4)
(o)	Plant Nutrient Conversion		
2.29	P	P_2O_5	0.437
1.20	K	K_2O	0.830
1.39	Ca	CaO	0.715
1.66	Mg	MgO	0.602

[1]Reprinted with permission from ASA-CSSA-SSSA-Publications Handbook and Style Manual, American Society of Agronomy, Crop Science Society of America, and Soil Science Society of America, Madison, Wisc., 1988.

Table A.3 Conversion Factors for Chemical Concentrations.[1]

Element and reported species (1)	F_1 (2)	F_2 (3)
Aluminum (Al^{3+})	0.11119	0.03715
Ammonium (NH_4^+)	0.05544	0.05544
Antimony (Sb)	—	0.00821
Arsenic (As)	—	0.01334
Barium (Ba^{2+})	0.01456	0.00728
Beryllium (Be^{2+})	0.22192	0.11096
Bicarbonate (HCO_3^-)	0.01639	0.01639
Boron (B)	—	0.09250
Bromide (Br^-)	0.01252	0.01252
Cadmium (Cd^{2+})	0.01779	0.00890
Calcium (Ca^{2+})	0.04990	0.02495
Carbonate (CO_3^{2-})	0.03333	0.01666
Cesium (Cs^+)	0.00752	0.00752
Chloride (Cl^-)	0.02821	0.02821
Chromium (Cr)	—	0.01923
Cobalt (Co^{2+})	0.03394	0.01697
Copper (Cu^{2+})	0.03147	0.01574
Fluoride (F^-)	0.05264	0.05264
Hydrogen (H^+)	0.99216	0.99216
Hydroxide (OH^-)	0.05880	0.05880
Iodide (I^-)	0.00788	0.00788
Iron (Fe^{2+})	0.03581	0.01791
Iron (Fe^{3+})	0.05372	0.01791
Lead (Pb^{2+})	0.00965	0.00483
Lithium (Li^+)	0.14407	0.14407
Magnesium (Mg^{2+})	0.08229	0.04114
Manganese (Mn^{2+})	0.03640	0.01820
Mercury (Hg)	—	0.00499
Molybdenum (Mo)	—	0.01042
Nickel (Ni)	—	0.01704
Nitrate (NO_3^-)	0.01613	0.01613
Nitrite (NO_2^-)	0.02174	0.02174
Phosphate (PO_4^{3-})	0.03159	0.01053
Phosphate (HPO_4^{2-})	0.02084	0.01042
Phosphate ($H_2PO_4^-$)	0.01031	0.01031
Potassium (K^+)	0.02558	0.02558
Rubidium (Rb^+)	0.01170	0.01170
Selenium (Se)	—	0.01266
Silica (SiO_2)	—	0.01664
Silver (Ag^+)	0.00927	0.00927
Sodium (Na^+)	0.04350	0.04350
Strontium (S^{2+})	0.02283	0.01141
Sulfate (SO_4^{2-})	0.02082	0.01041
Sulfide (S^{2-})	0.06238	0.03119

(continued)

Table A.3 Conversion Factors for Chemical Concentrations.[1] (continued)

Element and reported species (1)	F_1 (2)	F_2 (3)
Thorium (Th)	—	0.00431
Titanium (Ti)	—	0.02088
Uranium (U)	—	0.00420
Vanadium (V)	—	0.01963
Zinc (Zn^{2+})	0.03059	0.01530

[1]From Hem, J.D. 1985. *Study and Interpretation of the Chemical Characteristics of Natural Water*, 3rd ed., USGS Water-Supply Paper 2254, 263 pp.

Note: Milligrams per liter × F_1 = milli-equivalents per liter; milligrams per liter x F_2 = millimoles per liter (based on 1975 atomic weights, referred to carbon-12)

Table A.4 Other Useful Conversion Factors and Formulas

TDS (mg/l) ≈ EC (dS/m) × 640	for EC between 0.1 and 5.0 dS/m
TDS (mg/l) ≈ EC (dS/m) × 800	for EC > 5.0 dS/m
TDS (lbs/ac-ft) ≈ TDS (mg/l) × 2.72	
TDS (tons/ac-ft) ≈ TDS (mg/l) × 0.00136	
sum of cations/anions (meq/l) ≈ EC (dS/m) × 10	for EC between 0.1 and 5.0 dS/m
log cations/anions ($mmol_c$/l) ≈ 0.955 + 1.039 log EC (dS/m)	
log total soluble salts ($mmol_c$/l) ≈ 0.990 + 1.055 log EC (dS/m)	
ionic strength (mol/l) ≈ EC (dS/m) × 0.0127	
osmotic pressure (atm) ≈ EC (dS/m) × 0.40	for EC between 3 and 30 dS/m

Index

Acid extractants, 147–148
Acids, 419–420
Adjusted sodium adsorption ratio, 59–61
Adsorbed ions, 411
Adsorbed sodium, 411
Adsorption, of trace elements, 67–72
Adsorption isotherms, 70
Aerial photography, 201
Aggregate size distribution, 212–213
Aggregate stability, 91–92, 107–108, 212–213, 444
Agricultural drainage regulations, 451
Agricultural drainage water, 551
Agricultural salt problems, concerns over, 11–15; California, 5; extent of, 3–5
Agronomic practices, 379–382, 384
Air pollutants, 266–267
Alamo River, 436
Alfalfa, 116, 163, 240–241, 242, 266, 283, 316, 331, 378, 380–381, 384, 387, 434, 436, 437, 443, 474, 538, 545, 569
Alkali and alkaline earth metals, 74–75, 262
Alternate row irrigation, 415
Aluminum, 67, 209, 234, 423
Anion analysis, 209
Anion exclusion, 53–54
Anions, 47
Anthropogenic activities, 44, 75, 77
Antimony, 138
Applied water, distribution of, 314, 317–321, 351–352, 353, 355, 356, 511
Applied water. See also Irrigation water.
Aquatic life, 79, 458, 566. See also Fish
ARIMA models, 329, 331, 333–334, 339
Arizona, 13, 402, 407, 421, 453, 564
Arkansas River Basin, Colorado, 399–400, 434
Arsenate, 71
Arsenic, 77, 138, 220, 442
Artesian pressure, 362
Artichokes, 442
Asparagus, 161, 267
Aswan Dam, 584
Atmospheric salt deposition, 44
Auger holes, 205
Australia, 372, 400, 402, 561
Autocorrelation, 327–328, 331–333, 341
Avocados, 126, 163, 288, 478

Barium, 74–75
Barley, 13, 118, 122, 123, 124, 125, 126, 161, 164, 165, 169, 170, 172, 175, 266, 283, 288, 297, 382, 384, 434–435, 442, 569
Basin irrigation, 351
Beans, 119–120, 123, 124, 127, 128, 147, 171, 172, 175, 266, 295, 539–540
Beets, 172, 266
Bermuda grass, 267
Beryllium, 138, 209
Bicarbonate, 206, 208, 210, 221, 295, 445, 470, 476
Bioenergetics, 173–174
Biological wastewater treatment, 454–456, 457
Birds, 552. See also Waterfowl
Bird's-foot trefoil, 267
Border irrigation, 319, 350, 359
Border leaching, 415
Boron, 71, 77–78, 138, 142, 148–150, 155, 197, 220, 289, 294, 436, 451; concentration, 149–150, 211, 229–231, 410; deficiencies, 149–150, 229–230; toxicity, 21, 29–30, 149–150, 206–207, 212, 229–230, 262, 289, 294, 423–426, 440
Boron adsorption model, 423
Boron-affected soils, 423
Brine disposal, 453–454

Broadview irrigation district, California, 242, 562
Bromegrass, 267

Cabbage, 147, 442
Cadmium, 79, 138, 146, 148, 153, 155
Calcareous soils, 99, 101, 420–421
Calcite, 49–50, 67, 227; precipitation and dissolution, 492–493
Calcium, 29–30, 119, 122–123, 206, 209–210, 211, 221, 225, 227–229, 262, 375, 386, 410, 415–416, 418, 421, 445, 470, 474–476, 494 479
Calcium carbonate, 206, 512, 514
Calcium chloride, 420
Calcium distribution, 92–93
Calcium sulfate. See Gypsum
California, 5, 239, 436, 564, 565, 574
California Department of Water Resources, 453
California State Water Resources Control Board, 451, 552, 591
California Toxic Pits Cleanup Act, 552, 591
Canada, 379, 400
Cantaloupes, 437, 443. See also Melons; Muskmelons
Carbonate, 206, 208, 210, 445, 470
Carbonate alkalinity, 47
Carbonic acid, 492
Carrots, 147, 266, 442
Cation analysis, 209–210
Cation exchange capacity (CEC), 52, 53–54, 211, 493–495
Cations, 47
Cauliflower, 283
CEC. See Cation exchange capacity
Cereal crops, 267
Celery, 288
Chemical mass action, 48–50. See also Mass action
Chemical wastewater treatment, 456
Chloride, 29, 117, 197, 206, 210, 241, 250, 262, 288–289, 375, 440
Chlorine, 138, 220, 221, 225, 445
Chromatographic models, 421–422
Chromium, 75–76, 138
CIRF model. See Conceptual Irrigation Return Flow (CIRF) model.
Citrus, 75, 115, 170, 288, 295, 478–479, 539
Clay mineralogy, 98
Clay minerals, 52–53, 67, 423
Clay-sand mixtures, 96–98
Clays, swelling and dispersion, 19, 92, 93–95, 96–97, 99, 411
Clean Water Act. See Water quality regulations
Climate, 21, 43, 266–267, 443–444
Clover, 124, 125, 164, 240, 251, 266, 267–268, 442
Coal seam seeps, 373
Cobalt, 138, 154, 457
Coconut, 113
Cokriging, 329–330, 336–339
Collector ponds, 407
Colligative properties, changes in, 114–115
Colloidal clay, 92–95
Colorado, 44, 399, 574
Colorado River, 239, 436, 472–473, 476, 477, 587
Colorado River Basin, 44, 399, 401–402, 408, 463, 562, 586
Composite sampling, 193–194
Conceptual Irrigation Return Flow (CIRF) model, 507–528
Confined aquifers, 362
Constant capacitance model, 71
Continuous ponding, 414
Copper, 79, 138, 146, 220, 228–229
Corn, 120, 121, 126, 127, 143, 151, 171, 172, 240–241, 251, 252, 254, 255, 266, 267, 268, 288, 363, 384, 539–540
Corrugation irrigation, 350
Cotton, 115, 121–122, 123, 161, 164, 172, 249–250, 251, 266, 283, 313, 314, 317, 366, 434, 436, 437, 538–539, 545, 569
Cowpea, 267
Creeping bentgrass, 267
Cress, 147, 442
Crop management practices, 443–444

Crop rotation, 384, 435, 540–542. See also Crop-summerfallow rotation
Crop salt tolerance, 20–21, 47, 252–258, 262–298, 356–357, 364, 441–442; biological factors influencing, 267–268; data, sources of, 263; criteria, 263; herbaceous crops, 270; improving, 161; indices of, 168–169; measurement, 166–167; woody crops, 270
Crop selection criteria, 441–444
Crop stand establishment, 444–445
Crop water extraction patterns, 444
Crop water production functions, definition of, 305–307
Crop water requirement, 237
Crop yield, 445. See also Relative crop yield
Crop-summerfallow rotation, 372, 381–382, 384, 385–386, 387. See also Crop rotation
Crops, toxic elements, 442
Cross-correlation, 328
Crust formation. See Crusting
Crusting, 26–28, 102, 108, 423, 444, 445, 497
Cyclic dilution, 6
Cyclic irrigation, 435–437
Cyclic salinization, 6

Dallisgrass, 295
Date palms, 113
Davies equation, 47, 491
DDL. See Diffuse double layer
De-icing salts, 163
Dead Sea, 587
Deep percolation, 38, 535
Deep well injection, 408, 460–462
Depositional crust, 104, 107
Desalination, 453–454, 464
Desalting processes, 403, 407–408, 453–454
Deterministic control problems, 531–532
Deterministic models, 483
Diffuse double layer (DDL) model, 52–54
Dissolution, and trace elements, 64–66
Distribution canals, 35–36, 404–405
Ditch lining, 406
Dolomite, 67
Drainage, 25, 37, 238, 360–362, 379, 407, 586–589
Drainage effluents, 362, 462
Drainage water, 453, 551–553, 564–565; analysis, 441; availability, 441; discharge, 458; disposal, 398–399, 407, 433, 450–451, 457–462, 541–542; quality, 362, 557–558; salinity, 331–340
Drainage water reuse, 38–40, 121, 404, 407–408, 432–446, 451, 463–466; blending strategy, 39, 437–439, 441; cyclic strategy, 435–437, 438–439; long-term effects, 439–440; management of, 440; potential, 434
Drainage water treatment, 450, 453–457, 463
Drainage wells, 362
Drip irrigation, 35, 438
Drip/trickle irrigation, 193. See also Micro-Irrigation
DTPA-TEA, 145–147
Dynamic optimization concepts, 530–533

EC. See Electrical conductivity (EC)
Economic factors, 395, 441, 565–570
Effects of salinity, 308
Effects of salts on plants, 19
Effects of salts on soils, 18
Egypt, 163, 164, 584–585
Electrical conductivity (EC), 1–2, 46–47, 107, 206, 207, 221–224, 243–244, 377, 445, 472, 478–479
Electrical conductivity and ionic strength, 47
Electrical conductivity sensors, 203
Electrolytes, 95–107
Electromagnetic conductivity meters, 195–196
Electromagnetic induction sensors, 203–204
EPA. See U.S. Environmental Protection Agency (EPA)

EPP. See Exchangeable potassium percentage
Equilibrium chemical composition, calculating, 495
ESP. See Exchangeable sodium percentage
ESR. See Exchangeable sodium ratio (ESR)
Evaporation ponds, 362, 407, 453–454, 458–460, 464, 564, 588–589, 591
Evaporative salinization, 6
Evapotranspiration, 313–314, 316–317, 321, 362, 509
Excessive exchangeable sodium, 28
Exchangeable cations, 211
Exchangeable potassium percentage (EPP), 98
Exchangeable sodium percentage (ESP), 54, 58, 91–92, 99, 101–103, 107, 206, 207, 225
Exchangeable sodium ratio, (ESR), 58–59

Farm level analysis, 543–545
Federal agencies, 550–551. See also U.S. Environmental Protection Agency (EPA)
Federal farm bills, 568–569
Ferrous hydroxide, 456
Fertilizer salts, 241
Fertilizers, 264, 316, 384
Fish, 154, 393, 396, 586. See also aquatic life
Flocculation value (FV), 94, 98, 107
Flood irrigation, 35, 223
Flow injection analysis, 210
Foliar absorption, 444
Foliar injury, 20–21, 120–121, 224–225, 282–283, 286–287
Foliar salt adsorption, 282
Forage, 153–154, 440
Forecast modeling, 334–335
Fossil deposits, 44
Four-electrode salinity probe, 195–196, 203–204, 252
Fresh water imports, 403
Freundlich adsorption equation, 70
Fruit crops, 120, 172, 255, 268, 283, 288, 295, 357, 434
Furrow irrigation, 34, 35, 193, 223, 319, 350, 444
FV. See Flocculation value

Gapon constant, 56, 58, 494
Geologic maps, 378
Geologic outcrop seeps, 373
Germination, 121, 168, 267. See also Plant development
Gila River, 472–473
Glaciated Fort Union seeps, 373
Glenn Colusa Irrigation District, 516
Global water balance, 3–4
Glycophytes, 113, 115–116, 161, 172
Grain crops, 169
Grand Valley of Colorado, 250, 477
Grapes, 126, 170, 289
Gravimetric sampling, 196
Great Plains, 372–373, 375–376, 379, 382, 400
Great Salt Lake, Utah, 587
Great Salt Lake Basin, Utah, 43
Ground water depth, 247–251
Ground water quality, 205–206
Ground waters, chemical composition of, 50–52
Ground-water contamination, 562–565
Ground-water flow, 206
Ground-water overdraft, 562
Ground-water protection laws, 552–553. See also Water quality regulations
Groundwater management, 363–368
Gypsum, 19, 50, 108, 206, 211, 375, 416–419, 422, 423, 444, 470, 472, 479–480, 512–514, 523–524; precipitation and dissolution, 492–493

Halophytes, 113, 115, 121, 161, 172
Heavy metals, 79–80
Herbaceous crops, salt tolerance, 270
High-frequency irrigation, 223–224
History, 11–13, 584–585
Humidity, 172, 266
Hydraulic conductivity, 213–214, 373–375
Hydraulic conductivity, effect of electrolyte solutions on, 95–105

Hydraulic gradient, 206
Hydrologic submodel, 507–511
Hydrosalinity models, 504–528
Hydrostatic pressure seep, 373
Hypersaline waters, 1, 6

Imperial Valley, California, 407, 436, 574, 585
India, 164, 372, 400, 433
Indigenous soil salts, 25
Indus Plain, Pakistan, 43
Industrialization, 44
Infiltration, spatial variability, 486
Infiltration rate, 27, 101–103, 140, 196, 214–215, 318–319, 350; effect of electrolyte concentration on, 101; effect of ESP on, 101
Infrared thermometry, 202
Inorganic elements, 220
Institutional and legal constraints. See Legal and institutional factors
Interbasin water transfer, 403
Interceptor drains, 360, 379
Intermittent ponding, 414
International treaties, 401–402
Interseasonal irrigation management, 538 540
Intervention analysis modeling, 339–340
Ion accumulation in plants, 171–172
Ion chromatography, 210
Ion demixing, 59
Ion effects, toxicity, 120–121
Ion exchange, 55–59
Ion restriction in plants, 169–171
Ion toxicity, 287
Ion transport in plants, selective, 170–171
Ionic balances, 228
Ionic composition, 46–47
Ionic strength, 47, 66–67
Ions, 47
Iran, 372, 400
Iraq, 12
Iron, 67, 72–73, 138, 146, 228–229, 234–235, 423
Irrigation, 566; and salinity control, 349–360; management, 34–37, 246, 356–358, 538–543; methods, 266, 444–445; practices, 396- 397; return flows, 482; scheduling, 36–37, 316, 357–358, 365–368, 533–538, 544; uniformity, 318–321, 511. See also Applied water, distribution of
Irrigation water, 10–11, 44, 314; boron in, 230; quality of, 197, 205, 220 (See also Water quality); salinity, 221–225, 316; sodicity, 225; trace elements, 231; management, 358, 405–406, 444–445; saline, 286
Israel, 254, 438, 465

Judgement sampling, 187, 197

Kaolinite, 95, 100
Kenaf, 121, 124
Kesterson National Wildlife Refuge, California, 552, 556, 564, 566, 586
Kesterson Reservoir, California, 14, 78, 450, 457, 558, 562, 586, 587, 591
Kohlrauch's Law of independent migration of ions, 46
Kriging, 329–330, 343

Land leveling, 35, 351
Landscape effects, 43
Langmuir equation, 70–71
Lateral flow, 360
Leaching, 24, 30–32, 37, 140, 241, 358–360, 411–415, 444, 476, 482, 539; methods, 414–415; minimizing, 38
Leaching fraction (LF), 222–224, 230, 314, 444, 511, 525–527
Leaching requirement, 243–247; reducing, 161
Lead, 73, 79, 138, 146
Leaf growth, 123–124
Leaf injury. See Foliar injury
Leaves, salt balance, 125–127
Legal and institutional factors, 400–402, 549–558, 576–577
Lettuce, 127, 143, 164, 268, 442
LF. See Leaching fraction
Lime, 49–50, 206, 211, 375, 470
Lithium, 75, 138
Livestock, 153–154

Local regulations, 553
Lysimeter studies, 248–250, 474

Magnesium, 100–101, 206, 209–210, 211, 221, 225, 228–229, 262, 375, 386, 423, 445, 470, 476, 494
Magnesium carbonates, 59
Mancos shale, 44
Manganese, 67, 72–73, 138, 146, 228–229
Mangroves, 117, 170
Mass action, 55–56, 490–491. See also Chemical mass action
Melons, 436, 440, 443. See also Cantaloupe; Muskmelons
Mesopotamia, 12–13
Mercury, 79–80, 138
Mexico, 399, 401–402, 408
Micro-irrigation, 355–356, 357, 358, 360. See also Drip/trickle irrigation
Microwave sensors, 203
Mineral dissolution, 42–43, 65–66
Mineral nutrition, 115
Mineral saturation index, 65
Mineral weathering, 42–43, 44–45, 52, 99, 228, 239, 411
Miscible displacement models, 421–422
Model calibration, 497, 516–518
Model validation, 497–498, 518–523
Model verification, 497–498, 516–518
Mole drains, 379
Molecular biology, 174
Molybdenum, 76, 138, 153–154, 234, 440, 442, 451
Monitoring wells, 206
Montana, 375, 386–387, 400, 564
Montmorillonite, 93–94, 96, 100
Multispectral scanners, 202–203
Mung beans, 128
Muskmelons, 163, 168. See also Cantaloupe; Melons

Natural resources, 561–565
Nevada, 564, 587
New Mexico, 421
Nickel, 79, 138, 146, 457
Nile Delta, 585
Nitrate absorption, 117–118
Nitrogen, 265, 384, 415, 457
Nonmetals, 77–79
Nonpoint source pollution, 79
Nutritional imbalances, 287

Oats, 297, 382, 442
Observation wells, 205–206, 385
One-dimensional leaching, 412–413
Onions, 172, 266, 442
Onsanger-Fouss equation, 46–47
Optimal irrigation scheduling, 533
Organic osmolytes, 115–116
Organic solutes, 172
Ornamentals, 282, 296–297
Osmotic adjustment of plants 172–173
Osmotic effects, 115–117
Osmotic potential, 48, 114, 115
Osmotic pressure, 221
Osmotic stress, 289
Oxidation-reduction processes, 72–73. See also Redox reactions
Ozone, 21, 267

Pakistan, 43, 164
Panoche Drainage District, California, 456, 516
Parsnip, 442
Particle size analysis, 213
Passive root membrane filtration, 170
Peanuts, 254, 438
Peas, 123, 297–442
Pecos River, Texas, 241, 434, 474
Peppers, 128, 255, 283, 286, 288, 445
Perennial crops, 255
Permeability. See Soil permeability
Peru, 13
Pesticides, 13, 77, 80, 589
pH, 48–50, 68, 100–101, 141–142, 206, 210, 227–228, 423–424, 492
Phosphate, 71, 120–121
Phosphorus, 209, 384
Physical wastewater treatment, 456
Phytotoxicity. See Plant toxicity
Piezometer tubes, 206
Pipelines, 406
Plant breeding, 162–175

Plant development, 121–122, 168–169, 267
Plant nutritional requirements, 228; deficiencies, 139–151; imbalance, 28–30, 287–297; requirements, ion effects, 117–120
Plant response to salinity, 114–115, 169–174, 495–496 . See also Crop salt tolerance
Plant sampling, 197, 211–212
Plant tissue analysis, 151, 211–212
Plant tissue culture, 174
Plant toxicity, controlling, 33–37
Planting practices, 357
Plants, trace element toxicity, 28–30, 139–152, 287–297
Plum trees, 255–258
Point source pollution, 79, 550–551
Pore size distribution, 497
Porterville well water, 478–480
Post-leaching operations, 415
Potassium, 206, 209, 221, 228–229, 265
Potassium transport, 117
Potatoes, 143, 283, 438, 445
Pothole Seep, 374
Power plant cooling, 463
Precipitation, and trace elements, 64–66
Production functions, 305; models, seasonal, 313–314; models, transient, 307–313; nonuniformity of, 317; statistical estimation of, 315
Public policy, 570–577

Radishes, 147, 172, 266, 442
Rainfall, 103, 238–239
Rainfall simulator, 102
Random sampling, 187–189, 190, 194–195, 197
Rapeseed, 384, 442
Recharge areas, 378
Reclamation, 385–388, 410–426. See also Soil reclamation
Red fescue, 163
Redox potential, 141
Redox reactions, 43, 72. See also Oxidation reduction processes
Regional management programs, 393–397
Regional reuse, 433
Relative crop yield, 263–265, 268–270, 308–310, 313, 316. See also Crop yield
Relief drains, 361
Remote sensing, 201–203
Research needs, 525
Reverse osmosis (RO), 453
Rhizosphere chemistry, 59–61
Rhodes grass, 294
Rice, 125, 128, 141, 153, 163, 164, 251, 268, 288, 569
Root extension, 251
Root extraction term, 311
Root zone, 237, 241, 507–511; distribution of salts in, 444; salt accumulation, 37; salt concentrations in, 473–477; soil solution composition, 470
Root zone salinity, 19–20, 22–24, 241, 310–313, 314, 470–472, 482–498; controlling, 33–37; seasonal average, 310
Root-shoot interactions, 127–128
Runoff, 27, 423
Russia, 434
Rye grass, 442

Sacramento-San Joaquin Delta, California, 252–253
Safflower, 121, 283, 381, 382, 384, 437
Saline drainage water, 398–399, 404, 412, 432–434
Saline fields, nonirrigated, 361–362
Saline irrigation waters, 282–283, 285–287, 313, 314, 356–357
Saline seeps, 361–362, 372–388, 400, 589
Salinity, and drainage, 15; constituents, 1; control of, 356–358; control and drainage, 360–363; control measures, selecting, 33–40; criterion, 2; effects, 308; effect on crop yield, 445; measurement of, 2, 45–47, 203–204, 377; modeling, 40; monitoring, 40–41; parameters, 1–2; pollution of water resources,

37; and sodicity, comparison of, 113–114; spatial distribution, 190–191
Salinity submodel, 511
Salt, absorption, 282–286; accumulation, 11–13, 363–364; dilution, 421; distribution, 363–364; effects on soils, 18–19; effects on plants, 19–21; minimizing, 34; patterns, 352, 355, 356; precipitation, 22, 239–240; problems, diagnosing, 21–30; recovery, 465; removal by crops, 240; root zone, 37, 538 (See also Root zone salinity); sources of, 2–3, 42; stress in plants, 175; transport efficiency, 412–413
Salt affected soils, 47–61, 411
Salt balance, 473–477; equations, 506; in leaves, regulation of, 125–127; models, 307
Salt flows, 10
Salt grass, 266
Salt induced proteins, 175
Salt tolerance of plants. See Crop salt tolerance
Salt tolerant crops, 34, 163, 463. See also Crop salt tolerance
Salton Sea, California, 407, 585, 587
Samples, number of, 190–191
Samples, volume-variance relationships, 192
Sampling, objectives, 186
Sampling, Spatial variability, 191–193
Sampling intervals, 191–192
Sampling strategies, evaluation of, 194–195
Sampling techniques, 187–190
San Joaquin River, California, 79, 451
San Joaquin River Basin, California, 450, 451, 464
San Joaquin Valley, California, 13–15, 78, 121, 234, 242, 249–250, 255–258, 397–399, 407, 421, 433, 436, 437, 440, 442, 453, 454–455, 462, 478–508, 553–558, 562, 585, 586
San Luis Drain, California, 14, 453, 589
SAR. See Sodium adsorption ratio
Saturated soil paste, 207–209
Saturation extracts, 45–46, 50–51
Secondary saline deposits, 44
Sedimentary rocks, 74
Seed beds, 34, 444
Seedlings, 34, 267
Selenite, 71
Selenium, 13, 14–15, 72, 78, 138, 154, 155, 234, 433–434, 440, 442, 450, 451
Sensitivity analysis, 518, 520–522
Sesame, 283
Shallow ground water, 363–368, 375; analysis, 441
Silicates, 68, 71, 479
Silicon, 138
Silver, 138
Slope change seep, 373
Smectites, 92–93, 96, 497
Sodic soil, 18, 26; electrolytes, 95–101; hydraulic conductivity of, 95–101
Sodicity, 26 445
Sodicity, controlling, 33–37
Sodium, 117, 120, 125, 138, 206, 220, 221, 225, 228, 262, 287–288, 375, 386, 410, 411, 451–418, 421, 445, 476, 494; distribution, 92–93
Sodium adsorption ratio (SAR), 19, 58, 59–60, 206, 207–208, 225–228, 445, 470, 473–474, 478–479; sodicity and, 28
Sodium/calcium clays, 93–94
Sodium-affected soils, 415–423
Soil, aeration, 251; atmosphere, 492; chemical analysis, 206–211; chemical extractants, 81; chemical interactions, modeling, 490- 495; depth, 241–242, 252–254; deterioration, 561–562; fertility, 140, 265, 384; hydrological properties, 486–487; intake rate, 26; permeability, 18, 19, 26–27, 32, 445, 497; physical characteristics, 212–215; physical condition, 264; reclamation, 30–33; reclamation models, 421–423
Soil lime, 479
Soil moisture content, 141

Soil moisture probes, 378
Soil profile, 264–266, 416, 421
Soil salinity, 439; changes in, 254–258; chemical characteristics, 141; chemical precipitation, 141; chemistry of, 47; correlation and spectral analysis modeling, 343; diagnostic criteria, 23; field assessment of, 377; identifying, 40; management of, 365; monitoring, 254, 439; measurements, 22–23; potential causes of, 23–26; salt-affected, 47, 411–415; sampling, 195–196; spatial variability, 22–23, 340–346, 543; visual indications, 22
Soil samples, 45–46
Soil sampling, 195–197, 204
Soil sodicity, 225–228
Soil solution composition, calculating, 469–472
Soil solution extracts, 209
Soil strength, 26
Soil structure interactions, 91
Soil structure stability, 497
Soil survey maps, 378
Soil texture, spatial variations, 213
Soil types, spatial variation, 414
Soil water content, 2, 196
Soil water deficit, 238
Soil water system, 7, 10, 490
Solar gradient ponds, 464–465
Solute transport modeling, 487–490
Sorghum, 123, 127, 146, 169, 241, 283, 288, 434, 435, 535, 539–540
Soybeans, 120–121, 126, 143, 147, 170, 171, 268, 288
Spain, 517, 523
Special interest groups, 551
Spectral analysis, 328, 343–346
Spinach, 147
Sprinkler irrigation, 20–21, 35, 224–225, 228, 252, 282–283, 286, 318–319, 352–355, 356–357, 359, 444–445
Sprinkling, 414
Squash, 147, 266
State agencies, 551. See also California State Water Resources Control Board
State regulations, 552
Statistical techniques, 315–317, 327–346
Stochastic control problems, 532
Stochastic models, 483
Strawberry, 266
Subirrigation, 35, 252–253
Subsurface drainage, 251, 361
Subsurface return flows, 505
Sudangrass, 240
Sugar beets, 121, 124, 161, 168, 172, 283, 363, 436
Sugarcane, 161
Sulfate, 120, 206, 375
Sulfur, 419–420
Summerfallowing. See Crop-summerfallow rotation
Supersaturation, 65
Surface chemistry, 52–54
Surface flushing, 415
Surface irrigation, 104, 228, 319, 349–352, 414
Surface return flows, 505
Surface topography, 215
Surface waters, 458; chemical composition of, 50; contamination, 458, 562–565
Surface water diversions, 565
Surface water supplies, irrigation, 10
Systematic sampling, 189, 194–195

Tailwater ditches, 407
Tall fescue, 266
TDR. See Time domain reflectrometry.
TDS. See Total dissolved solids
Tellurium, 138
Tensiometers, 206
Test holes, 378
Texas, 240, 248,, 400, 421, 434, 465, 564
Textural change seep, 373
Tile drains, 414
Tillage, 32, 357, 421
Tilth, 18, 19, 26
Time domain reflectrometry (TDR), 204
Time series analysis, 327–329
Tin, 234–235
Titanium, 209, 234–235

Tobacco, 128, 175
Tomatoes, 115, 121, 128, 147, 163, 168, 171–172, 175, 243, 245–246, 254, 255, 266, 283, 288, 436–437, 440, 443, 545
Total concentration of soluble anions (TSA), 1
Total concentration of soluble cations (TSC), 1
Total dissolved solids (TDS), 1, 45, 333–334, 336, 451, 524
Toxicity, ion effects on, 120
Trace element deficiencies, 144. See also Plant nutritional deficiencies
Trace elements, 141–142, 231–235, 564, 588–589; accumulation of, 152–154; adsorption of, 67–72; concentrations, 142–143; concentrations, processes affecting, 64–73; concentrations in soils, inorganic, 80–83; effects on animals, 152; extractants, 145–151; geochemical behavior of, 73–80; in irrigation water, 234; solution-phase speciation, 66–67; spatial distribution, 81; toxicity, 144, 221
Tradable discharge permits, 575
Transition metals, 75–77
Trifoliate orange, 171
TSA. See Total concentration of soluble anions
TSC. See Total concentration of soluble cations
Tulare Basin, California, 437, 555–556, 564
Tungsten, 234–235
Turkey, 372, 400
Turnips, 147, 442
Two-dimensional leaching, 413–414

U.S. Environmental Protection Agency (EPA), 451, 460, 550–551. See also Federal agencies
U.S. Resources Conservation and Recovery Act (RCRA), 460
Under irrigation, 27
Underground Injection Control (UIC) Program, 460
Upward flow, 238, 247–248, 362
Uranium, 442
Utah, 43, 44, 564

Vanadium, 76–77, 209
Vegetative growth, 121–128
Vegetative growth and water deficit, 124
Videography, 202
Vine crops, 126, 268
Violada Irrigation District, Spain, 517, 520–521, 523–525
Volume-variance relationships, 192

Wastewater management, 407–408
Wastewater treatment processes, 453–457
Water allocation, 566–568
Water analysis, 441
Water and solute movement, 311
Water control structures, 405
Water deficit and vegetative growth, 124
Water delivery systems, 404–405
Water flow, 193; modeling, 483–485; in unsaturated soils, 307
Water marketing, 571–574
Water penetration rate, 104. See also Infiltration rate.
Water pollution discharge permits, 575–576
Water pricing, 566
Water quality, 375–376, 477–480. See also Irrigation water, quality of
Water quality control acts, 400–401, 451, 550–553, 575–576, 591
Water quality guidelines, 234
Water quality standards, 400–402
Water rights, 401, 566–567, 571–574
Water sampling, 197, 205
Water sodicity, 225–228
Water supplies, 402
Water supply blending, 437
Water table, 238, 247; control, 363; depth, 205–206, 238, 247–248, 361, 364–365, 385
Waterfowl, 77–78, 154, 564–565
Watershed management, 403
Weather modification, 403
Weathering. See Mineral weathering

Well water, 205–206
Wellton-Mohawk Salinity Control Project, 402, 407–408, 476
Wenner Array, 195–196
Wetlands, 396, 590, 565
Wheat, 13, 121, 124, 143, 147, 164, 169, 172, 251, 266, 267, 268, 288, 382, 384, 387, 434, 436, 437, 442, 443, 569
Wheat grass, 120, 163, 174
White lupin, 125
Wildlife, 393, 396, 552, 564, 565, 590
Woody crops, salt tolerance, 270, 280–283
World water chemistry, 5–6
Wyoming, 44, 564

Zinc, 79, 138, 145–146, 228–229, 457
Zirconium, 209
Zucchini, 147